Date Due

Global Biomass Burning

Global Biomass Burning
Atmospheric, Climatic, and Biospheric
Implications

edited by Joel S. Levine

The MIT Press
Cambridge, Massachusetts
London, England

This book was set in Times Roman by Circle Graphics and
was printed and bound in the United States of America.
Editorial and production services were provided by Editorial
Services of New England, Inc.

Library of Congress Cataloging-in-Publication Data

Global biomass burning : atmospheric, climatic, and
  biospheric implications / edited by Joel S. Levine.
      p.   cm.
    Includes bibliographical references and index.
    ISBN 0-262-12159-X
    1. Burning of land—Environmental aspects—
Congresses.   2. Fuel wood—Burning—Environmental
aspects—Congresses.   3. Atmospheric chemistry—Congresses.   4. Biogeochemistry—Congresses.   5. Climatic
changes—Congresses.
  I. Levine, Joel S.
TD195.B56G57   1991
363.73′87—dc20   91-19742
                                              CIP

√93

Contents

62

63

Contributors

Luc Abbadie
Ecole Normale Supérieure
Paris, France

Dilip R. Ahuja
The Bruce Company
Washington, D.C.

Funso Akeredolu
Obafemi Awolowo University
Ile-Ife
Oxo State, Nigeria

Edward Anders
University of Chicago
Chicago, Illinois

Kenneth J. Andrasko
U.S. Environmental Protection Agency
Washington, D.C.

Meinrat O. Andreae
Max Planck Institute for Chemistry
Mainz, Federal Republic of Germany

Michael J. Apps
Forestry of Canada
Northwest Region
Edmonton, Alberta
Canada

Bénédicte Ardouin
Centre National de la Recherche Scientifique
Commissariat à l'Energie Atomique
Gif-sur-Yvette, France

Ronald E. Babbitt
U.S.D.A. Forest Service
Missoula, Montana

Jean Baudet
Université d'Abidjan
Abidjan, Ivory Coast

Christophe C. Boissard
Centre National de la Recherche Scientifique
Commissariat à l'Energie Atomique
Gif-sur-Yvette, France

Bernard Bonsang
Centre National de la Recherche Scientifique
Commissariat à l'Energie Atomique
Gif-sur-Yvette, France

Michael M. Bradley
Lawrence Livermore National Laboratory
Livermore, California

Marie-Pierre Brémond
Centre National de la Recherche Scientifique
Commissariat à l'Energie Atomique
Gif-sur-Yvette, France

Ernst-Gunther Brunke
Council for Scientific and Industrial Research
Faure, South Africa

Jean Michel Brustet
Université Paul Sabatier
Toulouse, France

Hélène Cachier
Centre National de la Recherche Scientifique
Commissariat à l'Energie Atomique
Gif-sur-Yvette, France

Donald R. Cahoon, Jr.
NASA Langley Research Center
Hampton, Virginia

Daniel P. Y. Chang
University of California
Davis, California

Raymond L. Chuan
Femtometrics
Costa Mesa, California

Catherine C. Chuang
Lawrence Livermore National Laboratory
Livermore, California

Wesley R. Cofer III
NASA Langley Research Center
Hampton, Virginia

Michael T. Coffey
National Center for Atmospheric Research
Boulder, Colorado

Vickie S. Connors
NASA Langley Research Center
Hampton, Virginia

Bernard Cros
Laboratoire de Physique de l'Atmosphère
Brazzaville, Congo

Paul J. Crutzen
Max Planck Institute for Chemistry
Mainz, Federal Republic of Germany

Elen C. Cutrim
Cooperative Institute for Meteorological Satellite
Studies
Madison, Wisconsin

Robert A. Delmas
Université Paul Sabatier
Toulouse, France

Jane Dignon
Lawrence Livermore National Laboratory
Livermore, California

Joëlle Ducret
Centre National de la Recherche Scientifique
Commissariat à l'Energie Atomique
Gif-sur-Yvette, France

Leslie L. Edwards
Lawrence Livermore National Laboratory
Livermore, California

Wayne Einfeld
Sandia National Laboratories
Albuquerque, New Mexico

William R. Emanuel
Oak Ridge National Laboratory
Oak Ridge, Tennessee

Philip M. Fearnside
National Institute for Research in the Amazon
(INPA)
Amazonas, Brazil

Jack Fishman
NASA Langley Research Center
Hampton, Virginia

Jacques Fontan
Université Paul Sabatier
Toulouse, France

Michael Garstang
University of Virginia
Charlottesville, Virginia

Annie Gaudichet
URA-Centre National de la Recherche
Scientifique
Creteil, France

Steven J. Ghan
Lawrence Livermore National Laboratory
Livermore, California

Iain Gilmour
University of Chicago
Chicago, Illinois

Johann Georg Goldammer
University of Freiburg
Freiburg, Federal Republic of Germany

Gerald L. Gregory
NASA Langley Research Center
Hampton, Virginia

David W. T. Griffith
University of Wollongong
New South Wales, Australia

John Hallett
Desert Research Institute
Reno, Nevada

Wei-Min Hao
Max Planck Institute for Chemistry
Mainz, Federal Republic of Germany

Colin C. Hardy
U.S.D.A. Forest Service
Seattle, Washington

Patrick W. Heck
Lockheed Engineering and Services Co.
Hampton, Virginia

Dean A. Hegg
University of Washington
Seattle, Washington

Curtis W. Heisey
Atmospheric and Environmental Research Inc.
Cambridge, Massachusetts

C. Ross Hinkle
The Bionetics Corporation
Kennedy Space Center, Florida

Peter V. Hobbs
University of Washington
Seattle, Washington

Brent N. Holben
NASA Goddard Space Flight Center
Greenbelt, Maryland

Richard A. Houghton
Woods Hole Research Center
Woods Hole, Massachusetts

James G. Hudson
Desert Research Institute
Reno, Nevada

A. O. Isichei
Obafemi Awolowo University
Ile-Ife
Oxo State, Nigeria

Bryan M. Jenkins
University of California
Davis, California

Veena Joshi
Tata Energy Research Institute
New Delhi, India

Prasad S. Kasibhatla
Georgia Institute of Technology
Atlanta, Georgia

Yoram J. Kaufman
NASA Goddard Space Flight Center
Greenbelt, Maryland

Anthony W. King
Oak Ridge National Laboratory
Oak Ridge, Tennessee

Volker W. J. H. Kirchhoff
Instituto de Pesquisas Espaciais (INPE)
São Jose dos Campos
São Paulo, Brazil

Malcolm K. W. Ko
Atmospheric and Environmental Research Inc.
Cambridge, Massachusetts

Albert M. Koller, Jr.
NASA Kennedy Space Center
Florida

Thomas A. Kuhlbusch
Max Planck Institute for Chemistry
Mainz, Federal Republic of Germany

Werner A. Kurz
Environmental and Social Systems Analysts Ltd.
Vancouver, British Columbia
Canada

Jean-Pierre Lacaux
Centre de Recherches Atmospheriques
Lannemezan, France

Cong Lai
Institute of Environmental Medicine
New York University
Tuxedo, New York

Gérard Lambert
Centre National de la Recherche Scientifique
Commissariat à l'Energie Atomique
Gif-sur-Yvette, France

Daniel A. Lashof
Natural Resources Defense Council
Washington, D.C.

Don J. Latham
U.S.D.A. Forest Service
Missoula, Montana

Krista K. Laursen
University of Washington
Seattle, Washington

François Lavenu
Laboratoire d'Etudes et de Recherche en
Télédétection Spatiale (LERTS)
Toulouse, France

Steven W. Leavitt
University of Arizona
Tucson, Arizona

Peter J. LeBel
NASA Langley Research Center
Hampton, Virginia

Marie-Françoise Le Cloarec
Centre National de la Recherche Scientifique
Commissariat à l'Energie Atomique
Gif-sur-Yvette, France

Brigitte Lefeivre
Centre de Recherches Atmosphériques
Lannemezan, France

Jacqueline Lenoble
Université des Sciences et Techniques de Lille
Flandres Artois
Villeneuve d'Ascq, France

Joel S. Levine
NASA Langley Research Center
Hampton, Virginia

Hiram Levy II
Geophysical Fluid Dynamics Laboratory
National Oceanic and Atmospheric Administration
Princeton, New Jersey

Chungcheng Li
Yunnan School of Meteorology
Xiba, Kunming
People's Republic of China

Shao-Meng Li
National Center for Atmospheric Research
Boulder, Colorado

Jurgen M. Lobert
Max Planck Institute for Chemistry
Mainz, Federal Republic of Germany

Jennifer A. Logan
Harvard University
Cambridge, Massachusetts

Philippe Loudjani
Laboratoire d'Etudes et de Recherche en
Télédétection Spatiale (LERTS)
Toulouse, France

Thomas E. Lovejoy
Smithsonian Institution
Washington, D.C.

Jamie H. Lyons
University of Washington
Seattle, Washington

Dewey M. McLean
Virginia Polytechnic Institute and State University
Blacksburg, Virginia

Peter J. McNamee
Environmental and Social Systems Analysts Ltd.
Vancouver, British Columbia
Canada

Katherine Manissadjan
Université Paul Sabatier
Toulouse, France

William G. Mankin
National Center for Atmospheric Research
Boulder, Colorado

Monica A. Mazurek
Brookhaven National Laboratory
Upton, New York

Jean-Claude Menaut
Ecole Normale Supérieure
Paris, France

W. Paul Menzel
National Oceanic and Atmospheric
Administration/National Environmental Satellite
Data and Information Service
Madison, Wisconsin

James E. Miller
NASA Langley Research Center
Hampton, Virginia

Patrick Minnis
NASA Langley Research Center
Hampton, Virginia

Luiz C. B. Molion
Instituto de Pesquisas Espaciais (INPE)
São Jose dos Campos
São Paulo, Brazil

William J. Moxim
Geophysical Fluid Dynamics Laboratory
National Oceanic and Atmospheric Administration
Princeton, New Jersey

J. David Nance
University of Washington
Seattle, Washington

Raymond Nelson
Lockheed Engineering Services Corporation
Houston, Texas

Dominique Nganga
Laboratoire de Physique de l'Atmosphere
Brazzaville, Congo

Jozef M. Pacyna
Norwegian Institute for Air Research (NILU)
Lillestrom, Norway

Jack Paskind
California Air Resources Board
Sacramento, California

Joyce E. Penner
Lawrence Livermore National Laboratory
Livermore, California

Marcos C. Pereira
Instituto de Pesquisas Espaciais (INPE)
São Jose dos Campos
São Paulo, Brazil

Alain Podaire
Laboratoire d'Etudes et de Recherche en
Télédétection Spatiale (LERTS)
Toulouse, France

Wilfred M. Post
Oak Ridge National Laboratory
Oak Ridge, Tennessee

Ronald G. Prinn
Massachussetts Institute of Technology
Cambridge, Massachusetts

Elaine M. Prins
Cooperative Institute for Meteorological Satellite
Studies
Madison, Wisconsin

Stephen J. Pyne
Arizona State University
Phoenix, Arizona

Otto G. Raabe
University of California
Davis, California

Lawrence F. Radke
University of Washington
Seattle, Washington

Rei A. Rasmussen
Oregon Graduate Research Center
Beaverton, Oregon

Henry G. Reichle, Jr.
NASA Langley Research Center
Hampton, Virginia

Robert P. Rhinehart
NASA Langley Research Center
Hampton, Virginia

Allen Riebau
Bureau of Land Management
Cheyenne, Wyoming

Phillip J. Riggan
U.S.D.A. Forest Service
Riverside, California

Patricia D. Roberts
Hampton University
Hampton, Virginia

Jennifer M. Robinson
Pennsylvania State University
University Park, Pennsylvania

Alan Robock
University of Maryland
College Park, Maryland

Jose M. Rodriguez
Atmospheric and Environmental Research Inc.
Cambridge, Massachusetts

C. Fred Rogers
Desert Research Institute
Reno, Nevada

Glen W. Sachse
NASA Langley Research Center
Hampton, Virginia

Eugenio Sanhueza
Instituto Venezolano de Investigaciones Cientificas
(IVIC)
Caracas, Venezuela

Dieter H. Scharffe
Max Planck Institute for Chemistry
Mainz, Federal Republic of Germany

H. Eckhart Scheel
Fraunhofer Institute for Atmospheric
Environmental Research
Garmisch-Partenkirchen
Federal Republic of Germany

Paul A. Schmalzer
The Bionetics Corporation
Kennedy Space Center, Florida

Daniel I. Sebacher
ST Systems Corporation
Hampton, Virginia

Shirley Sebacher
ST Systems Corporation
Hampton, Virginia

Wolfgang Seiler
Fraunhofer Institute for Atmospheric
Environmental Research
Garmisch-Partenkirchen
Federal Republic of Germany

Alberto W. Setzer
Instituto de Pesquisas Espaciais (INPE)
São Jose dos Campos
São Paulo, Brazil

Ralph Seuwen
Max Planck Institute for Chemistry
Mainz, Federal Republic of Germany

Brian J. Stocks
Forestry Canada
Ontario Region
Sault Ste. Marie, Ontario
Canada

Daniel O. Suman
Rosensteil School of Marine and Atmospheric
Science
University of Miami
Miami, Florida

Ronald A. Susott
U.S.D.A. Forest Service
Missoula, Montana

Nien Dak Sze
Atmospheric and Environmental Research Inc.
Cambridge, Massachusetts

Roger L. Tanner
Desert Research Institute
Reno, Nevada

Didre D. Tanre
Université des Sciences et Techniques de Lille
Flanders Artois
Villeneuve d'Ascq, France

John A. Taylor
Australian National University
Canberra, A.C.T.
Australia

Steve Teague
University of California
Davis, California

Geoffrey M. Tennille
NASA Langley Research Center
Hampton, Virginia

Dennis A. Tirpak
U.S. Environmental Protection Agency
Washington, D.C.

Scott Q. Turn
University of California
Davis, California

Stephanie A. Vay
NASA Langley Research Center
Hampton, Virginia

Jean Bruno Vickos
Université Paul Sabatier
Toulouse, France

John J. Walton
Lawrence Livermore National Laboratory
Livermore, California

Darold E. Ward
U.S.D.A. Forest Service
Missoula, Montana

Peter Warneck
Max Planck Institute for Chemistry
Mainz, Federal Republic of Germany

Catherine E. Watson
NASA Langley Research Center
Hampton, Virginia

John G. Watson
Desert Research Institute
Reno, Nevada

Timothy M. Webb
Environmental and Social Systems Analysts Ltd.
Vancouver, British Columbia
Canada

Raymond E. Weiss
Radiance Research
Seattle, Washington

Robert B. Williams
University of California
Davis, California

John W. Winchester
Florida State University
Tallahassee, Florida

Steven M. Winnett
U.S. Environmental Protection Agency
Washington, D.C.

Edward L. Winstead
ST Systems Corporation
Hampton, Virginia

Wendy S. Wolbach
University of Chicago
Chicago, Illinois

Charles A. Wood
University of North Dakota
Grand Forks, North Dakota

David C. Woods
NASA Langley Research Center
Hampton, Virginia

Tommy W. Yip
NASA Langley Research Center
Hampton, Virginia

Véronique Yoboué
Centre de Recherches Atmospheriques
Lannemezan, France

Barbara Zielinska
Desert Research Institute
Reno, Nevada

Patrick R. Zimmerman
National Center for Atmospheric Research
Boulder, Colorado

Introduction

Global Biomass Burning: Atmospheric, Climatic, and Biospheric Implications

The History of the Atmosphere to 1990

When future generations reflect on the year 1990, it may be that the first year of the final decade of the second millenium proved to be the turning point in our attempt to save the earth's atmosphere. Over the last two decades it has become increasingly apparent that gases produced and released into the atmosphere by human actions and activities are significantly disturbing the composition and chemistry of the global atmosphere and increasing the concentration of the atmospheric greenhouse gases that control the climate of our planet.

For billions of years there has been, and still is today, a strong coupling between the atmosphere and the biosphere—the collection of the myriad forms of microscopic and macroscopic life on our planet. The relationship between the atmosphere and life began about 4.5 billion years ago, when the sun, the earth, the other planets, and the balance of the solar system formed out of an interstellar cloud of dust and gas, called the *primordial solar nebula.* Soon after its formation, the atmosphere and oceans of our planet formed as a result of the release of gases originally trapped within our planet during its formation. The atmosphere and oceans were born as a result of extensive and widespread volcanism that may have lasted 100 million years during the very early history of the earth. The volcanic gases, mostly water vapor, carbon dioxide, and nitrogen, contained the key elements necessary for life—carbon, oxygen, nitrogen, and hydrogen. Water vapor and carbon dioxide are greenhouse gases that trap heat energy and keep the mean temperature of the earth above the freezing point of water. Without these greenhouse gases, our planet would be a frozen, inhospitable world, incapable of producing and sustaining life.

The strong coupling between the atmosphere and the soon to be produced biosphere began shortly after the atmosphere and ocean formed, as atmospheric lightning and solar radiation struck the water vapor, carbon dioxide, and nitrogen in the early atmosphere, resulting in complex atmospheric molecules, such as formaldehyde and hydrogen cyanide. In the early ocean, these molecules of formaldehyde and hydrogen cyanide formed long-chained organic molecules of increasing complexity, eventually forming amino acids, the precursors of living systems.

Sometime prior to about 3.5 billion years ago the first living system appeared in the early ocean. The ocean protected the newly formed life from biologically lethal ultraviolet radiation emitted by the sun. Not a single molecule in the early atmosphere had the ability to absorb ultraviolet radiation emitted by the sun. Soon after the first living system appeared, one species of life developed the unique ability to produce its own food by combining carbon dioxide and water vapor in the presence of sunlight. This process, *photosynthesis,* forms carbohydrates utilized by the organism for food and releases a gas, oxygen. as a reaction by-product. Soon photosynthetic oxygen began to accumulate in the atmosphere. Accompanying the evolution of oxygen in the atmosphere was the buildup of another oxygen compound, ozone. Ozone is formed naturally from oxygen via atmospheric chemical reactions in the upper atmosphere or stratosphere, which contains about 90% of the total ozone in the atmosphere. Ozone is important since it is the only gas in the atmosphere that can absorb biologically deadly ultraviolet radiation emitted by the sun. In addition to forming atmospheric ozone, the buildup of oxygen had other important implications for the atmosphere and life. The presence and evolution of oxygen led to the development of the metabolic process—respiration in living systems. Atmospheric oxygen also led to fire on our planet. Respiration and fire ultimately became important sources of atmospheric gases, particularly carbon dioxide. In the 1990s, the burning of fossil fuels and biomass are the major sources of this important greenhouse gas.

About 600 million years ago, atmospheric oxygen reached about one-tenth of its present atmospheric level. At that point, ozone was formed in sufficient concentrations to absorb solar ultraviolet radiation and completely shield the surface of Earth (Figure I.1). For the first time in history, life could leave the ocean, which provided shielding from ultraviolet radiation for billions of years, and climb onto land. The opening of land was a major milestone in the evolution of life. Once on land, life flourished, both in numbers and biological diversity. Plant and animal life soon became significant controllers and regulators of the composition of the atmosphere through the biogeochemical cycling of elements and the production and emission of atmospheric gases through respiration, nitrification, denitrification, and methanogenesis. Almost every atmospheric gas, with the exception of the chemically inert noble gases, is cycled through or produced by microscopic and macroscopic plants and animals. Gases of biological origin include nitrogen, oxygen, carbon dioxide, carbon monoxide, methane, hydrocarbons, hydrogen, nitrous oxide, nitric oxide, ammonia, and hydrogen sulfide. In addition to geological processes, the composition and chemistry of the atmosphere and climate of the planet now were controlled and regulated by the natural biological cycling of elements and the biogenic production of gases.

For most of its 4.5-billion-year history, ozone increased in the earth's atmosphere (Figure I.1), but this trend was suddenly reversed in the 1980s with the discovery of the Antarctic ozone hole and direct indications that stratospheric ozone currently is decreasing.

The next major controlling force of atmospheric

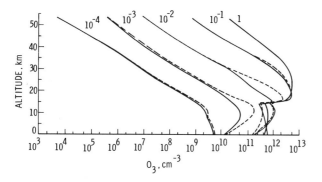

Figure I.1 The vertical distribution of ozone as a function of the buildup of oxygen. The level of atmospheric oxygen is expressed in units of present atmospheric level of oxygen—that is, 1 refers to the present atmospheric level of oxygen, 10^{-1} to $1/10$ of the present atmospheric level, etc. The solid line ozone profiles include the effects of natural sources of chlorine (volcanoes and sea salt spray) on ozone distribution; the broken line profiles do not include the effects of natural chlorine (Levine, 1982).

composition and global climate did not make its appearance on the planet until relatively recently—only about 150 years ago with the beginning of industry and technology as global forces. Today atmospheric gases are produced and released in a number of industrial and technological activities, including (1) the burning of coal, oil, and natural gas for energy production and for transportation systems (the burning of fossil fuels produces carbon dioxide, carbon monoxide, nitric oxide, and sulfur dioxide), (2) the increasing practice of land clearing and conversion of forests, grasslands, and agricultural stubble after harvesting by biomass burning (producing carbon dioxide, carbon monoxide, methane, hydrocarbons, and nitric oxide), (3) the widespread and increasing application of man-made fertilizers to agricultural fields (producing nitrous oxide and nitric oxide via microbial metabolic processes), (4) the increase in the world's rice fields and cattle population to feed our growing population (sources of methane via microbial metabolic processes), and (5) the direct emission into the atmosphere of many man-made chemicals from a variety of sources, including the gaseous propellants used in aerosol spray cans and chemicals used in air conditioning and refrigeration systems, fire extinguishers, and cleaning solvents (sources of chlorofluorocarbons and hydrochlorofluorocarbons).

The Events of 1990

The production and emission of atmospheric gases from industrial and technological activities and their adverse impact on the atmosphere and climate are major national and international concerns. In October 1990, the springtime ozone hole over the South Pole was larger than ever: More ozone had been destroyed through chemical reactions initiated by man-made chlorofluorocarbon chemicals, which produce chlorine, which leads to the chemical destruction of ozone (Figure I.2). Also in 1990 atmospheric greenhouse gases—carbon dioxide, methane, nitrous oxide, chlorofluorocarbons, and tropospheric ozone (which accounts for only about 10% of the ozone in the atmosphere and is produced by atmospheric chemical reactions involving nitric oxide, carbon monoxide, methane, and hydrocarbons as precursor gases)—continued to build up in the atmosphere at a rapid rate (Figures I.3 and I.4).

Yet 1990 was an important year in beginning to correct these global atmospheric problems. A number of separate events aimed at fostering greater awareness of and insight into these problems took

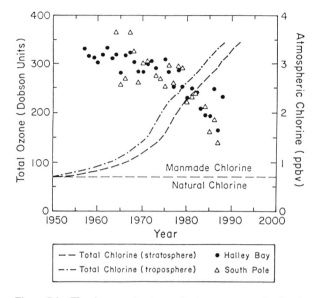

Figure I.2 The decrease in stratospheric ozone over the South Pole resulting from the buildup of man-made chlorine for the period of 1958 to the present (Solomon, 1990).

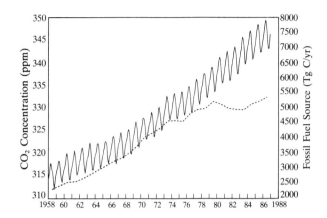

Figure I.4 The increase in the concentration of atmospheric carbon dioxide (represented by the solid oscillating curve) and carbon dioxide emissions from the burning of fossil fuels (represented by the broken line). The annual oscillation in the concentration of atmospheric carbon dioxide is due to the seasonal cycle of photosynthetic activity. The atmospheric carbon dioxide measurements were obtained at Mauna Loa, Hawaii (Lashoff and Tirpak, 1989).

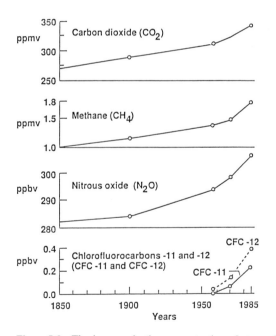

Figure I.3 The increase in the concentration of atmospheric greenhouse gases, carbon dioxide, methane, nitrous oxide, CFC-11 (CCl_3F), and CFC-12 (CCl_2F_2) from 1850 to the present (World Meteorological Organization, 1985).

place in 1990. These events were directed at three distinct groups—members of the general public, who demand that corrective actions be taken, and be taken immediately; the leaders of the world, who direct these corrective actions through scientific understanding and legislation; and the scientists of the world, who provide the scientific insight and understanding of these problems and the scenarios for their solutions.

On 22 and 23 April millions of people participated in the twentieth Earth Day celebration in the United States and abroad and sent an important message to the leaders of the world: Those living on this planet and breathing its atmosphere are vitally concerned about the future of the atmosphere and the planet. In June, the countries of the world reconvened the so-called Montreal Protocol and significantly strengthened the provisions of the earlier Montreal Protocol of 1987, which aimed at a global reduction and phaseout of the chlorofluorocarbons used in aerosol spray cans, which lead to the destruction of ozone. In August, the Intergovernmental Panel on Climate Change (IPCC) released the final report of its Working Group I, the Scientific Assessment of Climate Change. The IPCC was established by the World Meteorological Organization and the United Nations Environment Program in 1988. The objective of Working Group I was to assess available information on climate change. One hundred and seventy scientists from 25 countries contributed to the report. The report's Executive Summary states that "emissions resulting from human activities are substantially increasing the atmospheric concentrations of the greenhouse gases: carbon dioxide, methane, chlorofluorocarbons (CFCs) and nitrous oxide. These increases will enhance the greenhouse effect, resulting on average in an additional warming of the Earth's surface. The main greenhouse gas, water vapor, will increase in response to global warming and further enhance it.''

In October, the U.S. Public Broadcast System (PBS) telecast the 10-part series, *Race to Save the Planet.* This series, which was viewed by millions in the United States and abroad, highlighted the problems of stratospheric ozone destruction and the buildup of atmospheric greenhouse gases resulting from human activities and actions. Also in October the U.S. Congress passed the Clean Air Act, which amended the Clean Air Act of 1970, the year of the first Earth Day. Amendments to this bill were aimed at reducing atmospheric emissions resulting from human activities and actions. In November government officials and representatives from more than 130 nations participated in a United Nations Conference on Global Warming in Geneva, Switzerland. The conference was concluded with a declaration that emphasized the need for urgent action to cope with the global climate change challenge and called for negotiations to begin without delay on a worldwide treaty to control greenhouse gases resulting from human activities and actions.

Another important event addressing the impact of human activities and actions on the atmosphere and climate took place in 1990. More than 200 scientists, government officials, and environmental policy planners from 20 countries participated in the Chapman Conference on Global Biomass Burning: Atmospheric, Climatic, and Biospheric Implications, held in Williamsburg, Virginia, 19–23 March. The Chapman Conferences, named after the American geophysicist Sydney Chapman, are sponsored by the American Geophysical Union, the world's largest professional organization of geophysicists and earth scientists, with a membership in excess of 25,000. The conference was cosponsored and coorganized by National Aeronautics and Space Administration (NASA) Headquarters, the NASA Langley Research Center, the National Science Foundation, the U.S. Environmental Protection Agency, the U.S. Department of Agriculture Forest Service, the International Geosphere Biosphere Program, and the International Global Atmospheric Chemistry Project.

Biomass burning includes burning forests and savanna grasslands for land clearing and conversion,

burning agricultural stubble and waste after harvesting, and burning biomass fuel. The objectives of the Conference on Global Biomass Burning were to assess the extent of biomass burning as a global phenomenon, to quantify the production of atmospheric gases and particulates by the global burning, and to assess the effect of these gases and particulates on the composition and chemistry of the atmosphere, on global climate, and on the biosphere itself. The Chapman Conference was the first meeting to address these questions from a global perspective. The burning of the tropical rainforest and its effect on the biology and ecology of the region has long been a topic of great concern. The tropical rainforest is our planet's most productive and biologically diverse ecosystem, and its destruction would result in the loss of thousands of species of plants and animals. However, the thrust of the papers and discussions at the Chapman Conference was the *global* atmospheric, climatic, and biospheric impacts of the gaseous and particulate emissions resulting from biomass burning. The 63 papers collected in this volume were originally presented at the Chapman Conference and were authored or coauthored by a total of 160 scientists—104 from the United States and 56 from abroad—whose contributions offer important testimony to the truly international interest and concern associated with biomass burning.

These are some of the important findings reported at the Chapman Conference and documented in this volume:

1. On a global scale, the total biomass consumed by annual burning is about 8680 teragrams (Tg) (1 teragram = 1 million tons = 10^{12} g) of dry material (dm). The estimated total biomass consumed by the burning of savanna grasslands (3690 Tg dm/yr) exceeds the other components of biomass burning—the burning of agricultural waste (2020 Tg dm/yr), the burning of forests (1540 Tg dm/yr), and the burning of fuel wood (1430 Tg dm/yr).

2. Biomass burning is not a phenomenon restricted to the tropics. Large fires are a common feature of the world's boreal forests. For example, the Chinese–Soviet Union fire of 1987 destroyed more than 12 million acres of boreal forests in only 21 days.

3. The bulk of biomass burning is human initiated and appears to be increasing with time. Gaseous and particulate emissions resulting from biomass burning may have increased by as much as 50% since 1850. Photographs taken from space by astronauts have documented the growth of global biomass burning.

For example, the area of smoke palls over the Amazon increased from several hundred thousand km^2 in the 1970s to several million km^2 by the late 1980s.

4. In his paper in this volume, "Biomass Burning and the Disappearing Rainforest," Dr. Thomas Lovejoy writes, "The graph of increasing atmospheric CO_2 concentrations measured from the top of a volcano in the Hawaiian Island could be the symbol of this conference." Dr. Lovejoy is referring to the increase of atmospheric CO_2 concentrations as measured at Mauna Loa, Hawaii, since 1958 (Figure I.4). Inspection of this figure indicates a good correlation between the increase in atmospheric carbon dioxide (represented by the solid oscillating curve that increases with time) and carbon dioxide emissions from the burning of fossil fuel (represented by the broken line). The apparent correlation between these two curves suddenly stops in the early 1980s. Are we seeing a decrease in the ability of carbon dioxide sinks, such as the ocean or biosphere, to sequester atmospheric carbon dioxide, or are we seeing an escalation in the production of carbon dioxide due to other sources, such as biomass burning? What is certain is that biomass burning is indeed a significant source of atmospheric carbon dioxide. Estimates indicate that the burning of 8680 Tg dm/yr of biomass produces about 3,500 Tg of carbon per year (3,500 Tg C/yr) in the form of CO_2. Hence, biomass burning may produce 40% of the world's annual production of CO_2.

5. The type (such as oxidized or reduced carbon and nitrogen species) and quantity of gaseous and particulate emissions resulting from burning are a strong function of both the ecosystem and the phase of burning—that is, flaming versus smoldering.

6. Satellite measurements clearly indicate significantly increased tropospheric concentrations of carbon monoxide and ozone associated with biomass burning. Levels of ozone resulting from biomass burning in pristine tropical ecosystems are approaching levels lethal to native plant and animal life. Levels of carbon monoxide in these ecosystems are greater than in highly industrialized areas. Estimates indicate that on an annual basis, biomass burning may lead to the chemical production of about 38% of the ozone in the troposphere and produce about 32% of the world's carbon monoxide, about 39% of the particulate organic carbon, and more than 20% of the world's hydrogen, nonmethane hydrocarbons, methyl chloride, and oxides of nitrogen.

7. Burning appears to significantly enhance the microbial production and emission of nitrous oxide and nitric oxide from soils and methane from wetlands. It is believed that these enhanced microbial emissions are caused by changes in soil and wetlands chemistry and microbial ecology following burning.

8. The consensus of scientists and policy makers at the conference was that global biomass burning must be and can be reduced significantly over the next decade. A plan for the reduction of global biomass burning is suggested in the paper by Andrasko et al. in Chapter 55. Reduction will begin with global awareness and understanding about the adverse atmospheric effects of biomass burning. Indeed, awareness and understanding of biomass burning were the goals of the conference.

At the conclusion of the conference on 23 March 1990 the Public Broadcast System telecast a 90-minute live program over the PBS network and abroad via the Westar IV communications satellite. The program, *Biomass Burning and Global Change,* explored the subject of biomass burning and its effect on the atmosphere, climate, and the biosphere itself and summarized the findings of the week-long conference in Williamsburg with an international panel of conference scientists. Viewer response to the program and the subject material was strong.

Organization of This Book

Historical and geographical reviews of biomass burning and its effect on global atmospheric chemistry and global climate open this volume. The importance of global biomass burning as a significant source of atmospheric gases and particulates is evidenced by the priority given this area of research. The Impact of Biomass Burning on the World Atmosphere is a research activity of the International Global Atmospheric Chemistry (IGAC) Project, which is a core project of the International Geosphere-Biosphere Program (IGBP). IGAC and IGBP research activities dealing with biomass burning are described in this chapter. Space offers a unique perspective and vantage point to study and assess the geographical and ecosystem distribution of biomass burning. Measurements of biomass burning from Earth-orbiting platforms, including direct photography by astronauts and imagery from the NOAA, GOES, and Landsat satellites, are also presented in Part I.

Biomass burning in tropical ecosystems—particularly the Brazilian Amazon and Southern Africa—and its historic, present, and future trends are discussed in Part II. The chapter also includes case studies of gaseous and particulate emissions from burning in tropical ecosystems.

Biomass burning in temperate and boreal ecosystems is much more widespread and extensive than previously believed and is the subject of Part III. This chapter covers the extent of burning in these ecosystems, the gaseous and particulate emissions, and the impact of burning on soil chemistry, microbial ecology, and the biogenic production and emission of nitrous oxide, nitric oxide, and methane.

Most studies of biomass burning have involved field experiments and satellite measurements. Recently laboratory experiments in biomass burning have been developed and are discussed in Part IV.

The role of biomass burning in the global budgets of carbon and nitrogen species is discussed in Parts V and VI, respectively. The role of global biomass burning as a source of atmospheric carbon dioxide is considered. Other papers treat the effect of nitrogen oxides produced by biomass burning on the global atmosphere using simulations with general circulation models. Finally, observational evidence is presented to indicate that biomass burning may not be a significant source of nitrous oxide. New evidence suggests that as in the case of fossil fuel combustion samples, the nitrous oxide measured is really a sampling artifact.

Biomass burning is a significant global source of atmospheric particulates. The characteristics and optical properties of these particulates are presented in Part VII.

The impact of biomass burning on global climate is discussed in Part VIII. Papers consider how greenhouse gases and particulates produced by biomass burning affect climate and present policy options for managing biomass burning to mitigate global climate change.

Biomass burning over historic and prehistoric times is discussed in Part IX. Did global wildfires occur at the Cretaceous-Tertiary boundary at the time of the extinction of the dinosaurs some 65 million years ago? This possibility is debated in the first two papers. Other papers deal with the history of fire in the United States and Central America.

The year 1990 may indeed be the year we finally started to do something about global atmospheric problems. In 1990 important events activated general public awareness about global atmospheric problems, the awareness and interest of the governments of the world, and new insights within the scientific community. In 1990 people worldwide participated in the twentieth anniversary of Earth Day and viewed the PBS series *Race to Save the Planet.* In 1990 the governments of the world assessed global atmospheric environmental problems at the second Montreal Protocol meeting and at the United Nations Conference on Global Warming, and the Intergovernmental Panel on Climate Change released its long-awaited report. The U.S. Congress passed the Clean Air Act of 1990. And in Williamsburg, Virginia, scientists and environmental policy planners from all over the world met for the first time to discuss and assess a new and emerging atmospheric threat—global biomass burning. Global biomass burning—a significant source of atmospheric gases that are altering the composition and chemistry of the atmosphere and climate of our planet—is the latest chapter of a story that began some 4.5 billion years ago with the formation of the atmosphere by primordial volcanism.

It seems appropriate to conclude the introduction to this volume with the opening line of Chapter 61, written by Stephen Pyne:

We are uniquely fire creatures on a uniquely fire planet, and through fire the destiny of humans has bound itself to the destiny of the planet.

It may turn out that the destiny of our planet is also bound to global burning through its atmospheric, climatic, and biospheric implications.

It is a pleasure to acknowledge the assistance of the following colleagues in the planning and running of the conference: M. O. Andreae, M. Averner, W. R. Cofer III, P. J. Crutzen, J. Hoffman, C. O. Justice, V. W. J. H. Kirchhoff, A. M. Koller, Jr., J. D. Lawrence, Jr., H. Levy II, J. A. McSorley, L. C. B. Molion, J. L. Moyers, J. S. Pinto, R. G. Prinn, J. D. Rummel, A. W. Setzer, B. J. Stocks, and D. E. Ward.

I

Biomass Burning: Remote Sensing and Global and Geographical Distribution

Biomass Burning: Its History, Use, and Distribution and Its Impact on Environmental Quality and Global Climate

Meinrat O. Andreae

Pollution from biomass burning was one of the first occupational hazards: early humans must have been exposed to large amounts of carcinogenic compounds from the smoke of open fires in caves, huts, and tents. Today we are frequently confronted with the evidence of pollution from biomass burning in our daily lives: In North America, many communities have had to issue ordinances regulating domestic burning in fireplaces and wood-burning stoves to control atmospheric pollution in winter. In the tropics, airports like Santarém, in the middle of the Amazon Basin, have to be closed frequently during the burning season because of poor visibility resulting from the smoke from gigantic fires hundreds to thousands of kilometers away. Such large fires and the resulting smoke plumes are a common sight to air travelers in the tropics and even to the crews of the Space Shuttle (Figure 1.1).

Despite the magnitude of the problem posed by the emissions from biomass burning, little quantitative information is currently available on the amounts of pollutants emitted and their impact on the environment. In contrast, the effects of fossil fuel burning for heating, energy production, transportation, and so forth have been thoroughly investigated and quantified. This difference may largely be explained by the fact that the production and distribution pathways of fossil fuels are to a large extent under the control of relatively few organizations, either governments or large corporations, and are therefore well known and documented. Biomass burning, on the other hand, takes place to a large extent in the developing countries at the hands of individual farmers, ranchers, and housewives who do not keep records of amounts burned. In this review, I discuss the historical development of biomass burning and its role in agriculture and society and then provide current estimates of the type and amounts of pollutants emitted and their environmental effects.

The History of Fire

Fires must have existed on Earth ever since the evolution of land plants, around 350 to 400 million years

ago. The production of plant matter on dry land first made possible the accumulation of "fire potential" in the form of large amounts of combustible organic material. (Some of this accumulated, unburned fire potential from geological times is still present in the form of coal deposits.) Once the production of organic matter by plants had supplied the potential for wildfires, climatic and ecological parameters (length and intensity of dry seasons, lightning frequency, etc.) established a natural fire frequency. The intensity of the resulting wildfires is determined by the combination of the accumulation rate of fire potential and the fire frequency: frequency and intensity are inversely related, a fact that today is well known to foresters and which forms the basis of fire management through prescribed burning.

With the advent of herbivorous organisms, the rate of natural fire potential accumulation became controlled by the interplay of phyto-production on one hand and fire frequency and herbivory on the other (Schüle, 1990). In particular, hypsodont mammals (those with high-crowned teeth, e.g., horses, goats, and camels) are very efficient in removing vegetation in the lower layer of wooded savannas, where fires would normally propagate. Since these grazers are able to consume almost anything, especially small tree seedlings and shoots, they tend to act in the same way as fire in stabilizing savanna-type landscapes. A high rate of grazing by large reptiles (dinosaurs) may have contributed to the predominance of savanna-like vegetation on Earth during the late Cretaceous (about 70 million years ago), which would be accompanied by a high fire frequency and low fire intensity. High concentrations of black carbon (soot and charcoal) in the Cretaceous-Tertiary boundary sediments suggest that the end of the age of dinosaurs was accompanied by large fires which may have perturbed Earth's climate and thereby contributed to the dinosaurs' demise (Wolbach et al., 1985, 1988). In the early Tertiary, almost the entire land surface on Earth was covered with humid forests, where few and relatively ineffective grazers were present. A combination of climate change and the evolution of very large and effective hypsodont mammalian grazers led to

Figure 1.1 Biomass burning in Mozambique, Africa, as seen from the Space Shuttle. The picture was taken on 9 October 1984 by a Space Shuttle astronaut using a hand-held camera (Hasselblad, 70 mm lens). The mouth of the Zambezi River is visible in the center. Large plumes from savanna fires are entrained by easterly winds. Blackened areas from recent fires cover large areas. (Photograph by Space Shuttle Earth Observation Office, NASA)

the spread of savannas and steppes, which became the dominant land vegetation by the end of the Tertiary age. Fire frequency must have changed from very low to very high during this time. These ecological changes contributed to the evolutionary pressures that brought about the development of savanna-living monkeys on one hand and hominids on the other. Eventually, humans overcame the fear of fire common to all other primates and learned to use wildfires, initially as an accidental source of food in the form of insects, rodents, snakes, etc. caught and killed by the fires and later (about 1.5 to 2 million years ago) as an active tool of food preparation, hunting, and landscape control (James, 1989). It is likely that man-made fires began to make an ecological impact in the African savanna at this time.

Early humans thus had to develop quite sophisticated fire-management strategies, examples of which survived up to this century among the Aborigines of Australia (Jones, 1979), where *Homo sapiens* arrived

about 40,000 years ago. At about the same time, pollen records show a shift from pyrophobic vegetation (e.g., *Araucaria* spp.) to pyrotolerant and pyrophilic species (e.g., *Eucalyptus* spp.), and charcoal particles in sediment cores increase by three orders of magnitude. This coincidence suggests the possibility that human use of fire profoundly affected that continent's ecology (Schüle, 1990). More information on the relationship between fire and human evolution can be found in the reviews by James (1989) and Schüle (1990).

On a quantitative level, the history of biomass burning through geological and historical time is difficult to establish, since there is little physical evidence other than charcoal particles buried in sediments. Measurements of charcoal in dated sediment cores have shown some clear correlations between the rate of burning and human activity—for example, the increase of charcoal in Australian sediments coincides with human settlement mentioned above and the shift from wood burning to fossil fuel burning around the turn of the century is recorded in the morphology and number of charcoal and soot particles in Lake Michigan sediments (Griffin and Goldberg, 1979, 1983). Charcoal layers in Amazonian soils also document widespread catastrophic fires during periods of drought in Amazonia in prehistoric times (Sanford et al., 1985). However, much research remains to be done to clarify the relationship between changing human population densities and agricultural-industrial activity on one hand and biomass burning rate on the other.

The Role of Biomass Burning in Agriculture and Economy

In agriculture and economy, biomass burning serves a variety of purposes:

1. clearing of forest and brush land for agricultural use

2. control of brush, weeds, and litter accumulation on grazing and crop lands

3. nutrient regeneration in grazing and crop lands

4. control of fuel accumulation in forests

5. production of charcoal for industrial and domestic use

6. energy production for cooking and heating

Finally there is the category of wildfires, which are not started and controlled by people, but which often have large economic and ecological consequences.

Clearing of Forest and Brush Land
for Agricultural Use

Two approaches to forest clearing for agricultural use may be differentiated: shifting agriculture, which allows the land to return to forest vegetation after a relatively short period of use, and permanent removal of forest, which replaces forest with grazing or crop land. In both instances, the clearing and burning follows initially the same pattern: trees are felled, the vegetation is left to dry out in order to obtain better burning efficiency, and the material is then set on fire, often after bulldozing it together into large piles. The efficiency of this first burn is variable; it often does not exceed 10% to 30% (Fearnside, 1985, 1990). This apparent low efficiency is due to the large fraction of rainforest biomass residing in the tree trunks, only a small portion of which tends to be consumed in the first burn. The remaining material is then moved together into piles and left to rot or dry. It is usually set on fire again when it is judged to be dry enough. As a result of this practice, some 40% of the aboveground carbon from forest clearing enters the atmosphere through combustion, while the rest is released through rotting—that is, microbial decomposition.

In shifting agriculture, which is practiced by some 200 million people worldwide (Seiler and Crutzen, 1980), the cleared area is used for agriculture for a few years and abandoned when yields are felt to be declining; then a new area is cleared. Shifting agriculture is not expected to grow much beyond its present extent, which covers some 300 to 500 million hectares (ha) (Wong, 1978; Sommer, 1976; Seiler and Crutzen, 1980), because virgin forest lands which could be brought into this type of cultivation are becoming more scarce. Instead, lands presently under shifting agriculture are likely to become used for permanent agriculture. Often, however, because of poor soil conditions or land management, the areas become wasteland unsuitable for any type of agricultural use.

The permanent removal of rainforest for agricultural use is rapidly progressing as expanding populations require additional food and living space and as large-scale resettlement programs and land-speculation tactics are implemented. The global rate of deforestation is unclear. Based on work by Lanly (1982), Houghton et al. (1985), and Detwiler et al. (1985), a global deforestation rate of about 22 million ha/yr during the 1970s was estimated by Hao et al. (1990). It must be emphasized that this rate applies to the 1970s and is likely to have increased during the 1980s. Of these 22 million ha/yr, about 7.5 million ha/yr are cleared in virgin rainforest, amounting to an annual destruction of about 0.6% of the approximately 1200 million remaining hectares. Estimates of the fractions of cleared land which go to shifting agriculture and to permanent land use range from 1.9 to 3.4 million and 4.1 to 5.1 million ha/yr, respectively (Lanly, 1982; Detwiler et al., 1985; Houghton et al., 1985; Hao et al., 1990). The discrepancies between the estimates of the fate of land cleared in secondary tropical forest are even greater. Detwiler et al. (1985) estimated that 18 million ha/yr are cleared, all for shifting cultivation. Houghton et al. (1985), on the other hand, calculated that 13.5 million ha/yr were cleared in second-growth forest, of which 10.1 million ha were cleared permanently. This discrepancy remains to be resolved.

Biomass Burning in Tropical Savannas
and Brush Land

Tropical savannas cover an area (1530 million ha) roughly similar to that of the tropical forests (1440 million ha) (Lanly, 1982). These savannas typically consist of a more or less continuous layer of grass with interspersed trees and shrubs. They are burned periodically at intervals which may range from one to three years (Eiten, 1972; Sarmiento and Monasterio, 1975; Lacey et al., 1982; Menaut, 1983). This frequency may be increasing in some regions as a result of growing population and more intensive use of rangeland (Menaut, 1983). While lightning may start some fires in savannas, most investigators are convinced that almost all savanna fires are set by humans. Burning has several objectives: foremost probably is the control of weeds, shrubs, tree seedlings, and litter accumulation. Without burning, the grassy vegetation tends to be overgrown rapidly by shrubs and brush and under many conditions progresses to a chaparral or forest unsuitable for grazing, which is the predominant agricultural use of the grassy savannas. Only the grass and small plants are consumed in the fires; the larger trees are fire-adapted species which suffer little damage and actually thrive under conditions of periodic burning.

Burning is also perceived as a form of pest control because it reduces the population of insects, snakes, and so on. Another reason for burning is the belief that it is an effective way to return the nutrients—stored in the dry vegetation which accumulates during the dry season—to the soils and thereby promote regrowth of fresh grass during the following rainy season. This belief may be fallacious; it is discussed in the section on atmospheric emissions from fires.

Many of the tree savannas and open forest formations, particularly dry deciduous and semideciduous forests, are burned annually to facilitate access and harvest of nonwood forest products (Goldammer, 1988; Stott et al., 1990). Finally, fires are also set to drive game for hunting, a practice which may be one of the earliest uses of fire by humans. The savanna area burned each year is very large: Hao et al. (1990) estimate about 750 million ha, much more than the area burned in tropical rainforests. This applies especially to Africa, where about one-half of the global amount of savanna biomass burning is concentrated.

Fuel Wood and Charcoal
Fuel wood is a major source of energy in the developing countries, and its gathering often represents a large part of the workload of populations in Africa and parts of Asia. About 50% of fuel wood is used for cooking, about 30% for domestic heating, and the remaining 20% for various purposes, such as metalworking and pottery making. Because of the highly distributed nature of firewood gathering and use, the annual amount of wood burning is difficult to estimate. The number given by the Food and Agriculture Organization (FAO) (FAO, 1989) for 1987, 1050 Tg dm/yr (teragrams of dry matter per year; 1 teragram = 1 million tons = 10^{12} grams), is probably an underestimate because it does not cover the use of wood which is not marketed. Table 1.1 shows an estimate based on population and per-capita use statistics, which may be more reliable than the FAO estimate: For the tropical regions, the 1985 population

tion numbers are multiplied by an annual fuel wood consumption of 475 kg per person, based on the discussion in Seiler and Crutzen (1980). This value is actually a conservative estimate of biomass fuel consumption; rural populations consume as much as 1000 kg biomass fuel per person annually (Scurlock and Hall, 1990). Since much of this material is agricultural waste, however, I have included it in that category. These population-based estimates and the FAO statistics agree reasonably well for all continents except Asia, where the domestic use of coal in China and the burning of agricultural wastes may be replacing some fuel wood use. Throughout the developing world, population growth is increasing the demand for firewood, which produces scarcity in some regions. The ensuing calculations adopt a best-guess estimate based on the mean between the FAO and population-based estimates for the developing countries in the tropics and on the FAO data for the developed countries.

Charcoal production for domestic and industrial use has become an important alternative to the direct use of wood as fuel. Transportation of wood from the forest to consumers in the urban areas of the developing countries is cumbersome; therefore, wood is often replaced by charcoal, which has a higher energy density and is thus cheaper to transport. In recent years, large charcoal production plants have been established in countries like Brazil in order to supply charcoal for industrial use, especially for the smelting of pig iron. Often, the wood used in this type of charcoal

Table 1.1 Burning of fuel wood, charcoal, and agricultural waste in tropical and extratropical regions

Region	Population (1985, mill.)[a]	Fuel wood[b]	Fuel wood[c]	Fuel wood (best guess)[d]	Charcoal[c]	Agricultural waste[e]
				Tg dm/yr		
Tropics:						
America	430	180	150	170	7.5	200
Africa	550	240	240	240	9.3	160
Asia	2820	1200	490	850	3.3	990
Oceania	25	11	6	8	0	17
Total tropics	3830	1640	890	1260	20	1360
USA and Canada	260	—	80	80	0.5	250
Western Europe	380	—	40	40	0.2	170
USSR and Eastern Europe	390	—	50	50	0.2	230
World total	4840	—	1050	1430	21	2020

a. FAO (1986).
b. Based on population, a fuel-wood consumption of 475 kg per person per year, and a burning efficiency of 90%.
c. FAO (1989).
d. Mean of population-based and FAO estimate.
e. Based on total production of crops (see text).

manufacture is derived from fast-growing pine plantations. In some of the large projects in the Amazon, however, such as the Grande Carajas Program in the eastern part of the basin, it is anticipated that the wood required for charcoal production will be obtained by clear-cutting the surrounding rainforest. A precedent exists in the state of Minas Gerais, where pig-iron production consumed nearly two-thirds of the state's forests.

Charcoal production consists of the controlled partial combustion (pyrolysis) of dry wood. During this process, a mixture of water, carbon monoxide, methanol, tar, and other volatile products is distilled off, which in principle can be captured and processed into marketable by-products. It is likely, however, that in most of the charcoal kilns in the developing countries these volatile compounds are released to the atmosphere. Depending on the efficiency of the charcoal kiln, the ratio of carbon input (in the form of wood) to carbon output (in the form of charcoal) varies between about 2.0 and 1.2 (Ayres et al., 1987). For the purpose of our estimates of the amount of carbon released into the atmosphere due to charcoal production and use, we have therefore multiplied the amount of charcoal burned by a factor of 1.4 to account for losses during production.

Prescribed Burning and Wildfires in Forests

Wildfires are most common in temperate and boreal forests, since undisturbed tropical forests are usually too moist to allow the propagation of wildfires. According to Brown and Davis (1973), most wildfires are the result of human activities, such as sparks from railroad engines, with only a small fraction (10% to 30%) initiated by lightning. While individual wildfires may be very large and lead to smoke plumes visible over great distances, the area burned per year and the burning efficiency are relatively low. Seiler and Crutzen (1980) estimate that about 3.8 million ha of temperate and boreal forest are subject to wildfires each year.

Prescribed fires are a commonly used tool of forest management, particularly in North America. They serve to reduce the accumulation of dry, combustible plant debris, which is a major cause of destructive wildfires. They are also used to eliminate shrubby vegetation, which competes with tree crops for nutrients. In some regions, prescribed burns are used to maintain particular types of game habitat. Because the practice of prescribed burning is limited to North America and Australia, it has little impact globally but can lead to serious local and regional air pollution.

The area involved has been estimated to be on the order of 2 to 3 million ha/yr (Seiler and Crutzen, 1980).

Burning of Agricultural Wastes

A type of burning which is extremely difficult to quantify because of its distributed nature and because no material of direct economic value is involved is the burning of agricultural wastes, e.g., straw and stubble from grain crops. A very conspicuous source of air pollution in many tropical regions is the burning of sugar-cane fields before harvesting to facilitate the processing of the sugar canes (Kirchhoff et al., 1989). Based on an extrapolation from waste-burning practices in the United States and making some reasonable assumptions about the fraction of agricultural waste burned in developing countries, Seiler and Crutzen (1980) have derived an estimate of 1700 to 2100 Tg dm/yr. Table 1.1 presents a new estimate for this source, based on the FAO crop production statistics for 1985 (FAO, 1986). We have assumed that the amount of agricultural waste is the same as the amount of crops produced, which appears to be reasonable based on the discussion in Seiler and Crutzen (1980). We then assume that 80% of the waste is burned in developing countries (much of it as domestic fuel) and 50% in developed countries, and that the combustion efficiency is 90%. The resulting global estimate is 2020 Tg dm/yr, near the high end of the range suggested by Seiler and Crutzen (1980). It may be important to attempt a careful reexamination of this agricultural waste source, since it appears to be one of the major contributions to atmospheric emissions from biomass burning, and since it has the potential of being reduced substantially by alternative agricultural practices.

Geographical Distribution of Biomass Burning

The geographical distribution of biomass burning is discussed in some detail in the review by Seiler and Crutzen (1980). For the tropics, Hao et al. (1990) have estimated the amount of biomass burning and the resulting emissions in each 5° latitude by 5° longitude grid cell. The results obtained by these authors are summarized in Tables 1.2 and 1.3. Most biomass burning is taking place in the developing countries of the tropics (Tables 1.1 and 1.3), adding an emission of 28 Tg C/yr (from Table 1.1: 20 Tg × 1.4) from charcoal production and combustion to the tropical emissions of 3410 teragrams of carbon per year (Tg C/yr) from forest and savanna burning, fuel wood use, and

Table 1.2 Global estimates of annual amounts of biomass burning and of the resulting release of carbon to the atmosphere

Source	Biomass burned (Tg dm/yr)			Carbon released (Tg/yr)[d]
	Seiler and Crutzen[a]	Hao et al.[b]	Andreae[c]	
Tropical forest	2420	1260	1260	570
Savanna	1190	3690	3690	1660
Temperate and boreal forest	280	—	280	130
Fuel wood	1050	620	1430	640
Charcoal	—	—	21	30
Agricultural waste	1900	660	2020	910
World total	6840	6230	8700	3940

a. Seiler and Crutzen (1980).
b. Hao et al. (1990).
c. This paper: Tropical forest and savanna from Hao et al.; temperate and boreal forest from Seiler and Crutzen; fuel wood, charcoal, and agricultural waste from Table 1.1.
d. Based on a carbon content of 45% in dry biomass. In the case of charcoal, the rate of burning (based on FAO production statistics) has been multiplied by 1.4 to account for losses in the production process (see text).

Table 1.3 Biomass burning in the tropical regions

Region	Forest[a]	Savanna[a]	Fuel wood[b]	Agricultural waste[b]	Region total	
	Tg dm/yr				Tg dm/yr	Tg C/yr
America	590	770	170	200	1730	780
Africa	390	2430	240	160	3210	1450
Asia	280	70	850	990	2190	980
Oceania	—	420	8	17	450	200
Total tropics	1260	3690	1260	1360	7580	3410

a. From Hao et al. (1990).
b. From Table 1.1.

the incineration of agricultural waste, we obtain a total emission of 3440 Tg C/yr from the tropics. This represents 87% of the global emissions from biomass burning (3940 Tg C/yr; Table 1.2). As we will discuss later in the chapter, this is consistent with the observation that the impact of burning on the atmospheric environment appears to be concentrated in the tropics.

An interesting difference between the estimates of Seiler and Crutzen (1980) and Hao et al. (1990) is the relative importance of biomass burning in tropical forests and savannas. In the older study, forest burning was predominant, whereas the more recent work by Hao et al. suggests that savanna burning releases about three times as much carbon to the atmosphere as forest burning. This is especially evident in Africa (Table 1.3), where savanna fires account for almost 90% of the emissions from savanna and forest burning and for almost one-third of the global emissions from

biomass burning. A large proportion of savanna burning in Africa takes place within 15° north and south of the equator. The importance of savanna fires in Africa derives from the large area covered by this landscape type in Africa and the high frequency at which African savannas are burned (about 75% per year). In tropical America, forest burning is concentrated in the Brazilian states of Pará, Maranhão, Goiás, Mato Grosso, and Rondônia (Setzer and Pereira, 1991; Hao et al., 1990). Relatively little forest and savanna burning occurs in tropical Asia, with the exception of Indonesia, Thailand, Malaysia, and India, where rapid deforestation is ongoing (Lanly, 1982; Hao et al., 1990).

Seasonality of Biomass Burning
Frequent mention is made in the literature of "burning seasons," and it is often assumed that practically no burning occurs outside of these periods. Indeed, a

large fraction of the burning of forests and savannas takes place during the dry season, and therefore burning is most intensive in the Northern Hemisphere from December to March and in the Southern Hemisphere from June to September. However, experience in tropical countries shows that burning can be observed almost whenever and wherever there is plant material dry enough to burn. This applies to cooking and heating fires, burning of garden and agricultural wastes, burning of the areas surrounding living quarters to control insects, weeds, and snakes, and often burning just for the pleasure of watching the flames and smoke. Thus, even during the wet season in Amazonia, fires are started along the roads and rivers during short dry spells between rainy days. At the end of the rainy season, many farmers wait for cleared areas to become dry enough to burn, leading to an outbreak of burning in late May and early June in the Southern-Hemisphere part of Brazil. The large-scale forest-clearing burns are usually started later in the dry season in order to have drier fuel and a better burning efficiency. Based on a 1987 satellite survey in Amazonia, Setzer and Pereira (1991) found the highest rate of burning to occur in August and September at the southern perimeter of the Amazon Basin.

Emissions to the Atmosphere from Biomass Burning

Carbon Dioxide

The burning of organic materials produces water vapor and carbon dioxide as the primary products, according to the reaction

$$(CH_2O) + O_2 = CO_2 + H_2O$$

where (CH_2O) stands for the average composition of biological materials. This reaction is essentially the same as the photosynthesis-respiration reaction (photosynthesis from right to left, respiration from left to right). In this sense, biomass burning can be seen as an abiotic equivalent of the respiratory catabolism of biological material, returning the products of photosynthesis back to the atmosphere as CO_2. On a long enough time scale, therefore, biomass burning does not influence the atmospheric CO_2 budget but merely returns to the atmosphere CO_2 that had been removed by plants some time before. This, however, is not as comforting a balance as it at first seems: when biomass is burned and is not rapidly replaced by regrowth, CO_2 is added to the atmosphere and remains there until it is removed by some other process. It can

then contribute to the CO_2 greenhouse effect in the atmosphere and to global climate change.

We must therefore distinguish between gross CO_2 emission from the burning of savannas and forests and the net CO_2 release from deforestation. While prominent in the emission of trace gases to the atmosphere, the periodical burning of savannas has almost no influence on the CO_2 greenhouse effect because the CO_2 released by burning is reincorporated into savanna vegetation during the next growth cycle—i.e., within about one year. In contrast, the clearing of tropical forests by burning contributes directly to the CO_2 greenhouse effect, since the CO_2 emitted from the oxidation of the large amount of biomass stored in a rainforest (as much as 600 tons per hectare, ton/ha) cannot be taken up again by the vegetation which will grow on the same site, usually grass or an agricultural crop. Furthermore, in the case of deforestation, not only does the CO_2 emitted directly by burning contribute to the atmospheric burden, but in most cases the CO_2 released during the decay of the unburned above-ground and below-ground biomass will be greater than that from the burning itself.

The various estimates of the net amount of CO_2 added to the atmosphere from deforestation fall into a relatively narrow range: Houghton et al. (1985) give an estimate of 900 to 2500 Tg C/yr, Detwiler and Hall (1988) suggest a range of 400 to 1600 Tg C/yr, and Hao et al. (1990) calculate 700 to 2000 Tg C/yr. This agreement should not instill undue confidence in the accuracy of these estimates, however. It mostly reflects the fact that all authors base their estimates on the same data: The estimates on land-use conversion by Myers (1980) and by Lanly (1982) and the biomass densities given by Whittaker and Likens (1975), Brown and Lugo (1984), and Detwiler et al. (1985). Hao et al. conclude that almost half of the net CO_2 released comes from the conversion of forest to permanent agriculture and cattle holdings. Other major contributions come from conversion of forests to shifting agriculture, fuel-wood production, and the conversion from virgin to secondary forests. Their calculations suggest that only some 15% of the net CO_2 emission is released immediately during the clearing burns. The remainder is released slowly over the following years either during repeated burning of residual material or through its microbial decomposition. Detwiler and Hall (1988) propose that an additional net CO_2 release of about 25% may result from the oxidation of organic matter in cleared forest soils due to intensive cultivation. Incorporation of this component results in an estimate of 1800 ± 800 Tg C/yr for the release of CO_2 from land clearing in the tropics. This is a significant amount compared with the present annual emission of CO_2 from fossil fuel burning—5200 Tg C/yr. We can therefore conclude that the destruction of tropical forests results in about 25% of the global CO_2 greenhouse effect.

Other Trace Gases

In addition to CO_2 a large variety of other gases and particles are emitted from the fires (Table 1.4); they are the products of incomplete combustion of carbon compounds, such as carbon monoxide (CO), methane (CH_4), and other hydrocarbons, and of compounds containing other nutrient elements, such as nitric oxide (NO) and sulfur dioxide (SO_2), from the nitrogen and sulfur in amino acids and proteins. Particulate matter (aerosol) in the smoke consists of organic matter, black (soot) carbon, and inorganic materials such as potassium carbonate and silica. In fact, the element potassium derives its name from having been isolated from wood-burning ash (potash), where it is present in the form of potassium carbonate. The emission ratios in Table 1.4 are the molar ratio between the species of interest and CO_2 in the smoke, with the exception of the aerosol emission ratios, which are expressed as grams per kilogram (g/kg) $C(CO_2)$. (The ratios have been multiplied by 1000 to make the numbers easier to read.)

Dry plant biomass consists of about 45% carbon (by weight), most of the remainder being hydrogen and oxygen. On a mass basis, the nutrient element contents are relatively low: about 0.3% to 3.8% nitrogen, 0.1% to 0.9% sulfur, 0.01% to 0.3% phosphorus, and 0.5% to 3.4% potassium (Bowen, 1979). Consequently, the emissions from biomass combustion are dominated by the oxides of carbon—CO_2 and CO. The fraction of CO emitted depends on the characteristics of the fire: hot, fast fires with a good supply of oxygen produce relatively little CO, whereas smoldering fires emit a large fraction of this product of incomplete oxidation (Crutzen et al., 1979; Ward, 1986). Pyrolysis and incomplete oxidation in oxygen-deficient fires are also responsible for the emission of CH_4, nonmethane hydrocarbons (NMHC), hydrogen gas (H_2), and various partially oxidized organic compounds, such as alcohols, aldehydes, ketones, and organic acids (Greenberg et al., 1984). Consequently, these compounds are also released preferentially during the smoldering stages. Figure 1.2 shows the sequence of the emission of CO_2 (maximum in the

Table 1.4 Emission ratios for trace gases and aerosols from biomass burning, based on field and laboratory studies

Species	Field studies (range)[a]	Laboratory studies (range)[b]	Best guess[c]
Gases[d]			
CO	65–140	59–105	100
Methane	6.2–16	11–16	11
NMHC[e]	6.6–11.0	3.4–6.8	7
Nitrous oxide	0.18–2.2	0.01–0.05	0.1
NO_x	2–8	0.7–1.6	2.1
Ammonia	0.9–1.9	0.08–2.5	1.3
RCN[f]	–	0.24–0.93	0.6
SO_x[g]	0.1–0.34		0.3
Carbonyl sulfide	0.005–0.016		0.01
Methylchloride	0.023–0.033	0.02–0.3	0.05
Hydrogen	33		33
Ozone[d,h]	4.8–40		30
Aerosols[i]			
TPM[j]	12–82		30
POC[k]	7.9–54		20
EC[l]	2.2–16		5.4
Potassium	0.24–0.58		0.4

a. Based on data from Andreae et al. (1988a, 1990a); Cofer et al. (1988, 1989); Crutzen et al. (1979, 1985); DeAngelis et al. (1980); Delmas (1982); Greenberg et al. (1984); Hegg et al. (1987); Hegg et al. (1988); Khalil and Rasmussen (1984); Miner (1969); Ward (1986); Winstead et al. (1990).
 The range shown is a range of means from these studies, not a range of individual measurements.
b. Based on data from Crutzen et al. (1990) and Rasmussen et al. (1980). The range shown is from individual burns of different types of biomass (savanna grasses, pine straw).
c. Based on a more or less subjective evaluation of the information in columns 1 and 2.
d. Results are expressed as moles of substance X emitted per 1000 moles of CO_2 emitted.
e. Nonmethane hydrocarbons (C_2–C_{10}).
f. $(CN)_2$ + HCN + CH_3CN + C_2H_5CN + C_2H_4CN.
g. SO_2 + aerosol sulfate.
h. Photochemically produced ozone in the burning plumes.
i. All data in g/kg $C(CO_2)$.
j. Total particulate matter.
k. Particulate organic carbon (including elemental carbon).
l. Elemental (black-soot) carbon.

flaming stage), CO (maximum in the early smoldering stage), and the organic acids (maximum in the late smoldering stage) from an experimental fire (D. Scharffe, W. Hartmann, M. O. Andreae, unpublished data).

Nitrogen is present in plant biomass mostly as amino groups (R-NH_2) in the amino acids of the proteins. During combustion it is released by pyrolytic decomposition of the organic matter and then partially or completely oxidized to various volatile nitrogen compounds. NO is the single most abundant

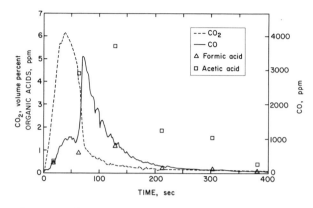

Figure 1.2 Concentrations of CO_2, CO, formic acid, and acetic acid in the smoke from an experimental biomass fire as a function of time. The flaming stage in this fire lasted for about 60 seconds. Concentrations are in volume percent for CO_2 and in volume mixing ratios (ppm) for the other species (1% = 10,000 ppm).

species emitted, but it represents only some 10% to 20% of the nitrogen initially contained in the fuel. Other nitrogen compounds (NO_2, N_2O, NH_3, HCN, organic nitriles and nitrates) account for another 10% to 20% of the fuel nitrogen, but some 60% to 70% of these compounds are released in as yet unknown forms, possibly as molecular nitrogen (Crutzen et al., 1990). Almost none of the fuel nitrogen is left in the ash. This is different for sulfur, about one-half of which remains in the ash (Delmas, 1982), probably due to the difference in the speciation of sulfur and nitrogen in plants. In contrast to nitrogen, which is almost exclusively present as amino groups in proteins, only part of the sulfur in plants is organically bound in the form of sulfur-containing amino acids in proteins (50% to 90%) (Turner and Lambert, 1980), and only this organic sulfur fraction is likely to be volatilized during combustion—mostly to SO_2 but to a small extent also to carbonyl sulfide (COS). The rest of the plant sulfur is in the form of sulfate and small amounts of sulfate esters that are pyrolyzed to sulfate during combustion. Sulfate is not volatile at the temperatures prevailing in biomass fires and is either retained in the ash or, to some extent, incorporated into smoke particles. The same applies to phosphorus, which is also present largely as phosphate esters that are broken down to nonvolatile phosphate during combustion. The ratio between the sum of identifiable nitrogen gas emissions and sulfur emissions (Table 1.4) is 14, very close to the N/S ratio in plant tissue, which is about 15 and quite constant from species to species (Thompson et al., 1970). This is

roughly consistent with the observations that about two-thirds of nitrogen is volatilized in still unidentified forms and that about one-half of the sulfur is left in the ash.

Table 1.5 combines the estimates of biomass burning rates in the tropical regions and worldwide (Tables 1.2 and 1.3) with a best estimate of the emission ratios (Table 1.4) to derive regional and global rates of pyrogenic emissions. The uncertainty of these estimates is about 50% in the case of CO_2, where the emission amount is relatively well constrained, and about a factor of 2 for most of the other gases, where additional uncertainty is contributed by the limited data base on emission ratios.

In spite of these uncertainties in the quantitative estimates, it is quite evident from this compilation that biomass burning results in globally important contributions to the atmospheric budget of several of these pollutant gases (Table 1.6). Since much of the burning is concentrated in relatively limited regions, such as the African savannas, and occurs over a limited time, it is not surprising that it results in levels of atmospheric pollution that rival and sometimes exceed those in the industrialized regions of the developed nations.

This applies especially to a group of gases that are the main ingredients of "classical" smog chemistry in the atmosphere—hydrocarbons, carbon monoxide, and nitrogen oxides (NO_x). As is discussed later, these species interact to form other pollutants, particularly ozone (O_3). About 20% to 30% of the global emissions of these compounds is attributable to biomass burning (Table 1.6). This budget ignores the natural emissions of the very short-lived terpenes and isoprene. Since only about half of the total source flux of these species is from anthropogenic emissions (the rest from natural sources), biomass burning accounts for roughly one-half of the anthropogenic sources of atmospheric hydrocarbons, CO, NO_x, and tropospheric ozone.

Methane (CH_4) and nitrous oxide (N_2O) contribute to the atmospheric greenhouse effect and are long-lived enough to enter the stratosphere and contribute to the ozone cycle there. Their pyrogenic emissions are on the order of about 10% of the global source flux. The emission ratios for both of these gases are uncertain, however, and it is quite possible that the pyrogenic fraction could be as high as 15% for methane. In the case of N_2O, on the other hand, recent evidence (Crutzen et al., 1990, Winstead et al., 1991) suggests that the emission ratios previously published may have been too high due to an experimental arti-

Table 1.5 Estimates of tropical and global emissions from biomass burning (Tg element/yr)

Species	Emission ratio	Tropical regions of				Tropics total	World total
		America	Africa	Asia	Australia		
Carbon burned[a]		780	1450	980	200	3410	3940
Gases[b]							
CO_2[c]		690	1270	860	180	3000	3460
CO	100	70	130	90	20	300	350
Methane	11	8	14	10	2	33	38
NMHC[d]	7	5	9	6	1	21	24
Nitrous oxide	0.1	0.16	0.30	0.20	0.04	0.70	0.81
NO_x	2.1	1.7	3.1	2.1	0.4	7.4	8.5
Ammonia	1.3	1.0	1.9	1.3	0.3	4.6	5.3
RCN[e]	0.6	0.4	0.9	0.6	0.1	2.1	2.4
SO_x[f]	0.3	0.6	1.0	0.7	0.1	2.4	2.8
Carbonyl sulfide	0.01	0.02	0.03	0.02	<0.01	0.08	0.09
Methyl chloride	0.05	0.10	0.19	0.13	0.03	0.44	0.51
Hydrogen	33	4	7	5	1	16	19
Ozone[b,g]	30	80	150	100	20	360	420
Aerosols[h]							
TPM[i]	30	21	38	26	5	90	104
POC[j]	20	14	25	17	4	60	69
EC[k]	5.4	3.7	6.9	4.7	1.0	16	19
Potassium	0.4	0.9	1.7	1.1	0.2	3.9	4.5

Note: Based on the emission ratio estimates from Table 1.4 and the estimates of the regional and global rates of biomass burning from Tables 1.2 and 1.3. All emission ratios are given in Tg element per year—e.g., in Tg N(NO_x) or S(SO_x) per year. The uncertainty of these estimates is about ±50% in the case of CO_2, about a factor of 2 for the other species.

a. From Tables 1.2 and 1.3.

b. Emission ratio expressed as moles of substance X emitted per 1000 moles of CO_2 emitted.

c. These values represent the actual amount of CO_2 emitted from the fires. They are 88% of the amount of carbon burned, since about 12% is emitted as CO, methane, and NMHC. These compounds will be eventually converted to CO_2 by photochemical oxidation in the atmosphere, however.

d. Nonmethane hydrocarbons (C_2–C_{10}).

e. $(CN)_2$ + HCN + CH_3CN + C_2H_5CN + C_2H_4CN.

f. SO_2 + aerosol sulfate.

g. Photochemically produced ozone in the burning plumes.

h. Emission ratio in g/kg C(CO_2).

i. Total particulate matter (emission rate in Tg/yr).

j. Particulate organic carbon (including elemental carbon).

k. Elemental (black-soot) carbon.

fact. Biomass burning may thus play only an insignificant role in the global budget of this species. Additional information on both the pyrogenic and the other sources of N_2O is needed before the budget of this species can be assessed with a reasonable amount of accuracy (Andreae and Schimel, 1989).

The global atmospheric budget of ammonia (NH_3) is not well known. Based on the discussion in Andreae et al. (1989) we estimate global emissions to be in the range of 20 to 80 Tg N/yr, with a best-guess estimate of 44 Tg N/yr. Microbial release from animal excreta and soils makes up the largest fraction of these emissions, while biomass burning contributes only some 12%.

In contrast to the nitrogen species, where py-

rogenic emissions are very important, only relatively small amounts of sulfur dioxide and aerosol sulfate are emitted. Biomass burning contributes only about 1.3% to the total atmospheric sulfur budget and represents about 3% of the anthropogenic emissions (Andreae, 1991). Still, since most of the natural emissions are from the oceans and most of the anthropogenic emissions are concentrated in the industrialized regions of the temperate latitudes, biomass burning does make a pronounced impact on the sulfur budget over remote continental regions, such as the Amazon and Congo basins (Andreae et al., 1988a; Andreae and Andreae, 1988; Bingemer et al., 1990). Here, sulfur

Table 1.6 Comparison of global emissions from biomass burning with emissions from all sources (including biomass burning)

Species	Biomass burning (Tg element/yr)	All sources (Tg element/yr)	Biomass burning (percent)	Reference for "all sources" estimate
CO_2 (gross from combustion)	3500	8700[a]	40	Bolin et al. (1986)
CO_2 (net from deforestation)	1800	7000[b]	26	Bolin et al. (1986)
CO	350	1100	32	WMO (1985)
Methane	38	380	10	Cicerone and Oremland (1988)
NMHC[c]	24	100	24	Ehhalt et al. (1986)
Nitrous oxide	0.8	13	6	Bolle et al. (1986)
NO_x	8.5	40	21	Logan (1983)
Ammonia	5.3	44	12	Andreae et al. (1989)
Sulfur gases	2.8	150	2	Andreae (1990)
Carbonyl sulfide	0.09	1.4	6	Khalil and Rasmussen (1984)
Methyl chloride	0.51	2.3	22	WMO (1985)
Hydrogen	19	75	25	Conrad and Seiler (1986)
Ozone	420	1100	38	Crutzen (1988)
TPM[d]	104	1530	7	Peterson and Junge (1971)
POC[e]	69	180	39	Duce (1978)
EC[f]	19	<22	>86	Turco et al. (1983)

a. Biomass burning plus fossil fuel burning.
b. Deforestation plus fossil fuel burning.
c. Nonmethane hydrocarbons (excluding isoprene and terpenes).
d. Total particulate matter (Tg/yr).
e. Particulate organic carbon (including elemental carbon).
f. Elemental (black-soot) carbon.

fluxes to the vegetation may be enhanced by as much as a factor of five due to tropical biomass burning.

Methyl chloride, CH_3Cl, is also released by biomass burning and has actually been proposed as a tracer for pyrogenic emissions (Rasmussen et al., 1980). This gas is of great significance for the stratospheric ozone budget; it is the second-largest source of stratospheric chlorine. The pyrogenic source corresponds to about 0.5 Tg Cl/yr (0.7 Tg CH_3Cl/yr), which is about 22% of the global emissions.

Emission of Aerosol Particles (Smoke)
Even though the smoke from biomass fires is one of the most obvious emissions, often visible in the tropics as a continuous pall hanging in the air for days or weeks, quantitative estimates on the amounts of particulate matter released are still highly uncertain. In Table 1.5 the emission of total particulate matter (TPM) is estimated as 104 Tg/yr, most of which is organic matter. Of the particulate organic carbon (POC) released (70 Tg/yr), about one-quarter (20 Tg/yr) is in the form of black, elemental (soot) carbon (EC). Based on these estimates, POC from biomass burning accounts for over one-third of the organic

carbon aerosol released globally (Duce, 1978). These estimates, however, are much lower than those of Seiler and Crutzen (1980), who suggested an emission rate of 200 to 450 Tg/yr for TPM, of which about 90% is assumed to be carbon compounds, and 90 to 180 Tg/yr for elemental carbon.

This discrepancy results from the large difference in the emission ratios used: Seiler and Crutzen's estimate is based on early work by Ward (personal communication to Seiler and Crutzen, 1979), who had suggested emission ratios of 24 to 180 g TPM/kg $C(CO_2)$. The value of 30 g TPM/kg $C(CO_2)$ is based on our work in Amazonia (Andreae et al., 1988a), the work by Ward (1986), and the data of Radke et al. (1988). In Amazonia, we sampled plumes that were quite "old" and had traveled for about one day following the burn and obtained apparent emission ratios of 6 to 25 g/kg $C(CO_2)$. Due to the delay between emission and sampling in our work, it is likely that a substantial amount of the aerosol in particles larger than a few micrometers in diameter had already fallen out. The emission measurements of Ward (1986) and Radke et al. (1988), on the other hand, were conducted in the immediate vicinity of the fires and were therefore not influenced by significant amounts of fallout. Ward reports TPM emission ratios in the range from 2 to 46 g/kg $C(CO_2)$, with average values of 22 and 37 g/kg $C(CO_2)$ for flaming and smoldering conditions, respectively. Based on 71 measurements from 10 different fires, Radke et al. (1988) derive a mean emission ratio of 44 ± 31 g/kg $C(CO_2)$ for aerosol particles of less than 2 μm diameter. As some of the particles sampled very closely above the fire are large enough to fall out almost immediately, it appears that a value of 30 g/kg $C(CO_2)$ is a reasonable estimate for the emission of particulate matter that may become subject to atmospheric transport.

Our measurements in Amazonia showed that the carbon content of this aerosol is about 66%, which is consistent with the aerosol particles consisting mostly of partially oxygenated organic matter. The content of black elemental carbon in smoke particles from biomass burning is highly variable. In smoldering fires it is as low as 4% (weight percent carbon in TPM), while in intensively flaming fires it can reach 40% (Ward, 1986; Patterson and McMahon, 1984; Cachier et al., 1989). We are using a value of 18%, based on our work in Amazonia (Andreae et al., 1988a). From this and the estimate for global TPM emissions in Table 1.5 (104 Tg/yr), we obtain a source estimate for black carbon aerosol of 19 Tg/yr. This accounts for a very large fraction of the global emissions, which have been estimated to fall into the range of 3 to 22 Tg/yr (Turco et al., 1983).

Charcoal Formation and Carbon Storage
In their 1980 paper, Seiler and Crutzen drew attention to an unexpected consequence of biomass burning: They argued that the formation of charcoal during burning would lead to the long-term sequestering of carbon in soils and sediments and therefore provide a sink for atmospheric carbon. In the absence of burning, photosynthesis and respiration in terrestrial ecosystems form a closed cycle in which just as much CO_2 is released to the atmosphere during the biochemical oxidation of the plant matter as was taken up during its production by photosynthesis. Thus, there is no net flux of CO_2 to or from the atmosphere. During combustion, however, some of the plant matter is turned into charcoal, which may either remain on the ground after the burn or be transported away in the smoke. Once formed, charcoal and soot carbon do not appear to be reoxidized back to CO_2, even over geological time scales. This process therefore provides a long-term sink for atmospheric CO_2 and, inci-

dentally, a net source of oxygen to the atmosphere. While this mechanism appears very plausible on a qualitative basis, its quantitative importance remains controversial.

The amount of charcoal and soot that becomes airborne during biomass fires can be estimated using the data on the elemental carbon component of the smoke aerosol given in Tables 1.4 and 1.5. The emission ratio estimate of 3.7×10^{-3} corresponds to about 1.4 g C per kg of burned biomass. Since only a fraction of the above-ground biomass is burned (20% to 80%, depending on vegetation type), the final fraction of biomass that becomes airborne as charcoal or soot particles is very small, probably less than 0.1%. This is consistent with the results of a study on the cycle of charcoal from biomass burning by Suman (1984, 1988), who estimated that only 5% of the charcoal produced in a coastal watershed reached sediments in the adjacent Gulf of Panama. Most of this charcoal is transported by rivers; only about 3% is carried by wind.

The main mechanism for charcoal storage must therefore be its incorporation into terrestrial soils and sediments. Seiler and Crutzen (1980) present a review of the literature and show that charcoal layers do indeed occur frequently in tropical and temperate soils. They estimate that 20% to 30% of the carbon present on the ground after a biomass fire is in the form of charcoal. At the burning efficiency of 25% assumed by Seiler and Crutzen, this would correspond to conversion of 15% to 22% of the preburn biomass into charcoal. Fearnside et al. (1990) measured the amount of charcoal produced in clearing fires in Amazonia and found a much lower rate of charcoal production—only about 3.6% of preburn above-ground carbon, which is about 20% of the value estimated by Seiler and Crutzen. As a result of these uncertainties, the range of estimates for the global rate of charcoal formation and the resulting sequestering of CO_2 is rather large—about 200 to 1700 Tg C/yr. If the actual value falls near the high end of this range, it could imply that as much as 20% of the approximately 7000 Tg C added per year to the world atmosphere may be removed by charcoal formation.

Outstanding Uncertainties
One of the most intriguing aspects concerning emissions from biomass burning is the fate of fuel nitrogen. It is well established from both field and laboratory studies that most of the nitrogen contained

in the biomass fuel is volatilized. However, only some 30% to 40% of the nitrogen released is in the form of chemical species (such as NO, NO_2, NH_3, and HCN) which are readily deposited and recycled into plant nutrients. Most of the remainder appears to be molecular nitrogen (Lobert et al., 1990), and therefore biomass burning may cause a substantial loss of nitrogen from tropical ecosystems.

The pyrogenic emission rate of N_2O, a trace gas whose atmospheric budget is very inadequately understood, also needs to be better defined. This is especially important since N_2O is a contributor to the greenhouse effect and a precursor of stratospheric NO_x, which in turn is important in regulating stratospheric ozone.

Another important issue is the production rate of charcoal during biomass combustion. Given the wide range of estimates, this process may be either a trivial sink of atmospheric CO_2 or large enough to almost compensate for the amount of CO_2 released due to deforestation.

Finally, much uncertainty persists regarding the rate of particulate emissions from biomass burning. Since aerosols can be observed and measured from satellites, they provide a potentially useful tool for the estimation of global rates of pyrogenic emission. It is therefore essential to have accurate data on the emission ratios and optical characteristics of smoke aerosols from biomass burning.

Transport and Photochemistry in the Smoke Plumes

Long-Range Transport of Smoke Plumes

The hot gases from the fires rise buoyantly in the atmosphere, entraining ambient air. Often, a cloud begins to form on the smoke plume. If this cloud produces rain, some particulate matter and soluble gases may already be scavenged and removed at this stage. In most cases, however, the clouds will dissolve again without rainfall, and the plume will start to drift horizontally with the prevailing winds. The height to which the smoke plumes can rise is limited in the tropics by the trade-wind inversion, which is present at an altitude of some 3 to 5 km. When the rising smoke plumes have become trapped at an inversion layer and start spreading horizontally, they will tend to become stretched out into relatively thin layers of great horizontal extent. Such thin haze layers can now travel very far from their point of origin without losing their identity. An example of one such layer, as ob-

served by Lidar in the middle of the Amazon Basin, some 1000 km from its source, is shown in Figure 1.3.

The further fate of the smoke-laden air masses depends on the large-scale circulation of the continent from which they originate. In tropical Africa, the plumes will usually travel in a westerly direction and toward the equator. As they approach the Intertropical Convergence Zone (ITCZ), vertical convection in the troposphere intensifies, destroys the layered structure, and results in the distribution of the pyrogenic emissions throughout the lower troposphere. Finally, in the ITCZ region, smoke and gases from biomass burning may be injected into the middle and upper troposphere, maybe even into the stratosphere. Air masses from the biomass burning regions in South America are usually moving toward the south and southeast, due to the effect of the Andes barrier on the large-scale circulation. Here, again, they may become entrained into a convergence zone, the seasonal South Atlantic Convergence Zone (SACZ). This convergence zone becomes established in austral spring, when biomass burning is abundant. The CO measurements from the Measurement of Air Pollution from Satellites (MAPS) satellite instrument tended to show a tendency for CO buildup in the middle to upper troposphere in the regions near the ITCZ and the SACZ (Reichle et al., 1986; H. G. Reichle, personal communication, 1989).

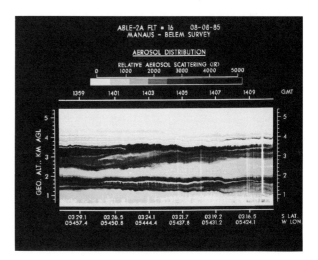

Figure 1.3 Vertical profile of haze layers from biomass burning as seen by Lidar over the central Amazon Basin. The vertical scale indicates the height above ground level; the horizontal scales, flight time (top) and geographical coordinates (bottom). Aerosol concentration based on infrared (IR) light scattering is indicated by a grey scale from low (white) to high (black). (Source: E. V. Browell, NASA/LARC)

Evidence for the efficiency of these transport mechanisms is abundant. The smoke plumes from the fires in South America and Africa are easily seen from space and have been the subject of investigations from the Space Shuttle. They have been mapped using various space-borne sensors, in particular the Advanced Very High Resolution Radiometer (AVHRR) instruments on the National Oceanic and Atmospheric Administration (NOAA) series satellites (Figure 1.1; Kaufman et al., 1990; Setzer and Pereira, 1990). At the ground, haze and smoke from fires some thousand kilometers away are still so dense that airports in the middle of the Amazon Basin are frequently closed during the burning season.

Less obvious than the visible smoke, but even more indicative of the large extent to which the emissions from burning are transported around the globe, are the results of chemical measurements from satellites, aircraft, and research vessels. Soot carbon and other pyrogenic aerosol constituents have been measured during research cruises over the remote Atlantic and Pacific (Andreae, 1983; Andreae et al., 1984). Ozone maxima in the ITCZ region over the Atlantic Ocean may be attributed to the presence of biomass burning plumes (Winkler, 1988). Carbon monoxide and ozone originating from fires in Africa were consistently observed during the NASA/CITE-3 research flights off the coast of Brazil (Andreae et al., 1990a). On an even larger scale, high levels of ozone and CO are seen from satellites over the tropical regions of Africa and South America, and large areas of the surrounding oceans (Fishman and Larsen, 1987; Reichle et al., 1986).

Smog Chemistry in the Smoke Plumes
The gases emitted in the biomass fires contain essentially the same constituents as the mixture which forms the starting material for urban smog: carbon monoxide, hydrocarbons, and nitrogen oxides. Once such a mixture is exposed to sunlight, the same chemical reactions take place, whatever its origin: Hydrocarbons are oxidized photochemically first to various peroxides, aldehydes, etc., then to CO. This CO is added to the amount directly emitted from the fires and is finally oxidized to CO_2. In the presence of elevated levels of NO_x, as will be the case in the smoke plumes, the oxidation of CO and hydrocarbons is accompanied by the formation of ozone (Crutzen, 1987). The efficiency of ozone formation— that is, the amount of ozone formed per molecule of hydrocarbon oxidized—depends on the proportions of hydrocarbons, NO_x, and O_3 present in the reaction

mixture and thus on the history of transport and mixing of the air mass involved (Chatfield and Delany, 1990).

Environmental Impacts

Ozone Pollution of the Troposphere
In view of the large amounts of hydrocarbons and NO_X emitted from biomass fires, it is not surprising that very high concentrations of ozone are produced in the plumes, often exceeding values typical of industrialized regions (Crutzen et al., 1985; Delany et al., 1985; Kirchhoff and Nobre, 1986; Logan and Kirchhoff, 1986; Cros et al., 1987, 1988; Andreae et al., 1988a, 1990a, 1991). The highest concentrations, typically in the range from 50 to 100 ppb, are usually found in discrete layers at altitudes from 1 to 5 km (Figure 1.4). Usually, the concentrations at ground level are substantially lower. They typically show a pronounced daily cycle, with a minimum at night and a maximum near midday. This cycle is controlled by the balance of O_3 sources and sinks: At night O_3 consumption by deposition to vegetation and by the reaction with NO emitted from soils reduces the concentration of ozone near the earth's surface; during the day, it is replenished by photochemical ozone

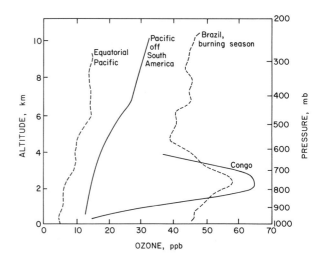

Figure 1.4 Vertical profiles of ozone in the tropical troposphere. The profile over the equatorial Pacific shows no influence from biomass burning (Fishman et al., 1987), while the profile over the Pacific off South America (Seiler and Fishman, 1981) suggests ozone enhancement due to long-range transport from the tropical continents. The ozone profiles over Brazil (Crutzen et al., 1985) and the Congo (Andreae et al., 1989a) show high ozone concentrations between 1 and 4 km altitude due to photochemical production in biomass burning plumes (figure modified from Fishman, 1988).

formation and downward mixing of ozone-rich air from higher altitudes. Over an unpolluted tropical rainforest, such as in the central Amazon Basin during the wet season, O_3 levels are very low—below 10 ppb at midday, and below 3 ppb at night (Gregory et al., 1990; Kirchhoff et al., 1990). Under somewhat more polluted conditions, e.g., in central Amazonia during the dry season, nighttime values are still very low, but daytime O_3 concentrations reach 10 to 30 ppb at the level of the tree crowns. Finally, in tropical Africa, where biomass burning currently has the strongest impact, surface ozone concentrations in excess of 40 ppb are frequently measured, especially during the dry season (Cros et al., 1988). These values are similar to those observed in the highly polluted regions of the eastern United States and of Central Europe (Logan, 1985, and references therein). Studies in temperate forest regions have linked such levels of ozone pollution to the damage to trees and vegetation that has become widespread in Europe and North America. In view of the sharp increase of O_3 with altitude frequently observed in the tropics, the likelihood of vegetation damage by ozone is especially high in mountainous regions, where the ground surface intersects the levels at which ozone concentrations above 70 ppb are encountered. Ozone episodes with ground-level concentrations of 80 to 120 ppb must be expected to occur in the tropics, particularly during the dry season, when photochemically reactive air becomes trapped under a subsiding inversion layer.

The increase of tropospheric ozone concentrations is a cause for concern even beyond the likelihood of plant damage by this gas: Ozone is also an efficient absorber of infrared radiation and thus acts as a "greenhouse gas" in the same way as CO_2. In fact, the global increase in tropospheric O_3 concentrations contributes about 15% to the overall greenhouse warming (Ramanathan et al., 1985). The increase of O_3 over the tropics contributes strongly to this effect: Photochemical ozone production from pyrogenic emissions accounts for about one-third of the total input of ozone into the troposphere (Table 1.6), about the same amount as produced from fossil fuel emissions in the industrialized regions of the Northern Hemisphere (the rest is due to stratospheric inputs). The ozone soundings from Natal (Kirchhoff and Nobre, 1986) and from the Brazilian savanna region (Delany et al., 1985) show that the effect of ozone production from biomass burning is quite pronounced even in the middle and upper troposphere, where it contributes most significantly to greenhouse warming. Consequently, biomass burning may be responsible for up to one-half of the greenhouse warming due to increasing tropospheric ozone (Figure 1.5).

Perturbation of Oxidant Cycles in the Troposphere

The large-scale change of tropospheric ozone levels that is being observed in the tropics and is expected to increase in the future is an indication of a fundamental change in the way the troposphere behaves chemically. Many gases, particularly hydrocarbons, are continuously emitted into the atmosphere from natural and anthropogenic sources. A buildup of these gases in the atmosphere is prevented by a self-cleaning mechanism, whereby these substances are slowly "combusted" photochemically to CO_2. The key molecule responsible for this oxidation process is the hydroxyl radical (OH). The reaction chains involved are such that ozone and OH are consumed when the concentration of NO_x is low. This is the "normal" condition of the unpolluted troposphere. Based on the observed increase of methane and CO in the atmosphere, Crutzen (1987) suggested the possibility of a global decrease in OH and O_3, leading through a feedback mechanism to a further increase in CH_4 and

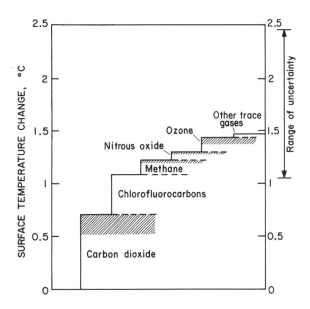

Figure 1.5 Cumulative equilibrium surface temperature warming ("greenhouse effect") due to increasing trace gas concentrations from A.D. 1980 to 2030 as computed by a one-dimensional climate model (after Ramanathan et al., 1985). Shaded areas indicate the effect attributable to deforestation and biomass burning. Due to climate feedback mechanisms the expected changes in global temperatures are 0.8 to 2.6 times the values given in this figure.

CO, and a possibly unstable condition of the atmosphere. Injection of large amounts of NO_x from biomass burning may reverse this process, since hydrocarbon oxidation in the presence of elevated amounts of NO_x *produces* additional amounts of O_3 and OH. The model calculations of Keller et al. (1991) predict a tenfold increase in regional OH concentrations to result from deforestation and biomass burning in the tropics. The consequences of driving a large fraction of the world atmosphere from its "natural" O_3-consuming mode of operation into the opposing, O_3-producing one are unforeseeable at this time.

Climate Change

We have already discussed the fact that the clearing of the tropical rainforests is responsible for about 25% of the greenhouse warming due to increasing CO_2 and that increasing tropospheric ozone as well adds to the greenhouse effect. But biomass burning also releases other gases that contribute to greenhouse warming, particularly methane and nitrous oxide. In the case of the latter two gases, biomass burning accounts for only about 10% of the global sources, but probably for a much greater fraction of the *increase* in global emissions, since many of the other sources, such as natural wetlands, are constant or actually shrinking. Based on the estimates of the temporal trends of methane source strengths from 1940 to 1980 given by Bolle et al. (1986), the pyrogenic contribution to the increase in methane emissions over that time period is 21%. The current uncertainty about the atmospheric mass balance of N_2O makes it very difficult to estimate the contribution of biomass burning to its increasing tropospheric concentration: It may be as high as 30%. The fraction of greenhouse warming attributable to biomass burning and deforestation is indicated as shaded areas in Figure 1.5.

The climatic effect of the smoke aerosols is beyond current understanding due to the complex nature of the interactions involved. Because of the limited lifetime of aerosol particles (on the order of days) compared to the lifetime of most of the "greenhouse gases," their climatic effects act initially more on a regional than a global scale. Since the equatorial regions, particularly the Amazon Basin, the Congo Basin, and the area around Borneo, are extremely important in receiving solar energy and in redistributing this heat vertically through the atmosphere, any change affecting the operation of these "heat engines of the atmosphere" must be viewed with great concern. Aerosols can influence climate directly through

changing the earth's radiation balance. They can reflect sunlight back into space and thus reduce the amount of heat the earth receives. Black-soot aerosol, on the other hand, may absorb sunlight and result in heating of the atmosphere. There appears to be a weak consensus by climatologists that these direct effects are not large at this time relative to other agents of climate change. A greater source of concern is the influence of pyrogenic particles on the behavior of clouds. Cloud droplets form on aerosol particles, called cloud condensation nuclei (CCN). The properties of the cloud depend on the number of available CCN: The more CCN, the more droplets form, which results in a smaller droplet size, given a constant amount of available water. Clouds made up of smaller droplets are whiter and reflect more sunlight back into space but are less likely to produce rain. Since clouds are one of the most important controls on the heat balance of the earth, any large-scale modification of cloud properties is likely to have a strong impact on global climate. The limited body of data that is available to date on the production of CCN by biomass fires suggests strongly that pyrogenic aerosol particles are very effective as CCN and that large numbers of such particles are released during burning (Eagen et al., 1974; Desalmand et al., 1982; Desalmand, 1987; Hallett et al., 1989; Rogers et al., 1990). Aging for up to 20 hours increased the percentage of aerosol particles that were effective as CCN to almost 100% (Hallett et al., 1989).

The potential changes in precipitation efficiency caused by increased abundance of CCN add to the perturbation of the hydrological cycle in the tropics caused by changes in the characteristics of the land surface brought about by deforestation and desertification. Tropical forests are extremely efficient in returning water that has fallen on them as rain back to the atmosphere in the form of water vapor. There it can form clouds and rain again, and the cycle can repeat itself many times. A region like the Amazon Basin can thus retain water (which ultimately comes from the ocean and will return there) for a long time and maintain a large "standing stock" of water. If the forest is replaced by grassland or, as often is the case, converted into an essentially unvegetated surface by erosion and loss of topsoil, water runs off quickly and returns through streams and rivers to the ocean without much chance of recycling. Beyond the unfavorable consequences that such large-scale changes of the availability of water will have on human activities, this modification of the hydrological cycle may itself perturb climate. The atmosphere makes use of the potential for storing energy in the form of water vapor and releasing this energy again when the vapor condenses into cloud droplets as an important mechanism for the redistribution of heat. If less water is available for this purpose, the heat absorbed at the ground has to be moved away by radiation and dry convection, leading to higher surface temperatures and a change in the vertical distribution of heat. Such perturbations of the heat and water balance of the atmosphere must be viewed with great concern.

Acid Deposition

After acid rain became a notorious environmental problem in Europe and North America, scientists were somewhat surprised to learn that it was widespread in the tropics as well (Table 1.7): Acid rain was reported from Venezuela (Steinhardt and Fassbender, 1979; Galloway et al., 1982; Sanhueza et al., 1989), from Brazil (Andreae et al., 1988b), from Africa (Lacaux et al., 1987, 1988; J. Lacaux, personal communication, 1989), and from Australia (Ayers and Gillett, 1988). In all instances, organic acids (especially formic and acetic acids) and nitric acid were shown to account for a large part of the acidity, in contrast to the situation in the industrialized temperate regions, where sulfuric acid and nitric acid

predominate. It was originally thought that the organic acids were largely derived from natural, biogenic emissions, probably from plants (Talbot et al., 1988; Andreae et al., 1988b; Keene et al., 1986). However, more recent evidence shows that they may be derived to a large extent from direct emission of acetic acid from biomass burning and to the photochemical formation of formic and acetic acid in the plumes (Helas et al., 1991). In regions affected by biomass burning, pyrogenic NO_x emissions overwhelm by far the natural sources of NO_x, e.g., soil emissions and NO formation in lightning bolts. Nitric acid is formed photochemically from the NO_x emitted in the fires. A modeling study simulating the effects of biomass burning and a moderate amount of additional pollution, mostly connected with the activities related to logging, etc., suggests that pH values of about 4.2 can be expected in the tropics due to the formation of nitric acid alone (Keller et al., 1991). The addition of organic acids to precipitation can be expected to drive pH down to values well below 4. Such values have indeed been frequently observed in Africa (J. Lacaux, personal communication, 1989). For comparison, the mean pH measured during the period 1963 to 1982 at the Hubbard Brook site in New Hampshire, a classical site for the study of acid rain,

Table 1.7 Rainwater pH and acid deposition at some continental tropical sites and in the eastern United States

	pH		Rainfall (cm)	Deposition (kg H^+/ha/yr)	Reference
	Mean[a]	Range			
Venezuela					
San Eusebio	4.6	3.8–6.2	158	0.39	Steinhardt and Fassbender (1979)
San Carlos	4.8	4.4–5.2	—	—	Galloway et al. (1982)
La Paragua	4.7	4.0–5.0	—	—	Sanhueza et al. (1989)
Brazil					
Manaus, dry season	4.6	3.8–5.0 ⎫	240[b]	0.29[b]	Andreae et al. (1990b)
Manaus, wet season	5.2	4.3–6.1 ⎬			
Australia					
Groote Eylandt	4.3		—	—	Langkamp and Dalling (1983)
Katherine	4.8	4.2–5.4	—	—	Galloway et al. (1982)
Jabiru	4.3		—	—	Ayers and Gillett (1988)
Ivory Coast					
Ayame	4.6	4.0–6.5	179	0.41	Lacaux et al. (1987)
Congo					
Boyele	4.4		185	0.74	Lacaux et al. (1988)
Eastern USA	4.3	3.0–5.9	130	0.67	Barrie and Hales (1984)

a. Volume weighted.
b. Annual average.

was 4.2 (Likens et al., 1984), and the mean pH in rain sampled throughout the eastern United States in 1980 was 4.3 (Table 1.7; Barrie and Hales, 1984).

Acidic substances in the atmosphere can be deposited to plants and soils either by rain and fog (wet deposition) or by the direct removal of aerosols and gases at the surface (dry deposition). In the humid tropics, wet deposition accounts for most of the deposition flux, whereas in the savanna regions, especially during the dry season, dry deposition dominates. Acid deposition has been linked to forest damage in Europe and the eastern United States. It is thought that such damage is usually caused by a combination of factors, including acid deposition, ozone damage, and the concurrent exposure to other sources of stress, such as drought or insect infestation (Bormann, 1985). Acid deposition can act on an ecosystem by two major pathways: direct damage through the deposition of acidic aerosols and gases on leaves, or soil acidification. The danger of leaf injury is serious only at pH levels below 3.5, which is rarely encountered in the tropics (McDowell, 1988). However, in view of the presence of the highly soluble gaseous nitric and organic acids from biomass burning, it is quite likely that fog and dew can reach such pH levels during the dry season in many regions. Tropical forests may be inherently more sensitive to foliar damage than temperate forests due to the longer average leaf life in the tropics, where the leaves of many trees are shed only after two or more years. The resulting longer exposure of individual leaves may make cumulative damage more pronounced (McDowell, 1988).

Many tropical soils are likely to be relatively resistant to rapid acidification due to their high sulfate adsorption capacity (McDowell, 1988). However, there are also large areas where soils are quite susceptible to acidification, such as in Venezuela (Sanhueza et al., 1988). Furthermore, even in relatively resistant soils, chronic exposure to acid deposition will eventually lead to soil acidification and associated problems, like leaching of aluminum, manganese, and other cations. Microbial processes in the soils are also at risk of being perturbed. Nitrogen cycling is likely to be influenced, both by decreasing pH and by the addition of nitrate and ammonium ions from wet and dry deposition.

Many species of tropical animals have stages in their life cycles where they live in rainwater collected in bromeliads or similar plants. Other species, like frogs and salamanders, depend on water collected between dead leaves, mosses, etc., to keep their skin moist. Acid deposition poses a serious risk to these animals, which are essential components of the forest ecosystem. This applies particularly to the insects, which are essential to many rainforest species for pollination (Baker et al., 1983).

Disruption of Nutrient Cycles and Soil Degradation

Tropical ecosystems, both natural and agricultural, are frequently deficient in nitrogen or sulfur or both (Sanchez, 1976). When an area is burned, a substantial part of the nitrogen and sulfur present in the ecosystem is volatilized. As long as these elements are deposited again relatively nearby, this would imply no net gain or loss. However, in the case of nitrogen, only some 40% of the fuel nitrogen can be accounted for in the emissions. If the remainder is indeed emitted as molecular nitrogen, a significant overall loss of nutrient nitrogen results. This possibility needs to be investigated further. The interactions between fire and grazing in grasslands add a further complication (Hobbs et al., 1991). The nitrogen loss due to burning is much greater from ungrazed than from grazed areas, since the latter have less above-ground biomass exposed to fire. This effect is large enough that grazing may control whether a burned grassland gains or loses nitrogen.

On the other hand, ecosystems that are not burned, such as remaining areas of intact rainforest, will receive an increased nutrient input. Studies of rainwater chemistry in the central Amazon Basin suggest that as much as 90% of the sulfur and nitrogen deposited there is from sources outside the forest ecosystem, with long-range transport of emissions from biomass burning playing a major role (Andreae et al., 1990b). The effects of such increasing inputs of nutrients to the rainforests are not known. It must be remembered, however, that the increased nutrient inputs are accompanied by an increase in acid deposition and ozone concentration. The complex effect of environmental perturbations on tropical forest ecosystems will be difficult to analyze.

Besides the immediate volatilization of nitrogen during the burns, long-term changes in the microbial cycling of nitrogen in the soils appear to result from the use of burning as a tool of agriculture (Anderson et al., 1988; Levine et al., 1988). NO and N_2O emissions from soils at experimental sites that were burned were about twice as high as from soils at unburned sites. This effect persisted for at least six months following the burn. Other studies have also shown enhanced fluxes of nitrogen oxides from soils following clearing of forest and conversion to grazing land (Matson et al., 1988), but in these studies the effect of burning was not isolated explicitly. Levine et al. (1990) found an increase in biogenic CH_4 and NO emissions from wetlands following vegetation burn-off. It is clear that much research still needs to be done in order to elucidate the effects of biomass burning on nutrient cycling and especially on nitrogen volatilization. The few studies available already do suggest, however, that deforestation and biomass burning contribute to the inputs of nitrogen oxides and the resulting perturbations of air quality and climate well beyond the direct emissions from the fires.

Perturbation of Stratospheric Chemistry and the Ozone Layer

Many of the trace gases released by biomass burning are involved in the chemical reaction cycles that control the stratospheric ozone layer. Methyl chloride, about 20% of which comes now from biomass burning, is the second-largest source of stratospheric chlorine. N_2O, for which biomass burning is also a possibly significant source, is long-lived enough to diffuse into the stratosphere, where it is broken down to reactive nitrogen species. Methane and hydrogen gas are oxidized in the stratosphere, forming CO_2 and H_2O. The fact that some biomass burning occurs in close vicinity to the Intertropical Convergence Zone is particularly relevant: In this region of intensive vertical convection, even relatively short-lived gases like nonmethane hydrocarbons may be injected into the stratosphere. COS is the only sulfur gas which is able to escape tropospheric oxidation and diffuse to the stratosphere. Here it is oxidized to form sulfuric acid aerosol particles, which scatter sunlight, and participate in the reaction cycles which control the abundance of stratospheric ozone (Oppenheimer, 1987). It is impossible at this time to predict what the overall effect of biomass burning on the fluxes of all of these species into the stratosphere and on the chemical reactions there will be. However, the potential for significant disruption of stratospheric chemical cycles is considerable enough to warrant further investigation by modeling and experiment.

Biomass Burning Studies and the International Global Atmospheric Chemistry (IGAC) Project

Ronald G. Prinn

Burning is a locally, regionally, and globally important biospheric phenomenon. Through burning, the chemical elements in vegetation are cycled back to the atmosphere and soil in chemical and physical forms and proportions which make biomass burning a distinct biogeochemical process. The process occurs naturally instigated by lightning strikes in dry vegetated regions. Today, however, the process is driven increasingly by human activity, which includes the burning of forested areas to facilitate land clearing, the burning of harvest debris to maintain cleared land, the extensive burning of natural grassland (particularly savannas) to sustain nomadic agriculture, and the burning of wood for heating.

Biomass burning has significant local and regional environmental effects, as depicted in Figure 2.1. A wide range of chemically reactive gases are produced during biomass burning, including carbon monoxide (CO), nitric oxide (NO), nitrogen dioxide (NO_2), many hydrocarbons, and ammonia (NH_3). These reactive gases influence strongly the local and downwind concentrations of the major atmospheric oxidants ozone (O_3) and the hydroxyl radical (OH). For example, the CO, NO, and NO_2 emissions drive the reactions

$$CO + OH \rightarrow H + CO_2$$
$$H + O_2 + M \rightarrow HO_2 + M$$
$$NO + HO_2 \rightarrow OH + NO_2$$
$$ultraviolet + NO_2 \rightarrow NO + O$$
$$O + O_2 + M \rightarrow O_3 + M$$

which convert CO and two O_2 molecules to CO_2 and O_3. The ammonium (NH_4^+) and nitrate (NO_3^-) compounds produced directly during burning or by subsequent oxidation of NO and NO_2 serve after deposition to fertilize the soil. The burned landscape has a distinctly different albedo and evapotranspiration rate than the preburn landscape, with consequent feedbacks to local and regional climate.

Biomass burning also has global impact. Andreae (1990) estimates the total biomass burned annually to be about 8.7×10^{15} grams of dry material (gm dm).

From this burning, annual emissions of about 3.5×10^{15} gm carbon as carbon dioxide (CO_2) and 4×10^{13} gm carbon as methane (CH_4) result (Andreae, 1990; Cicerone and Oremland, 1988). These emission rates can be compared to the 5×10^{15} gm carbon emitted annually as CO_2 from fossil fuel combustion and the 4×10^{14} gm carbon which must be emitted annually as CH_4 to balance its atmospheric accumulation plus chemical loss rates (Blake and Rowland, 1988; Prinn et al., 1987). Disturbance of soils resulting from biomass burning leads to global annual emissions of about 10^{12} gm nitrogen as nitrous oxide (N_2O) (Luizao et al., 1989; Anderson et al., 1988) compared to the total global annual emissions of about 1.3×10^{13} gm nitrogen as N_2O needed to balance the N_2O atmospheric accumulation plus loss rates (Prinn et al., 1990). The emissions from disturbed soils may also be the growing tropical N_2O source needed to help explain the observed N_2O trend and latitudinal gradient (Prinn et al., 1990). Because CO_2, CH_4, and N_2O are very long-lived, contribute significantly to the greenhouse effect, and influence very significantly the ozone layer (through stratospheric cooling or ozone destruction), the approximately 10% contributions of global biomass burning to the atmospheric inputs of N_2O and CH_4 and approximately 40% contribution of such burning to the total combustion input of CO_2 are clearly important. Note, however, that a significant fraction of the CO_2 from biomass burning is reincorporated into vegetation during regrowth.

The perturbations to local and regional atmospheric chemistry caused by biomass burning also have global significance. This is because several long-lived natural or anthropogenic atmospheric species such as CH_4, CH_3CCl_3 (an industrial solvent), $CHClF_2$ (a refrigerant), and CH_2FCF_3 (a proposed refrigerant) are destroyed predominantly by reaction with OH in the warm tropical lower atmosphere, which is the recipient of much of the emissions from biomass burning.

More than a decade has passed since the subject of biomass burning as not just a local but a global source

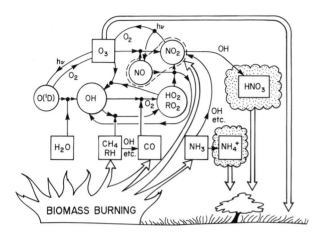

Figure 2.1 Schematic illustrating the local and regional atmospheric chemistry driven by emissions of hydrocarbons (CH_4 and RH where R is an alkyl, alkenyl, or aromatic free radical typically with 2 to 10 carbon atoms), carbon monoxide (CO), nitrogen oxides (NO, NO_2), and ammonia (NH_3) from biomass burning. Subsequent deposition of ammonium ions (NH_4^+), HNO_3, and O_3 has both beneficial and detrimental effects for the receptor ecosystems. The symbol $h\nu$ denotes a near-ultraviolet photon.

of trace gases was first discussed quantitatively by Crutzen et al. (1979). However, the paucity and difficulty of measurements in burning and burned regions, and the complexity of the ecosystem and combustion dynamics involved in biomass burning and its subsequent effects, has so far prevented acquisition of a clear understanding of the overall process. In light of this incomplete understanding, the biomass burning process and its effects on the atmosphere and biosphere have been recognized as one of several major research challenges facing atmospheric chemistry today. These several challenges are being met by an international group of atmospheric chemists and ecosystem biologists in an ambitious decadal global program called the International Global Atmospheric Chemistry Project (IGAC).

International Global Atmospheric Chemistry Project

The atmospheric concentrations of several trace gases (e.g., CH_4, N_2O, CO_2, and the chlorofluoromethanes $CFCl_3$ and CF_2Cl_2) are observed today to be increasing over the globe at rates that projected into the future will lead to major changes in both the chemical and radiative properties of the global atmosphere. The current understanding of these increases (particularly for CH_4, N_2O, and CO_2) is limited due to the chemical complexity of the atmosphere and its con-

nections to the oceans, the solid earth, and, most important, the biota. There is widespread agreement that therefore we need a major research program aimed at understanding quantitatively the chemical, physical, and biological processes that determine atmospheric composition, and we must use the knowledge gained from this program to address the past and future evolution of the earth's atmosphere.

The International Global Atmospheric Chemistry (IGAC) Project has been created recently by scientists from over 20 countries in response to the growing international concern about the above rapid atmospheric chemical changes and their potential impact on mankind. This project, while emphasizing atmospheric composition and chemistry, recognizes that the earth's atmosphere, oceans, land, and biota are interacting parts of a global system that determines the global environment and its susceptibility to change. The IGAC Project is therefore being carried out as a joint project of the Commission on Atmospheric Chemistry and Global Pollution (CACGP), which is an international organization focused on the atmospheric chemical part of this system, and the International Geosphere-Biosphere Program (IGBP), which is a broad-ranging interdisciplinary international undertaking that addresses the overall interactive system including the biota. The IGAC Project is intended to be a vital contributor to the broader interdisciplinary study of global environmental change, providing the important atmospheric chemistry component and recognizing its linkages with the biosphere and human activities.

The overall goal of IGAC is to develop a fundamental understanding of the natural and anthropogenic processes that determine the chemical composition of the atmosphere and the interactions between atmospheric composition and biospheric and climatic processes. A specific objective is to accurately predict changes over the next century in the composition and chemistry of the global atmosphere. The IGAC Project is broad and encompasses several urgent environmental issues, including the increasing acidity of rainfall, the depletion of stratospheric ozone, the greenhouse warming due to accumulation of trace gases, and the biological damage from increased oxidant levels.

The IGAC Project both utilizes and builds on existing national programs and begins new activities. IGAC specifically provides the international organization whereby essential scientific endeavors can be accomplished that involve large demands for physical efforts, technology, geographic coverage, or monetary resources beyond the capability of any single nation.

Atmospheric composition and chemistry and biosphere-atmosphere interactions are dependent on a number of climatic (temperature, pressure, humidity, cloud cover and opacity, precipitation, etc.) and biospheric (oceanic and terrestrial ecosystem sources and sinks, industrial sources, etc.) variables. The IGAC overall plan takes account of the fact that the spatial and temporal distributions of these variables are not uniform and for some (e.g., clouds and phytoplankton) the distributions are very patchy. There are also significant correlations between some of the patchy climatic and biospheric variables (e.g., green vegetation with precipitation, sea surface temperature with phytoplankton concentrations, etc.).

As a result of these patchy and sometimes correlated phenomena, it is essential to study a variety of regions over the globe with the full complement of the necessary measurements and associated theory being applied in each region. For this reason there are five major IGAC research areas (called "Foci") which address the boreal, midlatitude, tropical, marine, and polar regions (see Figure 2.2). These *regional* Foci are accompanied by a *global* Focus addressing global trends, distributions, sources, atmospheric transport, transformations, sinks, and modeling of trace gases and aerosols. This Focus includes the study over the globe of cloud condensation nuclei and their gaseous precursors and the role of these nucleii in determining cloud properties. A final Focus is on the *fundamental* experimental and theoretical work essential to all the other Foci—namely, measurement calibration, intercalibration, and intercomparisons; laboratory studies of fundamental molecular properties; and new instrument development. At the heart of each IGAC Focus are so-called Activities, which comprise achievable endeavors addressing the major IGAC goals and objectives. These Activities are proposed, planned, and implemented by a group of scientists, each expert in the area addressed by the Activity. The current Activities in IGAC are summarized in Table 2.1. The membership of the overall IGAC Project Steering Committee and the location of the IGAC Project Office are given in Table 2.2. Another facet of IGAC is cooperation where appropriate with other complementary IGBP or World Climate Research Programme (WCRP) Projects.

Marine Focus

The Marine Focus recognizes that the oceans are both major source and sink regions for various atmospheric

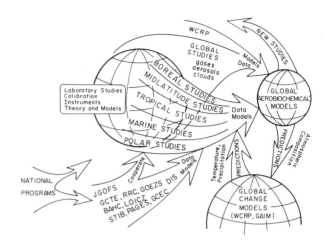

Figure 2.2 Schematic illustrating the major elements of the International Global Atmospheric Chemistry (IGAC) Project. Exchange of information and/or cooperation with several other existing or proposed projects of the International Geosphere-Biosphere Program (IGBP) are included in the IGAC plan. The other IGBP Projects are JGOFS = Joint Global Ocean Flux Study, GCTE = Global Change and Terrestrial Ecosystems, BAHC = Biological Aspects of the Hydrologic Cycle, STIB = Stratosphere-Troposphere Interactions and the Biosphere, RRC = Regional Research Centers, LOICZ = Land-Ocean Interactions in the Coastal Zone, PAGES = Past Global Changes, GOEZS = Global Ocean Euphotic Zone Study, GCEC = Global Change and Ecological Complexity, DIS = Data and Information Systems, and GAIM = Global Analysis, Interpretation and Modelling. The five regional IGAC foci (boreal, midlatitude, tropical, marine, polar), the IGAC global focus, the fundamental studies focus (laboratory, calibration, theory), and global chemical transport models developed through cooperation between the World Climate Research Program (WCRP) and IGAC will in sum provide the data and submodels for development of global aerobiochemical models. The aerobiochemical models predict atmospheric composition for use in global climate and biospheric change models under development by WCRP and GAIM. The global climate change models in turn predict future environmental variables (temperature, precipitation, etc.) for use in the regional and global aerobiochemical models (IGBP, 1990a,b; CACGP, 1989).

carbon, nitrogen, sulfur, and halogen compounds that influence climate, cloud microphysical properties, and atmospheric ozone. The chemistry of the marine atmosphere can be strongly affected by natural and anthropogenic substances carried from continental sources. In fact, the marine atmosphere constitutes a suitable environment to study the atmospheric chemical processes in "background air," isolated from the large and patchy features of the continental chemical inputs but still influenced by the integrated effects of these inputs. Furthermore, the marine environment can be substantially altered

Table 2.1 Current IGAC foci and activities and the leaders (convenors) of the groups of scientists involved in the activities

Focus	Convenor	Activity
Marine	F. Fehsenfeld, USA, and S. Penkett, UK	North Atlantic Regional Study[a]
	B. Huebert, USA	Marine Gas Emissions, Atmospheric Chemistry, and Climate[a]
	H. Akimoto, Japan	North Pacific Regional Study[b]
Tropical	C. Johannson, Sweden, and P. Matson, USA	Biosphere-Atmosphere Trace Gas Exchange in the Tropics[a]
	G. Ayers, Australia	Deposition of Biogeochemically Important Trace Elements[b]
	M. O. Andreae, Germany	Impact of Tropical Biomass Burning on the World Atmosphere[a]
	P. Crutzen, Germany	Chemical Transformation in Tropical Atmospheres/Interaction with the Biosphere[c]
	R. Cicerone, USA	Exchanges of Methane and Other Trace Gases in Rice Cultivation[c]
Polar	R. Schnell, USA	Polar Atmospheric Chemistry[b]
	R. J. Delmas, France	Polar Air-Snow Experiment[b]
Boreal	H. Schiff, Canada	High Latitude Ecosystems as Sources and Sinks of Trace Gases[a]
Midlatitude	W. Chameides, USA	Midlatitude Ecosystems as Sources and Sinks for Atmospheric Oxidants[c]
	J. Mellillo, USA, and K. Smith, UK	Exchanges of Trace Gases Between Midlatitude Terrestrial Ecosystems and Atmosphere[c]
	G. Robertson, USA	Midlatitude Ecosystems as Net Carbon Dioxide Sinks[c]
Global	R. Prinn, USA	Global Tropospheric Ozone Network[a]
	D. Ehhalt, Germany	Global Atmospheric Chemistry Survey[a]
	R. Charlson, USA	Chemical and Physical Evolution of CCN as Controllers of Cloud Properties[b]
	T. Graedel, USA	Development of Global Emission Inventories[a]
		Global Integration and Modelling[b]
	C. Keeling, USA	Global Tropospheric Carbon Dioxide Network[b]
Fundamental	F. Fehsenfeld, USA, P. Fraser, Australia, and C. Keeling, USA	Intercalibration/Intercomparisons[a]

a. Activity already begun.
b. Activity in detailed planning phase (should begin in one to three years).
c. Activity in conceptual phase (should move to detailed planning phase in one to three years).

when substances of continental origin are deposited on the ocean surface. Under the Marine Focus three Activities presently exist: the North Atlantic Regional Study and the East Asian/North Pacific Regional Study, which are devoted to the study of the influence of continental emissions on marine atmospheric chemistry and biogeochemistry, and the Marine Aerosol and Gas Exchange Study, devoted to understanding the physical and biogeochemical mechanisms controlling ocean-atmosphere exchange of trace species. The IGAC Marine Activities will include cooperative endeavors with the Joint Global Ocean Flux Study (JGOFS), which, like IGAC, is a project of IGBP.

Tropical Focus
The Tropical Focus recognizes that anthropogenic disturbances of the chemistry of the tropical atmosphere are especially important, as the removal of many atmospheric trace gases is determined by the high atmospheric temperatures and concentrations of hydroxyl radicals in the tropics. Large natural emissions of biogenic trace gases occur in tropical forests

and savannas, and these emitted gases play a large role in the regional photochemistry of the boundary-layer atmosphere in these ecosystems. These emitted gases or their photochemical products can also be transported rapidly to the free troposphere by moist convection and subsequently carried to other regions of the world, thus influencing extratropical atmospheric chemistry as well. In addition, tropical regions are experiencing rapid changes due to expanding agricultural and other activities often accompanied by the burning of large quantities of biomass. Continued change, especially due to population growth, will produce changes in the biogeochemical cycles of the affected tropical ecosystems with further repercussions for the emissions of trace gases. Under the Tropical Focus are five Activities (Table 2.1) devoted respectively to biosphere-atmosphere trace gas exchange, deposition of biogeochemically important trace species, biomass burning, chemical atmospheric transformations, and trace gas emission from rice cultivation. The tropical Activities will be carried out in close coordination with the IGBP Global Change and Terrestrial Ecosystems (GCTE) Project. I discuss

Table 2.2 IGAC project steering committee
and core project office

Chair	R. Prinn, Massachusetts Institute of Technology, USA
Vice-chair	P. Crutzen, Max-Planck Institut fur Chemie, Germany
Members	H. Akimoto, National Institute for Environmental Studies, Japan P. Buat-Menard, Centre des Faibles Radioactivities, France J. Dacey, Woods Hole Oceanographic Institute, USA R. Duce, University of Rhode Island, USA D. Ehhalt, Kernforschungsanlage Julich, Germany G. Farquhar, Commonwealth Scientific and Industrial Research Organization, Australia I. Galbally, Commonwealth Scientific and Industrial Research Organization, Australia V. Kirchoff, Instituto de Pesquisas Espacias, Brasil P. Liss, University of East Anglia, UK P. Matson, NASA Ames Research Center, USA H. Rodhe, University of Stockholm, Sweden E. Sanhueza, IVIC, Venezuela S. Wofsy, Harvard University, USA G. Zavarzin, Institute of Microbiology of Academy of Sciences, USSR
Project office	IGAC Project Office MIT, 54-1312 Cambridge, MA 02139 USA Tel: 617-253-4902 Fax: 617-253-0354

the important biomass burning Activity further in a later section.

Polar Focus

The Polar Focus recognizes that the Arctic and Antarctic glacial and oceanic regions are important in the global energy budget (through their influences on the planetary albedo) and on the global carbon dioxide budget (through cold, relatively dense CO_2-rich water sinking to great depth in the Norwegian and Weddell seas). Also, human activities in the Arctic and transport of polluted air from Europe and Asia are affecting the radiative properties and chemistry of the Arctic atmosphere, producing the "arctic haze" phenomenon. In addition, the massive Greenland and Antarctic glaciers contain a record of past atmospheric composition in trapped air bubbles going back 150,000 years or more. It is these trapped air bubbles which have yielded the evidence for the remarkable correlation between atmospheric temperature (as determined from the isotopic composition of the ice) and three major long-lived natural greenhouse gases

(CO_2, CH_4, N_2O). By understanding the manner in which atmospheric species are transferred to and incorporated into these glaciers, we can usefully extend these analyses to short-lived gases and particles and interpret them in terms of past atmospheric chemical states. Two Activities are currently planned in this Focus (Table 2.1). The first is devoted to the study of the chemistry and radiative properties of the present-day polar atmospheres and the second to the study of the transfer processes from the atmosphere into snow and glacial ice.

Boreal Focus

The forests, lakes, and wetlands, covering vast areas of North America, Scandinavia, and the Soviet Union, have significant influences on global climate and atmospheric composition and are the subject of the IGAC Boreal Region Focus. Climate modeling implies that warming due to increases in greenhouse gases will increase toward the poles, which could strongly enhance CO_2 and CH_4 emissions from northern wetlands and permafrost regions. These regions also emit other chemically active compounds and act as significant recipients for trace gas and aerosol species transported from midlatitudes. Emissions of trace gases like CH_4 are sensitive to changes in soil temperature and the amounts of water and organic matter in the soil. Under this Focus, one Activity is underway devoted to understanding the role of these regions as sources and sinks of gases and the ecosystem dynamics controlling these fluxes. The field component of a major joint Canada-U.S. effort (the Northern Wetlands Study) carried out in northern Canada as a part of this Activity has already been completed.

Midlatitude Focus

The Midlatitude Focus addresses the temperate region, particularly of the Northern Hemisphere, which is densely populated, with most of its ecosystems subject to strong past and present human disturbances, including conversion of forests to grasslands and agricultural lands. These midlatitude ecosystems experience extreme atmospheric chemical conditions such as high ozone concentrations and acid deposition due to industrial emissions. Strong chemical interactions occur between gases of industrial and biological origin (such as NO_2 and isoprene and terpenes), which lead to enhanced ozone concentrations with detrimental influences on the affected ecosystems. These processes are already being studied in some industrialized countries, but the investigations need to be expanded to other regions and extended to include their impact on overall nutrient cycling and trace gas exchange in the relevant ecosystems. In addition to the disturbances caused by atmospheric inputs, the roles of various agricultural activities (including heavy use of fertilizers) on trace gas emission and uptake need to be better determined.

Global Focus

The differences in the composition of the atmosphere over the globe and the short- and long-term variations in this composition are the net effect of several atmospheric and biospheric processes: biospheric emissions, atmospheric circulation, atmospheric chemical transformations, and finally deposition and its feedbacks, if any, to emissions. The rates of all of these processes can also be affected by climate changes. Thus the global distributions and trends of chemically, radiatively, and biologically important atmospheric species are signatures of the global distributions and changes in these controlling processes. The Global Focus in IGAC addresses this important globally integrating research area. Five Activities currently exist under this Focus, which are devoted, respectively, to implementation of a global tropospheric ozone network, development of a global chemical climatology for the major species involved in fast atmospheric chemistry (e.g., NO, NO_2, hydrocarbons, etc.), global studies of the chemical and physical evolution of cloud condensation nuclei, development of global industrial emissions inventories, and finally, global integration and modelling. A sixth Activity devoted to a global network for carbon dioxide (CO_2) is under consideration. The various observational Activities under this Focus will utilize fixed sites, aircraft, balloons, and satellites as observational platforms.

Fundamental Focus

All of the above foci of the IGAC Project have important common elements which form a Fundamental Focus for IGAC. The first element is theory and modeling, where a major challenge is the development of accurate chemical transport models which IGAC will carry out through cooperation with WCRP. The second element is laboratory determinations of fundamental molecular properties, including absorption cross-sections, rate constants, and homogeneous and heterogeneous reaction mechanisms. The third element is new instrument development, where major challenges are provided by the need to accurately measure highly reactive free radicals and a wide variety of key species at very low concentrations. The

fourth element is education, both to increase the number of scientists contributing to the Project and to inform the public of the rationale, goals, objectives, and accomplishments of the program. Thus far, one important Activity needing international cooperation has been formed under this Focus. This Activity is devoted to development of the most accurate calibrations for atmospheric measurements and the adoption of compatible measurement systems through a series of intercalibrations and intercomparisons. Of particular concern here are calibrations for several hydrocarbons and halocarbons, and carbon dioxide.

Impact of Biomass Burning on Atmospheric Chemistry and Biogeochemical Cycles

One of the major Activities in IGAC is a study of the Impact of Biomass Burning on Atmospheric Chemistry and Biogeochemical Cycles being carried out by a multinational science team headed by Professor M. O. Andreae of the Max Planck Institute for Chemistry. This Activity is a series of tasks aimed at (1) characterizing the production of chemically and radiatively important gaseous and aerosol species during biomass burning, (2) assessing the consequences of biomass burning on regional and global atmospheric chemistry and climate, (3) determining the short-term and long-term effects of fire on postfire exchanges of trace gases between terrestrial ecosystems and the atmosphere, and (4) understanding the biogeochemical consequences of atmospheric deposition of products of biomass burning.

As a part of this Activity, efforts are underway to construct worldwide regional inventories of the areal extent of burning and the quantities of biomass burned both at present (using in situ and remote sensing techniques) and also in the past (using techniques such as lake sediment analyses). Also, the quantities of biomass which potentially can be burned in the relevant terrestrial ecosystems will be determined. This inventory data together with existing and future measurements of the chemical and isotopic composition and yields of gases and particulates from burning at representative sites over the globe is expected to provide the necessary accurate estimates of the contributions of emissions from biomass burning to the global budgets of each species.

The above emissions measurements relevant to determining global budgets will also provide the basis for development of realistic fire dynamics and combustion chemistry theories and models. These are essential to explain the measurements, aid the global integration exercise, and provide predictive capability for future emissions due to biomass burning.

The emissions from fires alter significantly the subsequent chemistry of the affected air masses. These altered air masses need to be studied as they move away from the burning regions to elucidate their chemical evolution, their delivery of trace species to the global atmosphere through mixing, and their inputs to downwind ecosystems through deposition. Initially ground-based and aircraft measurements will play an important role in these studies with measurements of CO, CH_4, and perhaps other trace gases from satellites expected to play a major role later in this decade. One IGAC experiment planned for the 1992 to 1993 time frame is the Southern Tropical Atlantic Regional Experiment (STARE). Using aircraft and surface measurements, STARE will investigate the Brazilian and African biomass burning sources and the subsequent atmospheric transport and chemical processes leading to the observed enhanced ozone levels over the southern tropical Atlantic ocean. This experiment, which will also investigate the trace species sources and sinks in the relevant preburn and unburnt ecosystems, will be carried out jointly with the IGAC Activity on Biosphere-Atmosphere Trace Gas Exchange in the Tropics.

The effects of fires on subsequent (multiyear) ecosystem nutrient dynamics and accompanying emissions of trace species is of particular interest for N_2O and other nitrogen oxides. There is a need for longterm measurements in various ecosystems with different fire histories. These measurements together with experiments in artificially burned sites can be used to develop understanding and ultimately predictive models for the relevant ecosystem nutrient dynamics leading to the postburn trace gas emissions which are often significantly enhanced relative to the preburn values.

This Activity, like the other IGAC Activities, is being conducted by forming groups of scientists interested in participating in both the planning and implementation of the scientific investigation. Initial support for the field work will come primarily from funding agencies in the developed countries represented on the science team, but mechanisms for additional international support to enable further expansion of the science teams, particularly in the developing countries, is being actively sought.

Concluding Remarks

Research, particularly over the past decade, has led to the conclusion that atmospheric chemicals present in very small amounts but increasing at significant rates can have major importance for global climate and environmental changes over the next century. The need for accurate predictive capability is apparent, but realization of this capability is hampered by a lack of sufficiently accurate quantitative information about many of the atmospheric chemical and biospheric processes controlling atmospheric composition. While significant national programs addressing this subject exist in some countries, the scope and nature of the needed research demands resources, personnel, and global access that no one nation possesses. The IGAC Project has been created in response to this demand. Biomass burning is an example of an important biospheric process with considerable significance for atmospheric chemistry. The IGAC Activity devoted to the study of biomass burning and its effects on the atmosphere requires a cooperative effort involving many scientists working in field programs in several countries and over the downwind ocean regions and is an excellent example of the potential of international cooperation in global environmental research.

Acknowledgments

I thank G. Rodriguez for help in manuscript preparation. My research in atmospheric chemistry is supported by grants from the National Science Foundation and the National Aeronautics and Space Administration.

Astronaut Observations of Global Biomass Burning

Charles A. Wood and Raymond Nelson

It is a traditional practice in cultures all over the world to burn forests, farm stubble, pastures, and grasslands to clear land, improve pasturage, destroy weeds, aid hunting, and reduce fire hazards (Bartlett, 1956). Only in the last decade, however, has the extent of such deforestation and biomass burning been an important scientific concern because of potentially deleterious effects to the environment and reduction of species diversity (e.g., Nordin and Meade, 1982; Morgan, 1987). One of the most fundamental inputs for understanding and modeling possible effects of biomass burning is knowledge of the size of the areas burned. Because the burns are often very large and occur on all continents (except Antarctica), comprehensive and consistent measurements can come only from observations from space. Low-resolution weather satellite images and occasionally high-resolution Landsat images have been used to monitor deforestation, especially in the Amazon region (Malingreau and Tucker, 1988; Malingreau et al., 1987; Nelson and Holben, 1986). We present here information on another method for monitoring biomass burning, including immediate (e.g., smoke plumes and deforestation) and long-term (increased erosion) effects. Examples of astronaut photography of burning during one year give a perspective of the widespread occurrence of burning and the variety of biological materials that are consumed. The growth of burning in the Amazon region is sketched over 15 years using smoke as a proxy for actual burning, and possible climate effects of smoke palls are also discussed.

Astronaut Observations of the Earth

The first systematic remote sensing of Earth from space was accomplished by astronauts during the *Mercury* and *Gemini* spacecraft programs from 1961 to 1966 (Lowman, 1966). The scientific value of astronaut hand-held photography was recognized immediately and led ultimately to the development of the Landsat satellite remote sensing program. Although most remote sensing from space is now done with Landsat and other satellites, the U.S. space program has continued to include provisions for astronauts to document photographically the phenomena that they witness from Earth orbit. About 65,000 hand-held photographs of Earth have been taken by U.S. astronauts since Project *Mercury,* with about 85% having been acquired by Space Shuttle astronauts since 1981.

As has been documented elsewhere (Wood, 1989; Helfert and Wood, 1989), the value of astronaut photography is not simply in its resolution or areal coverage (which are comparable to Landsat MSS and TM images), but in the ability of the human photographer to select and isolate specific regions or phenomena of interest and capture their characteristics on film in real time. Thus, astronauts commonly document dynamic phenomena such as volcanic eruptions, dust storms, plankton blooms, and meteorological anomalies that are all short-lived and can be only accidentally imaged by satellites programmed by ground controllers. The value of the orbiting observer in documenting biomass burning lies in the human ability to (1) recognize and photograph previously unknown regions of burning, (2) find a hole in clouds to photograph deforested ground, and (3) realize that smoke palls are just as significant as a deforested landscape in documenting burning. Perhaps the most important contribution of the astronaut photographs, however, is in dramatically conveying the reality of a burning planet to scientists and laypeople alike, so that the phenomena can gain the attention necessary to foster a careful evaluation of its significance.

Examples of Biomass Burning Photography

Biomass burning can be observed by astronauts, not as glowing fires or thermal anomalies but as smoke plumes or, where individual plumes coalesce, as smoke palls. Prior burning can be detected by burn scars, which show up as sharply bounded albedo changes. A variety of plumes and burn scars are illustrated by Helfert and Lulla (1989) and Wood et al. (1989). Figure 3.1 shows the commonly observed continuum from individual smoke plumes to massive smoke hazes and smoke palls, with examples from Ethiopia, the USSR, Australia, and Zaire. Burn scars are most spectacular in desert scrub regions where burning makes a pronounced albedo marking. Figure 3.2 shows a picture of a fire scar in eastern Mongolia, approximately 50 km by 8 km, taken 31 October 1985. Other burn scars are visible—both recent dark ones and faint albedo features from older scars. Such scars provide excellent opportunities to measure the surface areas of biomass burns.

Global Burning in 1985: A First Estimate

No accurate information exists for the numbers, sizes, or locations of biomass burns on Earth, but an evaluation of one year's astronaut photography provides an indication of its widespread occurrence. We have searched through the photography for 1985, during which nine Shuttle flights occurred (Table 3.1). During the 70 days when Shuttle astronauts were in Earth orbit, they took 135 photographs of different fire plumes in 31 countries (Table 3.2). In some areas where fires commonly have been observed in other years, and probably did occur in 1985 (e.g., the Sahel), none were photographed in 1985; thus this listing is not a complete cataloging of all areas of burning in 1985.

The majority of the observed fires occurred in Latin America, with many fires also photographed in Africa (Figure 3.3). Note that any photograph may show a single fire, dozens of fire plumes, or continental scale plumes (Figure 3.4); thus we have no estimate of the number or area of fires. Table 3.3 lists the number of photographs showing fires on each continent during the 70 days of astronaut missions. These statistics are biased, however, by the fact that only three of the nine Shuttle missions overflew regions of the earth with latitudes higher than 28.5°N or 28.5°S; thus there was unequal opportunity to photograph fires in high latitudes. To correct for this, Table 3.3 also includes an adjusted list of fire photographs where the numbers of pictures taken in latitudes higher than 28.5° are multiplied by three. These data indicate that roughly one-third of the burning photographed in 1985 took place in Latin America, followed by Africa (23%), United States (18%), Europe (12%), and Asia (11%). These figures should be considered only order-of-magnitude values because of the difficulties of equally monitoring various parts

of the globe. For example, in Asia, where widespread deforestation is reported (Malingreau et al., 1987), two extraneous factors severely limit astronaut photography: persistently cloudy weather, and Shuttle mission profiles that typically limit observation of eastern Asia (including Indonesia and Papua New Guinea) because of crew sleep requirements.

Biomass Burning in the Amazon, 1973 to 1988: Where There's Smoke, There's Fire

Astronaut photography permits examination of the development of biomass burning in single regions over time, complementing the quasi-global snapshot for 1985. The best example is for Brazil, Bolivia, Paraguay, and Argentina, where *Skylab* photography from 1973 provides a temporally distant baseline, and Shuttle images document changes during the 1980s (Table 3.4). On the map in Figure 3.4 we have plotted the areas of smoke identified on astronaut photographs.

1973

In four days of astronaut observation of the Amazon and nearby regions in the burning season of 1973 (Figure 3.5a) only 10 photographs show fires. The burning is in southern Mato Grosso and Mato Grosso do Sul states of Brazil and further south near the Argentine-Paraguay border. Examination of the photographs shows that the burning occurs largely in agricultural fields, not as a result of deforestation. The area occupied by the smoke is estimated (by plotting boundaries of the smoke-covered areas on 1:1,000,000 maps) to be about 290,000 km^2. This and subsequent estimates are probably accurate within 10% to 15%.

1983

Unfortunately, after the 1973 information, no astronaut photography of the Amazon region exists until the eighth mission of the Space Shuttle in late August to early September 1983. At this time many fires were seen in eastern and southeastern Brazil and northern Argentina (Figure 3.5b). The 49 photos taken during seven days reveal that most fires represent the burning of agricultural stubble; the combined area of smoke plumes is about 340,000 km^2.

1984

Exactly one year after the 1983 photography, the extent, location, and nature of burning in South America had changed dramatically (Figure 3.5c). The

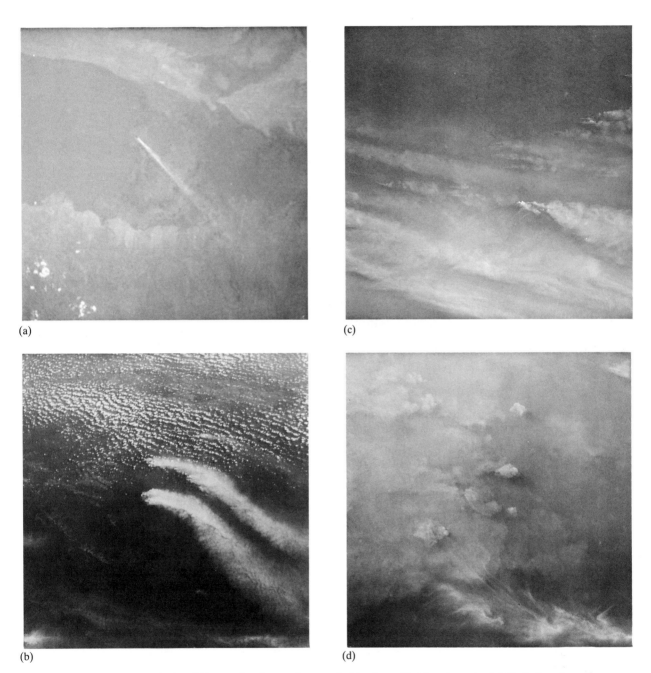

Figure 3.1 Illustrations of varieties of biomass burning as photographed by Space Shuttle astronauts. (a) Single fire plume from burning of presumed swamp grass near the tip of the Omo River delta, which is prograding into Lake Turkana/Rudolf in northern Kenya. Photo number STS28-78-96, August 1989. (b) Plumes from three groups of fires northeast of Novosibirsk, USSR, near 56.1°N, 77.3°E. Photo number STS28-84-90, August 1989. (c) Strong winds blow smoke from more than a dozen fire groups in Arnhem Land, Australia, toward the Indian Ocean. Photo number STS28-100-22, August 1989. (d) Massive smoke haze almost completely obscures the land surface in the south Congo basin (spacecraft was over 10.3°S, 24.1°E when photo was taken) in September 1988. Individual large fires causing the haze are visible as roughly circular concentrations of smoke. Photo number STS26-43-008.

Figure 3.2 Example of a large fire scar photographed from space: a 50 km long scar in Eastern Mongolia, near 47°N, 111°E. STS61A-32-046.

Table 3.1 Shuttle flights in 1985

Mission	Dates	Inclination	Number of fire photos
51C	Jan. 24–27	28.5°	7
51D	April 12–19	28.5°	12
51B	April 29–May 6	57.0°	11
51G	June 17–24	28.5°	25
51F	July 29–Aug. 6	49.5°	7
51I	Aug. 27–Sept. 3	28.5°	42
51J	Oct. 3–7	28.5°	7
61A	Oct. 30–Nov. 6	57.0°	12
61B	Nov. 27–Dec. 3	28.5°	12

26 photographs from 1984 include some individual fire plumes, but broad oblique views reveal that by early September an 1,880,000 km² smoke pall had formed. The location of burning had shifted from settled agricultural regions in eastern Brazil and Argentina to forest clearing in Rondonia and western Mato Grosso states in Brazil, as well as in Bolivia and Paraguay.

1985

The largest smoke pall ever photographed in South America, and probably in the world, occurred in late August to early September 1985. The 3,525,000 km² pall extended from Rondonia into Bolivia, Paraguay, Argentina, and out over the south Atlantic Ocean (Figure 3.5d). About 50% of the burning was for forest clearing, and the remaining was for agricultural and pastoral purposes.

1988

Because of the hiatus in Shuttle flights following the *Challenger* accident the next available astronaut photography is from late September to early October 1988. Once again a continental scale (3,000,000 km²) smoke pall was observed (Figure 3.5e), with an estimated 65% of the fires being to clear forests, with the remaining for agricultural purposes. Mission parameters for Shuttle flights did not permit photography

of the Amazon region during the burning season in 1989.

The drastic increase in area of smoke coverage from 1983 to 1984 in the Amazon and nearby regions of South America (Table 3.3; Figure 3.6) probably stems from the opening up of western Brazil for colonization; 1984 marks the paving of the famous highway BR-364 from Cuiaba in southwestern Mato Grosso to Porto Velho in northeastern Rondonia (Fearnside, 1989). Extensive deforestation is well known along this highway and from its offshoots (Malingreau and Tucker, 1988; Helfert and Lulla, 1989).

An interesting correlation (Figure 3.7) is hinted at between the area of observed smoke plumes or palls discussed above and the area of actual deforestation, based on data from Malingreau and Tucker (1988). The surprise is that the area of smoke observed on a few days in a given year should correlate with the area of deforestation, which is cumulative from the onset year of burning. The most likely explanation is that because the rate of deforestation is nearly exponential (Fearnside, 1989), the area of new deforestation each year dominates the cumulative area. This relation suggests that the more readily observed smoke area can be used for rough estimates of deforestation in regions where detailed measures have not yet been made.

Speculations on Biomass Burning and Climate and Weather Changes

Climate models have been run to study possible effects due to the deforestation of the Amazon. A general circulation model (GCM) run by Lean and Warrilow (1989) found that the change of large areas from rough, low albedo forests to smoother and

Table 3.2 Locations of biomass burning: Space Shuttle photography, 1985

Africa	Asia	Latin America	USA	Europe
Zaire	India	Mexico	USA-FL	Hungary
Sudan	Indonesia	Honduras	USA-MT	Greece
Zambia	China	Nicaragua	USA-GA	Italy
Tanzania	Japan	Venezuela	USA-LA	Turkey
Angola	Australia	Paraguay	USA-NY	USSR
Mozambique		Bolivia	USA-OR	
Madagascar		Brazil (27)	USA-CA	
Uganda		Peru		
Republic of South Africa		Ecuador		
Central African Republic		Argentina		

Note: Underlined: five to seven different fire photos.

Figure 3.3 Map showing the distribution of Space Shuttle astronaut photographs taken during 1985 that depict biomass burning fires. Dots represent individual fire photograph locations, and dark pattern highlights nations with photographed fires.

(a)

(b)

Figure 3.4 Examples of 1985 fire photos: (a) Northern Corsica, July 1985; STS51F-40-082. (b) Angola, June 1985; STS51G-46-77.

Table 3.3 Frequency of biomass burning: Space Shuttle photography, 1985

Africa	Asia	Latin America	USA	Europe
Observed (only 70 days in orbit)				
39	13	62	14	7
Corrected (9 Shuttle flights in 1985, only 3 at latitudes > 28°)				
39	19	62	32	21
23%	11%	36%	18%	12%

Table 3.4 Areas covered by Amazon smoke

Year	Smoke area (km^2)	Material burned
1973	290,000	Agricultural stubble
1983	340,000	Agricultural stubble
1984	1,880,000	Forest
1985	3,525,000	Forest (50%), agricultural (50%)
1988	3,000,000	Forest (65%), agricultural (35%)

brighter pastures results in reductions of evaporation, rainfall, and runoff, and increases in temperature. Similar results were independently obtained, including a lengthening of the dry season, by numerical modeling by Shukla et al. (1990).

These and other published models are based on the effects of replacing forest with fields or pastures, where the albedo increases from about 12% to 18% to 20%. We propose that another more extreme climate effect may be driven by the seasonal replacement of forest and fields by giant smoke palls with albedos of perhaps 50%. The existence of continental-scale smoke palls (e.g., Figure 3.8) with lifetimes of two to four months is a new phenomenon in the world, existing only since about 1984. GCMs need to be run to assess whether these palls will lead to even greater seasonal decreases in rainfall and increases in temperature than predicted by the current models.

Acknowledgment

We thank Mike Helfert for introducing us to the problems of deforestation and for sharing information and references.

Figure 3.5 The areas of smoke plumes and palls photographed over the Amazon and nearby regions over 15 years. (a) Locations of center points (black dots) of photos showing areas (pattern) of smoke photographed by Skylab 3 astronauts on 8 August and 3 and 4 September 1973. (b) As above for 30 August to 5 September 1983 based on photographs taken by Shuttle astronauts. (c) As above for 30 August to 5 September 1984. Three-sided boxes represent field of view of high oblique photographs. (d) As above for 27 August to 3 September 1985. (e) As above for 29 September to 2 October 1988.

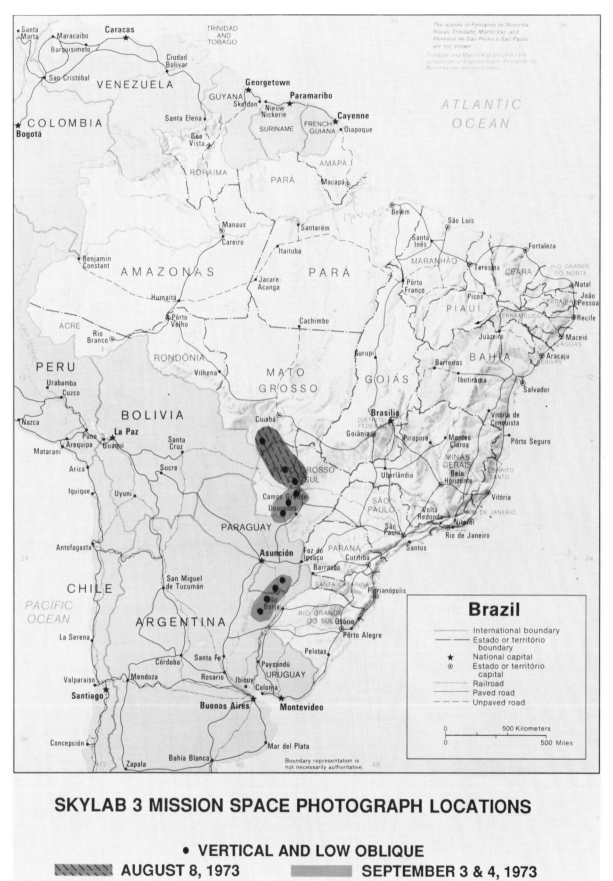

SKYLAB 3 MISSION SPACE PHOTOGRAPH LOCATIONS

- VERTICAL AND LOW OBLIQUE

▨▨▨ AUGUST 8, 1973   SEPTEMBER 3 & 4, 1973

Figure 3.5   (a)

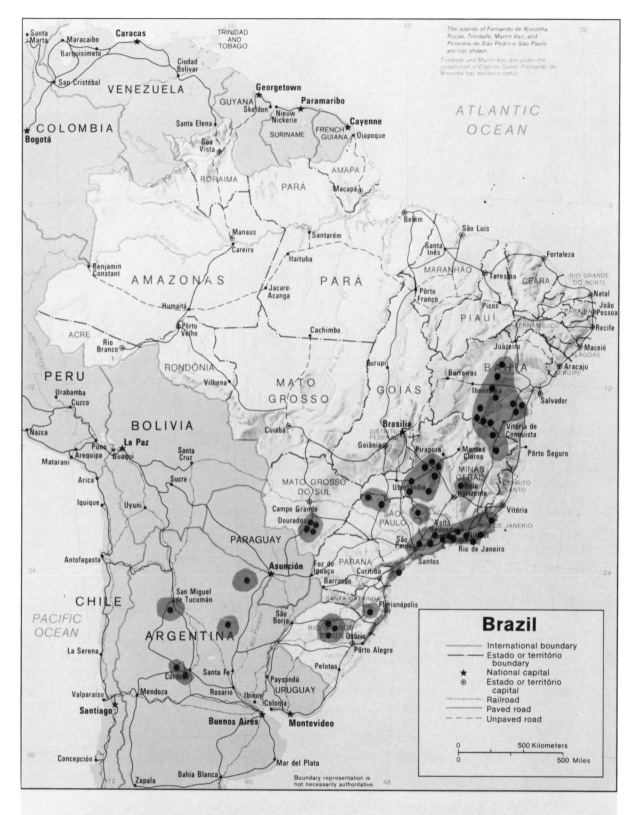

STS 8 MISSION SPACE PHOTOGRAPH LOCATIONS

• **VERTICAL AND LOW OBLIQUE**

Figure 3.5 (b)

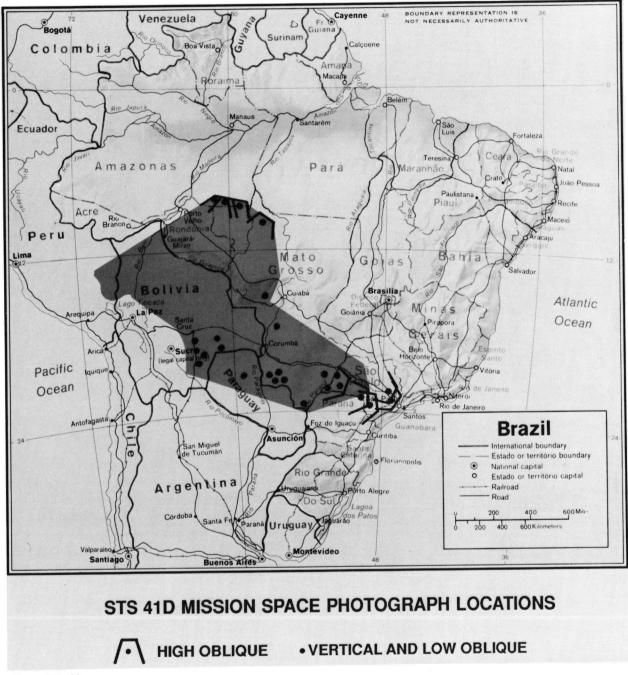

STS 41D MISSION SPACE PHOTOGRAPH LOCATIONS

 HIGH OBLIQUE   • VERTICAL AND LOW OBLIQUE

Figure 3.5 (c)

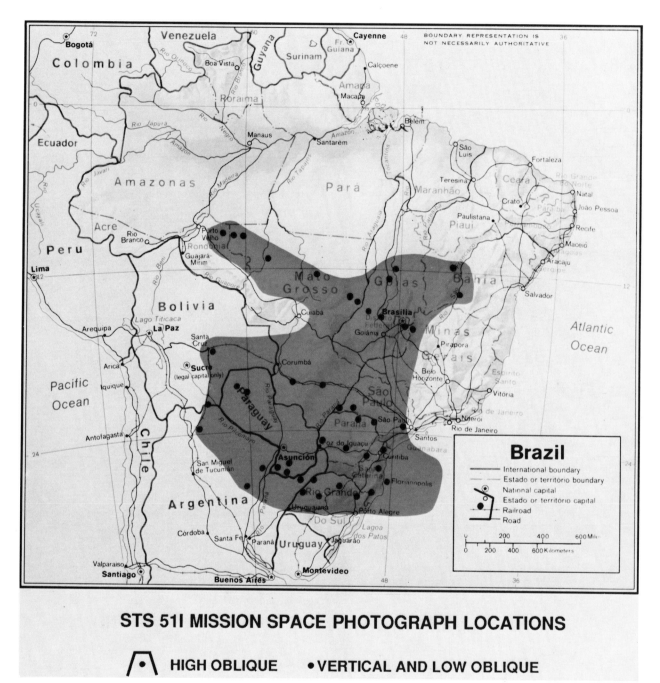

STS 51I MISSION SPACE PHOTOGRAPH LOCATIONS

⋀• HIGH OBLIQUE   •VERTICAL AND LOW OBLIQUE

Figure 3.5 (d)

STS 26 MISSION SPACE PHOTOGRAPH LOCATIONS

⚠ • HIGH OBLIQUE   • VERTICAL AND LOW OBLIQUE

Figure 3.5 (e)

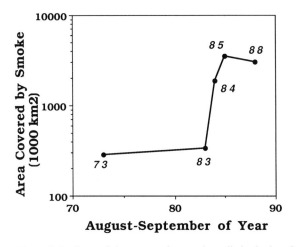

Figure 3.6 Area of Amazon-region smoke palls in the burning season from 1973 to 1988. Estimate uncertainties are probably less than 15%, which is minor considering the logarithmic scale.

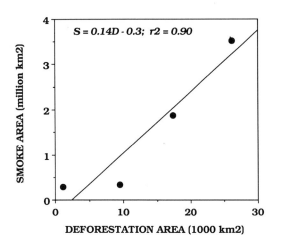

Figure 3.7 Areas of smoke in the Amazon and adjacent regions compared to estimates of areas of deforestation (the latter data from Malingreau and Tucker, 1988).

Figure 3.8 Part of the 3,000,000 km^2 giant smoke pall photographed over the Amazon by Shuttle astronauts in September 1988. The pall stretches nearly 1000 km to the high Andes Mountains in the background. STS26-38/38-014T.

Geostationary Satellite Estimation of Biomass Burning in Amazonia during BASE-A

W. Paul Menzel, Elen C. Cutrim, and Elaine M. Prins

During the past decade there has been increased human intervention in the tropics and subtropics, specifically in the Amazon Basin of South America. Each year intense biomass burning is associated with tropical deforestation in the selva (forest) and agricultural burning in the cerrado (grassland) regions of Brazil (Malingreau and Tucker, 1988; Setzer and Pereira, 1989; Booth, 1989). The implications of these activities are unknown, but some immediate concerns include loss of species, increased surface albedo and water runoff, decreased evapotranspiration, and higher concentrations of several greenhouse gases and large aerosols (Crutzen et al., 1985; Fishman et al., 1986; Matson and Holben, 1987; Andreae et al., 1988).

The most practical and economically feasible manner of monitoring the extent of burning associated with tropical deforestation and grassland management is through remote sensing. Considerable efforts have been focused on using the National Oceanic and Atmospheric Administration (NOAA) polar orbiting Advanced Very High Resolution Radiometer (AVHRR) with some success (Tucker et al., 1984; Matson and Holben, 1987; Malingreau and Tucker, 1988). Unfortunately, the AVHRR provides limited opportunity to monitor diurnal variations in the burning, and many of the fire observations saturate the sensor. This chapter presents the results of using Geostationary Operational Environmental Satellite (GOES) Visible Infrared Spin Scan Radiometer Atmospheric Sounder (VAS) infrared window (3.9 and 11.2 microns) data to monitor biomass burning several times per day in Amazonia. The technique of Matson and Dozier (1981) using two window channels was adapted to GOES VAS infrared data to estimate the size and temperature of fires associated with deforestation in the vicinity of Alta Floresta, Brazil, during the Biomass Burning Airborne and Spaceborne Experiment–Amazonia (BASE-A). Although VAS data do not offer the spatial resolution available with AVHRR data (7 km versus 1 km, respectively),

this decreased resolution does not seem to hinder the ability of the VAS instrument to detect fires; in some cases it proves to be advantageous in that saturation does not occur as often. VAS visible data are additionally helpful in verifying that the hot spots sensed in the infrared are actually related to fires. Furthermore, the fire plumes can be tracked in time to determine their motion and extent. In this way, the GOES satellite offers a unique ability to monitor diurnal variations in fire activity and transport of related aerosols.

Technique Description

As described in Matson and Dozier (1981), different brightness temperature responses in the two infrared window channels can be used to estimate the temperature of a target fire as well as the subpixel area it covers. As the temperature increases, the radiances for the shorter wavelengths increase much more rapidly than for the longer wavelengths. When the GOES VAS radiometer senses radiance from a field of view (FOV) containing a target of blackbody temperature T_t occupying a portion p (between zero and one) of the FOV and a background of blackbody temperature T_b occupying the remainder of the FOV, $(1 - p)$, the following equations represent the radiance sensed by the instrument at 4 and 11 microns, respectively (provided there is no atmospheric contribution or attenuation).

$$L_4(T_4) = p\, e_4\, L_4(T_t) + (1 - p)\, e_4\, L_4(T_b) + (1 - e_4)\, L_{4refl}$$
$$L_{11}(T_{11}) = p\, e_{11}\, L_{11}(T_t) + (1 - p)\, e_{11}\, L_{11}(T_b)$$

where

$L_j(T)$ is the integrated upwelling thermal radiance ($mW/m^2/ster/cm^{-1}$),

e_j represents the emissivity for channel j,

L_{4refl} is the reflected solar radiance in the shortwave channel.

Investigations suggest an emissivity for tropical rainforest of 0.96 in the 4 micron region and 0.97 in the 11 micron region (American Society of Photogrammetry, 1983). If T_b is known, these two nonlinear equations can be solved for T_t and p using numerical iteration techniques.

Since the radiance sensed by the GOES VAS instrument passes through the intervening atmosphere, corrections must be introduced to T_b, T_4, and T_{11}. Burning or smoldering fires are usually covered by clouds and smoke containing organic particles of varying sizes and shapes, necessitating a correction to the transmittance. Most of the smoke is composed of water vapor, but there are other constituents as well (Andreae et al., 1988). The 11 micron channel is more affected by atmospheric water vapor than the 4 micron channel. With *Nimbus*-2 data, it was found that the water vapor correction for a moist atmosphere is approximately 4°K at 300°K for the 11 micron window and 2°K at 300°K for the 4 micron window (Smith et al., 1970). By calculating a linear regression relationship between the brightness counts and brightness temperature in a variety of haze conditions (approximately 50) and extrapolating to clear sky conditions, the *Nimbus* corrections were found to be appropriate for the VAS data studied here (Prins, 1989).

The algorithm solution proceeds as follows. A fire is identified, and a nearby fire-free area is located. All observed brightness temperatures over the fire are adjusted for smoke attenuation; 2°K is added to the 4 micron brightness temperatures, and 4°K is added to the 11 micron brightness temperatures. T_4 and T_{11} represent the smoke corrected brightness temperatures over the fire for the two spectral bands. The background temperature, T_{b11}, is determined from the 11 micron brightness temperature in a nearby fire-free pixel (p = 0) that is adjusted for surface emissivity. Subsequently, the shortwave window reflected radiance for this same fire-free pixel, L_{4refl}, is solved. The input parameters T_{b11}, T_4, T_{11}, and L_{4refl} are now all in place; thus p and T_t can be solved.

Additional Considerations

Identifying Fires in the Satellite Image
The process of fire identification is strongly dependent on image interpretation. In this work, a fire was suspected if the haze corrected 3.9 micron observed brightness temperature showed a 4°K increase over the 3.9 micron background temperature and the haze corrected 11.2 micron observed temperature displayed a 1°K increase above the 11.2 micron back-

ground temperature. Hot spots were not considered fires unless they were accompanied by some indication of a fire in the visible channel, such as a smoke plume. The temporal resolution of the GOES VAS data proved to be extremely useful in two ways. First, a hot spot was often evident in the infrared channels at a certain time period, but the corresponding visible channel showed little or no indication of a fire, possibly due to obscuration by a plume produced by a fire located upwind. A visible image a half hour earlier usually clarified whether a fire was located there. Second, it is difficult to distinguish regular clouds from plumes in a single image. By looping a series of half-hourly visible images it becomes quite easy to identify the point sources of fires, since they remain at a constant location over time.

Spatial Resolution of the Satellite Sensor
The spatial resolution of GOES data is a limiting factor in this work; at nadir it is 0.9 km by 0.9 km in the visible, 6.9 km by 6.9 km in the 11.2 micron channel, and 13.8 km by 13.8 km in the 3.9 micron channel. With the noise limitations of the sensor in a typical scene (approximately 0.2°K for both infrared windows) (Menzel et al., 1983), GOES VAS data can resolve a fire with a target temperature of 450°K and area of 0.03 km^2 (for AVHRR the resolvable area is 0.00015 km^2). However, as mentioned earlier, the lack of horizontal resolution of the VAS also serves as an advantage in that the sensor does not saturate as often as the AVHRR does over fires. VAS saturates when a fire of 450°K occupies 5 km^2 of the field of view; for AVHRR this shrinks to 0.02 km^2. While VAS misses small fires that AVHRR senses, it measures the radiances of large fires that saturate the AVHRR.

The effects of sensor resolution were investigated for three cases of biomass burning observed at the same location at approximately the same time (within an hour) in NOAA AVHRR and GOES VAS images on 24 August 1988. Figures 4.1a and 4.1b show the AVHRR and VAS shortwave window (4 micron) images, respectively. The target temperature and area burned were calculated for three isolated fire areas in both images. Intercomparison was hindered by saturation in the AVHRR for large fires and poor detection by VAS for small fires; also, the time difference of one hour can reveal considerably different actual burning areas. The AVHRR estimates of burning range from several tenths of km^2 (using the algorithm in the Technique Description section) to several km^2 (assuming the saturated pixels are totally on fire). VAS falls between these extremes. Table 4.1 summa-

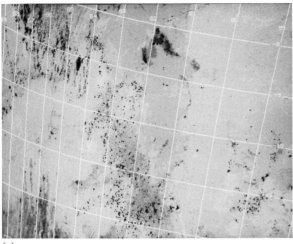

(a)

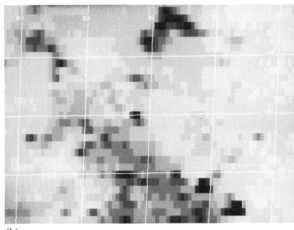

(b)

Figure 4.1 (a) AVHRR infrared channel 4 at 1942 UTC for 24 August 1988, 1 km resolution. The black pixels represent the hot spots; (b) GOES-7 infrared 3.9 micron window at 1830 UTC for 24 August 1988, 1 km resolution. Darker pixels show hot spots.

rizes the intercomparison. It is evident from the AVHRR 1 km data that the area covered by a VAS pixel includes a number of fires which act collectively to produce an average temperature and total area burned.

Cloud and Haze Contamination

The algorithm presented in the previous section is very sensitive to corrections for cloud and haze conditions. Figure 4.2 shows fire temperature and area as a function of smoke correction; as the correction ranges from 0 to 6°K, the area of the fire ranges from 0.1 to 2.2 km^2 while the temperature of the fire ranges from 635° to 440°K, respectively. Clearly, accurate determi-

Table 4.1 Comparison of AVHRR and VAS for selected fires in 1988

	AVHRR/burning area (km^2)		VAS/burning area (km^2)	
Fire	Number of saturated pixels	Estimate hot[a]	Estimate 450[b]	Estimate nominal[c]
A	7	0.30	1.76	1.69
B	6	0.19	1.39	1.21
C	11	0.27	1.61	0.92

a. "AVHRR estimate hot" is calculated using the technique of Matson and Dozier (1981) on only the hot pixels in the fire area.
b. "VAS estimate 450" is calculated assuming the fire temperature is 450°K.
c. "VAS estimate nominal" is calculated using the technique of Matson and Dozier (1981).

nation of the smoke correction is essential; the procedure outlined in the Technique Description section offers confidence within 1 K, which results in a relative error of roughly 50% in the area estimate.

To avoid the uncertainty associated with the smoke correction, a simplified algorithm has also been tested. It proceeds as follows. After identification of a fire, assume a fire temperature of 450°K. Determine the background temperature T_{b11} from a water vapor corrected 11 micron brightness temperature in a nearby fire-free (and smoke-free) pixel that is also adjusted for surface emissivity. Solve for L_{4refl} for this same fire-free pixel. Use T_4 and L_{4refl} to solve for p (assuming $T_t = 450°K$). The fire area now represents the area that a 450°K fire would occupy for the observed radiances; it is a relative measure of the fire area for a representative fire temperature and much less sensitive to the difficult smoke correction. Table 4.1 shows reasonable fire area results from the simplified 450 algorithm with respect to AVHRR estimates (the less saturation in the AVHRR, the better the comparison).

Results

The algorithm outlined in the section on Technique Description has been implemented on GOES VAS data for 1349 UTC, 1649 UTC, 1831 UTC on 4 September 1989 and 0149 UTC on 5 September 1989 during the BASE-A. The area considered covers 4 to 10S and 48 to 60W around Alta Floresta. The analysis provides a unique view of the diurnal variability of fire detection and burned area estimates for this region of the Amazon in 1989. Visible satellite images for 4 and 5 September 1989 are considered in conjunction with

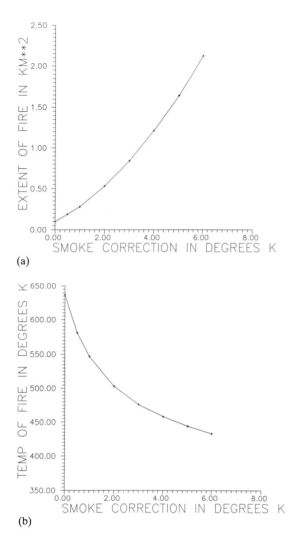

(a)

(b)

Figure 4.2 (a) Fire area as a function of smoke correction. (b) Fire temperature as a function of smoke correction.

infrared images for the South American continent in an attempt to determine the prevailing circulation and transport of constituents associated with biomass burning by tracking smoke plumes and clouds. Figures 4.3a to 4.3c show the visible, longwave infrared (11.2 micron) window, and shortwave infrared (3.9 micron) window images from 1830 UTC 4 September 1989. The fires are not apparent in the longwave window image but are very noticeable as dark spots in the shortwave infrared image.

Diurnal Variability

Fires detected on 4 and 5 September 1989 are presented in Table 4.2 and reveal a diurnal cycle in the number of fires observed; the maximum number was detected at 1649 UTC (one hour past noon) and the

(a)

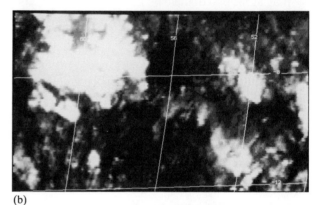

(b)

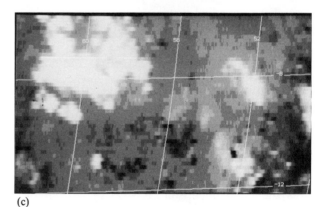

(c)

Figure 4.3 (a–c) GOES-7 visible, longwave infrared window and shortwave infrared window images, centered at Alta Floresta, from 1830 UTC 4 September 1989. Arrows on visible image indicate low-level circulation inferred from tracking smoke in sequence of GOES images.

Table 4.2 Fire temperature and burned area statistics within 4°-10°S and 48°-60°W around Alta Floresta (for one day during BASE-A in 1989)

Date	Time	Region	Number of fires	Mean fire temperature (°K)	Total area burned (km²)	Total area burned (km²) if $T_t = 450°K$
4 Sept. 89	1349	Selva	10	452	16.0	15.9
	1649	Selva	45	522	37.3	69.8
	1831	Selva	30	510	19.3	25.6
5 Sept. 89	0149	Selva	0	—	0.0	0.0

minimum at 0149 the following day. No fires were detected at 0149. This is due most likely a cessation in burning at the end of the day; the actual burning lasts only a few hours, so the fires disappear regularly in the GOES VAS image at 2131 UTC. This strong diurnal variation in detectability of fires was noted on many other days that were inspected. It appears that the optimum time to monitor biomass burning is between 1530 UTC and 1830 UTC (noon to early afternoon local time).

Burned Area Estimates

Table 4.2 also presents the fire statistics calculated from GOES VAS data. Not all fires detected were processed due to a variety of problems; on the average, temperature and area estimates were made for 90% of the detected fires. The mean fire temperatures are representative of low temperature or smoldering fires. The largest burned area in BASE-A was detected at 1649 UTC on 4 September 1989 and is estimated at 37.3 km². It is significant to note that this value is twice as large as those obtained at 1349 UTC and 1831 UTC. For those satellites which only offer one view of this region per day, it is imperative that these diurnal variations are considered.

The results from the simplified algorithm which assumes a fire temperature of 450°K are presented in the last column. Reasonable agreement is apparent; the same trends stand out, and the area of burning is usually within a factor of two. The simplified algorithm shows potential for routine application.

Aerosol Transport

Fires are evident throughout the BASE-A region (see Figure 4.3a). By tracking the smoke and haze in time, it is possible to infer the prevailing circulation and transport of aerosols. Winds at 1831 UTC on 4 September 1989 (see Figure 4.3a) were estimated from a half-hourly sequence of three visible and infrared (11 micron channel) images; a southerly flow through the Amazon Basin is indicated. This aerosol transport

Figure 4.4 GOES-7 visible image at 1301 UTC for 25 August 1988. Arrows indicate prevailing circulation inferred from tracking haze in sequence of GOES images.

was short-lived, as rain occurred in the area and the aerosols were washed out of the atmosphere. In drier conditions, it has been found (Prins and Menzel, 1990) that easterly winds in the northern portion of the Amazon Basin transport the aerosols westward. The Andes Mountains act to deflect the material south where westerlies transport the material over the Atlantic Ocean. Here they can be channeled to such distant regions as Antarctica (Andreae et al., 1988). Figure 4.4 shows a situation (from Prins and Menzel, 1990) where this circulation is prevalent. The haze transported from the continent over the Atlantic Ocean appears to be higher than the cirrus clouds, indicating aerosol presence high in the atmosphere (possibly as high as 10 km).

Conclusions

The GOES VAS is very useful for monitoring biomass burning. Comparison with AVHRR data reveals that the reduced spatial resolution of the GOES VAS does not severely hinder its ability to detect subpixel burning. The high temporal resolution of GOES VAS proves to be extremely useful in locating plumes associated with small fires. It also offers the unique opportunity to determine diurnal variability in the detection of biomass burning and the transport of related aerosols. Results indicate that the optimum time to monitor biomass burning is around 1630 UTC (1:30 P.M. local time). Burned area estimates determined from satellite measurements three hours earlier or later are a factor of two less. Under favorable conditions, aerosols associated with biomass burning are capable of being transported to distant areas and considerable heights. Convective activity or isentropic flow can transfer the material into the stratosphere and act to influence global atmospheric chemistry and climate.

The GOES VAS has demonstrated the capability to estimate the transport of biomass burning aerosols. However, the technique of Matson and Dozier (1981) applied to VAS data to estimate fire area and temperature is currently too cumbersome and time consuming to be used operationally; a reliable smoke correction must be determined for each fire. However, a simplified algorithm which assumes a fire temperature of 450°K shows some promise for routine application. Further work is suggested. With the improved thermal imaging capability anticipated with the next generation of GOES, modifications of these procedures in conjunction with data received from other remote sensing instruments offer unique opportunities to investigate the effects of biomass burning on the global habitat.

Remote Sensing of Biomass Burning in West Africa with NOAA-AVHRR

Jean Michel Brustet, Jean Bruno Vickos, Jacques Fontan,
Katherine Manissadjan, Alain Podaire, and François Lavenu

Savanna areas of the intertropical African zone are periodically under the influence of seasonal biomass burning, which, in most cases, is caused by fires deliberately lighted by humans (Jeffreys, 1951). Such a biomass combustion results in the emission of gases (CO_2, CH_4) (Seiller et al., 1980) that may affect the radiative balance of the earth or influence the atmospheric chemistry, on a local or large scale, through the formation of acids or oxidants.

Ozone and carbon monoxide, remotely sensed by satellite (Fishman et al., 1986, 1990), have revealed high concentrations in the intertropical area, over the African continent, and over the Atlantic Ocean under the trade winds. The measurements carried out as part of the Dynamics and Chemistry of the Atmosphere in Equatorial Forest (DECAFE) program have shown a strong acidity of the precipitations (Lacaux et al., 1987, 1988) and have revealed high ozone and hydrocarbons concentrations (Cros et al., 1987, 1988).

Biomass burning is generally observed during the dry season—namely, from November to March in the Northern Hemisphere and from June to October in the Southern Hemisphere. However, little is known about the frequency of these fires, their localization, and the extent of burned areas per year. Yet such information would be valuable for the design of realistic models of the transport and chemical evolution of airborne components emitted by biomass burning. Indeed, the intensity of these sources and their evolution through diurnal and seasonal cycles are important input data for these models.

Remote sensing measurements provide a valuable means of determining the extent of burning areas and estimating the overall distribution of the sources in time and space. The Advanced Very High Resolution Radiometer (NOAA-AVHRR) satellite is well adapted to a wide coverage of the large African savanna regions. It is necessary to watch the whole area even at times other than during the dry season, since two consecutive weeks without precipitation may be sufficient to allow the bushes to catch fire.

Some work has been done on the remote sensing of biomass burning, and we can cite Dozier (1981) and Dozier and Matson (1981) for their work on the identification of high temperature sources below the resolution limit (pixel); Malingreau et al. (1985) for a report on the remote sensing of forest fires in Kalimantan and in North Borneo; Malingreau et al. (1989) for a report on the deforestation of the Amazon; Matson et al. (1987) for the detection of fires from satellite images provided by NOAA-AVHRR; and finally, Matson and Holben (1987) for their investigation into satellite detection of forest fires in Brazil.

Geographic Localization of the Region Studied

The images examined in this chapter include the whole of West Africa—namely, within latitudes 5° and 14°N and 1° and 11°W. The study has been focused on a region that contains part of the Guinea territory, Mali, the Ivory Coast, and Burkina Faso.

Equipment and Methods

Satellite Data

The images have been recorded by the satellite NOAA-AVHRR. The spatial resolution of one pixel is of 1.1×1.1 km^2 and the swath width is of 2700 km. Two channels are used: Channel 3 is located in the middle infrared, between 3.55 and 3.93 μm. The luminance in this channel corresponds to the solar radiation reflected by the ground surface and the atmosphere, and to the thermal emission of the soils and hot sources, such as the fires we are looking for. Channel 4 is located in the longwave infrared within 10.3 and 11.3 μm. The luminance corresponds essentially to the thermal emission of soils or clouds. The area of a fire is not sufficiently wide to generate a signal distinguishable from the thermal emission corresponding to the whole pixel area.

Table 5.1 Fire area S_s, which causes saturation

$T_f(°K)$	400	450	500	550	600	650	700
p	0.05	0.017	0.007	0.0034	0.0019	0.0011	0.0007
$S_s(m^2)$	61,505	20,333	8498	4176	2312	1402	916

Methods

If T_3 is defined as the apparent equivalent temperature detected by a pixel in channel 3, the luminance should be expressed as follows, according to Dozier (1981):

$$L_3(T_3) = p \times L_3(T_f) + (1 - p) \times L_3(T_e)$$

where

$L_3(T_3)$ is the integral of the Planck function in channel 3 at the temperature T_3,

T_f is the apparent fire temperature,

T_e is the apparent temperature of the environment,

p is the part of the pixel corresponding to the fire.

Given that the apparent saturation temperature T_{3s} in channel 3 is about 320°K (Matson et al., 1987) and considering that the temperature of the environment is about 300°K, a simple calculation gives the fire surface necessary to saturate channel 3, which is a function of the fire temperature T_f, as evidenced by the following expression:

$$p = [L_3(T_{3s}) - L_3(T_e)]/[L_3(T_f) - L_3(T_e)]$$

The fire area S_s, which causes saturation, is the product of the total pixel area S_{pixel} by the p portion occupied by the fire:

$$S_s = p \times S_{pixel}$$

For a given environment temperature of 300°K, see the results in Table 5.1.

Thus, for a given environmental temperature of 300°K and for fire temperatures of at least 560°K (the most probable value estimated by Landsat; Brustet et al., 1991), the fraction p of a pixel that causes the saturation of channel 3 appears to be smaller than 0.4%. Furthermore, recent observations made by Landsat indicate that 80% of the fires exhibit an apparent area larger than 4000 m². Consequently, it appears that under the above temperature conditions, in most cases this prevents any calculation of the temperature and surface of a fire.

Results and Discussion

Figure 5.1 shows an example of luminance measured in channel 3. The albedo as well as the emission corresponding to the soil and vegetation are well separated from the emissions due to the fires that generally saturate channel 3.

For the whole scenario investigated (19 December 1986), there is a marked separation between the soil and fire luminances. Therefore, the fire luminance can be unambiguously separated from the environmental background. In such a case, a simple input signal threshold will be sufficient to sort out the pixels for which a fire has been detected.

In some cases, however, a threshold would be insufficient. This is illustrated by the scenario of 10 February 1987, displayed in Figure 5.2, where the environmental luminance appears to be more important than above, masking the borderline of fire emissions. Such an increase in the albedo and thermal emission is currently observable when the vegetation is very dry.

Figure 5.3 shows a two-dimensional histogram representing the emissions corresponding to channels 3 and 4. The pixels that are affected by fire emissions are seen to saturate channel 3. The environmental luminance corresponds to the dashed zone of the histogram. When the environmental signal increases in channel 3, it also increases in channel 4. The envi-

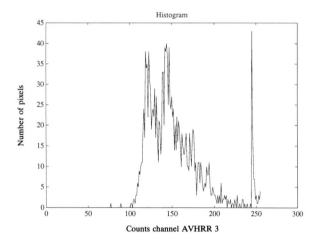

Figure 5.1 Histogram representing the numerical counts in channel AVHRR 3 in a typical example where the fire emissions are well separated from the environmental background (soil and vegetation). The pixels detecting fire emissions are saturated.

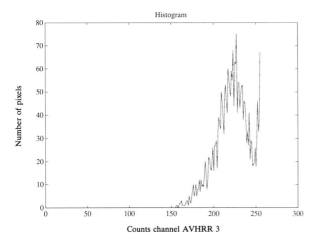

Figure 5.2 Histogram representing the numerical counts in the channel AVHRR 3 in a case where the environmental emission is important. An input signal threshold is insufficient in this case since it cannot selectively distinguish the environmental background.

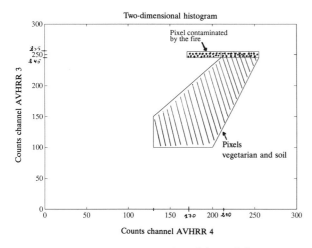

Figure 5.3 Schematic representation of the emissions corresponding to channels 3 and 4. The pixels detecting fire emissions are located in the dotted zone, whereas those detecting only environmental emissions are in the dashed zone. When the signal increases in channel AVHRR 3, it also increases in channel AVHRR 4. Consequently, the fire emissions and environmental emissions are located in the same zone.

ronmental and fire luminances are then located in the same zone.

In order to get rid of the environmental component, we have cut the signals of channels 3 and 4 below a given threshold. Some fires are then suppressed and the corresponding loss depends on the threshold value, determined from the two-dimensional histogram. This value will be proportional to the environmental luminance.

Figures 5.4a and 5.4b illustrate the problem of pixel separations within two-dimensional histograms by two examples: In the first case, displayed in Figure 5.4a, the fires are significantly out of the background. The density of the pixels contaminated by fire emissions increases toward high values of numerical counts in channel AVHRR 3. The most important ensemble of spots corresponds to the environmental component. An input signal threshold on channel AVHRR 3 is sufficient to short out the pixels detecting fire emissions. In the second one, displayed in Figure 5.4b, the signals corresponding to the fires are partially mixed together with those of the environment, the latter being apparent up to the saturation level in channel AVHRR 3.

Distribution of the Fires

Figure 5.5 represents a fire distribution within a 300 km² area, in a typical case where the fires can be well distinguished from the environment. Figure 5.6 represents the whole image, taken on 19 December 1986. For the convenience of this study, it has been divided into 300 km² units. The results have been obtained by setting up one threshold on channel 3. The fire area lies along a stretch from north to south and is about 500 km wide. The fire distribution within this area is rather homogeneous. The density of the sources, which is about 0.7%, rapidly decreases at both ends of this area. This corresponds to the forest zone in the southern limit and to the Sahelian regions in the northern one.

The fire distribution is such that the whole savanna area can be considered as a rather homogeneous source on a large scale. This is consistent with the observation that the zone mixing ratios over the forest are still high and constant, along horizontal layers (Cros et al., 1989).

The number of these fires is such that they can no longer be considered as sources of isolated plumes but have to be assimilated to a large planar source stretched along a large part of savanna regions.

For the same image, we have represented the case where a single threshold was set up on channel 3

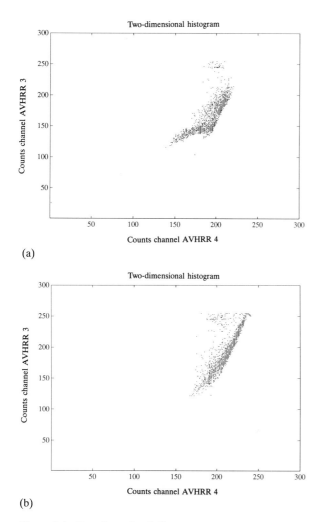

(a)

(b)

Figure 5.4 Two-dimensional diagrams representing the numerical counts in channels AVHRR 3 and 4. In the first case, displayed in Figure 5.4a, the fires are significantly out of the background. The density of the pixels contaminated by fire emissions increases toward high values of numerical counts in channel AVHRR 3. The most important ensemble of spots corresponds to the environmental component. An input signal threshold on channel AVHRR 3 is sufficient to short out the pixels detecting fire emissions. In the second case, displayed in Figure 5.4b, the signals corresponding to the fires are partially mixed together with those of the environment, the latter being apparent up to the saturation level in channel AVHRR 3.

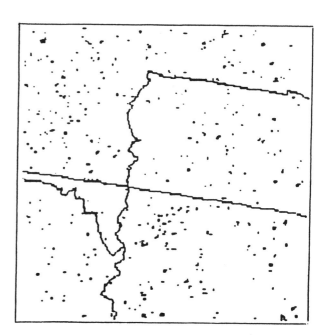

Figure 5.5 Distribution of the fires within a 300 km² area. These data are part of the image recorded on 19 December 1986, a case where the fires are well distinguished from the environment.

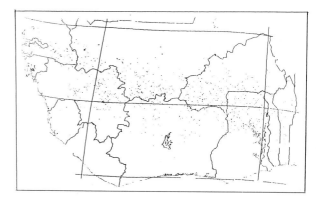

Figure 5.6 Distribution of the fires within the whole image taken on 19 December 1986. The fire distribution is homogeneous, and 0.7% of the pixels are saturated.

(Figure 5.7a) and the one where a double threshold was installed on channels 3 and 4 (Figure 5.7b). It appears that the pixel density is obviously too important in Figure 5.7a, whereas only the pixels detecting a fire are seen in Figure 5.7b. Clearly, the presence of an input signal threshold on both channels allows the elimination of most of the pixels for which the environmental luminance is very important.

Attempts to Obtain Biomass with GAC Images

Considering the importance of fire zones in Africa, we were prompted to investigate the possibility of using Global Area Coverage (GAC) images, which are obtained by degradation of Local Area Coverage (LAC) images, and exhibit an enhanced resolution of $4 \times 4 \text{ km}^2$. However, we have found that such a processing gives images where the fires become buried into the environmental background. Only 10% of the sources appearing on the original images are recovered after this operation. We conclude that GAC images are insufficient for a precise detection of active fires.

Evaluation of the Emissions

In the case of the image taken on 19 December 1986, 0.7% of the pixels corresponding to the active area, mentioned above, are saturated. A study of Landsat Thematic Mapper images has revealed that a fire generally activates two NOAA pixels (Brustet et al., 1991).

An investigation of the fire emissions by Landsat has led us to estimate the average emission of CO and NO for a fire of medium dimensions (Brustet et al., this volume). It can be deduced that the emission of the savanna area affected by the fires is 2.15×10^{13} mol cm^{-2} s^{-1} (1.23 kg C/s) for CO and 2.8×10^{11} mol cm^{-2} s^{-1} (0.042 kg NO/s) for NO. This evaluation does not take into account the separation between flaming and smoldering.

Lamotte et al. (1986) have determined the amount of CO_2 emission resulting from the combustion of 180,000 km^2 of savanna in the Ivory Coast. The values are respectively 11×10^6 tons in December, 67×10^6 tons in January, and 12×10^6 tons in February. Using the same emission factors as previously, the CO and NO emissions are given in Table 5.2.

The values previously obtained from the image recorded on 19 December 1986 are consistent with those obtained in January for the savanna regions of the Ivory Coast.

Our results are in full agreement with a recent numerical simulation by Lopez (1989, 1990) indicat-

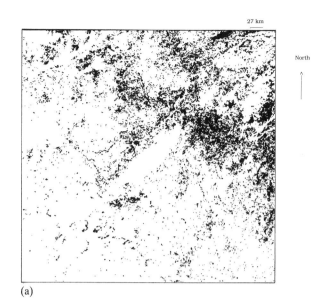

(a)

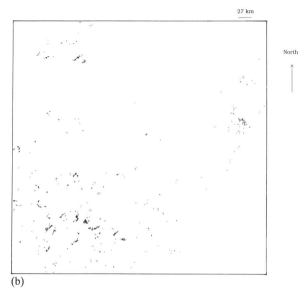

(b)

Figure 5.7 Two different treatments of the image are represented. In the first case (Figure 5.7a), a threshold has been set up on channel AVHRR 3, whereas in the second one (Figure 5.7b), an additional threshold has been installed on channel AVHRR 4. This procedure proves to be sufficient to separate the pixels contaminated by the fires from those corresponding to the environment.

Table 5.2 CO and NO emissions in mol. cm^{-2} s^{-1}

	CO	NO
December	2.6×10^{12}	4.4×10^{10}
January	1.6×10^{13}	2.7×10^{11}
February	2.8×10^{12}	4.8×10^{10}

ing that NO emissions of 5×10^{11} mol cm^{-2} s^{-1} originating from savanna regions are necessary to account for the high ozone concentrations detected over the forest in the North Congo.

Conclusions

The NOAA satellites are well adapted to an investigation of biomass burning on a large scale. The present study of West African regions indicates that active fires are usually detected on channel 3, except under particular conditions where the environmental luminance becomes too important. In such cases, an input signal threshold on channel 4 allows the environmental background to be eliminated. The results of our study have revealed a homogeneous fire distribution within a 500 km wide stretch where 0.7% of the pixels are saturated by fire emissions.

The saturation is reached with a fire surface of 4000 m^2 for an average environmental temperature of 300°K and a fire temperature of the 550°K. This saturation problem prevents any thorough quantitative study of the surface and temperature of fires. However, valuable complementary information may be obtained from high-resolution data of the Landsat Thematic Mapper type.

The daily cycle of NOAA satellites is convenient to determine temporal and spatial distributions of biomass burning in Africa. Furthermore, remote sensing can be also used by the governments of these wide regions as a means to control the evolution of bush fires.

Characterization of Active Fires in West African Savannas by Analysis of Satellite Data: Landsat Thematic Mapper

Jean Michel Brustet, Jean Bruno Vickos, Jacques Fontan,
Alain Podaire, and François Lavenu

Biomass burning is an important source of various gaseous compounds which are sent into the atmosphere. Once airborne, these compounds (CO, CH_4) may either trap the heat radiated by the earth and contribute to the greenhouse effect or influence the atmospheric chemistry on a local or large scale through the formation of acid compounds or oxidants (Seiler et al., 1979; Seiler and Crutzen, 1980; Stith et al., 1981; Greenberg et al., 1984; Crutzen et al., 1985; Cros et al., 1987; Kaufman et al., 1989).

Valuable experimental data concerning African regions have been obtained from Dynamics and Chemistry of the Atmosphere in Equatorial Forest (DECAFE), a program devoted to a thorough investigation of the dynamics and chemistry of the atmosphere over equatorial forests.

In particular, they have allowed the detection of strongly acidic precipitations (Lacaux et al., 1987, 1988) and have revealed abnormally high ozone concentrations (Cros et al., 1987, 1988).

Evidence for high ozone rates above the intertropical African regions of the Southern Hemisphere has been also obtained by Fishman et al. (1986, 1987, 1988, 1990) from satellite data recorded during the dry season.

These concordant observations are indicative of pollution, which is ascribed to the occurrence of bush fires in savanna regions. These fires take place during the dry season—namely, from November to March in the Northern Hemisphere and from June to October in the Southern Hemisphere. The highest ozone rates detected respectively over the North and South Congo are effectively corresponding to these periods (Cros et al., 1987, 1988).

The importance of biomass burning is such that it should be taken into account in the elaboration of physical and chemical atmospheric models. However, this requires a precise knowledge of the intensity of the sources, which depends on various parameters, such as the apparent temperature of the fire and its surface. To date, the means of providing access to these data are rather limited.

Satellite images constitute a major source of valuable information on the extent of areas being on fire at a given time. They are thus helpful for the characterization of biomass burning (Dozier, 1981; Matson and Dozier, 1981; Malingreau, 1984, 1985, 1989; Matson and Holben 1987).

The Advanced Very High Resolution Radiometer (NOAA-AVHRR) instruments are convenient for an investigation of fire distributions on a large scale (Grégoire et al., 1988; Brustet et al., this volume). However, the resolution of the images is limited by the pixel size, corresponding to an observation area of 1.1×1.1 km^2. It is thus difficult to obtain precise information on individual fires (surface, shape, temperature).

The high-resolution satellite Landsat-5, equipped with medium infrared channels, is more suitable for the characterization of vegetation fires. We have used its data in the present work, which deals with the observations of the national nature reservation, the Comoé, located in the northeast part of the Ivory Coast Republic.

Equipment and Methods

Satellite Data

We have analyzed the data recorded by the high-resolution detector Thematic Mapper, mounted on NASA's satellites Landsat-4 and Landsat-5, which were launched, respectively, in July 1982 and March 1984. These satellites have a rotation period of 16 days, and their orbit is at 715 km altitude. The observation corresponds to a ground surface of 170×180 km^2 with a spatial resolution of 30 m. The detection system involves seven spectral channels. Those used in the present work—namely, TM5 and TM7—are respectively locked on the wavelengths 1.65 μm and 2.2 μm. They are usually used to analyze the radiation reflected by soil and vegetation, but they are also sensitive to the emissions associated with biomass burning, which add to the reflected solar flux.

The images analyzed here were recorded on 9 January 1986 and 27 December 1986.

Methods

Dozier's Method

Dozier (1981) has developed an algorithm to estimate the temperature of hot sources and their surface. We have implemented the method in view of its use on the thermal channels AVHRR 3 and 4 of the NOAA satellite. The calculation is based on the following hypothesis: Given that p and $1 - p$ are, respectively, the parts of a pixel corresponding to the fire luminance and to the environmental background, the resulting signal will be a function of the sum of these two components. This is expressed as follows:

$$L_j(T_j) = p \times L^{cn}_j(T_f) + (1 - p) \times L^{cn}_j(T_b) \ (j = 3, 4)$$

where

j is the channel,

T_j is the apparent pixel temperature in channel j,

T_f is the apparent temperature of the hot source,

T_b is the environmental apparent temperature,

$L^{cn}_j(T)$ is the normalized integral of Planck function within the spectral window of channel j,

p is the part of the pixel corresponding to the fire.

In this equation, only the proper emission is taken into account.

Method Taking into Account the Smoke

Our method is derived from the algorithm designed by Dozier. The channels TM5 and TM7 used in the present work correspond to a shorter wavelength domain, for which important reflection phenomena are observed in addition to the emissions due to hot sources. Both these components are taken into account in this work.

The apparently homogeneous signal F1 detected by a pixel is a contribution of both the incident solar flux and the apparent reflection coefficient (albedo) on the ground surface and on the atmosphere (Figure 6.1a). Considering the emission of the soil as negligible, the expression of the signal is written as follows:

$$F1 = \rho^* \times E_{i,s}$$

where F1 is the signal detected by the pixel and ρ^* is the apparent reflection coefficient of this pixel and $E_{i,s}$ is the solar lightening.

Let us recall that each pixel is divided in two parts—

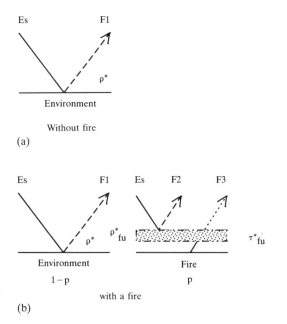

Figure 6.1 Detection of the solar radiation on a pixel. Figure 6.1a represents the interaction of the solar radiation with a pixel in the absence of fire. F1 is a function of the solar radiation E_s and of the apparent reflectance ρ^* of the pixel. Figure 6.1b represents the interaction of the solar radiation with a pixel in the presence of fire. The environment corresponds to the part $1 - p$, whereas the fire corresponds to the part p. The smoke is taken into account in the latter part. The luminance measured by the satellite and coming from part p is a function of the solar radiation E_s, the apparent reflectance of smoke, the smoke, and its apparent transmittance and the radiations emitted by the fire.

namely, part p corresponding to the fire area, and part $1 - p$ corresponding to the environment. The signal S coming from the pixel is then expressed as follows:

$$S = p \times (\text{fire signal}) + (1 - p) \times (\text{environment signal})$$

The p value is ranged between 0 and 1. The environment is assumed to have the same contribution for all pixels.

The emissions due to the environment are expected to be very different, depending on whether the vegetation is intact or has been already burned out. Accordingly, the two hypotheses have been considered in this work.

The smoke contains solid particles and liquid water that may mask the surface and reflect solar radiations. However, these effects have not been taken into account here. Due to the lack of information on the optical factor of the smoke, we have determined an apparent coefficient ρ^* equal to 0.60 obtained by an iterative method. Inconsistencies were found in testing other values: A smaller value, like 0.55, results in a p value greater than 1 for some pixels, whereas a large value, like 0.65, leads to a situation where a pixel can never be totally occupied by a fire. The proper emission factor of the smoke being considered as a layer over the fire τ^*_{fu} is expressed as follows:

$$\tau^*_{fu} = 1 - \rho^*_{fu}$$

It should be noted that this value cannot be considered as definitive, since the smoke depends on the amount of consumed biomass and on the wind intensity.

The resulting signal can be written as follows:

$S = p \times$ (fire emission through the smoke + solar reflection on the smoke) $+ (1 - p) \times$ (environmental luminance)

Using the notation of Figure 6.1b, this gives

$$S_i = p \times (F_2 + F_3) + (1 - p) \times F_1 \ (i = 5, 7)$$

where

F_1 represents the solar radiation reflected by the environment,

F_2 represents the radiation reflected by the smoke,

F_3 represents the radiation emitted by the fire through the smoke.

The fire is considered as a blackbody at the temperature T_f. Thence, the fire emission is defined by the Planck function L_λ

$$S_i = p[\tau^*_{fu}L^{cn}_i(T_f) + \rho^*_{fu}E_{i,s}] + (1 - p)\rho \times E_{i,s}$$

where

τ^*_{fu} is the apparent smoke transmission,

ρ^*_{fu} is the apparent smoke reflection,

$L^{cn}_i(T_f)$ is the integral of the Planck function in channel i,

$E_{i,s}$ is the solar lightening in channel i,

$E_{i,s}$ defined as the luminance observed on the neighbor channels is set equal to $\rho \times E_{i,s}$

S_i is the physical signal in channel i.

We thus obtain a system of two equations with two unknown terms:

$$S_5 = p[\tau^*_{fu}L^{cn}_5(T_f) + \rho^*_{fu}E_{5,s}] + (1 - p)\rho \times E_{5,s}$$

$$S_7 = p[\tau^*_{fu}L^{cn}_7(T_f) + \rho^*_{fu}E_{7,s}] + (1 - p)\rho \times E_{7,s}$$

The terms due to the atmospheric transparency are not taken into account, whereas the coefficients ρ^*_{fu} and τ^*_{fu} are supposed to be constant and identical in both channels.

For a given set of parameters ρ^*_{fu} and τ^*_{fu} the associated signals S_5 and S_7 allow us to define the parameter p and the apparent temperature T_f.

Results and Discussion

In Figure 6.2, the fire front appears as the clearest zone. Its apparent length here is 1.5 km, and its apparent width is 60 to 90 m.

The monodimensional histograms representing the luminances in channels TM5 and TM7 are shown in Figure 6.3.

There are two maxima: One corresponds to the reflectance of burned surface, and the other one to that of the vegetation.

The pixels contaminated by fire emissions show up clearly on the two-dimensional representations of TM5 and TM7. The luminance in each channel increases with the fire luminance. In Figure 6.4, zones 1, 2, and 3, respectively, represent a burned environment, an intact one, and an area where the fire is active.

From the system of two equations in the previous section we have built the network shown in Figure 6.5, which corresponds to values of p and T for two extreme kinds of environment—namely, a burned surface (Figure 6.5a) or a highly reflecting vegetation (Figure 6.5b). The fire temperatures range between $500°$ and $700°K$, and the pixel portions p are between 0 and 1. A single fire front has been considered in this example.

For the whole data collected on 9 January 1986, we have only considered the pixels contaminated by one fire. The histograms of the portions p contaminated by one fire are displayed in Figures 6.6a and b, which correspond, respectively, to the hypotheses of unburned and burned environments. Continuous distributions are obtained in both cases. It is also noteworthy that whatever the hypothesis is for the environment, more than one-half of the pixels exhibit a p value lower than 0.5. Also, a saturation peak is always observed in channel TM7.

If we use the preceding results to calculate the average p value in both cases, we get $p_m = 0.39$ in the case of a burned environment and $p_m = 0.3$ in the case of an intact environment.

In the same way, we have represented the histograms of apparent temperatures for the two hypoth-

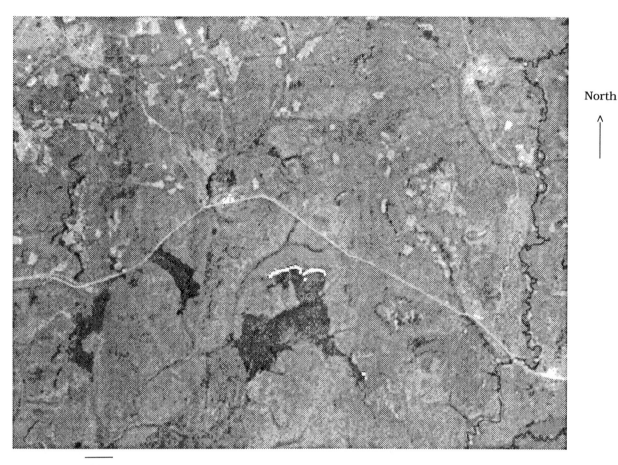

760 m

Figure 6.2 TM7 (2.2 μm) channel image. The area is 15 × 12 km². The fire front appears as the clearest (white).

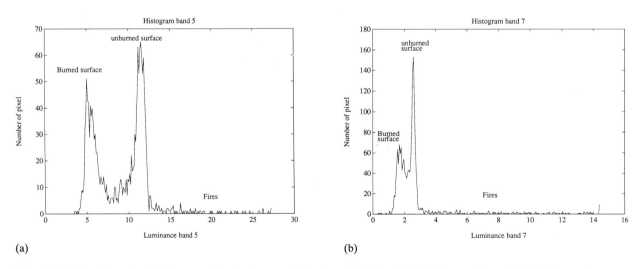

(a)   (b)

Figure 6.3 Monodimensional histograms of the luminances in the channels TM5 and TM7 (Figures 6.3a and 6.3b). Two maxima are observed—one corresponding to the reflectance of burned areas, and the other corresponding to the intact vegetation. Away from these maxima, the pixels exhibiting a more important luminance are detecting bush fires.

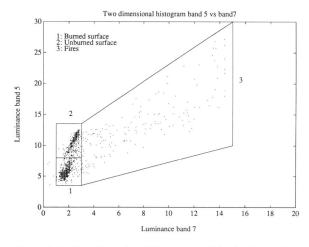

Figure 6.4 Two-dimensional histograms of the luminance in the channels TM5 and TM7. The pixels contaminated by fire emissions are well separated from those corresponding to burned and unburned environmental surfaces. The zones 1, 2, and 3 correspond, respectively, to burned environments, unburned ones, and active fires.

Table 6.1 Results of average p and T values

	Date	p_m	T_m(°K)	E-T
Burned environment	9 Jan.	0.39	568°	33.3
Unburned environment	9 Jan.	0.30	583°	45.7
Intermediate environment	27 Dec.	0.29	579°	38.7

eses in Figures 6.7a and 6.7b, respectively. The saturation peak is still visible on the two plots. We have calculated the average apparent temperature T_m = 568°K in the case of a burned environment and T_m = 583°K in the case of an intact environment.

We have analyzed the images of 27 December 1986 in the same way, except that an intermediate case has been considered for the environment. Figures 6.8a and 6.8b show that we still obtain a homogeneous distribution of the portions and of the apparent fire temperatures. The peak corresponding to the signal saturation in channel TM7 is also visible on each of the two distributions.

In this case, the average value of p_m is 0.29, and the average apparent fire temperature T_m is 579°K.

The preceding results of average p and T values are displayed in Table 6.1 along with the estimated standard deviation on T expressed as E-T.

Analysis of the Fire Areas
We have analyzed 20 fire fronts from the images recorded on 9 January 1986 and 71 others from those recorded on 27 December 1986.

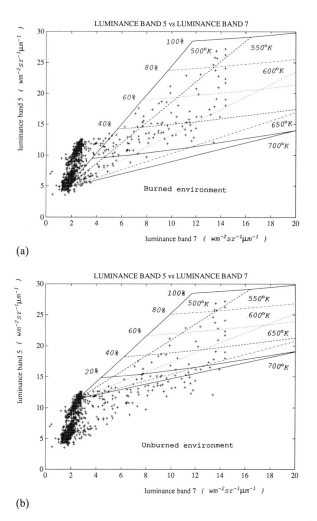

(a)

(b)

Figure 6.5 Network corresponding to temperature values ranged between 500° and 700 °K and pixel portions between 0 and 1. The environment is assumed to be already burned in Figure 6.5a and intact in Figure 6.5b.

These fires are very heterogeneous in size and shape. Indeed, the shape of a fire may be influenced by various factors, including the wind direction and its intensity, the relief (plateau, hillside, hollow), the various types of savanna vegetation (grass, bushes, trees, woods), and finally the way by which the fire has been ignited (from one point, or along a line). Figure 6.9 shows several examples of linear fire fronts as well as clusters corresponding to several fires that have merged together.

The grid pattern defines the NOAA-AVHRR pixels. It clearly appears that a fire contaminates several pixels (two on average).

The apparent size of the fire front can be determined from the enlargement of the image shown in Figure 6.10. Its apparent length (corresponding to 40

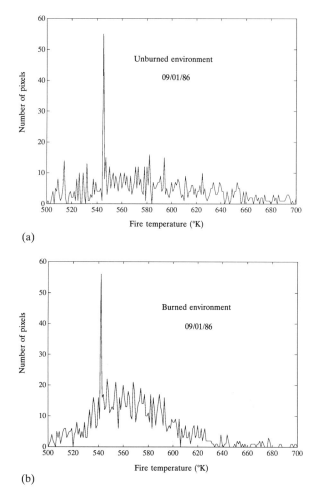

(a)

(b)

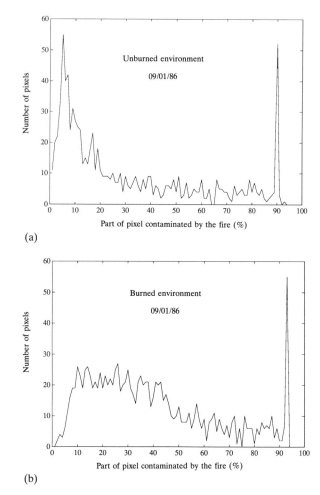

(a)

(b)

Figure 6.6 Histogram representing the part p of the pixels corresponding to the fires. All the pixels contaminated by a fire have been taken into account, and the two hypotheses for the environment have been considered. The average part p is 0.30 in the first case and 0.39 in the second one. The peak seen on the two figures corresponds to the signal saturation in channel TM7.

Figure 6.7 Histogram representing the apparent fire temperatures. All the pixels contaminated by fire emissions have been taken into account. The same hypotheses as in Figure 6.6 have been considered for the environment.

pixels) is 1200 m, whereas its apparent width (corresponding to 2 or 3 pixels on average) is on the order of 60 to 90 m. In fact, direct measurements of the width of the fire front carried out on the ground surface show that the effective value is on the order of 5 m. Thus, the width determined by spatial detection must include the active fire front, with the flames tilted by the wind appearing eventually larger, and the surrounding area that has just been burned out. The average length of the fire fronts detected on 27 December 1986 is 1.14 km. This value is obtained by dividing the total number of contaminated pixels by

the number of fire fronts and by the average width of a fire (2 pixels, on average).

Figure 6.11 shows the cumulative plot of fire front surfaces calculated from Landsat TM data and taking into account the two hypotheses considered above for the environment. Characteristic features such as the average fire front surface and the median of the distribution are provided in Table 6.2.

It clearly appears that the surface calculated from the pixel portions contaminated by the fire are overestimated. If we assume that the effective fire front width is 5 m, whereas its average length (obtained by remote sensing) is 1140 m, the surface of a typical medium fire is 5700 m^2.

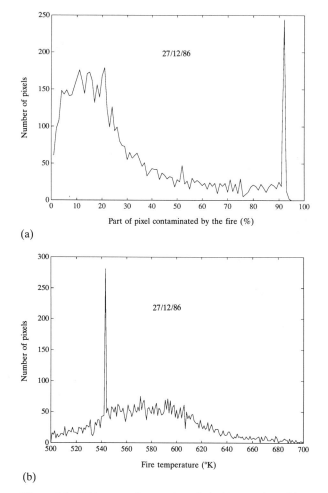

(a)

(b)

Figure 6.8 Histogram of p, portion of pixel contaminated by a fire and of fire temperature. These results are obtained with the 27 December 1986 image.

Table 6.2 Characteristic features of fire front surfaces

	Date[a]	S_m(°K)[b]	M(m²)[c]
Burned environment	1 Jan.	20,178	25,000
Unburned environment	1 Jan.	11,429	16,000
Intermediate environment	12 Dec.	20,216	30,000

a. The dates refer to the year 1986.
b. S_m represents the arithmetical average of the surface, in units of m^2.
c. M represents the median of the distribution.

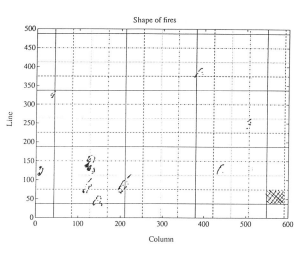

Figure 6.9 Shapes of fire fronts. These fires are heterogeneous in size and shape with some clusters corresponding to several fires that have merged together. The grid pattern defines NOAA-AVHRR pixels.

Estimate of CO and NO Emissions of a Representative Medium Fire

It is interesting to compare the emissions of a medium fire with those of an anthropogenic source such as automobile exhaust.

Considering the average surface calculated above (5700 m²), and given that the fire speed in the absence of wind is 300 meters per hour (Monnier, 1968), the surface burned per unit time can be evaluated at 95 m² per second. The amount of dry vegetation in Sudanian savannas is 5 tons per hectare on average, containing 45% carbon; 60% of this amount is burned out, and 80% is converted into CO_2 production of 10.26 kg C/s.

The corresponding amount of CO and NO emissions depends on the combustion type—namely, smoldering or flaming—and on the effective dryness of the biomass.

Considering that the CO/CO_2 ratio is on the order of 12% in volume, and that the NO_x/CO_2 ratio is 2 × 10^{-3} (Greenberg et al., 1984; Crutzen et al., 1985), the CO and NO_x productions are, respectively, 1.23 kg C/s CO and 0.02 kg N/s NO_x or 0.042 kg NO.

For comparative purposes, let us recall that a European car produces on average 20 g of CO per km and 2 g of NO (Joumard, 1987). Considering an average car speed of 60 km/h the emission of a medium vegetation fire per second is equivalent to the CO production of 3694 automobiles and the NO production of 1260 ones. These numbers give a rough idea of the magnitude of emission.

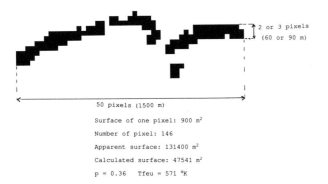

50 pixels (1500 m)

Surface of one pixel: 900 m²

Number of pixel: 146

Apparent surface: 131400 m²

Calculated surface: 47541 m²

p = 0.36   Tfeu = 571 °K

Figure 6.10 Enlargement of a fire. The length corresponds to about 40 pixels and the apparent width is 2 or 3 pixels.

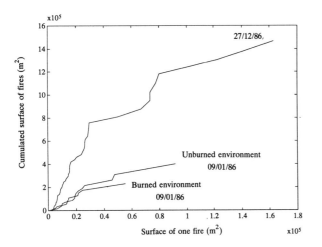

Figure 6.11 Surface distribution of fire fronts for three examples.

Conclusion

Landsat Thematic Mapper provides valuable information on biomass burning, such as the apparent temperature of a fire and its shape. However, the surface determined by remote sensing does not exactly correspond to the burning area, due to an artificial enlargement of the fire front width. This enlargement may have diverse origins. In particular, it is difficult to estimate the temperature of the areas that are behind the fire front and that have been just burned. Emissions from these areas may be detectable by Landsat channels, thus resulting in the observed enlargement of the fire front. Additional experiments including remote sensing by plane are necessary to allow a more complete understanding of these phenomena.

Biomass burning is an important source of atmospheric pollution on a global scale. This study indicates that a fire is a significant source of pollution on a local scale.

The Great Chinese Fire of 1987: A View from Space

Donald R. Cahoon, Jr., Joel S. Levine, Wesley R. Cofer III,
James E. Miller, Patrick Minnis, Geoffrey M. Tennille,
Tommy W. Yip, Brian J. Stocks, and Patrick W. Heck

On 6 May 1987 three fires were started in the boreal forest of the Helongjing Province of the People's Republic of China. The Helongjing Province borders the Soviet Union along the Amur River. These fires, resulting from human activities, quickly spread into one of the largest fires in Chinese history. The Chinese fires burned for three weeks and were officially declared out on 27 May, but flare-ups continued for at least another week. During the three-week period in May, the fires coalesced and destroyed a contiguous region of forest. This fire required the largest control operation in Chinese history, in which over 40,000 firefighters were mobilized.

Satellite imagery on 4 May, two days before the Chinese fires started, showed fire activity over an extensive region of the neighboring Soviet Union (Figure 7.1). This area extended across roughly 15° of longitude. The Soviet fires were not visible on the imagery before late April or after mid-June; therefore, they lasted four to five weeks. Because of the extensive area covered by these fires, most of them were likely to have been started by lightning strikes during frontal system passages, but human activity certainly cannot be ruled out as the cause of some fires. By the end of the month, the Soviet fires grew to almost three times the area of the Chinese fires.

The weather played a critical role in setting the stage for this wildfire event. Spring precipitation had been very light over the entire region, and there was little snowfall during the winter months. Snow cover disappeared by early April, and with low humidities experienced throughout the region, the combustibles on the forest floor dried out. These factors combined with strong gusty winds from several frontal passages led to the severe fire danger conditions, which persisted throughout most of the month. Rainfall near the end of the month finally brought assistance to firefighters in extinguishing the fires.

Over 200 images spanning a period of six months were examined in the study of this biomass burning event. The Advanced Very High Resolution Radiometer (AVHRR) on the NOAA-9 spacecraft produced all the imagery. Making use of the known geometric characteristics of the AVHRR instrument, the areal extent of the Chinese and Soviet fires was determined from the imagery. This chapter briefly reviews the AVHRR instrument, our imagery analysis techniques, and the methodology we employed in determining the extent of the burning.

The Advanced Very High Resolution Radiometer (AVHRR)

The AVHRR instrument has flown on several NOAA polar-orbiting satellites in sun synchronous orbit and provides digital imagery in the visible, near-infrared, and infrared wavelengths of the electromagnetic spectrum (Table 7.1). Of the five spectral channels available on the NOAA-9 instrument, only the first four were used in this study. Channel 1 measures reflected radiation in the visible wavelengths. Channel 2 measures reflected radiation in the near-infrared part of the spectrum. The reflectance of energy in the channel 2 wavelength is directly related to the spongy mesophyll leaf structure of vegetation (Justice et al., 1985). This characteristic of channel 2 provided contrast in the imagery between the boreal forest canopy and the damaged forest resulting from burning. Channel 3 and channel 4 are both thermal channels. The maximum blackbody radiance curve is skewed toward the channel 3 wavelengths and away from the channel 4 wavelengths at high temperatures (Matson et al., 1987). Therefore, the measured temperature of channel 3 will increase more rapidly than that of channel 4 for a given increase in terrestrial temperature. This characteristic makes channel 3 more sensitive to high temperature sources and makes this channel useful for forest fire detection.

The AVHRR imagery is archived in one of three forms. The highest resolution imagery is saved either in the High-Resolution Picture Transmission (HRPT) or the Local Area Coverage (LAC) data storage formats. HRPT is continuously transmitted directly to properly equipped receiving stations within the satel-

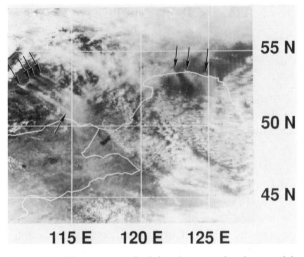

Figure 7.1 The arrows at the left point to smoke plumes originating from fires in the Lake Bakal area of the Soviet Union. The arrows at the right point to smoke plumes from Soviet fires along the Amur River (Chinese-Soviet border). (Taken 4 May, 1987, 6:20 UT NOAA-9, channels 1, 2, 4.)

Table 7.1 NOAA 9 characteristics

Channel	Band width (μm)	IFOV (milliradians)
1	0.58–0.68	1.39
2	0.725–1.10	1.41
3	3.55–3.93	1.51
4	10.5–11.3	1.41
5	11.5–12.5	1.3

Characteristics of each AVHRR spectral channel.

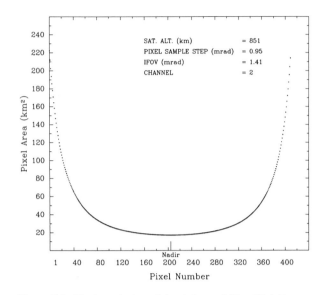

Figure 7.2 Pixel resolution of the Advanced Very High Resolution Radiometer Global Area Coverage product.

lite horizon. LAC imagery is archived by NOAA only by advance request. Global Area Coverage (GAC) imagery is degraded to a 4/15 sampling of the high-resolution imagery. This degradation is performed on the satellite before the imagery is transmitted to the ground. To produce the GAC product, every third scan is saved. The pixels within the retained scan line are processed in five pixel groups, where the first four pixels are averaged and saved as the GAC pixel and the fifth is discarded. This reduces the number of pixels from 2048 to 409 in a single scan.

There are 2048 observations (pixels in terms of imagery) taken along an AVHRR scan for each of the five channels simultaneously. The instrument scans ± 55.4° from nadir (almost 3000 km), and the center of each pixel is 0.95 milliradians apart. With the NOAA-9 satellite altitude at about 851 km in May 1987, and using the instantaneous field of view

(IFOV) of each channel (Table 7.1), the computed highest resolution of any channel at nadir is 1.2 km^2 (LAC) and 18.1 km^2 (GAC). The area of each pixel outward from nadir increases due to the earth's curvature (Figure 7.2). The pixel area was computed, neglecting atmospheric variability, by using the cross-track and along-track arc lengths and by approximating an ellipse.

Despite the reduction in resolution of the GAC imagery, there are several distinct advantages for its use. GAC imagery is readily available for the entire globe twice daily, day and night, from the NOAA archive. This makes it extremely useful to examine historical phenomenon with a high sampling frequency (twice a day with a single satellite, or four times a day with two satellites, which are offset by six hours). It should be noted that limited LAC imagery is sometimes available, but coverage cannot be relied on for past events. Because of the wide swath width and the reduced computer storage requirements as a result of the sampling, GAC imagery is very useful in studying large geographical areas over extended time periods. In contrast, Landsat imagery covers a much smaller geographical area with very high spatial resolution (30 m) and a sampling frequency of 16 days for any specific area.

Imagery Examination

Image processing techniques were used to examine the GAC imagery. To enhance smoke plumes used to

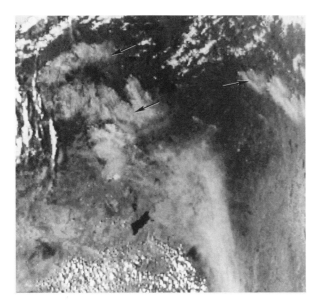

Figure 7.3 Arrows point to smoke plumes in the Soviet Union. The arrows pointing left show smoke in the Lake Bakal region and the arrow pointing right shows the Amur River area plumes. (Taken 6 May, 1987, 6:20 UT NOAA-9, channels 1, 2, 4.)

Figure 7.4 Areas of fire activity are indicated by the arrows. Brighter pixels in these areas contain fires. (Taken 6 May, 1987, 6:20 UT NOAA-9, channel 3.)

visually locate fires, channels 1, 2, and 4 were displayed through the red, green, and blue guns, respectively, of the display monitor (Chung and Le, 1984). Displaying the image in this manner takes advantage of the component of information available from each of the three channels. Smoke was seen in the visible and near-infrared channels, but not in the infrared channel—thus giving smoke a reddish-brown color because it has a red (channel 1) and green (channel 2) component and is different from clouds, which appear white due to components in all three channels. Otherwise, the shading between smoke and clouds overlaps in a single channel and can make differentiation between them very difficult. Figure 7.3 shows enhanced smoke plumes by combining channels 1, 2, and 4 as mentioned above. Even though the image has been printed in shades of grey from a color negative, the enhancement of smoke is still better than that of either channel individually. Figure 7.4 is the channel 3 image from the same time period. The brightest pixels are saturated and correspond to the fire locations. The map on Figure 7.1 and the use of the small lake on the lower-center portion of the images (near the Chinese, Mongolian, and Soviet border) will aid in geographical orientation and show the enormous area covered by fires. On 4 May 1987 satellite navigation showed that all the fires were on the Soviet side of the border.

The same display technique was used to enhance the saturated pixels (showing fires) in channel 3 and to show their relation to smoke in channel 1. For this purpose, channels 1, 3, and 4 were used. With this enhancement, the smoke appears as a very light red shade since it is present only in channel 1 (red). The locations of the fires are seen as bright pixels, and the relationship of the smoke and fires can easily be seen. The composition of channels 3 (green) and 4 (blue) clearly shows the areas which have been burned even through the smoke plume. Because of an approximate 3°K difference in brightness temperature between the forest canopy and the charred forest, these areas appear as a lighter shade compared to their surroundings and have been used to estimate the extent of burning during the time of the fires. This increase in temperature has also been observed in tropical forests and is possibly due to the absence of the evapotranspiration cooling effect within the canopy (Malingreau et al., 1985). Figure 7.5 shows both the smoke-enhanced image and an image showing the corresponding burning underneath the smoke plumes during the time of the Chinese fire. The lower large bright areas (b) under the smoke (a) show the spread of each Chinese fire before it had grown into a contiguous area. The active fires can be seen as bright pixels on the perimeter of the lighter (warmer) areas, showing the outward spread of the fire.

(a)

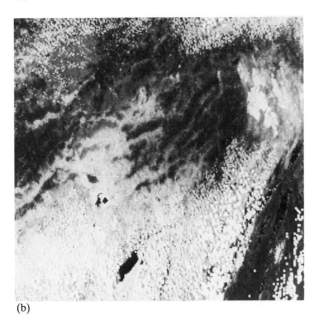

(b)

Figure 7.5 In the upper right corner of the two images, the Amur River region fires can be seen. The top image (a) is enhanced to show the smoke plumes. Three bright patches in the lower image (b) can be seen in the same area as the largest smoke plume. This corresponds to the Chinese fires. [Taken 16 May, 1987, 6:20 UT NOAA-9, (a) on channels 1, 2, 4; (b) on channels 1, 3, 4.]

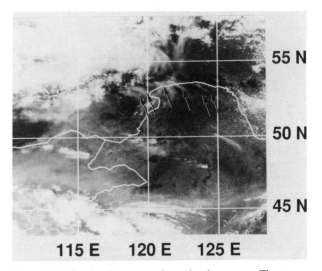

Figure 7.6 Smoke plumes are shown by the arrows. The two arrows at the right show Chinese plumes. The black area above the three arrows on the far right is the Chinese burn scar. (Taken 29 May, 1987, 6:20 UT NOAA-9, channels 1, 2, 4.)

After the fires had burned the forest, channel 2 was used to examine and estimate the area of the burn scars. The difference in reflectance between the vegetation of the forest canopy and the charred forest was pronounced and clearly delineated the burned area. These burn scars on the Chinese and Soviet side of the border, as shown in Figure 7.6, are the darkest areas. On the Soviet side, several smoke plumes can still be seen showing continued fire activity. On the Chinese side there are a couple of smaller plumes caused by flare-ups.

Figure 7.7 shows only the delineated burn scar resulting from the Chinese fire, and for comparison, Figure 7.8 is a Landsat mosaic taken over a two-week period near the end of the fire. In Figure 7.8, the burn scar is the darkened area, and the Amur River, the boundary between China and the Soviet Union, can be seen winding across the top of the scene. The difference in resolution between GAC (over 18 km^2) and Landsat (30 m) is apparent; however, the agreement in the outline of the burn scars is very good. Another good agreement between GAC and Landsat imagery is the indication of remaining vegetation in the burned areas. In the GAC image, the cloud contaminated pixels are the darkest pixels and are usually outlined in white, but the remaining changes in pixel shade are due to reflectivity changes from varying amounts of vegetation. In this image, the darker pixels show areas where there is more remaining vegetation. These pixels correspond to the lighter shaded

Figure 7.7 This is the outline of the Chinese burn scar as seen in the GAC data product.

areas in the Landsat burn scar where more vegetation remains. Overall, there is a very good agreement between the two sets of imagery.

Area Determinations

To determine the size of the burn scars, it was first necessary to understand how the area of a pixel changes along a scan. It was assumed that the geometry, given a nominal satellite altitude, does not change appreciably between scans, particularly over short segments of an orbit. Given this assumption, the problem of estimating the area of a geographical region can be reduced to totaling all pixels within some threshold by their position within a scan and multiplying that total by the appropriate pixel area for each scan position. This technique was used to determine the rate of spread of the Chinese fires during a three-day period and the total area covered by the Chinese and Soviet fires.

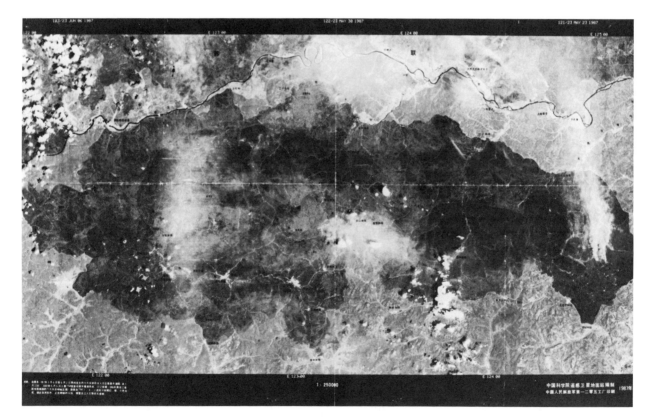

Figure 7.8 The Chinese burn scar as seen by Landsat. The burn scar is the darkest region. Some Soviet burn scars can be seen above (north of) the Amur River.

As mentioned previously, the Chinese fires were often intensified by dry frontal passages which increased wind speeds to fuel the fires but produced little or no precipitation to help extinguish the fires. The cloud cover from these dry frontal passages gave few opportunities to monitor the entire fire region. However, there was one opportunity to do this over three consecutive days—16, 17, and 18 May. Using channel 3 to penetrate the smoke plumes, temperature thresholds were determined to establish which pixels were in the burned areas for each of the three days. Based on the area estimates for each day, the burned region was determined to be increasing at a rate of about 15,000 hectares per day.

29 May 1987 was a predominantly clear day from which the Chinese and Soviet burn scars were visible in channel 2. A threshold was set and the burn scar pixels were tabulated for each scan position. The total area affected by the Chinese fire was 1.1 million hectares and by the Soviet fires, over 3.6 million hectares.

Summary

Biomass burning is a major source of atmospheric greenhouse gases. Even though biomass burning is most often thought of as a feature of the tropics, large areas in the northern boreal forests burn annually as well. The Chinese-Soviet fires destroyed over 5 million hectares of boreal forest in about a month, and climatology records indicate that dangerous fire conditions are present annually (Stocks and Jin, 1988). However, Asia is not unique in experiencing biomass burning in the boreal forests. Canada and Alaska have similar problems with boreal forest wildfires. These are all largely remote areas where biomass burning is best monitored and quantified using remote sensing techniques.

The use of AVHRR GAC imagery is a valuable resource in estimating the areal extent of biomass burning over large geographical regions. Because of the instrument characteristics, resolution, and relatively low computer storage requirements, this imagery is very suitable for this application. A visual comparison was made with Landsat imagery, and the difference in resolution can be readily seen, but it is also apparent that the GAC imagery did well in quantifying the burned region. An important reservation is that small burned areas will be missed, and hence some uncertainty exists with the estimations of area using GAC imagery.

The areal extent of burning determined with remote sensing techniques, combined with known trace gas emissions of the ecosystem in which the burning took place, can be used to determine the total trace gas emissions injected into the atmosphere by the fire (Levine et al., this volume, Chapter 34). The use of remote sensing in this manner can be used to determine the areal extent of biomass burning for each major ecosystem and to better quantify the global trace gas emissions from biomass burning.

Problems in Global Fire Evaluation: Is Remote Sensing the Solution?

Jennifer M. Robinson

Crutzen et al. (1979) and Seiler and Crutzen (1980) set the stage for subsequent analyses of global emissions from biomass burning to global atmospheric chemistry. They, and many after them, indirectly calculated mass emissions for a given region, r, and target species, s, using functions of the following form:

$$E_{r,s} = \Sigma_r(A_r \times B_r \times a_r \times b_r \times ef_s)$$

where

$E_{r,s}$ and ef_s are, respectively, the emissions (mass per time) and the emissions factor (mass of target species per mass CO_2) of species $s;$ and

A_r, B_r, a_r, and b_i are, respectively, area burned, biomass per unit area, fraction of above-ground biomass burned, and fraction of biomass that is above ground for a region r.

Alternate chain computations were used for other forms of combustion as appropriate, e.g., estimates of household fuel use were based on estimates of populations and fuel use per capita rather than on estimates of areas burned and biomass per area.

The accumulation of error through a calculation sequence may be computed via Taylor series expansion, making the assumption that the higher-order terms go to zero (Stuart and Ord, 1987). In chain multiplications (as in the above sequence), the higher-order terms in the expansion blow up if coefficients of variation of terms exceed around 30%—as is the case for many parameters in the above calculation (Robinson, 1989). Thus, the margin of error in estimates of emissions from biomass burning for given regions or species is very large but statistically undefinable. The likely, but difficult-to-specify case of intercorrelation among terms further complicates the statistical picture.

Estimates made in the above fashion are the only comprehensive global estimates we have on biomass burning. Given the large uncertainties there is a good probability that emissions estimates will have to be greatly modified for different regions and species, e.g., as has happened recently for N_2O due to the discovery of widespread measurement artifacts, and for emissions from tropical savanna after the report of new findings by Menaut (this volume, Chapter 17). Under the circumstances, independently derived estimates, such as that provided by isotopic analysis of CH_4 (Quay, 1990), are extremely valuable. Meanwhile, we have no other data that can be used in chemical models, and understanding the nature of uncertainties in chain calculations provides a useful framework systematically improving the precision of our estimates.

In this chapter I critically examine the prospects for reducing uncertainties over global biomass burning using remote sensing. First I consider the global temporal, spatial, and intensity distributions of fires and the remotely sensible signals they create and discuss the opportunities and problems that exist for matching available sensors to fire signal. Then I consider problems relating to instrumentation and to atmospheric interference.

Fire as Signal

Fires produce four forms of signal that are easily observed from space: direct radiation (heat and light) from active fires, smoke, postfire char, and altered vegetative structure (scar).

Heat and Light

Because fires radiate at temperatures much higher than the earthly background, and much lower than the sun, their emissions spectra, as defined by the Planck equation, fall in the energy void between the emissions spectra of solar radiation and that of terrestrial radiation. In the mid-IR window, therefore, the radiative anomaly produced by fires is thus amplified many thousands above the terrestrial background (Robinson, 1991).

This makes fires extremely bright in the 3.7 μm channel of Advanced Very High Resolution Radiometer (AVHRR)[1] or in equivalent channels for other sensors [e.g., the Geostationary Operational Environmental Satellite System (GOES)], and fires much smaller than a pixel can be detected. Matson and Dozier (1981) devised a procedure for computing the size and temperature of a sub-pixel thermal anomaly using two or more spectral channels (i.e., fire or snow). This inverts the Planck curve, and solves the resulting equations using information about background temperature deduced from surrounding pixels and specifications of parameters to define atmospheric interference and emissivities of the target and background.

Setzer and Pereira (1988) estimated fire emissions in Amazonia from counts of AVHRR channel 3 hot spots using the formula

$$E_s = H \times Ah \times B \times C_1 \times C_2 \times a \times b \times ef_s$$

where

E_s and ef_s are, respectively, the emissions (mass per time) and the emissions factor (mass of target specie per mass CO_2) of species s,

H is the number of hot spots counted,

Ah is the area per hot spot,

C_1 and C_2 are correction factors accounting for double counted and uncounted fires, and

B, a, and b are, respectively, biomass per unit area, fraction of above-ground biomass burned, and fraction of biomass that is above ground.

This procedure includes more terms than used by Seiler and Crutzen and thus expands the opportunities for error to accumulate. Because B, C, and Ah are poorly known, the procedure, at least in its present state of development, is probably less precise than the procedure used by Seiler and Crutzen (1980).

Light emitted by fires at night is also greatly amplified over background and may be detected using the low-illumination visible sensor onboard satellites in the Defense Meteorological Satellite Program (DMSP) (see Croft, 1977; Sullivan, 1984). This signal amplification is reduced by bright background near the time of full moon.

Smoke

Smoke plumes are much larger than the formative fire event. What we perceive when we see smoke is scattering and absorption by aerosols in the plume. Smoke scatters light effectively in the visible range because the size class distribution of smoke aerosols is close to visible wavelengths. The same principle makes smoke sensible to sensors operating in the

visible and near IR. Large plumes can be seen from relatively low-resolution (e.g., pixels of 4 to 8 km on a side) sensors such as meteorological satellites. Absorption and scattering of visible light commonly makes smoke apparent in the visible range of the spectrum.

A few of the gases in smoke, notably CO, CH_4, and O_3, have strong characteristic absorption lines that may permit monitoring of their presence in the troposphere. CO and CH_4 are likely targets because they are not abundant in the stratosphere, and because they are not soluble and do not suffer cloud scavenging. Unfortunately, however, no available platform monitors these gases.

Large fires in relatively clear atmospheres produce plumes that are easily tracked, but small, wispy plumes may be difficult to sense using any sensor. The cumulative degradation of atmospheric visibility by large smoke palls can be mapped and is quite spectacular in many tropical environments during the dry season.

Smoke is a nebulous signal that moves, gets diluted and thinned out, and changes chemically as it ages. The correspondence between remotely sensed signals from smoke and emissions by fire or between smoke aerosol and smoke gas is far from straightforward, and separation of the signal in the smoke from that of background terrestrial environments can be difficult. Attempts have been made to evaluate smoke by following the aerosol signal in the near-IR and visible channels of AVHRR (e.g., Kaufman et al., private correspondence), but the viability of the research approach has not yet been demonstrated in a concrete application, and many question whether the signal in smoke is strong enough to be read over the background noise of terrestrial landscapes.

Char and Scar

Fire-charred surfaces are black across the electromagnetic spectrum and stand out as high-contrast signals on most sensors. Char, unlike heat, is sensed at scale. Amplification does not permit the evaluation of sub-pixel scale events. Thus a charred spot can be evaluated only if the sensor has a smaller footprint than the fire. The photogrammetric rule of thumb is that a target of three to five pixels on a side or larger is required to quantitatively evaluate a signal. Thus, in fire regimes where expected fire size is less than about 10 to 30 square kilometers, high-resolution and high-cost sensors, such as Landsat, de Systeme Probatoire d'Observation de la Terra (SPOT), or the Marine Observatory Satellite (MOS), must be used for char monitoring. Coarser sensors (e.g., AVHRR Local Area Coverage, with nadir pixel size of 1.1 km) are useful for monitoring fire in tropical savanna and unmanaged boreal systems, and for occasional large fires in other environments. Even Landsat-TM cannot provide reliable measurements on char spots smaller than about 1 km.

By *scar,* I mean the alteration of vegetation caused by fire. Like char, scar is an unamplified signal. However, scar seldom provides the strong contrast against background that is expected for char. Scar behavior reflects local plant ecology and may be sensitive to precipitation, grazing, and other factors. Interpretation commonly requires the input of someone who knows the regional vegetation of the burn site. Automatic interpretation at the scale of large regions or continents is unlikely unless the computer can be programmed with local ecological understanding.

Despite these problems, scar can be a useful signal. In cases such as boreal forest and tropical forest, where fire effects are visible for years to decades after burning, study of canopy scars may be used to reconstruct historical patterns. For example, historical fires for the Canadian state of Ontario have been mapped by interpretation of Landsat imagery (Donnely and Harrington, 1978). On the assumption that tropical clearings are all burn scars, estimated rates of tropical deforestation made using Landsat are often used as surrogates for areas burned.

The Nature of Fire on the Landscape

Although fire is conspicuous from space, evaluating the global population of fire events from space is a logistical nightmare. Fire is heterogeneous in time, in space, and in intensity, on a multiplicity of scales, and opportunities for artifacts and missed observations abound. Temperatures commonly vary by hundreds of degrees between different locations in the same fire. Differences of 100°C or more may be found in the same location over a period of seconds as wind or hot pockets (patches of highly combustible fuel) cause flare-ups. High-frequency temperature variation is superimposed on a baseline trend in which the maximal heating rates achieved in the flaming stage of combustion are much higher than those of smoldering combustion, and smoldering combustion dwindles, unevenly, to nothing. Diurnal weather patterns, sometimes in combination with patterns of human activity, tend to synchronize hours of peak combustion, and seasonal and interannual patterns of precipitation and drought tend to synchronize days and months of high fire incidence.

Fire size, dynamics, and geometry are immensely variable. Fire in grasslands typically burns in a linear front. Open range fires and forest fires in remote regions commonly burn through several nights with more flaming than smoldering combustion. Fires used in the clearing of tropical moist forest are often lit from all sides and burn toward the center of the clearing. In such fires, flaming combustion rarely lasts more than a couple hours. In heavy fuels, the duration of smoldering combustion and completeness of combustion is sensitive to fuel moisture content and to fuel loadings and chemistry. I have seen slash fires in which the hollow centers of many large logs were still burning two days after the fire was lit. A hundred hectares of sugar cane may be burned in 15 minutes. Flame heights in a cane burn may exceed 10 m; the smoke plume is large and dark; it often produces a pyrocumulus cloud; it has little smoldering aftermath. The charred canes may be collected the following day.

Fires used for cooking, biomass burning for industrial purposes, and small fires used to eliminate vegetative debris are commonly *covert*—that is, they are either too small to be detectable by an affordable sensor, or they occur within a chamber or under a roof and therefore leave no signal other than general contribution to atmospheric turbidity.

Many tropical landscapes are a complex mosaic of pasture, wasteland, woodland, and mixed crops. Therefore one normally finds many types of fire side-by-side in the same landscape. For example, on a given day in the dry season in any agricultural colonization areas of Brazil, one may find grass fires, burning of woody slash on pastures, burning of primary forest, burning of secondary forest, and reburning of residuals from the first burn in forest clearing. The mix will vary considerably depending on where one is and on the part of the dry season. I once field checked an area in Chiapas, Mexico, where many fires were sighted by AVHRR, and concluded that the image included burning of agricultural waste, forest fire, burning of slash from primary forest conversion, burning of grazing lands, and burning of sugar cane—but that a very major part of the burning in the imaged area probably came from household fires and charcoal making, which would not have been detected by remote sensing (Robinson, 1987).

Sensor Problems

At-scale evaluation of char or scar is impractical where fires are small, and events smaller than about 50 hectares contribute significantly to the total amount of combustion. Small fires mean that higher-resolution sensors such as Landsat must be used, which becomes prohibitively expensive on a global scale. Moreover, char and scar have variable signal lifetimes. In tundra environments, a black cast may persist on the landscape for years. In agricultural burning, plowing may destroy the evidence in short order. In many ecosystems, vegetative regrowth does the same. In the case of Landsat, with a return observation interval of 15 days, it cannot be assumed that cloud-free imagery will be acquired before the char signal fades. Few data are available on signal duration in different environments, and for much of the world it is unclear how many images must be acquired to catch the fire activity of a full year.

Different problems arise when you try to base a fire evaluation program on the heat signals from fire. The amplification of thermal signals from fires is commonly sufficient to the AVHRR mid-IR channel (which was designed for calibration of sea surface temperature measurements and which has a dynamic range appropriate for that task). This makes size and temperature calculations impossible for larger or hot-ter fires. Figure 8.1 plots an envelope of resolvable fires—i.e., fires whose emissions would result in a channel 3 blackbody emission reading of 307°K or greater against a background temperature of 300°K for a sensor calibrated to saturate at 325°K.[2] Superimposed computed sizes and temperatures of 26 fires were analyzed by Matson and Holben (1987) in the Brazilian Amazon.

Of the 169 fires counted on the image from which Figure 8.1 is based, 143 (85%) saturated the sensor in the mid-IR. The saturation problem can be overcome by shifting to the coarser (8 km pixels) data from the GOES geostationary satellite, but only at a cost of losing sensitivity on the lower end of the range. How many fires would have been observed on the same day had the time of overpass been two hours later is an interesting question (Menzel et al., this volume, Chapter 4).

Temporal biases cause problems when using either mid-IR or nighttime visible signals. Sensors evaluate fires only at the time of overflight—i.e., at a fixed time of day or night. If the fire burns for many hours or days, the area burned may be quite different from

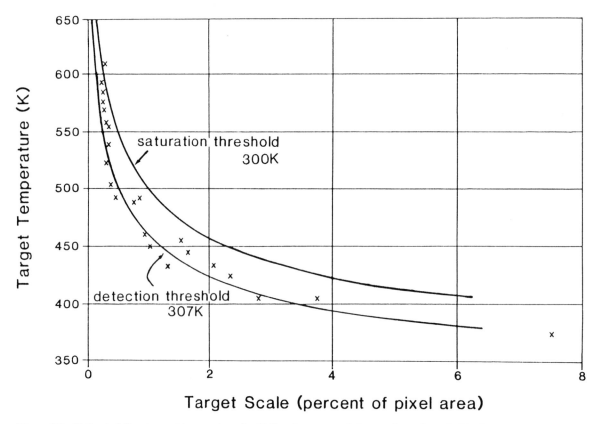

Figure 8.1 Estimated fire size and temperature for 26 fires in context of the envelope of resolvable fire events for an AVHRR-like sensor. Based on Matson and Holben (1987).

that evaluated at the instant of observation, and the same fire may be observed more than once. If the fire passes quickly, it may not be observed at all or may be evaluated during the long tail of smoldering combustion, when it will appear as a low-temperature event. Nighttime observation is problematic because fire activity is generally greatest near thermal noon, and some types of fires (such as those used in slash and burn agriculture) rarely burn at night. Furthermore, the DMSP was designed more for reconnaissance than for research, and insufficient information is available about its calibration to support quantitative analysis of radiant energy fluxes (J. Dozier, personal communication, 1985).

Atmospheric Interference

It is convenient to assume that burning is restricted to cloudless days. That is far from true. Many tropical regions where fire is intensively used in land management or as fuel are cloudy most of the time, even in the dry season. Figure 8.2, for example, shows percent of daylight hours with clear sky in September, October, and November—the principal season of burning in Amazonia.

Active radar is the only form of sensor that can penetrate heavy clouds. The imaging form of active radar (sidelooking airborne radar, or SAR), however, places such high demands on data transmission systems that SAR sensors have been flown only occasionally, on an experimental basis.

Thermal IR sensors penetrate some clouds and smoke, but signal strength is diminished, and separating the original signal after atmospheric interference is not a trivial problem, particularly when data are lacking on the depth and optical characteristics of smoke and clouds. Thus clouds and smoke interfere with fire evaluation to a variable and difficult-to-predict extent.

Satellites are almost useless for evaluating fire in places such as the "boiler box" of Southeast Asia, where persistent clouds make it difficult to study anything below the cloud deck using remote sensing. A more perverse problem may develop in Amazonia. The Brazilian government is now using AVHRR data to police regulations against burning. Helicopters are sent after suspicious hot spots, and heavy fines have been imposed on persons conducting illegal burns. We do not know how rapidly people will learn that they can burn with impunity under heavy cloud cover, but when they do, the correspondence between hot spots and fires will disintegrate.

Strategically speaking, the variances are as critical

as the norm. If mission planning presumes low cloud abundance, and high cloud abundance is encountered, data collection and interpretation encounter myriad problems. For example (see Figure 8.2b), the interannual variability of cloud cover in the western Amazon Basin is relatively high. The years 1987 and 1988 were both relatively dry with many days of clear-sky conditions; 1989 was quite cloudy. The BASE-A experiment of 1989 was designed assuming a repetition of the weather pattern of 1987 to 1988. We fell short of many of our objectives because cloud cover was too thick to permit measurements (personal observation). Similarly, it is difficult to tell, in Amazonian fire counts made with AVHRR, how much of the change in apparent fire frequency in 1987 and 1988 and in 1989 is real and how much is an artifact caused by greater cloud interference.

The presence of smoke further complicates signal interpretation. This subject is beyond the present scope. However, as discussed in various chapters in this volume, smoke optical density may become very high, and capping cumulus often forms over large smoke plumes.

Conclusions

Estimates of fire emissions based on remote sensing will be more precise than estimates like those of Seiler and Crutzen (1980) if and only if the relationship between the signals perceived by satellite and the total amount of biomass burned is less tenuous than the computation sequences used by Seiler and Crutzen. If, on the other hand, the remotely sensed data must be processed through a long string of calculations and numerous simplifying assumption, and if, furthermore, satellite data contain potentially large but difficult to quantify biases, remotely sensing is at best an alternative method of estimation. This alternative is not necessarily more precise than prior methods, and if satellite-based computations use the same data on biomass per unit area and fraction of biomass burned as previous calculations, they cannot be considered as independent from previous calculations.

Given the many vagaries of satellites, fires, and the atmosphere, and the large fraction of all fires that are below the detection thresholds of most conceivable sensors, the superiority of satellite-based estimates cannot be taken for granted. Particularly problematic are (1) regions where cloud cover is heavy and where a major fraction of the biomass that is burned is burned as small piles of agricultural waste, as house-

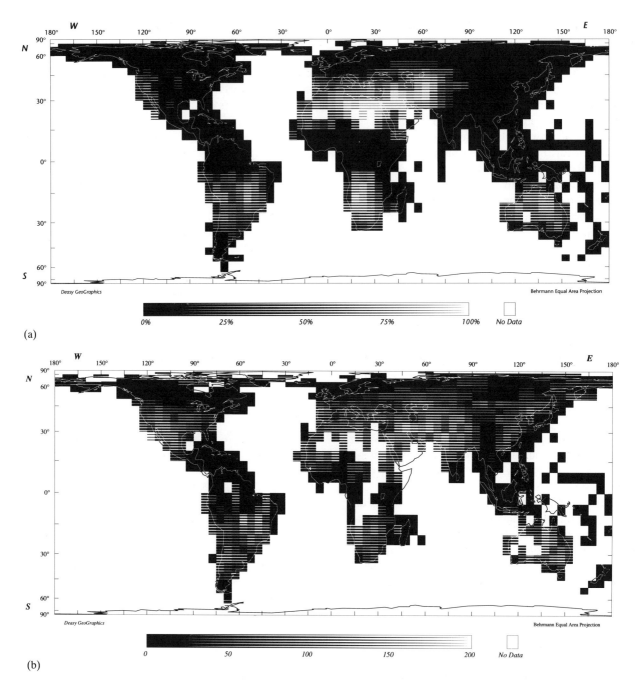

Figure 8.2 (a) Percent occurrence of completely clear sky. (b) Interannual variation in occurrence of completely clear sky for September, October, and November. Based on ground station data compiled by Warren et al. (1986). The same source provides data on a monthly basis and for various cloud types.

hold fuel, or in the making of charcoal, and (2) mixed fire regimes in which the sensor that is used is not equally appropriate to all fire types present in the landscape. I expect that between a third and a half of all biomass burning falls in one of these two categories.

Even where most fires give clear signals, unbiased satellite-based estimates cannot be made without an understanding of the spatial and temporal distributions and residence times of fire signals, and of events that might obscure them. The next generation of sensors may ameliorate some of the above problems, e.g., by providing routine observations of CO and CH_4. Other problems, e.g., imprecision over estimates of tropical deforestation, are likely to be ameliorated independently of fire-related research.

The above should be read as a plea for realism, balance, and candor. Remote sensing has a history of underestimating problems and promising more than it can deliver. In 1973, when Landsat became operational, it was widely believed that monitoring of land cover and cover change was at hand. Seventeen years later, general circulation models are parameterized based on vegetation maps that date to the 1920s (Wilson, 1984), we are still having trouble accounting for terrestrial sinks of CO_2 (Tans et al., 1990), and the characterization of land surface cover is rated as "F" by NASA's Earth System Science Advisory Committee (NASA, 1988).

I do not seek to discourage the use of remote sensing but rather to help nonremote sensors to appreciate the difficulties involved in the transition from images to quantitative information, and to encourage mission planning that is cognizant of problems at the outset and that accommodates them by foresight rather than patching them over in hindsight. Furthermore, in light of the many problems in satellite evaluation of fire, it is important that remote sensing of fire is not pursued to the exclusion of ground-based field measurements; that satellite-based studies be complemented by adequate ground-truth information; and that independent exercises, e.g., methane isotope studies, be used to check satellite-based results.

Acknowledgments

This paper owes a great deal to the input of D. S. Simonett and J. Dozier. Research was supported by NSF grant ATM-86-0921.

Notes

1. The imaging sensor carried by the NOAA polar orbiting meteorological satellites.
2. Calibrated saturation temperatures (in °K) for channel 3 on NOAA 6–11 have been, respectively, 352.3, 322.0, 324.2, 326.5 331.3, and 327.5 (unpublished data, James Silva, NOAA, 1987).

II

Biomass Burning in Tropical Ecosystems

Biomass Burning and the Disappearing Tropical Rainforest

Thomas E. Lovejoy

In Mexico's Sierra de Manatlán in the state of Jalisco, a dozen or so years ago, a new species of plant was discovered. Discovering a new species of plant is really a very ordinary event in tropical forest. So there had to be something quite special about this one for its discovery to be heralded by a front page story in the *New York Times*. This new species of plant turned out to be a wild relative of corn, the third most important grain in support of human society.

Wild relatives of corn had been known before. Yet in contrast to all but one of the other known wild relatives of corn, this particular plant species was perennial. That is, it springs anew from its root stock every year rather than having to start from seed as all annual plants do (including the domestic species of corn).

One other perennial species of corn had been discovered earlier in this century by a Smithsonian botanist. Unlike the present one, that species did not have the same chromosome number as the domestic variety, so the ability to introduce the characteristics of the wild form into the domestic variety had not been possible. With this discovery it suddenly became possible to take characteristics of the new perennial species of wild corn and begin to bring them into domestic corn and ultimately into corn agriculture.

Initially, the most important characteristics were the resistance of this perennial form to many of the virus diseases that affect corn, including one that knocked out a significant percentage of the U.S. corn crop in the early 1970s. Still somewhat elusive, but I think in the end, possible, is the hope of bringing the perennial characteristic into corn and agriculture. If a perennial corn agriculture were possible, it would represent a major benefit for human society: not having to plow every year, a savings in fossil fuel, and less soil erosion. All other things being equal, corn prices would be cheaper, as well as the myriad products which trace back to corn, including such abstruse things in our daily lives as the glue on the back of postage stamps.

In any case, the series of points that arise from this story, which I think of as the parable of the perennial corn, are as follows. First of all, this plant was found not in the United States but in another country. There is absolutely nothing about where a species occurs in this world that says anything about where it might prove to be of immense practical use. Indeed being in Williamsburg makes me think of Thomas Jefferson's statement that one of the greatest gifts a person can make to his country is to introduce a new species of plant. Not only was this plant found in another country, but it was also an endangered species. There is nothing about whether a plant or an animal is rare or abundant that says anything about whether it is going to have the kind of practical value the perennial corn represents.

Far more important than all of this is that the perennial corn makes an eloquent case for human society being best served by landscapes that are both domestic and wild. We are very lucky that the hillside on which the perennial corn grows had not been converted into an ordinary corn field, as is usually the case in Mexico. The parable of the perennial corn makes a strong case for the protection of the diversity of life on this planet.

If one concerns oneself with protection of the variety of life on Earth, one is immediately drawn to the tropical forest. Clustered on the 7% of Earth's dry land surface where tropical forests grow near the equator are somewhere between 50% and 90% of all plant and animals species. This is the biological treasury of the planet.

In contrast to climate in the north temperate zone, where temperature is the great variable throughout the year and rainfall is relatively constant, in tropical rainforests the situation is reversed. Temperature is relatively constant throughout the year, and rainfall varies. The months of October and November, the driest months of the year near the mouth of the Amazon at Belem, have rainfall roughly equivalent, for example, to an average month in Toronto, Canada.

The rainforests have year-round conditions of constant warmth and moisture, conditions highly favorable to life. In many senses it is not at all surprising that in regions of the tropics where there is sufficient moisture, essentially the greatest expression of life on Earth occurs: the tropical forest.

The numbers of species are always very interesting. For example, the tip of South America at Tierra del Fuego has only two species of ants. However, going north, just to the tropics in the vicinity of São Paulo, there are already 200 species of ants, with the equator and the equatorial forests still a long way off. The point this illustrates is that as one approaches the wet regions of the equator, species numbers rise dramatically. This repeats itself over and over again from one group of organisms to another, with obvious exceptions, of course; there are not more species of penguins at the equator than at the South Pole. But this pattern of species richness growing toward the equator is a general phenomenon.

Species richness expresses itself even on a very local basis. In southern New England forests there are 10 or 15 species of trees at most. Where the drainage is poor, or farther north, forests may drop to just a couple of species of trees, or even one. In the tropical forest we have richness almost beyond belief. In these kinds of forests, there are literally hundreds of species of trees. One 25-acre (10-hectare) plot in the Central Amazon contains well over 300 species of trees, and that number is low compared to some of the plots from the Western Amazon. This is an extraordinary richness in tree species. With huge numbers of species the almost automatic consequence is that many will occur at low densities with just a few numbers of individuals per unit area. That carries with it certain consequences of how these forests work.

In the temperate zone the dominant form of reproduction for trees is wind pollination, and it is a relatively efficient exercise. But if there are just a few individuals scattered over a large area, wind pollination would be an extremely inefficient approach. Therefore, in a tropical forest most of the trees depend on various animal species to do the pollinating for them, essentially moving from a shotgun to a rifleshot approach, where an individual animal will go directly from one individual of a tree species to the next. This is done by a variety of organisms, including bats, so if you are in the business of managing a tropical forest for whatever purpose, you are in the business of managing its bats.

These forests are also highly stratified. Literally, the species which occur at the top of the forest are so different from those which occur at the bottom of the forest that the only species the canopy and forest floor really share are the canopy trees themselves. Opportunities to get up into the forest canopy and really see what happens on top are a great treat for biologists.

At the end of the 1960s, when the world's most famous bird watcher, Roger Tory Peterson, visited the Amazon, he learned there was a tower providing access to the canopy. He then literally abandoned all his plans; nothing would do but that every morning before dawn he had to be up on that tower to see the bird life. Up in the canopy a Smithsonian entomologist, Terry Erwin, recently studied the kinds of insects which occur in the very tops of those trees. With those data the estimate of the numbers of species on Earth went up dramatically.

There are about 1.4 million described species of plants and animals. When I was a graduate student, the general idea was there were 3 or 4 million species total, so 1.5 or 2.5 million were yet to be described, and surely most of those were in the tropics. As time has gone on, people have taken new looks at tropical forests and the number of species estimated to occur in them has gone up. After Terry Erwin's look, the estimate of the total number of species on Earth went up to between 30 and 60 million species. That involves a lot of extrapolation, so the numbers may change. The important point is that just about every time a biologist goes into these forests and takes a serious fresh look, the estimate of the number of species in these forests goes up. The state of the tropical forest therefore becomes more critical every time a new look is taken at the number of species within it.

In most tropical forests the species are very poorly known. Birds are literally the best-known group of animals on Earth. Even so, I once caught a very distinctive little bird, the black-chested pygmy tyrant (*Taeniotriccus andrei*), but it took me two and a half years to identify it because it had never been known within 500 miles of where it was caught.

Something on the order of only one out of 20 species in the tropical forest has actually been seen by scientists and given a scientific name. Given that, it is easy to understand why the importance of those species—their practical value, what they mean to biology, and the rest—has only just begun to be understood. Many of these species occur in very local distributions, not throughout the Amazon or the Congo Basin or Southeast Asia. As the forest is cut and destroyed, basically those species with limited distributions become very vulnerable to extinction.

Most tropical forest trees have very shallow root

systems. The very same conditions of warmth and moisture so favorable to these incredibly exuberant vegetation formations are also highly favorable to decomposition. That sets in very rapidly, and the shallow root system is there to take the nutrients back into the living system. This ability to capture nutrients and pump them back into the trees is the secret of why tropical forests can in many instances, but not all, exist on some of the poorest soils on Earth. It is one of the great paradoxes and has fooled government decision makers many times.

Slash and burn agriculture, the biomass burning which the Chapman conference considered, involves cutting and burning trees to release the nutrients as ash in a pulse of fertilizer, which the peasant farmer can use for a short-lived agriculture of, say, three to five years. That was what the Amerindians did in those forests for generations. It works pretty well as long as too many people are not doing it.

The vast majority of the large number of species in any biological community are invertebrates. It is important for us to remember the world is really run by the plants and the wigglies and the squirmies. We vertebrates are just along for the ride. One of the important aspects of the insect community, with beneficial returns for ourselves, involves a continual struggle between the insect community and the plant community. There is constant grazing pressure on the vegetation from insects. The plant community evolves ways to resist this by producing certain kinds of biologically active molecules to discourage insect attack. They make the insects sick or even kill them. On the other side of this evolutionary struggle, the insects develop the ability to produce special enzymes to break down some of those compounds. And they occasionally evolve special strategies such as eating just a little bit of any kind at a time, which is a little bit like having arsenic for breakfast, strychnine for lunch, and cyanide for dinner.

The net results of these plant-insect struggles is that plant life of tropical forests is filled with these compounds; the tropical forests are essentially the greatest pharmaceutical factory on Earth. Small numbers of these are used in our hospitals in the industrialized world on a regular basis. When Ronald Reagan had his operation after the assassination attempt, he almost certainly was treated with curare, a major muscle relaxant used in any abdominal operation. Much of the modern pharmacopeia actually comes from plant origins, including the best-selling medicine of all time—namely, aspirin, originally a compound extracted from willow bark.

In the tropical forest the people who have the best knowledge of which plants are likely to have useful compounds are the indigenous peoples who have lived in close contact with the forest for millennia. They do not know them all but have found a number to be very effective. We have a great deal to learn from these people, now under terrible pressures, if only we can treat them with more respect than a World War II aviator's map which divided the tribes of the Amazon Basin into two kinds: hostile and unfriendly.

In many parts of the world the forest is being cut and burned at considerable rates. It is embarrassing for modern science not to be able to give a precise number, but in fact the funding for the kinds of analyses that are necessary has not been readily forthcoming, although this is beginning to change. Nonetheless, somewhere between 50 and 100 acres of tropical forest are destroyed every minute.

Among the most famous images of tropical forest destruction are those involving the BR364 highway in Rondonia, Brazil, where a highway was built with World Bank funding. In just five years (1982 to 1987), literally 20% of the forests of Rondonia were destroyed. This is one of the fastest rates of tropical deforestation anywhere in the world. This is not only a Brazilian problem. Tropical deforestation is a worldwide problem, and many of the Amazonian countries may have a relatively worse deforestation rate than does Brazil.

Almost anywhere in the world that forest destruction has taken place, whether in New England, Wisconsin, Europe, or São Paulo, the forest is not only destroyed but is also fragmented in the deforestation process. That creates an additional kind of problem: Once they become isolated, islands of habitat are unable to maintain the full variety of plant and animal species they contain.

Barro Colorado Island, located in Gatun Lake, created for the Panama Canal in the early part of the century, is a major biological station of the Smithsonian Institution, with long continuous records about its biology. In the 1920s there were 208 species of breeding birds described; 50 years later, at least five of those species had vanished, a fair portion of them because the island simply was not big enough. Something happens when a piece of forest is no longer a part of a larger continuous wilderness. The smaller the piece, the more rapid the change and the greater the loss of plant and animal variety. This has very serious consequences for protection of biological diversity of tropical forests or any kind of habitat, including in the United States, where some of our western national parks are suffering species loss for this reason.

That problem has led to a research project I have been involved in with the National Institute for Amazon Research (Instituto Nacional de Pesquisas da Amazônia—INPA) at Manaus, the World Wildlife Fund, and the Smithsonian Institution. The project takes advantage of the law in Brazil that 50% of any development in the Amazon has to remain in forest. We persuaded certain ranchers to arrange the geometry of their cutting to leave forest fragments of different sizes. We look at what happens in them so we can understand what it says about an ideal minimum size or how to manage a piece of forest to hold more than it otherwise might. The project is known as the Minimum Critical Size of Ecosystems or Biological Dynamics of Forest Fragments project.

An example of how the minimum-size problem works involves army ants. They occur in colonies of half a million or more and go through a three-week cycle during part of which they are in a quiescent bivouac state. During the other part they are in a very excited swarming state, advancing across the rainforest floor, going through the leaf litter and running up and down small tree trunks, looking for insect and other animal prey which normally sit still and are camouflaged. Birds like the white plumed ant bird make their living following army ant swarms and swooping down in front of them to steal some of those insects before the ants get to them.

What kind of area can support a system of such small organisms as this one? There must be enough army ant colonies so that on any given day, enough of them are swarming so that the birds which follow them can get enough food to support themselves, but they also must raise young and replace themselves in the population. The minimum area for a system of such small organisms is probably somewhere between 100 and 1000 hectares.

There are other consequences of this tropical deforestation. Because there is no longer protective vegetation, the rains cause devastating erosion. If deforestation proceeds over wide areas, it produces landscapes like that of the central plateau of Madagascar: Denuded of its native vegetation, the plateau has serious soil erosion, and the rivers hemorrhage to the sea with red soil. Nobody gains by that kind of treatment of landscape.

Other things can be disrupted, and one of them was very much in the news at the time of the Chapman conference—namely, the ability of the Amazon Basin and its forest to make about half of its own rainfall. The conventional geographer's wisdom is that vegetation type is caused by the climate. We are beginning to learn, however, that vegetation also can affect its own climate, and indeed that is the case in the Amazon.

Careful measurements made near Manaus and elsewhere by Brazilian scientists show what happens to rain that falls in a particular spot. Roughly half to three-quarters of it returns to the atmosphere. Indeed, if you fly in a small plane over the Amazon, you can see the rain clouds dropping rain here and there as they move westward toward the Andes in an almost unidirectional airflow. Not far behind, one can see plumes of moisture coming out of the forest canopy to create new clouds and more rain downwind.

A recent issue of *Science* includes a computer model analysis of what would happen to the climate in the Amazon were all forest to be replaced by pasture land of a particular African grass (Shukla et al., 1990). Rainfall is projected to fall 25%, average temperature goes up about 2°C to 3°C, the dry season extends, and even if all the species of the forest are not made extinct in the process of removing forest, the situation looks irreversible. It would be very hard to bring the forest back, and since some of the moisture in central Brazil leaks out of the Amazon Basin, there will be effects there as well. No one knows what it would do to global climate to change that much energy and water flow in a major chunk of the equator.

Somewhere along the apparently inexorable march of deforestation there is a threshold point at which an irreversible drying trend will be set in motion. Where that number is nobody knows; I would guess that it is between 30% to 40% deforestation. This is something we really need to understand because the problem could be beyond our control very suddenly.

Regarding the topic of the Chapman conference, the importance of biomass burning, in addition to the kind of atmospheric pollution caused by industrialized society, which is by far the greatest part of the global problem, there is a major contribution from burning of tropical forest and other biomass. The numbers are big, and they tie in inevitably to concerns about changing climate through the greenhouse effect and all its implications for society.

The greenhouse effect is a very old concept dating back to 1895. In fact, the normal temperature of Earth is set by the natural levels of greenhouse gases in the atmosphere; otherwise it would be some 60°F colder. The changes in atmospheric composition from burning fossil fuels, and the burning of biomass will inevitably bring some climatic change.

The controversy is not over whether there is going to be global heating; the controversy is about whether you can tell whether it has already started, how hot it will get, how fast it will happen, and what will happen in a particular place. It is very easy to get confused about press reports which "last" week show that a computer model predicted a 5°C warming with a doubling of CO_2 and "this" week involve only 3°C because the reporters changed some factors in the model. The headline then might read "Global Warming Cools." The bottom line is we have a serious problem, and society has to do something about it.

There are all kinds of things which will flow unhappily from climate change, and in the end the changes are going to happen relatively rapidly and be convulsive. They will be very disrupting to societies and economies. One of the most serious effects, which has gotten little attention as yet, is that on biological diversity. Biological diversity is going up in smoke in the tropics and contributing to the global climate change problem, but the global climate change problem is going to come back and hit biological diversity yet again.

In a time of natural climatic change like a swing between a glacial and an interglacial period, species would usually be able to migrate up in altitude or latitude to follow their ideal climate. The modern world, however, deals with fragments of habitat; most of our biological diversity is locked up in isolated parks and reserves. The rates of change suggested by some studies are 40 to 100 times faster than species have ever been known to track climate under natural conditions. Even if they could do that, there is an incredible obstacle course for them with the tremendous modification of the landscape. The point is, while there is a wave of extinction going on in the tropical forest, when climate change comes, it will create yet another wave of extinctions in the rest of the world.

There have been five major natural episodes of extinction in the course of the history of life on Earth. Most of these were about 10% reductions in the diversity of the time, with one big one at the end of the Permian on the order of 50%—a time some paleontologists say life itself was lucky to get through. Unless we change our ways, we are looking at extinction on the order of at least 20% to 25% of all plant and animal species. So that's the kind of problem we can bring down on ourselves and on society unless we begin to look at things very differently than we have been up to now.

What is causing all of this? In any particular place,

the problems are peculiar to it: A decade ago in Central America, slash and burn agriculture was the dominant cause, but there cattle ranching and lumbering were critical factors as well. Any place will have its own particular immediate causes. What lies behind those causes are big social vectors, with human population probably number one among them. We are adding 95 million more mouths to this planet every year. At the first meeting of the President's Council of Advisors in Science and Technology, the father of the "green revolution," Norman Borlaug, spoke eloquently about population and the need to get it under control.

I was really struck by something the rock star Sting once said: He said an Amazon Indian knows exactly where everything in his or her life comes from—precisely which tree, which plant, which stream—and those of us living in industrialized society simply don't have a clue. No wonder it is easy for us to get into environmental trouble. Our tendency to act as if we are only remotely connected with our environment is reflected in economic terms. Nothing represents this more in my view than the international debt between the industrialized North America and the developing Central and South America—about $1.4 trillion.

The graph of increasing atmospheric CO_2 concentrations measured from the top of a volcano in the Hawaiian Islands could be the symbol of this conference. It may well end up being the most famous graph of this century. While inexorably increasing, the line goes up and down every year. Each time the line goes down, something like 6 billion tons of carbon is removed from the atmosphere. It is removed by living things: springtime in the Northern Hemisphere as trees leaf out and the growing season begins and wood is laid down. That shows the immense power of biology, which we are lucky to have to help us get through the problem.

Indeed, we probably cannot manage to reduce our atmospheric pollution unless we begin to provide global management of forests. Cutting CO_2 contributions from deforestation and burning can reduce this equation by a billion tons of carbon. A massive reforestation program around the world, including in the United States, can remove another billion tons. The point is we at least have 2 billion tons of play just because of what trees can do if we manage them properly. That makes coping with the rest of the CO_2 excess by means of energy conservation or energy efficiency more attainable.

We have not only to think differently about energy use, but we also have to think differently about the international debt and whether there are ways to do debt-for-nature swaps for *reforestation,* conservation projects, research, and so forth. We also have to think differently about development and move toward sustainable development, which I think, in the end, has to be measured by whether biological diversity is able to survive in large units of landscape or not.

The industrialized countries which have tropical forests must get involved in some serious leadership by example. The Hawaiian Islands are the United States' most important tropical forest. Long before Captain Cook arrived and got his rather unhappy welcome, large numbers of plant and animal species had already gone extinct at the hand of the original Hawaiians. Forty-five of the 60 extinct species of birds disappeared prior to the arrival of Europeans in Hawaii, but subsequently a further 15 have, and many of the others are endangered. That pattern repeats for plant species. We simply have to do a much better job if anything we are to do to help in the rest of the world can be at all credible.

In ending, consider a South American pit viper. It's really hard to work up a sweet feeling about that snake. What could that possibly have to do with life in these United States? This snake has a rather unpleasant habit of injecting venom into its prey that causes their blood pressure to go to zero forever. Scientists studying how that worked discovered a previously unknown system of regulation of blood pressure in our own species, and knowing that system exists, scientists from the Squibb Company were able to build a molecule which plays on that system. That is the drug Capoten, the preferred prescription drug for hypertension in the United States and large parts of the world.

Literally, millions of people are living longer and healthier lives because of some knowledge based on the biology of a nasty snake in some far-away forest. This is an example of the power of biological diversity. It is the fundamental library for the life sciences, which in many senses, are the most important ones in support of human society. Allowing large chunks of diversity to go extinct is essentially a book-burning exercise. It is probably the greatest anti-intellectual act in our history. If people proposed doing this to the Library of Congress, they would not get beyond the first statement before people would be enormously upset. Somehow, because in this case the "books" are living things, people have accorded them lesser value.

We all know about species from the tropical forest that have affected our lives, like the rubber tree. Not one of us gets through the day without some use of rubber or imitation thereof. There was a time, of course, when rubber came only from the tropical forest of the Amazon, and the rubber barons lived very well indeed. They even built great opera houses like the one in Manaus which was just reopened.

Another species from the tropical forest that has profoundly affected the way society works is the giant water lily. The impressive pads have an underside of cellulose ribs and girders which inspired the first person who ever got it to bloom in horticulture, Joseph Paxton. He thought it would probably support a lot of weight and did an experiment that confirmed this. Being the kind of person who would not let a good thing alone, he thought it would be fun to design a glass house for these water lilies and built it according to structural principles you can take right from the underside of the giant water lily. That, in turn, became his basis for the design of the Crystal Palace of the great London exhibition of the 1850s, which is the acknowledged origin of modern metal beam architecture. Literally, half the buildings in the industrialized world stem from the underside of the giant water lily.

Tropical Wild-land Fires and Global Changes:
Prehistoric Evidence, Present Fire Regimes, and Future Trends

Johann Georg Goldammer

The impact of tropical wild-land fires and other bio-mass burning on the environment is receiving increasing attention by atmospheric sciences. Earlier models and calculations of the trace gases released by the combustion processes involved in forest conversion and other land-use practices (Crutzen et al., 1979; Seiler and Crutzen, 1980; Crutzen et al., 1986; Crutzen and Andreae, 1990) showed that these emissions contribute considerably to the total annual global release of greenhouse gases and a net carbon flux to the atmosphere. Contributions to this volume reflect the growing interest in this subject (Andreae, Chapter 1).

Wild-land fires, however, are not a recent phenomenon. As fire history reveals, northern boreal and circumpolar ecosystems and many of the temperate and mediterranean forest biomes have been shaped by fire since prehistoric times (Kozlowski and Ahlgren, 1974; Wright and Bailey, 1982; Wein and MacLean, 1983). Large-scale wildfires in Northeast Asia such as the conflagrations in Siberia and Northeast China during 1987 were reported early in this century and have been documented in fire history (Shostakovich, 1925; Goldammer and Di, 1990). Fahnestock and Agee (1983) additionally have emphasized the importance of smoke produced by prehistoric wildfires.

Recent synoptic and multidisciplinary approaches to tropical wild-land fire ecology prove that natural fires and widespread human burning practices within the tropics have also greatly influenced vast areas of vegetation throughout all tropical biota during historic and prehistoric times (Goldammer, 1990b).

This chapter points out that wild-land fires in the tropics have existed for a long time, that fire has shaped and changed the forest and other vegetation, and that fire regimes are undergoing changes—in the past, today, and, most likely, in the future. Wild-land fires are related to disturbances by nature and by humans. Natural disturbances may be characterized as of short duration, such as the effects of lightning, drought, and hurricanes; other natural disturbances

may be long-term processes, such as climatic fluctuations. Anthropogenic disturbances have not changed basically throughout human history. They are all related to forest conversion, slash and burn agriculture, hunting and grazing, and other wild-land uses (Bartlett, 1955, 1957, 1961; Goldammer, 1988). The difference between the past and today's fire scene, however, is the growing extent of fire-affected vegetation due to population pressure and accelerating degradation of vegetation cover. The future fire scenarios are largely determined by the human-induced feedback mechanisms between biosphere and atmosphere. Trace gases from tropical biomass burning are different from gases emitted by decay of plant life and play a significant role in the acceleration of change in the tropical biosphere and in the atmosphere.

Prehistoric Evidence

Global climatic changes throughout geological periods are drastically reflected by the fluctuations of the atmospheric CO_2 content and the temperature revealed by the Vostock ice core data for the past 160,000 years (Barnola et al., 1987, 1989). These climatic changes have largely influenced the development and biogeography of the tropical biotas which were subjected to migration both in area of distribution and altitude and to species extinction. Flenley (1979) suggests in his geological history of the rainforest that the tropical rainforests could not have developed as ecosystems undisturbed through extreme long periods as suggested by the classical school of tropical forest ecology (Richards, 1952; Whitmore, 1975); instead, large areas have carried savanna-type vegetation during the Quaternary (Pleistocene) due to cooler and more arid climatic conditions at that time (Prance, 1982).

Palynological reestablishment of tropical vegetation history in most cases has underestimated valuable fire history information, although valuable data on charcoal are generally available. New evidence about the ability of early hominids to use fire beginning about 1.5 million years ago (Brain and Sillen, 1988) suggests that fires must have played a similar role in the fluctuating or transition savannas of the Pleistocene, as could be shown for historic times. Australia's Prequaternary and Quaternary fire history seems to have received sufficient attention (Kemp, 1981; Singh et al., 1981), which is lacking in the lower latitudes in the Americas, Africa, and Asia.

The lack of radiometric data in the present tropical moist rainforest area can be explained by flooding and erosion processes (Colinvaux, 1989), by the change of river beds, and by the most likely rapid turnover of organic material, including charcoal in tropical soils. Occasional observations of charcoal under primary rainforests have been underestimated in their ecological significance until very recently. The ^{14}C dating of Amazon rainforest charcoal revealed abundant fires up to about 6000 years before present (B.P.) and reflect both human activities and climatic oscillations during the period of postglacial climate stabilization in the Holocene (Sanford et al., 1985; Saldarriaga and West, 1986).

Radiometric dates from modern primary rainforests in eastern Borneo, however, indicate that fire already must have occurred during the peak of the last glacial period (Wisconsin-Würmian glaciation) about 18,000 B.P. In their investigations in East Kalimantan (Indonesia), Goldammer and Seibert (1989, 1990) found charcoaled remnants of trees and dipterocarp seeds under primary dipterocarp rainforest. These fires presumably have been ignited by burning coal seams stretching along or near the forest surface. Thermoluminescence dating of burned clay on top of extinguished coal seams found in the intermix with active coal fires, revealed subsurface fires back to about 13,000 to 15,000 years B.P.; ongoing investigations (Goldammer, unpublished data) of burned clay tend to date the fires back to about 50,000 years B.P.

The evidence of prehistoric fires in today's rainforest biomes requires an interpretation beyond its immediate impact. The occurrence of fires may have fulfilled an ecological function (or task) during evolutionary or phylogenetic processes and related time scales. What was the role of fire in the development of the rainforest? Provided that the rainforest refugia (forest islands) theory has more substantial evidence than it has received criticism (Haffer, 1969; Simpson, 1972; Simpson and Haffer, 1978; Prance, 1982; Beven et al., 1984; Connor, 1986; Salo, 1987), one of the important functions of fire was between the refugia. The main postulate of the refugia theory is that diversification of species took place in rainforest patches isolated from each other. The forest refugia were separated by savanna vegetation and a seasonal climate with distinct dry fire seasons. The gene flow (seed dispersal, pollination) between the refugia was interrupted by "fire corridors," resulting in locally restricted diversification processes—one possible mechanism of the high species diversity in tropical rainforests.

In addition to that Darwinistic explanation of fire in evolutionary processes, another interpretation of the role of fire in diversification exists. In those primary rainforests where prehistoric charcoal was found, it must be distinguished whether the fuel affected by fire was originated in savanna biome or in a forest environment. The Borneo findings of Goldammer and Seibert (1989) showed that the analyzed charcoal was remnant from a dipterocarp forest. How and why could a forest burn which can be characterized as moist rainforest?

Fires in moist rainforests must have occurred because of brief climatic disturbances, not because of a climate change in the sense of thousands or ten thousands of years. Climatic oscillations of short duration (interannual climatic variability) are a possible explanation for periodic flammability of the moist rainforest biomes. The rainforest fires of 1982 to 1983 in Borneo, which affected a total of more than 5×10^6 hectares (ha) of primary dipterocarp forests, secondary and peat swamp forests in East Kalimantan (Indonesia), and the Malaysian territory of the island, are a striking example (Goldammer and Seibert, 1990). An extended drought triggered by the El Niño–Southern Oscillation (ENSO) event predisposed the rainforest to the fires set by shifting cultivators which largely spread into the surrounding forest lands. Such rainforest fires are not a recent phenomenon. Goldammer and Seibert (1990) showed that similar events had happened during ENSO droughts over the past 100 years. The collected fire data from prehistoric and presettlement times may be the result of a unique coincidence of periodic droughts (such as ENSO-related) and the permanent availability of fire sources (burning coal seams). In all other places lightning has been the main source of fire. Hurricane damage may also result in availability of flammable rainforest fuels. This has recently been demonstrated by the impact of Hurricane Gilbert on rainforests in Cancun (Yucatan, Mexico) in 1987 which were ready to burn after drying in 1989 (90,000 ha burned). Open questions are size (extent of burned area) and frequency of ancient rainforest fires. Prehistoric fire mapping is one of the upcoming challenges in tropical fire ecology.

The function of periodic fires which were limited in

size can be interpreted in two ways. The direct impact of lightning and lightning fires creates gaps (canopy openings) in the rainforest. These gaps are important elements for the regeneration and the dynamics of restructuring the rainforest (gap dynamics). In an undisturbed forest the pioneers and many other species would have been replaced by a few dominant *climax* species. Disturbances may be considered as basic elements for the theory of diversification through instability (Connell, 1978; Hubbell, 1979; Picket and White, 1985; Hubbell and Foster, 1986). Again, this theory is controversial to the Clementsian school of ecology (Clements, 1916). Clementsian climax postulates the explanation of species richness in tropical rainforests through stability over evolutionary time scales (Ashton, 1969, 1988; MacArthur, 1972; Whittaker, 1977).

One more evolutionary role of fire in rainforests may be added. Rainforest islands as diversification centers are mainly sought in climate-induced refugia or topographic features (isolated mountains) providing separation from neighboring gene pools. The process of forming of such islands within vast areas of closed rainforest may be performed by fire. Wildfires usually burn irregular patterns which create a mosaic of burned and unburned patches. This phenomenon is entirely explored in the temperate and boreal zones and explains the existence of diverse forest landscapes with diversity in stand ages and species composition. Similar effects of fire islands have been observed after the recent rainforest fires in Borneo (Goldammer and Seibert, 1990). This phenomenon becomes more visible in today's fire savannas of Western Africa where fire has shaped distinct, abrupt edges between rainforest patches and the surrounding pyrophytic vegetation. These fire-induced rainforest islands in a rainforest climate may have served, and may continue to serve, as centers of development of new varieties and even new species.

The role of fire in savanna and deciduous forest ecosystems is different from its role in rainforest biomes. Seasonal climate and seasonal vegetation are predisposing elements for the regular occurrence of fires. In many cases it is therefore difficult to determine whether natural fires or edaphic, orographic, and climatic conditions—or all of these elements combined—are the driving forces in shaping a savanna biome. Batchelder and Hirt (1966) suggested that the anthropogenic fire influence on tropical vegetation dates back to about 12,000 years B.P. Various documented examples on savanna formation support this assumption, such as the savannas of Rajastan (India), which were formed with the beginning of mesolithic grazing practices about 10,000 years B.P. (Jacobson, 1979). Maloney (1985) suggested that deforestation of landscape in Sumatra goes back 18,000 years. The savanna (monsoon) forests of Kampuchea are explained by the influence of fire on the dipterocarp forest for 2000 years (Wharton, 1966). On the other hand, as mentioned earlier, fire has been used by humans for more than 1.5 million years. Where and when was the beginning of the anthropogenic fire (and subsequently grazing) influence on tropical vegetation?

Paleofire regimes have varied with the influence of climate and man. Therefore a general picture cannot be drawn of the possible fire scene in the prehistoric past. However, a tentative prehistoric fire world during the last Pleistocene glaciation is shown in Figure 10.1. The map shows the postulated rainforest ref-

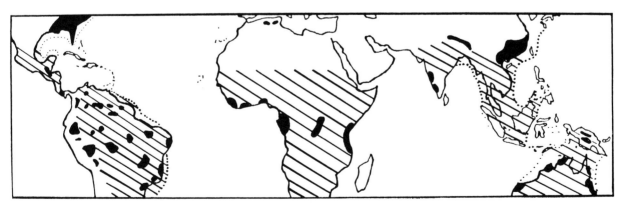

Figure 10.1 Suggested schematic distribution of fire-prone and of nonflammable vegetation types in the tropics during the peaks of glacial periods of the Pleistocene. Solid areas show tropical and adjoining forest refuges according to Prance (1982). Hatched areas between forest refuges illustrate the possible distribution of seasonal savanna-type vegetation characterized by short-return-interval fire regimes.

uges as compiled by Prance (1982), which are surrounded by vast areas of probably seasonal and drier vegetation subjected to fire in short return intervals.

Present Fire Regimes

After today's tropical climate became stabilized about 8000 to 10,000 years B.P., the rainforest reached its distribution and size as found by the early explorers and scientists. Alexander von Humboldt described the overall picture of the rainforest as damp, humid, and not flammable (Goldammer, 1990a). This impression was kept alive by tropical ecologists until very recently.

Recently, however, tropical fires have become visible through satellite imagery and have entered the calculations of climate modelers. In a satellite imagery such as the AVHRR a ground fire on a grassland with low fuel load, low emission and energy release may look completely similar to a slash and burn fire in which 50 to 100 times an amount of biomass is burned and a completely different bouquet of trace gases is released. Fires and fire regimes thus need to be classified and distinguished for a better understanding of the processes involved, ranging from the terrestrial ecological impact to the questions of trace gases and global carbon fluxes.

The characterization of fire and the general role and impact of fire in the environment can be classified as different fire regimes. A general model of tropical fire regimes proposed by Goldammer (1986b; see also Mueller-Dombois and Goldammer, 1990) is characterized by return interval, intensity and origin of fire, and main vegetation features. The main categories are related to anthropogenic and ecological gradients (sociocultural and ecological frame of landscape). From the seven fire regimes distinguished, four main types of tropical fire environments can be derived for the context of interest of this volume—rainforest fires, fires in deciduous forests and other open tree formations, tropical pine forest climax, and tropical savanna and grassland fires.

Rainforest Fires

Normally, fires do not regularly affect the tropical perhumid rainforest biomes. Natural disturbances locally may change the existing conditions which are unfavorable for fire—e.g., high humidity, fuel moisture, and lack of available surface fuels—eventually in return intervals up to hundreds of years. At present, human-caused fires are occurring more frequently and becoming a determining phenomenon of tropical

rainforest development. While the traditional system of slash-and-burn agriculture has been following a fallow cycle (corresponding to a fire return interval), present fire use overwhelmingly is related to conversion of rainforests to other permanent land uses (e.g., Malingreau and Tucker, 1988; Fearnside, 1990). Depending on weather conditions and burning skill, only a part of the above-ground biomass is consumed at the first burn. Subsequent burns usually follow, but sometimes the remaining biomass is left for decomposition and the agricultural crops are planted in between. In the case of conversion of forest to grazing lands (pastures) the same area will be reburned repeatedly, in many cases annually, until remnant woody vegetation, resprouting capability, and seed bank are eliminated (Watters, 1960, 1971; Peters and Neuenschwander, 1988).

The global extent of shifting cultivation fires and the rates of forest conversion are not known precisely. According to estimates of the Food and Agricultural Organization (FAO) (FAO, 1982, 1985; Lanly, 1985) about 500 million people were involved in shifting agriculture within the tropics during the early 1980s. It was estimated that shifting cultivation and its degraded forms at that time affected about 240×10^6 ha of closed forest and about 170×10^6 ha of open forest, totaling about 21% of the tropical forest area. The assessments on the extent of deforestation (in the sense of permanent forest conversion to other land use) in which fires are usually involved have a broad range, thus reflecting the lack of reliable information. The estimates of Crutzen and Seiler (1980) give a range of annually burned or cleared area of 21 to 62×10^6 ha due to shifting agriculture; other estimates are somewhat lower (Detwiler et al., 1985; Houghton et al., 1985; Myers, 1980, 1989). The most recent estimates of Myers (1989) still show an annually cleared area of 13.8×10^6 ha and a present net deforestation rate of 1.8% of rainforests in 28 investigated tropical countries.

A distinct feature of primary or closed rainforest fires is that they are used to remove woody vegetation from a specific and limited site. Because repeated disturbances result in increasing the flammability of rainforests and reducing them to savannas, the fires tend to escape from shifting cultivation plots and spread into the surrounding forest lands. Wildfires in vegetation characterized by uniformity of fuels, species, desiccating behavior, and climatic seasonality (fire season) tend to recur in shorter and more regular intervals. These fires shape uniform landscape patterns (larger fire mosaics) and are characteristic for the tropical deciduous forests and other open tree formations, including the savanna biomes.

Fires in Deciduous Forests
and Other Open Tree Formations

With increasing distance from the perhumid equatorial zone, the extent of regular or irregular droughts and wildfire occurrence is increasing. With decreasing precipitation and an increase of duration of drought periods, the tropical forest formations gradually develop toward semievergreen ecotones and finally to dry deciduous forests (Legris, 1963; Hegner, 1979). The distinction between these forest types and the general term *savanna,* which often includes open tree formations, is not clear (Hegner, 1979). Deciduous forests characterized by open understory and regular fire influence are sometimes designated as both forest formations and as savannas. One example is the *monsoon forests* of continental Southeast Asia, which are also referred to as *savanna forests* or *savannas* (Cole, 1986; Stott, 1988a, 1988b; Stott et al., 1990). This confusion of terminology has created problems in distinguishing and mapping tropical fire ecosystems and fire regimes which are the base of global models in trace gas emissions from biomass burning (e.g., Hao et al., 1990; other contributions in this volume).

In this chapter the term *forest* is used as long as trees are dominant landscape elements and are involved in the interactive processes between fire and vegetation. The main fire-related characteristics of these formations are seasonally available flammable fuels (grass-herb layer, shedded leaves) and adaptive mechanisms which allow the grass layer, other understory plants (shrub layer), and the overstory (tree layer) to survive and furthermore take advantage of the regular influence of fire. The most important adaptive traits are thick bark, ability to heal fire scars, resprouting capability (coppicing, epicormic sprouts, dormant buds, lignotubers, etc.), and seed characteristics (dispersal, serotiny, fire cracking, soil seed bank, and other germination requirements, etc.) (Goldammer, 1991). These features are characteristic elements of a *fire ecosystem.*

During the dry season the deciduous trees shed the leaves and provide the annually available surface fuel. In addition, the desiccating and finally dried grass layer, together with the shrub layer, adds to the available fuel, which generally ranges between 5 to 10 tons ha^{-1}. The fires are mainly set by forest users (graziers, nonwood forest product collectors). The forests are usually fired in order to remove dead plant material, to stimulate grass growth, and to facilitate or improve the harvest of other forest products (Goldammer, 1988). The fires usually develop as surface fires of moderate intensity (usually less than 400 kw m^{-1} according to Stott, 1988b) and tend to spread over large areas of forested lands. The tree layer is generally not affected by the flames, although crowning may occur earlier in the dry season when the leaves are not yet shed. In some cases fires may affect the same area twice or three times per year, e.g., one early dry season fire consuming the grass layer and one subsequent fire burning in the shed leaf litter layer (Goldammer, 1991). The size of these fires is usually larger than the intended area of impact. This is mainly due to the uniformity of available fuels.

Dry deciduous forests and moist deciduous forests are occurring on about 250×10^6 ha and 530×10^6 ha, respectively (Lamprecht, 1986; Windhorst, 1974). No reliable information exists on the extent of recurring fires in these areas. Goldammer (1986a) estimated that in Burma between 3 to 6.5×10^6 ha of forests annually are affected by fire. A recent report from Thailand contains similar figures for the predominating dipterocarp monsoon forests of about 3.1×10^6 ha per year (Royal Forest Department, Thailand, 1988). The analysis of historic information from British India reveals that during the last century and early this century almost all Indian deciduous forests were reported to burn every year (Goldammer, 1991). Figure 10.2 shows the global area of distribution of tropical deciduous and monsoon forests potentially burning in short-return intervals of between one and five years.

The ecological impact of the yearly fires on the deciduous and semideciduous forest formations is significant. The fire strongly favors fire-tolerant trees which replace the species potentially growing in an undisturbed environment. Many of the monsoon forests of continental Southeast Asia would be reconverted to evergreen rainforest biomes if the man-made fires were eliminated. Such phenomena have been observed in Australia where the aboriginal fire practices and fire regimes were controlled and rainforest vegetation started to replace the fire-prone tree-grass savannas (Ellis, 1985). The fire adaptations and the possible fire dependence of economically important trees such as sal (*Shorea robusta*) and teak (*Tectona grandis*) have been the focus of controversial discussions about the traditional fire control policy in British Indian Forestry for a long time (Pyne, 1990; Goldammer, 1991).

The fire climax deciduous forests are not neces-

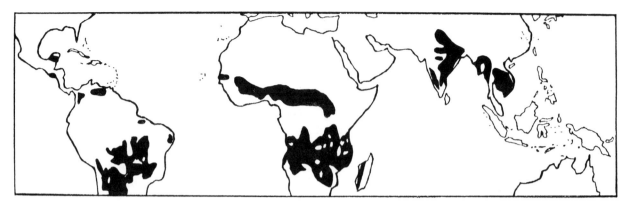

Figure 10.2 Present distribution of deciduous and monsoon forests in the tropics and adjoining regions which are affected by regular short-return-interval fires (solid areas). The boundaries given in this map are rough and may overlap with savanna biomes (Figure 10.4).

sarily in an ecologically stable condition. The long-term impact of the frequent fires is considerable erosion because of the removal of the protective litter layer just before the return of the monsoon rains. The erosion rates under standing deciduous forests regularly affected by fire may exceed 60 tons yr^{-1} (Goldammer, 1987).

Tropical Pine Forest Fire Climax
Approximately 105 species of the genus *Pinus* are recognized. From the main center of speciation in Central America and Southeast Asia some species extend into the tropics (Critchfield and Little, 1966; Mirov, 1967). The pines are largely confined to the zone of lower montane rainforest. They are usually found on dry sites and require a slight to distinct seasonal climate. Most tropical pines are pioneers and tend to occupy disturbed sites, such as landslides, abandoned cultivation lands, and burned sites. The fire ecology of tropical pines has been described for Southern Asia (Goldammer and Peñafiel, 1990) and for Central America (Munro, 1966; Koonce and González-Cabán, 1990); there are no pines occurring naturally between the tropics of Africa and in the whole of the Southern Hemisphere except Sumatra.

Besides the pioneer characteristics, most tropical pines show distinct adaptations to a fire environment (bark thickness, rooting depth, occasionally sprouting, flammability of litter) (Goldammer and Peñafiel, 1990). The tropical pure pine forests of Central America and South Asia in most places are the result of a long history of regular burning. Fire-return intervals have became shorter during the last decades and range between one to five years. These regularly occurring fires favor the fire-adapted pines which replace fire-sensitive broadleaved species. The increased frequency of human-caused fires has led to an overall increase of pines and pure pine stands outside of the potential area of occurrence in a nonfire environment (Munro, 1966; Kowal, 1966; Goldammer and Peñafiel, 1990). However, together with the effects of overgrazing (including trampling effects) and extensive illegal fuel-wood cutting, the increasing pressure of wildfires tends to destabilize the submontane pine forests resulting in forest depletion, erosion, and subsequent flooding of lowlands. Like in the tropical deciduous forests the fires are mainly set by graziers but also spread from escaping shifting cultivation fires and the general careless use of fire in the rural lands.

These tropical fire climax pine forests are occurring throughout Central America, the midelevations of the Himalayas, throughout submontane elevations in Burma, Thailand, Laos, Kampuchea, Viet Nam, Philippines (Luzón), and Indonesia (Sumatra) (Goldammer, 1987). The extent of the annually burned pine forest lands is not known. Figure 10.3 shows the area of natural distribution of pines within the tropics and the adjoining regions which are greatly influenced by the tropical climate and similar socio-ecological and cultural conditions.

Tropical Savanna and Grassland Fires
The various types of natural savanna formations are potentially of edaphic, climatic, orographic, or fire (lightning fire) origin and are influenced by wildlife (grazing, browsing, trampling). Together with anthropogenic influences (e.g., livestock grazing), fuel-wood cutting, and other nonwood product uses, most tropical savannas are shaped at present mainly by

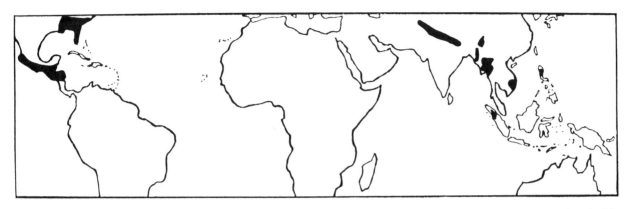

Figure 10.3 Area of natural and anthropogenic fire climax pine forests (*Pinus* spp.) within the tropics and adjoining regions (solid areas). These pine forests are characterized by occurrence of short-return-interval fires.

regularly occurring man-made fires. The impact of these fires became so dominant that only complete exclusion of fire and the other man-made influences would allow us to recognize the locally prevailing nonanthropogenic driver of savannization.

The role of fires in savanna ecosystem dynamics was recognized early. A publication of the German Colonial Service (Busse, 1908) describes the origin and impact of man-made forest fires and the importance of these fires for the rural savanna culture. The interactions of wildlife, man, and fire in the prehistoric climate and landscapes as highlighted by Schüle (1990) are of significant importance in the development of tropical savannas. Modern synoptic approaches toward an integrated savanna ecology have always considered fire as a major functional force (e.g., Tall Timbers Research Station, 1972; Huntley and Walker, 1982; Cole, 1986; Pätzold, 1986; these monographs contain numerous bibliographical sources on savanna fires).

There is a tremendous variety in physiognomy of the savannas occurring throughout the tropics of Africa; North, Central, and South America; and Asia. A common feature, however, is the grass stratum as the ground layer within the open savanna woodlands (tree savannas) or as the exclusive element in the grass savannas (grasslands) and in the ecotones between. From the point of view of fire ecology and biomass burning, the definition of a savanna ecosystem and its distinction from open forests should be based on the potentially available wild-land fire fuel. In this context savannas are defined as those ecosystems in which the grass stratum is the exclusive or predominant wild-land fire fuel; open deciduous forests, on the other hand, should predominantly be

characterized by available fuels from the tree layer (leaf litter).

The available fuel (biomass density) per hectare depends on the net primary phytoproduction and varies between 0.5 and 2.5 tons ha^{-1} in the dry savannas of the Sahel and up to 8 tons ha^{-1} in the moist savannas of Guinea (Menaut, this volume, Chapter 17). The susceptibility of savannas to extended fires depends on fuel continuity and density (Imort, 1989). Accordingly, the fire frequency varies between about one and five years depending on the spareness and required minimum accumulation of flammable plant material.

The total global area of tropical savannas annually affected by fire and the total global biomass burned are not known. Figure 10.4 shows the present extent of tropical savannas potentially affected by short-return-interval fires. The overlap of areas with the distribution of open deciduous forests (Figure 10.2) is due to uncertainties in definitions and boundaries and the ecotonal characteristic of vegetation. Another uncertainty is the continuously progressing transition of flammable savannas to nonflammable ecotypes (desertification), which reduces the extent of area affected by fire. In a pantropical savanna fire mapping we found that savannas occur on a total of 2.6×10^9 ha, of which up to 1.5×10^9 ha may be affected by fire each year. This upper limit of frequently burned-over savannas corresponds to a value of 7.9×10^9 g of biomass (dry matter) combusted annually and a prompt release of about 3.5×10^9 g of carbon to the atmosphere (Goldammer and Weiss, 1991). Thus savanna fires contribute to approximately 65% of the total of about 5.5 Pg of prompt annual carbon release (upper potential limit) to the atmosphere by tropical

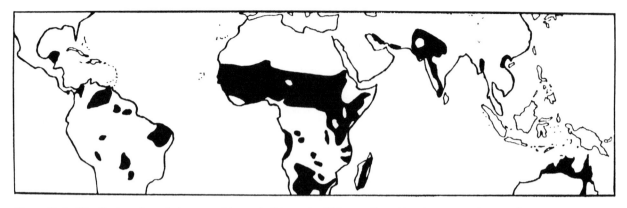

Figure 10.4 Distribution of tropical savanna biomes (solid areas) which are affected by short-return-interval fires. Rough map without distinction between the various savanna types (such as dry or moist savanna biomes).

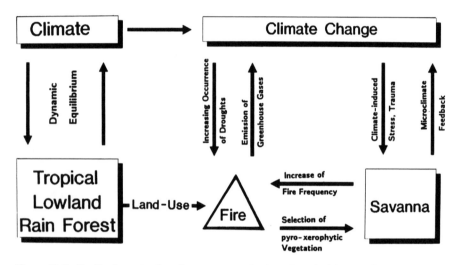

Figure 10.5 Feedback mechanisms between savannization of tropical forests, fire, and climate change (Goldammer, 1990a).

and subtropical vegetation fires (Goldammer and Weiss, 1991).

Future Trends

A future scenario of the extent of tropical fires and fire regimes can be derived from the suggested paleo-fire information and today's knowledge of tropical fire ecology. The process of expected change of the earth's vegetation cover induced by humans follows two different pathways. The first pathway is the large-scale dimension of savanna conversion, forest conversion, and depletion of vegetation cover (e.g., desertification) of the tropical biota through all kinds of human activities. These direct impacts induce the secondary pathway. Forest depletion, savannization, and the combustion processes involved result in a change of the physical and chemical environment.

Reduction and change of characteristics of vegetation cover result in change of evapotranspiration and albedo, thus consequently leading to changes in climate. The change of the water cycle, surface heating, and radiation regime may result in a drier and eventually warmer environment. Additionally the emission of trace gases through biomass burning contributes to the greenhouse effect and accelerates global warming (Figure 10.5).

Although the models on tropical greenhouse climate are still preliminary, there is sufficient evidence in trends to predict the future tropical wild-land fire scenario. Depending on political measures, the process of deforestation of the tropical rainforest may be finished soon due to either one of two reasons—because nothing will be left anymore (the worst-case scenario) or because legal restrictions will keep some of the rainforest and other closed forest formations

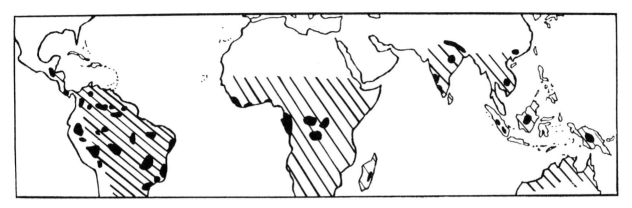

Figure 10.6 Possible distribution of fire-prone and of fire-protected tropical vegetation in a $2 \times CO_2$ climate. Hatched areas show large-scale degraded vegetation (savannized) regularly affected by short-return-interval fires. Solid areas tentatively show protected refuges (biosphere reserves), which may serve as gene-pool reserves and as source of forest expansion in a posthuman environment.

protected as reserves. Assuming that rainforest reserves will be maintained, a future tropical wild-land fire scenario in a $2 \times CO_2$ climate could look quite similar to the late Quaternary conditions (Figure 10.1 and Figure 10.6). The forest reserves (corresponding to the Pleistocene refuges) will be kept free from humans and fire. Between the refuges the non-agricultural vegetation will be largely degraded toward pyrophytic and xerophytic (tree-) savannas and characterized by short-return-interval fires.

However, there are differences between the paleo-fire scenario and the $2 \times CO_2$ fire scenario. One distinction is the available land mass bearing savanna fire climaxes. During the Pleistocene the total area of lowland savanna formations was larger due to the lowered global sea surface level and the exposure of the shelves. Under the $2 \times CO_2$ climate the available land mass may even be smaller because of a possible rise in sea level due to global warming. The other distinction is the fire climate. The tropical paleofire climate was cooler and drier than today, and the $2 \times CO_2$ climate will be warmer than today, with regional and local extremes in weather patterns (droughts, hurricanes).

Conclusions

Wildfires have been an integrated factor of the natural environment of Earth for millions of years. Evidence of ancient wildfires is given by fusain (fossil charcoal) embedded in the coal seams of the carboniferous period (Francis, 1961; Komarek, 1973). During geological time scales the role of fires in ecosystem processes was always interdependent with climate patterns and characteristics of the atmosphere. Car-

bon fixation and carbon release of the terrestrial biota occurred in and probably were triggers of the cyclic patterns of the glacial and the interglacial periods.

Since the beginning of burning fossil carbon sources in the mid-nineteenth century man has added a new driving force into global climate changes. Both the human-caused conversion of terrestrial biomass and the combustion of the buried fossil carbon storage will lead to an accelerated process of global change in the near future. The responses of the terrestrial tropical biota are not yet entirely predictable. However, from the past and present tropical fire regimes it can be concluded that wild-land fires will continue to burn in the future. Negative impacts of fires need to be encountered by integrated fire-management systems. Fire policies must be targeted to provide vegetational cover to protect the soil productivity and to maintain or increase the overall carbon storage of land biomass. Thus, forest or vegetation management strategies are needed which consider a global perspective in environmental policies.

Greenhouse Gas Contributions from Deforestation in Brazilian Amazonia

Philip M. Fearnside

The greenhouse effect is the sum of heat-absorbing actions of various gases emitted from a variety of human activities and natural processes in different parts of the world. Although carbon dioxide emissions from industrialized countries represent the largest single factor, other sources of greenhouse gases, such as tropical deforestation, also make significant contributions. Policies designed to control global warming must be based on an adequate understanding of the nature and magnitude of the gas sources, the cost and effectiveness of possible policy changes, and the benefits that are being derived from activities that now release greenhouse gases. The Brazilian Amazon, with the largest remaining area of tropical forest, is of central importance not only because deforestation in this region contributes a substantial amount of carbon to the atmosphere, but also because controlling deforestation is amply justified from the perspective of Brazil's own interests, independent of the question of global warming. Slowing forest loss is possible because the process of deforestation in Brazil is largely driven by factors that are subject to government decisions. Separate discussions have been published treating deforestation's causes in Brazil (Fearnside, 1987a), its meager benefits (Fearnside, 1985a, 1986a), heavy environmental costs (Fearnside, 1985b, 1988), and irrationality from the perspective of the long-term interests of the country (Fearnside, 1989a, 1989b). Measures that would help slow forest loss in Brazilian Amazonia have been reviewed both from the perspective of what the Brazilian government could do (Fearnside, 1989c) and that of possible contributions from other countries (Fearnside, 1990a). Potential impact on other countries makes Amazonian deforestation a focus of worldwide concern (Fearnside, 1989e).

The present and potential contributions to the greenhouse effect from deforestation in the Brazilian Amazon are uncertain because of the small amount and low reliability of data on several key components in the calculation. Brazilian Amazonia's great size and heterogeneity, combined with the relative pau-city of data, make these uncertainties a weak point in global carbon budget calculations.

The present contribution of deforestation is a function of the annual rate at which forests are being cleared, biomass of the forests, partitioning of biomass in above- and below-ground compartments, carbon content of the vegetation, fraction of above-ground carbon transferred to long-term pools such as charcoal, completeness of burning, reburning practices (including transformations to and from charcoal pools), rate of decomposition of unburned biomass, carbon stocks in replacement vegetation, and carbon stocks in soil under original and replacement vegetations. The ratio of gases released by deforestation affects contribution to the greenhouse effect. Calculation of potential release also requires knowing the total area for each vegetation type present.

All of these quantities are uncertain. The uncertainty of the overall result depends both on the uncertainty of each factor and on the sensitivity of the result to changes in that factor. Many uncertainties have multiplicative effects, rapidly degrading the reliability of the calculated releases (Robinson, 1989). Despite these limitations, it is essential that the best estimate possible be made from the available data. Where measurements are missing for needed quantities, such as the biomass of certain vegetation types, then guesses or assumptions based on similar vegetation elsewhere must be used. Use of such low-reliability values is preferable to extrapolating to the region from the few existing high-reliability biomass measurements: It is better to be approximately right than to be precisely wrong. Despite disagreements and conflicting data on such vital factors as forest biomass and deforestation rates, the conclusion remains inescapable that Amazonian deforestation makes a significant contribution to the greenhouse effect. More fundamental than disagreements about the magnitude of deforestation and biomass is lack of consensus over how the results should be interpreted in terms of policy changes.

Deforestation Rates

Controversy surrounds the existing estimates of the extent and rate of deforestation in Brazilian Amazonia. These controversies are analyzed elsewhere, and a "best estimate" is derived which calculates that 8.2% of the originally forested portion of the Brazilian Amazon had been cleared through 1988 (including old clearings), with new clearing in the forest (virgin and old secondary forest) area expanding at 20×10^3 km^2 per year (Fearnside, 1990d).

Much of the literature on the contribution of tropical deforestation to global warming has been based on the deforestation estimates of the Food and Agriculture Organization of the United Nations (FAO) for 1980 (Lanly, 1982). This survey is both out of date and unlikely to represent the true extent of deforestation even for the period it covers. The information it reports was obtained by a questionnaire sent to the government of each country, rather than from independent monitoring methods such as remote sensing. In the case of Brazil, the task of responding was given to the Superintendency for Development of the Amazon (SUDAM), the agency responsible for subsidizing and promoting large cattle ranches in the region. Much of the information available at the time (reviewed in Fearnside, 1982) is not reflected in the report.

The deforestation estimate adopted here (Fearnside, 1990d) uses as many as possible of the measurements on 1988 Landsat-TM images made by Brazil's National Institute of Space Research (INPE) (Brazil, INPE, 1989a, 1989b). In the state of Acre a discrepancy with previous results of the Brazilian Institute for Forestry Development (IBDF), now part of the Brazilian Institute for Environment and Renewable Natural Resources (IBAMA), led to using a projection from 1980 and 1987 data in this state. In the state of Rondônia the absolute value for deforestation was derived from the INPE Landsat results, but an unexplained jump relative to Landsat data interpreted by IBDF from the previous year (Brazil, IBDF, 1989) led to using an estimate for deforestation rate in this state derived from AVHRR results (Malingreau and Tucker, 1988; J. P. Malingreau, personal communication, 1988; D. Skole, INPA seminar, 1989; see Fearnside, 1990d). In all states the INPE data (Brazil, INPE, 1989a, 1989b) were used to estimate the area originally forested, but the alteration of *cerrado* (central Brazilian scrub savanna) was estimated using a number of assumptions regarding the

proportionality of alteration in different vegetation types, or continuation of previous trends.

By the "best estimate" calculation outlined above, the cleared area in the Legal Amazon totals 353×10^3 km^2, 268×10^3 km^2 (76%) of which is forest (Table 11.1). Of the original vegetation cover (Figure 11.1), 7.4% of the total and 6.4% of the forest had been cleared by 1988. These values do not include "old clearings" (clearings made prior to 1960, which the INPE/Our Nature Program measurements registered as 31,822 km^2 in Pará and 60,724 km^2 in Maranhão). These older secondary forests were not detected in the earlier Landsat-MSS studies (see Fearnside, 1982, 1986b) and so cannot be used in the present study for the purpose of establishing trends by comparison with older data. The INPE study's area values for old secondary forest have been included in the biomass and carbon release calculations by considering old secondary forest as a separate vegetation type. The area that has lost its original forest cover, including the old secondary forest area, is an area the size of Finland: 345×10^3 km^2, or 8.2% of the original forest area.

The average rate of deforestation can be conservatively estimated by assuming constant rates since the last available satellite measurement of cleared area (Table 11.2). This procedure underestimates the current rate of deforestation because the calculation averages deforestation over the period between the last two available satellite measurements, while all evidence indicates that areas cleared have, in general, been increasing every year. An exception to this trend may be clearing in 1989, mainly due to heavier rains during the dry season than in the two preceding years. The nearly constant increase in the rate of clearing renders obsolete the many greenhouse-effect calculations that have been based on deforestation estimates for 1980 or earlier.

Release of Greenhouse Gases

Available Estimates

Calculating the potential contribution of deforestation to the greenhouse effect requires comparison of carbon stocks present before and after clearing. Estimates of potential emissions have been evolving as better information becomes available. An estimate (Fearnside, 1985c) based on a seven-category classification of vegetation by Braga (1979) and biomass for dense forest based on the mean results from existing

Table 11.1 Original vegetation and best estimate of areas recently cleared in the Brazilian Legal Amazon from 1960 through 1988

State	Original vegetation (km²)[a]				Recently cleared area (km²)			Percent recently cleared		Source
	Forest	*Cerrado*	Humid savanna	Total original vegetation	Forest	*Cerrado*[b]	Total	Of forest	Of forest + *cerrado*	
Acre	152,589	0	0	152,589	8,634	0	8,634	5.7	5.7	(d)
Amapá	99,525	0	42,834	142,359	842	0	842	0.8	0.8	(e)
Amazonas	1,562,488	0	5,465	1,567,953	12,837	0	12,837	0.8	0.8	(e)
Maranhão	139,215	121,017	0	260,232	34,140	20,664	54,803	24.5	21.1	(e)
Mato Grosso	572,669	235,345	72,987[c]	881,001	67,216	134,277	201,493	11.7	24.9	(e)
Pará	1,180,004	22,276	44,553	1,246,833	91,200	1,722	92,922	7.7	7.7	(e)
Rondônia	215,259	27,785	0	243,044	30,634	989[f]	31,623	14.2	13.0	(e)
Roraima	173,282	0	51,735	225,017	2,187	0	2,187	1.3	1.3	(e)
Tocantins/Goiás	100,629	169,282	0	269,911	20,279	34,114	54,393	20.2	20.2	(e)
Legal Amazon	4,195,660	575,705	217,574	4,988,939	267,969	191,765	459,734	6.4	9.6	

a. Original vegetation in accord with the INPE map (Figure 11.1), with the savanna areas apportioned between humid savanna and *cerrado* in their approximate proportions in the savanna areas shown for each state. The forest in Tocantins/Goiás has been increased by 68,573 km² presumed to have been included in the INPE survey but not in the map of original vegetation. "Forest" includes both primary (virgin) forest and old secondary forests (from clearings prior to 1960 in Pará and Maranhão). Totals are areas of political units, including water surfaces, as in the INPE and IBDF reports (making the percentages underestimates). The area of Tocantins/Goiás is that used by Brazil, INPE, 1989a, b; it is at variance with the 235,793 km² used in previous INPE reports (e.g., Tardin et al., 1980) for the same geographical area.
b. *Cerrado* clearing, which was not measured in the INPE study (Brazil, INPE, 1989b), has been estimated assuming that this vegetation type is cleared in the same proportion as the forest within each state, with the exceptions of Rondônia (where proportionality is assumed excluding *cerrado* areas in Amerindian reservations) and Mato Grosso (where data exist for *cerrado* clearing in the western part of the state in 1983, and the ratio of *cerrado* to forest clearing observed there is assumed to apply to the entire state through 1988).
c. Pantanal (Mato Grosso humid savanna) area from IBGE data reproduced in Benchimol (1989). The remainder of the savanna area in Mato Grosso shown in Figure 11.1 (with correction for state area) is considered *cerrado*.
d. Linear projection from the last two years of available satellite data (see Fearnside, 1990c).
e. Brazil, INPE, 1989b, with corrections for state area and *cerrado* clearing (see text).
f. Rondônia *cerrado* clearing assumes that 6,946 km² of *cerrado* (25% of the 27,785 km² of *cerrado* in the state according to the INPE map) is exposed to clearing. The remainder is in an Amerindian reserve.

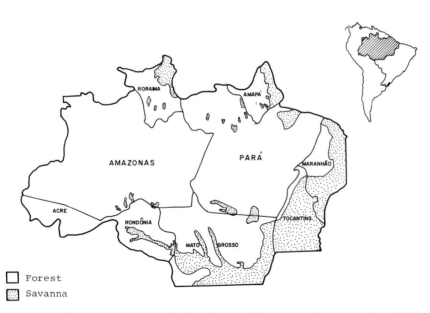

□ Forest
▨ Savanna

Figure 11.1 Forest and savanna in the Brazilian Legal Amazon (redrawn from Brazil, INPE, 1989a).

Table 11.2 Average clearing rates in the Brazilian Legal Amazon

State	Last previous data Year	Last previous data Source	Clearing total (km²)	Clearing total by 1988 (km²)	Average clearing rate in 1988 (km²/year) Forest	Average clearing rate in 1988 (km²/year) *Cerrado*	Average clearing rate in 1988 (km²/year) Total
Acre	1987	IBDF, 1988	8,133	8,634	501	0	501
Amapá	1978	Tardin et al., 1980	171	842	67	0	67
Amazonas	1978	Tardin et al., 1980	1,791	12,837	1,105	0	1,105
Maranhão	1980	IBDF, 1983a	10,671	54,803	3,437[a]	2,080	5,517
Mato Grosso	1980	IBDF, 1982	52,786	201,493	5,580	13,008	18,588
Pará	1986	SUDAM/IBDF, 1988	85,203[a]	92,922	3,788	72	3,860
Rondônia	1987	IBDF, 1989	22,913	31,623	3,916[b]	126	4,042
Roraima	1981	IBDF, 1983b	1,170	2,187	145	0	145
Tocantins/Goiás	1980	IBDF, 1983a	9,120	54,393	1,759	2,959	4,718
Legal Amazon				459,734	20,298	18,245	38,543

a. Pará and Maranhão clearing include reclearing in the area of old (pre-1960) secondary forest. Old secondary forest zones total 31,822 km² in Pará and 60,724 km² in Maranhão; of these an estimated 2,255 km² and 2,459 km² were cleared by 1986 and 1988, respectively, in Pará, and 10,369 km² were cleared by 1988 in Maranhão. Estimates in these states for years prior to 1986 had been unable to distinguish the old secondary forest from virgin forest, and the clearing in the old secondary forest region is therefore included without correction. For 1986 and 1988 in Pará and for 1988 in Maranhão the clearing within the old secondary forest area is assumed to have occurred in the same proportion as that in virgin forest.

b. Rondônia clearing rate assumed to follow the trend from the 1985 to 1987 period shown by AVHRR. Uncorrected deforestation values: 27,658 km² by 1985 (Malingreau and Tucker, 1988); 36,900 km² by 1987 (Jean-Paul Malingreau, personal communication, 1988); corrected for *cerrado* and 18% adjustment for pixel size effect (based on comparison made by David Skole, University of New Hampshire, Durham, NH, of 10 m resolution SPOT data with SPOT data degraded to 1.1 km resolution to simulate AVHRR): 24,195 km² by 1985 and 32,280 km² by 1987.

studies where direct measurements were made concluded that conversion of the Legal Amazon to cattle pasture would release 62 billion metric tons (1 billion tons = 1 gigaton = GT) of carbon. The biomass for the "upland dense forest" category used was 361.5 metric tons per hectare (MT/ha) dry weight total biomass, including live above-ground (251.7 MT/ha), below-ground (86.3 MT/ha), and litter and dead above-ground biomass (23.6 MT/ha). This biomass value from direct measurements is higher by a factor of two than the 155.1 MT/ha value for total biomass derived by Brown and Lugo (1984) from FAO forest volume surveys for "tropical American undisturbed productive broadleafed forests"—a value that has been used in recent global carbon balance calculations (e.g., Detwiler and Hall, 1988).

The Brown and Lugo (1984) forest volume estimate of 155.1 MT/ha is lower than biomass values derived using the same methodology for 15 of 16 locations for which volume information is given in the U.N. Food and Agriculture Organization (FAO) reports, making it unlikely that a mean value this low applies to dense forests in Brazilian Amazonia (Fearnside, 1986c). Revising the estimate of Fearnside (1985c), principally by incorporating FAO wood-volume information into

the dense forest mean and by using values for pasture biomass based on monitoring over an annual cycle at Altamira (Pará) and Ouro Preto do Oeste (Rondônia) (Fearnside, 1989d), yields an estimate of 49.7 GT as the potential release from conversion to cattle pasture (Fearnside, 1987b). The biomass calculations in the present paper yield an intermediate value of 51 GT (Tables 11.3 and 11.4).

The 16 locations in the FAO data set have a mean total (above- and below-ground) biomass of 226.1 MT/ha if calculated using the above-ground volume to biomass conversion factor derived by Lugo and Brown (1984) and the above- to below-ground ratio measured by Klinge et al. (1975) (see Fearnside, 1987b). Brown et al. (1989) have recently derived more reliable volume-to-biomass conversion factors, raising their estimate for mean above-ground biomass for undisturbed tropical American closed broadleaf forest by 28% to 47%. The mean above-ground biomass of 169.68 MT/ha (Brown et al., 1989) is equivalent to 222.3 MT/ha total biomass, using the Klinge et al. (1975) conversion factor of 1.31. This is in good agreement with the 226 MT/ha value used here for central Pará (Table 11.3), where the FAO surveys were concentrated. Both values are probably under-

Table 11.3 Approximate biomass and forest area by state

State	Forest type	Approximate area (km² × 10³)	Approximate biomass (MT/ha)	Area source	Biomass source
Acre	Bamboo	30	20	Sombroek, personal communication, 1989	
	Other low biomass	31	209	(25% of remainder)	
	Dense	92	418	(75% of remainder)	
Amapá	Mangrove	1	200	Braga, 1979	Guess
	Dense	92	354	Remainder	Jordan and Russell, 1983, for Jari
Amazonas	Flooded	30	216		
	Juruá/Purus	400	149		
	Western Amazonas	200	119	Guess	Commercial volume, 100 m³/ha (Sombroek, personal communication, 1989)
	Bamboo	30	20		
	Other low biomass	226	232	25% of forest on fragile soils (Sombroek, personal communication, 1989)	Assumed 50% of dense forest
	Dense	677	464		Mean from four locations around Manaus: Fazenda Dimona (327.7 MT/ha), Fearnside et al., nd-a; Fazenda Porto Alegre, Fearnside et al., nd-a; Reserva Ducke and environs (367.5 MT/ha), see Fearnside, 1987b, Klinge and Rodrigues, 1974; Reserva Egler (507.5 MT/ha), Klinge et al., 1975
Maranhão	Old secondary	61	100	Brazil, INPE, 1989a, 1989b	Guess
	Other	78	175		Guess based on 144.7 m³/ha trunk volume for forests in the Grande Carajas region of Brazil, SEPLAN/CODE-BAR/SUDA M, 1986
Mato Grosso	Northern	100	143	Guess	Based on 120 m³/ha merchantable bole found by Jaime Antonio Ubially and Edezio Cardoso Carvalho, Sombroek, personal communications, 1989
	Transition	473	83	Guess	Based on 70 m³/ha merchantable bole found by Jaime Antonio Ubially and Edezio Cardoso Carvalho, Sombroek, personal communication, 1989
Pará	Old secondary	32	100	Brazil, INPE, 1989a, 1989b	Guess
	Central	465	226	Guess	FAO forest volume surveys (mean of 16 localities; see review in Fearnside, 1986c, 1987b)
	West	249	356	Guess	Tucurui reservoir area; Cardenas et al., 1982
	North	158	354	Guess	Jari Project: Jordan and Russell, 1983
	Vine/low biomass	277	175	Guess	Assumed 25% of dense forest
Rondônia	Dense	215	418	Brazil, INPE, 1989a, 1989b	300 MT/ha above-ground biomass for Samuel reservoir; Brown, 1990; Martinelli et al., 1990

Table 11.3 (continued)

State	Forest type	Approximate area (km^2 × 10^3)	Approximate biomass (MT/ha)	Area source	Biomass source
Roraima	Montane	26	266	Braga, 1979	Seiler and Crutzen, 1980, for montane forest in general
	Other	147	119	Remaining forest	Assumed same as western Amazonas
Tocantins/Goiás	Transition	101	83	Assumed all forest reported in Brazil, INPE, 1989a, 1989b	Assumed same as transition forest in Mato Grosso
Legal Amazon	All forests	4,196	247	(Mean weighted by area present)	
			211	(Mean weighted by clearing rate)	
	Cerrado	576	70.7		

Table 11.4 Approximate carbon release from clearing in the Brazilian Legal Amazon

	Carbon release if all converted to pasture (GT)	Carbon release at current rate of clearing (GT/yr)
Forest biomass	47.3	0.196
Cerrado biomass	1.9	0.059
Soil (top 20 cm)	1.9	0.015
Total	51.1	0.270

estimates: the value used in Table 11.3 (from Fearnside, 1987b) for having used the lower (and less reliable) volume-to-biomass conversion (from Lugo and Brown, 1984) and the more recent estimate (Brown et al., 1989) for using a weighting scheme by forest type that results in a weighted mean volume lower than that found in 15 of the 16 localities that form the basis of the survey.

Land Use Transformations
The cattle pastures that replace forest last only about a decade before they cease to be productive. The vegetation that succeeds cattle pasture has a higher biomass than pasture, thus reducing somewhat the net release of carbon. However, degradation of soil under pasture, combined with rainfall changes expected should the scale of deforestation greatly expand, are likely to make low-biomass dysclimaxes, including grassy formations, the dominant land cover in a deforested Amazon (Fearnside, 1990b).

The rate of deforestation, together with the biomass of forest being cleared, affects the current (as opposed to potential) contribution of deforestation to the greenhouse effect. The rate of clearing was calculated for each state (Table 11.2) but must also be apportioned between various forest types within each state. This is done by assuming that within each state, each forest type is cleared in proportion to the area in which it occurs.

The areas of different forest types present and the biomass of each forest type are both uncertain quantities. In Table 11.3, the values listed have been derived from a variety of sources and have varying degrees of uncertainty. The area figures presented in Table 11.3 have been rounded off after carbon release calculations were made.

The factor most heavily influencing the total biomass present is the dense forest of the state of Amazonas. This has both the largest area and the highest biomass per hectare of any forest type. It also happens to be the unit where the largest number of direct biomass measurements have been made. This area represents approximately 37% of the total potential carbon release from conversion of the Legal Amazon to cattle pasture.

The Fate of Carbon Stocks

Biomass Carbon Char formed in burning is one way that carbon can be transferred to a long-term pool where it cannot enter the atmosphere. A burn of forest being converted to cattle pasture near Manaus resulted in 2.6% of above-ground carbon being converted to char (Fearnside et al., nd-a). This is substantially lower than the 20% assumed by Seiler and Crutzen (1980) when they identified charcoal formation as a potentially important carbon sink. Using the observed lower rate of charcoal formation would make global carbon-cycle models indicate a larger contribution of greenhouse gases from tropical deforestation than has been the case using the higher rates of carbon transfer to long-term pools (e.g., Goudriaan and Ketner, 1984).

The burning behavior of ranchers can alter the amount of carbon passing into a long-term pool as charcoal. Carbon budget calculations generally assume that forest is only burned once and that all unburned biomass subsequently decomposes (e.g., Bogdonoff et al., 1985). This is not the typical pattern in cattle pastures that dominate land use in deforested areas in the Brazilian Amazon. Ranchers reburn pastures at intervals of two to three years to combat invasion of inedible woody vegetation. Logs lying on the ground when these reburnings occur are often burned. Some char formed in earlier burns can be expected to be combusted as well. A typical scenario of three reburnings over a 10-year period would raise the percentage of above-ground carbon converted to charcoal from 2.6% to 3.6%, given the assumptions outlined in Figure 11.2 and Table 11.5, to be discussed later.

The remaining carbon would be released through combustion and decay; the relative importance of each affects the gases released. A one-burn-only scenario would release 27.5% of the preburn above-ground carbon through combustion and 68.9% through decay, whereas the scenario with three reburnings would release 40.6% through combustion and 54.8% through decay. Both combustion and decay release methane, 3.7 times more potent per ton of carbon in provoking the greenhouse effect than is carbon dioxide when the global warming potential over the lifetime of each gas is considered without discounting (Lashof and Ahuja, 1990).

Were a discount rate greater than zero applied, the importance of CH_4 relative to CO_2 would increase (and hence the impact of tropical deforestation relative to fossil fuel emissions). At discount rates of 1%, 2%, 3%, 4%, and 5%, respectively, methane provokes approximately 12, 17, 22, 25, and 28 times more global warming per ton of carbon than does carbon dioxide (Lashof and Ahuja, 1990). An alternative method of giving weight to short-term effects is to consider global warming potential without discounting up to a planning horizon, after which no effects are considered (Arrhenius and Waltz, 1990). Short planning horizons increase the relative impact of methane: Considering only the next 30 years rather than the 150-year lifetime of CO_2 raises the relative impact of CH_4 carbon from approximately 4 to 40 times that of CO_2 carbon.

Measurements of emission ratios of CH_4 to CO_2

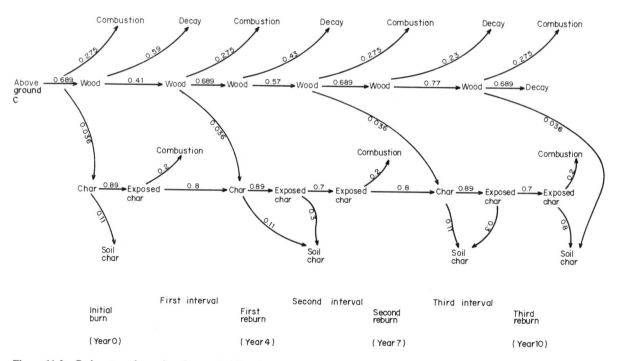

Figure 11.2 Carbon transformations in a typical burning sequence. See Table 11.5 for parameters.

Table 11.5 Parameters for carbon transformations

Parameter	Value	Units	Source	Comments
Total biomass	210.67	MT/ha dry weight	Table 11.2	Weighted mean for areas being cleared in 1988
Carbon content of biomass	0.50	Fraction of dry weight	Brown and Lugo (1984)	
Above-ground fraction	0.76		Klinge et al. (1975)	Near Manaus, Amazonas
Combustion efficiency in initial burn	0.28	Fraction of C released	Fearnside et al. (nd-a)	Near Manaus, Amazonas
Char C fraction in initial burn	0.04		Fearnside et al. (nd-a)	Near Manaus, Amazonas
Fraction of char on bio-mass following initial burn	0.89		Preliminary data from Fearnside et al. (nd-b)	Near Altamira, Pará
Exposed to soil char C transfer fraction during first interval	0.30		Guess	First interval = 4 years
Fraction surviving decay in first interval	0.41		Calculated from Uhl and Saldarriaga (1990)[a]	
Combustion efficiency in first reburn	0.275	Fraction of C released	Assumed equal to initial burn	
Fraction converted to char in first reburn	0.036		Assumed equal to initial burn	
Char C combustion frac-tion in first reburn	0.20		Guess	
Fraction surviving decay in second interval	0.57		Calculated from Uhl and Saldarriaga (1990)[b]	Second interval = 3 years
Combustion efficiency in second reburn	0.28	Fraction of C released	Assumed equal to initial burn	
Fraction of C converted to char in second reburn	0.04		Assumed equal to initial burn	
Fraction of char on bio-mass after first reburn	0.89		Assumed equal to initial burn	
Exposed to soil char C transfer fraction during second interval	0.30		Guess	
Char C combusted frac-tion in second reburn	0.20		Guess	
Fraction of char on bio-mass after second reburn	0.89		Assumed equal to initial burn	
Exposed to soil char C transfer fraction during third interval	0.30		Guess	
Fraction surviving decay in third interval	0.77		Calculated from Uhl and Saldarriaga (1990)[b]	Third interval = 3 years
Combustion efficiency in third reburn	0.28	Fraction of wood C released	Assumed equal to initial burn	
Fraction of C to char in third reburn	0.04		Assumed equal to initial burn	
Char C combustion frac-tion in third reburn	0.20		Guess	
Soil C release from top 20 cm	3.92 MT/ha		Fearnside (1985c, 1987b)	

(continued on next page)

Table 11.5 (continued)

Parameter	Value	Units	Source	Comments
Replacement vegetation biomass	10.67 MT/ha		Fearnside et al. (nd-a); Fearnside (1989d)	Pasture: average biomass throughout year at Ouro Preto do Oeste, Rondônia

a. Uhl and Saldarriaga (1990) report an average of 97.3 MT of above-ground dry weight biomass remaining three to four years after clearing a Venezuelan forest whose original above-ground biomass was believed to be 290 MT/ha based on estimates in the area by Stark and Spratt (1977). Assuming the combustion efficiency (0.275) and charcoal formation fraction (0.036) measured in Brazil (Fearnside et al., nd-a), the postburn above-ground biomass exposed to decay in Venezuela would be reduced to 200 MT/ha. Loss to decay over the 3.5-year interval (using the midpoints of the range of site ages) would therefore be 51%. Loss in a 4-year interval following the initial burn would be 59%.

b. Uhl and Saldarriaga (1990) report average biomass as 56 MT/ha for 6- to 7-year-old sites; 45.3 MT/ha for 8- to 10-year-old sites, 22.7 MT/ha for 12- to 20-year-old sites, and 7 MT/ha for 30- to 40-year-old sites. Assuming a linear decline in wood mass within each age interval (and using midpoints of age ranges as the limits of the intervals), the loss per year as a percentage of the wood mass at the beginning of each interval would be 14.7% for 0 to 3.5 years, 14.2% for 3.5 to 6.5 years, 7.6% for 6.5 to 9 years, 7.2% for 9 to 16 years, and 3.6% for 16 to 35 years. These loss rates have been used to calculate loss values for the intervals used in the present calculation (0 to 4 years, 4 to 7 years, and 7 to 10 years).

(expressed as percent volume) indicate values ranging from 0.5% to 2.3% with a geometric mean of 1.1% for samples collected from the ground near burning forest in the Brazilian Amazon (Greenberg et al., 1984) and ranging from 0.3% to 2.0% with a geometric mean of 0.8% when sampled from aircraft (Crutzen et al., 1985). The amount of methane released is heavily dependent on the ratio of smoldering to flaming combustion; smoldering releases substantially more CH_4. Aircraft sampling over fires (mostly from virgin forest clearing) indicates that a substantial fraction of combustion is in smoldering form (Andreae et al., 1988). Logs consumed by reburning of cattle pastures are virtually all burned through smoldering rather than flaming combustion (personal observation).

Termites are the major agent of decay for unburned wood (Uhl and Saldarriaga, 1990). A lively controversy surrounds the question of how much methane is produced by termites (Collins and Wood, 1984; Fraser et al., 1986; Rasmussen and Khalil, 1983; Zimmerman et al., 1982, 1984). Support for substantial emission potential from termites in deforested areas in the Amazon is provided by high population densities in fields in Pará where forest biomass remains present (Bandeira and Torres, 1985), and high methane emissions from termite mounds near Manaus (Goreau and de Mello, 1987). The billions of metric tons of wood that these insects would devour as Amazonia is deforested cannot help producing substantial contributions of methane regardless of which production rates prove to be correct.

The release of different greenhouse gases can be calculated based on available information from laboratory and field measurements. Low and high meth-ane release scenarios are shown in Tables 11.6, 11.7, and 11.8, using a range of available values for release from combustion and from termites.

In the low-methane scenario, 1550 grams of CO_2 per kilogram (g/kg) of fuel burned in mixed flaming and smoldering burns (i.e., initial burns) and 1400 g CO_2/kg fuel in smoldering burns (i.e., in reburns) (both values calculated by Kaufman et al., 1990, from Ward, 1986). Mixed combustion produces 5 g CH_4/kg fuel (calculated by Kaufman et al., 1990, from Ward, 1986). Smoldering combustion produces 7 g CH_4/kg fuel (calculated by Kaufman et al., 1990, from Greenberg et al., 1984). The carbon content of the fuel is assumed to be equal to that in the biomass being cleared (50%). Termites in the low-methane scenario release 0.2% of the carbon ingested as methane carbon (Seiler et al., 1984, cited by Fraser et al., 1986). The transformations in the low-methane scenario are summarized in Figure 11.3.

In the high-methane scenario, mixed and smoldering burns release the same quantities of carbon dioxide as in the low-methane scenario. Methane is produced at a rate of 6 g/kg fuel in mixed burns and 11 g/kg fuel in smoldering burns (calculated by Kaufman et al., 1990, from Ward, 1986). Termites release 7.8×10^{-3} molecules of CH_4 per molecule of CO_2 (Goreau and de Mello, 1987), or 7.9 g CH_4 carbon per kg fuel carbon, assuming that all carbon is released either as CO_2 or CH_4. The methane release from termites in the high-methane scenario is that measured in termite mound emissions near Manaus—a value only slightly lower than the emissions of the temperate zone species that led Zimmerman et al. (1982) to postulate massive global emissions from termites.

Table 11.6 Carbon release in the Brazilian Legal Amazon[a]

	Complete clearing of Legal Amazon (GT)					Annual net release in 1988 (GT/year)					Gross release per hectare (MT C/ha cleared) for complete clearing of the Legal Amazon			Gross release per hectare (MT C/ha cleared) for clearing in 1988		
	Carbon dioxide C	Methane C	Carbon monoxide C	Total C	CO_2 equivalent C	Carbon dioxide C	Methane C	Carbon monoxide C	Total C	CO_2 equivalent C	CH_4	CO_2	CO	CH_4	CO_2	CO
Low-methane scenario																
Forest	45.40	0.19	1.97	47.56	48.86	0.187	0.001	0.008	0.196	0.202	0.45	113.54	4.71	0.38	97.58	4.02
Cerrado	1.73	0.01	0.08	1.82	1.88	0.054	0.000	0.002	0.056	0.057	0.13	35.35	1.35	0.13	35.35	1.35
Total	47.13	0.20	2.05	49.37	50.74	0.241	0.001	0.011	0.253	0.259						
High-methane scenario																
Forest	45.25	0.39	2.49	48.13	50.18	0.187	0.002	0.010	0.198	0.208	0.92	113.18	5.93	0.79	97.27	5.07
Cerrado	1.72	0.02	0.10	1.84	1.93	0.055	0.000	0.003	0.058	0.059	0.26	35.25	1.70	0.26	35.25	1.70
Total	46.97	0.40	2.59	49.97	52.11	0.242	0.002	0.013	0.256	0.267						

a. Net release from biomass and soils. Gross releases would increase CO_2 carbon by 5.34 MT/ha, but would not affect other gases. For the low- and high-methane scenarios, respectively, gross release of CO_2 equivalent carbon would be 53.58 and 57.54 GT for clearing the Legal Amazon, or 0.283 and 0.341 GT/yr for annual release in 1988.

Table 11.7 Greenhouse gas emissions from deforestation of the Brazilian Legal Amazon (MT/ha)[a]

	CH$_4$	CO$_2$	CO
Low-methane scenario			
Forest			
Burning	0.44	115.45	11.77
Total	0.60	454.16	11.77
Cerrado			
Burning	0.12	33.10	3.37
Total	0.17	141.41	3.37
High-methane scenario			
Forest			
Burning	0.59	115.45	14.83
Total	1.23	452.73	14.83
Cerrado			
Burning	0.17	33.10	4.25
Total	0.35	140.99	4.25

a. Calculated using average biomass for forests in the Legal Amazon.

Table 11.8 Greenhouse gas emissions from complete deforestation of the Brazilian Legal Amazon (GT of gas)

	CH$_4$	CO$_2$	CO
Low-methane scenario			
Forest	0.25	190.55	4.94
Cerrado	0.01	8.14	0.19
Total	0.26	198.69	5.13
High-methane scenario			
Forest	0.51	189.95	6.22
Cerrado	0.15	59.16	1.78
Total	0.66	249.11	8.01

The effect of methane is to raise the impact of net carbon release from Amazonian deforestation by 14% to 18%, depending on whether the low- or high-methane scenario is used. The effect is slightly lower if gross carbon release is considered—the uptake of carbon by the replacement vegetation in the net release calculation affects only CO$_2$ since CH$_4$ does not enter photosynthetic reactions.

Carbon monoxide (CO) is also produced by burning (Tables 11.6, 11.7, 11.8). This gas contributes indirectly to the greenhouse effect by impeding natural cleansing processes in the atmosphere that remove a number of greenhouse gases, including methane. Methane removes hydroxyl radicals (OH), which react with CH$_4$ and other gases, including various chlorofluorocarbons (CFCs) that provoke stratospheric ozone depletion, in addition to the greenhouse effect.

For mixed flaming and smoldering combustion in the low-release scenario, 120 g CO result per kg of fuel (calculated by Kaufman et al., 1990, from Greenberg et al., 1984), while in the high-release scenario the equivalent figure is 150 g (calculated by Kaufman et al., 1990, from Crutzen et al., 1985). Assuming 50% fuel carbon, these values are equivalent to 0.096 and 0.12 kg CO carbon per kg of fuel carbon.

For smoldering combustion in the low-release scenario, 220 g CO is released per kg of fuel (Ward, 1986, cited by Kaufman et al., 1990), while in the high-release scenario the equivalent figure is 280 g (calculated by Kaufman et al., 1990, from Greenberg et al., 1984, and Ward, 1986). Assuming fuel carbon content

as above, these values are equivalent to 0.176 and 0.224 kg CO carbon per kg of fuel carbon, respectively. Complete clearing of the Brazilian Legal Amazon would release 5 to 8 GT of CO (Table 11.8). The global warming potential of a molecule of CO relative to one of CO$_2$ is 1.4 without discounting and rises to approximately 7 at an annual discount rate of 5% (Lashof and Ahuja, 1990). As with methane, the more conservative zero discount values have been used in computing CO$_2$ equivalents (Table 11.6).

Some carbon is released in other forms, such as nonmethane hydrocarbons (NMHCs) and graphitic carbon (soot). The data available are not sufficiently reliable to calculate emissions of these by difference. The carbon release from forest given in Table 11.4 corresponds to a gross release from biomass of 105.6 MT/ha, while the equivalent gross carbon release in the form of CO$_2$, CH$_4$, and CO totals 103.1 MT/ha (from Table 11.6). The implied difference of 2.5 MT/ha (2.3%) might be presumed to represent release in other forms, but uncertainties such as the carbon content of fuel used in deriving the gas emission relationships make this number highly uncertain. The implied difference is greater than the releases suggested by emission ratios from laboratory measurements on combustion of temperate-zone forest fuels. Using the ratios of particulates to methane and NMHCs adopted by Kaufman et al. (1990, based on Ward and Hardy, 1984, and Ward, 1986), the low- and high-methane scenarios imply NMHC releases of 0.29 to 0.39 MT/ha and 0.22 to 0.29 MT/ha for flaming and smoldering combustion, respectively, in forest of average biomass (using the combustion efficiency of 0.275 from Fearnside et al., nd-a; see Figure 11.2). The comparable releases of total particulates would be 1.47 to 1.97 MT/ha and 0.73 to 0.98 MT/ha; considering 7% of the total particulates to be

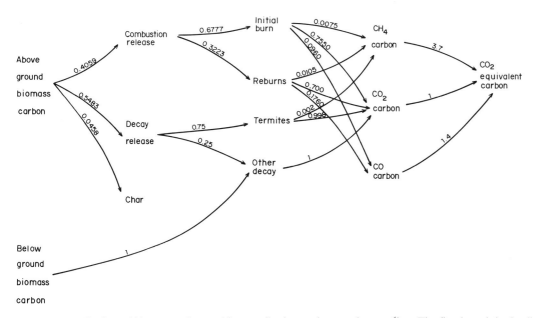

Figure 11.3 The fate of biomass carbon and its contribution to the greenhouse effect. The first branch in the diagram summarizes the results of Figure 11.2.

graphitic carbon (the fraction found over Amazonian fires by Andreae et al., 1988), the releases of graphitic C would be 0.10 to 0.14 MT/ha and 0.05 to 0.07 MT/ha.

Burning also releases some nitrous oxide (N_2O), which contributes both to the greenhouse effect and to the degradation of stratospheric ozone. A sampling artifact has made measurements prior to 1989 unusable. However, the amounts produced by biomass burning are substantially less than had previously been thought (Crutzen, 1990), so that ignoring the impact of N_2O from fire will not unduly bias the results of the present calculations. A greater bias may be introduced by ignoring the biological production of N_2O in the soil, which may be stimulated by deforestation. N_2O is released from soils in greater quantities in cattle pasture than in forest (observations in the dry season near Manaus by Goreau and de Mello, 1987; see also Goreau and de Mello, 1988). Burning in nontropical environments has been found to stimulate N_2O release from soils (Anderson et al., 1988, cited by Kaufman et al., 1990).

Soil Carbon Soil carbon in pasture is taken to be that in a profile equivalent to what is compacted from a 20 cm profile in the forest. It would not be fair to compare the amount of carbon (expressed in MT/ha) in the top meter of pasture soil to the top meter in forest soil, since soil under pasture undergoes compaction when exposed to sun, rain, and trampling of cattle. As the pores are crushed and soil bulk density increases, the amount of carbon in the top meter may increase as an artifact of including a greater weight of soil in the profile. The carbon in the top 20 cm of soil decreases from 0.91% to 0.56% by weight (see Fearnside, 1985c), based on soil carbon under forest and 10- and 11-year-old pastures at Paragominas (Pará) sampled by Falesi (1976). Considering the soil density as 0.56 g/cm^3 under forest at Paragominas (Hecht, 1981), the layer compacted from the top 20 cm of forest soil releases 3.92 MT/ha of carbon.

The 3.92 MT/ha release from the top 20 cm of soil represents 38% of the preconversion carbon present in this layer. This is higher than the 20% of preconversion carbon in the top 40 cm of soil that Detwiler (1986) concluded is released, on average, from conversion to pasture (based on a literature review). The difference is not so great as it might seem: Since carbon release is greatest nearest the surface, considering soil to 40 cm would thereby reduce the percentage released. One factor acting to compensate for any overestimation possibly caused by using a higher percentage of soil carbon release is the low bias introduced by having considered only the top 20 cm. If soil to one meter depth is considered (the usual practice), then the release would be increased to 9.33 MT/ha. The calculation to one meter depth considers that the top 20 cm of soil contains 42% of the carbon in a one meter profile (based on samples near Manaus: Fearnside, 1987b). Brown and Lugo (1982) have used a similar relationship to estimate carbon stocks to a depth of one meter from samples of the top 20 cm, considering 45% of the carbon in a one meter profile to be located in the top 20 cm.

Conversion of all forest and *cerrado* in the Legal Amazon to cattle pasture would release 1.9 GT of carbon from the top 20 cm of soil—about 4% of the total released from converting the region to pasture. Were the soil considered to a depth of one meter, and the assumption made that the proportion of carbon released remains constant with depth, the soil release would be 4.5 GT, or 8% of the total. Considering soil to one meter would add 0.014 GT per year to the 0.010 GT/yr release from the top 20 cm, given the 1988 rate and distribution of clearing.

Release of soil carbon would be expected when forest is converted to pasture because soil temperatures increase when forest cover is removed, thus shifting the balance between organic carbon formation and degradation to a lower equilibrium level (Cunningham, 1963; Nye and Greenland, 1960). A number of studies have found lower carbon stocks under pasture than forest (reviewed in Fearnside, 1980). For the same reason, naturally occurring tropical grasslands also have much smaller soil carbon stocks per hectare than do forests (Post et al., 1982). Lugo et al. (1986), however, have found increases in carbon storage in pasture soils in Puerto Rico, especially in drier sites, and suggest that tropical pastures may be a carbon sink. The present study treats soils as

a source of carbon when forests are converted to pasture. All carbon released from soils is assumed to be in the form of CO_2.

Global Contribution of Tropical Deforestation

Global carbon emissions from deforestation are uncertain, in part because of the uncertainty associated with Brazil's large contribution to the total. One estimate places the global annual total at 1.67 GT, of which 0.80 GT are ascribed to Brazil (Goldemberg, 1989). The Brazilian contribution of more than double the current estimate of 0.27 GT is probably due to using the AVHRR thermal infrared burning estimates from 1987 (Setzer et al., 1988) as the rate of deforestation. The global total implies that 0.87 GT of carbon are released annually from non-Brazilian deforestation and that the global total using the current estimate for Brazil would be 1.14 GT. Brazil's present contribution to the global total from deforestation would be 24%. Assuming a 5 GT/yr global total release from fossil fuels, deforestation in the Brazilian Amazon contributes 4.4% of the combined total from fossil fuels and deforestation. Using the fossil fuel release as the standard of comparison, as is the usual practice, Brazil's annual rate of deforestation in Amazonia represents 5.4% (Table 11.9). Using emission estimates for individual gases produces a similar result, since the loss of some carbon in forms not contributing to the greenhouse effect is compensated for by the greater impact of carbon in the form of meth-

Table 11.9 Carbon release scenarios from the present rate of clearing in the Brazilian Legal Amazon given different assumptions concerning average forest biomass

Average forest biomass (MT/ha)	Biomass carbon release[a] (MT/ha)	Carbon release			
		From forest clearing[b] (GT/year)	% of 5 GT global fossil fuel release	Total from Legal Amazon[c] (GT/year)	% of 5 GT global fossil fuel release
262.6[d]	120.1	0.252	5.0	0.318	6.4
252.0	115.2	0.242	4.8	0.308	6.2
225.0	102.9	0.217	4.3	0.283	5.7
222.5	101.7	0.214	4.3	0.281	5.6
200.0	91.5	0.194	3.9	0.260	5.2
174.0	79.6	0.169	3.4	0.236	4.7
115.1[e]	70.9	0.152	3.0	0.218	4.4

a. Assumes that the replacement vegetation is cattle pasture (10.67 MT/ha dry weight biomass; see Fearnside, 1987b); carbon content of vegetation 0.50 (after Brown and Lugo, 1982, 1984).
b. Includes 3.92 MT/ha carbon release from the top 20 cm of soil.
c. Includes release from *cerrado* (average biomass 70.7 MT/ha) and for soils assumed equal to forest release. *Cerrado* carbon release at current clearing rate is 0.059 GT/year (exclusive of soil release).
d. Value derived from FAO forest volume estimates and from available direct measurements (Fearnside, 1987b).
e. Value derived from FAO forest volume estimates for tropical American productive closed broadleaf forests (Brown and Lugo, 1984).

ane. Using CO_2 equivalent carbon release of 0.259 to 0.267 GT (for the low- and high-methane scenarios in Table 11.6), the contribution represents 5.2% to 5.3% of the global fossil fuel total.

Discussion and Conclusions

Deforestation in Brazilian Amazonia already makes a significant contribution to the greenhouse effect, and continuation of deforestation trends could lead to an even greater potential contribution to this global problem. Uncertainties concerning clearing rate, biomass, and other factors do not change the basic conclusion regarding the significance of deforestation. This can be seen by examining a series of hypothetical examples (Table 11.9): Were the average biomass of 210.7 MT/ha found to be incorrect, biomass values from other sources would result in contributions that, expressed as percentages of a 5 GT global total fossil fuel release, range from 2.8% to 4.6% if only the forest is considered, or 3.3% to 5.1% if the entire Legal Amazon is considered. The conclusion that the effect is significant is therefore quite robust.

Brazil emits 100×10^6 MT of carbon annually from burning fossil fuels (Goldemberg, 1989). This contribution to the greenhouse effect is balanced against the benefits of the country's industry and transportation powered by oil and coal, all domestic use of natural gas, etc. In contrast, each year's clearing of forest and *cerrado* in the Brazilian Amazon is now contributing to the atmosphere 270×10^6 MT of carbon—almost three times as much as Brazil's use of fossil fuels (Table 11.4). The benefits of deforestation, however, are minimal: It leaves in its wake only destroyed rainforests and degraded cattle pastures.

The contrast between costs and benefits of biomass burning and fossil fuel combustion are also tremendous on a per capita basis. Brazil's 140×10^6 population emits 714 kg of carbon per person per year from fossil fuels. A single rancher who clears 2,000 ha of forest (with an average biomass of 210.7 MT/ha; see Table 11.3) is emitting as much carbon as a city of 280,000 people burning fossil fuels (calculation patterned after Brown, 1988). Even a small farmer who clears one hectare per year is releasing 100 MT of carbon, the equivalent of 140 people in Brazil's cities. The gulf between the costs and benefits of deforestation compared to fossil fuel use makes slowing forest loss an obvious place for Brazil to start reducing its contribution to global warming.

Immediate action is needed to reduce emissions of greenhouse gases in order to minimize the global warming that continuation of current trends would provoke. While research and monitoring efforts must be fortified and continued, ample scientific evidence is already in hand to justify strong measures by governments throughout the world. Reducing fossil fuel burning and slowing the rate of tropical deforestation are areas that can be readily identified as targets for such measures. Governments must not wait for the availability of more research results nor for the appearance of observable temperature changes before taking action, or the opportunity will be lost to avert the most damaging impacts of the greenhouse effect.

Acknowledgments

Studies on burning in Altamira were funded by National Science Foundation grants GS-422869 (1974 to 1976) and ATM-86-0921 (1986 to 1988), and in Manaus by World Wildlife Fund–U.S. grant US-331 (1983 to 1985). I thank J. M. Robinson and S. V. Wilson for comments on the manuscript.

Biomass Burning in the Brazilian Amazon Region: Measurements of CO and O₃

Volker W. J. H. Kirchhoff

Figure 12.1 shows the general location of Brazil in South America, and the two sampling stations of interest to this work, Cuiabá and Natal. Ozonesondes have been launched from Natal regularly since 1978 (Kirchhoff et al., 1981, 1983; Logan and Kirchhoff, 1986), and a few rocket campaigns have been made (Barnes et al., 1987). Cuiabá belongs to the Legal Amazon complex. It is actually located in a *cerrado* (open forest, savanna) environment but receives air masses from the north, which bring with them the combustion gases from the heavy burning region near 10°S. Preliminary results are discussed in Kirchhoff et al. (1989).

Table 12.1 lists several sampling stations in operation. Natal, as shown in Figure 12.1, is a marine station. Surface ozone and CO measurements were started in September 1987. At Natal the prevailing winds blow constantly from the east. It is therefore a truly marine site. Nearby Fortaleza also belongs to the marine category. CO measurements were started in November 1987 and O₃ measurements in December 1988. Ozone measurements in Cuiabá were started in October 1985, and CO in September 1987. Cuiabá is the geographic center of South America. In two other stations in the open forest region of central Brazil measurements have started only recently: In Goiânia and Brasília measurements for O₃ started in September and June 1989, respectively, and CO measurements in Brasília started also in June 1989. The region near Alta Floresta and Campo Grande is presently under consideration to become part of the Brazilian sampling network.

The data availability for the Cuiabá station is shown in more detail in Figure 12.2. In April 1986 in the wet season, a lightning discharge destroyed completely the ozone sensor and data acquisition system. Measurements started again only in July 1987. Additional climate characteristics for Cuiabá and Natal are shown in Table 12.2.

Several sporadic field campaigns have also been useful to better understand biomass burning in Amazonia, especially the NASA-GTE-ABLE missions near Manaus (Kirchhoff, 1988; Kirchhoff et al., 1988, 1989; Browell et al., 1988; Kirchhoff and Rasmussen, 1989; Kirchhoff and Marinho, 1989).

Results

Figure 12.3 shows a three-year sequence of O₃ and CO data for Cuiabá and Natal. The basic consistent feature is the seasonality of the dry season peak (September) and the minima in the wet season (March; but note also the difference between dry and wet seasons from Table 12.2). In the wet season, in the absence of biomass burning, both stations have practically the same concentrations of O₃ and CO.

Figures 12.4 and 12.5 show the monthly means for O₃ and CO at both stations. Note that the seasonal peak at Natal is not negligible, either for O₃ or CO. This has led to speculations that possibly Natal might have a secondary influence of biomass burning (Kirchhoff and Nobre, 1986; Logan and Kirchhoff, 1986).

A different way of showing the influence of biomass burning on the O₃ and CO data in Cuiabá is given in Figures 12.6 and 12.7. Figure 12.6 shows the wet season histograms of daily O₃ in intervals of 5 parts per billion by volume (ppbv). Figure 12.7 shows the dry season

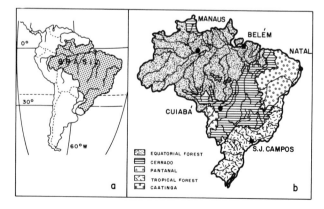

Figure 12.1 The geographic position of Brazil in South America (a), and the location of Cuiabá and Natal in Brazil (b).

Table 12.1 Operating sampling stations in Brazil

Name	Type of environment	Latitude south	Longitude west	Start O_3	Start CO
Natal	Marine[a]	5.9	35.0	Sept. 1987	Sept. 1987
Fortaleza	Marine	3.0	38.0	Dec. 1988	Nov. 1987
Cuiabá	Amazonia[b]	16.0	58.0	Oct. 1985	July 1987
Goiânia	*Cerrado*[c]	16.7	49.2	Sept. 1989	—
Brasília	*Cerrado*	15.0	48.0	June 1989	June 1989
Campo Grande	Pantanal[d]	20.4	54.5	Planned	Planned
Alta Floresta	Amazonia[e]	10.0	56.2	Planned	Planned

a. Used for comparison with continental sites.
b. Cuiabá is located in the so-called Brazilian Legal Amazon region. It is a *cerrado* environment but very close to severe deforestation in the north.
c. *Cerrado* is a type of open forest, the Brazilian savanna, where severe burnings take place yearly.
d. Seasonally flooded lands, Paraguai river system, area 145,000 km².
e. Sporadic measurements have been made in Belém and Manaus.

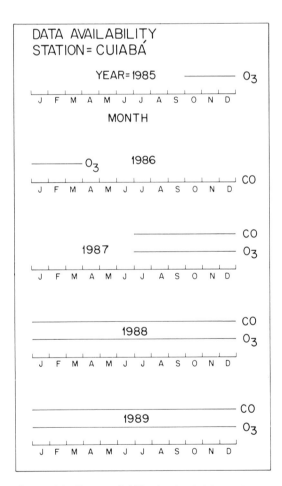

Figure 12.2 Data availability for the Cuiabá station.

months and the large dispersion that occurs for the O_3 concentrations maximizing in September.

The variations monitored on the ground appear to occur also in the upper tropospheric levels. Figure 12.8 shows ozone increases in the dry season at all levels below 5 km, compared to the wet season O_3 profile. This feature is also observed at Natal (Logan and Kirchhoff, 1986).

**Emissions of Carbon to the Atmosphere:
The Global Contribution**

The global contribution of biomass burning from the tropics is not well known, but most authors seem to believe that its contribution is very large, of the order of three times the fossil fuel emissions by all of Western Europe.

There are basically two different groups that estimate carbon emissions to the atmosphere: atmospheric chemists, physicists, and meteorologists, who calculate emissions based on burnt biomass; and oceanographers, geographers, forest engineers, and ecologists, who calculate emissions on the basis of deforestation rates, assuming that sooner or later all organic carbon will be transformed into atmospheric CO_2. Despite their different approaches, the estimates for the carbon emissions are surprisingly similar: between 2 and 3 billion tons of carbon per year, in the form of CO_2.

Both groups need two fundamental pieces of information for their estimates: the *mass* density of the biomass being either deforested or burned and the total *areas* affected. These fundamental quantities,

Table 12.2 Climatic characteristics for Cuiabá and Natal[a]

	Cuiabá (16°S, 58°W)	Natal (6°S, 35°W)
Yearly average temperature (centigrade)	24°C	26°C
Warmest month	Sept. (35°C)	Nov. (30°C)
Coldest month	June–July (16°C)	May (24°C)
Highest temperature	42°C	36°C
Lowest temperature	0°C	16°C
Rainfall rate (millimeters per year)	1500 mm/yr	1400 mm/yr
Three-month maximum rate	Dec.–Jan.–Feb. (50%)	April–May–June (50%)
Dry period	July–Sept.	Sept.–Dec.
Elevation (meters)	165 m	42 m
Prevailing wind from	North	East

a. Adapted from Nimer (1989).

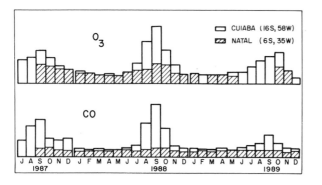

Figure 12.3 Carbon monoxide and ozone concentrations shown in the form of monthly averages for 1987, 1988, and 1989, for Natal and Cuiabá.

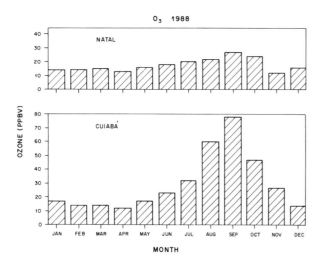

Figure 12.4 Monthly average ozone concentration observed at noon local time, for 1988, at Natal and Cuiabá.

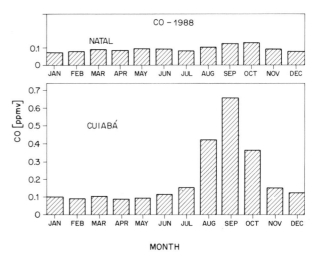

Figure 12.5 Monthly average carbon monoxide concentrations observed at noon local time, for 1988, at Natal and Cuiabá.

mass and area, are only precariously known in tropical countries. Mass densities have been measured by only a few field experiments in just a few spots of the immense tropical area of about 39.5 million square kilometers. Thus, one has to extrapolate and guess areas a lot. The exact areas affected in the different tropical countries are not well known, and data bases are probably manipulated and stored by bureaucrats that are unaware of scientific methods.

The Direct Measurement: A Proposal

The correct emission of CO_2 to the atmosphere could be calculated directly if enough measurement stations

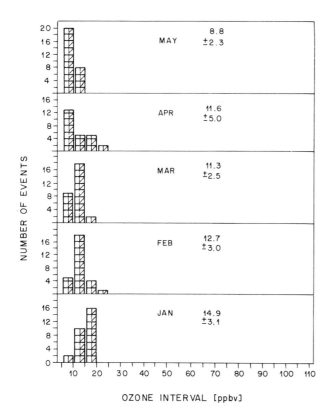

Figure 12.6 Histogram of daily average ozone concentrations during the wet season.

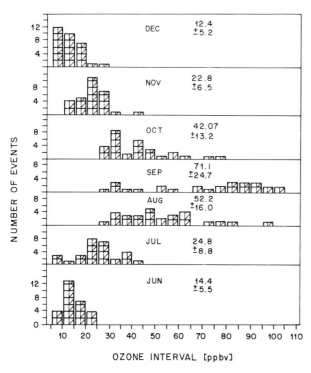

Figure 12.7 Histogram of daily average ozone concentrations during the dry season.

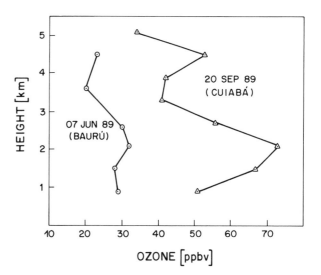

Figure 12.8 Vertical ozone profiles in dry and wet season periods.

were available. Emission rates will be meaningful if the following criteria are observed:

1. The area under study must be divided in *n* elementary subareas.

2. An elementary subarea is a region in which the measurements obtained at any point are essentially equal, such that one single measurement sensor is enough to characterize the elementary subarea.

The criteria above are evidently theoretical, but from our experience in Brazil it seems quite possible to implement a program following these principles. The coordination of such work should have an international flavor and be supervised by associations such as the International Global Atmosphere Chemistry (IGAC) Project or the International Geosphere Biosphere Programme (IGBP).

The establishment of elementary subareas assumes, evidently, that the measurements obtained represent the average behavior of that region whose surface area is well known. Thus, the temporal variations of the *n* sensors will define exactly the contribution of the total area. In order to ensure that only effects due to biomass burning are taken into account, all results must be subtracted from a comparison station in which the contribution of burning is known to be negligible.

Calculation of the Emission Rate: An Example

Presently, there are not enough measurement stations to apply the previous concepts for the calculation of an average emission rate. There is only one station (Cuiabá) in the biomass burning area and only one comparison station (Natal). This should be enough, however, to determine an upper limit, if it is assumed that Cuiabá is representative of the whole tropics.

The major CO removal mechanism from the atmosphere is its reaction with OH. Thus, in order to observe a concentration CO_{ex} (the difference between Cuiabá and Natal) there must be an emission $E(CO)$ given by

$$\int [OH]\, K[CO_{ex}]\, dh = E(CO)$$

where K is the reaction rate ($2.2 \times 10^{-13}\ cm^3 s^{-1}$). In practice, the integration is limited to 12 km. The lack of knowledge of an exact vertical distribution of the CO concentration is another limitation of the present calculation. However, the vertical CO profiles that have been measured in the area suggest an almost linear distribution with height. The exact values that have been adopted are shown in Figure 12.9, where the surface concentrations for 1988 have been used (the largest, compared with 1987 and 1989; the calculation will provide an upper limit). The numerical values for [OH] are taken from the literature. Table 12.3 shows the monthly contributions to the emission rate.

The resulting emission rate is thus 2.9×10^{12} $cm^{-2}s^{-1}$. This is the average rate that must be provided by biomass burning in order to accumulate in the atmosphere an excess CO_{ex} concentration as observed. The yearly emission rate in all tropics is thus

$$E(CO) = 62.8 \times 10^{12}\ g\ C/yr$$

That is, 62.8 million tons carbon per year are emitted to the atmosphere from biomass burning. This upper limit is about one-quarter of the value quoted in the literature. It represents about 7% of all CO sources.

It is possible to extrapolate this result to the CO_2 contribution. Assuming an average CO/CO_2 ratio of 10%, the CO_2 emission would be

$$E(CO_2) = 0.628 \times 10^{15}\ g\ C/yr$$

That is, the emission of C in the form of CO_2 would be 0.6 billion tons C per year. In Table 12.4 we summarize this result, comparing it with others. Assuming that for 1986 the global CO_2 emission from fossil fuel burnings was 5.6×10^{15} g C, the biomass burning contribution calculated above would represent 11.2% of that total. It is without any doubt a significant contribution, but not a major one. Again, the results in Table 12.4 must be used with caution. Higher or lower figures may be obtained depending on the specific assumptions used. But it seems that the above result is the best that can be obtained presently with the data available.

The International Cooperation Perspective

International cooperation can be separated in personal, scientist-to-scientist cooperation, which in general has been very good, and in cooperation through

Table 12.3 Monthly contributions to the emission rate of CO from biomass burning

Month	OH ($10^5 cm^{-3}$)	S ($10^{18} cm^{-2}$)	K[OH]S ($10^{11} cm^{-2} s^{-1}$)
July	6.0	0.9	1.2
August	6.0	4.7	6.2
September	7.0	8.1	12.5
October	7.5	4.8	7.9
November	9.0	0.8	1.6
Total			29.4

Table 12.4 Emission rates of CO and CO_2 from biomass burning

Source	CO (10^{12} g C/yr)[b]	CO_2 (10^{15} g C/yr)[c]
Crutzen et al. (1985)	342.8	2.0–3.3
Houghton et al. (1987)	—	1.0–2.6
This work[a]	62.8	0.6

a. Upper limit.
b. 10^{12} g = million tons.
c. 10^{15} g = billion tons.

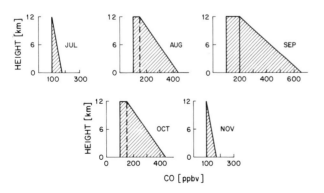

Figure 12.9 CO profiles adopted for the calculation of carbon emissions to the lower atmosphere. The surface CO variations are from the data of Figure 12.3.

scientific unions or agencies, which has been very bad, except for NASA. Scientific unions or agencies are more interested in political collaboration on paper than in real technical and operational cooperation. It is time that international cooperation be taken more seriously, especially on issues that have global importance.

Because of high costs, scientific problems are often tied to economic problems. Countries where biomass burning occurs extensively are in general poor countries. They can do most of the operational work, but there is a great need for serious and continuous international cooperation.

Table 12.5 lists the gross national product (GNP) for the Amazonian countries. Even if we take Brazil, for example, with the largest GNP, difficulties for doing science are large. In February 1990 Brazil had a monthly inflation of 70%. To import equipment under these circumstances is almost unthinkable. Difficulties in the other tropical countries are probably the same.

Table 12.6 shows the debt flows in developing countries. Note that from 1984 on there has been no new money available. This has, of course, a strong negative influence on science as well. How can we expect, under these circumstances, developing countries to

make investments in biomass burning studies, knowing so well the many social problems that plague these countries? The conclusion is simple: If the international scientific community and perhaps even government agencies and scientific unions want to keep alive work that is presently being done, or better, *increase* scientific activity on studies of biomass burning and the greenhouse effect in tropical countries, a much more aggressive approach must be adopted in terms of real international cooperation.

Table 12.5 Gross national product (GNP) for Amazonian countries for 1986 (billion of dollars)

Country	GNP
Brazil	$270.1
Venezuela	47.9
Colombia	31.3
Peru	24.5

GNP data from the World Bank, *World Debt Tables* (1987–1988).

Table 12.6 Long-term debt and financial flows in developing countries from 1983 to 1988 (billions of dollars)

	1982	1983	1984	1985	1986	1987	1988
Debt disbursed and outstanding	$562.5	$644.9	$686.7	$793.7	$893.8	$996.3	$1020
Debt service	98.7	92.6	101.8	112.2	116.5	124.9	131
Principal payments	49.7	45.4	48.6	56.4	61.5	70.9	72
Interest payments	48.9	47.3	53.2	55.8	54.9	54	59
Net flows	67.2	51.8	43	32.9	26.2	15.8	16
Net transfers	18.2	4.6	−10.2	−22.9	−28.7	−38.1	−43

Source: World Bank.

Ozone Concentrations in the Brazilian Amazonia During BASE-A

Alberto W. Setzer, Volker W. J. H. Kirchhoff, and Marcos C. Pereira

Ozone measurements in the Brazilian Amazon Basin have been conducted recently at ground level and with soundings (Kirchhoff, 1988; Kirchhoff et al., 1988) and in aircraft (Browell et al., 1988). Continuous surface data exist since 1985 for a site close to Cuiabá, Mato Grosso (MT), in the southern fringe of the Amazon forest (Kirchhoff, 1990), and strong effects of biomass burning have been detected in ozone concentrations during the dry season, from June through October (Kirchhoff et al., 1989b; Kirchhoff, 1990). The dry season coincides with the burning season, for centuries a time when fire has been used to clear areas where forest was recently cut and let to dry, or to renew pastures and agricultural land. The tropospheric ozone which results is produced by complex and multiple photochemical reactions between different compounds emitted by biomass burning (Fishman et al., 1979).

Ozone monthly averages in southern Amazonia have ranged from a maximum of almost 80 parts per billion (ppb) in September during the peak of the burning to 10 to 20 ppb in the wet season, from December to April. The low values are comparable to those found year round in nonpolluted sites such as Natal, at the Brazilian northeast coast (Kirchhoff, 1990).

Detection of fires in the Brazilian Amazonia with band 3 (3.55μ to 3.93μ) thermal images of the Advanced Very High Resolution Radiometer (AVHRR) on board the meteorological NOAA series satellites was developed by Pereira (1988) and is in operational use (Setzer and Pereira, 1990a, 1990b, 1991a, and 1991b). Coupling of pixels containing fires detected in such images and atmospheric contamination hundreds of kilometers downwind in Amazonia has been reported (Andreae et al., 1988; Kirchhoff et al., 1989). Total number of fire pixels detected in the dry season in the Brazilian Amazonia were about 315,000 in 1987, and 210,000 in 1988.

In this chapter results of ozone measurements made on board the Brazilian Institute for Space Research (INPE) airplane during the first week of Sep-

tember 1989 are presented and analyzed in relation to the temporal and geographical location of fires detected by the satellite before and during the sampling period.

Methodology

Ozone measurements were made with an ultraviolet photometric analyzer producing continuous recordings on a paper chart. Readings are in parts per billion by volume (ppbv), and precision is 1 to 2 ppbv. Calibration was made prior to the experiment as recommended by the manufacturer of the instrument, and dial adjustments before each flight. Air intake at 2 liters per minute was through a special port at the side of the fuselage, close to the cockpit to avoid contamination from engine exhaust. The airplane used was a two-engine Embraer Bandeirante EMB-101 also loaded with other research equipment for the Biomass Burning Airborne and Spaceborne Experiment–Amazonia (BASE-A) experiment as described by Kaufman et al. (1990).

Figure 13.1 shows INPE's airplane trajectory and Figure 13.2 the flight levels during BASE-A. Flight 1 from S. J. Campos, São Paulo (SP) to Brasilia, Federal District (DF), on 1 September was over land of varied uses but mainly agricultural and with sugar cane plantations that are burned before harvesting (see Kirchhoff et al., 1989a, for ozone associated with such burnings). Flight 2, from Brasilia to Porto Nacional, Tocantins (TO), on 2 September, was over savanna-type vegetation (*cerrado*) where open pasture is the predominant feature and where fire is regularly used to renew the grasses. Flight 3 on 2 September began with the transition from *cerrado* to forest and ended in Alta Floresta, MT, an area of fast forest conversion to large ranches. Flight 7 on 6 September covered forested areas where some deforestation is taking place, until Santarem, Pará (PA). Flight 8 on 6 September to Manaus was over relatively untouched forest. Flight 9 on 7 September to Porto Velho, Rondônia (RO), saw an increase in deforestation and fires

Figure 13.1 Approximate trajectory of the flights during BASE-A.

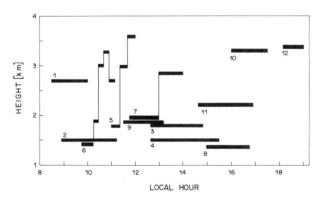

Figure 13.2 Height (km) and local times of flights shown in Figure 13.1.

toward Rondônia. Flight 10 on 8 September was over a region where extensive deforestation and burning took and was taking place, in the vicinity of the BR-364 road, and over *cerrado* areas close to Cuiabá, MT, also dominated by the use of fire. Flight 11 on 9 September to Uberlandia, Minas Gerais (MG), covered areas of intense agriculture, mainly soybean plantations, where fire was being used to burn grasses and weeds before plowing. The area covered in flight 12 on 10 September, the last one, was similar to the first one. Vertical profiles up to about 4 km were obtained over Alta Floresta, Santarem, and Cuiabá. Figure 13.2 shows altitudes and times of the flights.

The fire detection technique is based on the work originally presented by Matson et al. (1984). Band 3 (3.55μ–3.95μ) AVHRR thermal images were recorded daily by INPE at Cachoeira Paulista, SP, and screened for "hot" pixels with nominal radiometric brightness temperature between 316°K and 320°K, the former being the saturation limit of the sensor. As shown by Pereira (1988) and verified by field work, this range indicates major fires occurring at the time the image was produced by the NOAA satellite, around 2 to 4 P.M. local time. Morning or night satellite passes are not used, since fires are usually lit after noontime and die a few hours later. Fires with fronts Smaller than about 100 m or fires under dense canopies have not been detected by the threshold used. The thermal images and results used by BASE-A were the same produced by INPE in its operational program (Setzer and Pereira, 1990a, 1990b, 1991a, 1991b) to detect fires in Brazil in near-real time.

Results and Discussion

The number of satellite image pixels with fires detected in AVHRR band 3 during BASE-A and for two preceding days for rectangles surrounding Brazilian Amazon states and other states where flights took place was obtained. In 1989 rains were above normal for Central and North Brazil, which contributed to a reduction in the biomass burning activity compared to previous years; August and September had 50 to 100 mm over the normal (INEMET, 1990; INPE, 1990). Figure 13.3 shows ozone concentrations in ppbv for the transit flights, and represents data after flight altitude and ozone readings were stable. Figure 13.4 shows vertical profiles of ozone for Santarem, Alta Floresta, and Cuiabá.

The general pattern observed is that ozone concentrations were higher in *cerrado* and deforestation areas. Flight 1 measured about 40 ppbv, possibly caused by diverse agricultural fires in the region and by downwind combustion products brought by prevailing westerly and northwesterly winds from fires in Goiás, Mato Grosso do Sul, and Mato Grosso. Fire pixels for the previous day (31 August) in the areas limited by rectangles surrounding the states, and therefore including areas of a few states, numbered 184 for São Paulo, 9344 for Minas Gerais, 7266 for Goiás, 45 for Mato Grosso do Sul, and 1098 for Mato Grosso. Ozone ranged from 50 to 65 ppbv between Brasília and Porto Nacional during flight 2 and would probably have been higher for a flight in the afternoon when ozone-producing photochemical reactions

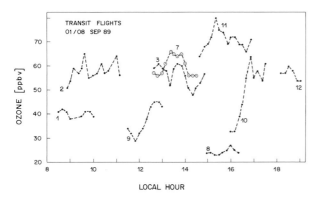

Figure 13.3 Ozone concentrations (ppbv) for flights of Figure 13.1.

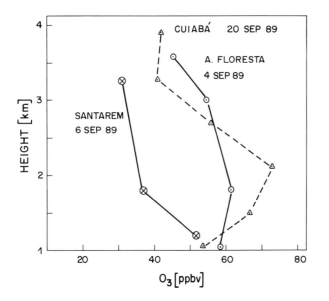

Figure 13.4 Ozone vertical profiles for flight 6 over Alta Floresta, Mato Grosso (MT), and flight 7 to Santarem, Pará (PA). The Cuiabá, MT, profile for 20 September was obtained after the experiment, with 1270 fire pixels for the Mato Grosso rectangle.

are more efficient. Fires for the rectangles in the previous day (1 September) were 4187 for Goiás. Ozone was also in the 50 to 60 ppbv range during flight 3 to northern Mato Grosso in the afternoon, and the fires numbered 8479 in the Tocantins rectangle for that day (2 September). Flight 7 would have kept the same range of ozone if it had not crossed the plumes from fires in South Pará carried by easterly winds, which caused an increase to about 65 ppbv; fires in the Pará rectangle in that day (6 September) were 514. Flight 8 had the minimum ozone readings of the experiment, in the 25 ppbv range; the Amazon state rectangle had 326 fires on that day (6 Septem-

ber), but most of them in parts of Rondônia and Acre are included in the rectangle and do not affect the ozone readings. Flight 9 shows an increase from 30 to 50 ppbv as the plane approached the Rondônia state, where biomass burning is common. Flight 10 shows a steep increase in ozone, from 35 to 60 ppbv as the plane moved into areas of more intense deforestation and burning in general; fires in the Rondônia rectangle that day (7 September) numbered 1142, and in Mato Grosso, 2922. The highest concentrations, in the 60 to 80 ppbv range, were in flight 11, which measured effects not only of regional fires but also from upwind fires in Mato Grosso and Rondônia; numbers of fires in the rectangles for the flight that day (8 September) in Goiás were 1277, 1640 for Mato Grosso, and 282 for Rondônia. The last flight, number 12, had concentrations of about 60 ppbv due to regional as well as upwind fires.

Without considering the size of fires, type of vegetation burned, combustion efficiency, or transport of pollutants, the above data indicate that ozone concentrations varied from an 80 ppbv range in a region with one or more fires per 1000 km^2 to 20 ppbv with less than 0.5 fires per 1000 km^2. Therefore, the larger the extent of biomass burning on a synoptic scale, the higher the ozone tropospheric concentration.

Conclusion

Tropospheric ozone is known as a biomass burning indirect by-product. This work has shown through in situ measurements of ozone and satellite detection of fires that on a synoptic scale concentrations rise sharply in regions of more intense burning.

Acknowledgments

The authors wish to thank Alfredo C. Pereira, Pedro C. D. Santos, Luiz F. Ribeiro, Luiz M. N. L. Monteiro, and the DOP/INPE-C.P. operators for their collaboration in different parts of this work. Support was partially provided by Fundação Banco do Brasil through "Our Nature" and INPE's Amazonia Program.

A Comparison of Wet and Dry Season Ozone and CO over Brazil Using In Situ and Satellite Measurements

Catherine E. Watson, Jack Fishman, Gerald L. Gregory, and Glen W. Sachse

Many researchers have examined the effects of biomass burning on the concentrations of ozone and carbon monoxide (Crutzen et al., 1979, 1985; Seiler and Crutzen, 1980; Delany et al., 1985; Logan and Kirchhoff, 1986; Andreae et al., 1988). It has been hypothesized, and is now generally accepted, that biomass burning is as large a source of pollutants as industrial activities. As such, regional biomass burning can have global consequences.

Several field experiments have measured the regional effects of biomass burning. Two such experiments, designed to understand the chemistry of the Amazon rainforest during both the wet season and dry season, were conducted in the Amazon Basin. The first experiment, ABLE-2A (Amazon Boundary Layer Experiment), took place from July to August 1985, the early dry season, when biomass burning was just beginning. The second experiment, ABLE-2B, took place during the wet season, from April to May 1987, when little biomass burning was occurring.

Comparing ABLE ozone data with tropospheric ozone concentrations derived from satellite data, using the method described by Fishman et al. (1989, 1990), shows a strong correlation between the direct measurements and the derived ozone concentrations, as well as a direct correlation of both to biomass burning. This comparison gives credence to the use of space-based platforms to monitor global chemistry and, in this case, the regional effects of biomass burning.

An examination of wind fields derived from National Meteorological Center u and v wind components allows the determination, to a first approximation, of the direction of transport of these enhanced values of ozone and CO. It appears that, during the early dry season, much of the enhancement in ozone seen over Brazil is due to transport from Africa, whereas the enhancement in the later dry season is due more to increased local biomass burning.

Data

ABLE Missions
During both ABLE-2A and ABLE-2B, ozone data were collected along the aircraft's flight path by an ozone sensing package which uses both C_2H_4 chemiluminescence and NO chemiluminescence; this instrument is discussed in detail in Gregory et al. (1983, 1987). CO data were obtained using the Differential Absorption CO Measurement (DACOM) instrument, also aboard the aircraft, described in Sachse et al. (1987). In this chapter we briefly examine the seasonal differences seen in the vertical profiles of CO and ozone obtained from aircraft measurements, as well as differences found in integrated concentrations of ozone from ozonesonde measurements taken north of Manaus, Brazil (3°S, 60°W).

Tropospheric Residual
The tropospheric residual data is derived from six years (1979 to 1981 and 1985 to 1987) of TOMS (Total Ozone Mapping Spectrometer) and SAGE (Stratospheric Aerosol and Gas Experiment) and SAGE II data. The stratospheric portion of the total column of ozone, represented by integrated SAGE profiles, is subtracted from the total column concentration, represented by TOMS measurements; the resulting value is called the tropospheric residual. This method is discussed in detail in Fishman et al. (1989, 1990). Table 14.1 shows how many tropospheric residual data points were used in each of the seasonal figures which follow.

Seasonality of Ozone and CO

Figure 14.1a shows averaged ozone and Figure 14.1b shows CO profiles during three ABLE survey flights from Belem (2°S, 48°W) to Manaus (3°S, 60°W). The wet season flight on 23 April 1987 is represented by

Table 14.1 Number of tropospheric residual points used in climatologies

Figure	Time frame	Points
14.2a	April–May 1987	51
14.2b	April–May climatology	241
14.3a	July–August 1985	66
14.3b	July–August climatology	227
14.4a	Expanded April–May climatology	361
14.4b	Expanded July–August climatology	344
14.4c	Expanded Sept.–October climatology	218

Note: The number of tropospheric residual data points used in the respective figures is given. Figures 14.2a and 14.3a contain data for their respective years only. Some years have more usable data than others, hence the different climatological totals in Figures 14.2b and 14.3b. Figures 14.4a, 14.4b, and 14.4c cover larger geographical areas.

Table 14.2 Intercomparison of Manaus and Natal, ozonesonde and Dobson measurements

Manaus intercomparison (ozonesonde measurements vs. tropospheric residual)

	ABLE	Satellite
Wet (April–May)	24	21
Dry (July–Aug.)	29	28

Natal intercomparison (Dobson measurements vs. tropospheric residual)

	Logan and Kirchhoff (1986)	Satellite
Wet (March–April)	28	24
Dry (July–Aug.)	41	34
Late dry (Sept.–Oct.)	48	39

Note: (top) Integrated ozonesonde measurements made in the forest north of Manaus during the ABLE missions are compared with seasonal trophospheric residual climatologies derived from six years of TOMS and SAGE data (1979 to 1981 and 1985 to 1987). (bottom) Tropospheric residual climatologies are compared with Dobson spectrophotometer measurements from Logan and Kirchhoff (1986). All concentrations are in Dobson Units (one Dobson Unit = 1 DU = 2.69×10^{16} molecules cm^{-2}). This figure shows how well the trophspheric residual depicts the seasonal cycles of ozone over Brazil.

Belem/Manaus Survey Flights

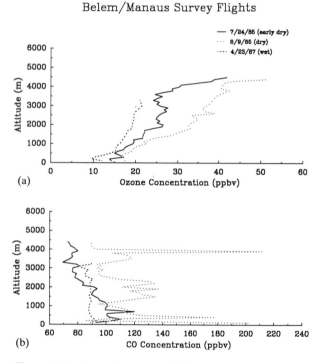

(a)

(b)

Figure 14.1 Averaged ozone and CO vertical profiles obtained during three survey flights from Belem (2°S, 48°W) to Manaus (3°S, 60°W), Brazil, during ABLE-2A and ABLE-2B. The wet season flight is represented by the dashed line, the early dry season flight by the solid line, and the flight in which several layers of smoke and haze were encountered is represented by the dotted line. The change in concentrations with the onset of dry season and increased biomass burning is evident for both CO and ozone.

the short dashed line, the early dry season flight on 24 July 1985 by the solid line, and the flight on 9 August 1985, during which the aircraft flew through several smoke plumes, by the dotted line. An increase from wet season to dry season is evident for both ozone and CO.

Ozone

In Figure 14.1a ozone increased throughout the lower troposphere from the wet season to the early dry season. On the 9 August flight, when the aircraft flew through several layers of haze caused by local biomass burning, ozone increased again at all altitudes.

In Table 14.2, the derived tropospheric residual climatology (1979 to 1981, 1985 to 1987) is compared with in situ measurements taken in the Manaus and Natal (6°S, 35°W) regions; all concentrations are in Dobson Units (one Dobson Unit = 1 DU = 2.69×10^{16} molecules cm^{-2}). The ozonesonde measurements (top) were taken at a site in the forest north of Manaus as part of both ABLE missions. It is evident that both the ozonesonde measurements and the tropospheric residual increase with the onset of the dry season and increased biomass burning; the ozonesonde measurements increased by 21% while the tropospheric residual increased by 33%. Fishman et al. (1990) found that the tropospheric residual agreed to within 10% with tropical integrated ozonesonde values, similar to the difference seen in Table 14.2 (top).

In the bottom half of Table 14.2, the tropospheric residual climatology is compared with three years of Dobson spectrophotometer measurements (1000-100 mbar) reported in Logan and Kirchhoff (1986). Both the tropospheric residual and the Dobson measurements increase with the onset of the dry season and continue to increase into October. The Dobson measurements increased 46% from wet to dry season and 17% from dry to late dry season while the tropospheric residual increased 42% from wet to dry season and 15% from dry to late dry season. The increases in Dobson values are closer to the tropospheric residual than the integrated ozonesonde measurements taken near Manaus. From Table 14.2, a spatial gradient is also evident, with ozone concentrations greater at Natal than Manaus.

The comparisons in Table 14.2 indicate that the derived tropospheric residual is an accurate representation of the seasonal changes occurring in Brazilian ozone values.

In Figure 14.2a, the average tropospheric residual is depicted for the timeframe of the ABLE-2B mission (April to May 1987). The contour lines are in Dobson Units with 2 DU between each contour. The M stands for Manaus, the B for Belem, and the N for Natal. Across most of Brazil there is a tropospheric residual concentration of 22 DU. Also notable are the latitudinal gradient in the Northern Hemisphere and the longitudinal gradient to the west of Brazil. An April-May climatology (Figure 14.2b), derived from the six years of TOMS and SAGE data, has the same features present in 1987 (Figure 14.2a), with gradients in the Northern Hemisphere and to the west of Brazil, and a concentration of 22 DU over most of Brazil.

In Figure 14.3a, the time frame of ABLE-2A (July to August 1985), the relatively constant wet season concentration of 22 DU (Figures 14.2a and 14.2b) has been replaced with an average concentration of 30 DU, an increase of 36%. The latitudinal gradient in the Northern Hemisphere is gone, and the longitudinal gradient west of Brazil has been replaced by a longitudinal gradient increasing toward the east, into the Atlantic. Much like the gradient seen in the Table 14.2 comparisons, tropospheric residual concentrations are greater at Natal than Manaus. In Figure 14.3b, a July-August climatology, the same pattern seen in the

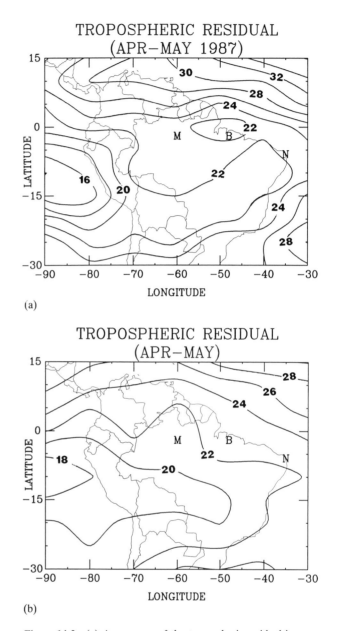

(a)

(b)

Figure 14.2 (a) An average of the tropospheric residual in Dobson Units during April and May 1987 (ABLE-2B) is shown. Across Brazil there is an average tropospheric ozone concentration of 22 DU. (b) An average of six years (1979 to 1981 and 1985 to 1987) of April and May tropospheric residual data is shown. The pattern for 1987 is very similar to the six-year climatology, indicating that 1987 was probably a typical wet season.

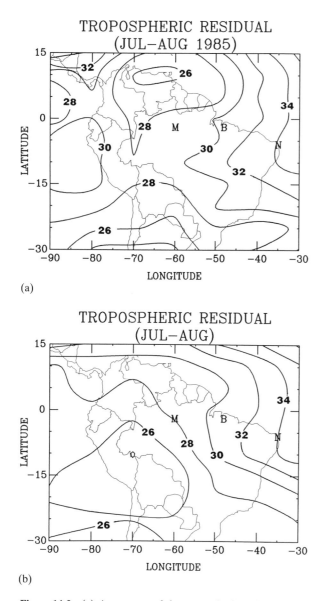

(a)

(b)

Figure 14.3 (a) An average of the tropospheric residual in DU during July and August 1985 (ABLE-2A) is shown. Across Brazil there is an average tropospheric ozone concentration of 30 DU where there was 22 DU during the wet season (Figures 14.2a and 14.2b). (b) An average of six years (1979 to 1981 and 1985 to 1987) of July and August tropospheric residual data is shown. The pattern for 1985 is very similar to the six-year climatology, showing that the increase in tropospheric ozone due to biomass burning is an annual event.

1985 data is also present. These four figures and Table 14.2 show that the derived tropospheric residual is giving an accurate representation of both the spatial and temporal distributions of tropospheric ozone over Brazil and that the seasonal increases in tropospheric ozone are annual events.

In Figure 14.4a, the climatology in Figure 14.2b has been expanded to include the South Atlantic and Africa. The contours are still in DU with 2 DU between each contour. In Figure 14.4b we see this same area in July-August and in Figure 14.4c in September-October. In Brazil the largest amount of biomass burning occurs in September and October. As in Figures 14.2b and 14.3b, Figures 14.4a, 14.4b, and 14.4c reveal a distinct seasonality in tropospheric ozone throughout the southern tropics. Relatively low values of tropospheric ozone are evident in the wet season climatology (Figure 14.4a) with increasing values as the dry season progresses and biomass burning increases (14.4b and 14.4c). The longitudinal gradient in Figure 14.3b is still present in Figures 14.4b and 14.4c and appears to be part of a much larger pattern of high tropospheric ozone values centered off the coast of Africa.

In July and August there is an area of high ozone concentrations off the coast of central Africa, slightly north of the maximum seen in September and October. Cros et al. (1988) presented vertical profiles taken near Brazzaville, Congo (4°S, 15°E), from November 1983 to October 1986. Each year ozone begins to increase in May, decreasing slightly in June, then increasing again, with the maximum values occurring in September and October; one specific maximum in October was attributed to transport of dry haze from biomass burning in southeast Africa (Cros et al., 1987). From these observations the slight difference in the location of the South Atlantic tropospheric ozone maximum (Figures 14.4b and 14.4c) may be due to the contribution of ozone from biomass burning in southeast Africa. However, the shift in the maximum could also be due to meteorological changes as well as the addition of South American ozone transported to the area by upper- and lower-level westerlies.

Carbon Monoxide

In Figure 14.1b CO has the same relative integrated value in both the wet season and early dry season, though the lower boundary layer is beginning to show an increase on the 24 July flight. On the 9 August flight, with several haze layers present, CO concentrations increased at almost all altitudes. Kirchhoff

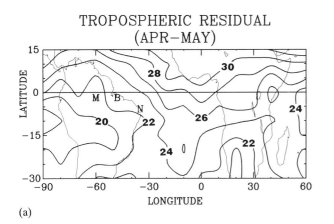

(a)

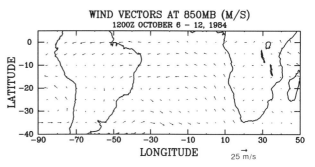

Figure 14.5 The derived wind fields from gridded NMC data are shown at 850 mb for 6–12 October 1984. The South Atlantic is dominated by an anticyclonic circulation. Vector size is directly proportional to wind speed, and the highest wind speed at each level is found in the lower right corner of the figure. During the dry season, air from Africa is transported toward Brazil at low latitudes and returned to Africa at high latitudes. The pattern of the wind field is very similar to the longitudinal pattern seen in the tropospheric residual (Figure 14.4c).

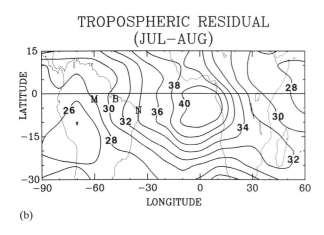

(b)

and Marinho (1988) found that surface CO concentrations in the Amazon rainforest were three to four times greater in the dry season than the wet season. At Natal, Kirchhoff et al. (1989) found maximum surface CO concentrations in September and minimum CO concentrations in February through April during 1987 and 1988.

CO mixing ratios obtained during the October 1984 Measurement of Air Pollution from Satellites (MAPS) experiment aboard the Space Shuttle (Reichle et al., 1990) also showed a correlation between areas of biomass burning and enhanced CO concentrations. High CO mixing ratios were measured over Brazil and southeast and central Africa, as well as over the South Atlantic off the coast of central Africa (coinciding with the tropospheric ozone maximum seen in Figure 14.4c). Coincident high values of both CO and ozone would preclude a stratospheric source, implicating other causes for these regions of enhanced CO and ozone.

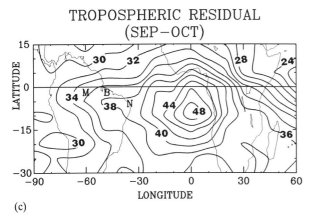

(c)

Figure 14.4 The climatologies in Figures 14.2b and 14.3b are expanded geographically to include the South Atlantic and Africa. (a) Wet season ozone values are relatively low throughout the southern tropics with a substantial increase (b) occurring as the dry season begins and biomass burning increases. (c) During the period of greatest burning in both Brazil and southern Africa, the tropospheric residual reaches a maximum, with the largest values off the coast of central Africa, stretching toward Brazil.

Transport

Figure 14.5 shows wind vectors derived from seven days of National Meteorological Center (NMC) *u* and *v* wind components during the October 1984 MAPS mission. The flow field is dominated by an anticyclone in the South Atlantic. Winds in the lower latitudes are easterly, bringing air from the coast of central Africa across the Atlantic to South America. Examining the longitudinal gradient in July through October (Figures 14.4b and 14.4c), it appears that the high ozone concentrations over Brazil may be due in part to the transport of ozone from Africa. Logan and Kirchhoff (1986) and Kirchhoff and Nobre (1986) hypothesized that the high values of ozone found at Natal were due in part to ozone coming from Africa. Kirchhoff and Nobre (1986) surmised that it would take 15 days, on average, for air leaving Africa to reach Natal. They also surmised that, because of the dominant subsidence in the Atlantic, air would not mix with higher altitudes and would retain its high concentrations of ozone as it was transported across the Atlantic.

To further examine the possible seasonal effects of long-range transport on local ozone concentrations, a graphical subtraction of the April and May climatological ozone concentrations from the July and August climatological concentrations is shown in Figure 14.6a. The gradient seen in Figure 14.4b is still evident, with a maximum off the coast of central Africa. The pattern is also consistent with the wind field in Figure 14.5. In Figure 14.6b, the September and October climatology minus the April and May climatology, the longitudinal gradient is still present, with an overall increase in ozone values. It appears from Figures 14.6a and 14.6b that during the dry season, ozone from central Africa is being transported to Brazil at low latitudes and being returned, enriched with ozone from Brazil, at high latitudes. This closed circulation may account for the maximum off the coast of central Africa, or it may be that Africa is producing more ozone than South America during the dry season.

In Figure 14.6c, July and August (early dry season) concentrations have been subtracted from September and October concentrations. The longitudinal gradient present in the dry season figures seen so far is not evident. There is little, if any, gradient present in the southern tropics. From Figure 14.6c, it would seem that local source regions are the dominant difference between the early dry season and the time of greatest burning in both Brazil and central and southeastern Africa. As the wind fields have not changed greatly from July and August to September and October, there is probably still as much transport occurring in September and October as there was in July and August. Whereas in Brazil in July and August the major difference from the wet season appears to be due to the transport of ozone, and likely CO, from Africa, in September and October the enhancement in ozone concentrations appears to be due to a mixture of both transport and increased local biomass burning, with a dominance by local burning.

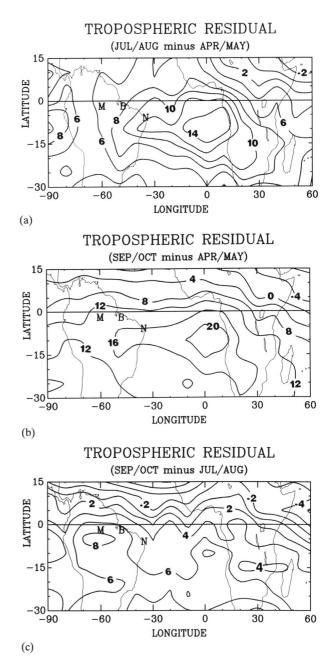

(a)

(b)

(c)

Figure 14.6 (a) The climatology shown in Figure 14.4a has been graphically subtracted from the climatology shown in Figure 14.4b. The strong longitudinal gradient seen in Figure 14.4b remains. (b) The climatology of Figure 14.4a has been subtracted from Figure 14.4c leaving a larger longitudinal gradient than seen in (a). Both subtractions indicate that some of the enhancement in ozone concentrations over Brazil during the dry season are due to transport from Africa. (c) The early dry season climatology (Figure 14.4b) is subtracted from the period of greatest biomass burning (Figure 14.4c). The longitudinal gradient has been replaced by regional sources over both Brazil (6 to 8 DU) and southern Africa (4 to 6 DU), indicating the dominance of local sources over transport in the late dry season.

Summary and Discussion

A comparison of direct measurements of Brazilian ozone and CO concentrations with space-based measurements shows a strong correlation between the two methods. This comparison also reveals a seasonality in both ozone and CO concentrations in Brazil. Dry season increases appear to be due both to the transport of ozone and CO from the African continent and increased local biomass burning.

The large values of both ozone and CO found off the coast of southern central Africa is most likely due to biomass burning in the African interior as well as an undetermined amount transported from South America. The coincident high values of both CO and ozone suggest photochemical sources as opposed to a stratospheric source for ozone. However, a better understanding of the dynamics occurring in the southern tropics throughout the year is necessary to fully understand the large values off the coast of Africa and the transport of these enhanced concentrations to Brazil and vice versa.

Likewise, knowledge of the chemical composition of the southern troposphere is also necessary to understand the complex interactions between chemistry and dynamics. With more measurement campaigns, such as those suggested by the IGAC (International Global Atmospheric Chemistry) Program, our data base will continue to expand, as will our understanding of this important region.

Acknowledgments

Part of the work for this study was performed under Contract NAS1-19000 to Lockheed Engineering and Sciences Corporation, Hampton, Virginia. This study was also supported in part by the NASA Global Tropospheric Chemistry Program.

Effects of Vegetation Burning on the Atmospheric Chemistry of the Venezuelan Savanna

Eugenio Sanhueza

Biomass burning in tropical savanna and rainforest regions is an important factor in the chemical composition of the atmosphere (Crutzen, 1987). On the global scale, burning of savanna grass produces three to four times greater emission of trace gases than deforestation processes of tropical rainforest (Hao et al., 1990).

As part of a comprehensive study of the Venezuelan savanna atmosphere, measurements of gases and particles, chemical composition of rain, and biogenic soil emission were made during burning and nonburning periods at several rural savanna sites. A review of the most significant findings is presented in this chapter, and their regional and global implications are discussed.

Venezuelan Savanna Climatic Region

The Venezuelan savanna region is located between the Amazon forest and the Caribbean Sea; it covers about half the country (Figure 15.1). Vegetation is diverse, ranging from semideciduous forest and woodland to scrub land and grassland. Two well-defined precipitation periods occur—a dry season from December to April and a rainy season from May to November. The amount of rain ranges from 800 to 1600 millimeters (mm) per year. Most of the vegetation burning occurs toward the end of the dry season and the beginning of the rainy period (between March and May), forming a clear seasonal pattern. No important variations in average temperature are observed between seasons.

On a regional scale, northeasterly and east-northeasterly trade winds are most frequent; therefore, a large amount of surface air entering the Venezuelan savanna region comes from the ocean. Wind speeds are lower during the wet season, when the Intertropical Convergence Zone (ITCZ) is over the central part of Venezuela; during this period, regional winds coming from the south also occur. It is important to note that the wind blows from the savanna region toward the Amazon forest, with stronger velocities during the

dry-burning period. Temperature profile measurements show that during the night the lower portion of the atmosphere becomes very stable, and periods of low dispersion capacity occur almost every day in the savanna region (Octavio et al., 1987).

The Venezuelan savanna region is affected very little by urban or industrial sources of air pollution (Sanhueza et al., 1988). Measurements of typical "anthropogenic" metals (i.e., Pb, V, Ni, Cd) show that the monitoring sites (Figure 15.1) were not affected by any significant local source of pollutants other than the emission from vegetation fires during the burning period (Sanhueza et al., 1986; Morales et al., 1990).

Complete data on the areas or types of vegetation burned in Venezuela do not exist. It is likely that about 155,000 km^2 of savanna grass is burned every year and that 50% is trachypogon grass (Sanhueza et al., 1988).

It is important to point out that measurements were always made away from fresh plumes; therefore, concentrations should be representative of relatively well-mixed atmosphere.

Chemical Composition of Rain during Burning Periods

The first rains of the season occur toward the end of the dry season, when vegetation burning still occurs throughout the savanna region, and they are heavily loaded with several compounds (Sanhueza et al., 1989; Sanhueza et al., 1990a). Table 15.1 shows pHs and concentrations of selected ions in rains collected during burning and nonburning periods. Concentrations of nitrate, ammonium, and phosphate are much larger during burning, showing that the atmospheric budget of these nutrients is enhanced during this period. The implication of this situation is discussed in the section on airborne particles later in the chapter.

In addition, higher concentrations of HCHO, HCOOH, and CH_3COOH were observed during burning periods. The results in Table 15.1 show that concentrations of HCOOH are higher than

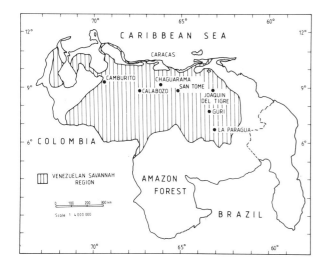

Figure 15.1 Venezuelan savanna region and location of monitoring sites.

Table 15.1 Chemical composition of rain during burning (B) and nonburning (NB) periods (units: $\mu eq\ L^{-1}$)

	La Paragua	J. Tigre	Chaguarama	Guri
pH				
B	5.6	4.7	5.0	4.7
NB	4.7	5.1	5.4	4.6
NO_3^-				
B	14.8	10.4	16.5	19.7
NB	3.2	3.7	4.6	2.7
NH_4^+				
B	60.0	9.8	81.1	42.6
NB	<1.9	2.6	13.4	3.1
PO_4^{3-}				
B	12.6	13.6	7.0	—
NB	<1.0	3.0	7.2	—
HCOOH				
B	30.0	30.4	46.4	25.4
NB	9.3	8.3	8.4	6.6
CH_3COOH				
B	18.6	—	34.2	17.2
NB	9.5	8.5	7.8	4.3
HCHO				
B	25.0	—	28.0	
NB	5.0	—	8.0	

Source: Sanhueza et al. (1990a).

CH_3COOH. According to Talbot et al. (1988), biomass burning emits 10 times more acetic than formic acid. Therefore, it is likely that the amount of these acids in rain is controlled by atmospheric processes, rather than by direct emission from fires. A good correlation has been observed between HCHO and HCOOH in Venezuelan rains (Sanhueza et al., 1990b), suggesting that formic acid in rain is mainly produced through cloud oxidation of formaldehyde.

No significant difference is observed in the rain acidity between both periods. The greater formation of HNO_3 and H_2SO_4 from photooxidation of NO_x and SO_2 emitted by the fires seems to be neutralized by a large direct emission of NH_3. As in the case of the rainy season, free acidity is dominated by organic acids (Sanhueza et al., 1990a).

Atmospheric Concentrations of Gases

A pioneer research effort on the seasonal variation of tropical ozone was made in the eastern part of the Venezuelan savanna region (Sanhueza et al., 1985). Monthly averages of the daily maximum clearly showed that higher ozone levels were produced during the vegetation-burning period. The annual mean O_3 "boundary layer" concentration was 17 ± 2 parts per billion (ppb), with lower values between June and August 1982, and a peak in March 1983 of ~31 ppb. The authors suggested that the higher O_3 levels observed during the burning period were produced photochemically by reactions between precursors emitted during the fires.

Gaseous nitric acid and ammonia were sampled with annular denuders in a woodland savanna site (Chaguarama) from April to December 1987 (Rondon and Sanhueza, 1990). At present, these are the only available measurements showing the seasonal variation of HNO_3 and NH_3 in the tropics. Higher concentrations of both gases were recorded during the vegetation-burning period.

A significant amount of NO_x is emitted to the atmosphere during the burning season in the Venezuelan savanna region (Sanhueza et al., 1988). Because of the high reactivity of the tropical atmosphere during this period, the NO_x should be rapidly oxidized to produce higher levels of HNO_3 (Rondon and Sanhueza, 1990).

It is now well established that ammonia is emitted in relatively large quantities by biomass burning. Considering that soil or vegetation emission of NH_3 resulting from biological activity should be lower during the dry season, the higher concentrations of NH_3 recorded in the Venezuelan savanna region by Rondon and Sanhueza (1990) could be mainly produced by direct emission of NH_3 during the fires. The NH_3 emitted by the fires plays a significant role in the acid-base equilibrium of the atmosphere, preventing a further acidification of Venezuelan savanna rains during biomass burning periods (Sanhueza et al., 1990a).

High nitrous acid levels (up to 1 ppb for a 12-hour average) were observed at night during the 1988 vegetation-burning season (Rondon and Sanhueza, 1990). Rainy season concentrations were always below 0.05 ppb. According to the authors it seems likely that HONO is directly emitted during the fires. This large production of HONO must increase the reactivity of the tropical atmosphere, through the production of OH radicals by photolysis:

$$HONO + h\nu \ (\lambda = 290 - 400 \ nm) \rightarrow OH + NO$$

Until now no emission factor for OHNO production from vegetation burning has been obtained. Field and laboratory experiments to evaluate this source are needed.

Airborne Particles

Atmospheric particulate matter was collected at several rural Venezuelan savanna sites, using HiVol samplers equipped with five-stage cascade impactors. The results show that during burning periods the atmospheric levels of small particles (< 0.49 μm) collected in the back-up filter are approximately five times larger than in the nonburning periods (Sanhueza et al., 1987). An example of size distribution during burning and nonburning periods is given in Figure 15.2. The dramatic increase of fine particles could be due to (1) an increase of the emissions in primary particles produced during the combustion process and (2) the production of secondary particles formed by oxidation of gaseous emissions from the fires.

The study of the chemical composition of the particles shows that during burning periods higher concentrations of inorganic water-soluble ions (i.e., SO_4^{2-}, NO_3^-, PO_4^{3-}, NH_4^+, K^+, Ca^{2+}) (Sanhueza and Rondon, 1988) and also of several elements (i.e., Mn, Cu, Zn) (Sanhueza et al., 1986; Morales et al., 1989) are observed. Significant primary emission of K^+, Ca^{2+}, Zn, and possibly Mg^{2+} and Cu occur in the fires. NO_3^-, SO_4^{2-}, and NH_4^+ must be produced in the atmosphere as final products of reaction of gaseous NO_x, SO_2, and NH_3 emitted during burning. Also,

the results indicate that a phosphorous compound must be emitted during biomass burning; however, until now it is unknown whether phosphate is released directly or whether it is formed in the atmosphere by photooxidation of a reduced phosphorous compound emitted during fires (Sanhueza and Rondon, 1988).

During vegetation burning several ions and elements were concentrated in the fine particles. As an example, Figure 15.3 shows that during burning

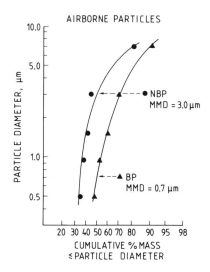

Figure 15.2 Particle size distribution (log-probability plots) of samples collected near La Paragua (Monte site) (Sanhueza et al., 1987).

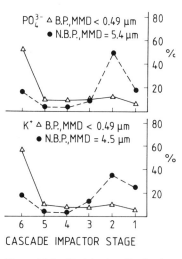

Figure 15.3 Particle size distribution of PO_4^{3-} and K^+. The 50% cutoff diameter for the CI are as follows: $1 > 7.2$ μm; $2 = 7.2 - 3.0$ μm; $3 = 3.0 - 1.5$ μm; $4 = 1.5 - 0.95$ μm; $5 = 0.95 - 0.49$ μm; and $6 < 0.40$ μm (Sanhueza and Rondon, 1988).

periods about 55% of PO_4^{3-} and K^+ are associated with particles smaller than 0.49 μm; it is important to note that during nonburning periods, coarse particles (>1.5 μm) predominate (Sanhueza and Rondon, 1988). Similar results, but less accentuated, were observed with Ca^{2+}, Zn, and Cu. On the other hand, SO_4^{2-}, NO_3^-, and NH_4^+ showed similar size distributions during burning and nonburning periods. This is consistent with the fact that these compounds are mainly produced in the atmosphere during both seasons by chemical reaction between gases.

Fine particles (< 1.5 μm) have a long atmospheric lifetime and can be transported long distances. Assuming a boundary layer of 2000 meters and an estimated dry deposition velocity of about 0.1 cm s^{-1} (centimeter per second), a residence time of about 12 days is calculated. With an average wind velocity of 10 km h^{-1} (hour = h), during dry periods particles can be transported over 3000 km before they are removed by dry deposition from the atmosphere. Transport of fine particles produced during vegetation burning should represent a loss in nutrients (i.e., N, P, K, Ca) for the savanna ecosystems, and probably a source of nutrients for the downwind Amazon rainforest. At present, not enough information (i.e., primary emission of particles from burning of savanna grass) is available to estimate the transfer of nutrients between savanna and forest regions, and the impact in both ecosystems should be evaluated.

Finally, the large input of cloud condensation nuclei (fine water-soluble particles, with a significant fraction of SO_4^{2-}) produced during vegetation burning into the savanna region could play a role in the rain pattern of the Amazon rainforest. This possible effect from savanna fires should also be investigated.

Soil Emission after Vegetation Burning

A large increase in the NO soil emission was observed from a burned plot in Chaguarama (woodland savanna) (Johannson et al., 1988). Figure 15.4 shows that NO fluxes increased by a factor of 10 after the burning and then slowly decreased over the following days. Therefore, vegetation burning, in addition to being a direct source of trace gases, also enhances biogenic production of NO from savanna soils. Given that much of the savanna region is burned every year, this effect is probably of regional significance. Until now, the reason for the stimulating effect of burning on NO emission from "dry" savanna soil was unknown.

Fluxes of N_2O from sandy soil did not change signif-

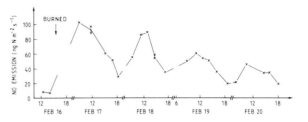

Figure 15.4 Soil emissions of NO after vegetation burning (Johansson et al., 1988).

icantly after the grass layer was burned at a grassland savanna site (Hao et al., 1988); however, preliminary measurements in soil with a high clay content show an increase in the N_2O emissions after vegetation burning (unpublished results). Additional measurements are needed.

Biomass Burning in Africa:
An Assessment of Annually Burned Biomass

Robert A. Delmas, Philippe Loudjani,
Alain Podaire, and Jean-Claude Menaut

It is now established that biomass burning is the dominant phenomenon that controls the atmospheric chemistry in the tropics. Africa is certainly the continent where biomass burning under various aspects and processes is the greatest. Three different types of burnings have to be considered—bush fires in savanna zones which mainly affect herbaceous flora, forest fires due to deforestation for shifting agriculture or colonization of new lands, and the use of wood as fuel. The net release of carbon resulting from deforestation is assumed to be responsible for about 20% of the CO_2 increase in the atmosphere because the burning of forests corresponds to a destorage of carbon from the biospheric reservoir.

The burning of annual plants that takes place in tropical savanna does not account for the CO_2 increase as the savanna recovers each year, taking the carbon from the atmospheric reservoir by photosynthesis; savanna burning contributes only to modulate CO_2 variations in the atmosphere. However, savanna fires, owing to the considerable amount of biomass concerned, should have an important influence on atmospheric chemistry. Many trace constituents, greenhouse gases (CH_4, N_2O), as well as reactive compounds (NO_y, NMHC) are emitted in larger proportion during the burning of vegetation than during the decay of organic matter. These latter compounds are ozone precursors, and thus biomass burning has recently appeared as an important cause of photochemical ozone formation in the troposphere, especially over the equatorial Atlantic downwind of Africa; this underlines the importance of the African continent with respect to this phenomenon (Fishman et al., 1986; Fishman and Larsen, 1987).

The amount of reactive or greenhouse gases emitted by biomass burning is directly proportional, through individual emission factors, to the biomass actually burned. This chapter evaluates the biomass annually burned on the African continent as a result of the three main burning processes previously mentioned.

Assessment of Biomass Burned Annually in Africa

Deforestation

The impact of humans on the natural vegetation of Africa is very longstanding, but it has greatly increased in the last 20 years because of economic development and population growth. The present rate of population increase is 3.3% per year. To assess the amount of biomass burned as a result of deforestation we used the data given by the Food and Agriculture Organization (FAO) for the periods 1976 to 1980 and 1981 to 1985 (Lanly and Clement, 1979; FAO, 1988). All African countries with closed forest are not affected to the same extent by clearing for agriculture. It is important to distinguish between closed forest and open forest formations, since both the processes and effects of deforestation are quite different in the two cases. Most of the deforestation occurs in a small number of countries such as Cameroon, the Ivory Coast, Nigeria, Madagascar, and Zaire (Table 16.1), but rates of deforestation are very different. In Zaire, where primary closed forest covers more than 1 million km^2, the rate of deforestation is rather low, about 0.1% every year; while in West African countries like Nigeria and the Ivory Coast, it consumes some 10% of the remaining forest area every year. In the Ivory Coast, deforestation has taken place throughout the last 20 years, and 70% of the forest area existing at the beginning of the century has already been cleared. For tropical Africa as a whole the closed forest areas which have disappeared annually during the last five years are estimated at 1.32 million hectares (ha) (Table 16.2).

The biomass burned annually as a result of deforestation depends on three parameters—the area annually cleared, the average phytomass, and the burning efficiency. The phytomass of tropical closed forest is quite variable depending on the region concerned. We will consider an average value of 400 tons dm ha^{-1} (tons dry matter per hectare) (Rodin et al., 1975; Whittaker and Likens, 1975). The fate of the cleared biomass is still much more uncertain, and thus

Table 16.1 Humid tropical forest areas in Africa in 1980 and annual rates of deforestation for the periods 1976 to 1980 and 1981 to 1985

	Hectares ($\times 10^3$)		
	1980	1976–1980	1981–1985
Angola	2,900	40	44
Benin	47	1.5	1.2
Cameroon	17,920	80	80
Central African Republic	3,590	5	5
Congo	21,340	22	22
Ivory Coast	4,458	310	290
Equatorial Guinea	1,295	2.5	3
Gabon	20,500	15	15
Ghana	1,718	27	22
Guinea	2,050	36	36
Guinea Bissau	660	15	17
Kenya	690	11	11
Liberia	2,000	51	46
Madagascar	10,300	165	150
Nigeria	5,950	285	300
Reunion	100	—	—
Senegal	220	—	—
Sierra Leone	740	5.8	6
Tanzania	1,440	10	10
Togo	304	2	2.1
Uganda	750	10	10
Zaire	105,650	165	180
Total	204,622	1,269	1,250

Source: FAO (1988).

Table 16.2 Annual deforestation and annually burned biomass in tropical Africa

	Annual deforestation ($\times 10^3$ ha)	Annually burned biomass ($\times 10^6$ tons dm)
Northern savanna regions	2	0.2
West Africa	720	72
Central Africa	349	34.9
East Africa and Madagascar	247	24.7
Tropical Africa	1318	131.8
Total for world	6113	611.3

Note: The average biomass considered is 400 tons dm/ha and the burning efficiency is 25% (data from FAO, 1988).

Table 16.3 Household consumption of fuel wood and commercial energy for selected African countries in 1980

	Household consumption of fuel wood	Commercial energy consumption
	($\times 10^6$ tons of oil equivalent)	
Sudan	9.6	1.3
Ethiopia	3.7	0.5
Niger	0.7	0.2
Senegal	1.4	1.4
Nigeria	14.6	9.6

Note: 1 m^3 of wood = 0.3 ton of oil equivalent. From Anderson and Fishwick (1988).

the burning efficiency is highly speculative. In the absence of accurate data we will adopt the analysis and the value (25%) given by Seiler and Crutzen (1980). Considering the above-mentioned area of closed forest cleared every year in tropical Africa, the biomass burned annually as a result of deforestation would be 130 MT dm (megatons dry matter) per year.

In mixed forest-grassland tree formations the human impact is more progressive than in closed forest, leading to a reduction of the wood component. This destructive action is going on under the combined action of humans, cattle, and fire. The increasing demand for firewood is the main cause of the disappearance of these open forest formations.

Use of Wood as Fuel

Compared with other sources, fuel wood is not a significant source of energy in industrialized countries, but in most developing countries it constitutes the main source of energy. The household consumption of fuel wood and the commercial energy consumption for selected African countries are shown in Table 16.3 (from Anderson and Fishwick, 1988). In Sudan, for example, fuel wood consumption represents 9.6 MT of oil equivalent and is seven times higher than commercial energy consumption, including oil, coal, and hydroelectric energy. This is the case in most African countries except in Nigeria, which is an important oil-producing country. Basically, fuel wood is the main source of energy in Africa; this will pose an important economic problem over the next decades. The reason is an increasing firewood demand following the increase in population, and a subsequent decrease in resources in many countries, especially in the Sahel zone. In this region the volume of fuel wood collected is already in excess of the annual wood productivity. In tropical countries many programs are sponsored by national and international organizations such as the Food and Agricultural Organization or the World Bank, to improve both the management of resources and the efficiency of wood-burning stoves (World Bank–PNUD, 1987; Anderson and Fishwick, 1988).

Firewood can be used directly for cooking or heating or it can be carbonized and transformed into charcoal. Generally, in small villages people burn firewood directly, but in large cities fuel wood is mainly used as charcoal to reduce transportation costs. In the Ivory Coast, for example (Table 16.4), the total fuel wood consumption is 8.2 MT per year for a population of 11 million inhabitants. In Abidjan (2 million inhabitants) household energy is mainly provided by charcoal. To extrapolate from the data to the whole of tropical Africa we consider a fuel wood consumption of 0.8 cubic meter (m^3) per capita for rural population, 0.6 m^3 per capita for urban population, the consumption of charcoal being 83 kilograms per year (kg yr^{-1}) per capita for urban population (Anderson and Fishwick, 1988). With these values and the 1988 FAO data for African countries the total consumption of fuel wood would be 230 MT dm per year (Table 16.5). Half of this quantity is transformed into charcoal, the conversion efficiency being about 10% in traditional charcoal ovens (Eimer and Ndamana, 1987; Vernet, 1988). Charcoal making is obtained by pyrolysis of wood, and this process leads to important emissions of methane (Delmas et al.,

Table 16.4 Firewood, charcoal, and total fuel wood consumption in the Ivory Coast and in the cities of Korhogo and Abidjan

	Population ($\times\ 10^3$ inhabitants)	Firewood consumption ($\times\ 10^3$ tons)	Charcoal consumption ($\times\ 10^3$ tons)	Total fuel wood consumption ($\times\ 10^3$ tons)
Ivory Coast	11,000	4100	410 (4100 tons wood equivalent)	8200
Abidjan	2143	264	341 (3410 tons wood equivalent)	3675
Korhogo	146	60.7	1.07 (10.7 tons wood equivalent	71.4

Note: Data extrapolated to year 1990. The total fuel wood consumption is calculated assuming a conversion efficiency of 10% for wood carbonization (data from the Ministry of the Environment of the Ivory Coast).

Table 16.5 Global estimate of firewood and charcoal consumption in tropical Africa

	Total population ($\times\ 10^6$)	Rural population ($\times\ 10^6$)	Urban population ($\times\ 10^6$)	Total wood consumption including charcoal production				Charcoal ($\times\ 10^6$ tons)
				Rural ($\times\ 10^6\ m^3$)	Urban ($\times\ 10^6\ m^3$)	Total ($\times\ 10^6\ m^3$)	Total ($\times\ 10^6$ tons dm)	
Northern savanna regions	37.213	30.654	6.559	24.523	3.935	28.458	20.85	0.54
West Africa:								
Without Nigeria	48.953	31.134	17.819	24.907	10.691	35.598	26.05	1.48
With Nigeria	105.438	69.037	36.401			50.000	36.60	3.02
Central Africa	60.063	39.161	20.902	31.328	12.541	43.869	32.20	1.73
East Africa	207.506	157.399	50.107	125.919	30.064	155.980	114.20	4.16
Total tropical Africa	459.173	327.385	131.788			314	230	11 MT

Note: Wood consumption rates are 0.8 $m^3\ yr^{-1}$ per capita for rural population, 0.6 $m^3\ yr^{-1}$ per capita for urban population, charcoal consumption being 83 kg yr^{-1} per capita for urban population.

1990) and probably also of other reduced trace gases, such as NMHC, NH_3, and H_2S; it can constitute a significant source for these gases.

Savanna Fires

Savanna fire is a widespread phenomenon in Africa. These fires extend from approximately 15°N to 25°S latitude. The area potentially submitted to fire is in the order of 10 million km^2. The savanna biomass varies in relation to annual rainfall from 1 ton dm ha^{-1} to as much as 25 tons dm ha^{-1} (Menaut, 1979), with pronounced latitudinal gradients but also with local variations due to vegetation types, orography, and rainfall patterns. To estimate the biomass burned during vegetation fires in savanna regions, several parameters have to be determined in each savanna type—the integrated above-ground biomass, the burned areas, and the burning efficiency. It is still difficult to give accurate answers to these questions since no extensive study of savanna fires in Africa has been made. Several research programs are presently under way—Savanna on the Long Term (SALT), Dynamics and Chemistry of the Atmosphere in Equatorial Forest (DECAFE)—but they cannot yet provide general results allowing global estimate of the phenomenon.

Our estimate of the biomasses of the different savanna zones was deduced from a model of conversion of solar energy into net primary production (NPP) (Monteith, 1972). The spatial extension of the model was made using Advanced Very High Resolution Radiometer (NOAA/AVHRR) satellite data. The NPP values were fitted to ground estimates using a statistical relationship (Loudjani et al., 1988). The total NPP estimated by the satellite model (Figure 16.1) was then converted into aerial NPP assuming that there is a migration of 50% of the dry matter produced to the root system (Table 16.6). To derive the above-ground biomass from the aerial NPP we use a conversion coefficient which takes into account grazing, death of plant organs, and root stocking for pluriannual plants. According to Menaut (1983) NPP is 25% higher than herbaceous biomass in the Guinean zone, where annual plants are dominant. In the Sahelian zone, where perennial plants are dominant, this coefficient is about 10%; in the Sudanian zone intermediate values were used (Table 16.6). All the coefficients are derived from field experiments in West Africa. The calculation procedure was extended, but without calibration, to the whole African continent on the basis of a numerical vegetation map of Africa developed by the Institute of International Vegetation

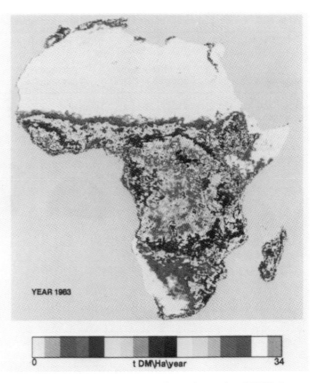

Figure 16.1 Net primary production estimate (total NPP) from satellite data.

(ICIV) in Toulouse on behalf of FAO (Lavenu et al., 1987). In this map five main savanna zones are defined (Table 16.6). The correspondence with the usual savanna classification is (1) Guinean savanna, zones 4 and 8; (2) Sudanian savanna, zones 5 and 6; and (3) Sahelian savanna, zone 7.

The data obtained from satellite determination of vegetation index corresponds to the biomass potentially submitted to fires. To derive the biomass actually burned we have to apply correction coefficients, giving first the burned area for each zone and then the biomass assumed to be burned taking into account the average burning efficiency. In the absence of general values for the whole African continent, we use the coefficients given by Menaut (1990) derived from field observation in the West African savanna. From this data the proportion of savanna burned annually would be, every year, 5% to 10% in the Sahelian zone, 25% to 50% in the Sudanian zone, and 60% to 80% in the Guinean zone; while average burning efficiency, taking into account both vegetation heterogeneity and water content variations in plants, would be 80%.

This description of the global estimate of the bio-

Table 16.6 Biomass annually burned in savanna fires in tropical Africa

Vegetation class	Number of pixels	Area (km^2)	Average NPP (tons dm/ha^{-1})	Total NPP (GT dm)	R_A	R_B	Above-ground biomass (GT dm)	R_{PB}	R_{EB}	Annually burned biomass (GT dm)
Woodland and tree savanna	13,523	3,042,675	19.8	6.02	0.5	0.75	2.25	0.7	0.8	1.26
Tree savanna or woodland savanna and shrub savanna intermingled	15,177	3,414,825	16.6	5.67	0.5	0.80	2.27	0.4	0.8	0.78
Tree and shrub savanna steppe or herbaceous savanna steppe	7,480	1,683,000	8.5	1.43	0.5	0.85	0.61	0.4	0.8	0.20
Shrub or herbaceous savanna steppe	4,766	1,072,350	9.9	1.06	0.5	0.90	0.48	0.1	0.8	0.04
Moist savanna under moist climate	3,949	888,525	15.3	1.36	0.5	0.75	0.51	0.7	0.8	0.29
Total for tropical Africa	44,895	10,101,375		15.54			6.12			2.52

Note: Area of one pixel: $15 \times 15 = 225$ km^2; units: GT dm = gigatons (10^9 tons) of dry matter.

mass burned annually in savanna fires can be summarized by the following relationship:

$$BB = NPP \times R_A \times R_B \times R_{PB} \times R_{EB}$$

where

BB = burned biomass,

NPP = total net primary production (roots + aerial) (see Figure 15.1),

R_A = aerial to total NPP ratio,

R_B = above-ground biomass to aerial NPP ratio,

R_{PB} = area annually burned (for a given area),

R_{EB} = burning efficiency.

All these data combined lead to an average biomass burned annually in African savanna of about 2.5 GT dm (gigaton = GT) per year (Table 16.6). Guinean and Sudanian savannas contribute 61% and 37%, respectively, of the total burned biomass. Taking into account the accuracy of the estimates the contribution of Sahelian savanna can be neglected. These values include both dead and living material whose percentages for each zone are given by Menaut (1990).

Biomass Annually Burned in Africa

The annually burned biomass associated with the three burning processes studied is shown in Figure 16.2. In spite of rather large uncertainties in this rough evaluation, two conclusions seem to be clear:

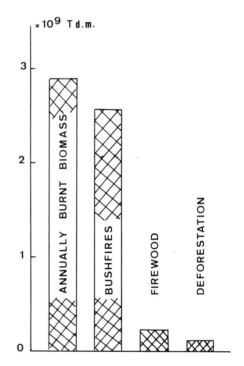

Figure 16.2 Biomass burning in tropical Africa: annually burned biomasses associated with forest fires, firewood, and savanna burnings. Total annually burned biomass (in gigatons of dry matter per year).

(1) The total amount of biomass burned every year is considerable: 2.9 GT dm (this represents, for the African continent alone, a value comparable with the annual world consumption of oil (5 GT); (2) in comparison with the other types of burning (forest fires—130 MT, use of wood as fuel—230 MT) savanna fires (2.5 GT) are totally dominant. This is a typical feature of the African continent; in South America, for instance, the role of forest fires is certainly much more important.

It is clear that biomass burning in Africa should have a great influence on atmospheric chemistry. Considering CO_2 emission factors of 950 g/kg for wood combustion and 1350 g/kg for savanna plant combustion (Delmas et al., 1990), the CO_2 release to the atmosphere would be $3.7 \, GT \, (CO_2)y^{-1}$. However, as stated previously, the majority of this CO_2 input is removed from the atmosphere by the subsequent regrowth of the savanna, and thus the net contribution to the CO_2 increase remains limited. It corresponds only to the biomass burned in forest fires and as household fuel, since the wood component of the biomass burned in savanna fires is slight. This contribution can be estimated in the order of 0.4 GT per year, which is small but not negligible.

The main consequence of these extensive fires concerns atmospheric minor constituents such as CO, CH_4, NMHC, NO_y, NH_3, CH_3Cl, and particulate carbon. Emission factors (EF) for these gases as well as for CO_2 depend on the relative importance of flaming and smoldering phases in the combustion and thus on the type of fuel and on its water content. For most of these gases emission factors have not yet been determined in Africa, so it would be somewhat premature to give global emission estimates for each trace gas species. However, recent studies of the tropospheric ozone field over tropical Africa have clearly shown the influence of biomass burning. It was first suggested by satellite measurement of total ozone and tropospheric ozone by Total Ozone Mapping Spectrometry (TOMS) (Fishman et al., 1986; Fishman and Larsen, 1987), which showed large increases in tropospheric ozone over equatorial Africa and the tropical Atlantic, during the dry season in the Southern Hemisphere, which the authors attributed to the influence of biomass burning in Africa. This was then confirmed by surface measurements of ozone in the Congo (Cros et al., 1987), which showed large increases in ozone at ground level with average values exceeding 70 parts per billion (ppb). Then a three-year study of ozone variations at Brazzaville (5°S) showed a typical seasonal cycle with enhanced con-

centrations reaching monthly average values of 35 to 40 ppb during the dry season in the Southern Hemisphere (Cros et al., 1988). An ozone layer, located at between 1 and 3 km altitude with concentrations of 100 ppb, was identified during this period by vertical profile measurements in the lower troposphere. The same phenomenon was also observed in the Northern Hemisphere during the TROPOZ I experiment (Marenco et al., 1990) and the 1988 DECAFE experiment (Helas et al., 1988). The influence of biomass burning on the photochemical ozone formation over Africa is now established. Quantitative estimates of the spatial extension and temporal evolution of the phenomenon are still to be made; they will be one of the aims of the future DECAFE, TROPOZ, and TRACE-A experiments (Delmas, 1990).

Seasonal Distribution of Fires

The seasonal distribution of biomass burning on the African continent is dominated by a preponderance of savanna fires. The occurrence of fire in savanna regions follows a seasonal cycle with a pronounced maximum in the middle of the dry season in both hemispheres. However, the duration of the dry season is not the same everywhere. It varies from eight to nine months in semi-arid regions to four to five months in humid regions close to the forest zones. At present it is not easy to define fire periods from objective data. Direct field observations are too scarce and scattered, and satellite determination of fires and burned areas are still not able to give a global view of the phenomenon throughout the entire year and for the whole continent. Adequate atmospheric chemistry data, such as continuous recording of CO in both hemispheres, would provide a good criterion. Unfortunately these measurements do not exist. The only objective data that can be used are variations of ozone, a secondary product of biomass burning, measured over yearly periods at Brazzaville (5°S) in the Southern Hemisphere and Enyelle (2°30′N) in the Northern Hemisphere.

The variations of the monthly mean of the daily maximum in surface ozone concentrations are shown in Figure 16.3. Because of convective mixing these values are representative of ozone concentrations of the African boundary layer (Cros et al., 1988). Rainfall variations indicate the respective dry seasons. As stated previously, increases in ground-level ozone concentration can be linked to biomass burning. The temporal distribution of burnings is thus characterized by two main burning periods—December to

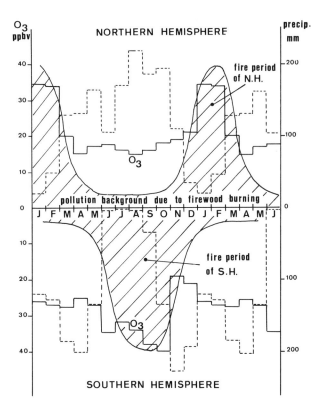

Figure 16.3 Fire periods in Northern and Southern hemispheres in tropical Africa characterized by monthly variations of ozone and precipitation. SH: O₃ and precipitation at Brazzaville (5°S); NH: O₃ at Enyelle (2°30′N), precipitation at Bangui (5°N). The curve representing fire periods in both hemispheres is qualitative; amplitudes on vertical axis have no significance.

ence on the atmospheric chemistry of the global troposphere. The biomasses associated with the three main burning processes (forest fires, firewood, and savanna fires) were studied separately because emission factors for atmospheric constituents can be different depending on the biomaterial consumed. Our results show that firewood combustion and forest fires are far less important than savanna fires. We chose to give just average values of burned biomasses because, in many cases—fuel wood consumption, burning efficiency in forest fires, burned areas in savannas—it would have been speculative to give an uncertainty range. However, the values obtained are certainly fairly well representative of the biomasses involved in the different burning processes. This first study is still approximative and general; detailed investigation is now required to provide data that could be used in global atmospheric chemistry models. Research programs are being undertaken in the framework of DE-CAFE and SALT (Delmas, 1990) to determine

1. Emission factors for the main trace gas species, taking into account vegetation heterogeneity and variability of plant water content.

2. The biomass actually burned for the whole African continent, on a monthly basis, with a spatial resolution of 15 km × 15 km, by means of remote sensing studies of biomasses and burned areas.

The final product of these combined programs should be numerical maps of trace gas fluxes, which could be used directly as source terms in models.

Acknowledgments

We would like to thank Dr. Doryane Kermel-Torres from the CEGET in Bordeaux and Professor Jean Baudet from Abidjan University for their help in finding data on firewood burning in Africa.

March in the Northern Hemisphere and June to November in the Southern Hemisphere—with maximum fire occurrence in January and February in the former and September and October in the latter. This figure is consistent with common field observations. However, it is quite probable that a continuous background of pollution is present throughout the year due to firewood burning and scattered fires lit during short periods of drought occurring during the rainy season. The only period during which the influence of biomass burning can be expected to remain limited corresponds to April and May. So it appears that biomass burning is certainly the dominant feature of the atmospheric chemistry of tropical Africa.

Conclusion

Our data as a whole show that, owing to the importance of the biomass annually burned (2.9 GT y⁻¹), biomass burning in Africa should have a great influ-

Biomass Burning in West African Savannas

Jean-Claude Menaut, Luc Abbadie, François Lavenu,
Philippe Loudjani, and Alain Podaire

The effects of savanna burning on vegetation and land use practices in the tropics have been described at length in a number of papers since the early 1900s. It has only been recently mentioned that biomass burning in tropical systems could play a major role in climate by releasing huge amounts of greenhouse gases (Wolbach et al., 1985, 1988; Crutzen, 1988; Crutzen and Andreae, 1990).

No general agreement has yet been found on the definition and extent of tropical savannas. They more or less encompass the tropical grasslands and woodlands described by Olson et al. (1983) and Atjay et al. (1987). According to these authors, savannas should cover about 28×10^6 km^2 (square kilometers) and their overall production of about 37 GT dm/yr (gigatons dry matter per year) largely exceeds that of tropical forests (about 26 GT dm per year, for 14×10^6 km^2). Long et al. (1989) and Hall and Scurlock (1990) have claimed that the global values given for savanna biomass and productivity had been seriously underestimated in the past (up to five times, including roots). They say that as much as 2.4 to 4.2 GT of carbon per year (gross flux) could be released by burning, which nearly doubles the value given by Hao et al. (1990). Whatever might be the accuracy of the data, there is little doubt that savannas, burning included, significantly account for regional and global carbon budgets (Hall, 1989; Hall and Rosillo-Calle, 1990). This is particularly true for West African savannas, which cover about 3×10^6 km^2, all in one block. In most other parts of the world, savannas are distributed in smaller units: They develop on marked reliefs or are intersected with other vegetation types, such as rainforests or arid scrub lands.

This chapter approaches the influence of West African savanna ecosystems on the regional climate by giving, as precisely as possible, the amount of volatilized elements (e.g., carbon, nitrogen, and sulfur) annually released by bush fires into the atmosphere. In spite of the relative functional similarity of West African savannas, fire behavior and effects vary with the different bioclimatic and phytogeographic zones of the region (Menaut, 1983; Menaut et al., 1985):

Guinea or humid zone:

- Precipitation (P) > 1000 mm $\times$ yr^{-1}, dry months < 2, P/PET (potential evapotranspiration) > 0.7
- Dense and high grass layer dominated by scattered trees; grass production is high and annual burnings are severe

Sudan or mesic zone:

- 500 < P < 1200, 2 < dry months < 8, 0.2 < P/PET < 0.7

- Higher tree density; lower grass production, but annual burnings may be severe according to drought

Sahel or arid zone:

- P < 600, dry months > 8, P/PET < 0.25
- Open vegetation types; low production of annual grasses; accidental burnings

In order to reach an acceptable accuracy, results are given for each of the zones described above (Figure 17.1) and summarized for West Africa.

Major Characteristics of Fire in West Africa

Historical Perspective: The Origin of Fire

Savannas display the basic ingredients for fire in nature (alternation of humid and dry weather, lightning strikes, and fuel), and have certainly been burned since they appeared (Komarek, 1972). In West Africa, pollen grains of grasses can be traced back to the Eocene, from which carbonized grass cuticles have been collected (Germeraad et al., 1968). The high density of fulgurites (sandy grounds fused by lightning strikes) found in Chad testifies to the frequent occurrence of lightning strikes in ancient times (Komarek, 1972). Fire is thus quite an ancient phenomenon, largely antedating the appearance of humans. Since their appearance in West Africa, there is little doubt that intentional burning has been practiced

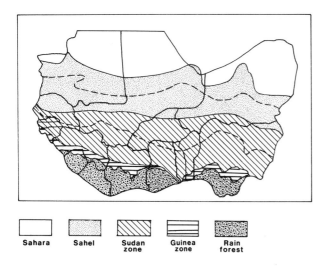

Figure 17.1 Phytogeographic and bioclimatic zones of West Africa. Broken lines separate the northern and southern parts of the Sudan and Sahel zones.

(Rose-Innes, 1972). Humans have increased the fire hazard tremendously, causing an extension of the open types of savanna to areas previously covered with scrub, woodland, or dry forest (Van der Hammen, 1983). However, the remoteness of the phenomenon, as a driving force in savannas, is still debated. Sanford (1982) held the view that burning was relatively rare before the human transition from hunting-gathering to farming (about 10,000 B.P.) and sporadic for a considerable time afterwards.

Reasons for burning are numerous and have been listed, for West Africa, by such authors as Rose-Innes (1972) or Monnier (1981). Some reasons especially apply to the Sudanian zone, where cattle breeding takes an increasing importance: Fire is set to get rid of the unpalatable stubbles and to initiate an off-season regrowth of succulent shoots of hemicryptophytes whose mature leaves are coarse and poorly nutritive. Such a type of fire seldom occurs in the Guinean zone (no significant cattle breeding) or in the Sahel (where herders prevent it from destroying the weak cover of annuals). In the Sudanian and Guinean zones of West Africa, fire is also set to clear ground for agriculture, to establish fire breaks around settlements, to get rid of parasites, to drive game out of hiding, and finally to improve communication. Accidental fires are frequent, not to mention the psychological motivations for burning the savanna (Guillaume, 1959).

Undoubtedly, natural fires still occur. However, the extent and frequency of man-made fires prevent their spread. They are now difficult to identify, and very few indisputable cases have been reported from West

Africa. Fires due to lightning have been observed in the field in Guinea by Tournier (1948) and in Nigeria by Jones (1963); some others have been detected by remote sensing (Vorphal et al., 1970). It also seems that, in areas where fire did not occur during the dry season, the thick herbaceous litter begins to ferment and fire might spontaneously burst out (Sillans, 1958).

Location and Extent of Fire

Phillips (1965) has provided valuable information on fire distribution and frequency in Africa. Unfortunately, the scale and the complexity of the accompanying map (1/20,000,000) make it difficult to use. However, it is quite clear that man-made fires run over West Africa. Depending on motivations and local ecological conditions, a number of fire sources can be permanently perceived from the end of a rainy season until the beginning of the following. Fire movements and effects on vegetation display a large spatial and temporal heterogeneity, which renders them difficult to survey. Large-scale fires are more likely to take place in areas having a climate moist enough to promote the production of a large amount of grass fuel, and seasonally dry enough to allow dried plant material to catch fire (Gillon, 1983). In West Africa, the savannas which experience the most frequent and severe burnings are the moist Guinean savannas (high grass biomass) and the South Sudanian savannas (lower but drier dead standing crop) where the desiccating effects of the "harmattan" wind in the dry season enhance fire-favorable conditions. The grass layer of these savannas is constituted of bunches of hemicryptophytes providing a thick and continuous fuel load to the fire. In the northern Sudanian savannas and in the Sahel, the sparse ground vegetation often constitutes a natural protection against fire spreading (Grosmaire, 1958). The even lower grass cover in the Sahel generally prevents fire occurrence (Cissé and Breman, 1982). Most herders also take care not to burn the fire-sensitive annuals in order to protect the seed pool and to preserve the maintenance of some edible plant material for the cattle till the dry season. Accidental, localized fires are nonetheless frequent, and Imort (1989) has even reported the annual occurrence of burning in some South Sahelian savannas of Mali.

Fire propagation is clearly related to local ecological conditions (climate and state of the vegetation). If fire spreads over most of the area every year, recent remote sensing studies have shown that, each year, considerable surfaces may remain unburned (Verger,

1980; Kouda, 1981; IBM-IRAT, 1983). However, these unburned areas do not remain constant over years and change in size and location with the prevailing ecological conditions. Natural fire breaks, such as rivers, forest galleries, or cornices of lateritic outcrops, may impede fire propagation (Jaeger, 1956). These natural fire breaks are rarely totally effective, and localized conditions may enable fire (fierce or creeping) to cross them.

Burning Regime

The extensive literature on fire in West Africa does not provide any quantitative information on the burning regime in "natural" conditions on a large scale. It is generally accepted by field ecologists that most savannas burn each year (Monnier, 1968, 1981). As seen above, local variations are frequent, but all savanna land (Sahel excepted) should burn within the time lag of two to three years.

Fire may appear at the end of the rainy season (early burnings) and remain until the next rainy season begins (late burnings). However, most fires occur in the heart of the dry season—i.e., in December and January, when the herbaceous cover has completely withered and is fairly dry. In areas of cattle migration, burning occurs with the arrival of the herds, which sometimes leads to off-season fires. Fire can even be set twice a year in localized areas. On the scale of West Africa, multiple burning has a negligible effect on the carbon budget and will not be taken into account in this study.

Differences in fire regime are observed between the various bioclimatic zones (Figure 17.2) At the local level, fire is considered to occur each year in humid to mesic savannas; its frequency decreases in drier savanna types with rainfall and available fuel. The period during which extensive burnings may occur increases with the length of the dry season, from the humid Guinean savannas to the dry Sudanian savannas. In the Sahel, burnings are more erratic and soon stop when the standing dead plant matter trampled by cattle disappears.

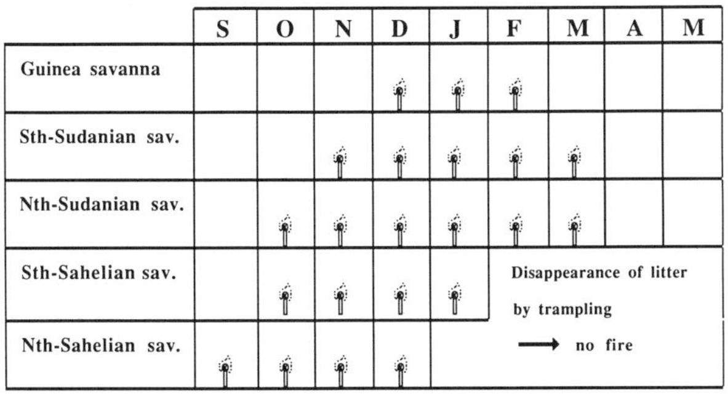

Fire behavior is extremely heterogeneous in all of its spatial and temporal aspects. Apart from detailed local studies, it is hardly possible to distinguish burning frequency from burning extent. In the present study, fire will be considered to be set up once a year in all savannas. Its behavior will be integrated as percentage of burned areas and biomasses and as functions of vegetation type and state.

Burning as a Function of Plant Cover

The degree of burning depends on the amount of herbaceous cover (production related to the climate of the elapsed year and to tree density) and on its level of desiccation (composition, phenology, and date of the last rain).

Type of Fire

The type of fire conditions the extent of burning and the chemical composition of the fumes. In the absence of a superficial humic layer, no ground fires occur. Crown fires developing through tree canopies seldom exist. Canopies are often not contiguous, and the rather low density of leaves in the dry season does not favor fire propagation. When fire occurs, many species have already shed most of their leaves but still display a number of green ones (persistent or new regrowing ones). Contrary to Mediterranean vegetation, savanna tree leaves have a low or null terpene content and thus a low degree of flammability. Tree burning mainly depends on the accumulation of fuel at its base. Tree cover usually depresses grass production underneath, and only dying or dead trees allow the development of a high grass biomass, the combustion of which may sometimes ignite them. The woody cover can be severely affected by fire but seldom actually burns.

Nearly all fires are surface fires. They burn the herbaceous layer, leaving more or less of it, depending on its degree of moisture. Tree leaf litter is in general completely burned. Though highly seasonal, leaf fall time and duration may considerably vary between years: 10% to 75% of total leaf fall can be present as litter when fire occurs (Hopkins, 1966; Menaut, 1974). Because tree leaf litter may amount to several tons per hectare and is different in chemical composition from grass litter (poorer in carbon, richer in other nutrients), such a variation bears significant implications for the chemical composition of

	S	O	N	D	J	F	M	A	M
Guinea savanna				🕯	🕯	🕯			
Sth-Sudanian sav.			🕯	🕯	🕯	🕯	🕯		
Nth-Sudanian sav.		🕯	🕯	🕯	🕯	🕯	🕯		
Sth-Sahelian sav.			🕯	🕯	🕯	🕯	Disappearance of litter by trampling → no fire		
Nth-Sahelian sav.	🕯	🕯	🕯	🕯					

Figure 17.2 Burning period in the different bioclimatic zones of West Africa.

the fumes. From Hopkins (1966), Menaut (1979), and Menaut and César (1979), it can be considered that, from year to year, tree leaf litter varies from 6% to 25% of the herbaceous biomass. Only very partial data on wood litter can be found in the literature. Wood litter seems to vary considerably with time and space, and it cannot be extrapolated to West African savannas as a whole. The woody biomass which is annually burned is certainly not negligible. Wood has a fairly different chemical composition from that of leaves and grasses. Moreover, logs may smolder on the ground for weeks after fire and produce a variety of important greenhouse gases. Wood burning remains an important missing component of the study.

Burning Efficiency

Burning efficiency is a direct function of fuel moisture. Disregarding the moistening effect of a recent rain, burning efficiency depends on the water content of plant material. Water content varies within plant species. In the dry season, grass leaves, which make up most of the herbaceous layer, have a fairly low water content (20% to 30%). Forbs have a higher water content (about 50%), but a much lower biomass; they are not addressed in this study.

Water availability affects the phenological stage of the vegetation and determines the relative proportions of live and dead matter, hence the average water content of the potential fuel. Before fire, the proportion of dead matter is significantly higher in dry savannas than in humid ones (Table 17.1). In any

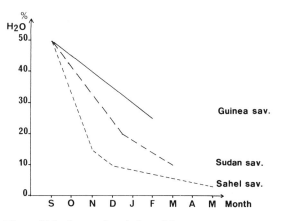

Figure 17.3 Seasonal evolution of the average water content of the grass layer in the Guinea, Sudan, and Sahel savannas.

bioclimatic zone, it also appears that grass dead matter increases with tree density, as an effect of competition for water. Considering the relative proportions of live and dead matter and of water in both, the weighted average of the water content of the grass layer steeply decreases from the time of maximum biomass (about 50%) to the heart of the dry season (about 30%) (Table 17.2). Early fires will consume a lower amount of fuel than late fires will. In addition, the combustion will not be as good, and gas emissions will notably differ between both types of fire. Figure 17.3 summarizes the above information for the three major types of savanna. Given that extensive fires are not likely to occur in systems having a water content above 40%, the risk of burning with the advance of the dry season can be easily deduced.

Proportion of Burned Biomass

Burning efficiency determines the ratio of burned to total biomass in the field. Early fires ignite the fairly humid grass cover with difficulty and burn only 20% to 25% of the grass cover. Middle to late fires destroy most of the existing biomass, up to 100% when fierce. However, due to spatial heterogeneity, only 65% to 95% of the biomass is consumed—85% on the aver-

Table 17.1 Proportion of dead matter in different savanna types of the humid (Guinea) and dry (Sudan) zones

	Dead matter (%)	
	Guinea savanna	Sudan savanna
Grass savanna	30	42
Shrub savanna	50	53
Dense savanna	65	68

Table 17.2 Weighted average of water content of the grass layer

	Water content in vegetative period (%)			Water content before fire (%)		
	Live	Dead	Mean	Live	Dead	Mean
Grass savanna	50	35	47	30	8	23
Shrub savanna	55	26	49	45	16	30
Dense savanna	60	40	53	50	18	31

age. In Ivory Coast, a study has shown that 12% of the biomass was burned in December or even before, that 75% burned in January (heart of the dry season), and that the remaining 13% burned in February (Lamotte et al., 1985). A comparable distribution, extended to the length of the dry season, should be valid for the other bioclimatic zones.

Occurring when vegetation has already reached a high degree of desiccation, middle and late fires consume the same amount of biomass. However, late fires occur when regrowth has already started and have a strong effect on vegetation regeneration and structure and thus on the fuel amount that will be available to the next fire.

Proportion of Burned Area

Seasonal variation Deshler (1974) provided an overview of the progression of fire sources with time and space in arid and mesic savannas along the southern border of the Sahara. Using remote sensing data, he observed that a large number of sources could be detected at the beginning of the dry season; only 20% of the land surface that will have burned at the end of the dry season was then actually burned. At the middle of the dry season, most sources had coalesced, and 80% of the surface which will be burned at the end of the dry season was burned. Burning ended with the dry season. Comparable data were found in Ivory Coast by Lamotte et al. (1985). In November 1984, 15% to 20% of northern, 5% to 10% of central, and 0% to 5% of southern Ivory Coast were burned. In January, these figures, respectively, reached 50% to 60%, 40% to 50%, and 20% to 50%. In February to March, burning was completed. All over West Africa, sources move southward, from dry to humid savannas, with the onset of the dry season.

Geographic variation The burned area at the end of the dry season, related to total land surface, also displays a clear spatial pattern. Only 5% to 15% of the surface of the Sahel burns every year. These values reach 25% to 50% in the Sudan zone, and 60% to 80% in the Guinea zone. Using Landsat imagery during several years on the Sudanian zone of Ivory Coast, Lavenu (1982, 1984) showed that only 25% to 49% of the total area was actually burned each year. Some areas were almost permanently (e.g., open savannas) or never burned (e.g., forest groves and galleries); in the other savanna types, pattern and area of burning changed over the years.

Calculating C and CO_2 Release from Savanna Burning

Local-Scale Experiments

Table 17.3 exemplifies the method of calculating the amount of carbon and carbon dioxide emitted by burning an experimental plot in a savanna of Ivory Coast. The standing biomass on the plot, distinguishing live from dead matter, was measured before burning. The carbon content of both compartments was determined. It has to be noticed that dead matter was poorer in carbon than live was, due to translocation of carbohydrates during withering to other plant parts (mainly to the root system). If dead matter totally burned, only 58% of live matter did. This enabled us to calculate the amount of carbon which was processed through fire. From this amount was deducted the quantity of carbon remaining in the ashes, in order to get the best possible estimate of carbon released in the fumes. Calculations made at a larger scale neglect the ashes; their amount is below the accuracy of biomass estimates.

Table 17.3 Method of calculating the amount of carbon and carbon dioxide emitted by burning an experimental savanna plot in Ivory Coast

	Mass before fire (tons/ha)	Burning efficiency (%)	Burnt mass (tons/ha)	C content (%)	C mass (tons/ha)	Ash mass	% C in ash	C released	CO_2 released
Live	0.12	58	0.05	45	0.023	7.5% of mass before fire	10		
Dead	0.23	100[a]	0.23	41[b]	0.094				
Total	0.35	80	0.28		0.117	0.026 (tons/ha)	0.003 (tons/ha)	0.114 (tons/ha)	0.42 (tons/ha)

a. 100% of the dead matter is burned.
b. Dead matter contains 41% C.

Table 17.4 Parameters for calculating carbon release

| | Sahel | Sudan zone | | Guinea zone |
		North	South	
Burnt area	5–15%	25%–50%		60%–80%
Maximum biomass	0.5–2.5 tons/ha	2–4 tons/ha	3–6 tons/ha	4–8 tons/ha
Maximum biomass before fire	95%	85%		90%–100%
$\dfrac{\text{Living}}{\text{dead}}$	$\dfrac{20}{80}$	$\dfrac{45}{55}$		$\dfrac{55}{45}$
$\dfrac{\text{Living}}{\text{dead}}$ C		$\dfrac{0.45}{0.41}$		
Burning efficiency		65%–95%		
Tree-leaf litter		6%–25% of herb biomass		
Burning efficiency of litter		100%		

Large-Scale Calculation

All parameters necessary to calculate carbon release have been discussed above and are presented in Table 17.4. A problem arises with the estimation of biomass before fire, information which is seldom readily available. No correlation has ever been looked for between the amount of dead matter and any climatic parameter. It cannot be easily estimated from satellite data during the dry season when plant dead matter and soil reflectances are mixed. Biomass before burning must be derived from maximum standing biomass, data commonly given in the literature for a number of sites and years in West Africa. Total biomass before fire is deduced from maximum biomass, using a coefficient close to 100% for early fires, progressively decreasing until fire occurs later in the dry season. Values given in Table 17.4 correspond to fires occurring in the heart of the dry season, apart from the Sahel, where fire occurs early and where there is almost no difference compared to maximum biomass. The difference is also low in the Guinea zone, where dead matter accumulates after flowering and has little time to decompose before fire occurs. In the Sudan zone, the difference increases: Part of standing dead matter disappears with animal foraging and trampling.

Using about 150 sites per year of maximum biomass based on the literature on West Africa, Loudjani (1988) found a fairly good correlation with annual rainfall (Figure 17.4). It is used in this chapter to estimate the range of maximum biomass in the different bioclimatic zones. Better correlations (simple or multiple) had been found with other climatic parameters at local scale; they did not improve, and sometimes decreased, the accuracy at larger scales. To obtain the variations in biomass with years, it should

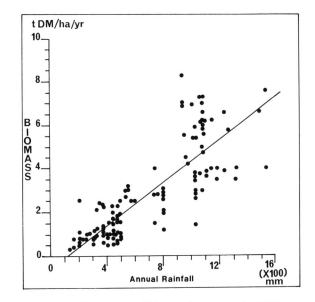

Figure 17.4 Estimation of biomass from annual rainfall ($Y = 4.9 \times 10^{-3}X - 0.58$; $R^2 = 0.7$).

be better to make use of the estimation of biomass (or production) by satellite imagery once calibration problems have been fully solved (Figure 17.5).

Using the values given in Table 17.4, Figure 17.6 shows how calculations are made, from herb and tree litter biomass before fire, to obtain the minimum estimate of carbon release per hectare in Guinea savannas; minimum, maximum, and average estimates are given in Table 17.5. The correlation displayed in Figure 17.4 enables one to obtain the same type of data for the different bioclimatic zones (Table 17.6).

Figure 17.5 Net primary production estimates in West African savannas, from satellite imagery of 1983 and 1984 (from Loudjani, 1988).

Table 17.5 Estimates of carbon release per hectare in Guinea savannas

	Minimum		Maximum		Average	
	C	CO_2	C	CO_2	C	CO_2
Herbs						
Living	0.35	1.28	1.5	5.51	0.86	3.16
Dead	0.40	1.47	1.2	4.40	0.80	2.94
Trees	0.04	0.15	0.5	1.84	0.17	0.62
Total	0.79	2.90	3.2	11.75	1.83	6.72

Note: All amounts are in tons per hectare.

Greenhouse Gas Emissions from West Africa

The method described above enables one to estimate carbon release and trace-gas emissions in West Africa. The area covered by the different bioclimatic zones has been derived from Figure 17.1. The northern part of the Sahel never burns (due to sparse vegetation and low biomass) and has been excluded. The area given in Table 17.7 covers only the savannas, which are likely to burn. On the average, 123×10^6 to C (tons carbon) and 438×10^6 to CO_2 are annually released into the atmosphere.

Hao et al. (1989) estimated that the average emission ratios of CO, CH_4, and C_2–C_4 hydrocarbons to CO_2, respectively, were 10%, 1.1%, and 0.9% by volume. Taking into account the volume-to-mass transformation (28/44 for CO and 16/44 for CH_4), Table 17.8 gives the emissions of CO, CH_4, and C_2–C_4 hydrocarbons in the different bioclimatic zones and in West Africa as a whole.

Given that the nitrogen content of grasses and tree-leaf litter averages 0.3% and 0.8%, respectively, the amount of nitrogen released from biomass burning has been calculated. From laboratory test fires, Lobert (1990) estimated the emission ratios of NO, N_2O, NH_3, and HCN–CH_3CN, respectively, to 12%, 0.7%, 3.8%, and 3.4% of the nitrogen content in the biofuel. On the basis of these data, the nitrogen species emission has been calculated and given in Table 17.9. The rest of the nitrogen either remains in the ashes (about 10%), may be present in smoke particles (about 5%), or is lost through "pyrodenitrification" (about 60%), according to Crutzen and Andreae (1990). It must be noted that such nitrogen losses seem of little importance to ecosystem functioning. They are easily made up for by nitrogen biological fixation and by nitrogen inputs (both organic and mineral) in the rain and dry deposition (Abbadie et al., 1990).

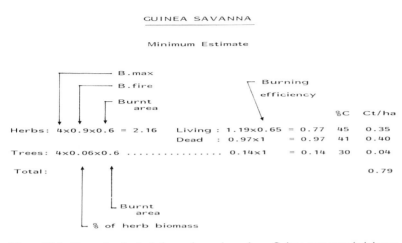

Figure 17.6 Example of calculating carbon release from Guinea savannas (minimum estimate).

Table 17.6 Estimates of carbon release per hectare for different bioclimatic zones (tons)

	Minimum		Maximum		Average	
	C	CO_2	C	CO_2	C	CO_2
Guinea savanna	0.8	2.9	3.2	11.6	1.8	6.7
Southern Sudan	0.2	0.9	1.3	4.7	0.7	2.4
Northern Sudan	0.2	0.6	0.9	3.2	0.4	1.5
Sahel	—	—	0.2	0.7	0.1	0.3

Note: All amounts are in tons per hectare.

Table 17.7 Estimates of carbon release and trace-gas emission in West Africa

	Area (10^6 ha)	C (10^6 tons)			CO_2 (10^6 tons) Mean
		Minimum	Maximum	Mean	
Guinea	16.9	13.5	54.1	30.4	113.2
Southern Sudan	80.6	16.1	104.8	56.4	193.4
Northern Sudan	77.1	15.4	69.4	30.8	115.7
Sahel	52.6	—	10.5	5.3	15.8
Total	227	45	239	123	438

Note: All amounts are either in 10^6 hectares or 10^6 tons.

Table 17.8 Emissions of hydrocarbons in West Africa

	C	CO_2	CO	CH_4	C_2–C_4
Guinea	30.4 (1.80 M)	113.2 (6.70 M)	7.2 (0.43 M)	0.45 (0.03 M)	1.02 (0.060 M)
Southern Sudan	56.4 (0.70)	193.4 (2.40)	12.3 (0.15)	0.77 (0.01)	1.74 (0.022)
Northern Sudan	30.8 (0.40)	115.7 (1.50)	7.4 (0.1)	0.46 (0.006)	1.04 (0.014)
Sahel	5.3 (0.10)	15.8 (0.30)	1.0 (0.02)	0.06 (0.001)	0.14 (0.003)
Total	123 (0.54)	438 (1.93)	27.9 (0.12)	1.75 (0.008)	3.94 (0.017)

Note: All amounts are in 10^6 tons, except for amounts between parentheses, which are mean values in tons per hectare. M is mean in t/ha.

Table 17.9 Nitrogen species emission in West Africa

	N burned	NO	N_2O	NH_3	HCN, CH_3CN
Guinea	270 (15.9 M)	32	1.9	10.3	9.2
Southern Sudan	480 (6.0)	58	3.4	18.2	16.3
Northern Sudan	290 (3.7)	35	2.0	11.0	9.9
Sahel	40 (0.7)	5	0.3	1.5	1.4
Total	1080 (4.7 M)	130	7.6	41	36.8

Note: All amounts are in 10^3 tons, except for amounts between parentheses, which are mean values in kilograms per hectare. M is mean in kg/ha.

The sulfur content of grass leaves has been measured in a number of savanna species; it averages 0.1%. According to Delmas (personal communication), sulfur volatilization is 30% to 60% of the initial sulfur content; the value of 40% has been used to calculate the emission of sulfur (Table 17.10).

Conclusions

The results of the study are significantly lower than those recently published. Both papers by Hao et al. (1989) and Hall and Scurlock (1990) provide a figure of about 2 tons C/ha/yr which are released by fire in tropical savannas. The same estimation for West Africa comes down to 0.54 tons (Table 17.8). Such values point out differences in calculation but have little significance. It seems rather misleading to extrapolate them to the world savannas. It has been shown that, for West Africa, the amount of carbon volatilized each year varies from 45 to 239 × 10^6 tons

Table 17.10 Emission of sulfur in West Africa

	Burnt mass	Sulfur content	Sulfur release
Guinea	74.2 (4400 M)	0.07 (4.4 M)	0.03 (1.8 M)
Southern Sudan	129.8 (1600)	0.13 (1.6)	0.05 (0.6)
Northern Sudan	78.6 (1020)	0.08 (1.0)	0.03 (0.4)
Sahel	9.5 (180)	0.01 (0.2)	0.004 (0.1)
Total	292.1 (1280 M)	0.29 (1.3 M)	0.1 (0.5 M)

Note: All amounts are in 10^6 tons, except for amounts between parentheses, which are mean values in kilograms per hectare. M is mean in kg/ha.

(Table 17.7), depending on the values of the parameters used for calculation (Table 17.4). Changing carbon emission on a global scale by a factor of 5, leads to somewhat different conclusions on the effects of savanna burning on the global carbon cycle and climate. Moreover, as already mentioned, the various authors who provide global values of carbon emissions make use of different and disputable figures for the total area covered by "tropical or burning savannas." Some authors (e.g., Hall and Scurlock, 1990) equate tropical savannas to the tropical grasslands defined by Lanly (1982). Some others (e.g., Hao et al., 1990), consider only the "humid savannas," which make up 60% of tropical grasslands. What about "tropical woodlands," as defined by Lanly, or the increasing cultivated areas, nested in the savannas, that may or may not burn? If they do, are their emissions (amount per hectare and composition) similar to savanna emissions?

A substantial improvement will be made only when estimations are performed, *for a given year,* at a regional scale, and then summed up at a larger one. It seems unrealistic to found such estimates on studies done at a scale larger than the basic bioclimatic and phytogeographic units of a region. For West Africa, the four units used here could not seriously be grouped into a lower number but should rather be detailed at a lower scale. In most other savannas of the world, it is probable that the basic units will have to be significantly smaller.

Table 17.4 clearly displays the parameters which will have to be more accurately estimated. No improvement in estimating total biomass and the relative proportions of live and dead matter before fire can be expected in the near future. Remote sensing studies can be done on a weekly or bimonthly basis from the time of maximum biomass but are not yet able to provide information on the amount of dead matter. When all calibration problems are solved, enabling an accurate perception of green biomass in thick and heterogeneous grass layers, the phenological evolution of the vegetation will have to be derived from field studies (relationships between climatic parameters and the proportion of dead matter for a given savanna type). Only field studies will allow estimates to be made of the amounts of grass beneath tree cover, tree-leaf litter on the ground, and dead wood which burns. Through satellite imagery burned areas can be determined and burning efficiency can be estimated (spectral responses are the same for savannas on which 50% or 80% of the biomass is burned, given

that the remaining material is dead and often blackened).

Whatever might be the accuracy of the present study, West African savannas clearly "play an important role in the global carbon cycle due to their large productivity, the potential interference of biomass burning with this productivity, and the pyrogenic formation of long-lived elemental carbon" (Crutzen and Andreae, 1990). Emissions of nitrogen and sulfur species are of similar importance and confirm that savanna burning causes a substantial effect on the atmospheric chemistry of the tropics and on the global climate (Crutzen et al., 1985; Andreae et al., 1988).

Tropospheric Ozone and Biomass Burning in Intertropical Africa

Bernard Cros, Dominique Nganga, Robert A. Delmas, and Jacques Fontan

It is now established that ozone occurs in layers of the tropical atmosphere when they are polluted by effluents coming from biomass burning, especially in the dry season. Several surface measurements account for this photochemical ozone production in the tropical troposphere. Most of these undoubtedly show a link between tropospheric ozone and biomass burning and were conducted in the southern tropics. In the northern tropics the seasonal surface ozone cycle does not clearly show the highest values in the dry season. Figure 18.1 summarizes some data available in northern and southern tropics of Africa and America. For the three southern sites—Sa da Bandeira (15°S) in Angola, Natal (6°S) in Brazil, and Brazzaville (4°20′S) in the Congo—the strong increase of ozone takes place from June to October during the dry season and therefore during the biomass burning period (Fabian and Pruchniewicz, 1977; Logan and Kirchhoff, 1986; Cros et al., 1988). On the other hand, the three stations in the northern tropics do not present a well-marked seasonal cycle—in Ndjamena (Chad) the maximum tropospheric O_3 values seem to occur in the rainy season (Fabian and Pruchniewicz, 1977; Chatfield and Harrison, 1977; Sanhueza et al., 1985).

Moreover, the seasonal cycles of total tropospheric ozone reported by Fishman et al. (1988), deduced from satellite data between 20°S and 20°N for Western Africa and most of South America, are consistent with the surface ozone cycle of the southern sites. Thus it would seem that the enhanced ozone is associated with widespread biomass burning of southern tropics, which is most prevalent from June to October.

Why is a comparable maximum not to be observed in the northern tropics as it is in the southern ones? In South America the strong continental asymmetry with respect to the equator gives a satisfactory reply. But in Africa this asymmetry is inversed and much less pronounced. Then it is difficult to have an opinion about the relative importance of southern and northern biomass burning on ozone production in Africa because few data are available. To bring some ele-

ments to a better understanding of tropospheric ozone's comprehensive behavior in the equatorial belt, we have undertaken surface ozone measurements in the northern Congo at Enyele (2°50′N, 18°E) during the one-year period from August 1988 to July 1989. We compared these results with those obtained at Brazzaville (4°20′S, 16°E) on the other side of the equator. Enyele is located in the heart of the forest and Brazzaville in a savanna area.

Results and Discussion

The results presented in Figure 18.2 depict the monthly variation of daily maximum ozone surface mixing ratios. These data clearly show a seasonal cycle with maximum values during the dry season, respectively in January and February in the northern tropics, from June to October in the southern ones. The surface ozone cycle in the African equatorial belt follows the equatorial rainfall regime (Figure 18.3).

Unfortunately we have not measured other precursor gases of ozone on these sites. But, as described by Kirchhoff (1988) for the Amazon Boundary Layer Experiment (ABLE-2) in Amazonia, NO concentration was estimated from nocturnal storm activity. During the night in a forest environment we can assume that the reaction between ozone and nitric oxide is the main sink for ozone. It causes the decay of its sudden increase due to the arrival of a thunderstorm over the site. The NO concentrations are presented in Table 18.1. The concentrations were generally high, and the highest values were obtained in March at the end of the dry season. Note that the value for February was measured by chemiluminescence during the Dynamics and Chemistry of the Atmosphere in Equatorial Forest (DECAFE) experiment in February 1988 and that it is consistent with the estimate of March.

The vertical ozone profiles obtained during the DECAFE experiment in February 1988 over the same region give some details on ozone increase in the dry season in northern tropics (Andreae et al., 1990). On

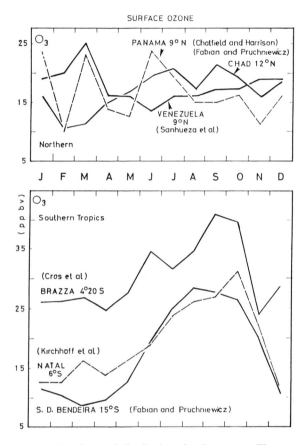

Figure 18.1 Seasonal distribution of surface ozone. The upper panel shows monthly values for three sites in northern tropics. Results for Panama are from Chatfield and Harrison (1977), those for Chad from Fabian and Pruchniewicz (1977), and for Venezuela from Sanhueza et al. (1985). The lower panel shows results in southern tropics at Brazzaville (Cros et al., 1988), Natal (Logan and Kirchhoff, 1986), and Sa da Bandeira (Fabian and Pruchniewicz, 1977).

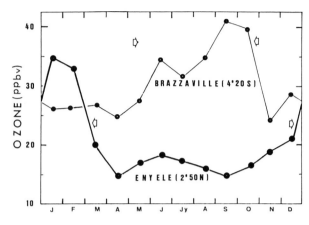

Figure 18.2 Comparison of surface ozone mixing ratio for two African sites on both sides of the equator. The regional dry season is indicated by the arrows.

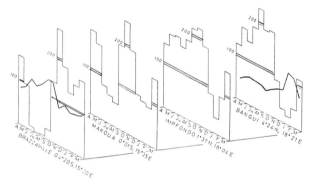

Figure 18.3 Evolution of surface ozone mixing ratio with the rainfall regime from the southern site to the northern one.

16 February in the afternoon (Figure 18.4a) the ozone layer centered on 2 km of altitude appears clearly. This rich ozone layer has mixing ratios of about 70 parts per billion volume (ppbv); it feeds the lower layer when vertical exchanges are well established. There is also in this layer an increase of Aitken nuclei, which is a good indicator of pollution by biomass burning over this remote area. Moreover the study of the trajectories of the different air masses over the measurement point on 16 February 1988 enabled us to trace their origins. This trajectography is made from European Center for Medium-Range Weather Forecasts (ECMWF) with the ASECNA network data. Just above the forest, we have the monsoon flow from west-southwest, which four days before was over the Atlantic Ocean. The polluted layer between 1.5 km and 3 km comes from northeastern savanna. Four

days before arriving over our site, it flowed just over bushfire areas, and three days before it climbed over the monsoon flow when it met the ITCZ (Figure 18.5).

In the southern tropics of Africa over a savanna region near Brazzaville a similar vertical behavior of ozone is found during the dry season. A typical vertical profile carried out in June 1988 is presented in Figure 18.4b. An ozone layer with mixing ratios over 100 ppbv appeared between 1 and 3 km of altitude just over the monsoon flow and below the trade wind inversion. Within this layer high densities of Aitken nuclei were recorded. They are characteristic of bushfire periods (Cros et al., 1981).

During the wet season, contrary to what was previously found in the dry season, it is clear that Aitken nuclei and ozone profiles are negatively correlated (Figure 18.6). However the ozone burden over the monsoon flow remains high. The biomass burning in

Table 18.1 NO concentrations estimated from nocturnal storm activity in Amazonia, 1988 (ppbv)

Jan.	Feb.	March	April	May	June	July	Aug.	Sept.	Oct.	Nov.	Dec.
—	360	389	—	224	193	—	—	—	94	198	—
—	~3[a]	3[a]	—	5[a]	3[a]	—	—	—	1[a]	4[a]	—

a. Number of nocturnal storms over the site. The value for February was obtained during the DECAFE experiment in February 1988.

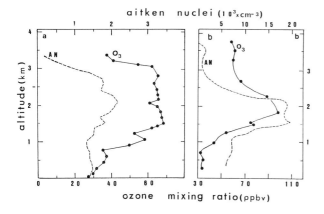

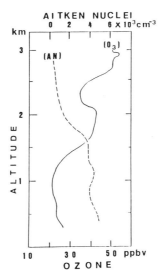

Figure 18.4 Ozone (solid line) and Aitken nuclei (dashed line) profiles measured during the dry season: (a) over the northern site at Impfondo near Enyele on 16 February 1988; (b) over southern site near Brazzaville on 20 June 1986.

Figure 18.6 Vertical profiles of ozone and Aitken nuclei measured near Brazzaville during the wet season on 7 January 1986.

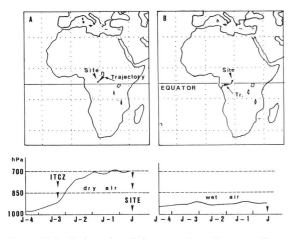

Figure 18.5 Trajectories of air masses from European Center Model (ECMWF) over Impfondo on 16 February 1988: (a) at 700 hectoPascal (about 3 km of altitude) over the site; (b) at 925 hectoPascal (about 750 m of altitude).

Africa is dominated by savanna fires (Delmas et al., 1990), which extend on both sides of the equatorial forest from 15°N to 25°S of latitude. The northeastern and southeastern flows, respectively, from northern and southern tropics play a major role in the transport of ozone and precursors over the continent. They, geographically and seasonally, widen the influence of fires on both tropics as described by Lacaux et al. (1990) from ground measurements of chemical content of precipitation in equatorial Africa. If we assume that the lower rich ozone layer represents the main impact of biomass burning on tropospheric ozone, we can quantify this effect from vertical ozone soundings. We find a contribution of 3 to 4.5 Dobson units (DU) in February and 3 to 8 DU from June to October. This layer contributes 10% to 25% of the total tropospheric ozone column. These values may explain in part the longitudinal evolution of tropospheric ozone over Africa reported by Fishman et al. (1988).

Conclusion

According to our measurements, the seasonal surface ozone cycle in the northern and southern tropics presents maximum values during the regional dry season: 35 ppbv in January to February at Enyele, 40 ppbv in September at Brazzaville (Figure 18.2). But evidence of maximum tropospheric ozone over equatorial Africa during the southern dry season is found only from analysis of satellite data. Assuming the accuracy of satellite measurements, it is important to obtain a better understanding of mechanisms which act on generation, consumption, and transport of tropospheric ozone on this region. They could explain an eventual disappearance or a non-appearance of a maximum of total tropospheric ozone during the northern dry season. To answer this question, multiplying surface and aircraft ozone measurements and ozone soundings over African tropics is necessary.

Acknowledgments

The authors wish to acknowledge the Forestière du Nord Congo (FNC) for its logistical assistance in the field, with special thanks to M. Bourdon, P. J. Naves, and J. L. Bonnefous. This work was supported in part by the French Department of Cooperation.

Savanna Burning and Convective Mixing in Southern Africa: Implications for CO Emissions and Transport

Vickie S. Connors, Donald R. Cahoon, Jr.,
Henry G. Reichle, Jr., Ernst-Gunther Brunke,
Michael Garstang, Wolfgang Seiler, and H. Eckhart Scheel

Convective mixing redistributes the boundary layer of air into and above the free troposphere. The observed vertical and horizontal gradients of conservative tracers have been explained by the rapid injection of surface air to higher altitudes by motions associated with convection (Pratt and Falconer, 1979; Newell et al., 1988; Dickerson et al., 1987; Ching and Alkezweeny, 1987; Lyons et al., 1987). The transport efficiency of cumulus clouds cannot be measured by the global meteorological network. However, short-term mesoscale networks have investigated how the trace gases are transported (Garstang et al., 1988; Scala et al., 1990).

Deep convection occurs during the rainy season in southern Africa. During October, the month of transition from the dry to the wet season, large subtropical anticyclones are positioned on either side of the continent. The surface winds provide southern Africa with clean air from onshore flow to both coasts. In the free troposphere, westerly flow dominates the circulation. The Great Escarpment, which is located 100 kilometers (km) from the east coast, rises nearly 2 km above sea level and acts as a barrier to the low-level easterlies. When the surface air temperature is warm enough, deep convection develops as the dry air from the west flows over the moist air along the scarp. Both subsidence from the oceanic anticyclones and downdrafts from nearby clouds might limit the development of cumulus clouds and the extent of the convective mixing. Yet, the net effect of the surface and cloud processes is to mix and transport surface air up to the free troposphere (Garstang et al., 1987).

The composition of the air in southern Africa is influenced greatly by the emissions from the vegetated surface and only slightly by industrial or technological activities. The southern African landscape includes vast grasslands, wooded savanna, and forest-savanna mosaics. Less extensive rainforest and desert regions are also present. The inhabitants of the region depend on the savanna, which they seasonally burn. The dry season persists from three to six months, and the burning period, which depends on both cultural and climatic factors, lasts for a shorter time. The time of burning in humid forests occurs following the felling or killing of trees. Shifting cultivators clear and burn in December or July, depending on whether hunting or gathering is important. Controlled burning is used to remove highly flammable ground cover. Grasslands are burned at the end of the dry season to maintain them; wooded savannas are burned early in the dry season to promote regeneration of grass. Shepherds and farmers burn during the mid-dry season, while nomads burn during any season. Warfare, pyromania, and incendiarism result in some uncontrolled fires. Within any given area, some burning is usually taking place during the dry season. Because new lands are being cleared for the increasing population, the effects of fire on the environment are occurring at an increasing rate (Batchelder, 1967; Newell et al., 1989).

The extensive burning releases gases and particulate materials into the atmosphere. Haze significantly reduces the visibility toward the end of the dry season over southern Africa. For example, Desalmand et al. (1985) reported that the mean concentrations of both Aitken and large particles in the lower troposphere increased three-fold within 24 hours following Abidjan bushfires. Carbon dioxide, carbon monoxide (CO), water vapor, hydrocarbons, nitrogen oxide, and smoke particles are emitted during biomass burning (Root, 1976). Fishman et al. (1985, 1986) have used imagery from meteorological satellites and columnar ozone measurements from the Total Ozone Mapping Spectrometer (TOMS) to link the production of tropical tropospheric ozone with biomass burning episodes in the Amazon Basin.

This study examines both the emission and the transport of CO from the surface to the free troposphere and the role of convection in redistributing this gas in the free troposphere over southern Africa. Upper-air soundings, the meteorological analyses from the European Center for Medium-Range Weather Forecasts (ECMWF), and the multispectral imagery from the European Space Agency's

Meteosat-2 satellite comprise the meteorological data base. The surface measurements of CO were measured at an atmospheric chemistry laboratory in Cape Point, South Africa. The CO in the middle troposphere was measured by the Measurement of Air Pollution from Satellites (MAPS) experiment flown on the space shuttle. This study will focus on the emissions and transport of CO from Africa south of the equator on 5–6 October 1984.

Carbon Monoxide

CO is a trace gas of global consequence because of its chemical coupling with radiatively important gases. The lifetime of CO in the troposphere depends on the altitude, latitude, and season. Because of its intermediate lifetime, CO is not uniformly mixed about the globe. In fact, Reichle et al. (1986, 1990) reported significant longitudinal gradients of CO in the middle troposphere. The inhomogeneity of CO depends on the distribution of sources and sinks of the gas. The vertical gradient of tropospheric CO is determined by the atmospheric stability. Schmidt et al. (1984) proposed a convective-upward and quasi-horizontal long-range transport scenario to explain the mid-tropospheric (6–11 km) CO maxima (>180 ppbv) they measured over Tenerife (25° to 28°N, 15°W) in May 1981. They suggested that large air parcels with high CO content could be transported from the source areas for long distances without dissolution.

A comprehensive review of the photochemistry and the sources and sinks of CO was provided by Logan et al. (1981). Inefficient combustion processes (biomass burning and fossil fuel use) and the oxida-tion of methane (CH_4) and other hydrocarbons are the primary sources of CO. Photochemical oxidation

$$CO + OH \rightarrow CO_2 + H$$

and soil uptake are believed to be the primary sinks. The production of CO by lightning in thunderstorms was estimated to be insignificant when compared to background values (Newell et al., 1988). Enhanced CO mixing ratios are associated with boundary layer air because there is no known upper tropospheric source of CO.

Although the sources and sinks for CO are believed to be known, the emission and uptake rates are not. Estimates of technological emissions of CO differ by orders of magnitude as a result of differing assumptions or techniques. Table 19.1 compares the best available CO emission estimates for the Southern Hemisphere and southern Africa. Logan et al. (1981) reported that biomass burning in the Southern Hemisphere contributes five times the CO emitted from technological sources; the emissions from biomass burning were estimated by the population density and average clearing rate per capita. Logan (1990) refined the estimate of CO emissions from biomass burning in the Southern Hemisphere to 550 million tons per year. The oxidation of methane and nonmethane hydrocarbons (NMHC), emitted from tropical rainforests and wetlands, represent the bulk of the other sources for the Southern Hemisphere (Gregory et al., 1986; Logan et al., 1981). Root (1976) estimated CO emission from the biomass burning in the African tropics as 134 million tons per year. He based his estimate on the Environmental Protection Agency (EPA) emission factors (0.04 tons CO per ton of fuel),

Table 19.1 CO emission estimates (million tons per year)

Location	Technology	Biomass burning			Other
		Savanna		Agriculture	
Southern Hemisphere	47[a]	100[a]	550[b]	120[a]	610[a]
African tropics (23°N–3°S)		134[c]			
Southern Africa (0°–35°S)	4.39[d]	206[e]			
South Africa (23°S–35°S)	0.34 (1984 only)[f]				

a. Logan et al. (1981).
b. Logan (1990).
c. Root (1976).
d. Logan (1987).
e. Griffiths (1972); Tucker et al. (1985).
f. Tyson et al. (1988).

838 million ha burned (the United Nations Food and Agricultural Organization's *Production Yearbook*), and the average weight of burnable vegetation (4 tons per hectare). Logan (personal communication, 1987) estimated the CO emission from technological sources in southern Africa to be an order of magnitude larger than that given by Louw (personal communication, 1987) for South Africa. Since South Africa is the most industrialized country in southern Africa, it emits the most CO from technological sources for that region. Tyson et al. (1988) reported the CO emissions from the technological sources in the Eastern Transvaal Highvelds (which is located in the "powerhouse" of South Africa) at 339,574 tons per year during 1984. Because the magnitudes of the technological emissions are neither well understood nor well defined, a global measurement system is needed for the critical trace gases.

The primary source of CO in southern Africa during October is the seasonal burning of the savanna. The area of tropical rainforest and wetlands is small (less than 15%) compared to that of the grasslands (nearly 56%). Therefore, the oxidation of locally produced CH_4 and NMHC is not an important CO source in southern Africa. Xeromorphic herbaceous plants (with grasses and sedges being dominant), scattered woody shrubs, and trees are found in the tropical savanna of southern Africa. The vegetation of the savanna is well adapted to its environment. The fibrous roots and underground plant organs enable its survival following the seasonal savanna burning; either the woody species have fire-resistant "cork" bark or they sprout new growth at the trunk base. The vegetation responds quickly to the rainfall that begins in late October and early November.

The tropical savanna is maintained largely by a delicate balance between climate, soils, topography, drainage, fire, animals, and human activities. The marked dry winter and wet summer seasons of the region favor the formation of the savanna landscape. Fire is a significant factor in maintaining most savannas, and the burning frequency depends on the current climate. During dry years the grasslands need to be burned only once every two or three years. However, most savannas in Africa, Asia, and Central America burn each year. Natural causes of fire, such as lightning strikes, account for fewer than 2% of tropical fires. Lightning occurs at the start of the rainy season when the fuel supply is low—either because the grasslands have already been burned or because they are wet. Man burns the savannas to drive game, shift agriculture, and maintain and create pastures.

Grazing is another principal factor in maintaining savannas: Heavy grazing results in regeneration of woody vegetation; light grazing enables the growth of a few large plants and the production of large amounts of fuel, resulting in hot fires which destroy much of the woody vegetation (Kesel, 1975).

The CO emission from the savanna grasslands of southern Africa will be estimated following procedures similar to Suman (1986), Crutzen et al. (1984), and Root (1976), where

$$\text{CO emission} = \text{Area burned} \times \frac{\text{Above-ground biomass}}{\text{Unit area}}$$

$$\times \text{Burning efficiency} \times \frac{\text{Emission factor}}{\text{Mass fuel burned}}$$

The area of the savanna in Africa south of the equator encompasses approximately 5.325 million km^2 or 56% of the southern landscape (Griffiths, 1972). Mathews (1985) determined that of the 724 $1° \times 1°$ elements in the study area, 372 were identified as grasslands, 239 as tropical-subtropical drought deciduous woodland, 92 as tropical rainforest, and 21 as desert. That is, the savanna grasslands comprised 51% of the vegetation in southern Africa. The drought deciduous woodlands accounted for 33%, the tropical rainforest 13%, and the desert 3% of the vegetation regimes. The relative prominence of the grasslands, as compared to the rainforests, is clear. Broadleaf tree savanna (2.1 kilograms per square meter) (kg/m^2), thorn tree tall grass savanna (1.3 kg/m^2), thorn tree desert grass savanna (1.2 kg/m^2), and velds (2.3 kg/m^2) comprise the range of grassland communities (Tucker et al., 1985). The total savanna biomass produced in southern Africa is 8,586 million tons per year. Of this, 80% is estimated to be above ground (Root, 1976) and 30% of that is grazed (Kesel, 1975). This leaves 4,843 million tons of fuel available to be burned. Only 50% of the fuel is actually burned (Logan et al., 1981). During the burning, CO, large aerosols, and Aitken nuclei, nonmethane hydrocarbons (NMHC), and NO are among the pollutants emitted (Crutzen et al., 1984; Desalmond et al., 1985). Emission factors depend on the fuel being burned and the temperature, duration, and intensity of the fire (Chandler et al., 1983). Root (1976) used 40 kg CO per ton of fuel (kg CO/ton) as the emission factor for several grasses. Logan et al. (1981) used a CO emission factor of 100 kg CO/ton. A more recent emission factor for CO is 85 kg CO/ton (Logan, 1990). Using this emission factor yields nearly 206 million tons of CO per year released during the biomass burning in southern Africa. This value compares well with Logan's current estimate of Southern Hemispheric CO emissions from biomass burning.

The period during which the CO is emitted into the atmosphere depends on the onset of the rainy season for a particular locale. Because the rainy season is variable, generalizations about CO emissions on less than an annual time scale are difficult to make. During 1983, the grasslands at the headwaters of the Zambezie River and in the Ethiopean highlands were in drought conditions (Justice et al., 1985). If the conditions in 1984 were similar to those of 1983, then the extent of the 1984 dry season would be underestimated from climatology. In Zaire, the dry season begins in May, and persists for about four months. In central Africa, the dry season lasts from five to seven months. The earliest thunderstorms start in Zambia in September and in Zimbabwe after mid-October. The rainy season begins in November in Mozambique, lasting from two to eight months; the dry season extends from May through October, typically. In South Africa, rainfall occurs in the summer from October to March and is primarily from thunderstorms and instability showers, particularly along the escarpment on the eastern high plateau (Griffiths, 1972; Garstang et al., 1987). The average dry season is four months long. Thus, the average emission rate is approximately 1.72 million tons CO per day during the dry season.

What is the fate of the CO-laden air? How much of the surface-generated CO reaches the high altitude of the free troposphere? These important questions will be considered by examining the distribution of CO during a burning period.

MAPS CO Data

The MAPS experiment was flown on the space shuttle during 5–13 October 1984. The radiometer, using the 4.67 μm fundamental absorption band of CO, measured the vertically averaged, middle tropospheric CO mixing ratios at one-second intervals between 57°N and 57°S during cloud-free conditions. Most of the signal is received from the CO in the tropospheric layer between 700 hectoPascals (hPa) and 200 hPa. Based on independent correlative measurements it appears that the inferred mixing ratios are from 20% to 40% low (Reichle et al., 1990).

The MAPS CO data set is presented in Figure 19.1. Several features are prominent in the data. The CO maxima are located over continental areas in subtropical and midlatitude locations. A broad CO minimum occurs over the western and central Pacific Ocean, and less extensive minima occur over Central America, southern South America, the plateau of Tibet, the Sahara Desert, and near Antarctica. The gas appears to be transported for long distances from the source locations (from the east coasts of Asia and Africa). The absence of enhanced CO mixing ratios over the maritime continent is puzzling and may be a result of differences in the vegetative emissions, convection, abundance of OH, or burning practices of the inhabitants.

The measurements of CO mixing ratios and the

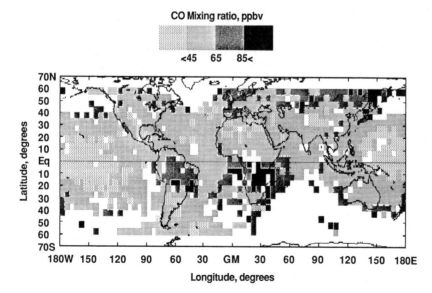

Figure 19.1 MAPS CO volume mixing ratios averaged into 5° longitude by 5° latitude grid squares during 5–13 October 1984.

meteorological situation over southern Africa on 6 October 1984 are the focus of this investigation of the net transport of CO from the region.

Meteorology

Over southern Africa on 5–6 October, the observed weather included haze (Zimbabwe, Malawi, Zambia), blowing smoke (South Africa and Mozambique), rain, and thunderstorms (Zimbabwe and Zambia). Calculations of the Lifted K-index, a predictor of atmospheric stability, from nearby upper air soundings indicated a greater than 60% to 80% probability of thunderstorms for Salisbury and Bulawayo, both cities in Zimbabwe. Haze was the most often reported weather condition for the region. A cloud band, extending across the Atlantic Ocean between the equator and 10°N, blended into a convective area over Africa between the equator and 26°S, and 20°E to 35°E. The clouds over southern Africa on 6 October at 12 UT are shown in the Meteosat-2 visible image in Figure 19.2. Light shades depict highly reflective surfaces (e.g., deserts, clouds); while dark shades depict surfaces of low reflectivity (e.g., lakes, oceans). Cloud Cluster A obscured Lake Tanganyika from view and covered eastern Zaire and western Tanzania. Cloud Cluster B was located over most of Zimbabwe and northeastern South Africa. Nelspruit, South Africa (25°S, 30°E); Harare, Zimbabwe (18°S, 31°E); and Bulawayo, Zimbabwe (20°S, 29°E) were all located under Cloud Cluster B. Maputo, Mozambique (26°S,

33°E), located on the east coast due west of the southern tip of Madagascar, had partly cloudy skies at that time. The upper air data from these four stations were used to characterize the atmospheric stability and the potential for thunderstorm development.

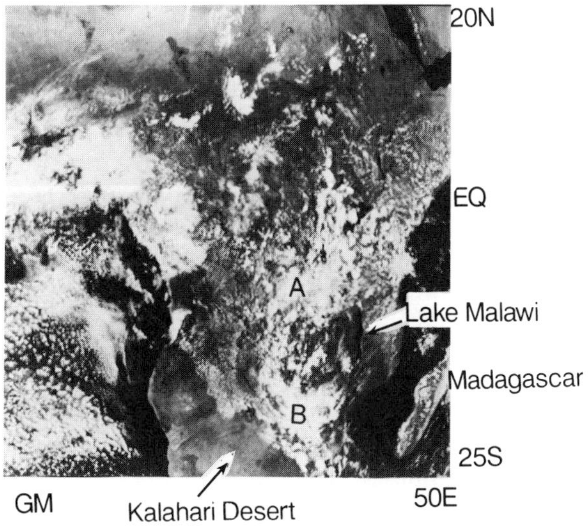

Thermodynamic Analyses

The thermodynamic analyses for Nelspruit (not shown) and Harare on 6 October (Figure 19.3) indicate that if the surface air was warm enough, the air parcels would rise freely to the tropopause. For this area, the warmest surface temperatures were between 25°C and 33°C on 5 October; and, on 6 October, the maximum surface temperatures ranged from 20°C to 28°C. These temperatures were warmer than that necessary for an air parcel to rise beyond the convective condensation level (CCL) up to the equilibrium level (EL). The altitude to which the buoyant air parcels would rise was verified by using a one-dimensional cloud model (that included entrainment) with the Nelspruit upper air data from 5–8 October as input data. For the 5 October case, the modelled air parcels rose to 9.3 km, while the sounding analyses (which did not include entrainment processes) resulted in the air parcels rising to the tropopause at 10.7 km. In both analytical results, the air parcels, representing the cloud tops over portions of Zimbabwe and northeastern South Africa (Cloud Cluster

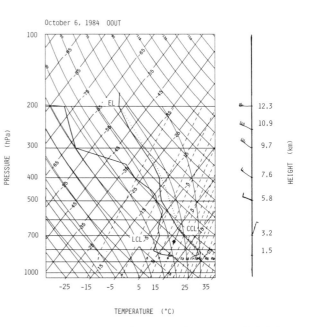

Figure 19.2 Meteosat-2 visible image over southern Africa at 12 UT 6 October 1984.

Figure 19.3 Thermodynamic analyses for Harare, Zimbabwe, on 6 October 1984.

B in Figure 19.2), rose to high altitudes (above 9 km or 300 hPa).

Another analysis tool that was used to examine the vigor and the vertical extent of the convective overturning was the vertical distribution of equivalent potential temperature (θ_e). The equivalent potential temperature, a conservative property under adiabatic conditions, is linearly related to the combined latent and sensible heating in the atmospheric column (Bolton, 1980). Consequently, θ_e profiles provide an excellent diagnostic for describing the convection that occurred on the previous day. Following the classification method of Aspliden (1976), deep penetrative convection is indicated when no distinct minimum occurs in the θ_e profile. In Figure 19.4, the θ_e profiles for the four locations on 7 October reveal that vigorous convection occurred in Harare and Bulawayo during 6 October, while essentially no vertical mixing above 4 km occurred in either Nelspruit or Maputo. In fact, on 6 October, Harare reported moderate thunderstorms and Bulawayo reported rain, while Maputo reported haze.

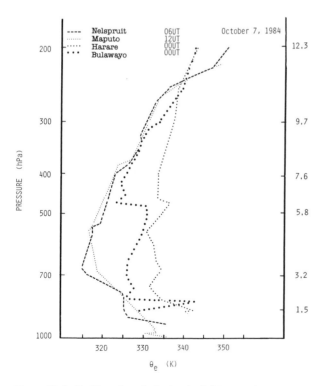

Figure 19.4 Profiles of equivalent potential temperatures over Nelspruit, South Africa; Maputo, Mozambique; Harare; and Bulawayo, Zimbabwe, on 7 October 1984.

Meteosat Image Analyses

The satellite imagery used to characterize the area and type of cloudiness over southern Africa was the Meteosat-2 imagery archived in the International Satellite Cloud Climatology Project (ISCCP). The Meteosat data used for the cloud characterizations were the ISCCP-B1 data. The B1 data are compressed versions of the full resolution imagery; the visible radiances are averaged to the resolution of the infrared radiances and then the matched radiances are sampled to the nominal 10 km spacing at three-hour intervals.

The cloud classification was accomplished following an approach similar to that developed by Desbois et al. (1982). Bispectral histograms were computed for all three combinations of the three Meteosat channels (infrared, water vapor, and visible). The image processing computer was then trained to recognize the scene type based on the results of the histograms. During the period from 5–7 October 1984 the area of active cumulus convection was generally small and variable. The greatest amount of convective activity was recorded (in the three-hour data) on 5 October at 12 Universal Time (UT); during these three days, the convective cloud area reached the maximum amount at 12 UT but the percent area of the total scene never exceeded 4% (about 11,000 pixels). The amount of middle cloud (or multiple layers of clouds) was steady through the period and occupied 6% to 8% of the scene; the amount of marine stratacumulus cloud was more variable and occupied 14% to 20% of the scene. The area of clear sky (over land and sea) maximized at 12 UT with values of 27% to 29% throughout the period; the clear scene area reached a minimum value of about 20% at 9 UT. The deep convection only covered a very small portion (1% to 4%) of southern Africa during the study period.

The vigorous convection in southern Africa was evident in the equivalent potential temperature profiles, satellite imagery, observed weather, and thermodynamic analyses. The boundary layer air was incorporated into the cloud layer and was then mixed to 530 hPa and higher during the growth stage of the cumulus clouds over Zimbabwe and South Africa on 5 October 1984. The MAPS CO data, like the θ_e profiles, diagnoses the history of an air mass. Because the CO source is near the surface, enhanced values of CO indicate that some vertical transport process has occurred. The primary transport mechanism in this case study was convection, with the daily evolution of the mixing layer as a secondary mechanism. The amount of CO that was transported from southern Africa into the global circulation is examined in the next section.

CO Mass Transport Calculations

The net mass transport of air or any conservative tracer can be determined by calculating the flux through each face of a cube situated over the desired area. The box model is oriented such that the north and south faces lie along the equator and 30°S, respectively, and the east and west faces lie along 10°E and 40°E, respectively. The flux was calculated by numerically integrating

$$\phi = \iint_s \rho\, V \times ds$$

where ds is the outward normal. Rewritten with pressure as the vertical coordinate, the mass flux equation becomes

$$\phi = \frac{M_{CO}}{M_{air}} \times \frac{1}{g} \int_p \int_x CO \times U \times dp \times dx$$

where M_{CO} is the molecular weight of CO, M_{air} is the molecular weight of air, g is the acceleration of gravity, CO is the CO mixing ratio by volume, and U is the wind component normal to the area over which the integration is being carried. It was assumed that no CO was destroyed in the volume nor was any transported through the tropopause. The limits of the integration were 1000 hPa and 100 hPa. The tropopause level was assumed to be 100 hPa everywhere over the region. The horizontal integration interval (dx) was 2.5° in latitude and longitude (ECMWF analysis grid spacing). The vertical integration interval was the spacing between the standard pressure levels of the ECMWF analysis. The data sources included the MAPS CO data on 6 October 1984, the ECMWF u and v wind analyses, and the African upper air data on 6–7 October 1984.

The following procedure was used:

1. The u and v wind components at 2.5° spacing along each edge of the box at each pressure level were extracted from the ECMWF analyses.

2. The average value of the appropriate wind component (u for the eastern and western faces; v for the northern and southern faces) was calculated as the average of the values at the four corners of the integration element (i.e., each $dp\ dx$ element).

3. The mass flux equation was solved by summing up the products of the CO mixing ratio and the averaged

wind component over each face, layer by layer, for each of the four faces. The sum of the fluxes through the four faces of the box yielded the net transport of CO from the volume.

For these calculations we assumed that the CO was uniformly mixed in the vertical. The CO fluxes as calculated above represent the CO emissions from within the box. Because the MAPS CO measurements are systematically low as compared to direct measurements, the MAPS data were recalibrated to concurrent CO measurements taken at Cape Point, South Africa. Comparisons between the Cape Point station data during unpolluted periods and shuttle overpasses showed that the MAPS data were 40% lower than the station data. Other comparisons between aircraft measurements of CO and nearly simultaneous MAPS measurements yielded similar results (Reichle et al., 1990) The MAPS CO mixing ratios, calibrated to the Cape Point data, are indicated interior to the surface area marked on the schematic of the box model shown in Figure 19.5.

The wind data used in the box model calculations were tested by comparing the measured and analyzed winds at the upper air sounding locations. The agreement was good for the nine stations located near the

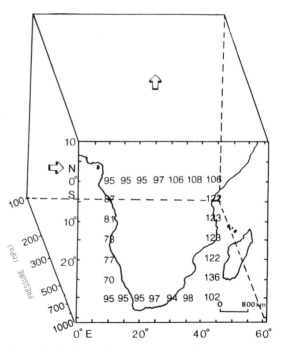

Figure 19.5 Schematic of transport model with adjusted MAPS CO mixing ratios along interior of box. Arrows indicate the positive directions of fluxes.

edges of the box. The standard deviations of the differences between the observed and analyzed winds were less than 30° in direction and 5 meters per second in speed, which is nearly 10% less than what was observed.

The overall validity of the box model was checked by calculating the total mass flux of air through the box. When the mass flux of air is averaged through the volume of the box and converted to a pressure change, the average barometric tendency was found to be $+1.2$ hPa per 12 hours for this day. The average value that was reported by the surface stations during 6 October 1984, was slightly less than $+1$ hPa per 12 hours. This good agreement between the modeled and observed pressure tendencies verified the mass conservation of the box model.

Results

Assuming that the CO was uniformly distributed in the vertical, we have calculated the mass transport of CO over southern Africa to be 365,000 tons per day. Recalling that the estimated average dry season, surface emission rate was 12,000 tons per day from technological sources and 1.72 million tons per day from savanna biomass burning, it appears that about 20% of the emitted CO was being transported away from Africa. It might be expected that a much larger fraction of CO produced would be transported away.

Overall, the model and its input data could account for some of the difference. The wind speeds are known to be in error by about 10%. Further, the assumption of a uniform 100 hPa tropopause height could introduce some error. The overall magnitude of these errors is checked by the previously described test of mass conservation. That test indicated errors on the order of 25%, which is far smaller that the observed factor of five difference between the calculated and estimated emissions. It is possible that some of the emitted CO is destroyed within the box, either as a result of photochemical processes in the air or as a result of processes at the surface. These effects are also thought to be small, however. The photochemical lifetime of CO is long relative to the residence time within the box, and the soil sink is weak (Logan et al., 1981). A major weakness of the calculation is the assumption that the CO was uniformly mixed in the vertical. This assumption was made because the MAPS instrument measured the CO in a single channel which is most sensitive to the CO in the middle and upper troposphere. In source regions the CO mixing ratio in the mixed layer is known to be higher than the free tropospheric value by factors of three or

four (see Crutzen et al., 1985). It is therefore possible that substantial amounts of the gas are transported out of the box at these lower altitudes, although it is not likely that this is sufficient to explain the large difference between the calculation and the estimate.

The previously described estimate is based on average data, and it assumes that the burning takes place at a uniform rate over the entire period of the dry season. It is more likely that the burning begins rather slowly as drying occurs, reaches a peak, and then tapers off near the end of the dry season. If this is correct, the average value would overpredict the short-term rates near the middle of the dry season. This study was carried out near the end of the dry season, and we therefore suspect that the seasonal average value is overpredicting the actual emission rate for the time period of this study. Further evidence of this is shown in the photographs taken by the astronauts aboard the space shuttle. The astronauts reported that many fires were burning in this part of Africa during the flight (K. Sullivan, personal communication, 1986), and this was confirmed by the photography. However, the photographs also show that the area previously burned (char) is vastly larger than the area that was undergoing active burning at the time.

Conclusions

These results show that satellite-derived data can be used to calculate CO emission rates when they are combined with global wind field data. At the present time the largest errors in these calculations probably result from the lack of a complete profile of the CO mixing ratio. It is likely that a substantial portion of the CO leaves the region in the mixing layer, and this region is not measured by the MAPS experiment. Errors in the wind field and the simplicity of the model used also contribute to the uncertainty, but these errors are probably significantly smaller. Since it is likely that the accuracy of the global wind field analyses will improve and since long duration satellite measurements of the vertical profile of CO are feasible, we feel that techniques similar to those described here will prove to be very powerful tools for the assessment of the strength of the various global sources of CO.

Light Hydrocarbons Emissions from African Savanna Burnings

Bernard Bonsang, Gerard Lambert, and Christophe C. Boissard

The burning of biomass in the tropics is a substantial source of reactive species which can play an important role in the chemistry and radiation balance of the troposphere. This source includes carbon monoxide and carbon dioxide; nitrogen species; and methane and nonmethane hydrocarbons (NMHC), particularly in the C_2–C_3 range (Crutzen et al., 1979; Greenberg et al., 1984; Crutzen et al., 1985; Hao et al., 1990). The subsequent photooxidation of NMHCs is a source of tropospheric ozone and of reactive species such as peroxy radicals, aldehydes, ketones, and peroxy acetyl nitrate.

The impact of nonmethane hydrocarbons on tropospheric ozone production has been intensively studied over Amazonia (Kirchhoff et al., 1988; Kirchhoff and Rasmussen, 1990). In the tropical belt of Africa, recent investigations have shown that a similar effect occurs and leads to an ozone enrichment at 2 to 3 kilometers (km) altitude, where its concentration reaches 60 parts per billion by volume (ppbv) (Cros et al., 1988; Marenco and Said, 1989). The Dynamics and Chemistry of the Atmosphere in Equatorial Forest (DECAFE) experiment undertaken in February 1988 in the equatorial forest of the Congo has demonstrated that the production of ozone can be accounted for not only by local emissions of isoprene by trees, but also by a photochemical oxidation of hydrocarbons originating from remote savanna areas (Rudolph et al., 1990). This assumption was founded on the significant high levels of NMHC and particularly of long-lived species such as acetylene observed in the boundary layer (Bonsang et al., 1988; Rudolph et al., 1990) and also on their increasing concentration trends versus altitude.

Experiment

An experiment was organized in West Africa in order to determine the background mixing ratios of NMHC during the dry season and to measure the compositions of savanna burnings. This experiment was conducted from 13 to 22 January 1989 in Ivory Coast at the experimental station of Lamto (06°13′N, 05°01′W) located at the border of the tropical rainforest and savanna (Figure 20.1), within a large reserve of 20 km². From December to February, the vegetation burned in this savanna region consists mainly of graminea (*Hyparrhenia displandra* and *Loudetia simplex*).

The dry season is characterized by an east to northeast flux of dry air coming from the Sahelian regions (Harmattan). It is associated with the presence of a haze layer of about 2 km thickness resulting from the advection of dust particles and smoke combustion aerosols. During the experiment, the temperature ranged between 30° and 31.5°C, and the relative humidity between 58% and 64%. These conditions were slightly drier than the average obtained in January

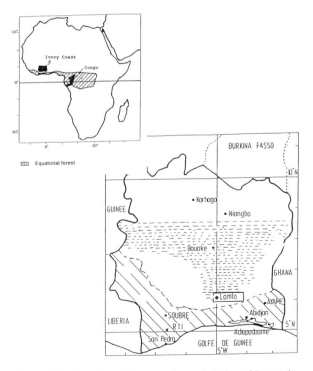

Figure 20.1 Location of the experimental station of Lamto in Ivory Coast.

Table 20.1 Concentrations of CO_2, CO, and hydrocarbons at Lamto and in background conditions, 1989

Sample number	Date	Sampling conditions	CO_2 (ppmv)	CO (ppmv)	CH_4 (ppmv)	C_2H_2 (ppbv)	C_2H_4 (ppbv)	C_2H_6 (ppbv)
1	13 Jan.	Aircraft 2400 m				0.100	0.829	1.210
2	13 Jan.	Aircraft 400 m			1.716	0.654	1.093	2.227
3	13 Jan.	Aircraft 200 m (fire plume)	352		1.670	2.323	7.819	3.851
4	21 Jan.	Lamto station soil			1.700	0.671	3.331	3.038
5	20 Jan.	Lamto station soil			1.649	1.024	6.620	2.791
6	21 Jan.	Lamto station soil			1.654	1.063	2.456	2.679
7	15 Jan.	Lamto station soil	374		1.680	1.072	1.587	2.686
8	14 Jan.	Lamto station soil			1.692	1.074	1.407	3.029
9	21 Jan.	Lamto station soil			1.680	1.076	1.527	2.873
10	20 Jan.	Lamto station soil			1.680	1.180	1.139	2.675
11	16 Jan.	Lamto station soil	352		1.656	1.400	3.388	3.221
12	18 Jan.	Lamto station soil	373		1.688	1.532	4.085	2.698
13	18 Jan.	Fires	394	6.58	1.974	4.739	46.13	37.32
14	16 Jan.	Fires	392	5.80	1.823	9.108	21.61	7.439
15	18 Jan.	Fires	443	14.46	2.384	28.627	88.90	50.90
16	18 Jan.	Fires	422	9.06	2.225	56.060	155.75	42.60
17	19 Jan.	Fires	762	23.60	2.718	134.22	356.25	60.75
18	16 Jan.	Fires	647	21.03	2.549	163.63	453.67	62.77
19	18 Jan.	Fires	556	36.63	3.310	174.51	625.53	149.60
20	19 Jan.	Fires	636	17.04	2.592	179.46	498.43	65.76
21	16 Jan.	Fires	731	35.20	3.560	297.40	913.40	133.30
22	16 Jan.	Fires	2109	197.20	15.000	2852.90	4811.55	752.00
Background at Lamto			352	0.18	1.68	1.08	1.50	2.86
Tropospheric background at 10°N			350	0.10[a]	1.64[a]	0.20[b]	0.06[b]	1.81[b]

a. From Marenco et al. (1989).
b. From Rudolph and Ehhalt (1981); Rudolph and Johnen (1990).

over the last 25 years (27.4°C, 68.7% humidity) (Lamotte and Tireford, 1988).

During an aircraft flight over the northern part of the country samples 1 to 3 were collected, respectively, at 2400 meters (m) height in the free troposphere, at 400 m height in the haze layer, and in a smoke plume at 200 m altitude. Samples representing the ground-level evolution of the local background (4 to 12) were collected at the station at 10 m height. Fires samples (13 to 22) were collected during two intensive experiments on 16 and 18 January at a short distance from the fires (between 1 and 5 meters).

Samples were collected by filling electropolished and previously evacuated stainless steel canisters. Analysis was performed within one month in the laboratory. NMHC were measured by gas chromatography, using a capillary column (AL_2O_3/KCl) and a flame ionization detector (FID) according to a technique previously described (Kanakidou et al., 1989). This method was improved in order to remove carbon dioxide during the injection procedure. CH_4 was measured by gas chromatography, with a packed carbosphere S column. CO and CO_2 were measured by FID gas chromatography with a methanizer. Typical analysis enables one to measure 28 different hydrocarbons including isomers of butenes, pentenes, dienes, and particularly isoprenes.

Results

Results of the whole experiment are reported in Table 20.1 for the different species measured. Samples collected during the aircraft flight are distinguished from samples collected at the station and those collected in the fire emissions. These last two sets of samples are sorted by increasing acetylene mixing ratios.

Background Conditions

At ground level, background methane concentrations are of the order of 1.645 ppbv, the range of concentrations for the main NMHC measured are for C_2H_6: 2.7 to 3.2 ppbv, C_2H_4: 1.13 to 6.62 ppbv, C_2H_2: 0.7 to 1.5

Table 20.1 (continued)

Sample number	C$_3$H$_6$ (ppbv)	C$_3$H$_8$ (ppbv)	Trans-2 C$_4$H$_8$ (ppbv)	1 C$_4$H$_8$ (ppbv)	Iso C$_4$H$_8$ (ppbv)	Cis-2 C$_4$H$_8$ (ppbv)	Iso C$_4$H$_{10}$ (ppbv)	N C$_4$H$_{10}$ (ppbv)	2 Me$_2$ Butane (ppbv)	1 C$_5$H$_{10}$ (ppbv)
1	0.486	0.330	0.025	0.141	0.204	0.020	0.025	0.127	0.015	0.094
2	1.044	0.689	0.042	0.098	0.302	0.028	0.136	0.150	0.056	0.073
3	2.109	1.183	0.136	0.563	0.647	0.108	0.167	0.542	0.037	0.312
4	1.346	0.939	0.127	0.396	0.461	0.113	0.244	0.315	0.023	0.185
5	2.072	0.919	0.171	0.550	0.527	0.154	0.255	0.415	0.027	0.274
6	0.878	0.559	0.081	0.171	0.286	0.067	0.116	0.152	0.028	0.062
7	0.615	0.555	0.066	0.210	0.153	0.071	0.053	0.112	0.016	0.067
8	0.512	0.595	0.019	0.122	0.169	0.017	0.052	0.120	0.008	0.055
9	0.552	0.581	0.065	0.177	3.895	0.049	0.153	0.381	0.059	0.056
10	0.528	0.602	0.046	0.100	0.284	0.041	0.164	0.285	0.060	0.048
11	1.148	1.020	0.079	0.196	0.611	0.072	0.393	1.485	0.000	0.000
12	1.160	0.783	0.069	0.271	0.334	0.058	0.106	0.209	0.014	0.119
13	19.62	9.047	1.348	3.584	2.971	1.000	0.526	2.124	1.055	1.475
14	3.466	1.589	0.095	0.698	0.307	0.076	0.105	0.519	0.035	0.515
15	28.61	9.167	1.724	4.456	3.683	1.286	0.672	1.815	0.591	0.984
16	29.24	7.046	1.574	4.987	2.913	1.162	1.077	2.875	0.700	1.050
17	51.30	8.834	1.752	7.620	4.385	1.365	0.485	1.470	0.523	1.373
18	53.02	8.442	1.494	7.374	3.929	1.159	0.495	1.637	0.480	1.305
19	130.10	22.410	4.844	19.210	10.160	3.615	0.979	3.920	1.527	3.937
20	68.59	8.371	1.989	10.520	4.915	1.501	0.498	1.197	0.657	1.986
21	124.80	15.240	3.492	17.070	7.281	2.650	0.551	1.913	0.785	2.775
22	664.10	67.410	15.260	85.150	33.400	11.910	2.065	7.923	3.511	12.090
Background at Lamto	0.563	0.575	0.042	0.166	0.161	0.044	0.052	0.116	0.012	0.061
Tropospheric background	0.020[b]	0.36[b]	—	—	—	—	0.053[b]	0.094[b]	—	—

ppbv, C$_3$H$_8$: 0.56 to 1.2 ppbv, and for C$_3$H$_6$: 0.51 to 2.1 ppbv. These concentrations are significantly greater than the background concentrations measured at the same latitude over the Atlantic Ocean by Rudolph and Ehhalt (1981) and Rudolph and Johnen (1990) (Table 20.1). Despite the fact that the ocean is also a source of NMHC (Bonsang et al., 1988), these high concentrations denote a significant influence of the emissions by the continent. Particularly, acetylene concentrations are about 10 times greater than the background at these latitudes; they are identical to the concentrations previously observed during the DECAFE experiment in the boundary layer of the equatorial forest of Congo (Bonsang et al., 1988; Rudolph et al., 1988). Samples 7 and 8 collected at the station before the period of local fires are averaged in order to determine the background concentrations at ground level. Considering the aircraft flight experiment, a very sharp decrease of concentration above the haze layer can be seen for most of the NMHC. Concentrations in the free troposphere drop to values very close to the background levels for this latitude

(i.e., 0.1 ppbv for C$_2$H$_2$, 1.2 ppbv for C$_2$H$_6$, 0.33 ppbv for C$_3$H$_8$, and 5 parts per trillion by volume (pptv) for isoprene. Concentrations in the fire plume (sample 3) are significantly greater than the concentrations in the haze layer. These observations are therefore consistent with an intense local production of NMHC in the boundary layer where the combustion products are localized.

Fire Samples

For the fire emissions (samples 13 to 22), it can be seen that the NMHC concentrations reached the parts per million by volume (ppmv) level. Simultaneously with NMHC, CO concentrations vary from 3.4 to 197 ppmv, CO$_2$ from 352 to 2100 ppmv, and CH$_4$ from 1.63 to 15 ppmv.

Comparing these fire samples with the samples representing the background level at the station, it appears that the concentrations far exceed the background values for typical species such as C$_2$H$_2$ and generally C$_2$–C$_4$ alkanes and alkenes. However, ratios of fire plume concentrations to background

Table 20.1 Concentrations of CO_2, CO, and hydrocarbons at Lamto and in background conditions, 1989 (continued)

Sample number	2 Me₁ butane	Iso C_5H_{12}	N C_5H_{12}	Cyclo C_5H_{12}	1 C_6H_{12}	N C_6H_{14}	Butane 2, 2 Dime	2, 3 Dime
1	0.017	0.125	0.117	0.006	0.059	0.070	0.015	0.010
2	0.015	0.955	0.109	0.015	0.034	0.043	0.015	0.015
3	0.097	0.445	0.254	0.012	0.206	0.087	0.005	0.005
4	0.058	0.086	0.142	0.009	0.139	0.092	0.010	0.013
5	0.059	0.278	0.322	0.036	0.195	0.235	0.110	0.015
6	0.029	0.103	0.088	0.007	0.032	0.045	0.015	0.012
7	0.018	0.027	0.057	0.041	0.035	0.026	0.017	0.013
8	0.012	0.025	0.048	0.039	0.015	0.020	0.038	0.001
9	0.034	0.790	1.117	0.077	0.029	0.366	0.166	0.019
10	0.035	0.540	0.853	0.086	0.022	0.489	0.016	0.086
11	0.000	4.115	3.320	0.434	0.000	1.177	1.133	0.483
12	0.030	0.056	0.155	0.084	0.092	0.102	0.026	0.033
13	0.627	0.465	1.198	0.278	0.985	0.879	0.092	0.060
14	0.051	0.472	0.672	0.249	0.172	0.297	0.163	0.047
15	0.546	2.750	0.730	0.069	0.818	0.670	0.170	0.193
16	0.600	6.911	5.209	0.290	0.997	2.355	0.438	0.317
17	0.587	6.239	0.674	0.345	1.364	2.084	0.177	0.467
18	0.565	7.668	0.557	0.146	1.338	1.867	0.175	0.514
19	1.780	16.760	2.348	0.248	3.937	7.981	0.997	1.286
20	0.879	9.034	0.577	0.001	2.363	5.887	0.158	0.682
21	1.224	16.670	0.713	0.035	2.079	3.722	4.865	0.321
22	5.725	110.400	3.645	0.449	14.380	85.480	40.750	2.560
Background	0.015	0.026	0.052	0.040	0.025	0.023	0.027	0.007

concentrations are significantly higher for acetylene and C_2–C_4 alkenes than for alkanes. The same kind of comparison indicates that some species do not seem to be produced during the combustion processes: cyclopentane, 2,3 dimethyl butane, 3 methyl pentane, and 1,2 butadiene. Taking into account that acetylene is a low reactive species with also a low tropospheric background, correlations with acetylene can be studied for the NMHC produced during the fire. Excluding the aircraft samples, which do not represent the initial composition of the emissions, high significant correlation coefficients are obtained for C_2H_4 (r = 0.994), dienes (propadiene and 1,3 butadiene), C_3–C_4 alkenes, isoprene and ethane (Figure 20.2). As a general rule it can be observed that mono- or polyunsaturated species are strongly related with acetylene and consequently are typical of fire emissions. On the contrary, isohexanes, C_4 and C_5 alkanes (except isopentane) are poorly correlated with acetylene. Mixing ratios of gases above the background level measured at Lamto are used to derive emissions factors relative to CO_2. Table 20.2 represents these emission ratios for CH_4, CO, and NMHC concentrations expressed in carbon. They are for $\Delta CO/\Delta CO_2$:

11.04% (standard deviation = 4.39%), for $\Delta CH_4/\Delta CO_2$: 0.54% (standard deviation = 0.24%), and for the total NMHC measured $\Sigma\Delta NMHC/\Delta CO_2$ = 0.81% (standard deviation = 0.38%).

Looking specifically at the NMHC composition of the fire emissions (Table 20.3), a relatively low variability may be observed, which denotes the existence of a typical signature. C_2H_4 represents the dominant hydrocarbon produced (44.1% ± 6.75%); acetylene is also produced in substantial amounts (16.5 ± 4.5%), then C_3H_6, C_2H_6, and C_3H_8 represent, respectively, 10.75 (±2.1), 9.05 (±3.35), and 2.07 (±1.2) of the total amount of NMHC.

It can be observed that nonmethane hydrocarbons represent all together a higher contribution than methane. Our emission ratios are slightly lower for CO and hydrocarbons than most of the figures already published for in situ experiments (Greenberg et al., 1984; Crutzen et al., 1985; Table 20.2) and for laboratory experiments (Hao et al., 1990). However, the dispersion of the emission ratios is important, and the difference remains within the range of variation. It is worth noting, moreover, that our data do not distinguish between the flaming and the smoldering

Table 20.1 (continued)

Sample number	Pentane		Propadiene	1, 2 butadiene	1, 3 butadiene	Isoprene
	2 Me	3 Me				
1	0.022	0.014	0.002	0.001	0.015	0.005
2	0.027	0.016	0.001	0.001	0.014	0.100
3	0.085	0.037	0.037	0.126	0.143	1.250
4	0.034	0.092	0.009	0.001	0.026	0.978
5	0.145	0.067	0.009	0.001	0.033	1.683
6	0.021	0.009	0.006	0.001	0.014	0.861
7	0.013	0.003	0.007	0.001	0.037	0.863
8	0.010	0.005	0.007	0.001	0.011	1.415
9	0.269	0.153	0.013	0.001	0.020	0.700
10	0.312	0.207	0.028	0.001	0.019	1.764
11	1.249	0.759	0.025	0.001	0.052	0.274
12	0.051	0.017	0.024	0.001	0.050	0.993
13	0.104	0.035	0.216	0.001	2.829	3.128
14	0.153	0.087	0.174	0.001	0.417	0.606
15	0.160	0.079	0.968	0.009	6.101	2.407
16	1.057	0.609	1.469	0.001	6.917	2.608
17	0.177	0.071	2.742	0.001	12.61	3.231
18	0.136	0.034	3.477	0.001	13.16	4.114
19	1.000	0.386	5.909	0.001	27.83	7.851
20	0.206	0.085	4.179	0.001	19.03	5.132
21	0.216	0.001	7.163	0.001	30.44	5.659
22	2.240	0.595	53.900	0.001	198.60	34.200
Background	0.011	0.004	0.007	0.001	0.024	1.139

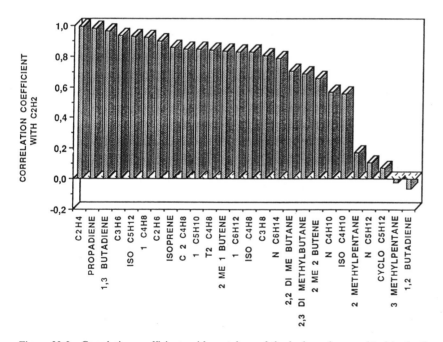

Figure 20.2 Correlation coefficients with acetylene of the hydrocarbons emitted in the fire plume.

Table 20.2 Emissions rate ratios compared to CO_2

	Arithmetic mean	Standard deviation	Range	Crutzen et al. (1985)		
				Arithmetic mean	Standard deviation	Range
$\Delta CO/\Delta CO_2$	11.04%	4.39%	5.7%–15.7%	15.4%	10%	6%–24%
$\Delta CH_4/\Delta CO_2$	0.54	0.24	0.25–0.78	1.2	1	0.3–2
$\Delta \Sigma NMHC/\Delta CO_2$	0.81	0.38	0.36–1.15	1.7	1.4	0.4–2.8

Table 20.3 Composition of the individual NMHC emission rate ratios

	Arithmetic mean	Standard deviation	Range	Crutzen et al. (1985) Arithmetic mean
$\Delta C_2H_4/\Delta \Sigma NMHC$	44.10%	6.75%	31.1%–51.7%	25.2%
$\Delta C_2H_2/\Delta \Sigma NMHC$	16.48	4.42	9.8–25.6	6.0
$\Delta C_2H_6/\Delta \Sigma NMHC$	9.05	3.35	6.7–17.1	12.0
$\Delta C_3H_6/\Delta \Sigma NMHC$	10.75	2.10	8.9–15.0	10.3
$\Delta C_3H_8/\Delta \Sigma NMHC$	2.08	1.18	0.90–4.59	6.5
$\Delta C_4H_8/\Delta \Sigma NMHC$	4.19	1.56	2.60–7.65	—

steps of the combustion processes, which are known to induce very different gas compositions, with a maximum of CO and hydrocarbon production during the smoldering stage (Hao et al., 1990).

Discussion and Conclusion

The agreement between this study in Africa and other published results for Amazonia—particularly concerning the emissions factors of CO, CH_4, and non-methane hydrocarbons relative to CO_2—indicates that a meaningful estimate of global emissions for savanna burnings could be made on the basis of in situ experiments. Using the figure of 1510×10^6 tons of carbon per year released as CO_2 by the biomass burned in savannas (Crutzen et al., 1990), the global corresponding NMHC emission would reach a value of 12×10^6 tons of carbon per year with our emission ratio, and twice this value when considering the Crutzen et al. data. In both cases, it represents a substantial fraction of the global NMHC source. It also appears very clearly that light alkenes are the major reactive NMHC produced during the combustion processes which have to be taken into account. The importance of these hydrocarbons in the local photochemistry can be deduced from the initial composition by considering the kinetic constant of oxidation with OH radicals. For this aim, R being the average composition of the emissions for a given

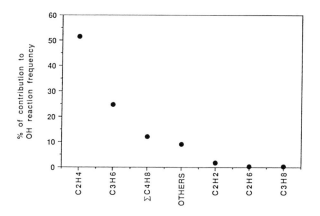

Figure 20.3 Relative contributions of NMHC to the OH reaction frequency.

NMHC (in volume per volume), its contribution to the OH reaction frequency is given by the ratio

$$^kOH\ ^R / \Sigma\ ^kOH\ ^R$$

It can be observed in Figure 20.3 that C_2H_4 contributes 50% of the reaction frequency with OH; this effect cumulated with those of C_3H_6 and butenes (ΣC_4H_8) reaches almost 90% of the reaction frequency with OH. It appears therefore that the effect of NMHC produced by biomass burning on the tropospheric photochemistry is limited to a few number of species—namely, C_2–C_4 alkenes.

Acknowledgments

This work was supported by Centre National de la Recherche Scientifique (PIREN) and Ministère de la Coopération. The authors thank the authorities of the Ivory Coast. Dr. J. Baudet from Department de Physique, Université d'Abidjan, Dr. J. P. Lacaux from Université Paul Sabatier, and Mr. J. L. Tireford from the station of Lamto are particularly acknowledged for the organization of this campaign.

Emissions of Carbon, Nitrogen, and Sulfur from Biomass Burning in Nigeria

Funso Akeredolu and A. O. Isichei

Savanna—321,550 square kilometers (km^2) of Sudan savanna, 376,070 km^2 of Guinea savanna, and 71,150 km^2 of Sahel savanna—occupies over 80% of Nigeria's 924,000 km^2 land area. Most of it is burned annually. It is estimated that swamp forest of the Niger Delta and coastal belt cover about 24,040 km^2, while the lowland rainforest proper covers about four times this area (89,630 km^2), although not more than 10% of the latter has escaped farming in recent times and no significant areas of high forest now remain outside the forest reserves (Onochie, 1979). The rest of the land area is made up of forest-savanna transition zones.

The rate of deforestation in Nigeria, estimated as 0.285 million hectares per year (ha/yr) (FAO, 1981), ranks ninth in the world (Woodwell, 1984) and proceeds at about 4.8% per annum (FAO, 1981). Existing subsistence farming practices, such as shifting cultivation, are largely responsible for the high rate of deforestation (Osemeobo, 1988). Biomass burning is an integral part of the shifting cultivation and slash and burn agriculture, which is common in many parts of West Africa. Burning takes place in the dry season between November and March annually in most of the forest and savanna zones.

There have been numerous studies on the effects of burning on vegetation in Nigeria (see Sanford, 1982), but the atmospheric implications have never been discussed. The objective of this chapter is to discuss some of those implications based on available data on biomass burning in Nigeria.

Extent of Biomass Burning by Geographical Area

Virtually any part of the country that experiences up to three months' dry season is affected by burning. What matters is the intensity of burning, which itself is dependent on the vegetation type and the time of year of burning. The amount of litter produced, especially leaf litter, is crucial in the burning of forest vegetation and plantations. Reported litter fall in savannas range from 1 to 4 tons per hectare per year

(tons/ha/yr), in forest from about 5 to 8 tons/ha/yr, and over 11 tons/ha/yr in plantations (Table 21.1). Okali and Ola-Adams (1987) reported standing woody biomass of 229.6 tons/ha in a primary forest and 91.1 tons/ha in a 28-year-old plot, both in a dry rainforest in southwestern Nigeria. Isichei (1979) reported between 10 and 55 tons/ha standing wood in various savanna formations in western Nigeria. Based on analysis of bole cores, these woody materials contained between 12 and 78 kg N/ha. When forests and savannas are burned, an estimated 10% of the woody plants, the small branches, are usually burned, while the main stems may be smoldered to various degrees. In savanna, standing dead herbaceous material is burned in addition to litter. Herbaceous biomass production varies widely, and Milligan and Sule (1982) reported a range of 1.5 to 18 tons/ha/yr, with the range of 2 to 6 tons/ha/yr being more typical (but see Egunjobi, 1974). The time of year when burning occurs is important. Egunjobi (1971) reported a 56% burn in December for a transition savanna in Nigeria and an 86% burn for an adjacent site in February and March. Higher burn figures could be obtained for the late dry season (see Isichei, 1979). The loss of C, N, and S will vary according to the mass of material burned and its elemental content at the time of burning.

Burning has for a long time been restricted to the savanna zone, but burning of forest and transition (between forest and savanna) areas is escalating. We have observed that most degraded forest vegetation with abundant *Chromolaena odorata* (formerly *Eupatorium odoratum*), which colonizes cleared forest lands, and almost all savanna lands are burned even if the burning may be spotty. Ajaiyeoba (1983) reported the following percentage losses of established pine plantations in five southern states in Nigeria at various times between 1976 and 1982: Anambra, 38% (90 ha); Bendel, 50% (32 ha); Ogun (percentage loss not available but 62.8 ha measured); Ondo, 36% (81 ha); and Rivers, 15% (3.5 ha). Kadeba and Aduayi (1985) reported litter production values in 7–9 and 9–11-year-old northern Nigerian pine plantations of 3

Table 21.1 Litter fall rates by vegetation types in Nigeria

Location	Vegetation type	Litter fall (tons/ha/yr)	Reference
Olokemeji	Forest	7.2	Hopkins (1966)
Olokemeji	Forest	5.6	Madge (1965)
Ibadan	Forest	6.9	Swift et al. (1981)
Olokemeji	Forest	5.9[a]	Isichei (1979)
Sapoba	Forest (mature)	8.7	Oguntala (1983)
Omo	Forest	4.6	Hopkins (1966)
Sapoba	Forest (young)	6.5	Oguntala (1983)
Olokemeji Plot A	Savanna	0.6[a]	Isichei (1979)
Olokemeji Plot B	Savanna	3.6[a]	Isichei (1979)
Igbeti	Savanna	1.3[a]	Isichei (1979)
Borgu woodland	Savanna	2.7[a]	Isichei (1979)
Borgu	Savanna	1.5[a]	Isichei (1979)
Mokwa	Savanna	2.4	Collins (1977)
Plantation	*Species*		
Sapoba	*Triplochyton*	8.2	Oguntala (1983)
Sapoba	*Nauclea*	6.5	Oguntala (1983)
Ore	*Gmelina*	10.5	Oguntala (1983)
Gambari	*Gmelina*	10.4	Oguntala (1983)
Gambari	*Melia composita*	5.2	Oguntala (1983)
Gambari	*Nauclea*	6.2	Oguntala (1983)
Ibadan	*Tectona*	6.5	Oguntala (1983)
Afaka	*Pinus* (9–11 yrs)	3.7	Kadeba and Aduayi (1985)
Afaka	*Pinus* (7–9 yrs)	3.0	Kadeba and Aduayi (1985)

a. Indicates leaf litter plus floral parts, etc.
b. Oguntala has indicated that most plantation litter is leaves.

and 3.7 tons/ha/yr, which was estimated to contain 15 kg N/ha. Assuming that only litter was burned in these southern plantations, and assuming the same nitrogen content, N emissions from them will be 4035 kg. It is estimated that about 2000 ha of forest zone plantations are affected by fire annually.

Based on Landsat images of the Kainji Basin of northwestern Nigeria acquired on 21 October 1986 (when the savanna was at its growth peak) and on 18 December 1984 (shortly after the annual savanna burning), Isichei (unpublished data) estimated that over 70% of the area had been burned. Although it is to be noted that burning is spotty, this areal extent of burning is applicable to most natural savanna formations in Nigeria. Bushfires are a common phenomenon in the West African region, and their causes are the same throughout this region (see Ampadu-Agyei, 1988). It is estimated that 20.7×10^6 ha of savanna and 2×10^6 ha of semi-deciduous forest are burned annually.

As with most developing countries, fuel wood remains an important source of energy for the rural population (which accounts for 70% of total) in Nigeria. It is estimated that 80 million cubic meters (m^3) of wood are consumed for this purpose annually (FAO, 1981).

Timber processing amounted to 11 million m^3 in 1982 (Alviar, 1983). Wood residue production (bark, sawdust, and shavings), estimated using a residue generation factor of 17.6 kilograms of dry wood per cubic meter wood processed (Environment Canada, 1975), is 193,600 tons per year. This residue is disposed of mostly by incineration.

Estimation of Emission Rates

Methodology
According to Seiler and Crutzen (1980), the amount of biomass burnt (M) is given by the following equation:

$$M = A \times B \times a \times \beta \qquad (21.1)$$

where

A = land area burned (m^2/yr),

B = biomass density (kg/ha),

a = fraction of biomass in the ecosystem that is above ground,

β = fraction of the above-ground biomass that is burned.

An emission factor was then applied to calculate the mass of CO_2 emitted from the burning of the biomass. By employing the molar ratios of given trace species to CO_2 (see Crutzen et al., 1985; Hao et al., 1989; etc.), the emissions of these trace gases are then calculated. As noted in the previous section, the biomass densities in Nigerian ecosystems cover a wide range. The biomass densities, fraction of above-ground biomass, and burning efficiency adopted (see Table 21.2) were employed to compute the emissions summarized in Table 21.3.

Published emission fluxes (see Delmas, 1982; Is-ichei and Sanford, 1980; and Robertson and Rosswall, 1986) were also employed to calculate the overall emissions from known land areas in different ecosystems. Closed forest, open woodland, and scrub (31×10^6 ha in total) receive some 460×10^6 kg N in the annual rainfall of 1200 to 2000 mm, assuming 15 kg/ha as the amount of N returned in rainfall. The rest of the country receives about 300×10^6 kg N per year, assuming 5 kg/ha as the amount returned in rainfall (Robertson and Rosswall, 1986).

Delmas (1982) obtained an emission value of 1.4 kg S/ha from savanna burning in Ivory Coast. Applied to Nigeria, this would result in a value of 29×10^6 kg sulphur emitted over 20.7×10^6 ha of savanna usually burned annually. For the 2×10^6 ha of burned semi-deciduous forest, the sulphur emission is estimated based on 0.2% sulphur content in woods (Bowen, 1966) and a biomass density of 80 tons/ha as 8×10^4 kg/yr.

Fuel wood burns almost completely. Assuming a burning efficiency of 90% and using the emission factors reported by Dasch (1982), the burning of fuel wood generates 5.84×10^8 kg/yr of particulate matter, 6.42×10^9 kg/yr of CO, 8.76×10^7 kg/yr of hydrocarbons, 4.09×10^7 kg of oxides of nitrogen, and 2.60×10^4 kg/yr of benzo(a)pyrene (see Akeredolu, 1989a).

Table 21.2 Values of parameters for equation 21.1

Biomass type	Biomass density (kg/ha)	Fraction above ground (%)	Burning efficiency (%)
Savanna			
Woody species	20	0.68[a]	25
Herbaceous species	6	0.47[a]	75
Forest	80	0.95[b]	25

a. Menaut and Cesar (1982).
b. Seiler and Crutzen (1980).

Table 21.3 CO_2 and trace gas emissions from savanna and forest burning in Nigeria

Species	Molar emission ratios relative to CO_2	Mass emission of species (Tg)	
		Savanna fires	Forest fires
CO_2	1	37	17.1
CO	0.1[a]	2.4	1.1
CH_4	0.012[b]	1.6	7.5E-2
NMHC[c]	0.01[d]	1.4E-1	6.4E-2
COS	10.8E-6[ce]	5.4E-4	2.5E-4
NO_x[f]	1.6E-3[g]	4.1E-2	1.9E-2
N_2O	5.0E-5[h]	1.8E-3	8.6E-4
NH_3	1.0E-3[i]	1.4E-2	6.6E-3
HCN/CH_3CN[ij]	6.0E-4[k]	1.4E-2	6.3E-3

a. Greenberg et al. (1984); Crutzen et al. (1989).
b. Hao et al. (1989); Greenberg et al. (1984).
c. NMHC = nonmethane hydrocarbons expressed as CH_4.
d. Crutzen et al. (1985); Bonsang (1990).
e. Nguyen (1990).
f. $NO_x = NO + NO_2$, expressed as NO.
g. Crutzen et al. (1985); Hao et al. (1989).
h. Hao et al. (1989); Crutzen et al. (1985).
i. Andreae et al. (1988).
j. HCN/CH_3CN, expressed as HCN.
k. Hao et al. (1989).

Impact on Biogeochemical Cycling of Elements

The monthly NO_3 concentration reported for Calabar showed a significantly higher value for January (more than twice the annual mean value) than for other months, which was attributed to bush burning emissions (Ette and Udoimuk, 1984). Bromfield (1974) also reported that a combination of low precipitation amount and high input of anthropogenic-S in the form of ash and gaseous-S released during bush burning by farmers was responsible for very high concentrations of S-SO_4 (about 10 times the annual mean) in rainfall samples for 11 sites in northern Nigeria. Ash and carbonaceous material were found on the gauges at the April collection. Jones and Wild (1975) concluded that the only source of S for the West African savanna soil apart from fertilizers is the atmosphere. Bromfield (1974) found 0.099 to 0.429 (mean 0.226) kg/ha of sulphur in dust and 0.49 to 1.89 (mean 1.14) kg/ha of sulfur in rain.

Nitrogen loss by burning from natural grassland ecosystems in savanna locations in western Nigeria was estimated to be 12 to 15 kg ha^{-1} yr^{-1} (Isichei and Sanford, 1980). Jones and Bromfield (1970) report the mean mineral N in rain in February (3.11 parts per million) (ppm), which is almost 10 times the annual average (0.37 ppm) and has a lower $NH_4-N:NO_3-N$ ratio (0.86) than the annual average ratio of 1.26. This supports the hypothesis of volatilization of NO_3-N through fire as the predominant source of N in February. Also in support of this hypothesis, Isichei et al. (1990) obtained a nitrate-nitrogen concentration of 0.8 milligrams per liter (mg/l) in March, the first month of the rainy season after the fires, 0.4 mg/l in April, and 0.2 mg/l for the rest of the season ending in October. Robertson and Rosswall, 1986 estimated that fire accounts for the bulk (about 87%) of all nitrogen lost from West Africa. According to those authors, approximately 8300×10^6 kg were lost by this pathway in 1978. About 67% of this loss occurred on clearing and burning forest vegetation, less than 4% via burning crop residues, and the remainder during near-annual burns of savanna, Sahel, and sub-desert systems.

It is of interest to know whether the inputs of the magnitudes estimated above have detectable impact on the lower troposphere of Nigeria. Elemental ratios are commonly used as tracers in source apportionment studies. The ratio of K/Fe in the ambient particulate matter has been used as a tracer for firewood emissions (e.g., Wolff et al., 1981; Courtney et al., 1980). Two different elemental ratios were calculated from the data reported for Bagauda, a rural location close to Kano, and Ile-Ife, both in Nigeria; Pelindaba, South Africa; and Chilton, England (see Table 21.4).

Most sources have K/Fe ratios of 0.35 or less (Dasch, 1982). The K/Fe ratio for wood smoke is between 15 and 230 (Watson, 1979, cited in Dasch, 1982). Therefore the ambient K/Fe will increase if wood smoke is a major source of air pollution. Crozat et al. (1978) found that zones with values of (Ca + K)/(Na + Mg) ratio higher than 3 correspond to those

where bush burning is widespread. It is clear that the elemental ratios for Chilton contrast with those for Bagauda, Ile-Ife, and Pelindaba. It also appears that high contributions from soil-derived Fe may have made the K/Fe ratios for the latter three locations to be low.

Discussion

It is estimated that savanna and forest burning generate 37×10^6 tons per year (37 Tg/yr) and 17×10^6 tons per year (17 Tg/yr) of CO_2, respectively. For a comparison, the gridded CO_2 emission data reported by Crutzen et al. (1989) when interpolated over Nigeria's latitude and longitude coordinates, yield CO_2 emission estimates for savanna and forest in Nigeria as 35 Tg/yr and 15 Tg/yr, respectively. The estimated CO_2 emissions estimated in this work is equivalent to a carbon flux of 550 kg C/ha/yr compared with 3600 kg/ha/yr estimated by Delmas (1982) for Ivory Coast.

The emission of nitrogen in all the species produced by biomass burning (including fuel-wood burning) was estimated as 2×10^9 kg N/yr. By comparison, prorating the estimated nitrogen emission from biomass burning obtained for West Africa (Robertson and Rosswall, 1986), on the basis of land area, nitrogen emission for Nigeria amounts to 1.4×10^9 kg N/yr. Thus, both emission estimates are in rough agreement. Greenberg et al. (1985) reported CO flux from fuel-wood burning in Kenya as $0.4 \text{ g cm}^2 \text{ day}^{-1}$. On this basis, the emission from fuel-wood burning in Nigeria is estimated as 1.3 Tg. By comparison, 6.3 Tg was calculated in this work.

Comparing these emission estimates with those for energy-related fossil fuel combustion in Nigeria (Akeredolu, 1989a), the ratios for biomass burning emissions expressed as percentages of fossil fuel emissions were 35% (for CO_2), 147% (for CO), and 5% (for NO_x). These values suggest that biomass burning generates a measurable impact on the cycling of carbon and nitrogen. With sulphur, no data were found with which to compare the estimates.

There are as yet no reported direct measurements of emissions of carbon, nitrogen, and sulphur in the Nigerian environment. Extrapolations here have been based on the composition of the sources before burning. The accuracy of the analysis of the chemical compositions of the vegetations is, however, not in doubt, so that at their face value, the estimates are fairly reliable. Most estimates have been pantropical (see Crutzen et al., 1989) and therefore involve extreme generalizations. The present estimates will help in arriving at better global estimates. There are

Table 21.4 Elemental ratios in some Nigerian aerosol samples

| Location | Ratio | | Reference |
	K/Fe	$\dfrac{Ca + K}{Na + Mg}$	
Bagauda, Nigeria	0.3	3.0	Beavington and Cawse (1978)
Chilton, England	1.6	1.6	Beavington and Cawse (1978)
Pelindaba, S.A.	0.4	4.1	Beavington and Cawse (1978)
Ile-Ife, Nigeria	0.1	3.6	Akeredolu (1989b)

however, some assumptions made which would need to be validated through further experimentation. First, the emission projections are based on the areal extent of the burned vegetation. Areal extent cannot be enough because, as we noted, burning could be spotty. Further, it is mainly the leaves that are burned both in litter form and in standing trees and other vegetation life forms. Field observations and on-the-spot measurements will give more accurate results.

There is also some promise of improvement in the estimates through the use of remote sensing. Remote sensing can accurately establish the areal extent of burning of vegetation and it can also be used to directly measure the concentrations of some pollutants in the atmosphere (e.g., Hamilton et al., 1978, for SO_2).

Acknowledgment

Funso Akeredolu acknowledges with thanks receiving a partial travel support from the National Science Foundation and NASA to attend the Chapman Conference on Global Biomass Burning, Williamsburg, Virginia, at which the material in this chapter was originally presented.

Influence of Biomass Burning Emissions on Precipitation Chemistry in the Equatorial Forests of Africa

Jean-Pierre Lacaux, Robert A. Delmas, Bernard Cros,
Brigitte Lefeivre, and Meinrat O. Andreae

The study of the precipitation chemistry and of its temporal and spacial variations enables us to tackle two different objectives. The first objective is linked with the quantification of the wet deposit of biogeochemically important elements. This deposition is the main sink of gases and particles emitted in the atmosphere and, as a consequence, is an important factor of the budget studies: It plays an essential role by limiting the concentration of these substances in the atmosphere and by providing nutrients necessary to the biological functioning of the plants. The second objective is the study of the physicochemical characteristics of the precipitations and of their evolution to determine the influence of the various gas and particle sources according to the meteorological conditions, by using the integrator character of the rain.

As part of the DECAFE program (Dynamics and Chemistry of the Atmosphere in Equatorial Forest), we carried out measurements on the precipitation chemistry in the forests of equatorial Africa. This chapter presents the analysis of the measurements, taken over a one-year period (November 1986 to October 1987), from two sampling sites of the equatorial forest of the Republic of Congo.

The combination of the physical and chemical characteristics of the precipitations, formed over the equatorial regions, can help to determine the seasonal influence of various sources of gases and particles emitted by the different ecosystems (savanna, forest, etc.). Indeed, the chemical composition of precipitation is the final result of the capture processes of atmospheric gases and aerosols by the liquid phase of cloud, fog, and rain. The typical dynamics (flux within the cloud) and microphysical characteristics (in particular, liquid water content) influence the chemical composition of atmospheric water. For example, fog forms in thin layers (about 0.1 g m^{-3}), while a highly convective cloud develops in atmospheric layers at about 10 to 15 km and is representative of the chemical content of thicker atmospheric layers. The liquid water content in clouds of the convective type is much higher (1 to 2 g m^{-3}), which leads to a dilution of the chemical content of precipitation.

We propose a comparative study of the chemical composition of rainwater from stratiform and convective clouds in order to identify and compare atmospheric sources of gases and particles (mainly biogenic sources and burning emissions by the vegetation). The chemical content of stratiform and convective precipitations represents, at the local and synoptic scales, the air chemical content of the planetary boundary layer and free atmosphere.

Biomass Burning Source of Gases and Particles in Africa

Recent studies have shown a heavy pollution of the atmosphere above the African equatorial forest, whatever the season, by gases and particles with radiative effects (Cachier et al., 1990; Bonsang et al., 1990; Delmas et al., 1990), oxidizing effects (increase of the tropospheric concentration of ozone and its precursors) (Cros et al., 1988), and acidifying effects (organic and mineral acidity of the air and precipitations) (Lacaux et al., 1987, 1988, 1990). This pollution comes from the biogenic emission of the soils and from the vegetation of the savanna and forest ecosystems, but above all from the vegetation burning, due to deforestation and to the seasonal burning of the savanna of both African hemispheres.

The Republic of Congo is located on both sides of the equator, and its latitude is between 3°N and 5°S. In this wooded area, a wet equatorial climate prevails. The migration of the Intertropical Convergence Zone (ITCZ) from its higher position (about 20°N) in July to its lower position (about 5°N) in January (Figure 22.1), indicates that the forest of Congo is submitted all year, in the low layers (altitude 1 to 2 km) to the southwestern monsoon flux. In the middle layers (2 to 4 km) the synoptical flux is a southeastern flux from June to September and a northeastern flux from December to March, while in the high layers (higher

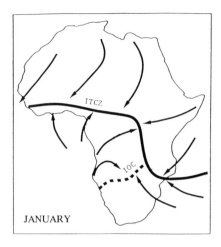

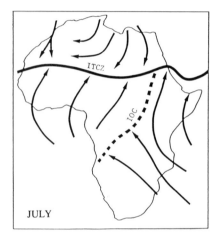

Figure 22.1 Position of the Intertropical Convergence Zone and Inter-Oceanic Influence in January and July.

than 4 km), the atmospheric circulation is generally an eastern direction.

The map of the African vegetation (Figure 22.2) shows that the forest zone ($2 \times 10^6 \, \text{km}^2$) is enclosed in the savanna zone ($10^7 \, \text{km}^2$) whose biomass quantity decreases with the pluviometry. During the dry season (December to April) the savanna zones of the Northern Hemisphere are burned and the northeastern flux carries the gases and particles produced to the forest zone; in the same way, the savannas of the Southern Hemisphere are burned during the dry season (June to September), and the southeastern atmospheric circulation carries the gases and particles to the forest zone. The combination of the atmospheric circulation and the burning of the African savannas influences the atmospheric chemistry above the forests almost throughout the year.

Table 22.1 gathers the estimates of Delmas et al.

(1990) and Menaut et al. (1990), adapted to the vegetation map of Figure 22.2, of the surfaces and quantities of biomass burned yearly in the African savannas. These estimates are deduced from the determination of the primary production by satellite detection and field calibration. The quantity of savanna dry matter burned each year is 2×10^9 tons.

The dry matter contains numerous chemical compounds; the most abundant elements are, in the African savannas, in percentage of dry matter, 46% for carbon, 0.2% for nitrogen, and 0.03% for sulfur (Delmas, 1982). The burning of the savannas thus produces gases and particles, with a more or less reduced chemical form according to the type of burning, which then evolve in the atmosphere toward higher degrees of oxidation:

• The carbon compounds: CO_2, CO, hydrocarbons, CH_4, acid, aerosols

• The nitrogen compounds: NO, NO_2, NH_3, HNO_3, PAN, aerosols

• The sulfur compounds: SO_2, H_2S, aerosols

During the ABLE-2A experiment in Amazonia, Andreae et al. (1988) revealed high concentrations of gases and particles in the plumes produced by vegetation fires (Table 22.2).

Precipitation Chemistry in the Equatorial Forests of Congo

Sampling Sites and Analytical Procedures

Two automatic precipitation collectors, avoiding the dry deposition before the onset of the rain, were located in the forests of Congo. The first one in the Mayombé massif at Dimonika (4°S, 12°30'E), and the second one in the heart of the evergreen forest at Boyelé in northern Congo (Figure 22.2). After each precipitation event, the collected samples are frozen and the collector is rinsed with deionized water, and thus prepared for a new set of collections. We have tested and chosen to preserve the rainwater samples by freezing them at −15°C, especially as the organic part of the chemical content is subject to a rapid degradation as a result of microbial activity.

Cl^-, SO_4^{2-}, NO_3^-, CH_3COO^-, and $HCOO^-$ were measured in our laboratory using a Dionex 10 liquid chromatograph with Dionex AS-1 and Dionex HPICE-AS2 columns, NH_4^+ with an HPIC-CS1 column. Na^+, K^+, and Ca^{2+} concentrations were determined using a Perkin-Elmer 305 atomic absorption spectrophotometer.

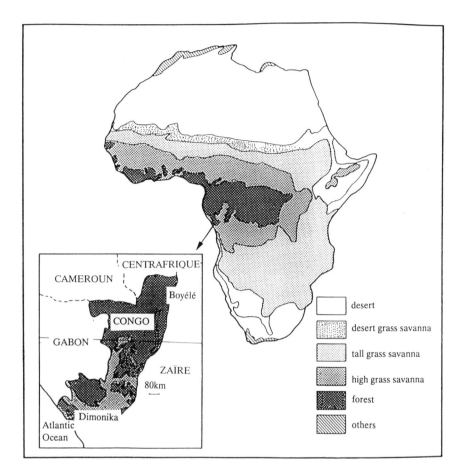

Figure 22.2 Vegetation map and sampling sites (Dimonika and Boyelé) in Congo.

Table 22.1 Estimation of yearly surface burned and biomass burned in the African savannas

Type of savanna	Area in km^2	Biomass above ground $\times$ 10^9 tons dry matter	% surface burned	Yearly burned biomass $\times$ 10^9 tons of dry matter
Desert grass	10^6	0.48	10	0.05
Tall grass	5 $\times$ 10^6	1.74	40	0.70
High grass	4 $\times$ 10^6	1.81	70	1.27
Total	10 $\times$ 10^6	4.00		2.00

Table 22.2 Concentrations of SO_2, HNO_3, and aerosols components in the biomass-burning plumes versus boundary layer in ABLE-2A

	Biomass burning plumes[a]	Boundary layer
SO_2	42	27 ± 10
SO_4^{2-} (aerosol)	302	129 ± 50
HNO_3	281	65 ± 47
NO_3 (aerosol)	328	106 ± 53
Formate (aerosol)	69	24 ± 12
Acetate (aerosol)	48	20 ± 12
Na	100	130 ± 65
K	271	103 ± 38
Cl	78	29 ± 24

Source: Andreae et al. (1988).
a. Arithmetic mean for four flights. All data in parts per trillion by mole (ppt).

Table 22.3 Mean chemical composition and acidity of the precipitations in μeq L^{-1} (November 1986 to October 1987)

	Dimonika	Boyelé
H^+	18.1	41
pH	4.74	4.39
Na^+	11.1	7.5
NH_4^+	6.4	15.7
K^+	2	3.3
Ca^{2+}	9.3	9.5
NO_3^-	8.6	19.4
Cl^-	13.3	8.5
SO_4^{2-}	10.5	10.1
$HCOO^-$	6.6 (5.9)[a]	23 (18.2)[a]
CH_3COO^-	3 (1.5)[a]	9.6 (2.7)[a]

a. Dissociated part.

Mean Chemical Characteristics in Precipitations

The mean chemical composition, weighted by the depth of the rainfall, for the two sites is presented in Table 22.3.

The hydrogen ion is the most abundant of all the ions and denotes the general character in acidity encountered in equatorial precipitation. The mean acidity of the precipitations in the two sites is very high and comes from a mixing of mineral acids (HNO_3, H_2SO_4) and organic acids (formic and acetic acids). If, for the study of the contribution of the sulfates to the acidity, we take into account the influence of the marine source, well characterized in the two sites by

Table 22.4 Contributions to acidity at Dimonika and Boyelé

	Dimonika		Boyelé	
	μeq L^{-1}	% acidity	μeq L^{-1}	% acidity
Dissociated organic acids				
Formic	5.9	35.8	18.2	46.6
Acetic	1.5		2.7	
Nitric acid	8.6	41.5	19.4	43.1
Sulfuric acid[a]	4.7	22.7	4.6	10.2

a. Taking into account the corrections of the marine and terrigenous sulfates.

Table 22.5 Temporal evolution of the acidity in Dimonika

	H^+	δ[a]	pH
Wet seasons			
November 1986 to February 1987	20.5	11.8	4.69
November 1987 to January 1988	17.5	11.5	4.76
Dry season			
June 1987 to September 1987	32.3	23.5	4.50

a. δ = standard deviation.

a Cl^-/Na^+ ratio of 1.20 in Dimonika and 1.13 in Boyelé, close to the marine ratio of 1.16, and the influence of the terrigenous source in gypsum (SO_4Ca) characterized by the content in Ca^{2+} of the precipitations, the following contributions to the acidity can be estimated as in Table 22.4.

The acidity of the precipitations with an organic origin is significant in both sites, with a higher percentage (47%) in the northern forest (Boyelé). The significant production of nitric acid in both sites has to be noticed too, with a percentage around 43%. Apart from the confirmation of the phenomenon of acidification of the equatorial ecosystem by the precipitations, the equivalent part of the chemistry of the compounds with a biogenic origin in the case of the organic acids and of the chemistry of the nitrogenous compounds for the nitric acid has to be emphasized.

The temporal evolution of the acidity in Dimonika shows that the acidity is a little higher during the dry season (see Table 22.5).

Biomass Burning Contribution to Precipitation Chemistry

To estimate the order of magnitude of the contribution by the vegetation fire source of some chemical compounds, we compared the chemistry of the rain in

Table 22.6 Mean chemical concentrations (μeq L^{-1}) and biomass burning contribution (%)

	Amazonia ABLE-2A (μeq L^{-1})	South Congo		North Congo	
		μeq L^{-1}	Biomass burning	μeq L^{-1}	Biomass burning
H$^+$	5.8	18.1	12.3 (68%)	41	35.2 (86%)
pH	5.24	4.74	4.91	4.39	4.45
NO$_3^-$	1.1	8.6	7.5	19.4	18.3
Organic acids (formic and acetic)	5.2	7.4	2.2 (30%)	20.9	15.7 (75%)
NH$_4^+$	1.9	6.4	4.5 (70%)	15.7	13.8 (88%)

Amazonia (Amazon Boundary Layer Experiment, ABLE) and in Congo (Table 22.6), supposing that the chemistry of the precipitations during the wet season in Amazonia is little influenced by the emissions of the biomass burning and that its chemical content is essentially due to the influence of the biogenic sources of the soils and the vegetation.

It can be noticed, with the two chemical chains chosen (nitrogen with NO$_3^-$ and NH$_4^+$ and organic acids with the formic and acetic acids) that the mean contribution of the vegetation burning source is about 80% of the chemical content. This result emphasizes the importance of the vegetation burning source on the atmospheric chemistry in Africa and confirms the increase of oxidizing, radiative, and acidifying effects due to the pollution of this source.

Nitrate Content of Rain as an Indicator of Biomass Burning

The chemical content of the rain is highly influenced by the nitrogenous compounds emitted by the vegetation burning: The contribution of the nitrates due to the combustion was estimated to be 87% and 94% of the total content in nitrates of the rain in southern and northern Congo, respectively.

Nitric oxide is the major nitrogen oxide released from biomass burning. Photochemical reactions take place, and a large fraction of the NO produced is oxidized to HNO$_3$ and organic nitrates. HNO$_3$ is very soluble in water, and this gas is easily captured by clouds. The analysis of the temporal evolution of nitrate concentration in the precipitation at Dimonika (Figure 22.3) confirms the following:

• During the dry season (June to September) the concentration is very high, especially in June and July when the influence of local fires and of the fires in the Southern Hemisphere is most pronounced, then the

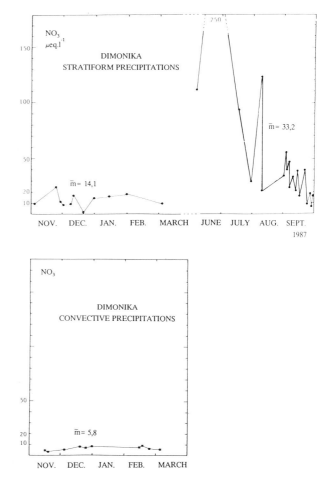

Figure 22.3 Nitrate concentration (μeq L^{-1}) in stratiform (dry and wet seasons) and convective (wet season) precipitations at Dimonika (southern Congo).

Table 22.7 Estimated nitrate content of the air (HNO_3 + NO_3 aerosol) from the nitrate concentration in rain

	Nitrate content of precipitations (μeq L^{-1})	Average liquid water content (g m^{-3})	Nitrate content of air (μg m^{-3})
Wet season			
Stratiform precipitation	14.1	0.5	0.44
Convective precipitation	5.8	1.5	0.55
Dry season			
Stratiform precipitation	33.2	0.5	1.03

concentration decreases evenly until it reaches values of the order of 10 μeq L^{-1} in September.

• During the wet season (November to March) the concentration fluctuates very little and is relatively constant around an average value of 14.1 μeq L^{-1} for rain from stratiform clouds and 5.8 μeq L^{-1} for rain from convective clouds.

From the mean nitrate concentration in stratiform and convective clouds and the liquid water content in the precipitation for every cubic meter (m^3) of air, we can estimate the average content in nitrates (gas and aerosol) of the air where this precipitation is formed (Table 22.7). We can assume a rapid capture of HNO_3 in the clouds because of its great solubility in water, and of nitrate particles which are generally carried by large aerosols (Savoie et al., 1984; Talbot et al., 1988).

The average liquid water content of the precipitation is 0.5 gram per m^3 of air in stratiform clouds and 1.5 gram per m^3 of air for convective clouds, which is representative of the liquid water content in tropical rain (Campistron et al., 1987). In the dry season, the nitrate content of the air, 1.03 μg m^{-3}, is indicative of the presence of gases and aerosols originating from the vegetation fires in the south, as in the concentration of 1.6 μg m^{-3} measured by Andreae et al. (1988) in smoke plumes in Amazonia (Table 22.2). The nitrate concentration during the dry season is remarkably even, of the order of 0.5 μg m^{-3} for both types of precipitation. Therefore it seems likely that this high value is due to a mixture of natural sources of nitrogen oxide from the soil of the savannas of Congo, and sources from vegetation fires in the Northern Hemisphere.

To generalize the results of nitrate content of the rain, we used numerical simulation by Keller et al. (1989), which models the increase in radiative and photochemically active gas emissions resulting from deforestation, colonization, and the development by humans of tropical forests. In particular, they made an estimate of the impact of vegetation fires on NO_x emission, which is the precursor of the nitrate content and the acidity in precipitation. The four scenarios representing the dry and wet seasons were simulated, two of them with emissions of nitrogen oxide from the soils, and two with additional emissions by the vegetation burning. The four simulated cases are summarized in Table 22.8.

From the average HNO_3 concentrations in the planetary boundary layer (altitude 2500 meters in the model) and from the amount of daily precipitation, we estimate the nitrate content of the rain in Congo and in Amazonia for each of the scenarios (Table 22.9). During the wet season the nitrate value of the rain measured was as given below:

• In Congo (7 μeq L^{-1}) it is of the same order of magnitude as that obtained in case C (2.8 μeq L^{-1}) when there is a strong source due to vegetation fires.

Table 22.8 Simulation of nitrogen oxide emissions in natural and biomass burning conditions during the wet and dry seasons (24h PBL average concentrations)

A—Wet Season
Soil emissions provide the source of NO_x
27% of NO_x is carried to the planetary boundary layer (PBL)
(NO) = 3 PPT → (NO_x) = 24 PPT → (HNO_3) = 44 PPT

B—Dry season
Soil emissions provide the source of NO_x
28% of NO_x is carried to PBL
(NO) = 9 PPT → (NO_x) = 85 PPT → (NHO_3) = 121 PPT

C—Wet Season
Soil emissions + biomass burning → NO_x
40% of NO_x is carried to PBL
(NO) = 46 PPT → (NO_x) = 352 PPT → (HNO_3) = 339 PPT

D—Dry Season
Soil emissions + biomass burning → NO_x
52% of NO_x is carried to PBL
(NO) = 132 PPT → (NO_x) = 787 PPT → (HNO_3) = 2960 PPT

Source: Keller et al. (1989).

Table 22.9 Nitrate content of rain in Congo and Amazonia for each of the scenarios defined in Table 22.8

| | HNO$_3$ (ppt) | | NO$_3^-$ in precipitation (µmole L^{-1}) | | Ratio of measured to modeled values |
			Modeled	Measured	
A—Wet season	44	Congo	0.3	7	23.3
Natural		Amazonia	0.4	1.1	2.8
B—Dry season	121	Congo	3.2	33.2	10
Natural		Amazonia	1.5	5.5	3.7
C—Wet season	399	Congo	2.8	7	2.5
NO$_x$ by biomass burning		Amazonia	3.2	1.1	0.4
D—Dry season	2960	Congo	77.1	33.2	0.4
NO$_x$ by biomass burning		Amazonia	35.6	5.5	0.2

Source: Congo, this study; Amazonia, ABLE (Andreae et al., 1989).

• In Amazonia (1.1 µeq L^{-1}) it lies between case A (0.4 µeq L^{-1}) and case C (3.2 µeq L^{-1}), indicating a smaller influence of the source of vegetation fires on the chemistry of rain.

The measured values for nitrates in rain during the dry season are as follows:

• The rain in Congo (33.2 µeq L^{-1}) is of the same order as that obtained in case D (77.1 µeq L^{-1}), the nitrate content at the beginning of the dry season, and those influenced by local fires, are close to those of case D.

• The rain in Amazonia (5.5 µeq L^{-1}) is intermediate between case B (1.5 µeq L^{-1}) and D (35.6 µeq L^{-1}) indicating a moderate influence of the source of vegetation fires during the dry season in Amazonia.

From the measured nitrate content in rain and from the model results, it is clear that rain at Dimonika in southern Congo is strongly influenced during all the seasons by vegetation fires, and that rain in Amazonia is less markedly influenced by fires in both wet and dry seasons. These results should be reevaluated for equatorial Africa using measurements of the nitrogen oxide flux naturally emitted by the soil and using natural hydrocarbons emitted naturally from vegetation, since the model used by Keller et al. (1989) uses only measurements made in Amazonia during the ABLE experiment.

Conclusion

The study of the chemical content of various precipitation types (convective and statiform rains)—the signature of the gases and particles content of the atmospheric layers in which they form—collected in the African equatorial forest enabled us to specify the influences of the generalized emissions of vegetation fires at continental scale.

The biomass burning source of gases and particles in Africa is very important, about 2×10^9 tons (dry matter) of savanna vegetation are burned each year during the dry seasons in both hemispheres. The products of this source are successively carried into the equatorial forest.

The ABLE comparison of the mean chemical composition of the rain collected in Africa with that of Amazonia shows that the influence of the gases and particles emissions by the savanna vegetation burning is preponderant. In particular, 90% of the nitrates, 80% of the ammonium, and from 30% to 75% of the organic acids (formic + acetic) are due to the savanna fire sources. The resulting acidity is very high, with a mean pH of 4.45 in the rain of northern Congo and a pH of 4.74 in southern Congo. The study of the temporal evolution of the nitrate content of the stratiform and convective rains of southern Congo shows the successive pollutions of the African forest by the savanna fires of the Southern Hemisphere (June to September) and of the Northern Hemisphere (December to April). This tendency is confirmed by the numerical simulation of the nitrate content of the precipitations during the dry and wet seasons, with nitrogen oxide sources due to the savanna fires.

Acknowledgments

We gratefully acknowledge the scientific and logistical collaboration of the University Marien Ngouabi of Brazzaville, with special thanks to Faustin Tondo at Dimonika and Cyrille Drago at Boyelé. This research was supported by the Programme Interdisciplinaire de Recherche sur l'Environnement, the French Ministry of the Environment (programme Grands Cycles), and the Ministry of Cooperation (programme Campus).

Biomass Burning Aerosols in a Savanna Region of the Ivory Coast

Hélène Cachier, Joëlle Ducret, Marie-Pierre Brémond, Véronique Yoboué,
Jean-Pierre Lacaux, Annie Gaudichet, and Jean Baudet

Biomass burning appears as a worldwide phenomenon which may rival industrial activities as a major source of atmospheric species (Delmas, 1982; Crutzen et al., 1985; Harriss et al., 1988; Newell et al., 1989). More than 80% of the biomass is burned in the tropical regions, primarily through human activities, and in this context, Africa is the world's leading contributor to atmospheric pollution. The major factor driving biomass burning on this continent is certainly the need for resources to support expanding population, which has led to important forest suppression for shifting agriculture and cattle grazing, and the use of fuel wood (Seiler and Crutzen, 1980). Among the various burning activities, clearing practices, agricultural burns, and the domestic use of charcoal and wood take place all the year long, whereas following a yearly repetitive pattern, extensive savanna fires rage during only three or four months of the dry season (Cachier et al., 1985).

Increasing savanna areas of the African continent are affected by fire each year due to deforestation and the accelerating turnover of the burns. The various combustion activities produce large amounts of atmospheric particles with potential importance for the radiative and chemical balance of the low-latitude troposphere (Penner et al., 1990). Due to the chemical nature of the fuels, these aerosols are primarily carbonaceous; they are also expected to include other trace elements preexistent in the vegetation or provided by the remobilization of local or distant soils. In order to characterize the biomass burning particulate emissions, we sampled aerosols at Lamto in the wooded savanna of the Ivory Coast, during periods when the atmosphere is primarily influenced by various prescribed nearby or distant fires.

We present here the results of parallel analyses which focus on the problem of tracing the biomass burning aerosols at different levels of investigation. Soluble ion measurements give evidence of enhanced levels of various cations (potassium, calcium) and anions (sulfate, nitrate, oxalate), and the appearance of detectable oxalate concentrations. Further indication is obtained by analytical transmission electron microscopy of the small individual particles focusing on their trace element content. At last, studies of the bulk carbonaceous content of the particles appear to provide primarily some possible indicators of the fire variability such as the isotopic composition ($\delta^{13}C$) or the relative importance of the black carbon fraction in the carbonaceous material (Cb to Ct ratio).

Experiment Procedure

Collection Site

The Lamto site (6°N, 5°W) is favorably situated between two vegetation zones, the tropical forest and the savanna in the southern part of the V Baoulé of the Ivory Coast (Figure 23.1). This transition region is moderately inhabited. Our sampling experiments took place during the dry season (January to April), a period of the year where the whole tropical northern Africa experiences various prescribed fires. Furthermore, in January, the Intertropical Convergence Zone (ITCZ) is at the latitude of the site (between 6°N and 7°N), its southernmost position, and Harmattan air masses may reach the Lamto region. At this period of the year, the Lamto atmosphere is likely to be fed by geographically very different sources and the aerosols to be representative of the biomass burning sources prevailing in tropical Africa.

Sampling

Aerosol collection took place in the dry seasons of the years 1980, 1982, and 1989. Filter holders were placed 5 meters (m) above the ground level. During the last experiment (the January 1989 experiment) savanna grass fires were ignited near the station, and particles collected at different distances from the field in order to characterize the particulate emissions of a typical burn prior to their atmospheric transport.

Soluble ion measurements and electron microscopy investigations were performed on the same filters. For this purpose, aerosol samples were size-segregated by means of a five-stage CRA-88 cascade impactor using

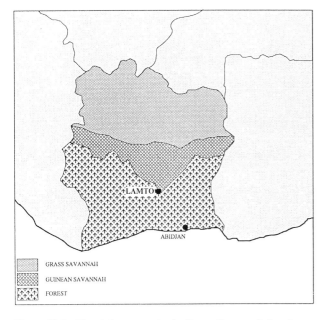

Figure 23.1 Vegetation zones in the Ivory Coast and situation of the sampling site.

nuclepore membranes (0.4 μm) as collection surfaces (Dinh and Pique, 1988). The mean flow rate was of the order of 1 cubic meter per hour (m³/hr) and the sample duration of 12 hours.

Filters devoted to carbon analyses were Gelman A/E or Whatman GF/F glass fiber filters, which had been previously cleaned by multisolvent extractions. These aerosol samples were kept frozen until analyzed. Bulk samples were obtained by low volume filtration (2 or 12 m³/hr); also, a few size-separated samples were collected with a Sierra 5-stage cascade impactor operating at 70 m³/hr.

Analysis Protocols

Microscopic investigations are performed on both bulk aerosols (less than 8 μm) and back-up filters of cascade impactors (less than 1 μm). The particles have to be deposited on nuclepore membranes and transferred directly onto electron microscope grids by dissolving under suction the filter substrate with chloroform (Gaudichet et al., 1986). Elemental analyses are made using a transmission electron microscope (JEOL 100C) fitted with a microanalysis system (energy dispersive spectrometer of X-rays TRACOR 420) allowing one to detect elements with an atomic number greater than 10.

The analysis of soluble ions is run on solutions obtained by ultrasonic extraction of filter aliquots with deionized water. The determination of cation contents (Ca^{2+}, Mg^{2+}, Na^+, K^+) is performed by atomic absorption with a Perkin Elmer 5000 spectrometer, whereas anion analysis is done with a Dionex 2001/SP ion chromatograph.

Carbon analyses necessitate the elimination of carbonates; this operation is achieved in HCl fumes. The carbon remaining on the filters is then referred to as atmospheric particulate total carbon. It includes both the organic and the black carbon components of the aerosol ($Ct = Co + Cb$). Black carbon is separated from the organics by means of a thermal treatment, and carbon contents are analysed by coulometric titration with a Ströhlein Coulomat 702C analyser. The isotopic composition of the carbonaceous matter ($\delta^{13}C$) is obtained on a Micromass 602 D mass spectrometer after oxidation into CO_2 and is expressed relative to the international standard PDB in percentages. All analytical protocols concerning the various carbon analyses have been detailed in previous papers (Cachier et al., 1985, 1989).

Results

Elemental Data

Results obtained from aerosols sampled continuously during the January 1989 experiment provide information on the mean composition of the aerosol found in the Lamto atmosphere during the fire season (Table 23.1). In our set of data, carbon is the most abundant element in the aerosol. Combustion emissions are likely to be one of the major sources of atmospheric particles as evidenced by the occurrence of important concentrations of black carbon aerosols. Furthermore, the ubiquitous presence of oxalate (Baudet et al., 1989) and the elevated potassium concentration levels which are assumed to trace the combustion of vegetation (Crozat, 1979; Cooper, 1980; Andreae, 1983) indicate overwhelming biomass burning inputs. A comparison with ABLE-2B data (Andreae et al.,

Table 23.1 Mean elemental composition of the bulk aerosol at Lamto, January 1989

Total carbon	13.6 μgC/m³
Black carbon	4.0 μgC/m³
K^+	4.29 μgK/m³
Ca^{2+}	2.46 μgCa/m³
S (as SO_4^{--})	1.40 μgS/m³
N (as NO^-_3)	0.72 μgN/m³

Note: Concentrations refer to the element.

1988), which shows less important concentration in haze layers sampled over Amazonia, stresses the extension and intensity of biomass burning practices in Africa during the dry season.

During the sampling period, individual important savanna grass fires could be identified. Figure 23.2 reports the relationship between the carbon concentrations in the bulk aerosol and the distance from fire areas; it clearly indicates that beyond an average distance of 500 m, individual fire inputs of particles cannot be distinguished. In this context, day and night data on the ionic content of individual samples collected at distances over 500 m from the fire plumes show systematically higher concentrations for each related species during the day (Table 23.2). Night inversion would have favored the inverse tendency; consequently, this feature is likely to be inferred to enhanced anthropogenic activities during the day.

Ions found in the Lamto aerosols are preexistent in detectable amounts in the vegetation and soil, and atmospheric concentrations are likely to result from the mobilization of these matrices by both combustion and wind. However, the particles emitted by the burning of the biomass display an important enrichment in trace species (Table 23.3), which may be due to the process of aerosol formation during the combustion (Cachier et al., 1985; Savoie et al., 1989).

This assessment is reinforced by the impactor data, which show enhancement of the main mode during visible fire events (Figure 23.3). It appears that biomass burning processes produce predominantly fine submicron particles (fifth stage and back-up filter of the cascade impactor). The sulfate and total carbon mass results show a striking mass distribution pattern with a sharp peak on the submicron mode; this could point out in the plume important conversions of gases emitted during the burn. The potassium data suggest a superimposition of aerosols of different origin; however, for this element, the enhancement of the fine mode concentrations during the day (Figure 23.4) confirms that the biomass burning production of po-

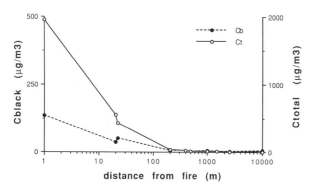

Figure 23.2 Black and total carbon concentrations in bulk aerosols sampled at different distances from fire areas. Beyond an average distance of 500 m, individual fire emissions cannot be distinguished from the biomass burning background aerosol burden.

Table 23.3 Soluble potassium, soluble calcium, and carbon (carbonates excluded) contents in plants, soils, and biomass burning aerosols at Lamto

	Plants	Soils	Aerosols[b]
K^+	0.30%[a]	0.04%[a]	9%
Ca^{2+}	0.18[a]	0.18[a]	5
Ct	42	4[a]	29

a. Data from Crozat (1978).
b. Assuming a conversion factor of carbon to organic matter of 2.

Table 23.2 Soluble ion concentrations for individual day and night aerosol samples at Lamto, January 1989

	Ca^{++} $\mu g/m^3$	K^+ $\mu g/m^3$	NO_3^- $\mu g/m^3$	SO_4^{--} $\mu g/m^3$	$C_2O_4^{--}$ $\mu g/m^3$
13 Jan. (day)	2.887	3.764	2.718	2.842	0.314
14 Jan. (day)	2.909	3.894	3.444	5.221	0.774
14 Jan. (night)	1.399	2.958	3.505	5.128	0.546
15 Jan. (day)	3.499	5.747	3.499	3.873	0.774
15 Jan. (night)	2.029	2.758	3.701	3.401	0.758
16 Jan. (day)	3.179	5.877	6.081	10.015	1.568
16 Jan. (night)	1.974	3.186	2.652	3.188	0.298
17 Jan. (day)	2.334	4.132	4.138	3.161	0.737
18 Jan. (day)	3.177	5.612	3.797	3.222	0.951
18 Jan. (night)	1.172	4.944	1.575	1.920	0.568

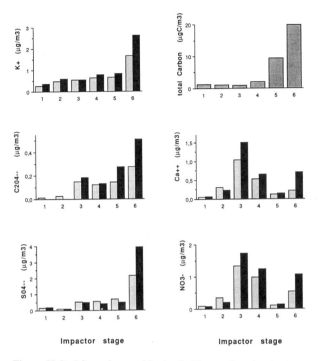

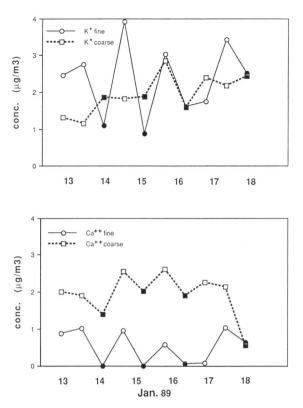

Figure 23.3 Mean size-partitioning in biomass burning background aerosols (10 samples). Mean K^+, SO_4^{--}, $C_2O_4^{--}$, Ca^{++}, NO_3^- data are from the January 1989 experiment; dark bars indicate data only from samples with visible fire events (n = 3) (CRA-88 five-stage cascade impactor with mean cutoffs at 50% collection efficiency, for stages 1 to 6 (μm): 13.60; 5.92; 2.50; 0.96; 0.50). Carbon data are from the January 1982 experiment and adapted from Cachier et al., 1985 (Sierra five-stage cascade impactor with mean cutoffs at 50% collection efficiency, for stages 1 to 6 (μm): >8.2; 3.5–8.2; 2.1–3.5; 1.0–2.1; 0.5–1.0; <0.5.)

Figure 23.4 Evidence for enhanced biomass burning activities during the day as reflected primarily in the coarse mode for soluble calcium concentrations and in the fine mode for potassium concentrations (black dots ■ • are for night samples).

tassium is predominantly controlled by the submicron fraction of the aerosol (Crozat, 1979). Calcium and nitrate ions exhibit similar mass distributions showing a predominance of the mass in the coarse mode (mean diameter of 2.5 μm); the other maximum observed for aerosols less than 1 μm is likely due to particle "bounce" effects during impactions. A possible explanation for the calcium and nitrate parallel patterns could be the production of calcium-rich aerosols by erosion of soils or vegetation surfaces accompanied by the absorption of acidic species such as HNO_3 originating from the abundant NO_x (Marenco, 1990) emitted during the burning of the biomass. Although the origin of the calcium aerosols may not clearly be understood, for this element too, Figure 23.4 gives evidence of enhanced production during the day.

Analytical Transmission Electron Microscopy

Electron microscopy of the bulk aerosol collected at Lamto during the dry season shows a mixture of various carbonaceous particles along with sparse atypical inorganic sulfates and soil dust-derived particles (Figure 23.5a). These investigations performed at the particle scale underscore the predominance of carbonaceous matter in the Lamto aerosol. Moreover, the morphology of the carbonaceous clusters appearing primarily as small sphere aggregates, indicates unambiguously the combustion origin of the aerosol. Sulfate particles are in the micron range; they are found either as pure potassium or mixed potassium and calcium sulfates, confirming a mobilization of sulfur, calcium, and potassium elements during the vegetation combustion and their fast attachment to the airborne particles. We also stress the prevailing unusual property of these sulfates, which were found very unstable under the electron beam; such particles

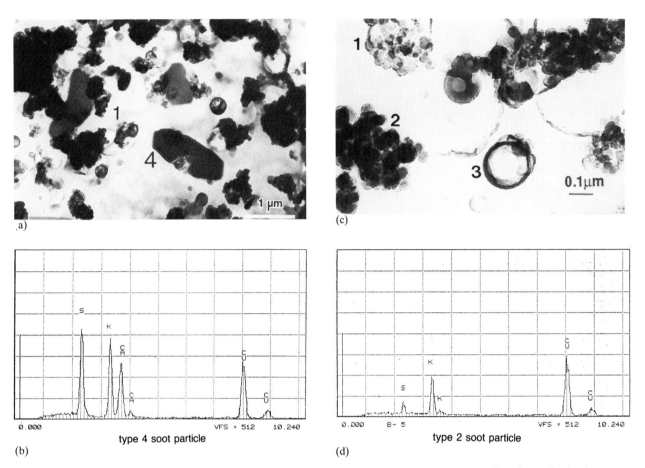

Figure 23.5 (a,b) Bulk aerosols: microsoot particles (1) and K-Ca particles (4); (c,d) impactor back-up filter (stage 6): the three types of biomass burning microsoots. K is an ubiquitous trace element of these particles; an additional S signal may be observed.

have never been observed in other types of aerosols (i.e., marine or crustal aerosols). Further indication is obtained by the observation of the submicron fraction of the aerosol (back-up filter of the cascade impactors, Figure 23.5b). The fine particles left on the filter appear to be primarily a black carbon (soot) aerosol and morphologically very similar to that found in urban atmosphere (Hallett et al., 1989). Most of them are chains or grapes of spherical black carbon. However, these clusters may be separated into two types, essentially on the basis of sphere size, which may have a diameter of either 20 or 30 mm. Beside these aggregates, some individual carbonaceous spheres are observed, showing an "empty" structure with a mean diameter of 200 to 400 mm. At last, X-ray analyses of all these carbonaceous structures permit detection of the presence of trace elements, primarily potassium and sulfur, a result which agrees with the above studies of size-partitioning of the main elements in the biomass burning aerosols.

An interesting facet of the electron microscopy data is the suggestion that biomass burning soot particles could be traced individually by their ubiquitous potassium content. Indeed, whatever the sample, potassium has been detected by X-ray analysis in each carbonaceous particle, sometimes as a single trace element and at other times as the major element in calcium- or sulfur-containing associations.

Carbonaceous Content of the Biomass Burning Aerosols

During the January 1989 experiment, the wooded savanna around the site experienced different prescribed burns, and aerosols were collected at variable distances from the fire. Results reported in Figure 23.6 show that the composition of the carbonaceous aerosol may undergo sizeable changes as indicated by the variability of the black to total carbon ratio (C_b/C_t). However, in this experiment, the aerosol coarse

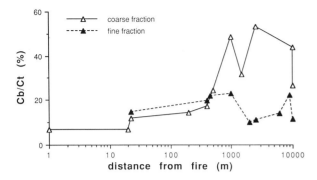

Figure 23.6 Variability of the composition (as indicated by the C_b/C_t ratio) in coarse and fine carbonaceous aerosols sampled at different distances from the fire.

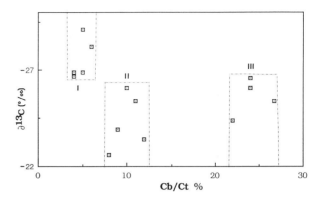

Figure 23.7 Isotopic composition ($\partial^{13}C$) and C_b/C_t ratio of biomass burning aerosols sampled at Lamto (from Cachier et al., 1989). Aerosols from group 1 are likely to originate from forest burns whereas those from groups 2 and 3 are produced by different savanna fire processes.

and fine (<1 μm) fractions do not exhibit similar behaviors. The coarse fraction displays a wide range of C_b/C_t values (from 7% to 53%; mean: 28% ± 13%) which can be partly related to the distance from the fire. On the other hand, the fine fraction shows a moderate range of C_b/C_t values (from 10% to 23%; mean: 15% ± 5%) with no distance dependence. The interpretation of these results must rely on the consideration of both the constancy of the C_b/C_t ratio of the fine particles and the variability of the coarse fraction. Our data suggest that the fine particles do not undergo major atmospheric transformations and therefore that the secondary organic aerosols, primarily derived from the oxidative transformation of organic gases emitted during the burn, are formed rapidly in the fire plume. The coarse particulate data, which exhibits growing importance of the black carbon component far from the plume, may indicate, on the other hand, a disappearance of the organic particles. Rather than a chemical evolution of the coarse particles, this pattern could be due to the preferential settling of the organic aerosols.

On the other hand, another determinant for the C_b/C_t variability of both size fractions of the aerosol could be the diversity of the biomass burning conditions related to the variability of the vegetation fuel, the spread, and the ventilation of the burn. Further insight on combustion processes may be gained with data obtained for aerosols previously collected in the 1980 dry season during periods with no proximate burns (Cachier et al., 1989). Their mean carbon composition is $C_b/C_t = 12\% \pm 7\%$. Segregated samples obtained simultaneously on cascade impactors show that at least 80% of the mass of the aerosols is attached to submicron particles. The low C_b/C_t value of 12% may thus be considered as a characteristic of the fine fraction aerosol and compares with that reported

for the 1989 fire experiment. This low mean value indicates that the major part of the particles emitted during the burning of the biomass is organic material. Furthermore, values as low as 4% are reported in this data set, indicating very incomplete combustions, a feature which we have never encountered in industrial combustion effluents (C_b/C_t range: 15% to 25%; Brémond et al., 1989). Samples with unambiguous dominant fire inputs were selected from this set of data. For these aerosols, the simultaneous use of the isotopic composition ($\partial^{13}C$) and the C_b/C_t data shows the emergence of three distinct groups of samples (Figure 23.7, Cachier et al., 1989). The isotopic composition which was shown in a previous work (Cachier et al., 1985) to trace the vegetation fuel (savanna grasses, which are C_4 plants against C_3 trees) clearly distinguishes between forest and savanna fires. But, on the basis of the C_b/C_t ratio, our data suggest that two very different patterns of burns are prevailing in savannas. The reasons for such a distinction are not yet clearly understood and deserve further studies.

Conclusion

Parallel investigations of the trace element contents of aerosols sampled during the dry season in the wooded savanna of western Africa show that the atmospheric particulate burden there is entirely dominated by biomass burning emissions. Our data suggest important fluxes to the atmosphere of particulate carbon (as black and organic aerosols), potassium, calcium, sulfur, and nitrogen (as soluble ions). The trace elements preexisting in both the vegetation and soils and found in the biomass burning aerosols may originate from the combustion of the vegetation tissues, the mobilization by the fire of local soils or dust deposited on plants (Hegg et al., 1990), or from Harmattan aerosols transported to the site; a clear source apportionment needs further investigations. These species are likely to undergo major oxidative gas-to-particle conversions in or by the fire plume, resulting in coarse aerosols for nitrogen probably by its neutralization on calcium-rich particles and submicron-sized aerosols for potassium, sulfur, and carbon. Due to their size and chemical state, these fine particulate emissions are likely to be exported in the tropical troposphere.

Focusing on the characterization of tropical biomass burning aerosols, it must be pointed out that besides important concentration levels of the above elements, detectable amounts of oxalate are always observed in such emissions. Also, it has been suggested that these combustion aerosols could be traced and differentiated from industrial effluents by their elevated potassium to black carbon ratio (Cooper, 1980; Andreae, 1983). We may propose here to characterize in the source region, the biomass burning aerosols of tropical Africa. Crozat (1978), after extensive work at the site of Lamto, reports about 20% of the atmospheric soluble potassium concentrations may be due to the combustion of the vegetation and the K/C_b ratio is of the order of 0.25, which compares with that reported by other authors (Andreae et al., 1984).

For the first time, to our knowledge, elemental and microscopic analyses are performed on the same biomass burning aerosol samples and shown to compare with chemical analyses for either the composition and the size-partitioning of the particles. Beside morphological indications on the soot particles, the electron microscopy proves to be a reliable tool for the discrimination of aerosols originating from the combustion of vegetation due to the ubiquitous presence of potassium in the carbonaceous clusters. Finally, the relative importance of black carbon in the fine particles as indicated by their black to total carbon ratio (C_b/C_t) is likely to give indication on the physical parameters of the combustion processes. In this respect, at Lamto, two different types of savanna fires may be observed. The first type with emissions displaying a mean composition of 10% may be related to low-temperature burns (backfires), whereas the second type could reflect large-scale savanna fires (C_b/C_t = 23%). These results point out the hazardous character of the interpretation of satellite data without indication on the physical parameters of the fires.

Acknowledgments

We are grateful for the financial support of the CNRS and the CEA and grants from the Programme Interdisciplinaire de Recherche sur l'Environnement, the French Ministry of the Environment (as part of the ATP Grands Cycles), and the Ministry of Cooperation (Campus Program). We thank the University of Abidjan for logistic help.

Long-Lived Radon Daughters Signature of Savanna Fires

Gérard Lambert, Marie-Françoise Le Cloarec, Bénédicte Ardouin,
and Bernard Bonsang

The combustion products injected into the atmosphere by biomass burning are not different from other aerosols and gas traces from various sources; however, their proportions are different. Moreover, their source strengths are not known with the desired accuracy, taking into account the important role they play in the atmospheric chemistry, particularly in Africa. For these reasons, it would be important to easily distinguish these combustion products from the other trace species present in the atmosphere. According to the work presented in this chapter, this could be done by measuring the long-lived natural radionuclides Pb-210, Bi-210, and Po-210, which are emitted from the plants during their burning.

Usually, the isotopes 210 of lead, bismuth, and polonium are produced in the air by the decay of the short-lived daughters of the atmospheric Rn-222, radon being injected into the atmosphere from the free surface of continental soils. Pb-210, Bi-210, and Po-210 are fixed at the surface of the atmospheric aerosols and subsequently removed from the atmosphere with residence times of about one week, so that the 20-year half-life of Pb-210 is not at all in radioactive equilibrium with its decay products Bi-210 and Po-210, whose half-lives are 5 days and 138 days, respectively. Therefore, typical activity ratios in continental atmospheres are 0.5 for Bi-210/Pb-210 and 0.1 for Po-210/Pb-210.

There are other natural sources of metallic aerosols in the atmosphere. It has been shown by several authors that high-temperature volcanic gases introduce into the atmosphere large amounts of metallic vapours, subsequently condensed on the volcanic aerosols (Buat-Menard and Arnold, 1978). These aerosols are characterized by very high enrichment factors, relative to the standard Earth composition (Zoller et al., 1974; Cadle et al., 1979; Mroz and Zoller, 1975; Lepel et al., 1978). More particularly, Lambert et al. (1976) found that volcanic aerosols are very much enriched in Bi-210 and Po-210 relative to Pb-210, with activity ratios Bi-210/Pb-210 and Po-210/Pb-210 varying between 10 and 100. All these results

are well accounted for by the differences of volatility of the different metals and/or their compounds (Pennisi et al., 1988). More especially for Po-210, Bennett et al. (1982) and Le Cloarec et al. (1984) showed that it is entirely volatilized from freshly emitted lavas. In effect, according to Le Guern et al. (1982), the temperature above which Po-210 can be considered as entirely in a gaseous state seems to be of the order of 350°C.

Similar results are expected for aerosols resulting from other high-temperature processes such as biomass burning. In effect, strong enrichments have been observed in fire plumes for a large number of metallic species (Artaxo, 1990). It was therefore reasonable to assume the same effect for Pb-210 and Po-210.

Measurements and Results

A short preliminary campaign was organized in this aim in January 1989 by J. Baudet, B. Bonsang, and M. P. Bremond at Lamto, Ivory Coast, just at the beginning of the regional fires. This station is situated at 6°20'N and 5°W, at the limit between savanna and dense forest, about 120 kilometers (km) to the north of Abidjan, in the middle of a 20 km^2 reserve.

Aerosols were collected by filtration of 3 to 10 cubic meters (m^3) of air. Their gross α and β activities were measured in the laboratory by classical methods, a few days after the samplings, and again several times remeasured in order to detect possible changes of the activity ratios Po-210/Pb-210.

The results are shown in Table 24.1, together with typical values relative to other continents, noncontaminated by fires. Several conclusions can be drawn from this preliminary study:

1. The samples collected to leeward of important fires are characterized by high activities of Pb-210 and Po-210. Moreover, these activities are practically equal. Consequently, the Pb-210 activity being up to 15 times its usual background, that of Po-210 reaches

Table 24.1 Pb-210 and Po-210 in air

Samples	Volume (m³)	Po-210 (pCi/m³)	Pb-210 (pCi/m³)	Po-210/Pb-210
Typical background	—	0.001–0.003	0.015–0.03	0.1
1[a]	4	0.022	0.041	0.54
7[a]	36	0.017	0.049	0.35
14[a]	89	0.010	0.041	0.24
4[b]	4	0.452	0.402	1.12
5[b]	4	0.270	0.273	0.99
6[b]	4	0.215	0.165	1.30
9[b]	4	0.195	0.217	0.90
13[b]	2.7	0.174	0.200	0.87
2	10	0.203	0.176	1.15
8	3	0.133	0.143	0.93
10	4	0.149	0.180	0.83
11	4	0.302	0.314	0.96
12	4	0.290	0.311	0.93

a. Samples collected out of local fires.
b. Small isolated fires.

up to 150 times its usual background. Such high activities enable using these radionuclides, and especially Po-210, as tracers of the combustion products.

2. On the contrary, in lack of local fires, the Pb-210 and Po-210 activities are considerably lower, even though they are significantly superior to their normal background concentrations: 0.02 picoCuries per cubic meter (pCi/m³) instead of 0.002 for Po-210. The activity ratio Po-210/Pb-210 is then intermediate between its normal background value and the radioactive equilibrium characterizing the combustion products.

All these results could be accounted for by long-range transport of combustion products from remote fires. They should be further analyzed by studying the concentration and emission of radionuclides in the main species of plants found in savanna.

Radionuclides in Grass

The concentrations of long-lived decay products of Rn-222, especially that of Po-210, in plants have not been very often studied, except in the case of tobacco leaves and edible vegetables (Marsden, 1964; Berger et al., 1965; Tso et al., 1966). Mean values of Po-210 in tobacco leaves vary within the range of 0.3 to 1.5 pCi/g of dry matter and are of the same order of magnitude in vegetables; the type of soil and the composition of the local atmosphere were reported by

these authors to have an influence on the amount of Po-210 in leaves. More recently, systematic analyses of the Po-210 concentration in plants as a function of the soil composition were performed (Simon et al., 1987) and concentrations as high as 25 pCi/g were reported in plants (mixed grasses, sagebrush, and forbs) grown in areas adjacent to a Wyoming U mine.

African grass grown in Lamto—*Loudetia simplex* and *Hyparrhenia diplandra*—which are the most commonly burned species, were analyzed for Po-210. Mean values of Po-210 activity were found to be within the range of 1.0 to 1.6 pCi/g of dry grass, which corresponds to 2.6 to 4.0 pCi/g of carbon, assuming a carbon content in dry matter of about 40% (H. Cachier, this volume, Chapter 23). As the plants were collected during the period of fires, some of them were washed to evaluate the contribution of activity from superficial deposition of Po-210 enriched aerosols; this contribution was found to be as high as 260% on stems of *Loudetia simplex*. Thus it seems that a good typical value of Po-210 in the most common grass found in savanna, in the Lamto area, is of the order of 1.2 pCi/g carbon. Assuming an annual flux of 2×10^9 tons carbon burned per year (Crutzen, 1989), this leads to a total emission of Po-210 into the atmosphere of the order of 2500 Ci per year—i.e., about 2.5% of its total source, half of which is due to the radioactive decay of the atmospheric Pb- and Bi-210, and the other half has been attributed by Lambert et al. (1982) to active volcanoes.

Table 24.2 Combustion products in plumes of intense fires

Samples	Po-210 (pCi/m^3)	CO$_2$ (ppmv)	CO (ppmv)	CH$_4$ (ppmv)	C$_2$H$_2$ (ppbv)	C$_2$H$_4$ (ppbv)	C$_2$H$_6$ (ppbv)
8	0.13	636	17.0	2.91	179	498	65.8
11	0.30	762	23.6	3.01	134	356	60.7
12	0.29	731	35.2	3.94	297	913	133.3
14	0.01	373	3.7	1.7	1.5	4.1	2.7

Table 24.3 Concentrations of combustion products per gram of carbon (for 1.2 pCi Po-210 per gram of C)

Samples	CO$_2$	CO	CH$_4$	C$_2$H$_2$	C$_2$H$_4$	C$_2$H$_6$
In mg/g C						
8	3372	106.2	5.43	1.32	3.96	0.54
11	2162	68.7	2.55	0.42	1.22	0.22
12	2057	116.7	4.50	0.98	3.26	0.50
In g C compound/g C(%)						
8	92	4.55	0.41	0.12	0.34	0.043
11	59	2.95	0.19	0.04	0.10	0.017
12	56	5.00	0.34	0.09	0.28	0.040

Comparison to Other Combustion Products

During the preliminary campaign of Lamto, several gaseous compounds of carbon were measured, more or less simultaneously with the aerosols filtrations: CO$_2$, CO, CH$_4$, and several C$_2$–C$_6$ NMHC. These gases were sampled in stainless steel cylinders previously evacuated; therefore, their sampling takes only a few minutes, instead of a few tens of minutes for the aerosols. Details of the gas sampling and measurement by gas chromatography are given in Bonsang et al. (this volume).

According to these authors, the most characteristic NMHC in combustion products are C$_2$H$_2$ and C$_2$H$_4$, which are presented here with C$_2$H$_6$, for comparison with alkanes. All the results relative to intense fires are shown in Table 24.2, together with the so-called background values, obtained in the same place, out of fires.

It may be seen in Table 24.2 that, except for CO$_2$ and CH$_4$ whose atmospheric residence times are very long, the values measured in fire plumes are much higher than the background ones. After subtracting the figures corresponding to sample 14, considered as typical of the local background, and assuming the above-mentioned Po-210 typical activity of 1.2 pCi/g carbon, it is possible then to relate the concentrations of the different substances to the initially available amount of carbon. The results are shown in Table 24.3, where they are expressed both in milligrams of gas, and in percentage of the initial carbon.

Despite the different timing of sampling for aerosols and gases, already mentioned, it may be observed that, for each species, the results are not very variable: a factor 2 for the chemically stable species (CO$_2$, CO, CH$_4$, and C$_2$H$_6$) and 3 for the more reactive C$_2$H$_2$ and C$_2$H$_4$. On the whole, the percentages are in agreement with most of the published values (Crutzen et al., 1979; Greenberg et al., 1984; Crutzen et al., 1985; Cofer et al., 1989) and particularly the results of the laboratory experiments of Hao et al. (1990), indicating 81% of carbon as CO$_2$, 6% as CO, 0.4% as CH$_4$, and 1.1% as NMHC. It is worth noting that the emissions of chemically reactive C$_2$H$_2$ and C$_2$H$_4$ are quite similar to that of CH$_4$.

Conclusion

The biomass burning is responsible for a significant injection of long-lived radon daughters, Pb-210 and Po-210, into the atmosphere.

The concentrations measured close to the fires are high enough, at least for Po-210, to enable using these radionuclides as tracers of the other combustion products, even far from remote fires, when the concentrations of the other combustion products are low. Closer to the fires, the lifetime of the aerosols being of the order of one week, it is then possible to use them as a reference for determining the residence times of shorter lived species.

Eventually, the Po-210 concentration being practically constant in grass, this nuclide can be used for calibrating the different combustion products relative to the carbon content of savanna regions. It would then be possible to evaluate the total production of several species through a global evaluation of the burned biomass.

Acknowledgments

This work was supported by Centre National de la Recherche Scientifique (PIREN) and Ministère de la Coopération. The authors thank the authorities of the Ivory Coast. Dr. J. Baudet from Department de Physique, Université d'Abidjan, Dr. J. P. Lacaux from Université Paul Sabatier (Toulouse), and Mr. J. L. Tireford from the station of Lamto are particularly acknowledged for the organization of this campaign.

Biomass Burning in India

Veena Joshi

Recent annual estimates of global emissions of greenhouse gases from anthropogenic sources are in the range of 22,700 teragrams (Tg) of CO_2, 1100 Tg of CO, 350 Tg of CH_4, and 5 Tg N_2O, etc., with an estimated 20% uncertainty. Of these estimated figures, 61% CO, 16% CH_4, and 44% N_2O is attributed to the burning of biomass from energy- and nonenergy-related sources. Of the world contribution from biomass combustion, it is estimated that biomass combustion in India, estimated as 496 Tg per year, contributes about 7% of the world's production of CO, CH_4, and N_2O (Ahuja, 1990). However, the estimates of global emissions from biomass burning suffer from major uncertainties due to lack of detailed information on areas burned, biomass loadings, fraction of biomass burned, etc. (Robinson, 1989).

In this chapter we have summarized the direct biomass burning practices in India. The review pertains to fire practices in forests, agricultural fields, grasslands, households, and industry. In forest land, extent of controlled burning for regeneration and fire prevention is estimated based on the forest statistics. The biomass burned annually due to accidental fires and for shifting cultivation is quantified based on a few earlier studies. In the case of household and small-scale industries, the biomass burned is quantified by extrapolating past data on energy consumption. In addition to wood and crop residues, the use of dungcakes and charcoal is also accounted for in calculating the total amount of biofuels burned annually. Wherever possible, regional and seasonal variations in the biomass burning practices are highlighted. This exercise has led to identification of data and information needs to improve the current estimates of biomass burned annually in India. The factors influencing the impact of National Programme on Improved Cookstoves (NPIC) in reducing the greenhouse gas emissions are discussed.

Fires in Forests

Forests in India can be broadly classified into four major groups: tropical, subtropical, temperate, and alpine. In India, as in other countries, the forest fires are a result of either controlled burning or accidents, genuine or deliberate. Forests are set on fire for inducing luscious growth of grass; for better grazing; for clearing land for agriculture; for collecting minor forest produce such as honey, edible and nonedible oil seeds, and flowers; and sometimes to catch wild animals. Fires are also started to destroy evidence of illegal felling. The available forest statistics are compiled to estimate the forest area burned and the biomass affected.

Forest Statistics

Table 25.1 is a compilation of basic statistics on forest area by its composition and by classification used officially. The classification of forest area is given in three categories: (1) reserved, (2) protected, and (3) unclassed. Although all areas are under the department of forest, the protected and unclassed areas are accessible to people more easily. The management of reserved forests is done through the working plans, which are documents containing silvicultural prescriptions for the management of a forest unit for a period of 10 to 15 years. Approximately 78% of the officially recorded reserved forest area in the country is covered by working plans—i.e., 316.8 thousand kilometers (km) in 1985 to 1986. The area regenerated naturally is given as an upper estimate of area affected by controlled fires for regeneration.

The National Remote Sensing Agency (NRSA) and the Forest Survey of India (FSI) have interpreted the Landsat imagery on a 1:1,000,000 scale for the period 1981 to 1983 and on a 1:2,500,000 scale for 1985 to 1987. This recent data on forest cover is given in Table 25.2. The data for 1985 to 1987 is closer to the truth due to improved techniques (bigger scale and better sensor) indicating that about 19.5% of the total geographical area is under forest cover. Although the NRSA data classifies forest area into dense and open forests based on crown cover, information on crown cover and standing biomass in different forests is not available. Hence the growing stock for the year 1982 to 1983 as reported by FSI, 4196 million meters (m)

Table 25.1 Statistical data on Indian forest area (1000 km²)

Year	Total forest area[a]	By composition		By classification			Natural regeneration area in existing tree forests
		Coniferous	Nonconiferous	Reserved	Protected	Unclassed	
1951–52	734.4	34	700.4	344.8	152.0	237.5	8
1960–61	689.5	44.3	645.2	316.1	240.5	112.1	4.4
1970–71	747.7[c]	39.2	699.2	360.2	212.7	115.1	13.6
1979–80[b]	736.6[d]	47.6	670.3	372.5	225.3	105.6	10.2
1985–86	752.3	40.6	711.2	406.1	215.1	131.1[e]	

Source: Teddy (1988); Mo and Mo (1987).
a. Area under forest departments.
b. Provisional.
c. Includes 6212 km² in 1965 to 1966 and 8566 km² in 1970 to 1971 for want of legal classification.
d. Also includes forest area that is not accounted for due to want of details.
e. 10,580 km² unclassed forest area outside Forest Department.

Table 25.2 NRSA and FSI assessment of forest cover

Category	1981–83		1985–87	
	Area (km²)	% of the total geographical area of the country	Area (km²)	% of the total geographical area of the country
Forest				
Dense forest (crown density 40% and over)	361,412	10.99	378,470	11.51
Open forest (crown density 10% to less than 40%)	276,583	8.41	257,409	7.83
Mangrove forest	4,046	0.12	4,255	0.13
Total	642,041	19.52	640,134	19.47
Scrub area: (tree lands with less than 10% crown density)	76,796	2.34	66,121	2.01
Uninterpreted area: (under clouds, shadows, etc.)	11,524	0.35	3,893	0.12
Nonforest: (includes tea gardens)	2,557,436	77.79	2,577,649	78.40
Total green cover	3,287,797	100.00	3,287,797	100.00

Source: FSI (1989).

(i.e., 6540 m³/km), is assumed as the above-ground biomass in our calculation. The net annual increment is 52 million m³ or about 1.24% of the growing stock (FSI, 1987). The growing stock in the year 1985 to 1986 would be 6850 m³/km, i.e., 4560 tons per km².

The spread and intensity of the forest in different parts of the country is uneven, as can be seen from Table 25.3. In the states belonging to the northeastern region less than 10% of reserved forest area is covered by working plans.

Controlled Burning for Natural Regeneration and Fire Prevention

In general terms, a fire is often considered a good preparation for regeneration but is inimical to existing regeneration. Various controlled burning practices are adopted to prevent accidental fires and to promote regeneration depending on the type of forests. These practices are briefly described below to enable estimates of areas affected by burning based on the statistics given above.

In the Chir pine forests a fire consumes the needle layers and the grass and provides an excellent seedbed. Here the practice is to burn the forest in the spring before seeds fall in areas due for regeneration. Burning in the sal forests is considered essential to thin out the underwood and replace it with light grass, to remove the toxic covering of dead sal leaves. However, burning of dry grassy sal forests is very harmful and justifiable only where the risk of hot-weather fires

Table 25.3 Status of Indian forests (area km²) by state

Study number	State/union territory	Geographical area	Recorded forest area (1985–86)	Forest cover assessment		Dense forest cover		Open forest cover		% area covered by working plan
				1987 assessment based on 1981–83 imagery	1987 assessment based on 1985–87 imagery	Based on 1981–83 imagery	Based on 1985–87 imagery	Based on 1981–83 imagery	Based on 1985–87 imagery	
1.	Andhra Pradesh	276820	63771	50194	47911	28580	25535	21119	21971	100
2.	Arunachal Pradesh	83580	51540	60500	68763	51096	54272	9404	14491	10
3.	Assam	78520	30708	26386	26058	18415	16688	7971	9370	54
4.	Bihar	173880	29227	28748	26934	13490	13412	15258	13522	100
5.	Goa	3698	1053	1285	1300	763	975	522	322	74
6.	Gujarat	195980	19320	13570	11670	7850	5259	5293	5999	41
7.	Haryana	44220	1699	644	563	43	130	601	433	100
8.	Himachal Pradesh	55670	21325	12882	13377	9908	7100	2974	6277	100
9.	Jammu and Kashmir	222240	20182	20880	20424	12978	10824	7902	9600	99
10.	Karnataka	191770	38645	32264	32100	16394	24749	15870	7351	100
11.	Kerala	38870	11218	10402	10149	8569	8312	1833	1837	79
12.	Madhya Pradesh	442840	155414	127749	133191	72174	91448	55575	41743	79
13.	Maharashtra	307760	64158	47416	44058	27244	26177	20032	17767	3
14.	Manipur	22360	15154	17679	17885	4670	5060	13009	12825	9
15.	Meghalaya	22490	8514	16511	15690	5749	3427	10762	12263	—
16.	Mizoram	21090	15935	19092	18178	2938	3883	16154	14295	2
17.	Nagaland	16530	8625	14351	14356	6379	4632	7972	9724	100
18.	Orissa	155780	59555	53163	47137	28573	27561	24391	19384	100
19.	Punjab	50360	2803	766	1151	96	97	670	1064	100
20.	Rajasthan	342210	31290	12478	12966	3048	2902	9430	10064	100
21.	Sikkim	7300	2650	2839	3124	1867	2410	972	714	100
22.	Tamil Nadu	130070	22370	18380	17715	10866	9759	7491	7909	95
23.	Tripura	10480	6298	5743	5325	330	1214	5403	4111	100
24.	Uttar Pradesh	294411	51337	31443	33844	18876	22632	12567	1121	100
25.	West Bengal	87850	11879	8811	8394	3512	3332	3223	2953	—
26.	Andaman and Nicobar Island	8290	7171	7603	7624	6807	6518	110	133	74
27.	Chandigarh	114	—	2	8		—	2	8	—
28.	Dadra and Nagar Haveli	490	206	237	205	187	149	50	56	100
29.	Daman and Diu	112	197	—	2	—	1	—	1	—
30.	Delhi	1490	—	15	22	—	12	15	10	—
31.	Lakshadweep	30	—	—	—	—	—	—	—	—
32.	Pondicherry	492	—	8	—	—	—	8	—	—
	Total	3287797	752273	642041	640134	361412	378470	276583	257409	

Source: FSI (1987, 1989); Mo and Mo (1987).

is exceptionally high. It is often recorded that natural regeneration of teak is more abundant in burned forests.

In Table 25.1 the area regenerated naturally is between 2.3% and 2.9% of the area covered by the working plan and gives an estimate of upper limit of controlled burning for regeneration.

Burning for Fire Prevention
Controlled burning is also carried out to prevent large-scale damage due to accidental fires. The usual method is to burn a strip around the periphery or to burn the adjoining forests outwards from the periphery. Isolation strips of more or less fire-proof evergreens can sometimes be put down to break up a large inflammable area, and where a suitable species is available, the practice is recommended. It is also sometimes recommended for plantations, as occurs with natural regeneration, to put a controlled ground fire through as soon as possible as a fire protection measure where the fire risk is great. However, it is now recognized that forest burning causes soil erosion.

There is no compiled and published data on the areas which undergo such controlled burning practices. These are marked in the working plans of various state forest departments. It is estimated that not more than 3% of the reserved forest area undergoes controlled burning as fire lines as a fire prevention measure (Chaturvedi, 1990).

Accidental Fires
Maithani et al. (1986) carried out an analysis of forest fire statistics in 13 states/union territories for the 10-year period 1968–1969 to 1977–1978. The fire seasons for different areas are different. However, it appears that most fires occur during January to June. Table 25.4 shows the incidence of fires in different years. The number of fires annually varied from 842 to 2958 with an average of 1903. After 1975 there seems to be a decrease in the number of fires. A study on the incidence of fire was carried out by the National Commission on Agriculture (NCA) from 1968 to 1973, and the data is given in Table 25.5 along with the more recent data for 1985 to 1988 from the Ministry of Environment and Forests. According to NCA data, the average number of fires annually was 3406, affecting an area of 2576 km^2 (i.e., 0.76 km per fire). The corresponding average for number of fires in the states covered by Maithani et al. (1986) is 2243. In 1986 Ministry of Environment and Forests compiled the data on fires for the period 1980 to 1985 (sixth plan) based on data provided by the states, and found that 17,852 fires occurred in the country, burning an area of 5,724 km^2. This amounts to an average of 3570 fires affecting 1,145 km^2 annually (i.e., 0.32 km^2 per fire). Based on the data for 1985 to 1988, it appears that the accidental fires affect about 10,000 km^2 of forest area annually. We have used this area as the current estimate of forest area affected by accidental fires. However, FSI inventory indicates that 33% to 99% of forest area in different states is affected annually by ground fires (FSI, 1987), which is about 80 times the area recorded as affected by accidental fires. But this estimate appears to include all types of fires in the forest and can serve as an estimate of maximum forest area affected by fire.

It appears that the existing method of reporting on fires in forests encourages an understatement of the incidence of fires and the area affected. The attempts to systematically quantify the area affected for each state annually has been initiated by the Ministry of Environment and Forests. The state of Orissa accounts for more than 90% of the area affected annually and needs further investigation.

Burning Due to Shifting Cultivation
Shifting cultivation, known as *jhum*, is practiced extensively in the northeastern hill regions and in parts of Orissa, Madhya Pradesh, Maharashtra, and Andhra Pradesh. The three different estimates on the area under shifting cultivation are summarized in Table 25.6.

The average fallow period based on the North-Eastern Council data is about 5.9 years; the average fallow period based on the Task Force data is about 4.3 years. The FAO estimate of area under shifting cultivation is 3 to 14 times higher except in the case of Orissa. The lower estimates of the Task Force compared to those of the FAO could also be due to large areas in the northeast being under the unclassed category or under private ownership. For the estimation of emissions we have taken 10,000 km^2 as the current annual area burned for shifting cultivation. If we consider 4.3 years as the fallow period, then based on the FAO estimate the annual area works out to be 22,023 km^2. This can be regarded as the current upper limit for the annual area under shifting cultivation.

Estimation of Above-Ground Biomass in Forests
The above-ground dry biomass is estimated using the following assumptions.

1. Growing stock in 1985 to 1986 = 4560 tons per km^2
2. Moisture content of growing stock = 10%

Table 25.4 Annual record of total number of forest fires in some of the states and union territories in India for the decade from 1968–69 to 1977–78

Period/Year	Andhra Pradesh (whole state)	Haryana (whole state)	Himachal Pradesh (whole state)	Jamu and Kashmir (whole state)	Uttar Pradesh (whole state)	Dadra & Nagar Haveli (whole state)	Gujarat (Vadodara circle)	Karnataka (Belgam circle, Bidar, and Humaabad Range)	Kerala (Malayattor divs. and central circle Perumbayoor)	Rajasthan (Udaipur and Jhalawar div.)	Total
1968–69	415	10	233	47	223	433	0	0	20	649	2030
1969–70	201	8	180	27	275	894	1	0	35	393	2014
1970–71	233	3	243	51	1009	558	7	0	43	691	2838
1971–72	302	3	166	25	153	296	2	56	52	320	1375
1972–73	226	26	121	51	573	465	0	4	11	1481	2958
1973–74	249	9	22	73	219	283	6	6	11	643	1521
1974–75	399	4	58	61	313	524	0	7	16	934	2316
1975–76	298	17	120	32	195	305	1	10	8	848	1834
1976–77	333	6	162	15	180	213	1	2	8	384	1304
1977–78	224	11	211	20	118	100	1	4	6	147	842

Source: Maithani et al. (1986).

Table 25.5 Data on accidental forest fires

Study number	State/UTs	Forest area (1000 km²)	1968–69 to 1972–73		Forest area (km²) burned during			FSI Inventory (1000 km²)
			Number of fires	Area burned (km²)	1985–86	1986–87	1987–88	
1.	Andhra Pradesh	63.77	275	688.78	12.76	8.70	5.12	—
2.	Arunachal Pradesh	51.54	—	—	8.44	4.94	3.73	479
3.	Assam	30.71	—	—	0	0	0	16
4.	Bihar	29.23	445	156.60	Only mild surface fires have occurred which have not caused any material loss.			19.5
5.	Goa	1.05	—	—	0	0	0	—
6.	Gujarat	18.78	400	296.20	293.42	80.95	50.50	9.5
7.	Haryana	1.69	40	24.40	14.99	1.46	6.18	—
8.	Himachal Pradesh	21.33	447	218.98	85.99	8.25	40.05	14.7
9.	Jamu and Kashmir	20.89	527	145.27	28.91	12.48	33.13	9.6
10.	Karnataka	38.64	—	—	—	—	—	17.3
11.	Kerala	11.22	—	—	5.46	8.91	16.95	—
12.	Madhya Pradesh	155.41	—	—	122.82	174.23	219.61	118.1
13.	Maharashtra	64.06	—	—	109.78	34.93	38.13	—
14.	Manipur	15.16	—	—	No assessment made as incidence of forest fires in the state occurred mostly in unclassed forests			15.0
15.	Meghalaya	8.51	—	—	0	0	0	8.0
16.	Mizoram	15.94	—	—	22.55	20.82	3.49	—
17.	Nagaland	8.63	—	—	0	0	0	7.5
18.	Orissa	59.56	550	509.88	8,938.41	9,241.84	9,786.39	55.9
19.	Punjab	2.80	41	22.20	8.12	2.85	5.85	—
20.	Rajasthan	3.12	—	—	32.71	14.39	10.72	—
21.	Sikkim	2.65	—	—	5.00	7.00	8.00	1.0
22.	Tamil Nadu	22.32	—	—	9.56	11.26	0	—
23.	Tripura	6.28	33	2.93	No assessment made			—
24.	Uttar Pradesh	51.27	614	507.92	138.71	116.66	117.43	29.7
25.	West Bengal	11.88	34	2.88	No major forest fire occurred during last three years and loss is insignificant			03.9
26.	Andaman and Nicobar Island	7.14	—	—	0	0	0	—
27.	Chandigarh	0.01	—	—	0	0	0	—
28.	Dadra and Nagar Haveli	0.20	—	—	.93	3.48	.29	—
29.	Daman and Diu	—	—	—	0	0	0	—
30.	Delhi	0.04	—	—	0	0	0	—
31.	Lakshadweep	—	—	—	0	0	0	—
32.	Pondicherry	—	—	—	0	0	0	—
	Total		3,406	2,576.04	9,838.56	9,753.15	10,345.57	804.7

Source: Gupta (1987); Chaturvedi (1990); FSI (1987).

3. The dry weight of growing stock represents 80% of total biomass standing in the forests on a dry basis (total standing biomass = 5140 tons/km²)

4. The biomass fallen on ground is 10% of total dry biomass = 51 tons/km²

5. Therefore, dry biomass above ground = 5.2 kg/m²

Another estimate is 12.5 kg/m² for nonconiferous forests and 9.6 kg/m² for coniferous forests (Hingane, 1990) with a weighted average of 12.3 kg/m². However, in our calculations we have used the estimate of 5.2 kg/m².

Burning Efficiency

The burning efficiency for different types of forest fires is assumed as follows (Chaturvedi, 1990): controlled fires, 10%; fires for shifting cultivation, 100%; accidental fires, 80%. The total biomass burned during 1985 to 1986 in forest fires is estimated in Table 25.7.

Agricultural and Grassland Fires

Agricultural residues in India are primarily used as fodder and fuel; they are also used as fertilizer and building material. However, in certain regions of the

Table 25.6 Statistics on shifting cultivation (km^2)

State	North Eastern Council, 1971 data, Ramakrishnan (1987) Annual	Total	FAO (1981), based on Landsat data Total	Task force on shifting cultivation (1983) Annual	Total	Fallow period (years)
Andhra Pradesh	NA	NA	4710	500	1500	3
Arunachal Pradesh	920	2480	7940	700	2100	3–10
Assam	700	4980	4160	696	1392	2–10
Bihar	NA	NA	NA	162	810	5–8
Madhya Pradesh	NA	NA	NA	125	1250	10–15
Manipur	600	1000	17770	900	3600	4–7
Meghalaya	760	4160	10240	530	2650	5–7
Mizoram	620	6040	16110	630	1890	3–4
Nagaland	740	6080	10970	192	768	4–9
Orissa	NA	NA	16580	5298	26490	5–14
Tripura	230	2210	6220	223	1115	5–9
Total	4570	26950	94700	9956	43565	
Number of dependent families	4,250,000			6,220,000		

NA = not available.

Table 25.7 Estimation of biomass burned in forest lands, 1985 to 1986 (dry biomass above ground = 5.2×10^{-3} Tg/km^2)

Category	Area (km^2)	Burning efficiency (%)	Amount burned (Tg)
Regeneration	7900	10	4.1
Prevention of fires	9500	10	4.9
Accidental	10000	80	41.6
Shifting cultivation	10000	100	52.0
Total			102.6

country the residues are also burned in the field. There is no comprehensive compilation on the nature and extent of agricultural fires. It is known that the crop residues are burned in Punjab, particularly in farms where a combined harvester and thresher is used. In the case of cotton, to avoid occurrence of pests in the subsequent crops, in some farms the residues are burned in the fields. After the sugarcane harvest, the remaining biomass is set on fire. In the hills, most of crop residues are burned after harvesting. However, there is no quantitative data either on the area subjected to these fires or on the amount of biomass burned in such fires. Hence, no attempt is made here to estimate biomass burned due to fires from agricultural fields. The use of crop residues as fuel is discussed in the next section.

The area under permanent pastures and other grazing grounds in India is 122,000 km^2; this is burned annually. The biomass burned annually is 24.8 teragrams (Tg), assuming a productivity of 0.198 kg per square meter (m^2). Another estimate on productivity of pasture lands is 0.84 kg/m^2 (Hingane, 1990).

Biofuels to Meet Energy Needs

The use of biomass as fuel accounts for about 50% of the total primary energy consumption in the country (Teddy, 1988). The major sources of biofuels are recorded and unrecorded firewood from forest land and trees on other land, crop residues, and dungcakes.

Biofuels are burned in a variety of cooking stoves made locally to suit the local diet. The most comprehensive data on the use of biofuels to meet energy needs in households and commercial establishments is based on a nationwide survey carried out in 1978 to 1979. There is a large regional variation in the proportion of various biofuels used in households (Natarajan, 1985). There is also a significant seasonal variation within a region (Bose et al., 1990). In Table 25.8, based on the population estimates for 1985 to 1986 (EDSG, 1986) and the actual per capita consumption in 1978 to 1979 (Natarajan, 1985), we have estimated the amount of biofuels burned.

Total biomass burned annually in households is 230 Tg. This calculation does not account for shifts in

Table 25.8 Estimation of biofuel burning, 1985 to 1986

Fuel	Annual norm (kg per capita)[a]		Annual consumption (million tons)		
	Rural	Urban	Rural	Urban	Total
Firewood	158.0	116.4	89.9	22.3	112.2
Crop residues	58.8	5.6	33.4	1.1	34.5
Dungcakes	133.0	35.6	75.7	6.8	82.5
Charcoal	0.2	3.8	0.1	0.7	0.8

Source: EDSG (1986); Natarajan (1985).
a. Population in 1985 to 1986: rural, 569 million; urban, 192 million.

Table 25.9 Annual consumption of biofuels in establishments, 1978 to 1979 (million tons)

Fuel	Rural	Urban
Firewood	2.68	0.96
Crop residues	0.31	0.17
Dungcakes	2.11	0.01
Charcoal	0.46	0.24

Source: Natarajan (1985).

domestic fuels in urban or rural areas. But the trends appear to be such that the total biomass consumption will increase due to an increase in population in areas using biofuels.

Biofuels are also burned in rural industries and services, whereas in urban areas they are used in service establishments. It is common to use rice husks in rice mills, bagasse in the sugar industry, coconut shells in copra, lime, and brick kilns, etc. For estimating the consumption for the year 1985 to 1986, we have assumed a decrease of 10% in the urban areas and an increase of 10% in the rural areas, except for charcoal, over the 1978 to 1979 consumption levels shown in Table 25.9. This is rather arbitrary; however it indicates the direction of the shift. The decrease in urban areas is due to a shift to commercial fuels, whereas in rural areas, the increase is due to increase in rural industries, the use of biofuels in the form of briquettes, etc., and the limited availability of commercial fuels. The biomass burned is about 9.0 Tg.

The total biomass burned from sources other than the burning of agricultural fields is 376.4 Tg. The contribution from the burning of biofuels in the household appears to be the largest—230 Tg per year.

Emissions Reductions Due to Improvements in Cooking Stove Efficiencies

The promotion of improved cooking stoves is a large-scale intervention in many developing countries for achieving multiple benefits at local, national, and global levels, including the potential for a reduction in the emissions of greenhouse gases. The significant contribution of the domestic biofuels to the total biomass burned in India makes conservation in the domestic sector an important option to reduce the emissions of greenhouse gases.

In India, there is a large-scale government program to promote more efficient cooking stoves. It is estimated that 6 million stoves were disseminated by March 1989 (DNES, 1989). The majority of these are two-port mud stoves with flues with a thermal efficiency of 20% or more in the laboratory (Garg, 1990). There are other small programs adopting different dissemination strategies. These programs—categorized as a designer's approach (Ravindranath, 1990; Sarin 1990), a disseminator's approach (Pal and Joshi, 1989), and a mass approach (Sadaphal et al., 1989)—have been evaluated recently.

Even in the best situations (i.e., the designer's approach, where stoves are designed for specific situations and disseminated through close monitoring) there is a reduction in fuel savings over time. On a large-scale dissemination there is a further reduction in the fuel savings achieved by the use of improved cookstoves (Ravindranath, 1989). In a disseminator's approach, where stoves are chosen based on local needs and available options, an insignificant reduction in fuel savings is observed (Pal and Joshi, 1989). In an evaluation of the national program in three states in India, it was found that the acceptability, defined as partial or complete use of the stove, varied widely within a region depending on the quality of dissemination. The improved stoves were used along with traditional stoves, and the designs of improved stoves were modified substantially to suit the requirements of a cook. An energy saving of 20% was observed (Sadaphal et al., 1989) in a village where the fuel consumption in traditional stoves was higher than that in other villages. Thus, any estimation of reduction in fuel consumption due to promotion of improved cookstoves would need to account for the scaling down in anticipated savings due to the above-mentioned field observations.

Subsequently, to calculate emissions, a knowledge of emission factors for different pollutants in small-scale burning is essential. In a study on metal stoves it was found that for most improved stove fuel combinations the emission factors for CO and TSP increase as compared to the emission factors from traditional stoves using different biomass fuels. These increases are offset by increases in efficiency (Joshi et al., 1989); offset in most cases is due to reduction in fuel consumption. It has been observed, however, that in the case of heavy mud stoves, commonly observed modifications in the field can increase the emission factors substantially (Joshi et al., 1990). The field observations in India on concentrations in kitchens show reduction in CO levels (Ramkrishna et al., 1989). A careful examination of all the factors influencing the field performance of cooking stoves seems to be essential to assess their potential to reduce the emissions of greenhouse gases.

Conclusions

The estimation of greenhouse gases due to biomass burning can be made more accurate by making the estimation of biomass burned more realistic and by measuring emission factors of greenhouse gases for different types of fires on a regional basis. In India, we find that

1. Most statistics are for fires in reserved forests. Although the productivity of protected and unclassed forests is lower, it would be useful if information on burning practices of shrubs, etc., were compiled.

2. The statistics related to the areas affected by controlled fires can be improved by an examination of working plans from different states. Here, the NRSA data may be found useful if crown density can be related to productivity.

3. The data on biomass loading can be improved by compiling information statewise and according to forest type. This again is probably feasible from the data collected by the forest department. However, additional information on the productivity of grasslands and other lands will be needed.

4. The estimates about burning efficiency can be improved by carrying out before-and-after measurements for different types of forest fires. This is possible, except probably in the case of accidental fires. In the case of accidental fires a measurement

after the fire can be used along with the data on biomass in the region to find out burning efficiencies.

5. In the case of agricultural fires, to generate the basic data on the areas affected by fire and amount of biomass burned, a baseline survey of burning practices in different agroclimatic regions would be required. The results of this can be used to identify areas affected and biomass burned.

6. A number of village-level surveys have been sponsored by the Department of Non-conventional Energy Sources, Advisory Board on Energy, and other government organizations. A smaller sample of the 1978 to 1979 survey has also been resurveyed. A compilation of these data would facilitate more accurate information on biofuel burning in households.

7. Recent data on burning of biofuels in commercial establishments in a few urban areas are collected by the Tata Energy Research Institute, New Delhi. However, in the context of global warming a more comprehensive survey will be required.

8. At present, data on emission factors of greenhouse gases from different sources in India does not exist, except for CO and TSP from biofuels burned in a few designs of cooking stoves. This data will be essential to reduce the uncertainty in the estimation of emissions from biomass burning.

Acknowledgments

The prompt service provided by Mr. B. Anil Kumar in gathering information from a variety of sources is gratefully acknowledged. The author is grateful to Mr. A. N. Chaturvedi and Mr. O. N. Kaul for the numerous discussions and for valuable information.

III

Biomass Burning in Temperate and Boreal Ecosystems

The Extent and Impact of Forest Fires in Northern Circumpolar Countries

Brian J. Stocks

Analyzing the extent and impact of forest fires in northern circumpolar countries essentially means documenting fires in Canada, the Union of Soviet Socialist Republics (USSR), Finland, Norway, Sweden, and Alaska—areas in which the boreal forest, or *taiga,* predominates as a vegetation type. An overwhelming number of large fires occur in this boreal forest zone, which stretches in two broad transcontinental belts across North America and Eurasia. Covering 12 million square kilometers (km²) (twothirds in the USSR and Scandinavia, and one-third in Canada and Alaska), and lying generally between 45° and 70°N latitude, the boreal zone contains extensive tracts of coniferous forest, which provide a vital natural and economic resource for northern circumpolar countries.

The boreal forest is composed of hardy species of pine (*Pinus*), spruce (*Picea*), larch (*Larix*), and fir (*Abies*), mixed, usually after disturbance, with deciduous hardwoods such as birch (*Betula*), poplar (*Populus*), willow (*Salix*), and alder (*Alnus*), and interspersed with extensive lakes and organic terrain. This closed-crown forest, with its moist and deeply shaded forest floor where mosses predominate, is bounded immediately to the north by a lichen-floored open forest or woodland, which in turn becomes progressively more open and tundra dominated with increasing latitude. To the south the boreal forest zone is succeeded by temperate forests or grasslands.

The overwhelming impact of wildfires on ecosystem development and forest composition in the boreal forest is readily apparent and understandable. Large contiguous expanses of even-aged stands of spruce and pine dominate the landscape in an irregular patchwork mosaic, the result of periodic severe wildfire years and a testimony to the adaptation of boreal forest species to natural fire over millenia. The result is a classic example of a fire-dependent ecosystem, capable, during periods of extreme fire weather, of sustaining the very large, high intensity wildfires which are responsible for its existence. This natural force has coexisted, somewhat uneasily, with in-creased human settlement and use of the resourcerich boreal zone over the past century. Organized fire suppression activities have been generally successful for the most part, but the initial concept that all fires should be suppressed at any cost has proven to be neither economically feasible or ecologically desirable. Fire is now managed, as are many other forest activities, in a multiple-use fashion where consideration is given to the natural ecological role of fire, the economics of fire suppression, and the priority of high-value resources.

The most recent forest fire statistics from northern circumpolar countries are presented and analyzed in this chapter. These statistics have been supplemented with older data gathered during earlier studies (Stocks and Barney, 1981). Although fire statistics are archived by administrative jurisdiction and not forest region, the vast majority of each country's large fires occur in the boreal forest. Canada and Alaska, despite having similarly progressive fire-management programs, still experience significant, resourcestretching fire problems often. Scandinavian countries, in contrast, do not seem to have major large fire problems, probably due to the easy access resulting from intensive forest management in those countries. Fire statistics for the USSR are scant and colloquial in nature, but indications are that major fire problems exist there, particularly in Siberia, where vast distances, a scattered population, and the lack of an effective infrastructure seem to mitigate against effective fire management. Canada has the longest record of detailed fire statistics available, and these statistics receive the most thorough examination in this chapter.

Forest Fire Activity

Alaska

Forest fire statistics are available for Alaska since 1940 and, over the past 50 years, the area burned in this northernmost U.S. state has decreased steadily

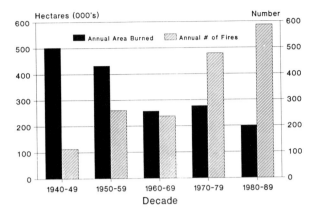

Figure 26.1 Annual forest fire statistics (area burned and fire occurrence) for Alaska, averaged by decade, 1940 to 1989.

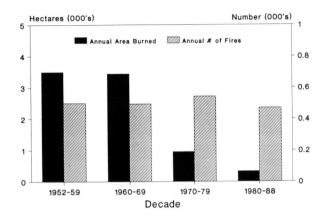

Figure 26.2 Annual forest fire statistics (area burned and fire occurrence) for Finland, averaged by decade, 1952 to 1988.

while fire incidence has increased fivefold (Figure 26.1). Both factors are the result of increased accessibility. Road and rail access meant both an increase in forest use, which resulted in increased fire occurrence, but also a corresponding enhanced detection capability and a shortened response time. Faster initial attack, particularly using smoke jumpers, coupled with aerial detection, are the major contributors to the reduction in area burned. Lightning fires, generally occurring in areas where response intervals are longer, account for a large percentage of the area burned in Alaska (38% of Alaskan fires are lightning-caused, and these fires account for 80% of the area burned). In addition, many fires in Alaska are fought on a priority basis, with extensive zones of limited protection, resulting in recent area-burned statistics being somewhat inflated as a result of selective fire suppression. An overall decrease in area burned, in view of this policy, reflects well on fire management activities in Alaska.

Scandinavia

Somewhat limited fire statistics are available for this region, with Finland having the only continuous records from 1952. Fire statistics from Sweden are available only from 1950 to 1980, when record keeping was discontinued. Norway fire statistics have only been recorded since 1980. Figures 26.2 and 26.3 present annual fire occurrence and area-burned statistics, averaged by decade, for both Finland and Sweden, respectively. Fire numbers and area burned in Sweden increased a small amount over the three decades of record, while in Finland fire occurrence remained constant and area burned decreased significantly over the past four decades. Fire activity levels in Norway

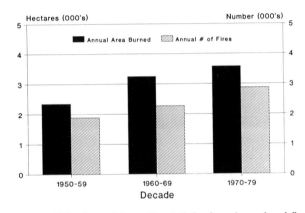

Figure 26.3 Annual forest fire statistics (area burned and fire occurrence) for Sweden, averaged by decade, 1950 to 1979.

and Finland appear similar during the 1980s and, assuming the trend evident in Sweden in the 1950 to 1970 period continued throughout the past decade, the annual area burned for Scandinavia as a whole would be slightly less than 5000 hectares (ha), resulting from 3800 fires. Unlike Canada, the USSR, and Alaska, Scandinavian countries do not appear to experience large forest fires. This would appear to be attributable to the high degree of accessibility in these smaller countries as a result of intensive forest utilization and management. In addition, lightning fires, a higher proportion of which tend to occur in remote areas, are not a major factor in Scandinavia where they account for less than 10% of all fires.

Union of Soviet Socialist Republics (USSR)

Although northern Russia and Siberia have long been noted as areas where extensive forest fire activity is common (Lutz, 1956), no documented statistics have ever been published by the USSR which would allow accurate quantification of the magnitude of the problem in that country. Fire statistics are collected in a highly localized fashion and have not been assembled to date in a manner that would permit a national or even regional summary. Documentary accounts from the early 1900s describe enormous forest fire losses covering thousands of square kilometers in Siberia, and giving the impression that it was difficult to find areas where evidence of recent fire was not present. In the particularly dry year of 1915, an estimated total of 14 million ha burned in Siberia (Shostakovitch, 1925). Periodically some qualitative accounts of the role of fire in the Siberian forests have been published, but these contain only partial statistics at best, which do not permit even rudimentary analysis. Coniferous forest types predominate, accounting for 82% of the forested area in Siberia, with larch and pine species being most common. Although some Russian literature seems to indicate that lightning is not a major cause of fires in that country, and that crown fires are uncommon, it seems likely that extensive, high-intensity crown fires must be a factor in the periodic severe fire years in Siberia in which vast areas are burned over. Further, it would seem that lightning would be a major causal agent in a region as large, remote, and underpopulated as Siberia.

1987 was a particularly severe fire year in Inner Mongolia and Siberia. The well-publicized Great China Fire burned over 1.1 million ha near the China-USSR border during the early spring of that year (Stocks and Jin, 1988). Satellite imagery from the NOAA-9 and NOAA-10 spacecraft showed that a much larger area was burning in central Siberia during the same period. Analysis of this low-resolution imagery revealed 40 to 50 fires, ranging in size from 20,000 to 2 million ha, had burned over a total of approximately 10 million ha in this part of the USSR. This may be an incomplete estimate, since portions of Siberia were not visible on the satellite imagery due to cloud cover. While the absolute accuracy of this estimate may be questionable due to the coarse resolution of the NOAA imagery, it still provides, in the absence of any official statistics from the USSR, a reasonable indication of the enormous forest fire problems that existed in this region in 1987. While fire activity in the USSR can be assumed to fluctuate dramatically from year to year, as is the case in other countries, the 1987 scenario is strong evidence that a major proportion of the earth's large boreal forest fires occur in Siberia. Higher resolution satellite im-

agery will be utilized in the next few years to better quantify forest fire losses in the USSR.

Canada

As in the USSR, the forested area of Canada is dominated by the boreal forest zone, which extends in a broad belt from the Atlantic to the Pacific oceans and lies immediately to the north of the heavily populated region along the Canada–United States border. Over the past century the use of the boreal forest zone in Canada, for both industrial and recreational purposes, has increased dramatically, and this has resulted in a concurrent increase in both forest fire incidence and the fire management capability mobilized to deal with this problem. Provincial and territorial agencies in Canada have progressed to the point where state-of-the-art centralized and highly computerized fire management systems are common, yet forest fires continue to exert a tremendous influence on the forest resource in this country. Periods of extreme short-term fire weather, in combination with a recognition of both the ecological desirability of natural fire and the economic impossibility of controlling all fires, have resulted in the realization that forest fires in Canada are a problem that cannot, and should not, be eliminated.

Detailed forest fire statistics have been archived since 1920 in Canada, and this extensive record permits a thorough analysis of trends in this country. Annual fire occurrence in Canada (Figure 26.4), without fluctuating greatly on a year-to-year basis, has increased rather steadily from approximately 6000 fires annually in the 1930 to 1960 period, to almost 10,000 fires during the 1980s. This is a reflec-

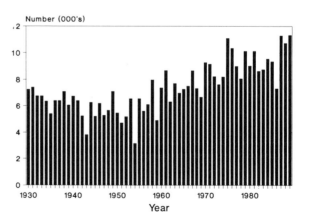

Figure 26.4 Annual forest fire occurrence statistics for Canada, 1930 to 1989.

tion of a growing population and increased forest use but is also due to an expanded fire detection capability. The area burned by Canadian forest fires (Figure 26.5) fluctuates tremendously on an annual basis, and the last decade is significant in this regard, due to significant losses in 1980, 1981, and 1989. Fire occurrence and area-burned statistics for the past 70 years in Canada, averaged by decades, are presented in Figure 26.6. While fire occurrence numbers were relatively constant over the 1920 to 1959 period and have increased steadily since that time, area burned actually decreased over the first four decades of record only to increase over the last three decades. The most dramatic increase occurred during the 1980s, primarily due to periods of short-term extreme fire weather in western and central Canada. Lightning accounts for 35% of Canada's fires, yet these fires result in 85% of the total area burned, due to the fact that lightning fires occur randomly and therefore

present access problems usually not associated with human-caused fires, with the end result that lightning fires generally grow larger, as detection and subsequent initial attack is often delayed.

The effect of Canadian fire management activities on area burned is evident from Figure 26.7, which shows that an overwhelming number of fires are controlled early, with the result that the 2% to 3% of the fires that grow larger than 200 ha account for 97% to 98% of the total area burned in the country. The evolution of increasingly improved fire management activities in Canada over time, as exemplified by the province of Ontario, is shown in Figure 26.8. The percentage of fires in smaller size classes has increased steadily over the past seven decades, and, correspondingly, larger fires represent a much smaller percentage. The increase in larger fires during the 1980s in Ontario is a function of "modified" or "selective" suppression, in which fires are prioritized

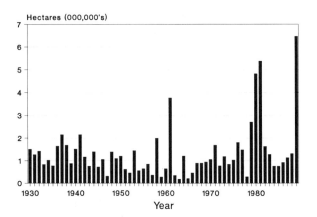

Figure 26.5 Annual area burned by forest fires in Canada, 1930 to 1989.

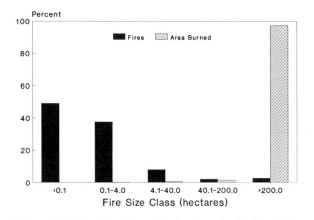

Figure 26.7 Canadian forest fire size class distribution, 1970 to 1985.

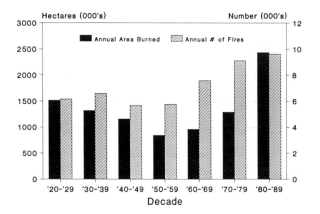

Figure 26.6 Annual forest fire statistics (area burned and fire occurrence) for Canada, averaged by decade, 1920 to 1989.

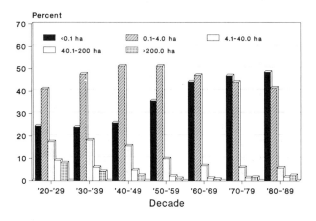

Figure 26.8 Ontario forest fire size class distribution, averaged by decades, 1920 to 1989.

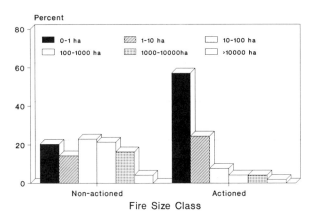

Figure 26.9 Forest fire size class distribution for actioned and nonactioned fires in northwestern Ontario, 1976 to 1988.

on the basis of values-at-risk, resulting in numerous fires in low-priority areas being permitted to run their course. This practice, a function of both economic realities and ecological considerations, is common in many jurisdictions in Canada. The effect of this approach in the province of Ontario, for example, is shown in Figure 26.9, where the size class distribution for recent actioned and nonactioned fires in a section of northwestern Ontario is presented. In the absence of active suppression activities, fire size distribution is somewhat normal, with large to very large fires common. Actioned fires, however, exhibit a negative exponential distribution, reflecting successful fire suppression activity.

Fire Emissions and Atmospheric Impact

In recent years there has been an increasing awareness within the scientific community of the impact of worldwide biomass burning on atmospheric chemistry, photochemistry, and climate warming (Crutzen et al., 1979; National Academy of Science, 1984). This information has been transferred to the public in general, with the result that worldwide attention has been focused on the tropics, where carbon released through biomass burning in developing countries is approaching the amount entering the atmosphere through the burning of fossil fuels in developed countries (Crutzen and Andreae, 1990). Research activity on the atmospheric impact of tropical biomass burning has correspondingly been accelerated through international cooperation, and many questions remain to be answered.

While the tropics remain the primary source of biomass burning emissions to the atmosphere, there is a strong need to quantify the contribution to global atmospheric trace gas budgets from forest fires in the Northern Hemisphere. Forest fires are constant and highly visible phenomena in northern countries where questions concerning the environmental impact of fires are being raised.

In recent years a cooperative Canada–United States study into the behavior and atmospheric environmental impact of large-scale, mass-ignition fires has been undertaken (Stocks, 1988; Stocks and McRae, 1991) in the boreal forest zone in Canada. Cooperating agencies include Forestry Canada, the U.S. Forest Service, the National Aeronautics and Space Administration, and the Ministries of Natural Resources and Environment from the province of Ontario. One of many results from this work has been a characterization of the smoke chemistry associated with fires in the boreal forest (Cofer et al., 1989, 1990; Hegg et al., 1990). Although this work is continuing, emission factors have been determined for these fires which, when combined with recent fire statistics, permit the estimation of the contribution of Northern Hemisphere forest fires to global atmospheric trace gas budgets. Approximately 89% of the carbon released to the atmosphere from boreal forest fires is in the form of carbon dioxide, 9% is carbon monoxide, while the remaining carbon is released as methane or nonmethane hydrocarbons (particulates). In general, the following emission factors can be agreed on for trace gases resulting from boreal forest fires: 445 grams per kilogram (g/kg) for carbon dioxide, 45 g/kg for carbon monoxide, and 4.55 g/kg for both methane and total nonmethane hydrocarbons (aerosols).

Average annual area-burned statistics for Canada, Alaska, Scandinavia, and the USSR are presented in Table 26.1. These statistics are averaged for the

Table 26.1 Fire statistics for northern circumpolar countries, 1980 to 1989 (annual average)

Country	Number of fires	Area burned (ha)
Canada	9618	2,437,717
Alaska	588	198,883
Finland	470	326
Sweden[a]	2875	3574
Norway	462	850
USSR[b]	20,000	3,000,000
Total	34,013	5,641,350

a. 1970 to 1979 data.
b. Estimated figures.

decade of the 1980s, the most recent and realistic data available. Due to the lack of official statistics for the USSR, an estimate of 20,000 fires and 3 million ha burned annually has been used for this country.

Assuming that total fuel consumption (ground, surface, and aerial fuels) averages 2.5 kilograms per square meter (kg/m^2), or 25,000 kg/ha, for boreal forest fires, that carbon accounts for 50% of this fuel weight consumed, and using the area-burned figures from the 1980s, it is possible to estimate total trace gas emissions from forest fires in northern circumpolar countries. Combining 5.64 million hectares burned with a fuel consumption of 25,000 kg/ha results in an annual total of 141 teragrams (Tg) of fuel, or approximately 70 Tg of carbon being consumed. This translates to 62.3 Tg of carbon dioxide, 6.3 Tg of carbon monoxide, and 0.7 Tg of both methane and total nonmethane hydrocarbons.

The latest estimate of biomass burning on a global scale (Crutzen and Andreae, 1990) indicates that between 4000 and 6000 Tg of carbon is released annually, with savanna burning and deforestation practices in the tropics being the primary contributors. While these figures are difficult to prove, it is clear that emissions from biomass and fossil fuel burning are roughly equal on a global scale. Assuming these estimates are accurate, forest fires in northern circumpolar countries contribute from 1.2% to 1.8% of the carbon released through biomass burning worldwide.

Conclusions

This analysis of the magnitude and atmospheric impact of forest fires in the boreal forest-dominated northern circumpolar countries leads to the following conclusions:

• Forest fires have always had, and continue to have, a significant and natural role in the world's boreal forest zone.

• There is a tremendous fluctuation in annual area burned in the boreal forest zone, primarily due to short-term extreme fire weather, but it also is a function of modified or selective fire suppression activities.

• The area burned in the boreal forest zone increased dramatically in the 1980s, particularly in Canada and most likely in the USSR.

• This level of fire activity in the boreal forest zone can be expected to continue, as fire management economics and the fire-dependent nature of the boreal forest will dictate such a result.

• The forest fire business in northern circumpolar countries would be affected quickly and significantly by a trend toward global warming, as fire management agencies currently operate with a narrow margin between success and failure.

• The best estimate at the present time is that forest fires in northern circumpolar countries contribute 1% to 2% of the carbon released globally through biomass burning.

Trace Gas and Particulate Emissions from Biomass Burning in Temperate Ecosystems

Wesley R. Cofer III, Joel S. Levine, Edward L. Winstead, and Brian J. Stocks

The growing awareness of the potentially large amounts of both trace gases and aerosols that are released into the atmosphere as a result of biomass burning (agricultural burning, wildfires, deforestation, etc.) has led to an increased emphasis on assessing the impacts of biomass burning on the global environment (NAS, 1984). Biomass burning is widespread in all ecosystems and may prove to be a substantial source of both radiatively and photochemically important trace gases and aerosols. It is generally believed that biomass burning (i.e., particularly for deforestation and agricultural practices) is increasing on the global scale, or at the very least, has been underestimated in the past. The advent and use of satellite imagery has contributed significantly to the current assessments of the extent of global surface burning. If global temperatures are warming, as often suggested, a higher incidence of wildfires globally would be a logical consequence. Clearly, a substantial increase in the understanding of the impacts of global biomass burning on the earth's atmosphere is presently needed.

Since Crutzen et al. (1979) suggested that biomass burning could be a significant contributor to the budgets of several important atmospheric trace gases (CO_2, CO, H_2, CH_4, N_2O, NO_x, COS, and $CHCl_3$), many efforts have been made to expand our understanding of trace gas production and resulting emissions from fires. Seiler and Crutzen (1980) have estimated the input of CO_2 to the atmosphere from biomass burning may be comparable to that of fossil fuel burning. Many gases of considerable photochemical importance are released into the atmosphere during burning (Darley et al., 1966; Evans et al., 1977; Stith et al., 1981; Westberg et al., 1981; Greenberg et al., 1984; Delany et al., 1985; Levine et al., 1985; Hegg et al., 1987; Andreae et al., 1988; Cofer et al., 1988a, 1989, 1990a, 1990b). Three of the major gaseous emissions from biomass burning are greenhouse gases: carbon dioxide (CO_2), methane (CH_4), and nitrous oxide (N_2O). All three significantly impact the climate of our planet (Kuhn, 1985;

Ramanathan et al., 1985). They are also known to be currently increasing in our atmosphere at rates ranging from about 0.3% per year for CO_2 and N_2O to about 1.0% per year for CH_4 (Rasmussen and Khalil, 1981; Keeling et al., 1984; Weiss, 1981; Rinsland et al., 1985). The causes of these increases are not explicitly known, but increased global biomass burning is thought to be a major contributor. Estimates of the rates of increase coupled with estimates of the contributions attributed to biomass burning for several of the environmentally significant gases are presented in Table 27.1.

During the combustion of biomass, large amounts of living and dead carbon-based fuels are chemically transformed into gaseous, liquid, and solid phases and released into the atmosphere. The chemical composition of biomass burning emissions depends primarily on the rate of energy release (intensity) or combustion. Combustion is strongly coupled to fuel moisture, fuel type, fuel size, fuel array, ignition pattern, terrain, and weather. Some of these parameters are not easily isolated from one another.

Many attempts have been made to experimentally characterize emissions from biomass burning. Most of the large field studies of trace gas and aerosol production from biomass burns have occurred in the tropics, in the western and southeastern United States, or in Canadian boreal forests (Crutzen et al., 1979; Delmas, 1982; Greenberg et al., 1984; Ward and Hardy, 1984; Crutzen et al., 1985; Andreae et al., 1988; Cofer et al., 1988a, 1989, 1990b). The limited sets of measurements of gas and aerosol emissions from these fires have been extrapolated to estimate emissions from biomass burning on a planetary scale (Seiler and Conrad, 1987). Tropical fires have received special emphasis because they are believed to constitute the most significant fraction of atmospheric emissions from biomass burning (Seiler and Crutzen, 1980). A summary of most of these results for the gases CO, CH_4, and N_2O appears in Table 27.2. As can be seen, there is a large spread in the reported emission ratios for most of these measurements.

Table 27.1 Radiatively active and chemically active gases increasing in the atmosphere: estimated annual contribution from global biomass burning

Gas	Rate of increase	Estimated contribution from burning	Estimated total production	Source
CO_2	0.3–0.4%	$2–4 \times 10^{15}$ g C	$7–10 \times 10^{15}$ g C	a.
CO	0.5–5.0	$110–450 \times 10^{12}$ g C	$150–400 \times 10^{13}$ g C	b.
CH_4	1.0–1.5	$60–100 \times 10^{12}$ g C	$300–560 \times 10^{12}$ g C	c.
N_2O	0.2–0.3	$1–2 \times 10^{12}$ g N	$7–20 \times 10^{12}$ g N	c.
NO	NA	$5–16 \times 10^{12}$ g N	$20–60 \times 10^{12}$ g N	d.

Sources: (a) Seiler and Crutzen (1980); (b) Logan et al. (1981); (c) Seiler and Conrad (1987); (d) Ehhalt and Drummond (1982).

Table 27.2 Summary of emission ratios ($\Delta X/\Delta CO_2$; V/V; in percent) for carbon monoxide, methane, and nitrous oxide

Location	CO		CH_4		N_2O		Source
	Range	Mean	Range	Mean	Range	Mean	
Brazil	5–30	11.5	0.2–2.4	0.8	Not measured		a.
Brazil	5–15	8.5	not measured		Not measured		b.
United States	11–25	17	1.0–3.4	2.2	0.2–0.5	0.22	c.
Flaming	5–11	8	0.2–0.7	0.3	Not measured		d.
Smoldering	12–48	25	0.5–2.0	1.0	Not measured		d.
Canada							
Flaming	3–8	5.8	0.3–0.9	0.6	0.01–0.02	0.015	e.
Smoldering	5–10	8.2	0.6–1.1	0.9	0.02–0.05	0.04	e.

Sources: (a) Greenberg et al. (1984); (b) Andreae et al. (1988); (c) Crutzen et al. (1979); (d) Ward and Hardy (1984); (e) Cofer et al. (1988a, 1989).

Emission ratios are determined for the trace gases of interest by normalization with CO_2. An enhanced level of CO_2 is determined for a particular smoke sample. This is the measured in-plume CO_2 mixing ratio less the ambient background level. The same is done for the other trace gases of interest. Individual trace gas (or aerosol) enhancements are then normalized ($\Delta X/\Delta CO_2$) to the enhanced CO_2 to provide emission ratios.

Since the factors governing biomass burning and emissions are typically complex, the large range in reported emission ratios in Table 27.2 is not surprising. The lack of discrimination between emissions from flaming and smoldering combustion in most of these data reported in Table 27.2 is probably the most significant cause of the spread.

About 3.6 teragrams (10^{12} g) of smoke particles have been estimated to be released into the atmosphere each year as a result of biomass burning in the United States alone (Ward et al., 1979). Penner et al. (1991) estimates about 85 Tg of smoke particles globally are released each year by biomass burning.

Knowledge of the total mass, size distribution, and chemical composition of the particulates produced from biomass burning is important for studies of air quality, visibility, atmospheric chemistry, global budgets, and because of the potential for these emissions to impact climate (Radke et al., 1978; Stith et al., 1981; Patterson and McMahon, 1984; Ward, 1984; and Patterson et al., 1986). For example, the emissions of nitrogen, carbon, and sulfur-containing particulates during a biomass burn are probably important components of these trace species budgets and should be included in the trace species balances for these elements.

It is interesting to note that few measurements of N_2O emissions from major biomass burns have been made. Nitrous oxide, chemically inert in the troposphere, diffuses into the stratosphere where it is chemically transformed into nitric oxide and is responsible for about 70% of the annual global destruction of stratospheric ozone via the nitrogen oxide catalytic cycle (Turco, 1985). Recently, Muzio and Kramlich (1988) have reported results revealing a

N_2O producing artifact in stainless-steel grab sample bottles containing combustion products from fossil fuels. Increases in N_2O levels in these bottles appear sufficiently fast to render the frequently used grab sampling techniques for N_2O analysis highly questionable. Unfortunately, previously reported results for biomass burn produced N_2O using grab sampling techniques are also questionable due to a similar (or the same) N_2O producing artifact (see Cofer et al., 1990b). Nevertheless, N_2O does appear to be a product of biomass burning.

Although numerous factors are known to influence combustion emissions from biomass fires (see Ward and Hardy, 1984; Ward, 1984), these research efforts (Cofer et al., 1988a, 1988b, 1989, 1990a, 1990b) have been focused on characterizing the emissions of large biomass fires from a variety of different vegetations, mostly in temperate ecosystems. There are some good reasons to expect that differences in chemical products would result as a consequence of parameters largely defined by ecosystems, such as fuel size and array. In this chapter, we attempt to demonstrate the importance of fundamental ecosystem properties in influencing combustion emissions. We will largely use data obtained from fires in boreal forest, chaparral, and graminoid wetlands to support our contentions.

Fires

Three prescribed fires in Canadian boreal forest were sampled—Thomas, Peterlong, and Hill. The 375 ha Thomas prescribed fire (48°23′N, 82°52′W) was conducted on 29 July 1988 and consisted essentially of tramped balsam fir, paper birch, white cedar, aspen, and black spruce fuels. Surface fuel consumption (SFC) averaged about 7 kilograms per square meter (kg/m^2), duff consumption (DC) about 3 kg/m^2. The Peterlong Lake fire (48°8′N, 81°27′W) occurred on 22 August 1988 and consumed 217 ha of slashed balsam fir, aspen, black spruce, birch, and jackpine. SFC and DC were both around 2 kg/m^2. The Hill fire (47°57′N, 83°36′W) consisted of 486 ha of balsam fir, black spruce, aspen, and jackpine that were burned on 10 August 1989. SFC was 7.5 kg/m^2; DC equaled 3.1 kg/m^2. The Thomas and Peterlong fires are described more fully in Cofer et al. (1990b). This is the first published data from the Hill fire.

Two graminoid (grassy) wetlands fires were sampled at the Kennedy Space Center (Merritt Island National Wildlife Refuge) on 9 November 1987 and on 7 November 1988. Both burn sites were located on the east central coast of Florida (28°40′N, 80°40′W). Ap-

proximately 500 ha were burned during the 1987 fire, and about 700 ha during the 1988 fire. These grasses were predominantly in standing water. Interspersed with the grasses were small amounts of scrub oak and saw palmetto. The Kennedy fires are described more fully in Cofer et al. (1990a).

Two prescribed chaparral fires (Lodi I and II) were conducted on 12 December 1986 and on 22 June 1987 in the Lodi Canyon (34°10′N, 117°47′W) located in the San Gabriel Mountains of southern California. These fires were reported in Cofer et al. (1988a, 1989) and represent a Mediterranean-type chaparral ecosystem.

Sample Collections

Bell 204B helicopters were used to collect gas and aerosol samples in each of the fires. A twin-engined Cessna 402 fixed-wing aircraft was used once in conjunction with one of the helicopters for higher altitude smoke sampling during the 1988 Kennedy Space Center fire. The sampling protocol required the collection of upwind background air samples. Helicopter sampling typically involved the visual selection of a portion of smoke plume associated primarily with flaming, mixed, or smoldering combustion, followed by sampling traverses through the targeted portion. Sometimes two helicopter passes were conducted through the same approximate portion of smoke to acquire enough gas sample to fill a 15-L Tedlar sampling bag, constituting one sample.

High-volume pumps (Hi-Vols) were used to draw air samples through a manifold sampling system. They were loaded with either 102 millimeter (mm) diameter Teflon or glass fiber filters. These Hi-Vols served three purposes: to regulate and maintain sampling flows, to collect smoke particles on Teflon filters, and to remove particulates (glass fiber filters) before filling the Tedlar bags with smoke samples. Sampling flows (velocities) were chosen to match forward aircraft speeds for particle collections (isokinetic sampling). The Teflon filters had nominal pore size diameters of 2 μm and were efficient (≥98% capture) for collection of particles down to 0.1 μm in diameter. After several passes through the smoke plumes the helicopter returned to the on-site laboratory (or temporary helopad) where the gases were transferred from the Tedlar bags to stainless-steel grab bottles and the filters were changed and the used ones packaged. The Cessna 402 was fitted with a grab sampling system almost identical to that used for the helicopter collections. Filtered smoke samples, however, were fed directly as collected into stainless-steel grab bottles.

Chemical Analyses

Chemical analyses for carbon dioxide (CO_2), carbon monoxide (CO), hydrogen (H_2), methane (CH_4), total nonmethane hydrocarbons (TNMHC), and nitrous oxide (N_2O) were performed within several days. Aliquots were withdrawn from the stainless-steel grab bottles and analyzed by techniques described previously (Cofer et al., 1988a, 1989). Briefly, thermal conductivity gas chromatography was used for CO_2 analysis; the hot mercury oxide reduction technique for CO and H_2 (detected as mercury vapor); Ni^{63} electron capture gas chromatography for N_2O; flame ionization gas chromatography for CH_4; and our TNMHC technique (Cofer and Purgold, 1981) for hydrocarbons. The precision of the analytical techniques was assessed in the field with calibration standards and determined to be within ±2%.

Filters were analyzed for soluble inorganic ions as follows. Water/methanol extractions were performed on quartered portions of the Teflon filters. Mass densities ($\mu m/m^3$) for soluble ions were calculated based on a premeasured 0.4 m^3 per minute flow and an estimated collection time. Aliquots of extract were injected into a Dionex ion chromatograph for analyses.

Results and Discussion

Mean CO_2-normalized emission ratios ($\Delta X/\Delta CO_2$; V/V; where Δ = above background) and their standard deviations for the trace gases CO, H_2, CH_4, TNMHC (total nonmethane hydrocarbons as methane), and N_2O, for the graminoid wetlands, chaparral, and boreal forest fires studied are presented in Table 27.3. Mean emission ratios determined for flaming phase combustion can be seen to be lower than for smoldering phase combustion for each ecosystem studied. This is both consistent with observations of other researchers (Gerstle and Kemnitz, 1967; Dash, 1982; Ward and Hardy, 1984) and reasonable, since flaming typically represents the early phase of combustion where the most easily ignitable and readily combustible fuels are oxidized. Generally these fuels are small and well arrayed (accessible to air), which contributes substantially to efficient combustion. Efficient combustion of wood fuels translates into very high levels of CO_2 and H_2O production. At lower combustion efficiency, the less than fully oxidized products (e.g., CO, HCs, etc.) are produced at the expense of CO_2 and H_2O. Thus, in a very real sense, the chemical composition of trace gases and aerosols released during fires are direct indicators of the efficiency of the combustion process for wood-based materials. A logical outcome of this is that the early flaming stages of combustion would be expected to produce the lowest emission ratios (ERs) for fires, regardless of the vegetation. This conclusion is supported by our mean ERs in Table 27.3. As with all generalities, however, there are conditions during flaming combustion that can produce copious amounts of partially oxidized gases, reduced gases, and carbon-based aerosols. Examples of this would occur when strong turbulent airflows, caused by the vast amount of energy and material being released upward by a fire, sweep into a combustion zone and thermally quench vaporized fuels before combustion can occur or be completed. Zones or pockets in which oxygen concentrations have been depleted—that is,

Table 27.3 Mean emission ratios ($\Delta X/\Delta CO_2$, V/V, in percent) for graminoid wetlands, chaparral, and boreal forest prescribed fires

Fire type	Number measured	Intensity	CO	H_2	CH_4	TNMHC	N_2O
Wetlands system	20	Flaming	4.7 ± 0.8	1.0 ± 0.5	0.27 ± 0.11	0.39 ± 0.17	0.021 ± 0.007
	10	Mixed	5.0 ± 1.1	1.1 ± 0.5	0.28 ± 0.13	0.45 ± 0.16	0.024 ± 0.011
	15	Smoldering	5.4 ± 1.0	1.3 ± 0.5	0.34 ± 0.12	0.40 ± 0.15	0.031 ± 0.012
Chaparral system	9	Flaming	5.7 ± 1.6	2.2 ± 0.6	0.55 ± 0.23	0.52 ± 0.21	0.014 ± 0.004
	9	Mixed	5.8 ± 2.4	2.3 ± 0.9	0.47 ± 0.24	0.46 ± 0.15	0.020 ± 0.010
	6	Smoldering	8.2 ± 1.4	2.5 ± 0.6	0.87 ± 0.23	1.17 ± 0.33	0.039 ± 0.008
Boreal system	18	Flaming	6.7 ± 1.2	2.1 ± 0.5	0.64 ± 0.20	0.66 ± 0.26	0.018 ± 0.005[a]
	14	Mixed	11.5 ± 2.1	3.1 ± 0.9	1.12 ± 0.31	1.14 ± 0.27	0.020 ± 0.005[a]
	13	Smoldering	12.1 ± 1.9	3.2 ± 0.5	1.21 ± 0.32	1.08 ± 0.18	0.021 ± 0.005[a]

a. Does not include data from the Thomas fire.
Note: Graminoid wetlands = Merritt Island I (9 November 1987), Merritt Island II (7 November 1988); Chaparral = Lodi 1 (12 December 1986), Lodi 2 (22 June 1987); Boreal fires = Thomas (29 July 1988), Peterlong (22 August 1988), Hill (10 August 1989).

where transport of oxygen cannot keep up with fuel release and therefore oxidation processes are limited—have also been suggested (Cofer et al., 1989; Hegg et al., 1990) to explain high levels of reduced gas emissions (including H_2) sometimes measured during flaming combustion. Both effects would be most likely to occur during the early intense stages of combustion. However, the highest efficiencies for biomass combustion would generally be expected during flaming combustion.

Mean emission ratios (Table 27.3) for mixed phase combustion products in the wetlands and chaparral fires do not differ significantly from those obtained during the flaming stages. In fact, ERs obtained for flaming, mixed, and smoldering combustion in the wetlands fires are essentially the same. We believe that these results provide strong evidence of the importance of the ecosystem in determining the nature of the biomass fire emissions. Since the wetlands fires (see Cofer et al., 1990a) consisted mostly of uniform grasses (small fuels that were well arrayed) in standing water, their combustion was efficient. The low ERs obtained for CO, H_2, CH_4, and TNMHCs would indicate this. That these ERs remained low throughout the fire suggest minimal smoldering combustion, consistent with grasses in standing water. Such grasses were observed to burn quickly down to water level and extinguish.

The chaparral ecosystem consisted of small (although larger than the wetlands grasses) well-arrayed fuels (see Cofer et al., 1988a). However, the chaparral system did not prohibit combustion of any accumulated surface litter. This, coupled with the larger diameter fuels, very likely contributed to the slightly higher mean ERs obtained for smoldering combustion in the chaparral. Nevertheless, ERs for the chaparral system are still on the low end of the spectrum of most reported results from biomass fires.

Significant differences can be seen in the mean ER results for the boreal forest system. ERs are higher for each respective phase of combustion, and mixed-phase combustion in the boreal system appears to be much more in line with the results for smoldering combustion. These results are consistent when combustion in the duff layer is taken into account. Duff is a loosely compacted organic layer of plant materials, including humus, that, when burning, can be a significant source of reduced trace gases and smoke aerosol (Ward, 1983; Sandberg, 1983). Duff layers are a common ecosystem feature in temperate forest regions where forest floor decomposition is slow and are usually a significant fraction of the biomass combustion.

While smoke aerosols typically comprise less than 3% of the carbon released into the atmosphere during biomass burning, they are usually of most concern to the public because of their obvious odors, deposition, and impact on visibility. The impacts from smoke aerosol, however, can extend much further, into atmospheric chemistry (Stith et al., 1981; Darley et al., 1966), weather, and climate (Robock, 1988; Radke et al., 1988). Usually, elemental carbon production is thought to largely occur during the intense stages of combustion (Patterson et al., 1986), with mostly organic carbon aerosol associated with the latter smoldering phases (Mazurek et al., 1991; Penner et al., 1991). The fraction of aerosol production versus CO_2 (like all products of less efficient combustion) is higher during smoldering. The increased aerosol ERs would be expected to consist largely of organic carbon aerosols.

Results expressed as ERs (based on mass) for the extractable inorganic ions chloride (Cl^-), nitrate (NO_3^-), sulfate ($SO_4^=$), sodium (Na^+), ammonium (NH_4^+), and potassium (K^+), released as aerosols during fires in the three systems, are shown in Table 27.4. With two exceptions, ERs obtained for the wetlands, chaparral, and boreal systems are in very close agreement. The notably high ERs for Cl^- and Na^+ for the wetlands samples quite likely are the result of surface sea salt coatings on the vegetation resuspended as aerosol during the "explosive" combustion process. Since the Kennedy Space Center is a coastal site, this would be reasonable. The other notable is the high ER obtained over the chaparral system for NO_3^-. Hegg et al. (1987) also observed very high NO_3^- levels for the Lodi Canyon fires and suggested revolatization of previously deposited nitrogen pollutants as an explanation. Certainly, such fires could additionally resuspend surface deposited pollutants as exemplified by our Na^+ and Cl^- results for the prescribed fires at the Kennedy Space Center.

In general, emissions for the reported soluble ions, however, are relatively consistent among the fires that we have studied. Potassium (K^+) and sulfate ($SO_4^=$) are usually the ions in largest concentration. The small levels of ammonium (NH_4^+) ion we measured would appear to be in conflict with the results determined by Andreae et al. (1988) over Amazonia. However, our NH_4^+ levels were determined early in the fire emission history, and much of the ammonia (NH_3) released during the fires that we studied (see LeBel et al., 1989, 1991) may not have had time to transform into ammonium ion in the atmosphere, possibly leading to higher NH_4^+ concentrations as measured by Andreae et al. (1988).

Table 27.4 CO_2-normalized emission ratios for soluble inorganic ions

System	Number of samples	$\Delta X/\Delta CO_2$; all $\times$ (E-4)					
		Cl^-	NO_3^-	$SO_4^=$	Na^+	NH_4^+	K^+
Wetlands	10	8.7 ± 1.5	0.9 ± 0.4	3.8 ± 0.8	5.7 ± 1.1	0.3 ± 0.2	5.6 ± 2.2
Chaparral	10	0.9 ± 0.4	3.4 ± 0.9	2.9 ± 0.9	1.0 ± 0.5	0.5 ± 0.3	5.7 ± 2.0
Boreal	20	0.8 ± 0.3	0.5 ± 0.2	2.7 ± 0.8	0.7 ± 0.3	0.2 ± 0.2	2.9 ± 1.3

Conclusions

Measurements of emissions from fires in graminoid wetlands, Mediterranean-type chaparral systems, and boreal forest ecosystems suggest that ecosystem relatable parameters (e.g., fuel size) generally influence combustion emissions in predictable ways. This point is most clear when results from the wetlands fires are compared to the burning emissions from boreal forests. Generally, the inorganic fraction of the particulate emissions is relatively close in composition regardless of ecosystem. Both aerosol and trace gas emissions are strongly influenced by the phase of combustion.

Particulate and Trace Gas Emissions from Large Biomass Fires in North America

Lawrence F. Radke, Dean A. Hegg, Peter V. Hobbs,
J. David Nance, Jamie H. Lyons, Krista K. Laursen,
Raymond E. Weiss, Phillip J. Riggan, and Darold E. Ward

In this chapter we describe the results of airborne studies of smokes from 17 biomass fuel fires, including 14 prescribed fires and 3 wildfires, burned primarily in the temperate zone of North America between 34° and 49°N latitude. The prescribed fires were in forested lands and logging debris and varied in areas burned from 10 to 700 hectares (ha) (over a few hours). One of the wildfires ultimately consumed 20,000 ha and burned over a period of weeks. The larger fires produced towering columns of smoke and capping water clouds. As an indication of scale, the prescribed fires were visible only as small features in meteorological satellite imagery, but one of the wildfires studied produced a persistent, visible plume more than 1000 kilometers (km) long. Details of these fires and their fuels are given in Table 28.1.

The measurements were made aboard the University of Washington's C-131A research aircraft. This twin-engined, 20,000 kilogram (kg), propeller-driven airplane carries instrumentation for measuring the size and nature of aerosol particles, trace gas concentrations, and meteorological parameters. Major portions of the aerosol system have been described by Radke (1983) and the trace gas instrumentation has been described by Hegg et al. (1987). Details concerning the analysis of data, in addition to those presented in subsequent sections, can be found in Radke et al. (1988) and Hegg et al. (1990).

Our studies have focused on factors that could impact global climate through alteration of the earth's radiation balance. These include emissions of trace gases and smoke particles from biomass burning, the optical properties of the smoke, and the interaction of the smoke particles with water clouds.

Particle Emission Factors

Particle (and other) emission factors for the fires were computed using the carbon-balance method of Ward et al. (1982) as adapted for aircraft sampling by Radke et al. (1988). This method requires in-plume measurements of all major carbon-containing products of combustion. These include CO_2 and CO, which are of primary importance, a less important subset of hydrocarbons (CH_4, C_2H_2, C_2H_6, C_3H_6, C_3H_8, and C_4 isomers), and the smoke particles (approximately 60% carbon). We assumed that all of the carbon in the fuel was released into the plume during the fire after being converted to the different carbonaceous products of combustion listed above. We further assumed that the concentration of particulate mass remained proportionally constant with reference to the concentrations of the gaseous products of combustion listed above for every Lagrangian parcel in the plume throughout the period of our measurements. This assumption is valid for particles with aerodynamic diameters less than 3.5 μm (a 3.5 μm cutpoint cyclone was used to remove larger particles before sampling on Teflon filters). The ratio of the mass concentration of carbon in the plume to the mass fraction of carbon in the fuel gives a measure of the mass of fuel required to produce the emissions contained in a unit volume of air in the plume after combustion. The reciprocal of this quantity, when multiplied by the corresponding concentration of particle mass, gives the particle emission factor (EF). This is summarized in the following equation:

$$ EF = \frac{\overline{P}C_f}{(\overline{P}C_p) + (\overline{CO_2}C_{CO_2}) + (\overline{THC}C_{THC}) + (\overline{CO}C_{CO})} $$

(28.1)

where $\overline{P}$, $\overline{CO_2}$, $\overline{CO}$, and $\overline{THC}$ are the excess (i.e., above background) mass concentrations of aerosol particles, CO_2, CO, and total hydrocarbons, respectively; and C_f, C_p, C_{CO_2}, C_{CO}, and C_{THC} are the fractional masses of carbon in the fuel, smoke particles, CO_2, CO, and total hydrocarbons, respectively. The derivation of equation (28.1) is given by Radke et al. (1988). For the current study we assumed a fuel carbon fraction (C_f) of 0.5 (Bryam, 1959; Susott et al., this volume, Chapter 32). C_p is measured using programmed ramped temperature heating of quartz filters exposed to smoke, and analysis of the carbon

Table 28.1 Fires examined in present study

Fire	Date	Location	Size (ha)	Type of fire	Fuel
Abee	22 Sept. 1986	Montesano, Wash.	40	Prescribed	Debris from Douglas fir and hemlock
Eagle	3 Dec. 1986	Ramona, Calif.	30	Prescribed	Standing black sage, sumac, and chamise
Lodi 1	12 Dec. 1986	Los Angeles, Calif.	40	Prescribed	Standing chaparral, chamise
Lodi 2	22 June 1987	Los Angeles, Calif.	150	Prescribed	Standing chaparral, chamise
Hardiman	28 Aug. 1987	Chapleau, Ont.	325	Prescribed	Debris from jack pine, standing aspen, and paper birch
Wheat	31 Aug. 1987	Rosalia, Wash.	~10	Prescribed	Wheat stubble
Myrtle/Fall Creek	2 Sept. 1987	Roseburg, Ore.	2,000	Wildfire	Standing pine, brush, and Douglas fir
Silver	17–19 Sept. 1987	Grants Pass, Ore.	20,000	Wildfire	Douglas fir, true fir, and hemlock
Satsop	19 Sept. 1987	Satsop, Wash.	40	Prescribed	Debris from Douglas fir and hemlock
Troy	8 Oct. 1987	Troy, Mont.	70	Prescribed	Debris from pine, Douglas fir, and true fir
Battersby	12 Aug. 1988	Timmins, Ont.	718	Prescribed	Jack pine, white and black spruce
Peterlong	22 Aug. 1988	Timmins, Ont.	217	Prescribed	Jack pine, white and black spruce
Carbonado	27 July 1989	Enumclaw, Wash.	40	Prescribed	Debris from Douglas fir and hemlock
Summit	1 Aug. 1989	Grangeville, Idaho	100	Wildfire	Debris from pine, Douglas fir, and true fir
Hill	10 Aug. 1989	Chapleau, Ont.	486	Prescribed	"Chained" and herbicidal paper birch and poplar
Wicksteed	12 Aug. 1989	Hornepayne, Ont.	700	Prescribed	"Chained" and herbicidal birch, poplar, and mixed hardwoods
Mabel Lake	25 Sept. 1989	Kelowna, B.C.	29	Prescribed	Debris from hemlock, deciduous, Douglas fir

released with the pyrolysis products (Johnson et al., 1981).

The derived emission factor is strictly valid only at the point of measurement. Any process that adds or removes a pollutant between the source and the location of the measurement will change the derived emission factor for that pollutant. Such changes are negligible if the measurements are made close to the fire, but at high altitudes or at long distances downwind, mechanisms such as cloud scavenging and gas-to-particle conversion may significantly change the measured values of the emission factors. The emission factors for particle mass given below were derived from measurements in smoke columns above the fire or at short distances downwind in stabilized smoke plumes. In some cases, however, the plume had been scavenged by capping cumulus clouds.

Many of the prescribed biomass fires studied exhibited marked temporal characteristics. Most of the fires began with intense combustion and long flame lengths as the small, dry fuels (leaves, needles, twigs, etc.) burned. Particle emission factors during this period were comparatively small and the smoke appeared rather dark (in some cases very dark) at visible wavelengths. Some of the smaller fires (those which ignited rapidly) then evolved naturally into a state characterized by combustion of the predominantly

"smoldering" type. Compared with the initial "flaming" phase of combustion, the "smoldering" phase was marked by higher emission factors and light gray to white smokes. This type of progression was observed at the Mabel Lake fire (Figure 28.1), where the particle emission factor increased by a factor of almost three as the fraction of the total fire characterized by smoldering combustion increased (it took most of the first 18 minutes after ignition for the smoke to rise from the surface to the 3 km sampling altitude of the aircraft).

A similar temporal progression was observed at the Abee prescribed burn (Figure 28.2), except that the particle emission factor decreased again near the end of the fire. Here, the fuels were similar to those at Mabel Lake, the most notable difference being that the Abee fire was in the Pacific coastal region of Washington while the Mabel Lake fire was in the interior of British Columbia. In spite of the fuel similarities, the emission factors measured at the Abee fire were much greater than those measured at Mabel Lake. The Abee fire had the highest emission factors, for a sustained period, of any fire in this study. The emission factor increased rapidly over the first 25 minutes of combustion. However, during the initial 10 to 15 minutes the smoke column was capped by a vigorous cumulus cloud (about 1.5 km in depth) that may

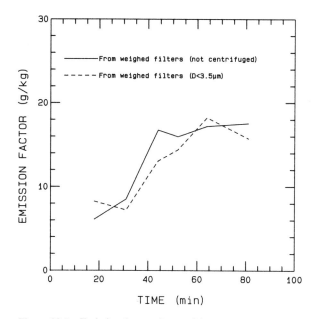

Figure 28.1 Emission factors for particle mass as a function of time after ignition for the Mabel Lake fire.

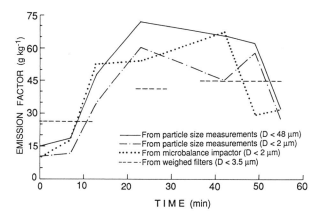

Figure 28.2 Emission factors for particle mass as a function of time from the beginning of observations on the Abee fire. A vigorous cumulus cloud capped the smoke column for the first 10 to 15 min. The measurements were made downwind of this capping cumulus in the stabilized plume of smoke (from Radke et al., 1988).

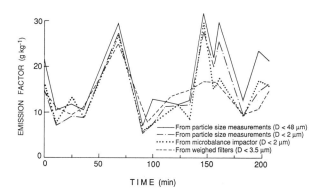

Figure 28.3 Emission factors for particle mass as a function of time after ignition for the Lodi 1 fire (from Radke et al., 1988).

have significantly scavenged the smokes (we have no quantitative estimates of scavenging by this cloud; however, as we will discuss in a later section, we observed no significant scavenging of accumulation mode smoke particles by any other cumulus cloud with a vertical depth less than 2 kilometers. Thus, we interpret this period of relatively low emission factors as being caused by some combination of efficient, predominantly flaming combustion and cloud scavenging.

Other temporal responses of this type were observed in an earlier study of smaller (<5 hectares) fires in Oregon (Radke et al., 1990) and by Ward and Hardy (in press) in their study of 38 prescribed test fires where the products of combustion (CO, CO₂, CH₄, nonmethane hydrocarbons, and particles less than 2.5 μm) were found to produce a continuum of data as a function of combustion efficiency. They define combustion efficiency as the percent of the carbon contained in the fuel that is completely oxidized to CO₂ during combustion. Combustion efficiency was found to be 90% to 95% for the flaming phase and 60% to 90% for the smoldering combustion phase in their study.

Most of the remaining fires listed in Table 28.1, particularly the larger ones, were characterized by multiple periods of ignition that produced complex temporal patterns. Lodi 1 (Figure 28.3) was such a fire.

Average particle emission factors for each of the fires are listed in Table 28.2. The three sets of values shown represent results derived from volume integrated size distributions for particles with diameters <2 μm and an assumed particle density, from smoke particle masses on weighed filters that were preceded by a centrifuge with an aerodynamic particle cut <3.5

Table 28.2 Particle emission factors (in grams of smoke per kilogram of biomass fuel burned) and single-scattering albedo and specific absorption for all fires studied

	Emission factor (derived from size distribution of particles) (<2 μm diameter)			Emission factor (from weighted filters containing particles) (<3.5 μm diameter)			Emission factor (derived from size distribution of particles) (<48 μm diameter)			Mean value of single-scattering albedo ($\bar{\omega} = \sigma_s/\sigma_E$)	Mean values of specific absorption B_A (m² g⁻¹)
	Mean	Standard deviation	Number of samples (N)	Mean	Standard deviation	Number of samples (N)	Mean	Standard deviation	Number of samples (N)		
Abee	35.1	20.3	7	37.4	9.8	3	44.5	23.1	7	ND	ND
Eagle	7.9	5.7	8	11.3	4.8	2	10.8	6.0	8	0.85 (N = 2)	0.67 (N = 2)
Lodi 1	14.3	7.3	16	13.5	4.4	13	17.6	7.9	16	0.89 (N = 13)	0.74 (N = 13)
Lodi 2	15.5	6.5	9	23.0	19.6	8	17.7	7.0	9	0.80 (N = 7)	0.68 (N = 1)
Hardiman	16.2	12.4	11	10.5	3.0	9	21.7	14.6	11	0.87 (N = 9)	0.79 (N = 5)
Wheat	43.8	—	1	ND	—	—	47.9	—	1	NA	ND
Myrtle/FallCreek[a]	19.5	12.1	10	6.1	3.1	8	29.3	16.5	10	0.84 (N = 4)	0.73 (N = 16)
Silver[a]	26.4	13.6	2	20.2	12.7	3	32.5	17.1	2	0.90 (N = 2)	0.36 (N = 2)
Satsop	24.6	7.1	2	12.0	—	1	34.3	10.8	2	0.86 (N = 1)	0.38 (N = 1)
Troy	17.1	12.4	5	9.7	3.0	5	30.2	19.5	5	0.82 (N = 3)	0.82 (N = 3)
Battersby	18.2	21.1	11	20.9	10.6	11	20.3	20.7	11	0.84 (N = 7)	0.89 (N = 4)
Peterlong	ND	ND	ND	16.9	5.1	4	ND	ND	ND	0.84 (N = 4)	1.39 (N = 2)
Hill	5.5	3.5	6	10.2	6.5	3	6.9	4.3	6	0.84 (N = 3)	0.59 (N = 4)
Wicksteed	12.9	9.1	10	10.8	4.7	5	13.3	9.1	10	0.60 (N = 6)	0.66 (N = 5)
Carbonado	10.7	7.6	7	ND	—	—	11.1	7.9	7	0.89 (N = 9)	ND
Summit[a]	17.6	4.1	2	ND	—	—	18.33	4.4	2	0.82 (N = 5)	ND
Mabel Lake	19.9	8.4	5	12.8	4.3	6	23.4	5.9	3	0.82 (N = 12)	0.41 (N = 14)
Averages	16.4	13.2	112	15.0	10.6	81	21.2	15.4	110	0.83 ± 0.11 (N = 83)	0.64 ± 0.36 (N = 73)

a. Wildfires.
NA = not analyzed.
ND = no data.

μm, and from volume integrated size distributions for particles with diameters <48 μm. The weighed filter approach is the simplest and most accurate.

As shown in Table 28.2, the emission factor for biomass smoke particles with aerodynamic diameters <3.5 μm averaged 15.0 grams per kilogram of fuel burned. Particles of this size contribute overwhelmingly to the light-scattering coefficient and visual light produced by smokes. They also have atmospheric residence times of a few days to weeks depending on the altitude to which they are lofted and whether they are scavenged by a capping cumulus cloud. During the flaming phase of combustion, a high rate of heat release can generate a very strong central convection column capable of lofting smoke particles to high altitudes (>6 km). However, lifting of this magnitude also has the potential to produce large cumulus clouds that are capable of scavenging substantial portions of the lofted smoke (see the section on cloud and precipitation scavenging of smokes). The net effect is still uncertain and needs further study.

Coarse and giant particles, which have diameters between 2.0 and 48 μm, averaged about 20% of the particulate mass. Electron microscopy showed these particles to be a mixture of coagulation products and fuel and soil debris.

It is presumed that much of the variance in the particle emission factors that we measured is due to the changing proportional contributions from different phases of combustion typical of biomass fires, although this is difficult to quantify from an airborne perspective. Because of this difficulty, it is not clear how much of the total variance remains unexplained. However, additional factors influencing particle emissions have been uncovered.

Stith et al. (1981) noted an increase in emissions with increasing fuel moisture. Cofer et al. (1989), Ward (1989), and Hegg et al. (1990) suggest that combustion efficiency in open fires also depends on the amount of oxygen delivered to the combustion zone by turbulent airflows. Thus, some fires, especially large ones, may be characterized by periods of oxygen depletion or perhaps by "thermal quenching" of oxidation reactions. In many of the fires studied the particle emission factor increased as the supply of oxygen to the combustion zone decreased. An example is the Battersby fire (Figure 28.4). Here we use the ratio of CO to CO_2, which gives a measure of the extent of oxidation in the plume, as an indicator of oxygen availability, with high ratios suggesting lim-

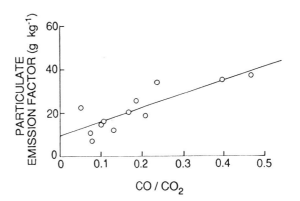

Figure 28.4 Emission factors for particles as a function of the CO/CO_2 concentration ratio in the plume from the Battersby fire. The CO/CO_2 ratio is a measure of the extent of oxidation in the plume, with higher values indicating less oxidation (from Hegg et al., 1990).

ited oxygen (further discussion of this phenomenon is presented in the section on trace gas emissions).

In view of the difficulty of accounting for all of the variability in particle emissions from biomass fires, our studies may not have bounded the full range of emission factors possible from open fires of biomass fuels. Fires with greater areal extent than those studied here could have significantly larger smoke particle emission factors. It should also be noted that the prescribed fires studied here occurred within a restricted set of meteorological and fuel moisture conditions designed to prevent fire escape yet still allow an acceptable fraction of the fuel to burn. Extreme conditions that give rise to rapidly moving wildfires may produce different smoke emission factors.

Particle Size Distributions

The size distributions of particles in the smokes displayed a high degree of consistency. The particle volume distributions in particular showed comparatively little variation in shape from one fire to another or during any one fire while near the source, although particle concentrations varied widely. Shown in Figure 28.5 are average number and volume distributions measured in the smoke plumes at distances about or <5 km from three fires. These size distributions are much like those of other urban/industrial pollution aerosol, with three distinguishable modes: a nucleation mode (diameters <0.1 μm), an accumulation mode (diameters from 0.1 to 2.0 μm), and a coarse mode (diameters >2.0 μm). For reasons probably related to different particle generation mecha-

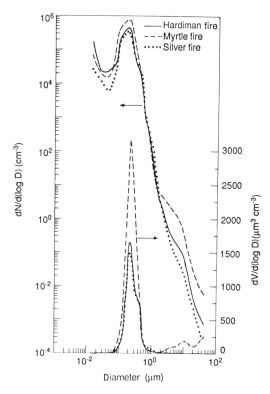

Figure 28.5 Average number and volume distributions of particles measured in the smoke plumes near (<5 km from) the sources of three fires (from Radke et al., 1988).

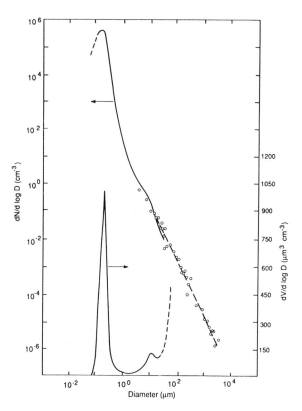

Figure 28.6 Mean number and volume distributions of particles measured in the ascending column at the Lodi 1 fire. Data points representing large, giant, and supergiant particles were measured with laser hydrometer cameras.

nisms for the accumulation and coarse mode aerosol, there is a pronounced minimum in the particle volume distributions in the region between 1.0 and 3.0 μm diameter that separates these two modes. Not only was this characteristic of the smoke from these fires, but it also seems to be generally true in the atmosphere. The accumulation mode often dominated both the number and volume distributions. Particles in the accumulation mode consisted primarily of tarry, condensed hydrocarbons that were typically spherical in shape. The fraction of accumulation mode particles that were nonspherical aggregates was rather small, especially when compared to smokes from fossil fuel fires (Radke et al., 1990b). The biomass smoke particles also contained some water-soluble inorganics (primarily $SO_4^=$ and NO_3^-; Hegg et al., 1987), which accounts, in part, for their activity as very efficient cloud condensation nuclei (Radke et al., 1978; Hallet et al., 1989).

The magnitude of the coarse particle mode showed considerable variation. For example, in the Lodi 1 fire (chaparral fuel type), there was a mode near 10 μm comprised mostly of condensed hydrocarbons aggregated with a significant fraction of soil particles (Cofer

et al., 1988). In addition to this mode, an examination of the data from our laser hydrometer cameras showed that particles of ash and soil debris up to and exceeding 1 millimeter (mm) in diameter were often present in concentrations greater than $10 \, m^{-3}$ (Figure 28.6). Although there were large variations in the concentrations of these supergiant particles, their presence was typical of all the large fires studied. Lofting of soil and ash at Lodi 1 has also been reported by Einfeld et al. (1989). The amount of coarse mode particles from fires of biomass fuels may be related to the rate of heat release and to horizontal and vertical wind velocities (Ward, 1990; Susott et al., this volume). On occasion, the mass in the coarse mode exceeded that in the accumulation mode. It is important to remember that coarse mode and especially supergiant particles are generally undercounted and, despite our efforts to improve the sampling efficiencies of these particles, were probably underestimated here as well.

The nucleation mode was the most variable mode in these smokes. It was occasionally the most promi-

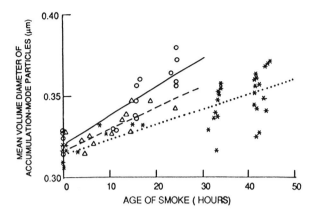

Figure 28.7 Geometric mean volume diameter of particles in the accumulation mode (0.2 to 2 μm diameter) as a function of the age of the smoke in the plume of the Silver fire. The measurements are grouped according to the value of the light-scattering coefficient due to dry particles as follows: stars and dotted lines: 2×10^{-4} m^{-1} $<\sigma_{sp}$ $<5 \times 10^{-4}$ m^{-1} ($r = 0.89$); triangles and dashed lines: 5×10^{-4} m^{-1} $<\sigma_{sp}$ $<5 \times 10^{-3}$ m^{-1} ($r = 0.77$); circles and solid lines: $\sigma_{sp} >5 \times 10^{-3}$ m^{-1} ($r = 0.81$).

nent mode in the number distribution, but more often it was much smaller than the accumulation mode, and it seldom contained significant particle mass. Its mode concentration is largely driven down by coagulation very near the fire (Fuchs, 1964) and up by continuing gas-to-particle conversion in the smoke plumes.

Coagulation also affects the mean size of the accumulation mode, although particle concentrations need to be rather large for a long period of time to obtain definitive measurements. The dependence of the coagulation rate on particle concentration is illustrated in Figure 28.7, where the smoke from a large multi-wildfire conflagration (Silver) was tracked for more than 1000 km (about 48 hours downwind from the source). The geometric mean volume diameter of the particles in the accumulation mode generally increased with age for all samples. The rate of increase was positively correlated with the light-scattering coefficient due to dry particles (σ_s), which is generally proportional to the mass concentration of particles in the accumulation mode (Waggoner et al., 1981). This result is to be expected for a process dominated by coagulation.

Intermode coagulation is also evident in the Silver fire data. Samples taken at various distances downwind of the fire showed a general decline with age in the volume concentrations of particles in the nucleation and accumulation modes, while the volume concentrations of particles in the coarse-particle mode

increased with age. This was, presumably in large part, due to coagulation, which is a sink for smaller particles and a source for larger particles, with mass being added to the small size end of the coarse mode faster than sedimentation can remove particles from the larger end. These smoke samples were normalized by the CO concentration (which we assumed was conserved in the plume) to compensate for the dilution effects of plume dispersion (for details, see Radke et al., 1990a).

Optical Properties

The optical properties of smoke particles are often described by the volume optical extinction coefficients (σ_e, σ_s and σ_a, where the subscripts indicate the form of extinction—e for total extinction, s for scattering, and a for absorption), the specific or mass normalized extinction ($B_i = \sigma_i/\rho$, where ρ is the mass concentration and i again represents the form of the extinction coefficient), and the albedo for single scattering ($\tilde{\omega} = \sigma_s/\sigma_e$). The extinction coefficients depend on the complex refractive index of the particles, their size distribution and concentration. The specific extinction and the albedo for single scattering are normalized to concentration and therefore depend only on the fundamental chemical and physical properties of the particles.

Direct, in-situ measurements of σ_e and σ_s over a range of wavelengths centered on 540 nm yielded σ_a by the relation $\sigma_a = \sigma_e - \sigma_s$ (Weiss and Radke, 1990). This permitted real-time measurements with a time resolution of about 1 second (about 80 meters of flight path). σ_e is measured with an optical extinction cell (OEC). The OEC is an enclosed, 6.4 meter path length photometer designed to measure the change in brightness of a regulated light source over a single pass of the cell. For smoke plume studies, an initial brightness measurement (I_o) was taken of background air prior to entering the plume and this was compared continuously to brightness of the light source (I) with plume air flowing through the cell. A reference light path was used to normalize all light measurements to constant lamp brightness. σ_e was calculated continuously from

$$\sigma_e = (1/L)\ln[I_o/I] \qquad (28.2)$$

where L is the length of the cell. In the aircraft, sample flow was supplied by ram air through a nearly isokinetic probe protruding ahead of the aircraft through the forward cabin wall. Near the entrance to the OEC, the sample flow was split to provide parallel flow to the nephelometer and the OEC.

Preflight calibration of the OEC/nephelometer system was accomplished by ensuring that $\tilde{\omega}$ did not deviate from unity for laboratory generated nonabsorbing particles (such as polyethylene glycol, NaCl, or $(NH_4)_2SO_4$) with an appropriate and known size distribution. This system has been used by us since December 1986 beginning with the Lodi 1 prescribed burn near Los Angeles.

The optical properties of smoke particles from biomass burning can be highly variable. They appear to depend on a number of factors, including the type of fuel burned, the intensity of the fire, flame height, and the phase of combustion (McMahon, 1983; Patterson and McMahon, 1984). In the early, flaming phase of combustion, when both high temperatures and oxygen deprivation prevail, conditions support pyrolytic production of particles consisting of high concentrations of both single spheres and sooty chains that are optically very absorbing. In the later, smoldering phase of combustion, smoke production is dominated by weakly or nonabsorbing liquid (waxy) and solid particles, consisting of organic materials with a range of volatility, and nonsooty inorganic materials.

Patterson and McMahon (1984) measured specific absorption of smoke particles collected on filters exposed to laboratory fires of pine needles. They found a substantial increase in the production of elemental (or sooty) carbon in the flaming phase of combustion relative to the smoldering phase, with the specific absorption coefficient ranging from 0.04 m^2 g^{-1} for the smoldering phase to 1.0 m^2 g^{-1} for the flaming phase. These were measurements on smokes in a laboratory environment where burning conditions were controlled. In larger fires of over 100 ha the smokes often consist of mixtures of particles from all combustion stages. However, we studied two prescribed fires where the transition from a mostly flaming phase to the smoldering phase was evident in the optical properties.

The two fires were the Troy fire (8 October 1987) and the Mabel Lake fire (25 September 1989). Both were small fires (70 and 29 ha, respectively) of downed debris consisting of mixed wood types in which there was intense burning in the initial stages of combustion and relatively uncomplicated meteorological conditions. In both fires the aerosol was optically dark during the early stages of intense burning and became less dark as the fires progressed to a predominantly smoldering phase. This result is shown

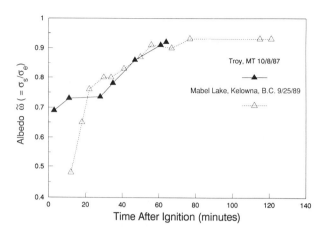

Figure 28.8 The time evolution of the single-scattering albedo $\tilde{\omega}$ for smoke from the Troy (solid triangles) and Mabel Lake (hollow triangles) prescribed burns.

in Figure 28.8, where it can be seen that the single-scattering albedo (measured at altitudes of 2 to 2.5 km) was initially less than 0.7 in both fires and approached a relatively stable value of about 0.9 during the smoldering phase. Mean values of specific absorption for both fires were approximately 0.4 $m^2 g^{-1}$, which is similar to the laboratory average values reported by Patterson and McMahon (1984).

Mean values of $\tilde{\omega}$ and B_a for each of the 17 biomass fires in this study are listed in Table 28.2. The average albedo was 0.83 ± 0.11 and the specific absorption was 0.64 ± 0.36 m^2 g^{-1}, indicating that biomass smokes are typically grey-white in visual appearance.

All of these fires produced periods (usually less than 10% of the fire's temporal extent) of dark smokes. One of the largest prescribed fires in the program (Wicksteed) produced the darkest plume cross-section (average albedo = 0.37). Since there is reason to believe that exceptionally large fires might be characterized by oxygen deficient flaming combustion, with large particle emission factors and low albedos, this result suggests that it would be desirable to take measurements in the smokes from fires larger than the largest in this study, or in a "blow up" or mass wildfire.

Finally, the visible wavelength albedo of biomass smokes appears to be modified by interactions with clouds. Smokes that pass through a cumulus cloud typically experience an increase in albedo of as much as 10% (smokes become whiter). Although the exact mechanism for increased albedo from cloud detraining smoke is not known, gas-to-particle conversion, chain aggregate collapse, coagulation, dilution and albedo-dependent scavenging may all play a role.

Biomass Smokes as a Source of Cloud Condensation Nuclei

Warner and Twomey (1967) and Hobbs and Radke (1969) noted (with some surprise) that agricultural and forestry biomass burning produced copious quantities of rather efficient cloud condensation nuclei (CCN). Further, it was found that these particles affected cloud droplet concentrations and size distributions, even for cumulus clouds with "continental" droplet concentrations. See, for example, the observations of Eagan et al. (1974) downwind of a small wildfire in coniferous forest lands in Washington state. Such measurements indicate the power of biomass combustion CCN to have a major impact on cloud microphysics (for further discussion see Radke, 1989).

Recent attempts to quantify the amounts of biomass burned globally (NAS, 1984, 1986) together with more quantitative work on smoke particle size, smoke mass yield per mass of biomass burned, and the ratio of CCN to total smoke particles (Radke et al., 1988; Hallett et al., 1989) encourage an estimate of the global CCN production from this source. Such estimates are important to the global change issue since global burning is far more extensive in the tropics and the production of CCN by industrial pollution may be less prominent.

There are a number of ways to estimate the CCN source strength, but, in keeping with the measurement uncertainties, a simple approach seems justified. If it is assumed that about 5×10^{12} CCN are produced for every gram of biomass burned (Warner and Twomey, 1967), and that about 7×10^{12} kg of biomass are burned every year (Seiler and Crutzen, 1980), then an annual average estimated global production of about 3.5×10^{28} CCN per year or about 10^{21} CCN s^{-1} from biomass burning is the result. An alternative estimate is obtained by observing that about 30%–100% of submicron biomass smoke particles serve as CCN (Hallett et al. 1989), that such submicron smoke is produced with an efficiency of about 15 g kg^{-1} of biomass burned (from Table 28.2), and that the particle mass median diameter of such smoke is about 0.3 μm (from Figure 28.5). Assuming that the density of a 0.3 μm diameter smoke particle averages 1.0 g cm^{-3} (Stith et al., 1981; Radke et al., 1990), one finds a production rate of about 1–3×10^{20} CCN s^{-1}. The uncertainties in both of these calculations are at least an order of magnitude. Thus, the result of this simple analysis is that the direct emission of particles from global biomass burning produces about 10^{19}–10^{22}CCN s^{-1}. This rate is similar to estimates of global CCN production from oceanic (Radke, 1989; Twomey and Wojciechowski, 1969) and industrial (Radke and Hobbs, 1976) sources of CCN. CCN production must have a seasonally magnified impact, especially in the tropics because of the dominance of biomass burning as a source of CCN and the seasonal nature of biomass consumption by fires. The same may apply to CCN production from biomass fires in the boreal and temperate regions. Here, however, the anthropogenic sources for CCN would be comparable to that from biomass fires.

Cloud and Precipitation Scavenging of Smokes

In the present sequence of experiments we have examined 17 biomass fires or fire complexes. Of these, 10 (Abee, Battersby, Carbonado, Hardiman, Hill, Myrtle/Fall Creek, Satsop, Troy, and Wicksteed) featured fire capping cumulus clouds and four (Hardiman, Battersby, Hill, and Wicksteed) reached cumulonimbus proportions. Of these four, the Hardiman fire produced only light to moderate precipitation and none may have reached the ground. At the other end of the intensity scale, the Hill fire produced a persistent, mature thunderstorm with heavy precipitation.

Our studies show that large fires can produce capping cumulus clouds that can substantially reduce total smoke emissions and that, contrary to the suggestion of Cotton (1985), biomass smokes are not resistant to prompt and efficient cloud and precipitation scavenging. Indeed, interactions with and removal by cloud processes appear to largely determine the residence time of smoke particles in the atmosphere since, for the accumulation mode at least, sedimentation and other forms of dry deposition are slow (Jaenicke, 1981).

Our approach to measuring scavenging efficiencies depends on a variety of sampling schemes, all of which face some difficulties for a single aircraft experiment. The choice of which approach to use is largely dependent on the size of the fire. The schemes that we have used are listed below, together with a brief critique of each:

1. *Aerosol deficit.* In principle, this is the simplest and most powerful approach. It requires measurements of an aerosol property (mass, size distribution, etc.) along with a conservative tracer believed to be proportional to the aerosol emitted (and largely insoluble in cloud water, e.g., CO) in the ascending smoke

column below cloud base and in the smoke exiting the capping cumulus cloud. The conservative tracer is used to correct smoke concentrations for dilution due to mixing with ambient air inside the cloud. The difficulty is that this method requires a period of steady-state emission.

2. *Emission factor.* We initially believed (based on ground observations) that the evolution of the particulate emission factor would be relatively slow as the phase of the fire moved from flaming to smoldering. Thus, a combination of emission factors measured below cloud base, interstitially (between cloud droplets) in the cloud, and exiting the cloud, would provide a useful measure of scavenging. Again, this requires a steady-state fire (which is uncommon); however, unlike the "aerosol deficit" method, a dilution correction is not required.

3. *Interstitial.* This requires measuring the ratio of the mass of smoke (or a surrogate for the smoke) interstitial to the cloud droplets and in the cloud droplets. After converting the amount in the cloud droplets to an equivalent air concentration (see Hegg et al., 1984), the ratio produces directly the scavenging fraction for the accumulation mode aerosol. If a surrogate for smoke is used (we used the sulfate accumulation mode aerosol coemitted integrally with the smoke), one must be satisfied that smoke is scavenged similarly, and that the surrogate is inert (i.e., it has no sources or sinks different or in addition to those for smoke).

4. *Cloud water.* This requires measuring in close time proximity the concentration of smoke particles going into the cloud and the amount of smoke in the cloud water. Using volumetric and cloud liquid water content measurements to yield equivalent air concentrations, the amount of accumulation mode aerosol scavenged is determined. This method suffers from the same problems as 3 above. In addition, errors will be introduced if the cloud is precipitating.

Our initial efforts focused on the "aerosol deficit" and the "emission factor" methods. Of these, only our work with "aerosol deficit" remains active since the "emission factor" method did not provide consistent results. The "aerosol deficit" method is most valid for small cumulus clouds (cloud depths <2 km), because the short time needed to sample just below cloud base and climb to measure the smoke escaping the cloud allows the period of required steady-state smoke emissions to be rather short.

Of the small cumulus cases examined, especially the cumulus capped wildfires (Myrtle/Fall Creek) and

the Troy prescribed burn, no significant scavenging of accumulation mode particles occurred. Significant removal of the supermicron aerosol was observed. For the large cumulus, the Battersby fire produced clear results by the "aerosol deficit" method. The aerosol property examined was the particle size distribution. Shown in Figure 28.9 are the percentage removal of smoke particles as a function of particle diameter for two sets of measurements. Here we see most of the supermicron aerosol removed (95% to 100% in one case), a removal minimum at about 0.35 μm in particle diameter, and increased removal in the center of the smoke accumulation mode mass peak near 0.2 μm. Using the same method, the Hardiman fire produced similar results, with about 80% supermicron removal, about 40% accumulation mode removal, and a removal minimum at about 0.3 μm diameter (Figure 28.10). These results are for scavenging by all mechanisms. The vertical dashed line in Figures 28.9 and 28.10 represents our theoretically predicted minimum particle size for nucleation scavenging, which was computed using the cloud droplet size distribution measured just above cloud base. Interestingly, the result of this calculation falls in the region of minimum scavenging efficiency. The mechanism responsible for the efficient removal from 0.1 to about 0.3 μm is uncertain. However, it is clear that significant smoke scavenging occurs with the onset of only

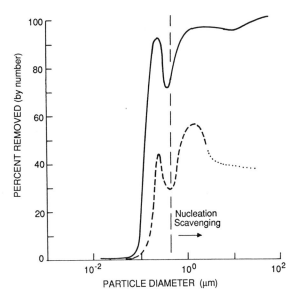

Figure 28.9 Two measurements of the percentage of the mass of smoke removed by cloud and precipitation scavenging as a function of smoke particle size for the Battersby biomass fire. The vertical line marks the calculated lower size limit of a nucleation scavenging mechanism.

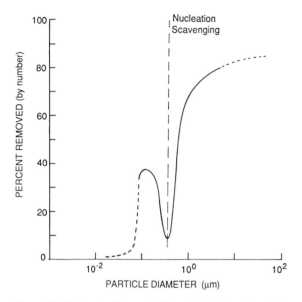

Figure 28.10 As for Figure 28.9 but for the Hardiman biomass fire.

Table 28.3 Mass scavenging efficiencies by cloud and precipitation processes for submicron diameter smoke particles

Fire	Method 3 ("interstitial")	Method 4 ("cloud water")
Hill	—	50 ± 30% 40 ± 20%
Wicksteed	75 ± 20%	80 ± 50%

Note: See text for description of these two methods.

slight precipitation and it appears to increase with precipitation intensity.

The large Hill and Wicksteed biomass fires, which produced considerable precipitation, allowed scavenging estimates by the "interstitial" and "cloud water" methods; the results are summarized in Table 28.3. The Hill fire, which provided good data for a calculation by the "cloud water" method, yielded scavenging efficiencies of 40% to 50% for accumulation mode particles. The same method applied to the Wicksteed fire showed about 80% removal. Additional measurements obtained at Wicksteed allowed the use of the "interstitial" method, which produced a nearly identical scavenging efficiency of 75% for accumulation mode particles.

The difference between the scavenging efficiencies computed for the Hill and Wicksteed fires may be related to the concentration of smoke particles within the smoke plumes. The rate of release of smoke parti-

cles at the Hill fire was approximately three times greater than that for the Wicksteed fire during the times when measurements were taken. If the accumulation mode particles were being mostly cloud nucleation scavenged, the theoretical studies of Jensen and Charlson (1984) suggest that some scavenging limits could be reached that are inversely proportional to the concentrations of the inputted aerosol. This is in accord with our observations. However, since it is not clear that nucleation scavenging was the primary route by which the accumulation mode smoke entered the liquid water phase (in fact, the Battersby data suggests that it is not), speculations on the issue are premature.

The four large Canadian prescribed fires that produced cumulus clouds have provided us with significant new experimental evidence suggesting that scavenging of biomass smokes by capping cumulus clouds can be quite efficient. The cloud scavenging picture which emerges is:

• Supermicron smoke particles are removed with considerable efficiency in all but the smallest capping cumulus clouds. These particles no doubt participate in the early production of precipitation-sized drops.

• Accumulation mode smoke particles (which represent the bulk of the smoke particles with potentially great atmospheric residence times) can enter cloud water with significant efficiency (40% to 80%) and are removed from the cloud with equal efficiency (30% to 90%) by precipitating cumulus with depths greater than 2 km.

Trace Gas Emissions

Emissions of trace gases from biomass burning are known to be an important source of several trace gases (such as CO_2, CO, and CH_4) in the atmosphere (Crutzen et al. 1985). In addition, local and regional air quality can be greatly affected by the emissions from fires (Radke et al., 1978). Nevertheless, quantitative assessments of the contributions of these emissions to the global budgets of certain trace gases are difficult to make. Biomass burning worldwide is not well quantified, either as to the area burned or to the proportion of the total mass of biomass consumed for a given area. In addition, emissions of trace gases are dependent on the fuel's chemical and physical properties. (Work is in progress at the Intermountain Fire Sciences Laboratory of the U.S. Forest Service to develop models that can be used for predicting the release of trace gases from biomass fires in fuels exhibiting different chemical and physical properties) (Ward, 1990). Despite these shortcomings, the emissions from various temperate zone biomass fires show considerable consistency.

The trace gas data gathered in our measurements on the smokes from biomass burning not only warrants attention in its own right but encourages its use for extrapolation to global scales. The contributions of emissions from fires to global trace gas fluxes can be most simply calculated using available estimates of the amounts of CO and CO_2 produced annually from biomass burning (e.g., Crutzen et al., 1985; Mooney et al., 1987). To make use of this information, the relative emission ratio of each of the trace gases we measured to CO (or CO_2) was calculated using the data listed in Table 28.4. This value was then multiplied by the estimated worldwide emission flux of CO (or CO_2) from biomass burning, to yield an estimate of the contribution of biomass fires to the global flux of each trace gas measured. Since our estimates for CO_2 emissions (see Table 28.4) are, in general, somewhat higher than published values, CO has been used as the "ratio species" in our estimations of global fluxes.

Table 28.5 lists the mean emission factor ratios for certain trace gas species, as well as the estimated fluxes from global biomass burning found using these ratios. The value for worldwide CO emissions due to biomass burning used in calculating the fluxes was 800 teragrams per year (Tg yr^{-1}) (Crutzen et al., 1985). By way of comparison, a calculation using Radke's (1989) estimate of about 10^4 Tg yr^{-1} of biomass burned globally and our emission factor for CO, yields an essentially identical global flux of 910 Tg yr^{-1} of CO.

Ozone, the first species listed in Table 28.5, is not emitted directly during combustion of biomass but is instead the product of the chemical interaction of reactive hydrocarbons and NO_x in the smoke plume (Evans et al., 1974; Radke et al., 1978). The correlation between available NO_x and the production of O_3 can be examined by studying the relationship between our emission factors for O_3 and NO_x. A linear regression of the O_3 emission factor onto the NO_x emission factor (based on the data in Table 28.5) results in an intercept of -2.2 Tg yr^{-1}, a slope of 1.4, and a correlation coefficient (r) of 0.8 significant at the 98% confidence level. This suggests that O_3 is produced in smoke plumes in a process similar to that involved in photochemical smog production. As has been discussed previously (see Dismitriodes and Dodge, 1983), the mechanism responsible for O_3 production

is apparently regulated by the availability of NO_x. If the amount of NO_x in the plume is insufficient to react with the available reactive hydrocarbons, O_3 will not be produced. Correspondingly, the production of O_3 increases with increasing NO_x.

The contribution of emissions from biomass burning to global ozone fluxes can also be assessed using our results. Total tropospheric O_3 production has been estimated to be on the order of about 4,000 Tg yr^{-1} (Hegg et al., 1990). This is two orders of magnitude greater than the value for the global flux of O_3 listed in Table 28.5 (32 Tg yr^{-1}). Consequently, biomass burning is a minor source of tropospheric ozone.

Based on our data, the estimated flux of NH_3 from global biomass burning is on the order of about 8 Tg yr^{-1}. This value represents roughly 50% of the total worldwide flux of the species, thereby substantiating a previous assertion (see Hegg et al., 1988) that emissions from biomass burning are a significant source of atmospheric NH_3.

A study of biomass fires in the Amazon Basin by Andreae et al (1988) yielded measurements of significant fluxes of particulate NH_4^+. The authors postulated that if these emissions also contained large amounts of gaseous NH_3 (which they did not measure), the contribution of biomass burning to the global flux of NH_3 would be quite substantial. Our global flux estimate for NH_3 of 8 Tg yr^{-1} is approximately twice the flux of NH_4^+ from biomass fires (as estimated by Andreae et al.). If the fluxes of NH_4^+ and NH_3 are assumed to be additive, one obtains a combined flux of about 10 Tg yr^{-1}. This flux would then be the most significant contributor to the atmospheric NH_3 reservoir (see Galbally, 1985).

The estimated worldwide flux of CH_4 listed in Table 28.5 (32 Tg yr^{-1}) agrees favorably with previously calculated values (for example, Crutzen et al., 1985, obtained a global flux of 40 Tg CH_4 yr^{-1}). Furthermore, measurements of CH_4 emissions from the 10 fires illustrate an important characteristic of biomass fires. As is discussed by Hegg et al. (1990), the ratio of CO to CO_2 in a smoke plume serves as an indicator of the extent of oxidation in a plume, with high ratios indicating only limited oxygen availability. In such an oxygen-limited environment, the amount of CH_4 produced would be expected to be higher due to greatly increased H_2 levels (the relationships between concentrations of O_2, H_2, and ratios of CO to CO_2 measured close to the ground at the Battersby fire are shown in Figure 28.11). This correlation between CO to CO_2 ratios and CH_4 emissions is illustrated by a linear regression of the CH_4 emission factor onto the

Table 28.4 Average emission factors (and standard deviations) for various trace gases from 10 biomass fires in North America

Fire	CO	CO_2	O_3	NH_3	CH_4	C_3H_6	C_2H_6	C_3H_8	C_2H_2	N-C_4[a]	N_2O	F12[b]	NO_x[c]
Lodi 1	74 ± 16	1664 ± 44	14 ± 13	1.7 ± 0.8	2.4 ± 0.15	0.58 ± 0.05	0.35 ± 0.12	0.21 ± 0.12	0.32 ± 0.05	0.11 ± 0.07	0.31 ± 0.14	0.045 ± 0.010	8.9 ± 3.5
Lodi 2	75 ± 14	1650 ± 31	0.19 ± 0.36	0.09 ± 0.04	3.6 ± 0.25	0.46 ± 0.03	0.55 ± 0.15	0.32 ± 0.12	0.21 ± 0.03	0.10 ± 0.05	0.27 ± 0.31	0.009 ± 0.003	3.3 ± 0.8
Myrtle Fall Creek	106 ± 20	1626 ± 39	−0.5 ± 0.2	2.0 ± 0.9	3.0 ± 0.8	0.7 ± 0.13	0.60 ± 0.13	0.25 ± 0.05	0.22 ± 0.04	0.02 ± 0.04	—	0.0025 ± 0.0015	2.54 ± 0.70
Silver	89 ± 50	1637 ± 103	4.7 ± 4.0	0.6 ± 0.5	2.6 ± 1.6	0.08 ± 0.01	0.56 ± 0.33	0.42 ± 0.13	0.19 ± 0.09	0.2 ± 0.1	0.27 ± 0.39	0.0	0.81 ± 0.69
Hardiman	82 ± 36	1664 ± 62	−0.5 ± 0.4	0.1 ± 0.07	1.9 ± 0.5	0.58 ± 0.09	0.45 ± 0.26	0.18 ± 0.13	0.31 ± 0.35	0.02 ± 0.04	0.41 ± 0.52	0.0004 ± 0.0003	3.3 ± 2.3
Eagle	34 ± 6	1748 ± 11	6.5 ± 2.9	—	0.9 ± 0.2	0.25 ± 0.06	0.18 ± 0.05	0.05 ± 0.02	0.08 ± 0.02	0.2 ± 0.08	0.16 ± 0.013	0.008 ± 0.003	7.2 ± 3.8
Battersby	175 ± 91	1508 ± 161	−0.9 ± 0.6	—	5.6 ± 1.7	0.9 ± 0.15	0.57 ± 0.45	0.27 ± 0.12	0.33 ± 0.06	0.07 ± 0.06	—	—	1.05 ± 1.33
Hill	90 ± 21	1646 ± 50	−0.29 ± 0.12	—	4.2 ± 1.3	0.65 ± 0.17	0.48 ± 0.17	0.15 ± 0.06	0.25 ± 0.05	0.04 ± 0.01	0.18 ± 0.06	0.0	—
Wicksteed	55 ± 41	1700 ± 82	−1.25 ± 1.16	—	3.8 ± 2.8	0.62 ± 0.40	0.51 ± 0.34	0.17 ± 0.12	0.22 ± 0.12	0.04 ± 0.03	0.22 ± 0.14	0.0017 ± 0.0022	—
Mabel Lake	83 ± 37	1660 ± 70	−0.3 ± 0.4	—	3.5 ± 1.9	0.46 ± 0.21	0.38 ± 0.21	0.11 ± 0.07	0.22 ± 0.06	0.03 ± 0.01	0.04 ± 0.05	0.0	—
Overall average emission factor	83 ± 16	1650 ± 29	2.2 ± 1.6	0.90 ± 0.43	3.2 ± 0.5	0.53 ± 0.08	0.46 ± 0.08	0.21 ± 0.05	0.24 ± 0.04	0.083 ± 0.028	0.23 ± 0.05	0.0074 ± 0.0051	3.9 ± 16

Note: With the exception of O_3 and NH_3, for which values are based on continuous measurements and filters, respectively, all values shown are based on the analyses of samples collected in steel canisters. Units are g kg^{-1}. From Hegg et al. (1990).
a. Straight chain paraffin with carbon number of 4.
b. CF_2Cl_2
c. NO_x = NO + NO_2

Table 28.5 Mean values of EF_x/EF_{CO} for biomass burning (where EF_x is the emission factor of trace gas species x and EF_{CO} the emission factor of CO) calculated from the data listed in Table 28.4[a]

Species	EF_x/EF_{CO}	Estimated flux from biomass burning worldwide (Tg yr^{-1})	Estimated contribution of biomass burning to worldwide flux of species (%)
O_3	0.04 ± 0.03	32	1
NH_3	0.01 ± 0.008	8	50
CH_4	0.04 ± 0.008	32	~7
C_3H_6	0.007 ± 0.001	6	?
C_2H_6	0.006 ± 0.001	5	?
C_3H_8	0.003 ± 0.0007	2.4	?
C_2H_2	0.003 ± 0.0006	2.4	?
$N-C_4$	0.001 ± 0.0007	0.8	?
N_2O	0.003 ± 0.001	2.4	16
F_{12}	0.0001 ± 0.00007	0.08	20
	$(0.00005 \pm 0.00003)^b$	$(0.04)^b$	$(10)^b$
NO_x	0.07 ± 0.04	56	40

Note: Also shown are estimates of the global fluxes of various trace gases from biomass burning based on the EF_x/EF_{CO} ratios and estimates of worldwide CO emissions from biomass burning from Crutzen et al. (1985). From Hegg et al. (1990).
a. The mean values of EF_x/EF_{CO} were obtained by first ratioing the values of EF_x to EF_{CO} for each of the fires listed in Table 28.4 and then averaging these 10 ratios.
b. A more conservative estimate obtained by eliminating results from the Lodi 1 fire.

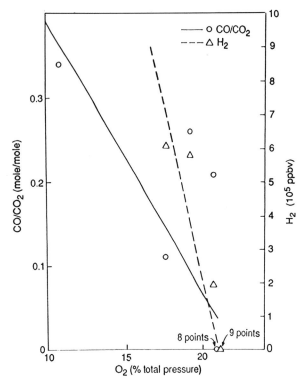

Figure 28.11 Concentration ratio of CO/CO_2 (circles) and H_2 concentration (triangles) versus O_2 in the plume from the Battersby fire. The data were obtained at or within 100 m of the ground. The lines shown are linear regressions (from Hegg et al., 1990).

ratio of CO to CO_2 emission factors using the data in Table 28.4. Such a calculation yields a correlation coefficient (r) of 0.75 significant at >98% confidence level, thus suggesting that biomass fires are often oxygen limited. A graphical presentation of this regression analysis is given in Figure 28.12.

Based on the data listed in Table 28.5, we obtain an estimated global flux of nonmethane hydrocarbons (NMHC) from fire emissions of about 15 Tg yr^{-1}. In view of the fact that only a portion of all the NMHCs have been incorporated into our data set, together with the uncertainties in the calculations, this value is in reasonable agreement with the result of about 30 Tg yr^{-1} obtained by Crutzen et al. (1985).

Measurements of N_2O emissions from combustion sources have recently come under scrutiny due to the discovery of an artifact in the use of sampling containers. Muzio and Kramlich (1988) have found that the storage of moist combustion products containing SO_2 and NO for relatively short periods can stimulate the production of high concentrations (on the order of several hundred parts per million, ppm) of N_2O in containers that previously contained no N_2O. Many of our trace gas measurements are based on steel canister samples. We do not, however, feel our N_2O measurements are artifacts. The concentrations of SO_2 required for N_2O formation in the containers are at least three orders of magnitude larger than the SO_2 concentrations we measured in the plumes from the fires in this study (Muzio and Kramlich stated that

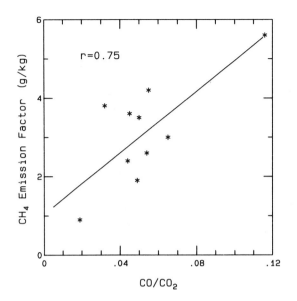

Figure 28.12 CH$_4$ emission factor versus the ratio of CO to CO$_2$ emission factors for ten of the seventeen fires studied. r is the correlation coefficient, and the line shown is a linear regression. Data is taken from Table 28.4.

SO$_2$ concentrations below 600 ppm resulted in negligible production of N$_2$O; typical SO$_2$ concentrations we have measured are only on the order of 1 to 15 ppb).

The estimated global flux of N$_2$O shown in Table 28.5 (2.4 Tg yr^{-1}) corresponds to an emission of about 1.5 Tg N yr^{-1}. This value is essentially the same as the estimate of 1.6 Tg N yr^{-1} by Crutzen et al. (1985). Our emission flux of 2.4 Tg N$_2$O yr^{-1} represents approximately 16% of the total worldwide flux of the species, thereby strengthening the contention that biomass burning contributes significantly to the atmospheric N$_2$O reservoir.

One persistent, unexpected, and not completely explained feature of our biomass fires is the presence of freons in the smoke. Because F12 cannot be produced by fires, the presence of this trace gas in the smoke plume must be due to the resuspension of F12 that had been previously deposited onto the fuel bed (Hegg et al., 1990).

Our estimated flux of F12 from biomass burning (0.08 Tg yr^{-1}) represents a significant fraction (about 20%) of the annual global emission of this species (this percentage value was calculated using the National Research Council (1983) emission estimate of 0.4 Tg yr^{-1} of F12). This suggests that the deposition of F12 may be significant globally. This hypothesis is contrary to the current belief that the only sink for F12 that is important in atmospheric budget calculations is loss by photodisassociation in the strato-sphere. While the actual method by which freons could be sequestered in biomass or soils is uncertain, we are confident that this result is not an artifact. We do, however, have some concern about the results from Lodi 1, which may be anomalous (see Table 28.4). These elevated emissions heavily influenced the average value of the F12/CO ratio used in calculating the flux of F12 due to biomass burning (for a detailed discussion of our concerns with the F12 emissions from the Lodi 1 fire, the reader is referred to the paper by Hegg et al., 1990). However, a calculation of the global flux of F12 that omits the Lodi 1 fire emission measurements yields a value of 0.04 Tg yr^{-1}. This result is still a significant fraction (about 10%) of the total global flux of F12, thus suggesting that surface deposition may indeed be an important sink for tropospheric F12.

Our estimated flux of NO$_x$ from fire emissions is 56 Tg yr^{-1} (Table 28.5). This value corresponds to about 19 Tg N yr^{-1} (Hegg et al., 1990). By comparison, a previous estimate of the global flux of NO$_x$ due to biomass burning yielded a value of 12 Tg N yr^{-1} with a range of 4 to 24 Tg N yr^{-1} (Logan, 1983). Even though our estimate of 19 Tg N yr^{-1} falls within Logan's wide range, our emission value suggests that the flux of NO$_x$ from fire emissions may be more important than has been assumed previously. Further evidence for the impact of biomass burning on the worldwide flux of NO$_x$ arises from the fact that our estimated flux value is about 40% of total NO$_x$ emissions. This indicates that NO$_x$ emissions from fires are comparable to emissions of this species from fossil fuel combustion (c.f. Crutzen et al., 1979; Logan, 1983).

A fraction of the high emission of NO$_x$ from biomass burning is evidently due to the revolatilization of NO$_x$ that had been previously deposited on the fuel bed (for further discussion of this process, the reader is referred to Hegg et al., 1988, 1990.) It appears therefore that resuspension by biomass burning of previously deposited urban/industrial NO$_x$ should be included in global flux estimates.

Summary and Conclusions

Our field studies of biomass burning indicate that rather different amounts of particles and trace gases are produced as a result of a complex interplay between fuel types (including surface and soil organics), fuel conditions (primarily moisture content), and combustion physics (meteorology, spatial distribution of the fuels, and the scale and character of the fire). In view of the variances generated by these poorly defined interactions, we hesitate to extrapolate our measurements and observations beyond the temperate zone in which they were obtained. Nevertheless, with respect to global change implications, the main features of our work may be summarized as follows:

• The particle emission factor (for particles <3.5 μm diameter) for all of the fires studied averaged 15 grams per kilogram of fuel burned.

• Biomass smoke particles have a size distribution characterized by prominent nucleation (<0.1 μm), accumulation (0.1 to 2 μm) and coarse (>2 μm) modes. The accumulation mode generally dominates smoke mass and visible light optical properties.

• The single scattering albedo of the smokes at visible wavelengths averaged 0.83 ± 0.11 and the specific absorption averaged 0.64 ± 0.36 m^2 g^{-1}.

• A large fraction (30% to 100%) by mass of the smoke particles produced in biomass burning are efficient cloud condensation nuclei (CCN).

• Estimates of the global production of CCN by biomass burning suggest a flux of about 10^{19} to 10^{22} CCN s^{-1}. This is similar to other globally important sources of CCN.

• Large biomass fires are often capped by cumulus clouds. When the cloud depth exceeds 2 km, a large fraction of the smoke particles are scavenged by the capping cloud.

• Biomass burning may account for as much as 50% of global emissions of NH$_3$.

• The combination of NH$_3$ and NH$_4^+$ emissions from biomass burning may dominate the global NH$_3$ cycle.

• Pollutants such as NO$_x$ and SO$_2$, which have been previously deposited on biomass fuels, may be volatilized and resuspended in the atmosphere by biomass burning. This effect is likely most important in urban regions.

• Resuspension of F12 during biomass burning suggests that surface deposition may be an important, and previously unrecognized, sink for F12.

• As fires increase in size, the combustion process may become oxygen deficient causing a more prolific production of smoke particles and saturated hydrocarbons (such as CH$_4$) than might otherwise be expected.

Acknowledgments

This contribution summarizes the results from a long period of research on biomass fires that would not have been possible without the able assistance of the technical, scientific, and aviation staff of University of Washington's Cloud and Aerosol Research Group. The work was supported by the following grants and contracts: Naval Research Laboratory Contract No. N0014-86-C-2246, Sandia National Laboratories Contract No. 57-0343, U.S. Department of Agriculture Forest Service cooperative agreement No. PSW-87-0020, Defense Nuclear Agency under Project IACRO 89-903, Task RA, Work Unit 00024, British Columbia Ministry of Forestry, and Intermountain Research Station, Forest Service, U.S. Department of Agriculture Agreement No. INT-89426-RJVA. We thank all of these organizations for their support.

Ammonia and Nitric Acid Emissions from Wetlands and Boreal Forest Fires

Peter J. LeBel, Stephanie A. Vay, and Patricia D. Roberts

Biomass burning has become recognized as a significant source of many trace atmospheric gases. As much as 5% of the earth's total land area experiences burning each year (National Academy of Sciences, 1984). Measurements have shown biomass burning to be a source of such important tropospheric gases as carbon monoxide (CO), methane (CH_4), carbon dioxide (CO_2), nitrous oxide (N_2O), and nitric oxide (NO) (Crutzen et al., 1979, 1985; Andreae et al., 1988; Cofer et al., 1989). As concern for the atmospheric effects of biomass burning increases, fires are becoming better characterized. Recently, elevated levels of nitric acid (HNO_3) (LeBel et al., 1988) and ammonia (NH_3) (LeBel et al., 1988; Hegg et al., 1988) have been measured in biomass burn smoke plumes. These gases play a significant role in the chemistry of tropospheric nitrogen species as well as in the cycling of nitrogen between the biosphere and the troposphere, but very little is known about their production during biomass burning.

Nitric acid, produced photochemically in the troposphere from NO and NO_2, is an atmospheric reservoir for the nitrogen oxide species which control production of ozone. Because it is removed from the atmosphere primarily by heterogeneous absorption into clouds and by wet or dry deposition, HNO_3 plays a critical role in acid precipitation (Finlayson-Pitts and Pitts, 1986). Ammonia, the primary gaseous alkaline species in the atmosphere, acts to chemically neutralize the acidity of atmospheric acids by the formation of aerosols. The wet and dry deposition of the particulate form of NH_3 and HNO_3 are significant sources of important soil nutrients ammonium (NH_4^+) and nitrate (NO_3^-).

NASA has developed a sensing technique involving collection of gases on a metal oxide denuder surface that is capable of simultaneous, sensitive measurements of HNO_3 and NH_3. This chapter presents the results of a study of the applicability of the denuder technique for measurements of HNO_3 and NH_3 in a smoke plume during biomass burning and discusses measurements made during three prescribed fires in temperate wetlands and boreal ecosystems.

Experiment Procedure

Sampling Method
A Bell 204B helicopter was used to collect gas samples from within smoke plumes at low altitudes. Low-altitude sampling permits collection in specific areas of a fire before large-scale mixing can occur and is the experimental method that has been used successfully during several prescribed fires (Cofer et al., 1989, 1990a). The sampling system consists of 1.75 centimeter (cm) diameter aluminum probes that protrude forward from the helicopter through a nose access hatch. Sampling velocities through the probes are chosen to match the typical forward speed of the helicopter, ensuring that rotor downwash is well behind the sampling inlets. A pumping system delivers outside air to a manifold for sampling by the NH_3/HNO_3 denuder system.

Measurement Technique
Simultaneous NH_3 and HNO_3 measurements were made using the tungsten oxide diffusion denuder technique described by LeBel et al. (1985, 1988), Braman et al. (1982), and McClenny et al. (1982). A denuder tube consists of a 35 cm by 6 millimeter (mm) O.D. quartz tube with a tungsten oxide (WO_3) coating vacuum deposited on its interior surface. During sampling, NH_3 and HNO_3 are collected by selective chemisorption on the WO_3 surface. A sample is thermally desorbed after collection is completed. This technique was developed for rapid, sensitive aircraft measurements of NH_3 and HNO_3 in the background troposphere (LeBel et al., 1988). For smoke plume measurements from the helicopter where space was limited, denuder tubes were used as grab samplers and were analyzed on the ground after the flights.

In the helicopter, a sample was collected by pumping air from the manifold through the denuder at 1.0

standard liter min^{-1} (SLM) using a micropump powered by a 6 volt direct current gel cell. Immediately after sample collection, the denuder tube ends were sealed. The tubes were analyzed in an on-site ground laboratory immediately after sampling. The analysis procedure involves heating a tube to 350°C with a carrier gas flowing. Collected HNO$_3$ is desorbed as NO and detected by chemiluminescent reaction of NO with ozone. Ammonia is then desorbed as NH$_3$, converted to NO by a gold-foil catalyst at 650°C, and is detected. The net integrated desorption signal is related to NH$_3$ (or HNO$_3$) concentration by calibration with a permeation tube calibration/dilution system.

The WO$_3$ denuder technique has a demonstrated sensitivity for NH$_3$ and HNO$_3$ of approximately 20 parts per trillion by volume (pptv) for a 10 minute sample. For these biomass burn experiments, sampling times for NH$_3$ and HNO$_3$ in the smoke plume ranged from 30 to 60 seconds. The precision for denuder field measurements of NH$_3$ and HNO$_3$ is approximately 10%.

Field Experiments

Three prescribed fires conducted in two diverse ecosystems (Table 29.1) were the basis for this study. A wetlands ecosystem located within the Merritt Island National Wildlife Refuge (MINWR) at the NASA John F. Kennedy Space Center, Florida (28°40'N, 80°40'W) was the site for two of the fires (KSC1, KSC2). The third prescribed fire (CAN) took place in a boreal forest in northern Ontario, Canada, in the vicinity of Peterlong Lake (48°08'N, 81°27'W).

The KSC1 fire in 1987 burned approximately half of a 1019 hectare (ha) site that had last burned three years earlier. Graminoid marsh covered 17% of the overall site and varied from areas dominated by *Spartina bakeri* to areas dominated by *Juncus roemerianus* or *Cladium jamaicense*. Mixed oak and saw palmetto scrub (*Quercus spp.* and *Serenoa repens*) were the predominant vegetation types covering 24% of the

land area within the site. Standing water, with an average depth of 10 to 20 cm, covered part of the site during the burn. Meteorological conditions during the 1987 fire included wind from the east southeast at 15 to 25 kilometers per hour (km h^{-1}), 80% relative humidity, an ambient temperature of 24°C and 40% to 60% cloud cover.

The KSC2 fire in 1988 burned approximately one third of a 2006 ha site last burned in late 1985. Although the composition of the plant communities in the 1987 and 1988 fires were similar, the proportions of each differed. In the KSC2 fire, graminoid wetlands made up 15% of the site, while the mixed oak and saw palmetto scrub covered only 6%, and cattail marsh (*Typha spp.*) dominated (30% of the total area). The site had less standing water, an average depth of 5 to 10 cm, than the previous year. Wind during the KSC2 fire was from the northeast at 6 to 10 km h^{-1}, the relative humidity was 40%, the ambient temperature was 22°C, and the sky was clear.

The northern boreal fire (CAN) consumed all of a 217 ha site that had been previously cleared of harvestable timber. Remaining trees were knocked down and allowed to dry. The fuel consisted of 38% balsam fir (*Abies balsamea*), 32% aspen (*Populus tremuloides*), 15% black spruce (*Picea mariana*), 11% birch (*Betula papyrifera*), 4% jack pine (*Pinus banksiana*), and an organic duff layer. In the week preceding the fire, 3.5 cm of precipitation fell on the site. Pockets of standing water were observed prior to the fire. During the fire, the wind was from the southeast at 17 km h^{-1}, the relative humidity was 35%, ambient temperature was 22°C, and the sky was clear.

Ambient background measurements for the KSC2 and CAN fires were made from the helicopter prior to ignition of the fire. Background measurements for the KSC1 fire, taken upwind of the fire, were found to be influenced by the smoke. For this reason, data from the KSC1 fire were reduced using preburn surface level measurements as background. During each fire, smoke plumes associated with flaming, smoldering,

Table 29.1 Characteristics of NH$_3$ and HNO$_3$ measurement experiments during boreal and wetlands fires

Date	Fire	Location	Site	Size (ha)	Sampling altitude (m)	Number of smoke plume measurements	Sampling time (sec)
November 87	KSC1	MINWR[a]	Graminoid wetlands	500	45–90	8	60
August 88	CAN	Northern Ontario, Canada	Boreal forest	220	20–45	13	30
November 88	KSC2	MINWR[a]	Graminoid wetlands	700	60	11	30

a. MINWR is Merritt Island National Wildlife Refuge, Kennedy Space Center, Florida.

or mixed combustion were targeted for sampling. The 60 second denuder sampling time required multiple passes through the same apparent portion of a plume during the KSC1 fire. For the KSC2 and CAN fires, the shorter 30 second sampling time required only a single pass through a plume. Sampling altitudes during the three fires ranged from 20 to 90 m with the lowest altitude passes being made over smoldering combustion.

Results

Average NH_3 and HNO_3 mixing ratios measured in the smoke plume as well as the background levels for the three prescribed fires are shown in Table 29.2. Note that in all experiments, significant emissions (i.e., concentrations much greater than background) of both NH_3 and HNO_3 were measured in the smoke plumes.

In Table 29.3, the average net (background subtracted) NH_3 and HNO_3 mixing ratios measured during each fire are categorized by the most prevalent type of combustion: flaming, mixed, or smoldering. It should be emphasized that this categorization is a qualitative assessment of the dominant type of com-

bustion present while a sample is being taken. Table 29.3 shows the average measured concentration as well as the standard deviation in each combustion category. The standard deviations tend to be large, in a few cases comparable to the average, illustrating the high degree of variability in the smoke plume at low altitudes over a fire.

We feel that differences between the two wetlands burn sites (see Levine et al., 1990, for more detailed site information) as well as in the meteorology account for the higher concentrations measured during the second wetlands fire. The KSC1 and KSC2 results are combined in Table 29.3 into a wetlands average for comparison with the boreal (CAN) results. For nitric acid, wetlands averages are comparable to the boreal concentrations. However, for ammonia, the concentrations measured over the boreal fire are much higher than the wetlands concentrations in all fire categories. Figure 29.1 illustrates these results. It shows boreal HNO_3 mixing ratios of the same magnitude as wetlands concentrations. Figure 29.1 also shows boreal NH_3 mixing ratios five to eight times greater than the wetlands NH_3 concentrations for all stages of combustion. The boreal fire was more intense than either of the wetlands fires. Our results are consistent with the suggestion by Cofer et al. (1989) that intensely burning fires may produce more reduced species than less intense fires. The NH_3 concentrations measured during the boreal fire were the highest NH_3 mixing ratios that we have observed. Our overall average of 32 NH_3 measurements in two different ecosystems was approximately 150 parts per billion by volume (ppbv), and our boreal forest average NH_3 concentration was approximately 350 ppbv. By comparison, Hegg et al. (1988) reported a much lower average NH_3 concentration (approximately 50 ppbv) from forest fires in the western United States.

Table 29.2 Average measured smoke plume and background NH_3 and HNO_3 concentrations during three prescribed fires

	Mixing ratio (ppbv)		
Fire	KSC1	CAN	KSC2
NH_3			
Smoke plume	20.9	355.1	78.9
Background	1.2	7.4	7.9
HNO_3			
Smoke plume	14.6	44.7	44.5
Background	0.1	0.1	3.8

Table 29.3 Summary of NH_3 and HNO_3 mixing ratios measured during boreal and wetlands fires

	Mixing ratio (ppbv)			
Fire	Flaming	Mixed	Smoldering	Mean
NH_3				
KSC1	28.6 ± 11.0	17.0 ± 7.3	4.7 ± 2.0	19.7 ± 13.1
KSC2	88.0 ± 86.4	73.6 ± 28.1	61.2 ± 39.2	71.0 ± 45.6
Wetlands average	54.0 ± 59.6	54.8 ± 36.6	45.0 ± 42.3	51.4 ± 44.3
CAN (boreal)	423.4 ± 210.0	319.7 ± 74.3	214.6 ± 80.0	347.7 ± 163.3
HNO_3				
KSC1	20.6 ± 19.8	7.8 ± 4.8	9.1 ± 2.1	14.6 ± 14.7
KSC2	83.3 ± 50.8	21.6 ± 7.3	31.8 ± 18.2	40.7 ± 34.1
Wetlands average	47.5 ± 46.7	17.0 ± 9.4	25.3 ± 18.5	30.8 ± 30.7
CAN (boreal)	65.9 ± 35.2	31.9 ± 20.6	16.6 ± 13.6	44.6 ± 33.1

Griffith (1990) has also measured high concentrations of NH_3 in the smoke plumes of burning biomass.

In all three fires, flaming combustion produced higher NH_3 and HNO_3 concentrations than smoldering combustion, both for the reduced species NH_3 as well as the oxidized species HNO_3. It is interesting to note that the ratio of the average smoldering concentration to the average flaming concentration for each fire, although less than 1.0, is greater for NH_3 than for HNO_3, suggesting that smoldering fire produces relatively more reduced species than oxidized species. (The exception to this is the KSC1 fire where the smoldering component was minimal.)

During these experiments, measurements were made by Cofer et al. (1990a, 1990b) of CO_2 mixing ratios concurrent with the NH_3 and HNO_3 measurements. We have normalized NH_3 and HNO_3 mixing ratios to mixing ratios of CO_2 in the smoke plume (i.e., ΔCO_2 = in-plume CO_2 minus background

CO_2). These NH_3 and HNO_3 normalizations or emission ratios ($\Delta X/\Delta CO_2$; X = NH_3 or HNO_3) are summarized by combustion category for the three fires in Table 29.4 and shown graphically in Figure 29.2. Nitric acid emission ratios are about the same magnitude in both boreal and wetlands fires (3 to 4×10^{-4}). By contrast, the NH_3 emission ratios from the boreal fire (an average of about 23×10^{-4}) are significantly higher than from the wetlands fires (approximately 8×10^{-4}). Our highest average NH_3 emission ratio, 31.4×10^{-4}, was observed in smoldering plumes during the boreal fire. We also note that smoldering combustion gives higher NH_3 emission ratios than flaming combustion. This is similar to results that have been observed from other reduced gases (Cofer et al., 1989, 1990b). (Again, the exception is the KSC1 fire, where there was a very limited smoldering component.)

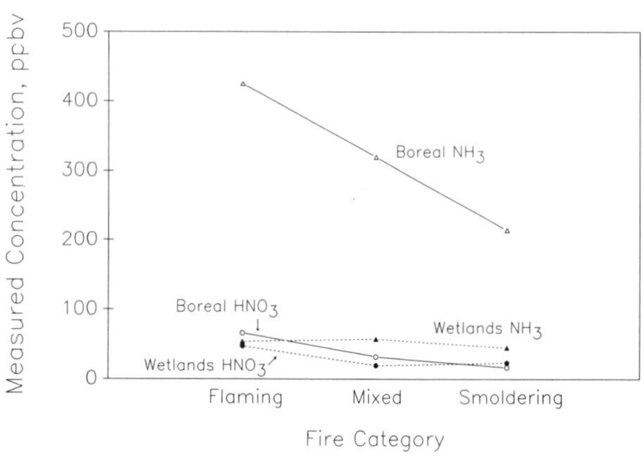

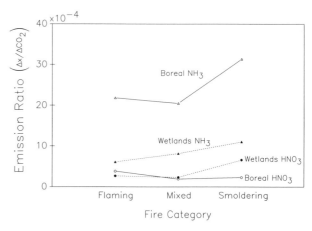

Figure 29.1 Ammonia and nitric acid mixing ratios versus fire category measured during boreal and wetlands fires.

Figure 29.2 CO_2-normalized NH_3 and HNO_3 emission ratios versus fire category for boreal and wetlands fires.

Table 29.4 Summary of CO_2-normalized NH_3 and HNO_3 emission ratios for boreal and wetlands fires

| Fire | Emission ratio, ΔNH_3 (or HNO_3)/ΔCO_2 ($\times 10^{-4}$) | | | |
	Flaming	Mixed	Smoldering	Mean
NH_3				
KSC1	6.7 ± 8.2	2.5 ± 1.1	0.7 ± 0.3	4.2 ± 6.1
KSC2	5.1 ± 6.1	11.0 ± 10.3	15.1 ± 9.0	11.2 ± 9.1
Wetlands average	6.0 ± 6.8	8.2 ± 9.1	11.0 ± 10.2	8.4 ± 8.6
CAN (boreal)	21.9 ± 6.8	20.5 ± 2.7	31.4 ± 4.9	23.1 ± 6.3
HNO_3				
KSC1	1.3 ± 1.3	1.1 ± 0.7	1.3 ± 0.3	1.2 ± 0.9
KSC2	4.6 ± 3.6	2.9 ± 0.5	8.7 ± 6.0	5.7 ± 4.8
Wetlands average	2.7 ± 2.9	2.3 ± 1.1	6.6 ± 6.1	3.9 ± 4.3
CAN (boreal)	3.4 ± 1.1	1.9 ± 0.6	2.3 ± 1.6	2.7 ± 1.2

Our average NH_3 and HNO_3 emission ratios, based on 32 measurements during prescribed fires in two ecosystems, are 13.6×10^{-4} and 3.5×10^{-4}, respectively.

Discussion

Applying our emission ratios to a new estimate by Andreae (1990) of global CO_2 production due to biomass burning of 3940 teragrams carbon per year (Tg C yr^{-1}), one can estimate a global flux of NH_3 and HNO_3 from biomass burning using the method first described by Crutzen et al. (1979). For NH_3, our average emission ratio (13.6×10^{-4}) gives a global estimate of 6.3 Tg N yr^{-1} from biomass burning. This is very close to the estimate by Andreae (1990) for NH_3 of 5.3 Tg N yr^{-1} due to biomass burning and would account for 14% of the total global atmospheric NH_3 budget (estimated as 44 Tg N yr^{-1} by Andreae, 1990). Hegg et al. (1990) estimate a global NH_3 flux of 8 Tg yr^{-1} based on oxalic acid filter measurements during forest fires in the western United States and Ontario, Canada.

Our nitric acid emission ratio is derived from measurements of HNO_3 and CO_2 at low altitudes in fresh smoke plumes. Since HNO_3 is closely coupled photochemically with tropospheric NO_x and is involved in heterogeneous chemistry in the lower troposphere, one would expect different results from measurements in an aged plume at high altitude. Andreae et al. (1988) reported HNO_3 concentrations in biomass burn haze layers at high altitude over Amazonia that were much lower (averaging less than 0.5 ppbv) than the HNO_3 levels from our low altitude smoke plume measurements (an average of about 39 ppbv). Our average HNO_3 emission ratio (3.5×10^{-4}) is not unlike the HNO_3 emission ratio (2 to 5×10^{-4}) derived from the measurements of Andreae et al. (1988). Based on our average emission ratio, we would estimate a global flux of HNO_3 from biomass burning of 1 to 2 Tg N yr^{-1}.

Conclusions

During three prescribed fires in boreal and wetlands ecosystems, we have demonstrated that the tungsten oxide denuder technique is capable of simultaneous measurements of NH_3 and HNO_3 from a helicopter flying at low altitude through biomass burn smoke plumes. In all experiments we saw emissions of HNO_3 and NH_3 from the smoke plumes that were consistently much higher than background levels. Boreal

and wetlands HNO_3 emissions were comparable (30 to 40 ppbv), although the two ecosystems were quite different. By contrast, NH_3 emissions were five to eight times greater in the boreal fire than in the wetlands fires. The average NH_3 concentration produced by the boreal fire, approximately 350 ppbv, was seven times greater than NH_3 concentrations previously reported from forest fires in the western United States. CO_2-normalized emission ratios for nitric acid averaged 3.5×10^{-4}. Ammonia emission ratios ranged from 8.4×10^{-4} in wetlands fires to 31.4×10^{-4} in smoldering plumes of the boreal fire. Our average NH_3 emission ratio of 13.6×10^{-4} relates to a global flux of NH_3 from biomass burning of 6.3 Tg N yr^{-1}, suggesting that biomass burning plays an important role in the tropospheric budget of ammonia.

Acknowledgments

These experiments were conducted in collaboration with Joel S. Levine and W. R. Cofer III of the NASA Langley Research Center; their support is greatly appreciated. We thank Randy Cofer for making his CO_2 measurement results available and Paul Schmalzer for detailed information on the ecology of the KSC sites. A special thanks to A. M. Koller and C. Ross Hinkle at Kennedy Space Center and Brian J. Stocks of Forestry Canada for their help in carrying out these experiments.

FTIR Remote Sensing of Biomass Burning Emissions of CO_2, CO, CH_4, CH_2O, NO, NO_2, NH_3, and N_2O

David W. T. Griffith, William G. Mankin, Michael T. Coffey, Darold E. Ward, and Allen Riebau

Global biomass burning plays an important role in the budgets of many species in atmospheric chemistry. Recent estimates suggest that 3 to 6 Petagrams (1 Pg = 10^{15} g) of carbon are burned globally as biomass each year (Crutzen and Andreae, 1990; Seiler and Crutzen, 1980), an amount similar in magnitude to that burned as fossil fuel. In contrast, however, biomass burning takes place less efficiently, and both fuel and fire characteristics vary widely, resulting in the emission of a far more complex mixture of species to the atmosphere. To understand the role of biomass burning in atmospheric chemistry, it is essential not only to identify and quantify these emissions but also to understand their dependence on the fuel and fire characteristics which determine them. These characteristics include such parameters as fuel composition and moisture content, fuel loadings and fire intensity, meteorology, and type of combustion (e.g., flaming, smoldering, or glowing). Only with such an improved understanding will we be able to make reliable assessments of future emissions and feedbacks from biomass burning in a world of changing climate.

A growing number of studies have made significant progress in characterising biomass burning emissions (e.g., Crutzen et al., 1979, 1985; Greenberg et al., 1984; Delany et al., 1985; Andreae et al., 1988; Cofer et al., 1989, 1990; Lobert, 1989; Hegg et al., 1990), resulting in a growing collection of data. The results are normally quoted as relative emission ratios, defined as the amount (in moles) of substance emitted divided by that of CO_2, CO, or of fuel burned; these emission ratios can then be multiplied by estimates of regional or global emissions of CO_2 or CO or fuel burned to arrive at regional or global estimates of biomass burning emissions. Natural variability in meteorology, biomass, and fire types causes these measured emission ratios to vary widely, by factors of 4 to 5 or more, and the detailed relationships between the emissions and their determining parameters remain unclear. In addition, different techniques may introduce systematic differences, especially where reactive or difficult-to-measure species (such as NO_x and NH_3)

are involved. All measurements to date have been made by conventional methods for atmospheric trace gas analysis, using either in situ instruments or grab sampling with subsequent laboratory analysis. In general, one technique can be used for the analysis of one species (or class of species); to obtain an overview of smoke plume composition, coordinated measurements from several techniques are required. This overview is, however, essential to the understanding of the many relationships between emissions and the controlling fire characteristics.

This work introduces remote sensing of biomass burning emissions using high-resolution Fourier transform infrared (FTIR) absorption spectroscopy over open paths in smoke plumes from biomass fires. There are several advantages to this type of smoke composition measurement, which address some of the disadvantages of previous measurements:

• Simultaneous measurements of a wide range of gas phase species in the smoke can be made with a single technique.

• Pseudo-continuous measurements may be made before, during, and after a fire.

• The technique is a remote sensing technique; no sampling is required, so there are no uncertainties due to loss or production of species in sample transfer lines or containers.

• Measurements are integrated over a long path through the smoke plume and therefore are not subject to small-scale local variations as may be the case with in situ and grab-sampling methods.

This technique provides an overview of the combustion products from burning not available from any other single technique used to date and can yield much valuable information on the gaseous emission products from biomass burning and the factors which control the balance of those emissions.

Using FTIR absorption spectroscopy over long, open paths, we have made measurements of the smoke composition from sagebrush and forestry slash fires in rural areas of Wyoming and Montana, and from various fuels in the stack of a large-scale combustion laboratory. In this chapter we report a preliminary analysis of the spectra obtained from the field studies and present simultaneous measurements of CO_2, CO, CH_4, CH_2O, NO, NO_2, NH_3, N_2O, and HCN in the plumes of four fires.

Experiment Procedure

In summary, the experiment consists of a Fourier transform spectrometer with suitable input optics collecting infrared radiation from a collimated IR source up to 200 meters (m) distant in field measurements, and 3 m distant in the laboratory stack. Smoke from field or laboratory fires passes through the infrared beam between source and spectrometer, and IR spectra of the smoke are recorded continually before, during, and after a fire. Subsequent analysis of the IR absorption spectra yields simultaneous concentrations of many gas-phase species, averaged over the optical path.

FTIR Spectrometer

The Fourier transform spectrometer consisted of a Michelson interferometer (EOCOM Corp.) with a Ge/KBr beamsplitter, a range of selectable optical filters, an InSb/MCT sandwich detector, and a controlling PDP-11/23 minicomputer (Mankin, 1978). Interferograms were written directly as collected onto one-half–inch magnetic tape for later processing. The interferometer had a maximum resolution of 0.06 cm^{-1} (apodized), but was used at 0.12 cm^{-1} in all measurements because absorption lines typically have widths of 0.15 cm^{-1} due to atmospheric pressure-broadening. Spectra were recorded alternately in the regions from 700 to 1900 cm^{-1} (MCT detector, no optical filter) and 1800 to 3900 cm^{-1} (InSb detector, 2.5μm short wavelength cutoff filter). A single scan required 5 to 10 seconds; spectra were collected and coadded in groups of 5, 10, 20, or 40 scans depending on the signal level and time resolution required. Time resolution for coadded scans was thus 0.5 to 4 minutes.

Field Measurements

The optical arrangement for long-path field studies is illustrated in Figure 30.1. The input optics for the interferometer consisted of a 300 millimeter (mm) diameter periscope (M1 & M2) to allow for beam height adjustment and pointing at the source, a 300

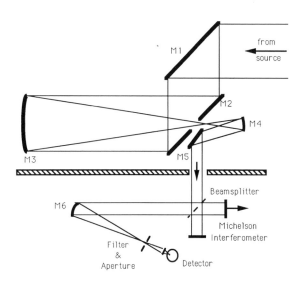

Figure 30.1 Schematic diagram of the FTIR long-path absorption spectrometer in field use.

mm diameter × 1 m focal length (f.l.) spherical mirror (M3) to focus and collect the incoming beam, and a 50 mm diameter × 200 mm f.l. spherical mirror (M4) to recollimate the collected beam into the interferometer.

The spectrometer assembly, computer, and tape drive were mounted in a 2 × 3 m enclosed trailer, with entry for the IR beam through rear double doors. Power was supplied by a 2.2 k VA Honda petrol generator. The source, located 25 to 200 m distant dependent on fire site and local topography, was a silicon carbide gas furnace igniter (Carborundum Corp., 28 V, ca. 150 Watt) powered by a small (400 VA) generator. The source was collimated by a 250 mm diameter Cassegrain telescope mounted on a tripod. When the source is focused by its telescope onto the input aperture of the receiver telescope, and the receiver telescope focuses the source telescope onto the input aperture of the interferometer, the throughput of the entire optical system is interferometer-limited at source-receiver separations less than about 140 m; at greater pathlengths light is effectively lost due to the lower throughput of the long path, and signal-to-noise ratios are degraded. In practice, under field conditions, although the two telescopes could be accurately and stably aligned with one another, the focussing was less precise, and some signal was lost. Signal was also lost through distortion of the beam by the two aluminised plate glass periscope mirrors, which were not optically flat.

In operation, a site was chosen over or upwind of the proposed fire site, and the trailer and source set up. Visual alignment of source and receiver telescopes was usually sufficient to obtain a signal through the interferometer, which could then be maximized by fine adjustment of the first periscope mirror and source telescope. The telescope alignment was then stable for several hours, although temperature changes of up to 15°C occurred through the day, necessitating some interferometer realignment. The trailer and source could usually be set up in one to two hours by two people. Spectra were recorded before ignition to determine background atmospheric composition, then continuously for the duration of the fire.

Laboratory Measurements

The laboratory experiments were carried out in the large-scale combustion laboratory of the Intermountain Fire Sciences Laboratory, U.S. Forest Service, Missoula, Montana. The combustion laboratory consists of a large room (about $12 \times 12 \times 20$ m L × W × H) with controllable temperature and humidity, a large "inverted funnel" with a 1.5 m diameter exhaust stack above the fuel bed, and an observation platform with sampling ports and windows in the stack 15 m above the fuel bed. The spectrometer and source were installed on the platform for measurement of cross-stack spectra. The IR source (the same as for field experiments) was collimated by a 75 mm diameter spherical mirror, traversed the stack twice (for a total smoke path length of 3 m), and was directed by a flat mirror into the interferometer. Signal levels were typically five times higher than in field experiments, resulting in spectra with good signal to noise ratios (>200:1) from 5 to 10 scans in 30 to 60 sec. Spectra were recorded before ignition and continuously until burnout. A typical burn time for 30 kg fuel was 1 hour.

In parallel to the IR measurements, stack temperature and flow and fuel bed weight were continuously measured, and samples were withdrawn continuously and monitored for CO_2, CO, NO, and NO_x using NDIR and chemiluminescence instruments. Filter samples for particulates were also collected.

More details of the laboratory measurements will be published separately; in this chapter, only preliminary results relevant to field measurements will be presented.

Description of Fires

Measurements were made during five prescribed burns of sagebrush scrub land (Wyoming) and coniferous forest waste (Montana). The individual fires are next described in detail.

- *F1:* 3 October 1989, Muddy Springs, 50 km south of Casper, Wyoming, on open high plains, elevation 2200 m. Vegetation burned was sagebrush scrub interspersed with thin grass. Weather was dry, sunny, and cool, with a light southerly wind following a cool change. Air temperature was below 0°C overnight, warming to 13°C and relative humidity 49% at noon. The spectrometer and source were located 107 m apart across a shallow (4 to 5 m) valley. The valley was burned from below; the fire front moved up-valley, through the IR beam to the top of the gully. Spectra were recorded from before the first arrival of smoke from below, through the flaming of sagebrush directly under the beam, and continued well through final smoldering. Ignition of sagebrush was fast, with flame heights reaching 2 to 3 m, and almost total combustion of above-ground biomass.

- *F2:* 5 October 1989, Curry Creek, 20 km south of F1 at Muddy Springs. Vegetation was as for Muddy Springs, but on a flat site. Temperature and relative humidity were 2°C and 69% respectively at 9:30, warming to 6 to 8°C during the burn. Wind was light northerly. The path was 170 m long and 2 m above ground over the flat terrain on the downwind edge of a 200×200 m plot which was burned starting from the upwind edge. Background spectra were recorded for 30 minutes before ignition, and continuously in smoke for 1.5 hours after ignition. During this fire, detailed measurements of meteorological and fire parameters and plume characteristics were also made (A. Riebau and M. Sestak, to be published).

- *F3:* 16 October 1989, Ward Canyon, Lolo State Forest, 100 km west of Missoula, Montana. Fires were of piled waste from forestry operations in mixed pine/fir/larch forest in hilly terrain. Slash piles were mainly made up of stemwood and branches 30 to 40 cm in diameter, typically 4×4 m × 2 to 3 m high, with little fine fuel, needles, or litter. The piles burned initially with good flaming combustion and little visible smoke. The 128 m optical path was set up parallel to and immediately upslope (downwind) of a line of three piles which were ignited together at 13:00. Weather was clear and dry, temperature 6 to 8°C. Spectra were recorded before and for three hours during the burning of the slash piles. Early spectra in the sequence, when flaming was intense, were complicated by emitted radiation from the hot flames and plumes.

- *F4:* 17 October 1989, site and conditions as for F3. Spectra were recorded directly of the IR emission from flames, plume, and coals of a slash pile fire similar to that described for F3. Due to the high temperatures of the IR-emitting plume gases, the spectra are more complicated than the simple absorption spectra. Analysis of these spectra will be reported separately.

• *F5:* 18 October 1989. Gold Creek, Lolo National Forest, 30 km northeast of Missoula, Montana. Weather was clear and dry. This fire was again a burn of waste from logging operations in pine/larch forest and in this case, a broadcast burn, in which the waste branches, small fuel, and some logs were left scattered on the ground, and the fire propagated across the ground through the site. The site was about 2 hectares of sloping hillside, ignited along the top edge, so that the fire spread slowly downhill, burning the whole site in about six hours. The spectrometer and source were set up across the uphill edge of the site, with a path length of 32 m which was totally immersed in the smoke from the fire below. No spectra were recorded of the background before ignition due to an instrument fault; spectra of smoke were recorded for two hours, beginning about two hours after initial ignition.

Analysis of Spectra

There are several possible approaches to the quantitative analysis of IR absorption spectra, ranging from the simple comparison of measured peak heights or areas with those from calibration spectra, to a more precise least-squares fitting of observed to calculated spectra. The preliminary analyses presented here were made by comparison of peak height or area measurements with calibration spectra. The "calibration" spectra are in this case *calculated* spectra for a wide range of compositions of the various absorbing species. The spectral calculation program for generating "calibration" spectra assumes a homogeneous path through mixtures of a given composition, pressure, and temperature. Absorption line parameters (positions, strengths, line widths, and temperature dependencies) were taken from the AFGL atmospheric line parameters compilation (Rothman et al., 1983, 1987). The calculated spectra include the effects of pressure and Doppler broadening (i.e., the Voigt profile) and apodization.

For each molecule of interest, absorption lines which were reasonably clear of other features in the spectrum were selected to minimize interferences and local baseline variations. The lines were also chosen to have absorptions of less than about 70% to avoid excessive nonlinearities. Spectra in each region containing selected absorption lines were then calculated for a range of concentrations of the molecule of inter-

est and included other absorbers in that spectral region. Peak heights and areas for the selected absorption lines were then measured from a local baseline, and a calibration curve of absorption vs concentration was generated for each line. From the calibration for each line used, the path-averaged concentration of absorber was retrieved from the corresponding peak heights in the experimental smoke plume spectra. Reported concentrations are the means of the retrievals from all lines used for each molecule. The absorption lines used in the analyses are tabulated in Table 30.1, and Figures 30.2 and 30.3 show two examples of measured and calculated spectra in the regions containing CO, CO_2, and NH_3 absorption.

The systematic errors associated with these calibrations primarily reflect the accuracy of the AFGL line parameters, typically 5% to 10% (Rothman et al., 1983, 1987). Other sources of systematic error should be small by comparison; these include the lineshape calculation in calibration spectra, pressure and temperature variations, and any systematic differences between the measured and calculated spectra due to instrumental distortions and nonlinearities, phase errors, etc. Random error is due principally to noise in the spectra, and the consequent errors in measuring local baseline and peak absorption. These errors vary

Table 30.1 Absorption lines used in quantitative analysis of spectra

Molecule	Lines (cm^{-1})	Detector	Precision (%)
CO_2	2047.5, 2056.7, 2062.8	InSb	10
CO_2	723.9, 725.5, 727.0, 733.2, 734.7, 736.2, 737.8, 739.3, 746.8, 749.8	MCT	25
CO	2050.8, 2055.4	InSb	5
CH_4	2917.6, 2926.8, 2938.2, 2948.4, 3038.5	InSb	7
CH_2O	2778.5, 2781.0	InSb	20
N_2O	2201.7, 2207.6, 2211.4, 2214.2, 2216.8	InSb	4
NO	1900.1, 1903.1, 1906.0	InSb	45
NO_2	1598.4, 1598.9, 1599.9	MCT	70
NH_3	867.9, 892.2, 951.8, 967.3, 992.7, 1007.5, 1027.0, 1046.4, 1103.4, 1140.6	MCT	15

Note: Alternate spectra were recorded using MCT and InSb detectors. The precision values are estimates of the relative errors in retrieved mixing ratios, determined from the observed noise levels in each region, and the variance in each set of determined mixing ratios for each species.

234

Chapter 30

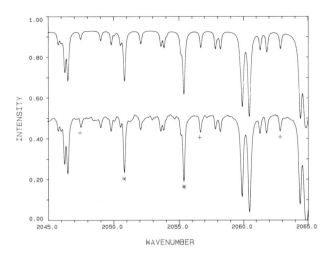

Figure 30.2 Typical spectra from 2045 to 2065 cm^{-1} showing lines used for CO_2 (+) and CO (*) analyses. Lower trace: measured spectrum beginning at 12:46 from fire F2 (InSb detector, 40 scans). Upper trace: calculated spectrum (path 177 m; temperature 7°C; pressure 0.8 atm; H_2O 4500 ppm; CO_2 690 ppm; CO 35 ppm).

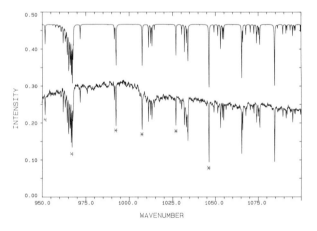

Figure 30.3 Typical spectra in the 1000 cm^{-1} region used for NH_3 analyses. Lower trace: measured spectrum beginning at 12:43 from fire F2 (MCT detector, 80 scans). Upper trace: calculated spectrum (path 177 m; temperature 7°C; pressure 0.8 atm; H_2O 4500 ppm; NH_3 1.3 ppm). NH_3 lines used in the analyses are marked with an asterisk (*).

from molecule to molecule and in different regions of the spectrum. Estimated percent errors are included in Table 30.1, and are based on both the observed signal-to-noise ratio and the measured line-to-line variation in retrieved concentrations for each molecule.

This method of analysis measures the total column amount of an absorber in the optical path. For all later discussion, these absorber column amounts are converted to path-averaged mixing ratios using the measured optical path length, atmospheric pressure, and temperature. Since the results are ultimately quoted as emission ratios relative to CO_2, any systematic error in calculating the path-averaged mixing ratios will cancel out. The actual mixing ratios in the plumes usually exceed the path-averaged values, since normally only a part of the optical path actually contains smoke at any one time.

Results

In all, 115 absorption spectra of the four prescribed burns were recorded. We present here detailed concentration and emission ratio chronologies of several species for fire F2, and averages for all four fires.

Analysis of the spectra as described above yields initially path-averaged mixing ratios for a number of species as a function of time with three- to five-minute resolution. Figure 30.4 shows the temporal changes of CO_2 and CO mixing ratios in the plume from the second sagebrush burn, F2. The coarse features of these time series can be interpreted in terms of visual observations of the fire behavior. The time marked A in the plot corresponds to the first arrival of smoke in the beam. This smoke originated from combustion about 200 m upwind. At time B, fuel directly under the beam began to burn near the source, and the spectra for the next 15 minutes correspond to smoke immediately above flames. At time C, more fuel immediately beneath the beam was ignited, after which the fire subsided to smoldering combustion.

Emission ratios relative to CO_2 are of more interpretive value than raw mixing ratios and remove some systematic errors. The emission ratio for species X is calculated as

$$\text{Emission ratio } (X) = \frac{\text{Excess } X \text{ over background (ppm)}}{\text{Excess } CO_2 \text{ over background (ppm)}}$$

and stated as mole fraction, mole percent, or parts per thousand (per mil, or mmol/mol). Figure 30.5 shows

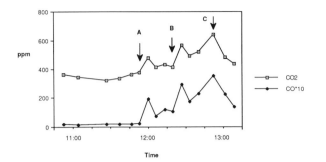

Figure 30.4 Path averaged mixing ratios of CO_2 and CO from sagebrush fire F2 as a function of time. The CO values have been multiplied by 10 for clarity. Times A, B, and C are discussed in the text.

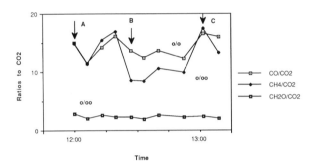

Figure 30.5 Emission ratios of carbon species CO, CH_4, and CH_2O relative to CO_2 as a function of time. For CO the units are mole percent, for CH_4 and CH_2O they are mmol/mol.

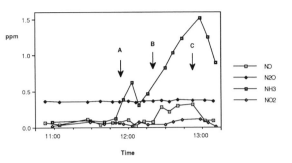

Figure 30.6 Path averaged mixing ratios of NO, NO_2, NH_3, and N_2O as a function of time for sagebrush fire F2.

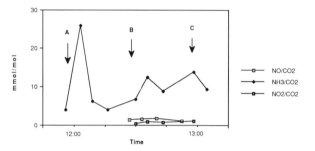

Figure 30.7 Emission ratios of nitrogen species NO, NO_2, and NH_3 relative to CO_2 as a function of time. All units are mmol/mol.

the emission ratios for CO, CH_4, and CH_2O as a function of time for fire F2. (Note change of time-scale from Figure 30.4.) In this fire, the CO/CO_2 ratios vary from 12% to 16% with a mean of 14.1 ± 1.7% (one standard deviation), with lower values during flaming combustion, and higher values during smoldering. The CH_4 emission ratio varies by a factor of 2 between flaming and smoldering (8 to 17, mean 12.7 ± 0.3 mmol/mol), while CH_2O shows a fairly constant emission ratio of 2.3 ± 0.3 mmol/mol.

Figures 30.6 and 30.7 show similar data for the nitrogen-containing species NO, NO_2, NH_3, and N_2O. For most of the duration of the fire NO and NO_2 were at or below the detection limits of about 0.1 and 0.05 ppm, respectively, and were emitted in measurable concentrations only in relatively thick smoke. Nevertheless, it is clear that ammonia is the major IR-absorbing nitrogen compound in the smoke. NH_3 is on average about five times more abundant than NO_x (NO + NO_2), and NO appears to be dominant over NO_2. Neglecting the one apparently high value of NH_3/CO_2 (see below), the mean emission ratios for

NH_3, NO, and NO_2 relative to CO_2 are of the order of 8, 1.4, and 0.8 mmol/mol, respectively. The precision of the data do not warrant further detailed analysis in terms of flaming/smoldering differences. In MCT spectra (NH_3 and NO_2) the precision in the determined emission ratios is mainly limited by the poor (25%) precision for the CO_2 measurements, which must be made from a weak band at the cutoff of the detector response. The high measured NH_3/CO_2 value at 12:03 is due to an apparently low CO_2 measurement from that spectrum. In addition, NO and NO_2, as pointed out above, are near detection limit and precision is also poor. In contrast to other techniques, however, the detection limit and precision for NH_3 are good.

The measurements of N_2O warrant special mention. As can be seen from Figure 30.6, there is no measurable change in the path-averaged N_2O mixing ratio before or during the fire. Given the precision for individual N_2O measurements of about 4% (and hence a precision for difference measurements from background of about 6%), we conclude that N_2O did not increase in the smoke at any time by more than 6%, or 20 parts per billion (ppb). For an excess CO_2 mixing ratio of 250 parts per million (ppm), this corre-

Table 30.2 Mean emission ratios relative to CO_2 for fires F1–F3, F5

Species	F1	F2	F3	F5
CO	181	141	183	196
CH_4	12.5	12.7	11.4	16.3
CH_2O	2.4	2.3	1.9	2.1
NO	0.8	1.1	<0.5	<1.5
NO_2	0.4	0.6	<0.4	<1.0
NH_3	8.4	8.3	1.4	3.3
HCN	0.3	0.1	<0.3	<0.2

Note: All values are in parts per thousand (mmol/mol), and are the average of all measurements above background for each fire.

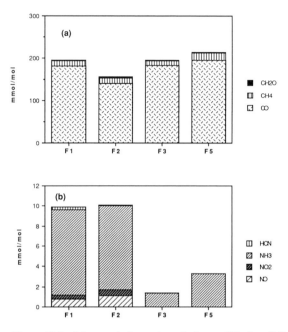

Figure 30.8 Mean emission ratios relative to CO_2 for all fires. (a) Carbon species. (b) Nitrogen species. All values are in mmol/mol, and are the average of all measurements above background for each fire.

sponds to an upper limit to the emission ratio relative to CO_2 of less than 8×10^{-5}.

Similar analyses were made for all fires, and the results are summarized as emission ratios relative to CO_2 in Table 30.2 and Figure 30.8. These data represent the averages of all measurements made above background for each fire. There are only minor differences between the relative average emissions of CO, CH_4, and CH_2O from the four fires, despite their being very different in nature. Figure 30.8b shows clearly that NH_3 is the major nitrogen-containing gas-phase species measured in all fires. Fires F1 and F2 had similar fuel and burning conditions and showed very similar emissions: a total N:C ratio of the order of 1%, with the NH_3:NO_x ratio of the order of 5:1. In F1 and F2, traces of HCN were also measured with concentrations of about 5% to 10% of NO_x. N_2O emission is insignificant by comparison to the other nitrogen species. Fires F3 and F4 showed significantly less nitrogen emission; in each case NO and NO_2 were below detection limits, and NH_3 was the only nitrogen-containing gas detected. This can be related to fuel nitrogen contents, which are discussed below.

Finally, we briefly report results from three laboratory fires which will bear on our later discussion. The fuel beds for these fires were chosen to consist of the same fuels as were burned in the field studies, and consisted of (1) sagebrush, (2) dead, dry pine needles, and (3) a mixture of needles, litter, and wood simulating the broadcast burn F5. In each case the flaming and smoldering combustion phases were well separated in time, and we have selected individual spectra recorded during each phase of each fire and analysed them for CO_2, CO, NO, NO_2, and NH_3. Figure 30.9 summarizes the results: for flaming and smoldering combustion in each fire, Figure 30.9a shows the emis-

sion ratios relative to CO_2 for CO and CH_4, and Figure 30.9b shows the emission ratios for NO, NO_2, and NH_3. The increase in CO and CH_4 emission ratios from totals of less than 3% during flaming to more than 12% during smoldering combustion is clear from Figure 30.9a. In Figure 30.9b, flaming emissions are dominated by NO and NO_2, while NH_3 is dominant during smoldering combustion. Both NO and NO_2 were below detection limits in the FTIR spectra in the smoldering phases of the pine-needle and broadcast-simulation fires. There is a suggestion that total detected nitrogen emissions are lower during the flaming phase, but in view of experimental error for these individual spectra this may not be significant.

Discussion

Technique

The study has demonstrated the utility of FTIR long-path spectroscopy for the measurement of many species in smoke from biomass burning. The principle advantages of the technique are its ability to measure many species simultaneously and pseudo-continuously in the same smoke sample, and the remote sensing advantage, i.e., that there are no uncertainties derived from the handling of smoke samples.

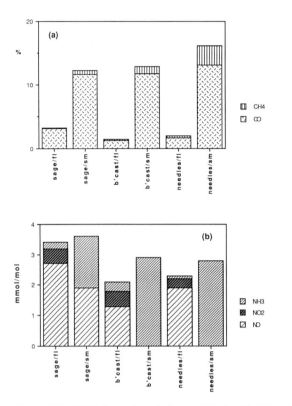

Figure 30.9 Emission ratios relative to CO_2 for (a) CO and CH_4 and (b) NO, NO_2, and NH_3 for three laboratory fires during the flaming and smoldering phases. Units are % for CO and CH_4, and mmol/mol for NO, NO_2, and NH_3. See text for details of fires.

Detection limits with the instrumentation used were of the order of 50 to 100 parts per billion (ppb) over a 100 m path for most molecules exhibiting infrared absorption spectra. The main species emitted by biomass burning which are not detected by FTIR are probably N_2 and H_2. The instrumentation could be set up in the field in one to two hours including all alignment procedures after transport to the site, could be realigned within a site in 20 to 60 minutes, and worked reliably under all field conditions encountered. In all, only one hour of data was lost due to instrument malfunction (and that turned out to be trivial!). The principle limitations are derived from the fact that the measurement is ground based and relatively immobile. It is therefore effectively restricted to prescribed burning measurements, where the fire site and approximate behavior, and wind conditions, are known in advance. Our measurements were restricted to smoke close to the ground and may therefore be biased against vigorous combustion which creates a buoyant, rapidly rising plume. The use of towers or steep terrain would mitigate this disadvantage.

There is considerable scope for improvement in both measurement precision and detection limits. Improvements are possible in both optical design and quantitative analysis of spectra. The long-path optical system used here was put together quickly using existing optical components that were not necessarily optimal for the task. The IR throughput of the entire system was about 20% of that without long-path optics, and with a more efficient design and better components, most of this loss of light should be regained, translating directly to a fivefold reduction in detection limits and a similar increase in precision. The major loss of throughput can be traced to the two large, flat mirrors used in the periscope for beam height adjustment and pointing. These mirrors were aluminized plate glass, and their lack of flatness over the large dimensions (300 × 420 mm) introduced distortion of the focussed beam. Optically flat mirrors of this dimension would be very expensive; a better solution would be to not use such mirrors at all and point the telescope directly. The second major source of throughput loss was the inability to easily focus the two telescopes under field conditions; this facility must be built into the mirror mounting designs. A further valuable improvement would be the use of two data channels in the FTIR spectrometer, permitting the recording of interferograms from both MCT and InSb detectors simultaneously instead of alternately. This would allow truly continuous measurements of all species at all times.

A second avenue for improvement lies in spectral analysis and quantitative retrieval of molecule concentrations from spectra. In the present analysis we have used the simplest option of comparing measured peak heights to those from calibration spectra. Better precision and detection limits should be possible using least-squares fitting procedures to match observed with calculated spectra. This improvement arises because many lines or a whole band are effectively used in the analysis for a given species instead of the few lines used here (Haaland and Eastman, 1980; Ying and Levine, 1989). The application of such fitting techniques to our existing spectra will be particularly useful in obtaining better excess CO_2 concentrations (which currently limit the precision of all measured emission ratios), lowering the detection limits for NO and NO_2, and giving a better measure of N_2O emissions. This analysis is underway.

Emissions from Biomass Burning

The fire-averaged results for emission ratios relative to CO_2 can be compared first to results of other studies. CO measurements range from 5% to 25%, with values below 10% being typical from aircraft-based sampling (e.g., Andreae et al., 1988; Cofer et al., 1989, 1990; Hegg et al., 1990) and values above 10% from ground-based sampling (e.g., Crutzen et al., 1979, 1985; Greenberg et al., 1984). In addition, our laboratory measurements, and those of Lobert (1989; also Lobert et al., 1990) indicate that flaming combustion generally results in $CO:CO_2$ emission ratios in the range 3% to 10%, and smoldering around 15% to 20%. These observations may be tied to the argument (Andreae et al., 1988) that airborne sampling is biased toward flaming combustion (and ground sampling toward smoldering) because vigorous, flaming combustion produces hotter, more buoyant plumes which rise in the atmosphere more than those from smoldering. In any case, our $CO:CO_2$ ratios are consistent with other ground-based measurements, although at the high end of the range, 14% to 19%, even under conditions of apparent flaming combustion. CH_4 measurements from other studies span the range 2 to 17 mmol/mol relative to CO_2, and also tend to cluster into lower values from airborne sampling and higher values from the ground. Again our own measurements, 11 to 16 mmol/mol, are at the higher end of (but well within) the range.

Our measurements of the emission ratio for formaldehyde show it to be a ubiquitous product of biomass burning, with a fairly constant emission ratio of about 2.3 mmol/mol relative to CO_2. Although aldehydes, ketones, organic acids, and other oxygenated hydrocarbons are known to be produced in combustion in general, we are not aware of other similar measurements of CH_2O for comparison. Formaldehyde, as a source of HO_2 radicals in the atmosphere (Warneck, 1988; Finlayson-Pitts and Pitts, 1986), will play an important role in photochemistry and ozone production in aging plumes from biomass fires.

The results for ammonia and nitrous oxide are perhaps of most interest. In all fires, NH_3 appears to be the major nitrogen-containing gas emitted, about five times greater than NO_x emissions. We measure an $NH_3:CO_2$ ratio of about 8 mmol/mol from sagebrush fires and 1 to 3 mmol/mol from forest slash. Hegg et al. (1988, 1990) report values for $NH_3:CO_2$ of 0.14 to 3.2 mmol/mol from a range of fires, LeBel et al. (1988) measured 0.4 to 2.2 mmol/mol from a Florida wetlands fire, and Andreae et al. (1988) infer 1 to 2 mmol/mol from aircraft measurements of NH_4^+ aerosol in the dry season over Amazonia. Our results thus appear to show significantly greater emissions of NH_3 than previous studies. The remote sensing technique

is, however, very suitable for ammonia measurements, because the absorption bands are strong, isolated, unambiguous, and easy to measure (see Figure 30.3) There is also no uncertainty in NH_3 measurements due to collection efficiency or losses during sampling and measurement. Ammonia is a notoriously difficult compound to sample and measure. A significant partitioning of fuel nitrogen into NH_3 instead of NO, NO_2, or N_2O following burning will have important ramifications for the cycling of nutrient nitrogen, because the NH_3 has a very different atmospheric chemistry than the oxygenated species. The NH_3 measurements will be treated in more detail in a forthcoming publication.

The spectra also indicate an upper limit for N_2O emission ratios of 8×10^{-5} relative to CO_2. This value is significantly lower than earlier measurements by other groups (Crutzen et al., 1979, 1985; Cofer et al., 1989, 1990; Hegg et al., 1990), but more in agreement with recent values (Elkins et al., 1990; Lobert, 1989). The discovery of artifact N_2O formation in canister samples from fossil fuel burning plumes (Muzio and Kramlich, 1988; Muzio et al., 1989) has lead to careful revision of measurements and techniques for N_2O from combustion sources, and it is recognized that this artifact may also occur in biomass burning samples. Recent measurements of $N_2O:CO_2$ emissions are of the order of $1–2 \times 10^{-5}$ and lower (Elkins et al., 1990; Lobert, 1989). A soil contribution to N_2O emission is also possible, and may help explain higher observed values (Cofer et al., 1990). We believe our upper limit of 8×10^{-5} for the $N_2O:CO_2$ emission ratio to be significant in that it is the only measurement made which is not subject to any doubt concerning sampling artifacts.

The emissions of all nitrogen species from the four field fires studied (Figure 30.8b) can be compared to the nitrogen contents of the fuels burned. Both sagebrush fires show total emissions of $N:CO_2$ of close to 1%, made up principally of NH_3, NO, and NO_2, with traces of HCN and negligible N_2O. Taking a CO_2:total C ratio of about 0.85, this corresponds to an overall N:C mole ratio of about 0.85% for all measured species. We have measured the elemental compositions of samples of grass, sagebrush, and litter from the F2 site: the samples were divided into separate "fractions" (leaves, twigs, wood, etc.), and the weighted mean compositions determined; the N:C ratios in mole percent were as follows:

Grass and twigs	1.2%
Sagebrush	1.0%
Litter	2.4%

Depending on the fractions of grass, sagebrush, and litter burned, the mean N:C ratio in all burned biomass thus probably lies around 1.3% to 1.5%, which can be compared to the observed ratio of about 0.85%. We can thus account for about 60% of the total fuel nitrogen with the sum of measured NH_3, NO, NO_2, N_2O, and HCN. Lobert et al. (1990) have found from laboratory biomass burning studies that the nitrogen budget cannot be balanced; in general, less than 60% of the total fuel nitrogen can be accounted for in the known nitrogen-containing emissions, ash, etc. They postulate N_2 to be the missing species by default; our results for the sagebrush fires, despite the increased amount of NH_3 observed, would support this conclusion. N_2 cannot be detected in the IR and in the presence of atmospheric N_2, and we have not yet seen any spectral features which might be attributable to other major nitrogen-bearing species.

For the two forest waste fires we have not measured fuel nitrogen contents, but they are expected to be much lower than those for sagebrush, of the order of 0.2% to 0.4% for wood and somewhat higher for bark, twigs, and litter. The total N:C emissions of about 0.1% for the slash pile fires (F3) and about 0.3% for the broadcast burn (F5) are in qualitative agreement with these estimates, but again the measured N:C ratios in emissions are lower than fuel N:C ratios. In the case of the slash piles, which consisted mainly of stemwood and large branches, there was very little small fuel, needles, or litter. In contrast, the ground layer of litter and the top layer of soil burn with the broadcast fire, with consequent higher emissions of nitrogen detected.

The simultaneous measurements of several species provided by FTIR long-path absorption spectroscopy also provide valuable insight into the relationships between biomass burning emissions and fire parameters. Our laboratory studies (Figure 30.9), and those of others (e.g., Lobert, 1989; Lobert et al., 1990), consistently show that flaming combustion produces more highly oxidized products such as CO_2 and NO, while smoldering leads to increased emissions of more reduced species such as CO, CH_4, and NH_3. Measurements of emission ratios for flaming versus smoldering combustion, and estimation of the proportion of flaming versus smoldering combustion in any one fire or fire type, are two of the most important quantities required for good estimates of local, regional, and global emissions of trace gases from biomass burning.

Figures 30.4 and 30.5, combined with visual observations as outlined above, demonstrate semi-quantitatively the variation of emission ratios with flaming and smoldering combustion. At time B the decrease of both CO and CH_4 emission ratios with the onset of flaming combustion can be clearly seen. For CH_4 it is particularly marked, with a decrease of a factor of two from 16 to 8 mmol/mol, while for CO the decrease is less marked, from 16% to 13%. The CO values are at the high end of the range compared to other measurements, and are discussed below together with results from other fires. In Figure 30.7, NH_3 appears to increase at the start of flaming combustion, but the precision in this case is lower and this observation may or may not be significant.

The results from the laboratory studies provide more insight into flaming-smoldering differences, since in the laboratory fires these two combustion phases are well separated in time. Figure 30.9 clearly shows the increase in CO and CH_4, and the shift from NO/NO_2 to NH_3 emission, in the smoldering phase compared to flaming. While NO and NO_2 are the dominant detectable nitrogen species during flaming combustion, in two cases they fall below the detection limits during smoldering and NH_3 is dominant. There is also a suggestion in the data that the total detected nitrogen emission is lower during flaming; this may be related to possible N_2 emission but may also be an artifact of the measurement precision. A more complete analysis of all spectra of the laboratory fires will shed more light on these issues. The analysis of these spectra is considerably more time consuming because of the variation in stack temperature from 30°C to 150°C during a fire and will be carried out in full using the spectral fitting techniques outlined in the previous section.

Acknowledgments

We have pleasure in thanking many people for their willing help and cooperation in making this project possible in a tight schedule: Jack Fox and the NCAR machine shop for construction and mounting of the optical components with the FTIR spectrometer; the operational fire staff of the Bureau of Land Management, Casper, Wyoming, and U.S. Forest Service, Superior and Missoula, Montana, for lighting the fires and stopping the trailer from burning down; and staff of the Intermountain Fire Sciences Laboratory (Ron Babbit, Lyn Weger, Pat Boyd, Ron Susott) for support in all Missoula measurements. David W. T. Griffith is grateful for a senior Fulbright fellowship and to the National Center for Atmospheric Research for financially supporting the study. The National Center for Atmospheric Research is sponsored by the National Science Foundation.

Aerosol Characterization in Smoke Plumes from a Wetlands Fire

David C. Woods, Raymond L. Chuan, Wesley R. Cofer III, and Joel S. Levine

Emissions from biomass fires are potentially a large and important source of trace gases and aerosols to the earth's atmosphere. Biomass burning may involve as much as 2% to 4% of the land area of our planet each year (NAS, 1984). The effects of biomass burning on atmospheric chemistry, weather, and climate have not yet been adequately assessed at either regional or global scales (Crutzen et al., 1985; NAS, 1984; Greenberg et al., 1984; Delmas, 1982; Seiler and Crutzen, 1980; and Crutzen et al., 1979). Smoke aerosols released from biomass fires are a complex chemical mixture of organic and inorganic material, simultaneously involving the gaseous, liquid, and solid phases. About 3.6 teragrams (1 Tg = 10^{12} g) of smoke particles have been estimated to be released into the atmosphere each year as a result of biomass burning in the United States alone (Ward et al., 1979). Penner et al. (1991) have estimated that about 85 Tg of smoke is released globally each year. Particulates released during biomass burning impact air quality, visibility, atmospheric chemistry, and the global budgets of some of the constituents. Because of the potential for these emissions to impact climate (Radke et al., 1978; Stith et al., 1986), knowledge of these emissions is important. Robock (1988) has recently documented the effect of forest fire smoke emissions in reducing surface temperatures in northern California.

In this chapter, we present results from airborne measurements of aerosol mass loading, size distribution, and elemental composition obtained in a smoke plume from the burning of vegetation at a Florida wildlife refuge. These are important parameters in assessing the impact of biomass burning on the atmosphere. Our results show that there was a high concentration of carbon-containing aerosols and salt crystals in the 0.1 μm to 0.2 μm size range, giving rise to a relatively strong fine particle size mode, during the hot flaming phase of the burning, compared to that during the smoldering phase, when a higher concentration of coarse particles were produced. We also found that the composition and morphology of the aerosols differed with size. We used the aerosol mass concentration along with CO_2 concentrations to calculate ratios of aerosol and CO_2, which we found to be higher for the smoldering phase than for the flaming phase of combustion.

Merritt Island Prescribed Fire

The prescribed burn was conducted by the U.S. Fish and Wildlife Service on 7 November 1988 at the NASA Kennedy Space Center (also the Merritt Island National Wildlife Refuge). The burn site was located on the east central coast of Florida (about 28°40′N, 80°40′W), in a subtropical climate (NASA, 1979). Approximately 700 hectares (ha), consisting mostly of graminoid (grassy) wetlands largely covered by *Spartina bakeri* and *Juncus roemerianus* vegetation was burned. About 5 to 10 centimeters (cm) of standing water on average covered the soils. Average elevation was about 1 meter (m) above sea level, and the burn was conducted in predominantly live fuel. Mixing layer height for the prescribed burn was about 1.5 km, and the smoke did not penetrate the mixing layer. Winds were out of the northeast at 6 to 10 kilometers per hour (km/h), temperature was 22°C with zero cloud cover, and the relative humidity ranged from 40% to 45%.

Measurement Procedure

The quartz crystal microbalance (QCM) multistage cascade impactor classifies aerosol particles by aerodynamic size and measures aerosol mass loading on each of several impactor stages, corresponding to different size bands, with piezoelectric crystals arranged in a microbalance circuit (Chuan, 1976). The aerosol particles captured on the impactor stages may be examined by scanning electron microscopy (SEM) and energy dispersive X-ray analysis (EDXA) to determine their elemental composition and morphology. The results of these measurements may be presented as aerosol size distributions (mass loading as a func-

Figure 31.1 Smoke plume from the biomass burning at the Merritt Island National Wildlife Refuge near Kennedy Space Center in Florida on 7 November 1988.

tion of particle aerodynamic diameter) and elemental composition as a function of size. The QCM has been used in this manner to characterize aerosols in volcanic eruption plumes (Woods and Chuan, 1982) and background stratospheric aerosols (Chuan and Woods, 1984).

The measurements in the smoke plume at the wildlife refuge were obtained by operating a QCM aboard a twin-engine Cessna 402 aircraft as it flew through various segments of the plume. The plume is shown in Figure 31.1 during the hot flaming phase of the combustion as the aircraft approached for sampling. It leveled off in an atmospheric inversion layer at about 1400 m forming the long thin layer shown spreading to the left in the photograph. Visual observations from the aircraft suggested the cloud was heavily concentrated with particulates and vaporized materials. During the flaming phase of combustion, the aircraft made five penetrations through the main part of the plume at altitudes ranging from 900 to 1370 m above the surface and then one pass downwind within the long thin inversion layer. An integrating nepholometer, which measures the aerosol scattering coefficient,

was used, primarily, in this case, to map the plume profile and to determine the boundaries. Its fast response gave immediate indications of when the aircraft entered and exited the plume as well as variations of aerosol concentrations within the plume. Wide variations in aerosol concentrations along a given flight path were observed.

After about 45 minutes of measurements the sampling was interrupted until the smoldering phase started. Before sampling was resumed the QCM crystals (impactor stages) were replaced with new ones so that the samples collected in the smoldering plume would be separate from those collected during the flaming phase. When we resumed sampling, the smoke plume was much closer to the ground, and its boundaries were much less well defined. We made several passes through the plume at altitudes between 245 and 150 m above the ground.

Figure 31.2 shows two size distribution plots of $\Delta C/\Delta \log D$ vs D, where C is the aerosol mass loading in $\mu g/m^3$ and D is the aerodynamic diameter of the aerosol particles in units of μm. The plot in Figure 31.2a is typical of the size distributions measured

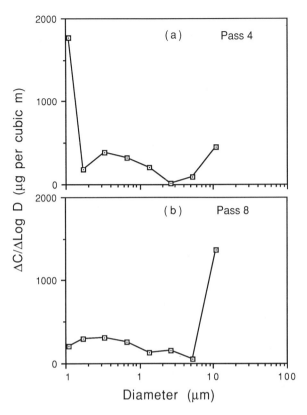

Figure 31.2 Aerosol size distributions in smoke plume from the Merritt Island Wildlife Refuge fire. (a) Measured during the hot flaming combustion. (b) Measured during smoldering.

Figure 31.3 SEM photograph showing particles collected from the smoke plume on the impactor stage which collects particles of 0.1 μm aerodynamic diameter with 50% efficiency. (a) Aggregates made up of smaller particles (about 0.1 μm) containing carbon. (b) Particles collected near the center of the impaction site similar in composition to those in (a) along with a considerable number of salt crystals.

during the hot flaming phase of the combustion. It was obtained during a penetration through the plume at 1130 m above the surface. It has a trimodal feature with a high concentration of fine particles peaking near 0.1 μm. These small particles were probably produced by the efficient burning at high temperature. The middle mode in the size distribution, in the range between about 0.2 and 2.0 μm, is relatively low in concentration and was probably formed by a combination of coagulation of smaller particles and inefficient burning. The large size mode (coarse mode) appears to peak beyond 10 μm and is probably made up of particles formed from incomplete burning mixed with some crustal particles blown up into the plume from the surface.

Figure 31.2b shows a typical size distribution measured as the aircraft flew through the plume during the smoldering phase of the burn. This measurement was obtained closer to the ground at about 150 m. The major difference between this size distribution and the one above is that there is no fine particle mode near 0.1 μm and there is an enhancement in the coarse mode. The decrease in the fine particle mode

and increase in the coarse mode in the smoldering plume are perhaps due to the lower temperature inefficient burning producing many condensible organic aerosols and few soot aerosols.

Aerosol samples collected in each impactor stage were examined by SEM and analyzed by EDXA. Figure 31.3 is a SEM photograph showing particles collected on the impactor stage, which collects particles larger than 0.1 μm aerodynamic diameter, corresponding to the large peak in the size distribution in Figure 31.2a. The particles in Figure 31.3a impacted near the outer edge of the circular impaction site.

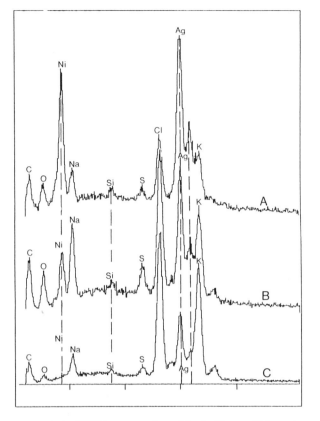

Figure 31.4 EDXA spectra of particles in Figures 31.3a and 31.3b. A is a scan of aggregates in 31.3a, B a scan of aggregates in 31.3b, and C a scan of one of the crystals in 31.3b.

amounts of Na and Cl or K and Cl. Spectra from these crystals were compared with laboratory samples of NaCl and KCl. Based on these comparisons and their shapes, these crystals are assumed to be NaCl and KCl crystals. They are presumed to be from sea salts which were absorbed into the plants from the marine atmosphere. They were most likely volatized during high temperature burning and subsequently recondensed forming crystals. We have observed similar processes in volcanic eruption plumes (Woods and Chuan, 1985).

Examples of particles in the accumulation mode are shown in Figure 31.5. Figure 31.5a shows particles collected on the impactor stage corresponding to aerodynamic diameters larger than 0.33 μm and Figure 31.5b is a higher magnification of one of these particles. The particles appear to be fluffy agglomerates made up of smaller particles. The agglomerates probably have a low mass density since their physical size is much larger than their aerodynamic size. The EDXA spectrum in Figure 31.6 (spectrum A) shows, in addition to C, O, Na, Cl, and K as found in the smaller size mode, traces of S, Mg, and Ca. The dark stains under the particles in Figure 31.5a are characteristic of acid stains (Woods et al., 1985) and are probably caused by H_2SO_4 or HCl. Examples of particles collected on the impactor stage corresponding to 1.3 μm aerodynamic diameter are shown in Figures 31.5c and 31.5d. The clusters appear to have been formed by coagulations, since several different types of particles are apparent. Some Si signals above background (spectrum B in Figure 31.6) were found in these particles in addition to the elements found in smaller ones in Figure 31.5a. There is evidence that Si is ingested by plants in this region (Lanning and Eleuterius, 1983).

The total aerosol mass loading over the size interval measured by the impactor (0.1 to >10 μm) was obtained by summing the mass loadings measured on each of the eighth impactor stages. These mass loadings were compared with CO_2 concentrations (Cofer et al., 1990) for each pass through the plume, and emission ratios (aerosols:CO_2) were calculated. Figure 31.7 shows a comparison of the mean values of the normalized emission ratios for flaming and smoldering combustion. The ratio during smoldering is much higher than that during flaming since the less efficient combustion during smoldering produces larger aerosols and more aerosol mass (see Figure 31.2b) relative to CO_2.

Small individual particles on the order of about 0.1 μm can be seen along with large clusters or chain-like aggregates made up of smaller particles. These clusters have the physical appearance of soot particles formed during high temperature burning. A scan of these particles with EDXA produced spectrum A in Figure 31.4. The peaks are highlighted by the vertical dash lines; Si, Ni, and Ag peaks are background signals from the substrate. The Si is from the quartz crystal, and the Ni and Ag are used as electrodes. The elements present in the sample are C, O, Na, S, Cl, and K. Spectrum B in Figure 31.4 was obtained by scanning the more closely packed particles in Figure 31.3b, which were near the center of the impaction site. The individual small particles and aggregates are similar in composition to the ones in Figure 31.3a, but there is also an abundance of cubic or crystalline shaped particles. These crystals were probed individually by EDXA. Spectrum C in Figure 31.4, which shows a moderate amount of Na and relatively high amounts of Cl and K, is from one of these crystals. Each crystal probed by EDXA either contained high

Figure 31.5 Particles collected from the smoke plume on impactor stages for aerodynamic sizes larger than 0.33 μm (a and b) and larger than 1.3 μm (c and d).

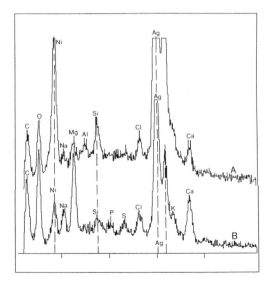

Figure 31.6 EDXA spectra of particles in Figures 31.5a (spectrum A) and 31.5c (spectrum B).

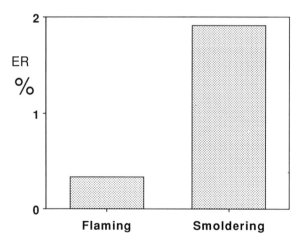

Figure 31.7 CO_2^- normalized emission ratios for total smoke aerosol ($0.1 - >10$ μm) determined for flaming and smoldering combustion.

The Measurement of Trace Emissions and Combustion Characteristics for a Mass Fire

Ronald A. Susott, Darold E. Ward, Ronald E. Babbitt, and Don J. Latham

Concerns increase about the effects of emissions from biomass burning on global climate. While the burning of biomass constitutes a large fraction of world emissions, there are insufficient data on the combustion efficiency, emission factors, and trace gases produced in these fires, and on how these factors depend on the highly variable chemistry and burning condition of the fuel. Measurements made by satellites or aircraft can be used to estimate areas burned or ratios of specific emissions, but measurements close to the ground are needed to estimate actual amounts of fuel consumed, the rate of fuel consumption, and carbon and heat release by the fire. Ground measurements also provide information on changes in emissions as rapid flaming combustion passes, and burning is reduced to a much lower intensity smoldering phase of the fire—information that frequently gets integrated and lost by sampling from platforms at higher altitudes. The work reported here describes our development of a self-contained monitoring package specifically for use in high-intensity biomass fires and how it can be applied to numerous emission problems. In addition to understanding fire processes and fuel chemistry affecting the release of emissions, there is a fundamental need for a better understanding of fire dynamics of large fires (induced winds, heat release, turbulence, etc.). This information is needed to help solve problems associated with major wildfires.

The control of large fires remains a major concern to both the United States and Canadian governments. More than 2 million hectares (ha) of wild land burned in 1988, one of the worst fire seasons in recent times in the United States. Large fires such as the Canyon Fire in Montana burned over 100,000 ha in less than 24 hours—an incredible rate. Even though fire behavior associated with surface fires is reasonably well understood and predictable (Rothermel, 1972, 1983), models for predicting the behavior of mass fires and crown fires are not available (Albini, 1984). A better understanding is needed of fire behavior, spatial distribution of gaseous fuels, temperature profiles, and fluid motions within and adjacent to large fires. Feedback mechanisms resulting in rapid increases in the rate of heat release and rate of spread are poorly understood, as are mechanisms associated with the formation of emissions—especially emissions important from an atmospheric radiation transfer standpoint.

Results are presented of a continuing study of research that was started in 1988. A new sampling system was designed to provide fire dynamics data from within the fire. This chapter describes the sampling system, the measurements it provided on one biomass fire, and some valuable parameters that can be calculated such as emission factors, combustion efficiency, and rate of fuel consumption. The large prescribed fire in Ontario, Canada, provided a practical test of this package that can be used to assess the application of the monitoring concept to a broad range of biomass fires. Measurements of wind vectors, temperature, and emissions are reported for a 40-minute period from ignition through the critical period of maximum release of heat to the near extinction of the smoldering combustion phase.

Studies of large fires have included measurements of the concentration of CO, CO_2, and O_2 at or near street level between the piles of brush that simulated houses along streets (Countryman, 1964). Work completed by Hegg et al. (1989), Einfeldt (1989), and Ward and Hardy (1989) for the Lodi tests in California suggests emissions of graphitic carbon to be correlated positively with the rate of heat release. Emissions of chlorofluorocarbons, oxides of nitrogen, and lead may be related to the deposition of anthropogenically produced air pollutants. Ward and Hardy (1984) found that emission factors for total particulate matter without regard to size will increase with an increase in fire intensity, but emission factors for particles less than 2.5 μm mean mass cut-point diameter tend to decrease as the rate of heat release increases. Emission factors for particles less than 2.5 μm diameter, particles collected without regard to size, CH_4, CO, and CO_2 are all correlated with combustion efficiency (Ward and Hardy, 1990). Among other parameters, this study tested for nonstandard O_2 conditions on emissions production.

Experimental Methods

Selection of the Site and Fire Information

Hill Township near Chapleau, Ontario, was selected as the site for the prescribed burn (PB). The 490 ha PB was located about 30 km northwest of Chapleau and about 200 km north of Sault Ste. Marie, Ontario. This site was selected because of the uniformity of fuels and the relatively flat terrain. The site consisted of tramped fuels of white birch (*Betula papyrifera* Marsh.) and poplar (*Populus tremuloides* Michx.) with some jack pine (*Pinus banksiana* Lamb.) and black spruce (*Picea mariana* [Mill.] B.S.P.). The area had been three times treated with herbicides prior to the prescribed burn on 10 August 1989. Fuel loading as determined by Forestry Canada (McRae, 1989) for the unit was about 19.4 kg/m². The moisture content for fine fuels less than 0.5 cm in diameter averaged 23.3%, with the duff fuel top 2 cm averaging 43.6% and the lower layer 85.9%.

Weather associated with the Hill PB was influenced by a large area of high pressure in Illinois eastward through Ohio. Winds were generally less than 5 m/s and carried moist, warm air from the central region of high pressure to the north across Ontario. The dewpoint temperature was about 14° to 15°C during the time of the Hill PB with a surface temperature of 26.0° to 26.9°C and relative humidity declining from 58% at 13:30 to 49% by 16:00. Surface winds were generally from the southwest at 3 to 4 m/s.

Our Fire Chemistry and Dynamics array of sensors and towers was in an area of intense study on the northwest side of the PB. A 25 m spacing was selected for the six towers with five of the towers placed in a pentagon around a central tower. Figure 32.1 shows the tower placement and scale, as well as the location of important site characteristics such as fuel wind rows, fuel breaks, and a road. Figure 32.1 also shows estimated fire front contours at selected times as the fire spread within the study area.

Equipment Development

Design of Fire Atmosphere Sampling System (FASS)

The operation of the FASS is controlled by a microprocessor-based computer designed for instrument applications. Special considerations in the design were portability, low power consumption, and ability to withstand exposure to the harsh fire environment.

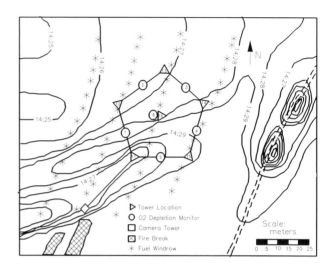

Figure 32.1 Pattern of fire spread through the matrix of towers from 14:25 to 14:30 hours on 10 August 1989.

The system collected information needed to characterize the atmosphere just above the flaming zone. This includes real time combustion gas concentrations, particulate matter production, and several fire dynamics parameters. The 40 kg packages of instruments were suspended, using standard television antenna towers, at a height of 12 m above the ground. The FASS network consisted of six systems deployed for the 1989 mass fire experiments. All systems started sampling simultaneously when signaled by a wire link connecting them that a fire was detected near the tower network.

Both glass fiber and Teflon filter mats were used to collect particulate matter during three consecutive periods (flaming, intermediate, and smoldering phases of the fire.) The FASS computer actuated valves to control the exposure sequence and combination of filters by combustion phase. Integrated samples of combustion gases were collected in sample bags for laboratory analysis. There were three bags for this purpose, one for each fire phase. Fire dynamics sensors consisted of wind and temperature measuring devices located on the tower (Figure 32.2). A three-dimensional wind measuring array, located on top of the tower (14 m), was constructed of three separate turbine type anemometers placed in an XYZ configuration. The vertical component of the air flow in the fire is used to calculate the flux of combustion products from the fire.

Methods for Real Time Measurements of CO_2, CO, and O_2

Combustion gases were drawn through filters by pumps for sampling particulate matter and the

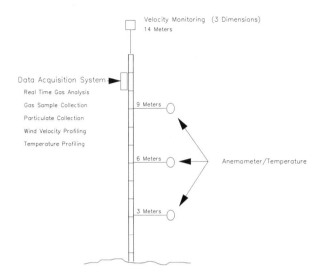

Figure 32.2 Position of the FASS sensors on the tower.

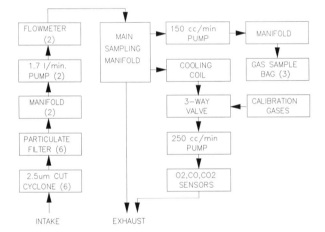

Figure 32.3 Diagram of the flow of the gases through the analyzers in the FASS packages.

filtered gases then analyzed for CO, CO_2, and O_2 (Figure 32.3). By minimizing volumes in the connecting tubing, the gas transit time between the external sample port and the sensors was kept to a few seconds. A valve also connected the sensor pump to a sample bag filled with calibration gas (CO_2 at 7310 parts per million, ppm, and CO at 500 ppm in air) to provide a check of the sensor span during the fire. The O_2 sensors were G. C. Industries Inc.[1] model 33–270 electrolytic cells, designed to measure O_2 in gas mixtures with CO_2 concentrations up to 20%. O_2 sensor

accuracy was specified to be 2% of full scale (or 0.5% O_2) at constant temperature, and 5% of full scale over the range of 0° to 50°C. The CO sensors were also electrolytic cells manufactured by G. C. Industries Inc. (model 44–300). Linearity was specified to be 1%, and accuracy depended on the CO concentration accuracy of a calibration gas mixture. The CO_2 sensors were nondispersive infrared gas monitors manufactured by Valtronics (Model 2015 BMC). Accuracy was specified to be within 3% of full scale. The three gas sensors were calibrated by standard gases with certified composition and zero grade air.

Wind Sensors Triaxial wind speed was measured by three turbine anemometers manufactured by Qualimetrics (model 24201). Calibration of one of these anemometers in the Intermountain Fire Sciences Laboratory (IFSL) wind tunnel was used for all 18 anemometers placed in the Hill PB. Wind speed accuracy is expected to be within about 0.25 m/s.

Thermocouples All thermocouples exposed to the fire environment were constructed of 24 gauge type K (chromel/alumel) wire protected by a braided stainless-steel sheath. Thermocouples within the sample packages were 30 gauge with a glass fiber sheath. Temperature accuracy is expected to be within 4°C from 0° to 400°C, with larger errors at higher temperatures.

Sampling Protocol

The system computer timed and controlled data acquisition events during the experiment. In general, the objective was to maximize the data collected during the 40 minutes following fire detection near the tower array. It was also necessary to record baseline data (at a lower sampling rate) prior to the fire as well as smoldering data for several hours after the fire front had passed. All computer clocks were set to real time (EDT) based on the Hill PB experimental clock.

Real Time Sensors On the day of the burn, the FASS was programmed to first calibrate the gas sensors at 12:00. From this time on the FASS could accept a trigger from a temperature-activated fire sensor. Every two minutes the program made a reading of background (no fire) data. These data provided an ambient air measurement for the gas sensors, prefire wind speed and direction, and air temperatures. The approaching fire was detected when a ground temperature sensor exceeded 67°C. These sensors were placed in piles of fuel up to 30 m west of the tower array, in the direction anticipated for fire arrival. All

[1]The use of trade or firm names in this chapter is for reader information and does not imply endorsement by the U.S. Department of Agriculture of any product or service.

FASS packages were linked and began real time measurements (two readings per second) when the fire was detected. The next 40 minutes were viewed as three fire phases: a 10-minute flaming phase, a 10-minute intermediate phase, and a 20-minute smoldering phase. A second calibration was made at the start of the smoldering phase to reduce the effects of sensor drift. Finally, the packages returned to taking one reading every two minutes, as in the prefire sampling phase. This extended smoldering phase continued until the tower package was turned off, or storage memory was full.

Integrated Filter and Gas Bag Samples The tower packages collected physical samples during three of the sampling phases, for later analysis. Particulate matter was collected for flaming, intermediate, and smoldering phases. Three gas sample bags were also filled in each tower package over the same time intervals as the three filter particulate samples.

Analytical Methods for Postfire Sample Analysis
The gas sample bags from the Hill PB were analyzed for CO_2 and CO composition using infrared gas analyzers (Horiba model PIR-2000). Bags of zero air and a span gas mixture were also analyzed to adjust the instrument to zero and to calculate bag composition. The accuracy of this method is about $\pm$ 1% composition for both CO_2 and CO.

The membrane and glass fiber filters were weighed at IFSL to determine smoke particle loading. Filters were conditioned at 50% relative humidity and 22°C for 24 hours for initial and final weighing on a Cahn model C-32 microbalance. Filter blanks were within 0.010 mg before and after the experiment. Membrane filters were also analyzed for trace metals using an X-ray fluorescence method at NEA Inc. of Beaverton, Oregon. The glass fiber filters were analyzed for graphitic and organic carbon fractions by Sunset Laboratories, using a ramped temperature protocol for carefully volatilizing the organic carbon fraction while retaining the graphitic carbon fraction.

**Calculated Parameters and Methods
of Adjusting Data**
Emission factors were calculated using the carbon mass-balance method (Ward et al., 1979; Radke et al., 1990). The method is based on the partial oxidation of fuel (typically $C_6H_9O_4$) to CO_2 and products of incomplete combustion. The assumption was made that the carbon is released from the fuel proportional to the mass-loss rate of the fuel consumed by the fire. This assumption was validated for laboratory fires by

Nelson (1982). For each gram of carbon released, approximately 2 grams of $C_6H_9O_4$ are consumed. By accounting for the carbon contained with the CO_2, CO, hydrocarbons, and particulate matter, the equivalent amount of fuel consumed can be evaluated. (In practice, the CO_2 and CO account for 95% to 99% of the carbon released from the fuel.)

An emission factor is computed by dividing the concentration of the emission (g/m^3) by twice the total carbon concentration (g/m^3). The FASS units were specifically designed to make the measurements needed by phase of combustion so that these calculations could be made for the flaming, intermediate, and smoldering combustion phases. In the case of particulate matter, the concentration was computed by averaging the mass of particulate matter collected on the glass fiber and Teflon filter mats and dividing this by the average volume of gas sampled.

Rate of fuel consumption calculations were made as in Ward and Hardy (1984) by multiplying the carbon concentration as measured using the CO_2 and CO real-time sensors (g/m^3) by the vertical component of smoke plume velocity (m/s), and the resulting number by two (accounts for the molar ratio of fuel to carbon). The error associated with not measuring the particulate matter and hydrocarbons in real time is less than 5%. Individual source strengths for emissions can be calculated by multiplying the concentration (g/m^3) by the vertical velocity (m/s).

The rate of heat release was computed by multiplying the rate of fuel consumption by the fuel heat of combustion corrected for the heat remaining in the products of incomplete combustion. The real time CO concentration was the only significant emission resulting from incomplete combustion and was used in making the adjustment to the heat of combustion. Based on oxygen bomb calorimeter measurements of the heat of combustion, a value of 20 kilojoules per gram (kJ/g) was selected (McRae, 1989) and 10.1 kJ/g was used for CO (Perry and Chilton, 1973). The resulting rates of heat release driving the convection column were not corrected for other forms of heat loss due to radiation and soil heating.

Presentation of Data

In this section, the general fire behavior and deviations from ambient conditions for each of the measured parameters are discussed. Because this was the first full-scale deployment of the FASS system, the performance of the sampling system is discussed. Concentration of combustion products and the "normalcy" of the data are presented. Comparisons are made of the concentrations of various combustion gases and the multiple measurements of these gases through grab sampling and real time techniques. General observations of the wind field parameters are presented.

Fire Behavior

The area of the fire ignited surrounding the FASS array was 97 ha. Ignition of this block commenced at 14:19 hours with approximately 8 ha involved by 14:23 hours. At 14:22 hours, a line of fire was ignited 50 m to the east of the array of towers (Figure 32.1). Approximately two minutes later, three segments of fire were ignited 60 m to the west of the tower array, and this line of fire terminated at the fire break test site just to the south. By 14:25:44, the fire had advanced from the central segment of fire to within 25 m of the base of FASS package tower 12. The triggering mechanism was activated in this location and the six FASS packages simultaneously began sampling the environmental parameters.

Fuels were distributed across the tower array non-uniformly. Wind rows of concentrated heavy fuels, resulting from the cable clearing of the area, extended north-northeast, and FASS tower 12 was on the easterly edge of one of these wind rows. FASS towers 16 and 13 were in a second wind row with a third wind row passing just to the east of FASS towers 14 and 15. The center lines of each of the three wind rows of heavy slash are shown by the lines of asterisks in Figure 32.1. The fire spreads along these wind rows more rapidly than between the wind rows so that the spread pattern actually tended to follow the wind rows and seemed to move forward from wind row to wind row in pulsing actions. Potentially, some of the pulsing periodicity of the fire could have been associated, on a micro scale, with the spread of the fire between wind rows of fuel.

The rate of spread of the fire from the line of ignition to the trigger location was calculated to be about 0.5 m/s. Flame lengths were on the order of 8 to 15 m based on video coverage of the burn area and from photographs taken from a helicopter observation point. The main fire front flanked along a line extending from FASS towers 12 to 13 toward FASS tower 16 in the middle of the sample array while a finger of fire was running through the array from FASS tower 11 toward FASS tower 15. The north flank of the finger along a line from FASS 11 to FASS 15 and the southerly flank spreading southward from FASS 12 and FASS 13 converged in the vicinity of

FASS packages 16 at approximately 14:29:00, or about 4.25 minutes after the FASS packages were activated.

The flame burnout time for the fire was on the order of one to three minutes depending on the location in or between wind rows. This affected the duration of the flaming phase emissions production at all of the FASS package locations.

There were three major convergence zones associated with the fire ignition pattern for block A that could have affected the measured patterns of airflow in the vicinity of the FASS packages.

1. The first of these occurred during the time that the center fire was converging to the east of the FASS array. This was believed to occur during the period from 14:26 to 14:30.

2. The second occurred soon after ignition of the area to the west of the fire break test area during the time that the lines of fire were converging at approximately 14:34.

3. The third occurred during the time that major segments of the fire on the southwestern flank of the block were converging to the south and slightly to the east of the FASS array.

The ignition of the 97 ha block was completed by 14:55.

Gas Concentrations

The concentrations of the primary gases needed for carbon mass balance calculations and emission factors were available from analyses of gas collected in sampling bags and the measurements of the real time sensors. Table 32.1 contains the results of the sample bag analysis for CO_2 and CO, by sampling phase. CO_2 is highest for the flaming phase, with the highest level measured for package number 14, at 4490 ppm. Inter-

Table 32.1 FASS sample bag gas concentrations in ppm

Box number	Flaming		Intermediate		Smoldering	
	CO_2	CO	CO_2	CO	CO_2	CO
11	3217	126	1937	152	1305	172
12	3420	142	1751	120	1353	164
13	4166	168	1969	175	1224	151
14	4490	154	2140	198	1264	170
15	3963	137	a	a	1248	170
16	3922	140	1961	174	1321	172
Average	3863	144	1951	164	1286	167

a. Valve did not open.

mediate phase values drop to about 50% of the flaming value, and the lowest levels (1224 ppm) occur during the smoldering phase on package 13. CO_2 levels were quite uniform across the sample array with deviations from the average of about 11% during flaming and only 3% during smoldering. Trends in CO concentrations with fire phase are less obvious, and these values tend to be uniformly between 120 and 200 ppm. Flaming phase CO averaged about 12% lower than the later phases, but the data are not consistent.

Continuous CO_2 and CO measurements were made during the three fire phases. The background concentrations of CO_2 and CO in air can be assumed to be 375 and 1 ppm, respectively. The average of the concentration curve over the time assigned to a phase should be equal to the sample bag concentration. Figure 32.4 shows the real time sensor concentrations for FASS package 11. The CO_2 levels in Figure 32.4a show the high variability of these levels, from near

ambient levels to 9000 ppmv during the 10 minutes of the flaming phase. The rapid response of the CO sensor (Figure 32.4b) emphasizes the variability even more. Comparison of Figures 32.4a and 32.4b in the first 200 to 300 seconds of flaming show the high efficiency of the initial flaming pulse, with little CO produced. The CO concentrations increase relative to CO_2 as the fire goes to the smoldering phase. The nearly constant region at about 14:46:40 is due to switching a flow of calibration gas to the sensors.

Table 32.2 provides a summary of real-time sensor results for all FASS packages, by phase of combustion. The second smoldering phase data were taken from the two hours of sampling following the initial 40 minutes of the fire. Figure 32.5 shows a comparison of average CO_2 and CO as a scatter plot of the real time results versus the sample bag measurements. Clearly, there is good agreement between the two methods. A linear regression of the CO_2 data gave a slope of 1.17

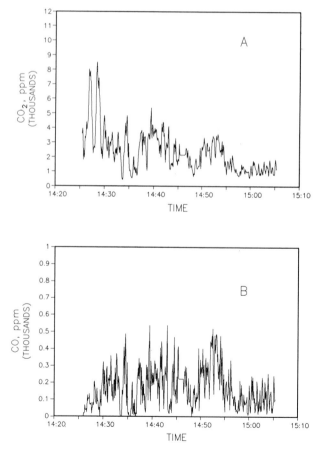

Figure 32.4 Example of real-time gas concentration measurements on FASS 11: (a) CO_2; (b) CO.

Table 32.2 FASS real-time combustion gas sensor summary

Box number	CO_2 (ppm)		CO (ppm)	
	Average	Maximum	Average	Maximum
Flaming phase (10 minutes)				
11	3518	8616	136	515
12	3595	11551	123	500
13	4737	12536	169	523
14	5467	11995	137	442
15	4772	12660	125	394
16	3848	8370	132	637
Average	4323	10955	137	502
Intermediate phase (10 minutes)				
11	2623	5604	189	767
12	2322	5233	128	664
13	2683	5798	170	648
14	2709	6602	161	545
15	3162	5219	197	468
16	2272	5514	155	808
Average	2629	5662	167	650
Smoldering phase 1 (20 minutes)				
11	1648	4666	163	595
12	1763	4659	173	493
13	1786	4501	164	555
14	1757	4889	157	524
15	1688	4772	165	577
16	1759	4628	161	839
Average	1734	4686	164	597
Smoldering phase 2 (2 hours)				
11	722	1830	40	222
12	641	1201	61	169
13	729	1313	52	159
14	717	1259	47	130
15	613	1409	64	115
16	830	1252	37	134
Average	709	1377	50	155

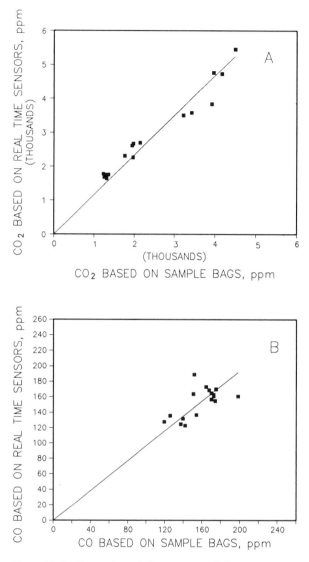

Figure 32.5 Comparison of the average real-time gas concentrations with measurements based on sample bags: (a) CO_2; (b) CO.

phase minima of about 90% to 92% of air level, or a decrease of about 0.02 volume fraction. Some broad decreases in O_2 level were seen in later-phase curves, but most of the detail was lost in noise and drift of the sensor. The O_2 sensors should be improved in resolution or accuracy for this application on relatively dilute combustion gases.

Particulate Matter Concentrations
The tower packages collected particulate matter by fire phase on two filter media (glass fiber and Teflon). Table 32.3 gives particulate matter concentrations by fire phase. The concentrations ranged from 15.1 to 39.9 mg/m³. These values are about five to 10 times lower than those of CO, and therefore make a smaller contribution to the carbon balance and combustion efficiency calculations. All packages were within about 25% of the average value for a phase. The smoldering phase has slightly higher concentrations and produces more total particulate matter in the 20-minute sampling. The average for the two filter media was used to calculate emission factors for PM5.0.

The glass fiber filters were analyzed for carbon content, proportioned between graphitic and organic carbon. The organic carbon percentage averaged lowest for the flaming phase (45.5%) and increased at later times (52.1% for intermediate and 53.6% for the smoldering phase). There is considerable variation during the flaming phase, reflecting the highly turbulent fire and entrainment of ash and soil particles. Graphitic carbon averaged 8.1% during the flaming phase and decreased dramatically in the intermediate (2.9%) and smoldering (1.5%) phases. The intermediate phase results are closer to those for the smoldering phase. Low graphitic carbon during the intermediate phase indicates a substantial decrease in flaming combustion. The total percentage carbon in the fine particles is approximately equal (53.5% to 55.1%) for all three fire phases.

The Teflon filters were analyzed for trace inorganic elements. Figure 32.6 shows a comparison of the Hill PB elemental profile to those from experimental burns in California and the Pacific Northwest. The calcium levels were high compared to other reports and are probably due to the soil chemistry at the burn site. Considerable levels of elements from aluminum to calcium were measured, along with manganese, iron, and zinc. Although there is considerable variation between tower boxes, the S, Cl, and K are highest during flaming, while Al, Si, Ca, and Fe tend to increase in the smoldering phases. These results are consistent with other studies.

with $r^2 = 0.93$ (the intercept is forced to zero). The high slope is attributed to remaining difficulties in calibrating the sensor and a slight decrease in bag CO_2 due to leaky valves. A plot of the CO data (in Figure 32.5b) gave a linear regression line slope of 0.966 with $r^2 = 0.28$. The relatively small range of concentrations measured leads to the poor fit, but the two methods give comparable concentrations. The agreement should improve as we improve the real time sensor calibration methods. Relative differences of 3% to 5% should be possible in future experiments.

Oxygen levels were measured in real time by the tower-mounted systems. The data show early flaming

Table 32.3 FASS filter particulate (PM 5.0) concentrations (mg/m³)

Box number	Flaming		Intermediate		Smoldering	
	Glass	Teflon	Glass	Teflon	Glass	Teflon
11	24.5	28.9	39.9	24.8	36.1	24.6
12	21.2	15.1	19.5	15.8	28.7	25.6
13	29.1	26.2	26.7	23.3	27.4	24.5
14	28.8	24.9	26.5	24.0	27.0	25.2
15	28.7	28.5	a	a	30.4	28.2
16	32.1	18.3	23.7	21.1	28.9	25.8
Average	27.4	23.6	27.3	21.8	29.8	25.7

a. Valve did not open.

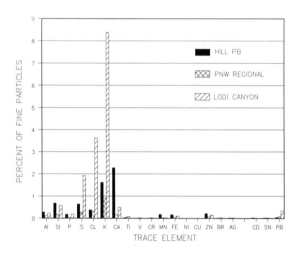

Figure 32.6 Inorganic composition of particulate matter compared to literature values for Pacific Northwest and Lodi Canyon, California, burns.

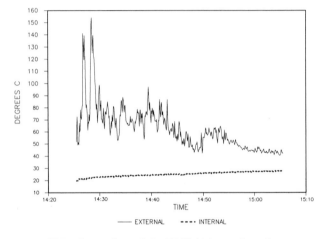

Figure 32.7 Comparison of the FASS 11 internal gas flow temperature to the external air temperature of the fire environment.

Temperature Measurements

Survival of the sensitive instruments in a harsh fire environment requires the internal temperatures to remain close to ambient. Thermal protection of the FASS depended on a low-density ceramic insulation and a highly reflective aluminum outer wrap. Figure 32.7 compares the internal gas flow temperature to the external fire environment temperature at the inlet, for FASS 11. While the outside temperatures reached over 150°C, the internal box temperatures increased only about 10°C. FASS 12 was exposed to air temperatures as high as 450°C, but average temperatures for the flaming phase ranged from 80° to 137°C. There was no damage to any of the FASS packages. The modest internal temperature changes suggest that the system designs can withstand much more severe fires.

Wind Speed and Direction

Wind speeds were measured in three dimensions by the set of turbine anemometers positioned above the FASS, at the top of the tower. The positive or negative direction sensors on these anemometers failed but the absolute values of the vector wind speeds were reliable. Other data indicate that the wind was always from the west and predominately slightly south of west. These results were consistent with other horizontal wind direction measuring systems located at the same tower locations.

Vertical wind directions should be mostly up, due to buoyancy of the hot combustion gases. The vertical component of wind velocity varied rapidly from 0 to over 6 m/s for brief peaks but averaged between 2 and 3 m/s for most of the fire. Generally, we expected the temperature and CO_2 concentrations to be correlated with the vertical velocity of the combustion gases. In fact, this was found to be a valid correlation and for the FASS tower location, we found no evidence during the flaming and intermediate sampling periods for significant down drafts. On occasion, the concentration of excess CO_2 dropped to near zero, and the vertical velocity was also near zero. We concluded that the errors in calculating the rate of fuel consumption and the rate of heat release from the carbon mass balance considerations would be minimal. The vertical velocity was used in conjunction with the real time gas sensors to calculate the mass flux of carbon from the fire site.

Discussion

In this section, we discuss the findings from the FASS data set. The rate of fuel consumption is used in evaluating the rate of emissions production for the fire over time as well as the total rate of heat release corrected for combustion efficiency. The emission factors are presented by combustion phase and weighted for the entire fire.

Rate of Fuel Consumption

The rate of fuel consumption is important for calculating source strength and the rate of heat release over

the life history of the fire. Here we use the data from the vertical wind velocity, CO concentration, and CO_2 concentration to calculate the rate of release of carbon from the biomass fuel. Four of the six FASS packages yielded satisfactory data for calculating rate of fuel consumption (FASS packages 11, 12, 14, and 16). The integrated fuel consumption is compared with the fuel inventory measurements of total fuel consumption by Forestry Canada (McRae, 1989).

Elemental carbon composition of the fuels is important to the accuracy of carbon mass balance methods. Fuels typical of those burned during the Hill PB were collected by Forestry Canada and shipped to the IFSL. Table 32.4 provides a list of the species and size ranges analyzed, along with the elemental composition of the fuel (C, H, N, and O estimated by difference). The moisture and ash component available from thermogravimetric analysis were used to reduce composition values to an ash-free, moisture-free basis. This basis allows easier comparison of just the organic component of the forest fuels. The woody component is the largest fraction of the fuel, and the carbon fraction of the woody material approaches 50%.

Generally ash is low (0.5% to 1%) in the woody material but is up to 7% of black spruce twigs. The

Table 32.4 Elemental analysis of Canadian fuels from the 1988 mass burn

Sample identification	C(%)	H(%)	N(%)	O(%)
Timmins duff	58.19	5.29	1.82	34.70
Timmins litter	57.54	6.33	1.26	34.87
White birch twig (0.0–0.49 cm)	54.39	6.30	0.85	38.47
White birch wood (3.0–4.99 cm)	49.86	6.01	0.10	44.03
White birch wood (7.0+ cm)	49.50	6.07	0.09	44.33
White birch bark (3.0–4.99 cm)	57.82	6.82	0.47	34.89
White birch bark (7.0+ cm)	58.74	7.08	0.49	33.69
Black spruce twig (0.0–0.49 cm)	56.45	6.37	0.61	36.57
Black spruce wood (3.0–4.99 cm)	52.02	6.10	0.18	41.70
Black spruce wood (7.0+ cm)	50.59	6.05	0.05	43.31
Black spruce bark (3.0–4.99 cm)	57.92	6.06	0.48	35.54
Black spruce bark (7.0+ cm)	53.34	5.77	0.29	40.60
Jack pine twig (0.0–0.49 cm)	55.03	6.26	0.57	38.13
Jack pine wood (3.0–4.99 cm)	51.38	5.97	0.05	42.60
Jack pine wood (7.0+ cm)	51.37	6.15	0.04	42.43
Jack pine bark (3.0–4.99 cm)	52.31	5.91	0.48	41.30
Jack pine bark (7.0+ cm)	54.34	5.92	0.24	39.50
Poplar twig (0.0–0.49 cm)	56.52	6.60	1.06	35.82
Poplar wood (3.0–4.99 cm)	50.71	6.09	0.15	43.05
Poplar wood (7.0+ cm)	49.25	6.07	0.05	44.63
Poplar bark (3.0–4.99 cm)	54.80	6.54	0.82	37.84
Poplar bark (7.0+ cm)	52.75	6.33	0.40	40.52

Note: Compositions are on a dry weight and an ash-free basis.

duff and litter samples have 10% to 20% ash. This ash component should be investigated in more detail as a source of potential hazardous emission or as a useful tracer for smoke from biomass burning. Because oxygen is the only other major component it can be estimated by difference. The propensity for soot or particulate matter emissions to form may be indicated by the oxygen content, which covers a range of 33.7% to 44.6% for the Canadian samples. Soot formation is expected to increase with lower oxygen content, making the white birch bark an important source of soot.

Measured Rate of Fuel Consumption The peak rate of fuel consumption occurred during the time when the fine fuels were consumed by the advancing fire. Peak rates of fuel consumption were measured at 40.8, 64.5, 61.4, and 36.1 g/m²/s for packages 11, 12, 14, and 16, respectively. Figure 32.1 shows the time of ignition for the fuel immediately surrounding each tower location. The maximum rate of fuel consumption occurred at an elapsed time from triggering of the FASS sample packages of 2.83, 1.00, 4.33, 2.5, and 4.33 minutes for packages 11, 12, 14, 16A, and 16B, respectively. (For FASS package 16, two peaks of fuel consumption were measured and are identified as 16A and 16B.) From Figure 32.1, the wind rows adjacent to and upwind from FASS towers 12 and 14 contributed to the high rates of fuel consumption. In addition to a wind row being immediately upwind of FASS tower 16, the fire converged immediately under FASS tower 16 and probably contributed to both the sustained high rate of fuel consumption and the double peak phenomena observed for the rate of fuel consumption.

The ambient winds were substantial, averaging 3.1 m/s at 14:00 (McRae, 1989), immediately prior to ignition of block A of the Hill PB. Undoubtedly, smoke drifted into the region of flux measurements because the fire was ignited in steps starting with the center fire and extending out past the FASS array in an ever increasing circle. This could have influenced the measurements more during the smoldering phase than during the initial ignition phase. It is likely that some of the emissions created in the vicinity of the FASS packages exited the area without being accounted for through the carbon mass-balance and flux measurements. The overall significance is not known but is thought to be no greater than the level of error associated with the fuel consumption measurements.

Total Fuel Consumption Total fuel consumption for each of the four sets of measurements was calculated by integrating the rate of consumption. The cumula-

tive fuel consumption over time is shown in Figure 32.8. Total fuel consumption averaged 11.46 kg/m² with a standard deviation of 1.65 kg/m². The range of fuel consumption of 9.76 to 13.61 kg/m² coincides with the areas of total fuel loading (Figure 32.1). The rate of fuel consumption was still decreasing at the end of the sample period and averaged 1.06 g/m²/s after 40 minutes of sampling. The fuel consumption rate continues to be significant through the end of the smoldering sample period. Figure 32.8 shows a steep rate of fuel consumption over the first four to six minutes following ignition, during the time when the fine fuel component is consumed.

The FASS data allow calculation of fuel consumption for the flaming, intermediate, and smoldering combustion periods by integrating the rate of fuel consumption curves over the limits of integration for those periods (Table 32.5). The average total fuel consumption of 11.46 kg/m² compares with the inventory value from data collected by Forestry Canada of

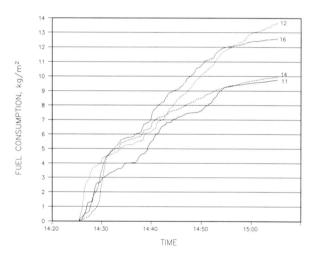

Figure 32.8 Cumulative fuel consumption for each of the four sets of data (packages 11, 12, 14, and 16) for Hill PB.

Table 32.5 Fuel consumption by phase of combustion and the total, for each of the four FASS packages (kg/m²)

FASS number	Flaming	Intermediate	Smoldering	Total
11	4.00	3.17	2.57	9.76
12	5.24	3.13	5.24	13.61
14	5.58	2.12	2.26	9.96
16	5.73	3.35	3.45	12.53
Average	5.14 ±0.68	2.94 ±0.48	3.38 ±1.16	11.46 ±1.65

10.63 kg/m^2 or about 8% greater than the ground inventory. Examining the fuel consumed for the first six minutes for each FASS package yields measured fuel consumption averaging 4.22 kg/m^2. This is about twice the fine fuel loading on the unit. However, an additional component of fuel including a portion of the duff and larger-diameter fuel fractions was consumed during the flaming phase.

An approximate mathematical expression was fit to the rate of fuel consumption data to simplify other source strength and heat release calculations for the entire area of the Hill PB. More rigorous modeling of the fuel consumption process is beyond the scope of this report. Ward and Hardy (1984, 1989) modeled the rate of fuel consumption using a similar set of carbon flux data for broadcast burns of logging slash on clear-cut units in Oregon and Washington. They used an equation that provides for an exponential growth and decay of the combustion process. We use a linear growth function to describe the buildup phase and an exponential decay function for the diedown phase.

The four sets of data for the rate of fuel consumption were averaged (Figure 32.9). The maximum rate of fuel consumption was approximately 20 g/m^2/s (corresponding to a relative rate of 1.0 in Figure 32.9), which occurred 4.5 minutes after the sensors were first challenged by the approaching fire. A linear fit with a slope of 0.074 g/m^2/s/s was used for the buildup phase. The diedown phase uses an exponential decay function stretched over two hours with a time-decay constant of 10 minutes. The resulting

model lines for growth and decay are also shown in Figure 32.9. The exponential decay function gave a reasonably good fit for the period from 12 minutes to the end of the sampling period. In addition, because of the indicated high rate of fuel consumption sustained at the end of the sampling period, the model was extended for another 80 minutes. While details of the model are not exact due to ignition pattern, fuel concentrations, and ambient winds, the model does give a good representation of the average trend in burning rate with time. The model fuel consumption values are easily calculated for any time interval and can be used to compute the source strength for particulate matter, CO, CO_2, and other compounds as a function of the rate of ignition for the unit.

Emission Factors and Emissions Released Per Unit Area Burned

Figure 32.10a shows the average CO emission factors compared to results from fires in Douglas fir and

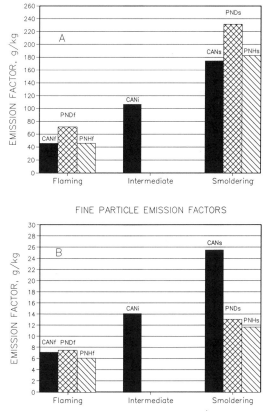

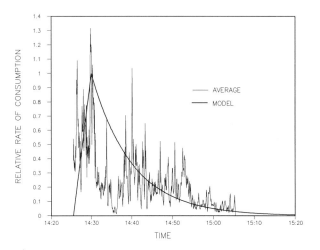

Figure 32.9 Fit of a simple model to the average rate of fuel consumption.

Figure 32.10 Comparison of emission factors measured for the Hill PB Canada (CAN) with data from Pacific Northwest Douglas fir (PND) and Pacific Northwest hardwood (PNH): (a) CO; (b) PM 5.0. Flaming, intermediate, and smoldering phases are indicated by *f, i, s.*

hardwood slash in the Pacific Northwest. There is excellent agreement with the hardwood slash, but the Hill PB has somewhat lower CO emission factors than for Douglas fir slash. Emission factors for CO increased from 46 g/kg during flaming to 174 g/kg during smoldering. Combustion efficiency decreased from 95% for flaming to 82% for smoldering as a result of the increased emission of incompletely oxidized combustion products.

Figure 32.10b shows the average fine particle emission factors from the Canada fuels compared to the Northwest United States fuels. There is good agreement during the flaming phase, but emission factors for PM 5.0 during the smoldering combustion phase for the Canada fuels are about twice those for the United States fuels. The average PM 5.0 increased from 7.17 g/kg during flaming to 25.6 g/kg during smoldering.

Emission factors are weighted based on the fuel consumed during each of the sampling periods for the flaming, intermediate, and smoldering combustion phases of 5.14, 2.94, and 3.38 kg/m^2, respectively. The emission factor data shown in Figure 32.10 for each sampling period is multiplied by this fuel consumed during the sampling period to give the total emissions by phase in Table 32.6. The totals in this table represent the total emissions produced by the fire on a square meter basis.

The total emissions released per square meter of area burned were 165, 1141, and 18,887 g/m^2 for PM 5.0, CO, and CO_2, respectively, based on our measured fuel consumption. By dividing the total emissions by the total fuel consumed (11.46 kg/m^2), fire-weighted emission factors are calculated for the Hill PB of 14.4, 99.6, and 1648 g/kg for PM 5.0, CO, and CO_2, respectively.

Application of Unit Area of Measurement to the Hill Fire

The total emissions released from the Hill PB can be calculated based on the fuel consumption and the fire-weighted emission factors. The total area burned of

Table 32.6 Total emissions released on a g/m^2 basis for block A of the Hill PB

Emission	Flaming	Intermediate	Smoldering	Total
PM 5.0	36.9	41.6	86.4	164.9
CO	236.4	313.7	590.5	1141
CO_2	8980	4813	5094	18,887

390 ha multiplied by the emissions released per square meter yields a total emissions production of 643, 4450, and 73,700 tons of PM 5.0, CO, and CO_2 for the Hill PB. Block A, the area of the intensive fire chemistry and dynamics study (97 ha), released about 25% of the emissions of the total Hill PB.

A source strength calculation requires information pertaining to the rate of ignition of the fuels and the rate of consumption during the burnout time for the fuels. The mathematical expressions for the rate of fuel consumption can be applied here, along with emission factors for the individual phase of the fire. The rate of ignition is not known accurately, but we estimate the helitorch and advancing fire ignited new area at a steady state of about 350 m^2/s 20 minutes after starting the center fire. Ignition of new area stopped about 50 minutes into the fire, and remaining fires burned together 10 minutes later. Figure 32.11 shows how this ignition pattern would generate fine particulate matter from block A of the Hill fire. The flaming phase produces most of the particles in the first 20 minutes, but smoldering dominates for the remainder of the fire. Other emissions can be computed accordingly. Maximum rates of production of PM 5.0, CO, and CO_2 are estimated as 0.047, 0.33, and 6.3 tons per second.

The FASS-measured CO production allows the calculation of an effective heat of combustion as the fuel is consumed. For example, if 10% of the original fuel carbon is converted to CO, the heat of combustion of that CO can be subtracted from the heat expected from the total fuel burned (20 kJ/g). Other products of combustion can be ignored as a first order approximation. The overall average heat of combustion was

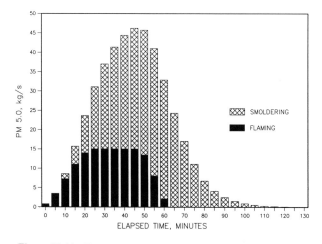

Figure 32.11 Source strength over time for particles less than 5.0 micrometers.

calculated to be 19.22 kJ/g, or about 96% of the total. This value changes slightly with phase of combustion as combustion efficiency changes. Peak heat release rates of about 500 kW/m^2 were calculated about five minutes after the fire reached the FASS array. The estimates of ignition rates in the previous paragraph result in a 75,000 megawatt (MW) heat release rate for the entire block A of the Hill PB. Some of this heat is lost to soil heating or is radiated away, but a majority heats the surrounding air and smoke to provide the driving force to build the strong convection column.

Summary and Conclusions

The Fire Atmosphere Sampling System (FASS) has been tested within an intense mass fire without major problems. The measurements and samples taken provide detailed information on emissions within the fire. Continuous measurements during a fire can provide needed insight on the flaming and smoldering of large fires, and temporal variations of emissions. The vertical measurements of air flow, combined with the concentrations of CO_2 and CO, accurately calculate the mass flux of carbon released. The measured fuel consumption based on the carbon mass flux agreed with ground inventory methods. These inventory methods are not possible in fuels common to global biomass burning, where the FASS can be easily applied. The detailed fuel consumption rate allows more realistic and accurate models of the local burning process to be developed.

The computer system incorporated in the FASS allows numerous possibilities for modifications to meet alternate needs. Emphasis in the current system was on release of carbon and the nature of particulate matter, but additions could easily be made to measure nitrogen gases or other greenhouse and hazardous emissions. The self-contained system can easily be transported to fires throughout the world to provide a ground truth addition to remote sensing studies.

Acknowledgments

The efforts of many groups and individuals aided the development, construction, and deployment of our experiment—we gratefully acknowledge their contributions. The fire research personnel of Forestry Canada, Ontario Region, provided needed support in site selection, fuels characterization, and tower assembly and were gracious hosts during our visit to Canada. Of special recognition are the space, accommodations, telephone, facsimile, and aircraft support provided by the Ontario Ministry of Natural Resources. This research was supported in part by funds provided by the Defense Nuclear Agency under project IACRO 89-903, Task RA, Work Unit 00024.

Carbonaceous Aerosols from Prescribed Burning of a Boreal Forest Ecosystem

Monica A. Mazurek, Wesley R. Cofer III, and Joel S. Levine

Carbonaceous particles in the tropospheric boundary layer can both scatter and absorb solar radiation. These particles are emitted from numerous sources, including combustion of vegetative matter. Although biomass burning is now recognized as a major and increasing source of particulate carbon to the global atmosphere (e.g., Duce, 1978; Seiler and Crutzen, 1980; Andreae, 1990), the impact of these smoke aerosol emissions on tropospheric chemistry and on global climate is not well known. Much of the uncertainty is linked to the limited inventories of nonvolatile carbonaceous aerosols that are emitted from biomass combustion processes, to the few measurements of smoke aerosol physical and optical properties, and to a narrow understanding of the fate and dispersal of smoke aerosol within the troposphere. Consequently, general circulation models (GCMs) have few real data on which to predict global climate change due to the changing chemical and radiative properties of the earth's atmosphere now linked, in part, to the increased emissions of carbonaceous aerosols from global biomass burning.

In the past, carbonaceous particles found in the smoke plumes of combusted biomass have been evaluated as bulk aerosol components. Two forms of smoke aerosol, organic carbon (OC) and elemental carbon (EC), are radiatively active components; OC particles scatter solar radiation (depending on composition) and EC particles scatter and absorb the same radiation. Total carbon (TC), OC, and EC have been identified in smoke from the tropics (Andreae et al., 1988; Cachier et al., 1989; Suman, 1988). Recent estimates of the global emissions of particulate organic carbon and elemental carbon due to biomass burning show major releases to the troposphere of these radiatively important carbonaceous particles (69 Tg/yr for OC; 19 Tg/yr for EC) (Andreae, 1990).

Detailed chemical measurements which relate aerosol composition to emission sources of the carbonaceous matter contained in smoke aerosols, especially those from large-scale biomass combustion, are few. Stable carbon isotope analyses ($^{13}C/^{12}C$ content)

are available for smoke aerosols collected from burning of savanna grassland and tropical rainforests (Cachier et al., 1986, 1989). Significant differences are noted in the ratio of $^{13}C/^{12}C$ present in the smoke aerosol from these types of biofuels, and such isotopic distinctions have been applied to source characterizations of carbonaceous aerosols in the remote marine atmosphere (Cachier et al., 1986). Standley and Simoneit (1987, 1990) have identified molecular tracers present in smoke aerosols sampled from prescribed burning of a coniferous forest. These organic compounds were traced directly to the coniferous biofuels and consisted of plant waxes, resins, and thermally altered wood biopolymers. Other analytical techniques have been applied to carbonaceous subfractions of whole wood smoke emissions, also for source reconciliation (Hildemann, 1990; Mazurek et al., 1990a, 1990b). In these applications, the solvent-soluble organic compounds isolated from wood smoke aerosol produce unique chemical profiles that serve as "fingerprints" to be used in distinguishing among different types of wood burned. Therefore, given the ability to measure key carbonaceous subfractions present in smoke aerosol which can be related chemically to the biomass fuel source, it is useful to apply these analytical methods to natural wildfire and prescribed burn emission studies.

The goal of this project consists of three objectives. First, evaluations are needed of the sampling methods and analytical techniques used in measuring smoke aerosol emissions from biomass combustion. Accurate quantitative measurements of particulate carbon emissions rely on sampling and analytical methods which have been evaluated fully in terms of quality assurance/quality control (QA/QC) parameters. Second, an accounting is necessary for the mass of total carbon present as smoke particles and for the subfractions which constitute the total particulate carbon aerosol mass. This chemical mass balance provides a quantitative basis for describing the mass distribution of radiatively important aerosols (OC and EC fractions) emitted during biomass burning. Finally, a third objective is to compile emission data from the combustion of diverse ecosystems, where measurements are made of carbonaceous particles, especially those that effectively scatter and absorb light. These emission data are needed to provide a global distribution of particulate carbon released to the atmosphere due to prescribed burns, agricultural land clearing, and natural wildfires.

As an initial step toward these objectives, smoke aerosol and background aerosol particles are collected from the controlled burning of boreal forest where vegetation species and relative mass distributions are known. Chemical mass balances are constructed for the total mass of carbonaceous aerosol particles emitted during the prescribed burn. A carbonaceous species inventory is developed for aerosol particles present under background, smoldering, and full-fire conditions; the production of OC and EC particles is noted for these two fire regimes. Distributions of the solvent-soluble organic components of the sampled aerosols are generated to identify molecular properties that can be traced to unburned and pyrolyzed materials present in the boreal forest fuels.

Methods

Smoke Aerosol Collection
A boreal forest prescribed burn (486 hectares) was conducted in the Hill township of northern Ontario, Canada, on 10 August 1989. The fuel source composition consisted of spruce (13%), balsam fir (25%), white birch/poplar (46%), decayed wood (12%), and miscellaneous wood (4%) (Stocks, 1990). The mass of unburned woody material was 9 to 10 kilograms per square meter (kg/m^2), of which 6.6 kg/m^2 was consumed.

A helicopter platform configured for smoke plume sampling was used (e.g., Cofer et al., 1988). Aerosol particles that were unsegregated with respect to particle diameter were collected on prefired quartz microfiber filters (102 millimeters, mm, diameter). Collection times ranged between 0.5 to 4.0 minutes per sample at flow rates of 360 liters per minute (lpm). Several burn conditions were sampled: (1) background aerosol (out-of-plume) during full-fire conditions; (2) full-fire burning of in-plume aerosol; and (3) smoldering in-plume aerosol. Several samples of the full-fire and smoldering in-plume smoke aerosol were collected to identify a range of ambient mass concentrations of the carbonaceous aerosol present during the two fire regimes. The background aerosol sample

was collected as a method of evaluating the performance of the sampling equipment under dynamic conditions and provided a check for sampling-derived artifacts, if present. A travel blank consisting of a prefired and unused filter accompanied the sampled filters. The travel blank was used to monitor storage and handling conditions that existed for the collected sample filters and, later, to assess the total analytical background of possible carbonaceous artifacts. All filters were stored and shipped in annealed borosilicate glass jars that were fitted with Teflon-lined caps. Once collected, the smoke aerosol samples and blank filters were frozen ($-27°C$) until analysis.

Analysis
The analytical objective for this study is to construct the following mass balance relationship for the smoke aerosols collected during the prescribed burn:

Total carbon (organic + elemental): organic carbon: elemental carbon: elutable organics

Several analytical methods are required for the above mass determinations: (1) total carbon analysis by a combined pyrolysis/combustion measurement technique (Johnson et al., 1981); and (2) quantitation of solvent-soluble organics (i.e., compounds having 6 to 40 carbon atoms) by high-resolution gas chromatography (HRGC) flame ionization (Mazurek et al., 1987). Measurement of these bulk carbon constituents present in smoke aerosol particles permits a quantitative framework for identifying the release of these carbonaceous species into the atmosphere from biomass combustion.

Figure 33.1 shows a flow diagram of the extraction, separation, and quantitation procedures used to determine the elutable organics fraction (i.e., solvent-soluble organic compounds that are detected by HRGC). These micromethods have been developed for the quantitative recovery of extractable organic matter contained in the atmospheric fine aerosol fraction (i.e., aerosol particles having nominal diameters <2.1 μm) (Mazurek et al., 1987, 1988, 1990b). Routine quality control procedures are incorporated into the methodology and include monitoring of sampling artifacts, analysis of procedural blanks, and an accounting of analyte recovery. The organics are extracted from the filters by ultrasonic agitation using successive additions of hexane (two-volume additions) and benzene/isopropanol (three-volume additions). The serial extracts are filtered and then combined. The total extracts are reduced to volumes of 200 to 300 μL. The neutral fraction of the elutable

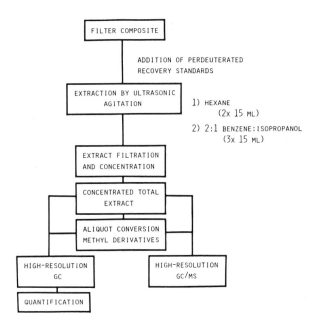

Figure 33.1 Flow diagram of the methods used for the extraction, quantitation, and characterization of elutable organic aerosol components.

organics is defined operationally as that fraction which elutes from the bonded phase (DB-1701) of the analytical column and is detected by the flame ionization detector (FID) of the HRGC without further derivatization. An aliquot of the total extract is derivatized by addition of diazomethane (Mazurek et al., 1987). This step converts reactive organic acids to the respective methyl ester or methyl ether analogues. Injection of the derivatized fraction onto the HRGC column produces chromatographic data for the acid plus neutral (acid + neutral) fraction. The mass of the acid fraction of the elutable organics is calculated by difference.

Quantitation of the elutable organics fraction is accomplished by computerized HRGC analyses that combine (1) area counts relative to a conjection standard (1-phenyldodecane); (2) relative response factors for the perdeuterated surrogate standard (n-$C_{24}D_{50}$); and (3) recoveries of the surrogate standard. Sampling and analytical procedural artifacts are monitored by high-resolution gas chromatography/mass spectrometry (HRGC/MS) analysis. This quality control/quality assurance step screens contributions of possible common sampling contaminants such as plasticizers, rubber stabilizers, silicone greases, lubricating greases, and oils (Mazurek et al., 1988). Overall, the analytical protocol for the elutable organics fraction is designed to provide quantitative

results which are adjusted for any losses associated with volatilization, incomplete extraction, or instrumental bias and for any additions of sampling or analytical organic artifacts.

Results and Discussion

Smoke Aerosol Bulk Carbon

Ambient mass concentrations for the total, organic, and elemental carbon components of the smoke aerosols are summarized in Table 33.1. Mass loadings per filter are given to allow comparison with the travel blank mass levels for the same carbon species. The travel blank contains low levels of TC (27 μg/filter); the carbon present is in the form of OC, and EC is not detected. The total carbon measured in the travel blank is small (2% to 4%) compared to the mass/filter loadings of total carbon found in the in-plume smoke aerosol samples. Comparison of the total carbon masses for the travel blank and the full-fire background aerosol shows a higher proportion of blank-to-sample loadings. However, it will be shown later that the major component of the full-fire background sample, the OC fraction, has a distribution profile that is distinct from that of the travel blank. Therefore, the carbonaceous material present in the full-fire background sample is composed mainly of compounds other than those found in the travel blank.

Different chemical properties for the smoke aerosol are seen in Table 33.1. Ambient mass concentrations for all the bulk carbon fractions are highest for the full-fire, in-plume smoke samples. These maximum ambient mass concentrations observed for TC, OC, and EC are 50 to 350 times the levels of TC and OC, and 1 to 40 times the levels of EC, that were measured for fine particulate matter (1982 annual averages) for the Los Angeles area (Gray et al., 1986). In comparison to a polluted urban atmosphere, the bulk carbon species concentrations are significantly higher for this prescribed burn event in rural Ontario. The distribution of OC and EC fractions within the total mass of aerosol carbon also is different for the biomass smoke aerosol and urban plume aerosol: EC accounts for less of the total carbonaceous aerosol mass in the prescribed burn aerosol studied here.

Important properties relating to the combustion conditions present during the prescribed burn can be tracked by the OC and EC ambient mass concentrations and by the OC:EC mass ratios. The vast propor-

Table 33.1 Ambient mass concentrations of bulk carbon species present as smoke aerosol from prescribed burning of a boreal forest (Ontario, August 1989)

Sample	Total carbon, $\mu g/m^3$ (μg/filter)	Organic carbon, $\mu g/m^3$ (μg/filter)	Elemental carbon, $\mu g/m^3$ (μg/filter)	Carbonaceous smoke aerosol as organic carbon (%)	Ratio of organic carbon to elemental carbon (OC:EC)
Full-fire background	244; (44)	233; (42)	11; (2)	95	21
Full-fire in-plume					
(Sample 1)	2280; (822)	2160; (780)	120; (42)	95	18
(Sample 2)	1720; (617)	1560; (560)	160; (57)	91	10
Smoldering in-plume					
(Sample 1)	1080; (1551)	1030; (1480)	50; (71)	95	21
(Sample 2)	575; (829)	569; (820)	6; (9)	99	95
Travel blank	—; (27)	—; (27)	—; (0)	—	—

tion (91% to 99%) of carbonaceous smoke aerosol is present as OC, regardless of the combustion temperature. Full-fire conditions produce the greatest ambient mass concentration of OC aerosol and are two to four times the ambient mass levels measured during smoldering fire conditions. Formation of EC particles is greatest during full-fire combustion relative to smoldering combustion. Elemental carbon ambient mass concentrations are similar for the two full-fire samples (120 to 160 $\mu g/m^3$), but vary over a wider range during smoldering conditions (6 to 50 $\mu g/m^3$). The relative mass distribution of organic carbon to elemental carbon within the smoke aerosol can be followed by the OC:EC mass ratios shown in Table 33.1. The OC:EC ratios have a range of 10 to 95 for all smoke aerosols sampled; full-fire smoke plume aerosol have lower OC:EC ratios (10 to 18) and smoldering in-plume aerosols have higher, more variable ratios (21 to 95). The OC and EC data show nonuniform production of these particles during various stages of the prescribed burn, and major differences in the type of carbonaceous aerosol that is generated (OC versus EC).

Smoke Aerosol Elutable Organics

Elutable organics are a subfraction of the OC smoke aerosol; further separation and analysis of this organic mixture by HRGC provides data on the chemical composition of these solvent-soluble compounds. Figures 33.2a to 33.2d show the HRGC plots of the elutable organics found in the aerosol samples and the travel blank. With the exception of the travel blank, the HRGC plots of the elutable organics are complex mixtures of individual organic compounds. Separate injections and analyses of an *n*-alkane standard mix-

ture run under identical HRGC conditions have shown that the elutable organics distributions reproduced in Figure 33.2 are comprised of organic compounds having 6 to 40 carbon atoms. In contrast, elutable organics present in the travel blank (plus analytical blank) show a very simple profile having few components that contain 6 to 25 carbon atoms. The species distribution of elutable organics present in the travel blank (Figure 33.2d) confirms the low background levels of the bulk carbonaceous materials (TC, OC, and EC as measured by combustion analysis) that are contributed from sources relating to sample handling and storage, and to sample analysis.

The ambient mass concentrations of elutable organics measured in the aerosol samples and the travel blank are given in Table 33.2. The full-fire and smoldering in-plume concentrations presented in Table 33.2 are two orders of magnitude greater than those for the urban Los Angeles ambient aerosol samples (1982 monthly averages) that were processed in a similar fashion (Mazurek et al., 1989; Hildemann, 1990). One-third of the total carbonaceous aerosol mass is composed of elutable organics (acid + neutral fraction) for both the full-fire and smoldering in-plume smoke aerosol samples. An even higher proportion of total carbonaceous aerosol mass (77%) is composed of elutable organics (acid + neutral fraction) for the full-fire background aerosol. A breakdown of the neutral and acidic species present as elutable organics for the full-fire and smoldering in-plume aerosol (measured by HRGC analysis) shows complex organic acids accounting for nearly 20% of the elutable organics ambient mass concentrations. The full-fire background aerosol and the travel blank sample are composed essentially all of neutral organic

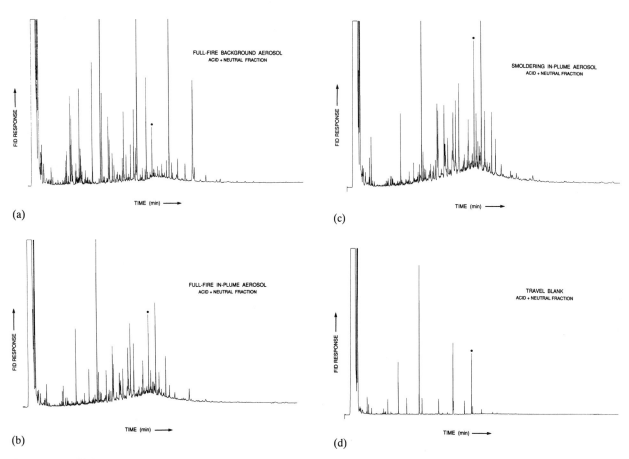

Figure 33.2 HRGC plots of the elutable organics from (a) full-fire background aerosol (acid + neutral fraction); (b) full-fire in-plume aerosol (acid + neutral fraction); (c) smoldering in-plume aerosol (acid + neutral fraction); and (d) travel blank (acid + neutral fraction). The solid circle indicates the peak corresponding to the surrogate standard (n-$C_{24}D_{50}$; retention time = 32 minutes).

Table 33.2 Ambient mass concentrations of elutable organics species present as smoke aerosol from prescribed burning of a boreal forest (Ontario, August 1989)

Sample	Neutral fraction, μg/m³ (μg/filter)[a]	Acid + neutral fraction, μg/m³ (μg/filter)[a]	Acid fraction, μg/m³ (μg/filter)[a]	Carbonaceous smoke aerosol as elutable organics (%)
Full-fire background	188; (49)	181; (47)	—; (—)	77
Full-fire in-plume[b]	495; (204)	601; (247)	106; (43)	30
Smoldering in-plume[b]	229; (377)	283; (466)	54; (89)	34
Travel blank	—; (8)	—; (8)	—; (—)	—

a. Represents the average μg/filter mass loading. Masses of the elutable organics (μg/filter) values reflect the additional masses of associated hydrogen, oxygen, nitrogen, etc., atoms. The masses of these atomic substituents are not accounted for in the OC bulk analysis, which measures just atomic carbon. Multiplication of the (μg/filter) values in Table 33.1 by a factor of 1.2 will yield the equivalent "organics" mass which takes into account the masses of the substituent atoms (Gray et al., 1984).
b. Filter samples 1 and 2 combined and extracted as a single composite.

compounds. These data for the prescribed burn studied indicate that massive concentrations of elutable organics are emitted during full-fire and smoldering fire conditions. Such large concentrations of organics (180 to 600 $\mu g/m^3$) may be important precursors in the production of high ozone levels which are commonly found during biomass burning (e.g., Evans et al., 1977; Stith et al., 1981; Andreae, 1990). So in addition to affecting atmospheric optical properties, emissions of elutable organics may also influence photochemical processes (oxidant production) within the earth's troposphere.

Conclusions

Carbonaceous particles are produced at nonuniform rates for the prescribed burn studied. The ambient mass concentrations for the particulate carbon smoke aerosol are highest for the full-fire burn conditions and do vary significantly throughout the burn. Therefore, collection strategies must define ranges in smoke aerosol concentrations that are produced. The mass distributions of the OC and EC fractions, which comprise the total mass of particulate carbon plume aerosol, also change with combustion temperature. Highest EC concentrations are observed under full-fire conditions. However, the vast proportion of smoke aerosol carbon is in the form of OC particles, regardless of the fire temperature. These OC and EC ambient mass concentrations show the relative mass distributions of light scattering (OC particles) versus light scattering *and* absorbing (EC particles) carbonaceous smoke aerosols. Formation of light scattering OC particles is a significant process for the prescribed burn studied.

Acknowledgments

This work was conducted under Contract No. DE-AC02-76CH00016 with the U.S. Department of Energy under the Atmospheric Chemistry Program within the Office of Health and Environmental Research.

Biomass Burning: Combustion Emissions, Satellite Imagery, and Biogenic Emissions

Joel S. Levine, Wesley R. Cofer III, Edward L. Winstead, Robert P. Rhinehart, Donald R. Cahoon, Jr., Daniel I. Sebacher, Shirley Sebacher, and Brian J. Stocks

Air chemistry and Earth's climate are controlled by the presence of trace gases present in very minute quantities—parts per million by volume (ppmv) (10^{-6}), parts per billion by volume (ppbv) (10^{-9}), and parts per trillion by volume (pptv) (10^{-12}). Many of these chemistry and climate controlling gases are increasing in concentration with time. Atmospheric greenhouse gases, carbon dioxide (CO_2), methane (CH_4), nitrous oxide (N_2O), and tropospheric ozone (O_3) are all increasing with time. Chemically active gases, nitric oxide (NO) and carbon monoxide (CO), are also increasing with time. These gases have varied sources of production, including the burning of fossil fuels (source of CO_2, CO, and NO), microbial metabolic activity $(CO_2, CH_4, N_2O,$ and NO), and chemical reactions (O_3). All of these gases are produced directly or indirectly (O_3) by biomass burning. Biomass burning includes the burning of the world's forests and savanna grasslands for land clearing and agricultural use, the burning of agricultural waste and stubble after harvesting, and the burning of biomass for fuel. Global biomass burning is more extensive than previously believed and appears to be increasing with time (Levine, 1990a, 1990b). It is also thought that the bulk of global biomass burning is human-initiated rather than naturally occurring, i.e., lightning induced fires (Levine, 1990a, 1990b). Biomass burning may be an important driver for global change in the atmosphere and climate.

This chapter deals with two different, but related, aspects of biomass burning. The first part of the chapter deals with a technique to estimate the instantaneous emissions of trace gases produced by biomass burning using satellite imagery. The second part of the chapter concerns the recent discovery that burning results in significantly enhanced biogenic emissions of N_2O, NO, and CH_4. Hence, biomass burning has both an immediate and long-term impact on the production of trace gases to the atmosphere. The objective of this research is to better assess and quantify the role and impact of biomass burning as a driver for global change. It will be demonstrated that satellite imagery of fires may be used to estimate combustion emissions and may in the future be used to estimate the long-term postburn biogenic emissions of trace gases to the atmosphere.

Estimates of Gaseous Emissions from Biomass Burning Using Satellite Imagery

On 6 May 1987 three fires resulting from human activities began in the extensive boreal forest of the Helongjing Province of the People's Republic of China. The Helongjing Province borders the Soviet Union along the Amur River. The Helongjing Province fires combined and continued to burn for about three weeks. More than 40,000 Chinese firefighters were called in to battle the fire. During the time that the Chinese fire was spreading, fire activity began across the Amur River in the Soviet Union. The exact cause of the fires in the Soviet Union is not known, but the possibility exists that they were related to the Chinese fires. The fires in China and the Soviet Union in May 1987 were very large events, perhaps the largest fires recorded in each country. Collectively, they may have been the largest fires recorded in history. These fires are referred to as either "the Great Chinese Fire of 1987" or "the Great Chinese/Soviet Fire of 1987." Due to the very large area and mass of boreal forest destroyed by this event, it was decided to study this fire using satellite imagery and to estimate the gaseous emissions released to the atmosphere by the fire. The areal extent of biomass burning by ecosystem which is deduced from satellite imagery provides an indication of the total amount of dry biomass burned within that ecosystem per unit time. The amount of dry biomass (M) (in units of grams of dry matter per unit time, or g dm/unit time) burned in a particular ecosystem is related to the total area (A) burned within that ecosystem by the following relationship (Seiler and Crutzen, 1980):

$$M = A \times B \times \alpha \times \beta$$

where B is the average organic matter per unit area in the individual ecosystem (g dm/m^2), α is the fraction of the average above-ground biomass relative to the total average biomass B, and β is the burning efficiency of the above biomass. Parameters B, α, and β vary with the ecosystem and are determined by assessing the dry biomass matter before and after a fire.

The complete and efficient burning of biomass produces water vapor and carbon dioxide as the primary combustion products, according to the reaction

$$CH_2O + O_2 \rightarrow H_2O + CO_2$$

where CH_2O represents the average chemical composition of biomass material. Dry biomass matter consists of about 45% carbon by weight, with the remainder being hydrogen and oxygen. In general, due to the incomplete and inefficient combustion of biomass matter, CO_2 accounts for about 90% of the gaseous carbon released, with CO accounting for about 10%, and CH_4 and nonmethane hydrocarbons (NMHCs) each representing about 1%. Dry biomass nitrogen ranges from about 0.3% to 3.8%, sulfur from 0.1% to 0.9%, phosphorus from 0.01% to 0.3%, and potassium from 0.5% to 3.4%. The total mass of carbon species ($M(C)$) produced by biomass burning is related to the mass of the burned matter (M) by

$$M(C) = 0.45M$$

To quantify the production of gases other than CO_2 and H_2O produced by biomass burning, we must determine the emission ratio (ER) for each gas. The emission ratio for each gas is defined as

$$ER = \frac{\Delta X}{\Delta CO_2}$$

where ΔX is the concentration of species X produced by biomass burning and equals $X^* - \overline{X}$, where X^* is the measured concentration in the biomass burn smoke plume and $\overline{X}$ is the background atmospheric concentration of the gas, and $\Delta CO_2 = CO_2^* - \overline{CO_2}$, where ΔCO_2 is the concentration of CO_2 produced by biomass burning, CO_2^* is measured concentration in the biomass burn plume and $\overline{CO_2}$ is the background atmospheric concentration of CO_2. Note that all species emission factors are normalized with respect to CO_2. CO_2 is the normalizing factor in species emission factors since the concentration of CO_2 produced by biomass burning may be related to the amount of dry biomass burned by simple chemical stoichiometric considerations, and the measurement of CO_2 in the

atmosphere and in a biomass burn smoke plume is a relatively simple and routine measurement. It is important to emphasize that the emission factor for the same species will vary depending on the ecosystem that is burning and the conditions of the fire. The species emission factor depends on the nature and chemical composition of the burning biomass material and the temperature (flaming vs. smoldering) of the fire. These parameters vary from ecosystem to ecosystem. Hence, it is critical to obtain measurements of species emission ratios from the major ecosystems of the world.

Using images obtained by the Advanced Very High Resolution Radiometer (AVHRR) aboard the NOAA-9 meteorological satellite, Cahoon et al. (1991) have estimated the areal extent of the 1987 fires in the boreal forests of China and the Soviet Union. Cahoon et al. (1991) estimate the fire covered at least 2.5 million acres in China and about 10 million acres in the Soviet Union during its 21-day lifetime in May 1987. Hence, the fires covered at least 12 million acres or 4.86×10^{10} m^2. According to Seiler and Crutzen (1980), the following numerical parameters are characteristic of a boreal forest: $B = 25 \times 10^3$ g m^{-2}, $\alpha = 75\%$, and $\beta = 20\%$. Using these parameters with the satellite-determined estimate for the total area of the fire, we find: $M = 1.8 \times 10^{14}$ g and $M(CO_2) = 0.45M = 8.1 \times 10^{13}$ g C. Cofer et al. (1990) have measured gaseous emissions from boreal forest fires and determined the emission ratios (ER) for boreal forest fires to be $CO/CO_2 = 12.3\%$, $CH_4/CO_2 = 1.26\%$ and $NMHCs/CO_2 = 1.03\%$. Using these emission ratios, we calculate the gaseous carbon emissions from the Chinese/Soviet fire during its 21-day lifetime to be: CO_2: 7.1×10^{13} g C; CO: 8.6×10^{12} g C; CH_4: 8.8×10^{11} g C; and NMHC: 7.21×10^{11} g C.

The calculation of gaseous emissions from the Chinese/Soviet fire of 1987 clearly illustrates that satellite imagery of fires may be used to determine gaseous emissions. To utilize this technique, we must know certain ecosystem dependent parameters, i.e., B, α, β, and the emission ratio for each gas under consideration. However, once these parameters are known for each of the world's major ecosystems, satellite imagery of fires will provide important information on gaseous emissions from these fires. The planned NASA Earth Observing System (EOS) will provide a unique platform for the routine global monitoring of burning and may provide definite information on the role of global burning as a source of trace gases to the atmosphere.

Microbial Biogenic Emissions

Microbial metabolic activity in the soils and wetlands of the world is a significant global source of trace gases to the atmosphere (Levine, 1989). Nitrifying and denitrifying microorganisms utilize ammonium and nitrate in soil to produce biogenic emissions of nitrous oxide and nitric oxide, and methanogenic microorganisms in wetlands, swamps, and rice paddies utilize carbon dioxide, acetate, and formate to produce biogenic emissions of methane. Biogenic emissions of N_2O, NO, and CH_4 due to microbial metabolic processes are very significant sources of these gases to the global atmosphere. Recent measurements indicate that burning results in significantly enhanced emissions of all three gases.

Biogenic Emissions of N_2O and NO

Measurement of biogenic emissions of NO and N_2O were obtained before and after two prescribed fires occurring in the San Dimas Experimental Forest, a chaparral ecosystem located in the Angeles National Forest, San Gabriel Mountains, about 60 km northeast of Los Angeles, California (Anderson et al., 1988; Levine et al., 1988). These controlled fires were planned and implemented by the Los Angeles County Fire Department as part of their fire-management and controlled burn activities at the San Dimas Experimental Forest. All flux measurements reported in this chapter were obtained using the flux chamber method. This technique and the detection instrumentation and calibration methods are discussed in detail elsewhere (Anderson et al., 1988;

Levine et al., 1988, 1990). The measurements from July 1986, two weeks after the fire, are summarized in Table 34.1 (Anderson et al., 1988). The measurements from the 22 June 1987 fire are summarized in Table 34.2 (Levine et al., 1988). Both sets of measurements indicate that NO and N_2O soil emissions from wetted sites exceeded those from dry sites and that the burned sites had significantly enhanced emissions. Wetting stimulates the metabolic activity of soil microorganisms and hence leads to the metabolic production of NO and N_2O. The artificial wetting performed in these experiments simulated natural wetting due to rainfall, i.e., 1 gallon (3.8 liters) of water was applied to the 0.4 m^2 area of the flux chamber. NO emissions from burned and wetted sites reached 60 ng N m^{-2} s^{-1}, which exceeded NO fluxes from agricultural fields measured after the application of fertilizers (Anderson et al., 1988). Comparable increases in NO emissions after burning were found in the tropical savanna of Venezuela (Johansson et al., 1988). N_2O emissions from burned and wetted sites also approached 60 ng N m^{-2} s^{-1}. N_2O emissions from unburned and wetted sites were below the detection limit of the measurement (2 ng N m^{-2} s^{-1}). On the three occasions that N_2O emissions were detectable (from the three burned and wetted sites), the ratio of the NO emissions to the N_2O emissions ranged from about 2.7 to 3.4. This observation may shed some light on the process responsible for the enhanced postfire emissions of both NO and N_2O as discussed later.

Over the years, researchers at the San Dimas Experimental Forest have measured the concentrations

Table 34.1 NO and N_2O fluxes from burned versus unburned San Dimas sites (July 1986)

Soil conditions	Mean flux NO (ng N m^{-2} s^{-1})		Mean flux N_2O (ng N m^{-2} s^{-1})	
	Unburned	Burned	Unburned	Burned
Dry	9.7 (4.5, 5)[a]	13.3 (1.1, 9)	NA	Not detectable[b]
Wetted[c] (1 day)	21.4 (3.5, 5)	60.7 (4.5, 14)	NA	53.9 (10.0, 7)
Wetted[c] (7–11 days)	NA	23.7 (2.6, 7)	NA	69.3 (12.0, 8)
Wetted[c] (180 days)	1.1 (1.1, 2)	22.3 (2.3, 8)	NA	Not detectable[d]

Source: Anderson et al., 1988.

NA indicates measurement was not attempted.

a. The first number in parentheses represents the standard error; the second number represents the number of measurements. The standard error = $s/\sqrt{n}$, where s is the standard deviation and n is the number of measurements.

b. Flux was not detectable with our measurement system. The minimum detectable flux was 4.8 ng N m^{-2} s^{-1} over a 30-min period at 298°K.

c. Most sites were irrigated with 1.3 to 2.5 cm of distilled water 30-min prior to making NO flux measurements. Under these conditions, N_2O was not detectable. For detection of N_2O, sites were flooded or continuously drip-irrigated.

d. Flux was not detectable with our measurement system. The minimum detectable flux was 1.6 ng N m^{-2} s^{-1} over a 30-min period at 298°K.

Table 34.2 Mean NO and N$_2$O fluxes from burned versus unburned San Dimas sites (June 1987)

| | Mean NO flux measurements (ng N m^{-2} s^{-1}) | | | | Mean N$_2$O flux measurements (ng N m^{-2} s^{-1}) | | | |
| | | After fire | | | | After fire | | |
	Before fire	Light burn	Moderate burn	Heavy burn	Before fire	Light burn	Moderate burn	Heavy burn
Grass, dry	11.15 (0.65)	10.15 (0.35)		3.5 (0.50)	BDL	BDL		BDL
Grass, wetted	30.5 (0.50)	27.25 (3.76)		58.15 (8.67)	BDL	8.65 (0.88)		21.85 (1.16)
Chaparral, dry	10.0 (0.50)		8.85 (0.35)		BDL		BDL	
Chaparral, wetted	15.5 (1.50)		45.05 (0.45)		BDL		12.77 (3.13)	

Source: Levine et al., 1988.

Note: Wetted measurements correspond to 1 gal. (3.8 L) of water applied to the 0.4 m^2 area of the flux chamber. The number in parentheses after the mean NO flux value represents the standard error of the mean ($s/\sqrt{n}$), where s is the standard deviation and n is the number of measurements (Arkin and Colton, 1953).

BDL indicates that the N$_2$O flux was below detection limit of our gas chromatograph. The minimum detectable N$_2$O flux with our instrumentation was 2 ng N m^{-2} s^{-1}.

of ammonium (NH$_4^+$) and nitrate (NO$_3^-$) in the soil of the chaparral forest before and after prescribed burns (DeBano et al., 1979; Dunn et al., 1979). De-Bano et al. (1979) concluded that burning significantly increases the concentration of ammonium and decreases the concentration of nitrate in the soil. DeBano et al. (1979) found that immediately after an intense fire, soil associated with chamise vegetation increased in ammonium concentration from 3.79 to 10.77 kilograms per hectare (kg/ha) and the nitrate concentration decreased from 0.58 to 0.47 kg/ha. They also found that immediately after an intense fire, the soil associated with Ceanothus vegetation increased in ammonium concentration from 1.69 to 12.13 kg/ha and the nitrate concentration decreased from 1.05 to 0.33 kg/ha (DeBano et al., 1979).

In summary, we found that NO emissions from moderately or heavily burned and wetted grass and vegetated sites reached 60 ng N m^{-2} s^{-1}. These fluxes exceeded NO fluxes measured from a fertilized corn field within days after fertilization and greatly exceeded NO fluxes from natural (unfertilized) grasses (Anderson and Levine, 1987). N$_2$O emissions from lightly, moderately, and heavily burned and wetted sites were also significantly enhanced. N$_2$O fluxes were not detected from these sites prior to the burn even after wetting (indicating an N$_2$O flux of less than 2 ng N m^{-2} s^{-1}). From burned and wetted sites, the flux of NO exceeded the N$_2$O flux by factors ranging from 2.7 to 3.4. Preburn and postburn measurements of soil ammonium and nitrate (DeBano et al., 1979) and laboratory studies of biogenic emissions of nitric oxide and nitrous oxide (Anderson and Levine, 1986) have provided important information on the micro-

organism and the metabolic pathway responsible for the postburn enhancement in nitric oxide and nitrous oxide. Laboratory studies showed that over a wide range of oxygen levels (from 0% to 20%), the ratio of NO to N$_2$O emissions for nitrifying bacteria ranged from about 1.0 to 8.5, while that ratio for denitrifying bacteria was less than unity (Anderson and Levine, 1986). This observation coupled with the measurements that show soil ammonium, the substrate for nitrification, increased after burning, while soil nitrate, the substrate for denitrification, decreased after burning (DeBano et al., 1979), suggests that the postburn enhanced emissions of nitric oxide and nitrous oxide most probably resulted from nitrification associated with enhanced postburn concentrations of soil ammonium.

Biogenic Emissions of CH$_4$

Our group obtained measurements of biogenic emissions of CH$_4$ before and after two prescribed wetlands fires at the NASA John F. Kennedy Space Center (KSC) within the Merrit Island National Wildlife Refuge in Brevard County, Florida (Levine et al., 1990). These controlled fires, planned and implemented by the U.S. Fish and Wildlife Service as part of their fire management and controlled burn activities at KSC, occurred on 9 November 1987 (KSC I), and on 11 November 1988 (KSC III). During these burns we obtained prefire and postfire measurements of CH$_4$ emissions from two different wetlands sites, a *Juncus roemerianus* marsh and a *Spartina bakeri* marsh with standing water.

Preburn and postburn measurements of biogenic

emissions of methane were obtained at two different standing water sites (a *Juncus roemerianus* marsh and a *Spartina bakeri* marsh) at both the 9 November 1987 (KSC I) and 11 November 1988 (KSC III) fires. These measurements, along with measurements of the temperature and pH of the standing marsh water, are summarized in Tables 34.3 and 34.4. Inspection of these tables indicates that the magnitude of the methane emissions increased significantly following the 1987 and 1988 fires. In both measurement sets the preburn and postburn emissions form the *Spartina bakeri* sites exceeded the emissions from the *Juncus roemerianus* sites. However, the magnitude of the methane emissions from both sites in 1987 was less than the magnitude of the emissions from both sites in 1988. We believe that the explanation for the lower methane emissions and the higher values of pH in 1987 is due to the fact that only six days before the 1987 fire (9 November 1987), both the *Juncus roemerianus* and *Spartina bakeri* marshes were very dry, i.e., not covered with any standing water. On 3

November 1987, a late season tropical depression produced 14.5 cm of rain in a 24-hour period. This storm provided enough rain for the marsh to be covered with standing water less than a week before the fire. In direct contrast, for the 1988 measurements, both marsh sites were flooded and continuously covered with standing water (about 18 cm) since August, about three months before the 11 November 1988 fire. We believe that the lower-valued methane emissions recorded in 1987, compared to 1988, were due to the fact that the marshes, due to the very recent rain and standing water, had not yet reached equilibrium as anaerobic environments. The higher values of pH for the standing water in 1987 are also attributed to the fact that very recent rains were responsible for the standing water.

After the 1987 fire, methane emissions in the *Juncus roemerianus* marsh increased from below the detection limit of the instrument (0.3×10^{-3} g CH_4 m^{-2} d^{-1}) to 5–6 $\times 10^{-3}$ g CH_4 m^{-2} d^{-1}. Measurements of methane emissions at our unburned, control *Juncus roemerianus* site were always below the detection limit of the instrument. Methane emissions in the *Spartina bakeri* site, which were below the detection limit of the instrument before the fire, increased to 34.5×10^{-3} g CH_4 m^{-2} d^{-1} after the fire. Measurements of methane emissions before and after the 11 November 1988 fire in both *Juncus roemerianus* and *Spartina bakeri* marshes showed a very similar trend. The main difference is that in the 1988 measurements, methane emissions were detected before the fire. We attribute the detectability of the methane emissions before the 1988 fire to the fact that standing water covered both marshes for at least three months prior to the start of the measurements, thereby establishing the anaerobic environment needed for the production of methane.

Measurements of biogenic emissions of methane from the two standing water sites, before and after the 11 November 1988 fire (KSC III), are summarized in Table 34.4. Both of these marsh sites were covered with about 18 cm of standing water for at least three months before the fire. The CH_4 emission histories from both sites following the fire is similar. Three days following the fire the methane emissions from both sites were somewhat below the prefire level. Five days after the fire the *Juncus roemerianus* site exhibited its maximum flux (about 100% above the prefire value), and three days after the fire the *Spartina bakeri* site exhibited its maximum flux (about 50% above the prefire value). We believe that the decrease in meth-

Table 34.3 Mean CH_4 flux measurements, water temperature, and pH: 9 November 1987 fire, NASA Kennedy Space Center

Date	Sites 1 and 4, *Juncus roemerianus* in standing water		Site 2, *Spartina bakeri* in standing water, burned
	Burned (Site 1)	Unburned (Site 4)	
Mean CH_4 flux measurements			
5 Nov.	<0.3[a]	<0.3	< 0.3
7 Nov.		<0.3	
9 Nov. (fire)			
12 Nov.	5.0 ± 1.1[b]		33.5 ± 6.7
13 Nov.		<0.3	
18 Nov.		<0.3	
19 Nov.	6.0 ± 1.3		34.5 ± 6.9
Water Temperature, °C/pH			
5 Nov.	22.5/7.1	—	23.4/7.0
7 Nov.		—	
9 Nov. (fire)			
12 Nov.	18.5/7.5		18.4/7.5
13 Nov.			
18 Nov.		—	
19 Nov.	21.3/7.5	—	21.3/7.5

Source: Levine et al., 1990.
Note: All flux measurements corrected for temperature effects; CH_4 flux units in 10^{-3} g CH_4 m^{-2} d^{-1}.
a. Below minimum detection limit of instrument which is 0.3 $\times$ 10^{-3} g CH_4 m^{-2} d^{-1}.
b. The first number represents the mean flux value based on at least three individual flux measurements. The measurement accuracy is based on the measured spatial variability within the site. The precision of the methane flux instrument is ± 0.1 $\times$ 10^{-3} g CH_4 m^{-2} d^{-1}.

Table 34.4 Mean CH$_4$ flux measurements, water temperature, and pH: 11 November 1988 fire, NASA Kennedy Space Center

Date	Site 1, *Juncus roemerianus* in standing water		Site 2, *Spartina bakeri* in standing water	
	Unburned	Burned	Unburned	Burned
Mean CH$_4$ flux measurements				
28 Oct.	27 ± 5[a]		114 ± 19	
2 Nov.	31 ± 4		146 ± 19	
8 Nov.	25 ± 3		143 ± 23	
11 Nov. (fire)		20 ± 4		53 ± 9
12 Nov.	25 ± 5	17 ± 3		56 ± 9
14 Nov.	25 ± 4	28 ± 6	100 ± 18	176 ± 34
16 Nov.	30 ± 5	52 ± 11	107 ± 18	121 ± 25
Water temperature, °C/pH				
28 Oct.	20.5/6.17		20.5/6.17	
2 Nov.	17.5/5.67		18.1/6.22	
8 Nov.	17.8/5.45		20.8/5.84	
11 Nov. (fire)		22.2/7.05		22.9/8.50
12 Nov.	21.4/5.66	21.4/6.64		23.0/6.55
14 Nov.	19.7/5.64	23.4/6.26	23.6/6.19	24.1/6.58
16 Nov.	20.9/5.88	23.1/6.21	22.1/5.61	25.2/6.37

Source: Levine et al., 1990.
Note: All flux measurements corrected for temperature effects; CH$_4$ flux units in 10^{-3} g CH$_4$ m^{-2} d^{-1}.
a. The first number represents the mean flux value based on four individual flux measurements. The measurement accuracy is based on the measured spatial variability within the site. The precision of methane flux instrument is ± 0.1 × 10^{-3} g CH$_4$ m^{-2} d^{-1}.

ane emissions within 48 hours of the fire is due to large quantities of methane purged from the marsh water due to the high temperature of the fire as well as due to the removal of methane-saturated flora, which serve as a conduit for methane emissions (Sebacher et al., 1985).

Methane emissions from different wetlands ecosystems are summarized in Table 34.5. Note that only two ecosystems, high-latitude, peat-rich terrain and Alaskan alpine fen, produced methane emissions that exceeded the maximum postburn emissions measured after the 1988 fire.

After the short-lived postburn decrease in methane emissions in 1988 the emissions began to increase, as they did in 1987. Methane is produced by the metabolic activity of methanogenic bacteria in a strictly anaerobic environment. Methanogens can only utilize a restricted suite of compounds for both growth and the metabolic production of methane. The compounds needed by methanogens for growth and methane production are carbon dioxide, carbon monoxide, acetate, formate, methanol, and methylated amines (Cicerone and Oremland, 1988). Of these six compounds needed for methane production, four are produced in large concentrations during biomass burning: carbon dioxide, carbon monoxide, acetate, and formate. Carbon dioxide and carbon

Table 34.5 Methane emissions from different wetland ecosystems

Ecosystems	CH$_4$ emissions, g CH$_4$ m^{-2} d^{-1}
High-latitude, peat-rich ecosystems (Svensson, 1980)	0.3
Alaskan alpine fen (Sebacher et al., 1986)	>0.250
Maximum postburn emission reported in this chapter (*Spartina bakeri*)	0.176
Preburn emission range reported in this chapter (*Juncus roemerianus* and *Spartina bakeri*)	0.025–0.146
Alaskan wetland ecosystems (Sebacher et al., 1986)	0.112
Sawgrass (Bartlett et al., 1985)	0.11
Swamp forest (Bartlett et al., 1985)	0.072
Florida Everglades (Bartlett et al., 1985)	0.013–0.03
Forested freshwater swamps, southeastern United States (Harriss and Sebacher, 1981)	0.0046–0.068
Great Dismal Swamp (Harriss et al., 1982)	<0.001

Source: Levine et al., 1990.

monoxide are the two most abundant gases produced during biomass burning (Cofer et al., 1990). Our chemical analyses of biomass burn particulates collected by airborne filters in the smoke plume indicate that these particulates contain large concentrations of both acetate and formate. Carbon dioxide is a very water soluble gas and acetate and formate are very water soluble compounds. In general, methanogens are dependent upon other microorganisms for providing them with the needed carbon dioxide, carbon monoxide, acetate, and formate (Cicerone and Oremland, 1988). However, biomass burn combustion products are a significant source of the substrates needed by methanogens for the production of methane. Hence we hypothesize that the postburn enhancement of methane emissions is related to the enhanced concentrations of carbon dioxide, carbon monoxide, acetate, and formate supplied to the methanogens by biomass burning. The carbon dioxide, acetate, and formate are dissolved into the marsh standing water and diffuse down to the marsh floor where they are utilized by the methanogens in the production of methane. The acetate- and formate-rich ash particles began falling to the marsh as soon as the fire began. We estimate that these particles diffused through the marsh standing water in only a matter of hours. It is possible that the postburn enhancement of methane emissions is also related to the massive destruction of marsh flora by the fire and the elimination of a user of nutrients, making more nutrients available to the marsh methanogenic bacteria.

Measurements of preburn and postburn biogenic emissions of methane from a wetlands indicate a significant postburn emission enhancement. We hypothesize that the postburn enhancement in methane is due to enhanced levels of carbon dioxide, formate, and acetate, all water soluble biomass burn combustion products, that readily dissolve into the standing water of the wetlands marsh and serve as the substrate for the production of methane by methanogenic bacteria. These measurements indicate for the first time that burning results in enhanced biogenic emissions of methane in the wetlands.

Biomass Burning: A Global Phenomenon

Recent estimates suggest that global burning is much more widespread and may cover 300 to 700 million ha (3 to 7 million km²) per year (National Academy of Sciences, 1984), which corresponds to 2% to 5% of the land area of our planet (1 hectare, ha = 2.47 acres). Much of this burning is human initiated and appears to be increasing with time (Levine, 1990a, 1990b). When we think of human-initiated burning, we usually think of only the burning of the tropical rainforest and the tropical savanna for land clearing. In Brazil alone, about 8 million ha of tropical rain forest were burned for land clearing in 1987 (Booth, 1989). Brazil possesses 357 million ha of tropical forest, which is 30% of the world's total and three times as much as Indonesia, which is second to Brazil in its tropical forest extent (Booth, 1989).

However, the burning of living and dead biomass for land clearing is not restricted to the tropics. For 9 out of 10 world regions there has been a significant transformation of natural forests and woodlands to grasslands, pasture, and croplands via burning for the period 1850 to 1980 (International Institute for Environment and Development and World Resources Institute, 1987). According to this study, the forests and woodlands area for each of the 10 world regions changed by the indicated amount from 1850 to 1980: Latin America, from 1420 to 1151 million ha, a decrease of 19%; tropical Africa, from 1336 to 1074 million ha, a decrease of 20%; Soviet Union, from 1067 to 941 million ha, a decrease of 12%; North America, from 971 to 942 million ha, a decrease of 3%; South Asia, from 317 to 180 million ha, a decrease of 43%; Pacific developed countries, from 267 to 246 million ha, a decrease of 8%; Southeast Asia, from 252 to 235 million ha, a decrease of 7%; Europe, from 159 to 167 million ha, an increase of 4%; China, from 96 to 58 million ha, a decrease of 39%; North Africa and the Middle East, from 34 to 14 million ha, a decrease of 60%. By combining all 10 world regions, we find that the total area of forests and woodlands has changed from 5919 to 5007 million ha from 1850 to 1980, a decrease of 15%.

It is interesting to note that several regions that have experienced the greatest loss of forests and woodlands are extratropical, i.e., the Soviet Union, North America, and China. In fact, two very large recent fires occurred in the temperate and boreal forests. In the autumn of 1988, nearly 0.6 million ha of Yellowstone National Park burned (Jeffrey, 1989). These fires covered two-thirds of the 0.9 million ha of Yellowstone, the world's largest temperate forest. Altogether in the hot and dry summer and autumn of 1988, 1.5 million ha in the western United States and 0.8 million ha in Alaska burned (Jeffrey, 1989). However, these fires were dwarfed by fires in China and the Soviet Union a year earlier in what may have been the largest fires ever recorded, as described in this chapter. These statistics indicate that fires, mostly human initiated, have transformed a significant amount of forests and woodlands into grasslands, pasture, and croplands over the last 130 years, and that these fires are not limited to the tropics (Levine, 1990a, 1990b).

The burning of the world's forests, grasslands, and agricultural stubble has several distinct impacts on the atmosphere and climate. Burning leads to the production of atmospheric greenhouse gases, carbon dioxide, methane, and tropospheric ozone (Levine, 1991). Burning also leads to the atmosphere buildup of carbon monoxide and nitric oxide, which forms nitric acid, the fastest growing component of acid rain. Burning of the world's forests destroys a major sink for atmospheric carbon dioxide and an important source of atmospheric oxygen due to photosynthetic activity. Burning also enhances the biogenic production of greenhouse gases, nitrous oxide, methane, and nitric oxide, the precursor of nitric acid (Levine, 1989). Global burning has significantly increased over the last century with the bulk of it human-initiated (Levine, 1990a, 1990b). The countries of the world must reexamine their biomass burning practices with the goal of a significant reduction. The role of human-initiated biomass burning as a driver of global atmospheric change must be reduced.

Changes in Marsh Soils for Six Months after a Fire

Paul A. Schmalzer, C. Ross Hinkle, and Albert M. Koller, Jr.

Wetlands are major sources of methane releases to the atmosphere (Cicerone and Oremland, 1988; Devol et al., 1988; Aselmann and Crutzen, 1989). Releases are seasonally variable in natural and cultivated wetlands (e.g., Crill et al., 1988; Schutz et al., 1989). Methane is an important trace gas in the atmosphere, involved in numerous chemical reactions (Cicerone and Oremland, 1988; Levine, 1989), and it is a greenhouse gas (Ramanthan et al., 1985; Dickson and Cicerone, 1986). Increasing atmospheric concentrations of methane (Rasmussen and Khalil, 1981a, 1981b; Khalil and Rasmussen, 1983, 1987; Blake and Rowland, 1988) have implications for global climatic change.

Fires occur naturally in many wetland systems (e.g., Cohen, 1974; Wade et al., 1980; Wilbur and Christensen, 1983; Izlar, 1984). Marsh burning for wildlife management is widely practiced in North America (Wright and Bailey, 1982; Smith and Kadlec, 1984; Thompson and Shay, 1985; Mallik and Wein, 1986). Marsh burning has a long history in the southeastern United States (Penfound and Hathaway, 1938; Lynch, 1941; Zontek, 1966; Vogl, 1973; VanArman and Goodrick, 1979; Hackney and de la Cruz, 1983). Nonetheless, there are only a few studies on changes in soil nutrients in wetlands following fire. Faulkner and de la Cruz (1982) examined fire effects on soils of irregularly flooded *Juncus roemerianus* and *Spartina cynosuroides* marshes. Wilbur and Christensen (1983) examined nutrient changes in the peat soils of a pocosin after fire. Reviews of fire effects on soils (Raison, 1979; Wells et al., 1979) are based nearly entirely on upland, well-drained soils. Chemistry of wetland soils differs greatly from that of well-drained soils (Ponnamperuma, 1972; Reddy and Patrick, 1984). Thus, differences in fire effects in wetlands compared to uplands may be expected.

Fire influences many ecosystem properties. Total nutrients, particularly nitrogen, may be reduced by volatilization of that in biomass, but available nutrients may be increased by deposition of ash and partially burned residue (Raison, 1979; Wells et al.,

1979; Faulkner and de la Cruz, 1982). Vegetation removal results in microclimatic changes at a given site, particularly soil temperature regimes. Nutrient and temperature changes have the potential to affect activity of the microbial community. Only recently have the effects of fire on methane production in wetlands been examined (Levine et al., 1990; this volume, Chapter 34).

In this study, we examined changes in soil nutrient levels in marsh systems after fire. These studies were conducted in conjunction with studies of particulates and gases generated from biomass combustion (LeBel et al., 1988; this volume, Chapter 29; Cofer et al., 1990; this volume, Chapter 27) and flux measures of methane and nitric oxide before and after the fire (Levine et al., 1990; this volume, Chapter 34). Here we present data through six months postfire, past the time during which flux measurements were made. These data indicate that changes in soil properties occur at different times after the fire and persist for different intervals, indicating the need for long-term postfire observations.

Study Area

This study was conducted at John F. Kennedy Space Center (KSC) on Merritt Island on the east coast of central Florida (Figure 35.1). KSC consists of 57,000 hectares (ha) of land and lagoonal waters. Areas not actively used by the space program are managed as Merritt Island National Wildlife Refuge by the U.S. Fish and Wildlife Service (USFWS).

Merritt Island and the adjacent Cape Canaveral form a barrier island complex of Pleistocene and Recent age (White, 1958, 1970). The topography is marked by a series of ridges and swales derived from relict dunes deposited as the barrier islands were formed. Erosion has reduced the western side of Merritt Island to a nearly level plain (Brown et al., 1962). Elevation ranges from sea level to about 3 meters (m) in the inland areas and to 6 m on the recent dunes. Soils of the area have been derived primarily from

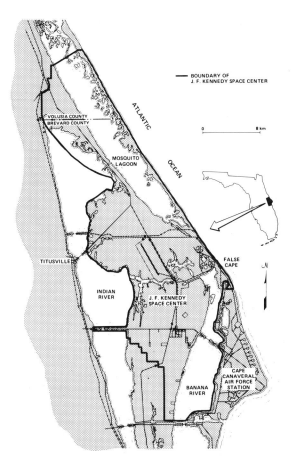

Figure 35.1 Location of John F. Kennedy Space Center.

Table 35.1 Plant communities and land use types within the controlled burn site

Type	Area (ha)	Percent
Cattail marsh	594.5	29.6%
Impounded waters	355.5	17.7
Cabbage palm savanna	313.2	15.6
Graminoid marsh[a]	294.8	14.7
Mixed oak, saw palmetto scrub	117.4	5.9
Cabbage palm hammock	82.6	4.1
Dikes	75.6	3.8
Red bay, laurel oak, live oak hammock	59.3	3.0
Live oak, cabbage palm hammock	36.8	1.8
Willow swamp	23.1	1.2
Mixed mangrove	10.9	0.5
Hardwood swamp	10.7	0.5
Ruderal	10.2	0.5
Salt marsh	5.2	0.3
Other	16.0	0.8
Total	2005.8	100.0

a. This type includes separate marsh sites dominated by *Juncus roemerianus, Spartina bakeri,* or *Caldium jamaicense.*

deposits of sand and sandy coquina but vary greatly with landscape position, drainage, and age of parent material (Huckle et al., 1974).

The climate is warm, humid subtropical. Annual precipitation averages 137 centimeters (cm) and ranges from 5.64 cm in January to 20.22 cm in September. Mean daily maximum temperatures are 22.3°C for January and 33.3°C for July; mean daily minimum temperatures are 9.5°C for January and 21.8°C for July (Titusville station climate records). Freezing temperature may occur in winter but do not persist. Thunderstorms are frequent in the summer with numerous lightning strikes that sometimes cause wildfires.

The flora of KSC consists primarily of temperate species with some subtropical elements (Poppleton et al., 1977). Vegetation patterns are complex (Schmalzer and Hinkle, 1985; Provancha et al., 1986) and vary with landscape position, drainage, soil characteristics, and other factors. Scrub, slash pine flatwoods, and hammocks are the most important upland

communities. Wetlands include hardwood swamps, and freshwater, brackish, and saline marshes of varying species composition. Most of the marshes fringing the lagoon systems (Banana River, Indian River, and the Mosquito Lagoon) have been impounded for mosquito control (Leenhouts and Baker, 1982). Controlled burning on KSC is conducted by the USFWS for habitat management and fuel reduction (Lee et al., 1981; Adrian et al., 1983).

The site for the specific study was a 2006 ha controlled burn unit previously burned in December 1985. Plant community composition is given in Table 35.1. Soil and vegetation studies were conducted in the graminoid marsh type that constitutes 14.7% of the area; fluxes of methane and nitric oxide were also measured in these marshes (Levine et al., 1990). Cabbage palm savanna (15.6% of the unit) is similar to the graminoid marsh type except for the presence of scattered cabbage palm (*Sabal palmetto*).

Methods

Within the fire management unit, we selected two representative marshes scheduled to be burned and one marsh that would be left unburned. At the time of the site selection (September 1988), all three marshes were flooded to a similar depth (about 15 to 30 cm). One marsh to be burned was dominated by *Juncus roemerianus* and the other by *Spartina bakeri*. The marsh that remained unburned was primarily *Juncus roemerianus* with other taxa present.

Vegetation Sampling

We sampled vegetation of the *Juncus* and *Spartina* marshes preburn with five 15 m transects each and determined vegetation cover by taxa in the >0.5 m and <0.5 m layers. Vegetation of the unburned marsh was sampled in November 1989, one year after the fire.

Biomass Sampling and Analysis

We sampled above-ground biomass of the *Juncus* and *Spartina* marshes preburn by harvesting 25 plots (0.25 m²) in each marsh. Samples were separated by taxa and by live and dead categories, weighed, dried at 100°C for 24 hours in a forced air drying oven, and dry weights determined. Dried samples were ground in a Wiley mill. For cations and phosphorus, 1 g of oven-dried material was dry ashed at 450°C in a muffle furnace and taken up in hydrochoric acid. Analyses for Ca, Mg, and K were performed on a Perkin-Elmer Model 6500/XR inductively coupled spectrometer (Wallace and Barrett, 1981); total phosphorus was determined on a Technicon Autoanalyzer (Technicon Industrial Systems, 1983d). Total Kjeldahl nitrogen (TKN) was determined by digesting a 0.25 gram (g) sample in 2 milliliters (ml) concentrated H_2SO_4, 2 ml 30% H_2O_2, and 4 ml of K_2SO_4 digestion mixture in a model BD-40 block digester and analyzing on a Technicon Autoanalyzer (Technicon Industrial Systems, 1983b). Standing crops of Ca, Mg, K, P, and TKN were calculated by multiplying the amount of each biomass category for a plot by its respective concentration and summing for the plot.

Soil Sampling and Analysis

We sampled soils of the *Juncus* and *Spartina* marshes that burned and of the unburned marsh. All sites were sampled preburn (18 days); postburn (3 days); and 1, 3, and 6 months postburn. Each sampling consisted of 25 soil samples, each composited from at least 5 soil cores (2 cm diameter); we collected samples from the 0 to 5 cm and 5 to 15 cm layers. Samples were randomly distributed throughout the marshes. Postfire sampling of the burned marshes was random within burned areas of the marshes.

We homogenized and then subsampled the soil samples. Subsamples were oven-dried at 50°C for 24

hours or until dry. The rest of the samples were air-dried. All chemical analyses except organic matter were conducted by the NASA/KSC Environmental Chemistry Laboratory.

Oven-dried samples were used for analyses of nitrate- (NO_3-N), nitrite- (NO_2-N), and ammonium- (NH_4-N) nitrogen. Exchangeable NO_3-N, NO_2-N, and NH_4-N were extracted in 2N potassium chloride (Keeney and Nelson 1982) and then analyzed on a Technicon Autoanalyzer (Technicon Industrial Systems, 1973, 1983a).

All other analyses used air-dried samples. pH was determined on a 1:1 soil-to-water slurry (McLean, 1982) using an Orion pH meter. Conductivity was measured on a 1:5 soil-to-water solution using a conductivity meter (Rhoades, 1982). Exchangeable cations (calcium [Ca], magnesium [Mg], and potassium [K]) were extracted in neutral 1N ammonium acetate (Knudsen et al., 1982; Lanyon and Heald, 1982) and analyzed by atomic absorption spectrophotometer (Perkin-Elmer Corporation, 1982). Available phosphorus was determined by extraction in deionized water (Olsen and Sommers, 1982) followed by analysis on a Technicon Autoanalyzer (Technicon Industrial Systems, 1983c). Total Kjeldahl nitrogen was determined by micro-Kjeldahl digestion (Schuman et al., 1973) followed by analysis on a Technicon Autoanalyzer (Technicon Industrial Systems, 1983b). Organic matter was determined by combustion (Nelson and Sommers, 1982) by Post, Buckley, Schuh, and Jernigan, Inc., Orlando, Florida.

Results

Vegetation

Preburn composition of the *Juncus* marsh was primarily *Juncus roemerianus* with some *Sagittaria lancifolia* (Table 35.2). The *Spartina* marsh was dominated by *Spartina bakeri* with some *Sagittaria lancifolia* and *Juncus roemerianus* (Table 35.3). *Juncus roemerianus* with various other graminoids and herbs dominated the marsh that remained unburned (Table 35.4).

Biomass

Juncus and *Spartina* dominated the biomass categories of their respective marshes (Table 35.5). Standing dead biomass equalled or exceeded live biomass. Biomass contained substantial quantities of major nutrients (Table 35.6) that could be released by fire. Live biomass was the more important pool of P and K, dead biomass was more important for Ca, while live and dead biomass pools of TKN and Mg were similar.

Table 35.2 Preburn composition of the *Juncus* marsh (means with standard deviations in parentheses) (N = 5)

Taxa	Cover (%)	
	>0.5 m	<0.5 m
Juncus roemerianus	91.5 (11.8)	—
Ludwigia peruviana	1.2 (2.7)	—
Open water	—	0.3 (0.6)
Sagittaria lancifolia	14.3 (8.7)	—
Salix caroliniana	1.3 (3.0)	—
Spartina bakeri	0.1 (0.3)	—
Urtricularia inflata	—	0.7 (1.2)

Table 35.3 Preburn composition of *Spartina* marsh (means with standard deviations in parentheses) (N = 5)

Taxa	Cover (%)	
	>0.5 m	<0.5 m
Erianthus giganteus	1.5 (2.0)	—
Eriocaulon spp.	—	5.1 (6.8)
Ipomoea sagittata	0.2 (0.3)	—
Juncus roemerianus	16.1 (19.4)	—
Mikania scandens	0.1 (0.3)	—
Open water	—	3.2 (2.1)
Polygonum spp.	—	0.1 (0.1)
Sagittaria lancifolia	16.7 (5.8)	—
Spartina bakeri	72.1 (13.7)	—

Table 35.4 Composition of the unburned marsh (means with standard deviations in parentheses) (N = 5)

Taxa	Cover (%)	
	>0.5 m	<0.5 m
Andropogon capillipes	0.7 (1.2)	0.1 (0.3)
Erianthus giganteus	8.3 (5.0)	—
Fuirena breviseta	0.4 (0.3)	0.1 (0.3)
Gratiola ramosa	—	0.7 (0.9)
Hyptis alata	0.3 (0.6)	—
Ipomoea sagittata	2.5 (1.3)	0.3 (0.4)
Juncus roemerianus	53.4 (13.5)	0.4 (0.9)
Ludwigia repens	—	0.3 (0.6)
Mikania scandens	1.1 (2.2)	0.6 (0.8)
Oxypolis filiformis	7.5 (6.6)	0.1 (0.3)
Panicum chamalonche	—	1.8 (3.1)
Panicum rigidulum	4.3 (7.6)	0.2 (0.4)
Pluchea rosea	5.1 (2.7)	8.8 (5.2)
Rhynchospora inundata	9.3 (9.5)	2.0 (3.8)
Rhynchospora microcarpa	6.3 (6.3)	1.5 (1.9)
Sacciolepis striata	0.1 (0.1)	—
Sagittaria lancifolia	—	0.8 (0.4)
Scleria reclinata	1.8 (3.6)	0.5 (1.2)
Spartina bakeri	2.9 (5.1)	—
Xyris spp.	0.1 (0.3)	0.1 (0.1)

Table 35.5 Major categories of biomass in the *Juncus* and *Spartina* marshes (means with standard deviations in parentheses) (N = 25)

Biomass type	Biomass (g/m²)	
	Juncus marsh	*Spartina* marsh
Live *Juncus*	692.6 (214.7)	19.0 (52.0)
Dead *Juncus*	883.1 (292.0)	59.5 (195.3)
Live *Spartina*	—	594.2 (372.0)
Dead *Spartina*	—	717.9 (336.2)
Live *Sagittaria*	115.5 (48.7)	112.9 (55.1)
Dead *Sagittaria*	113.4 (50.8)	186.2 (156.5)
Total live	812.6 (204.3)	772.6 (371.2)
Total dead	996.5 (294.4)	963.7 (366.5)
Total	1809.2 (436.2)	1736.2 (632.2)

Note: Minor species are omitted.

Soils

Comparison of the preburn composition of the soils of the three marsh sites indicated that they were similar but not identical (Table 35.7). In particular, the unburned reference stand was higher in pH and calcium but lower in magnesium and organic matter than the other marshes. The *Spartina* marsh soils had higher conductivity than the others and greater TKN levels in the surface 0 to 5 cm soils. Differences in other parameters were minor or were significant in one soil layer but not the other.

Soil parameters showed different patterns of change postfire. Hydrogen ion concentration decreased in the burned marshes immediately postfire in both the 0 to 5 and 5 to 15 cm layers but returned to preburn values by one month postfire (Figure 35.2). This change is equivalent to a pH increase of 0.16 to 0.28 units. The pH of the unburned site did not change postburn. After one month postburn, pH in burned and unburned sites changed similarly.

Organic matter increased at one month postburn in the surface layer of both burned marshes and remained elevated through six months postfire (Figure 35.3). Small increases occurred in the 5 to 15 cm layer.

Table 35.6 Standing crops of major nutrients in the marshes preburn

Biomass Category	*Juncus* X̄ (SD) (N = 25)	*Spartina* X̄ (SD) (N = 25)
Total Kjeldahl nitrogen (g/m²)		
Total live	6.86 (1.80)	4.59 (1.87)
Total dead	6.06 (2.57)	5.64 (3.11)
Total	12.92 (3.93)	10.24 (4.75)
Total phosphorus (g/m²)		
Total live	0.603 (0.210)	0.247 (0.072)
Total dead	0.341 (0.313)	0.134 (0.065)
Total	0.943 (0.400)	0.381 (0.112)
Total calcium (g/m²)		
Total live	1.064 (0.434)	1.054 (0.327)
Total dead	2.407 (0.795)	2.248 (1.286)
Total	3.471 (1.048)	3.302 (1.486)
Total magnesium (g/m²)		
Total live	0.675 (0.165)	0.468 (0.125)
Total dead	0.702 (0.235)	0.527 (0.288)
Total	1.377 (0.354)	0.995 (0.379)
Total potassium (g/m²)		
Total live	7.666 (1.945)	6.212 (1.989)
Total dead	1.191 (0.491)	0.860 (0.496)
Total	8.588 (2.151)	7.072 (2.264)

Table 35.7 Comparison of the soils of the three marsh sites preburn[a]

Soil horizon	pH[b] Juncus $\overline{X}$	Spartina $\overline{X}$	Unburned $\overline{X}$	Conductivity (µmhos/cm) Juncus $\overline{X}$ (SD)	Spartina $\overline{X}$ (SD)	Unburned $\overline{X}$ (SD)
0–5 cm (N = 25)	5.01	5.00	5.89	93.0 (13.6)	155.8 (58.3)	105.9 (16.7)
5–15 cm (N = 25)	5.12	5.17	6.28	54.6 (7.5)	92.4 (23.0)	74.9 (12.3)

Soil horizon	Potassium, K (mg/kg) Juncus $\overline{X}$ (SD)	Spartina $\overline{X}$ (SD)	Unburned $\overline{X}$ (SD)	Calcium, Ca (mg/kg) Juncus $\overline{X}$ (SD)	Spartina $\overline{X}$ (SD)	Unburned $\overline{X}$ (SD)
0–5 cm (N = 25)	47.16 (57.30)	43.05 (18.79)	24.26 (9.82)	410.70 (55.28)	614.02 (174.89)	782.38 (240.14)
5–15 cm (N = 25)	13.70 (10.63)	10.50 (2.79)	28.82 (91.56)	192.37 (34.93)	303.31 (35.94)	447.24 (78.89)

Soil horizon	Magnesium, Mg (mg/kg) Juncus $\overline{X}$ (SD)	Spartina $\overline{X}$ (SD)	Unburned $\overline{X}$ (SD)	Organic matter (%) Juncus $\overline{X}$ (SD)	Spartina $\overline{X}$ (SD)	Unburned $\overline{X}$ (SD)
0–5 cm (N = 25)	89.0 (14.96)	141.3 (40.41)	69.86 (17.08)	10.4 (5.7)	10.6 (3.5)	5.1 (1.1)
5–15 cm (N = 25)	40.39 (7.42)	55.70 (7.96)	36.38 (7.05)	1.8 (0.6)	2.3 (0.7)	1.7 (0.3)

Soil horizon	Phosphorus, PO_4-P (mg/kg) Juncus $\overline{X}$ (SD)	Spartina $\overline{X}$ (SD)	Unburned $\overline{X}$ (SD)	Nitrate and nitrite-nitrogen, NO_3-N and NO_2-N (mg/kg) Juncus $\overline{X}$ (SD)	Spartina $\overline{X}$ (SD)	Unburned $\overline{X}$ (SD)
0–5 cm (N = 25)	1.94 (0.57)	3.38 (1.99)	3.37 (3.61)	3.61 (2.93)	3.12 (1.49)	3.67 (0.68)
5–15 cm (N = 25)	0.71 (0.41)	0.98 (0.86)	1.51 (1.49)	2.38 (0.33)	2.00 (0.39)	3.00 (0.68)

Soil horizon	Nitrite-nitrogen, NO_2-N (mg/kg) Juncus $\overline{X}$ (SD)	Spartina $\overline{X}$ (SD)	Unburned $\overline{X}$ (SD)	Nitrate-nitrogen, NO_3-N (mg/kg) Juncus $\overline{X}$ (SD)	Spartina $\overline{X}$ (SD)	Unburned $\overline{X}$ (SD)
0–5 cm (N = 25)	0.09 (0.06)	0.10 (0.05)	0.18 (0.04)	3.53 (2.93)	3.03 (1.47)	3.49 (0.68)
5–15 cm (N = 25)	0.08 (0.08)	0.08 (0.04)	0.17 (0.05)	2.30 (0.33)	1.92 (0.39)	2.83 (0.66)

Soil horizon	Ammonium-nitrogen, NH_4-N (mg/kg) Juncus $\overline{X}$ (SD)	Spartina $\overline{X}$ (SD)	Unburned $\overline{X}$ (SD)	Total Kjeldahl nitrogen, TKN (mg/kg) Juncus $\overline{X}$ (SD)	Spartina $\overline{X}$ (SD)	Unburned $\overline{X}$ (SD)
0–5 cm (N = 25)	5.35 (1.48)	7.86 (4.44)	5.34 (1.44)	2986.9 (2024.6)	5508.2 (2655.5)	4115.8 (1074.1)
5–15 cm (N = 25)	3.00 (0.71)	3.09 (0.55)	2.66 (0.95)	562.6 (120.5)	664.8 (272.4)	646.5 (231.7)

a. One-way analysis of variance (ANOVA). For variables where the overall ANOVA was significant ($p \geq .05$), Duncan's multiple range test was performed.

b. Means and significance tests calculated on hydrogen ion concentration.

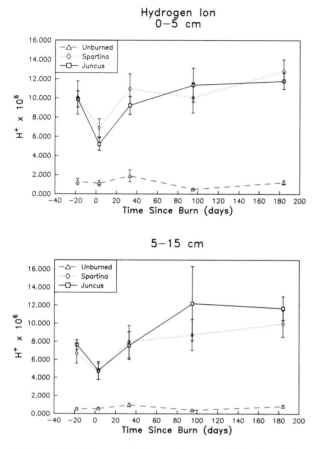

Figure 35.2 Changes in hydrogen ion concentration through 6 months postburn. Bars show 95% confidence intervals (N = 25).

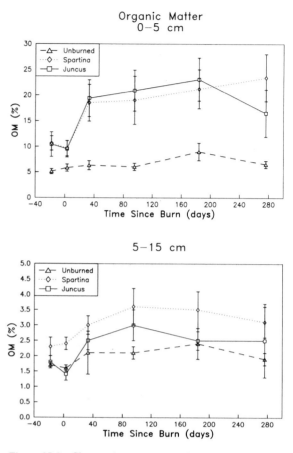

Figure 35.3 Changes in percent organic matter through 6 months postburn. Bars show 95% confidence intervals (N = 25).

Potassium did not change immediately postfire but increased in the 0 to 5 cm layer of both burned marshes by one month postfire and remained elevated at three and six months postfire (Figure 35.4). In the unburned marsh, potassium increased at six months after the fire. The 5 to 15 cm layer showed no definite patterns of change with time.

Calcium increased at one month postburn in the 0 to 5 cm layer of both burned marshes (Figure 35.5); one month postburn levels were more than twice those immediately postburn. Calcium increased in the 5 to 15 cm layer of the *Spartina* marsh, but the magnitude of the increase was less than at 0 to 5 cm. Levels in the 0 to 5 cm layer of the burned marshes remained high through six months postburn. In the unburned marsh, there were gradual increases in calcium with time.

Magnesium increased sharply at one month postburn in both the 0 to 5 and 5 to 15 cm layers of the burned marshes (Figure 35.6) and remained elevated

through six months postfire. The unburned marsh showed a modest increase at three and six months after the fire in the surface layer only.

Conductivity in all three sites increased at one month postfire, but the increase on the burned sites was greater than on the unburned (Figure 35.7). At three months postfire, conductivity on both burned sites was high relative to the unburned marsh, but at six months postfire conductivity remained high only in the *Spartina* marsh.

Phosphorus increased in the surface layer (0 to 5 cm) of the burned marshes at one month postfire (Figure 35.8); during this period, phosphorus levels in the unburned marsh declined. Phosphorus remained higher in the burned marshes than the unburned through six months postfire but declined between the three- and six-month samplings. In the 5 to 15 cm layer, phosphorus declined from preburn through six months postburn.

Total Kjeldahl nitrogen did not show clear re-

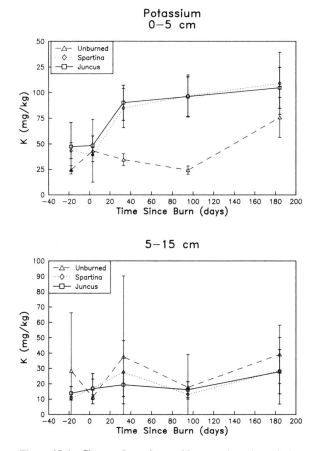

Figure 35.4 Changes in exchangeable potassium through 6 months postburn. Bars show 95% confidence intervals (N = 25).

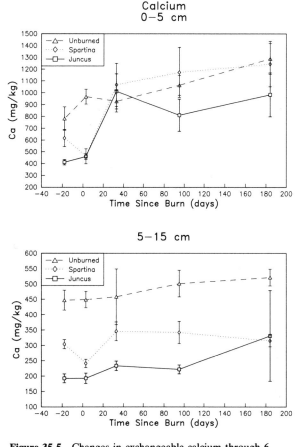

Figure 35.5 Changes in exchangeable calcium through 6 months postburn. Bars show 95% confidence intervals (N = 25).

sponses to fire through six months postburn. TKN in the 0 to 5 cm layer of the *Juncus* marsh increased immediately postfire while that of the *Spartina* marsh remained unchanged and the unburned marsh declined (Figure 35.9). At six months postburn, TKN levels in all marshes were greater than initial values in both the 0 to 5 cm and 5 to 15 cm levels.

Ammonium-nitrogen increased from preburn to immediately postburn and one month postburn in the 0 to 5 cm layer of all sites and in the 5 to 15 cm layer of the *Spartina* and unburned marshes but not in the *Juncus* marsh (Figure 35.10). After one month postburn, NH_4-N levels in the 5 to 15 cm layer of all sites and in the 0 to 5 cm layer of the unburned marsh declined. In the 0 to 5 cm layer of the burned marshes, NH_4-N levels remained high and exceeded the unburned site at six months postfire.

Nitrate-nitrogen in the 0 to 5 and 5 to 15 cm layers declined from preburn to postburn in both the burned and unburned marshes (Figure 35.11). From imme-

diately postburn to six months postburn, there were gradual increases in NO_3-N levels in the surface layer until concentrations exceeded preburn at six months after the fire.

Nitrite-nitrogen levels were always low (<0.18 mg/kg) and showed no pattern of change with fire (Figure 35.12).

Discussion

Vegetation

Composition of the *Juncus* and *Spartina* marshes sampled here is typical of these wetland types on Merritt Island (Schmalzer and Hinkle, 1985 and unpublished). *Spartina bakeri* occurs only in Florida and southeast Georgia (Godfrey and Wooten, 1979). *Spartina cynosuroides,* a related species, occurs on the Gulf Coast (Faulkner and de la Cruz, 1982). *Juncus roemerianus* is a more widespread species, ranging from Delaware to south Florida and west to

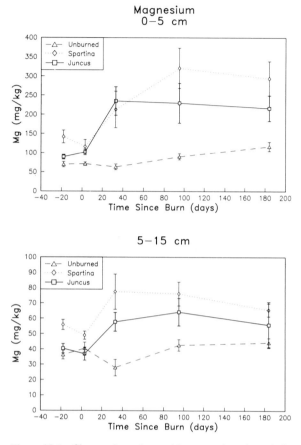

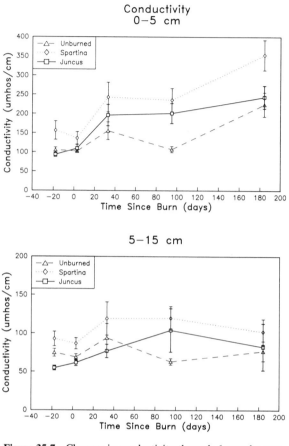

Figure 35.6 Changes in exchangeable magnesium through 6 months postburn. Bars show 95% confidence intervals (N = 25).

Figure 35.7 Changes in conductivity through 6 months postburn. Bars show 95% confidence intervals (N = 25).

Texas (Eleuterius, 1976). It is an important component of salt and brackish marshes in North Carolina (Williams and Murdoch, 1972), Georgia (Gallagher et al., 1980), Florida (Kurz and Wagner, 1957), Mississippi (de la Cruz, 1974; Eleuterius, 1972, 1984), and Louisiana (Hopkinson et al., 1978; White et al., 1978), occurring on diverse sites and soils with and without tidal flushing.

Biomass

Few data are available on biomass of *Spartina bakeri* marshes. Chynoweth (1975) reported mean live *Spartina* biomass of 429.1 g/m^2 and mean dead *Spartina* biomass of 699.1 g/m^2; our values are slightly higher. Our biomass data for *Juncus roemerianus* marshes are within the range reported elsewhere (Table 35.8). Live-to-dead ratios of <1.0 are common for *Juncus* marshes without regular tidal flushing (e.g., Hopkinson et al., 1978). Eleuterius (1984) noted that *Juncus* marshes accumulate standing dead material until re-

moved by tides or fire. Standing dead stems of *Juncus roemerianus* require up to eight years to decompose completely (Eleuterius and Lanning, 1987). Fewer data are available on nutrient standing crops than on biomass in *Juncus* marshes; our data are similar to that reported elsewhere (Table 35.9). No other nutrient standing crop data for *Spartina bakeri* marshes have been reported. For *Spartina cynosuroides,* Faulkner and de la Cruz (1982) reported standing crops of N (7.25 g/m^2), P (0.29 g/m^2), K (3.70 g/m^2), Ca (0.88 g/m^2), and Mg (1.04 g/m^2) similar or less than we found in *Spartina bakeri* (Table 35.6).

Biomass of these marshes contains substantial quantities of nutrients that can be transformed by fire into gases, particulates, or ash, and released into the atmosphere or deposited on the standing water surface of the marsh. Volatilization is most significant for nitrogen (Knight, 1966; Raison, 1979). Elevated levels of nitrous oxide (N$_2$O) (Cofer et al., 1990; this volume, Chapter 27) as well as nitric acid (HNO$_3$) and ammonia (NH$_3$) (LeBel et al., this volume) were de-

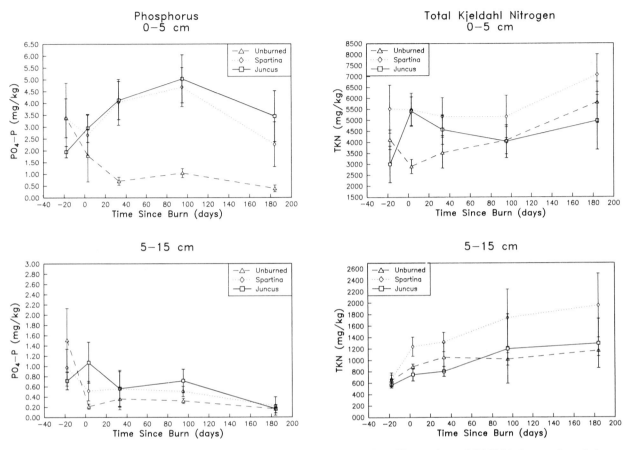

Figure 35.8 Changes in available phosphorus through 6 months postburn. Bars show 95% confidence intervals (N = 25).

Figure 35.9 Changes in total Kjeldahl nitrogen through 6 months postburn. Bars show 95% confidence intervals (N = 25).

Table 35.8 Comparison of biomass standing crops in *Juncus roemerianus* marshes

Location	Live biomass (g/m²)	Dead biomass (g/m²)	Reference
North Carolina	843[a]	1611	Williams and Murdoch (1972)
Georgia	800–1500	700–1500	Gallagher et al. (1980)
Mississippi	1531–4225	416–1632	Eleuterius (1984)
Louisiana	827,[b] 1240[c]	905[b]	Hopkinson et al. (1978)
Louisiana	1959[c]	800–1000	White et al. (1978)
Florida	$\overline{X}$ = 812.6	$\overline{X}$ = 996.5	This study

a. Includes dying.
b. Annual mean.
c. Peak live.

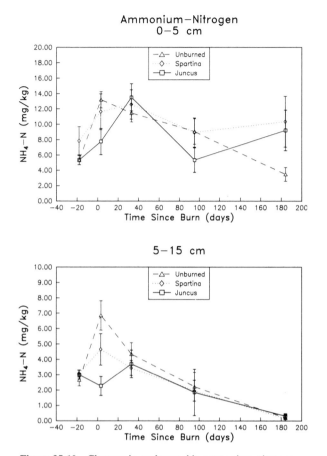

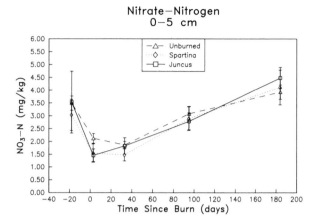

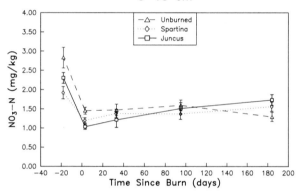

Figure 35.10 Changes in exchangeable ammonium-nitrogen through six months postburn. Bars show 95% confidence intervals (N = 25).

Figure 35.11 Changes in exchangeable nitrate-nitrogen through six months postburn. Bars show 95% confidence intervals (N = 25).

Table 35.9 Comparison of nutrient standing crops in *Juncus roemerianus* marshes

Location	Total Kjeldahl nitrogen (g/m²)	Phosphorus (g/m²)	Potassium (g/m²)	Calcium (g/m²)	Magnesium (g/m²)	Reference
Georgia	8–10	1.5	10–15	5–8	1.0	Gallagher et al. (1980)
Mississippi	8.76	0.22	3.88	1.44	1.94	Faulkner and de la Cruz (1982)
Florida	12.92	0.94	8.86	3.47	1.38	This study

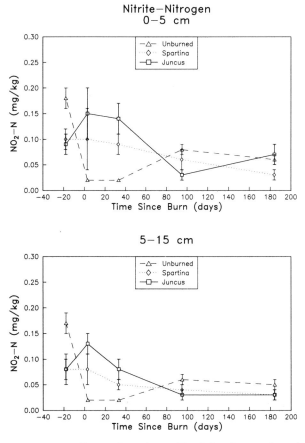

Figure 35.12 Changes in exchangeable nitrite-nitrogen through six months postburn. Bars show 95% confidence intervals (N = 25).

tected in the smoke plumes above this and similar fires. Faulkner and de la Cruz (1982) estimated that 70% of the nitrogen and 35% of the potassium in biomass of a *Juncus* marsh were lost to volatilization. Losses of similar magnitude are reported for other ecosystems (e.g., DeBell and Ralston, 1970; Christensen, 1977; Gillon and Rapp, 1989).

Soils

Effects of fire on soil may result from direct heating of the soil, redistribution and changed availability of nutrients, or the changed microclimate that results from vegetation removal (Raison, 1979). Changes may occur immediately postfire or subsequently due to vegetation changes or the activities of soil organisms. In these marshes, standing water limited or prevented direct heating of the soil. Vegetation cover and therefore shading of the water and soil surface was removed. Ash and partially burned residues were added to the water and soil.

Few data on responses of wetland soils to fire are available. The studies most comparable to ours are that of Faulkner and de la Cruz (1982) in *Juncus roemerianus* and *Spartina cynosuroides* marshes in coastal Mississippi and that of Wilbur and Christensen (1983) in North Carolina pocosins. Some differing aspects of these studies need to be mentioned. The marshes sampled by Faulkner and de la Cruz (1982) were irregularly flooded tidal marshes that were at low tide at the time of the fires. Ash was deposited directly on the soil surface in these fires. Fires were conducted in winter, January for *Spartina* and February for *Juncus*. Soil samples (0 to 2 cm and 2 to 10 cm) were collected 26 and 68 days postfire in the *Spartina* marsh and immediately (0.5 hour) and 29 days postfire in the *Juncus* marsh. The pocosin site studied by Wilbur and Christensen (1983) had hemist peat soils over muck. The site was not flooded when burned (March), and the fire consumed 0.5 to 6 cm of the surface peat. Soils (0 to 5 cm) were sampled periodically through 15 months postfire. Our sites were nontidal marshes where water levels varied with seasonal precipitation and evapotranspiration. They were flooded from before the November burn through three months postfire (February), but water levels receded below the surface by six months postfire (May). These were mineral soils with considerable organic matter in the surface (0 to 5 cm) layer.

The transitory increase in pH following fire in our study is similar to that observed by Faulkner and de la Cruz (1982). In their *Juncus* marsh, they found elevated pH immediately postburn, but it declined by 29 days postburn; their *Spartina* marsh had elevated pH 26 days postburn, and at 68 days postburn pH was still elevated compared to preburn. In the pocosin, Wilbur and Christensen (1983) found no pH response after fire; however, their first sampling was two weeks postburn and a very short-term response could have been missed. pH increases frequently occur after upland fires due to the input of basic cations in ash; magnitude and duration of these pH increases vary (Raison, 1979; Wells et al., 1979). Abundant organic acids in marsh soils may act to return pH to preburn values rapidly.

Neither Faulkner and de la Cruz (1982) nor Wilbur and Christensen (1983) reported organic matter data. In uplands, organic matter may decline postfire due to direct combustion or accelerated decomposition (Wells et al., 1979; Schoch and Binkley, 1986); however, light burning may result in no changes or increases in organic matter (Wells et al., 1979). Thus, the increase in organic matter after fire that we observed has not been reported in wetlands before.

Increases in cations (K, Ca, and Mg) frequently occur after fire in uplands (Raison, 1979; Wells et al., 1979) except where losses of organic matter reduce cation exchange capacity. In our burned marshes, K, Ca, and Mg increased at one month postburn. Faulkner and de la Cruz (1982) found that K increased immediately postburn in the *Juncus* marsh and remained high one month later; a small K increase occurred 26 days postburn in *Spartina* that was greater 68 days postburn. Calcium increases were greater immediately postburn in *Juncus* than one month later; *Spartina* marshes had high Ca values one and two months postfire. Magnesium increased with time postfire through one or two months in both marshes. Wilbur and Christensen (1983) found elevated K levels through the first growing season postfire, increased Mg for two growing seasons postfire, but no increase in Ca. Conductivity is a general measure of soluble salts. Its increase in soils here occurred at the same time as available cations (K, Ca, Mg).

Phosphorus also increased at one month postburn. Phosphorus increased immediately postburn in the *Juncus* marsh studied by Faulkner and de la Cruz (1982) and remained elevated at the later sampling; it also increased in their *Spartina* marsh. Wilbur and Christensen (1983) found increased PO_4-P through the first growing season postfire. Increased P levels are common in uplands after fire (Raison, 1979; Wells et al., 1979).

Both Faulkner and de la Cruz (1982) and Wilbur and Christensen (1983) found that increases in cations and phosphorus occurred primarily in the surface layer soils as we observed here. In uplands, this is also initially the case but leaching may occur and transport material down the soil profile (Raison, 1979; Wells et al., 1979).

The delayed increase of cations and phosphorus may be due to time required for materials to be leached or mobilized from ash and partially burned residue left by the fire. Christensen (1977) found that extractable N (ammonium or nitrate) and P in ash from pine-wiregrass savannas were much less than total N and P in ash, while extractable cations (K, Ca, Mg) were somewhat less than total cations in ash.

Nitrogen dynamics postfire are complex. In upland, well-drained soils total N may be reduced by combustion of organic matter in severe fires but light burning may increase N levels due to the addition of partially burned material or stimulated nitrogen fixation (Raison 1979). Most of the nitrogen in these marsh soils is in the organic form; thus, small changes may be hard to detect against background levels. Faulkner and de la Cruz (1982) found no changes in TKN levels in *Juncus* marsh immediately or 29 days postburn. There were slight declines in TKN in their *Spartina* marsh, although no sediment organic matter combustion occurred during their fires. Wilbur and Christensen (1983) did not report TKN data.

Ammonium-nitrogen increases postfire in many upland ecosystems (Raison, 1979), including chaparral (Christensen, 1973; Christensen and Muller, 1975; DeBano et al., 1979; Dunn et al., 1979; Anderson et al., 1988), coastal fynbos (Stock and Lewis, 1986), and coastal strand (Hinkle et al., 1990). Subsequent nitrification causes increases in NO_3-N (Christensen, 1973; Christensen and Muller, 1975; Stock and Lewis, 1986). Ammonium- and nitrate-nitrogen increases occur after logging and burning in tropical forests (e.g., Matson et al., 1987; Montagnini and Buschbacher, 1989). These trends are not universal. In pine-wiregrass savanna, Christensen (1977) found no significant changes in NO_3-N or NH_4-N; he did find seasonal differences with NH_4-N high in winter and spring and low in summer, and NO_3-N high in winter and low in spring and summer.

Ammonium-nitrogen increased postfire in both our burned and unburned sites; a clear difference in burned sites was not apparent until six months postfire. Wilbur and Christensen (1983) found that NH_4-N levels were higher in burned sites on some dates the first season postfire but not consistently so; they did not observe the large flush of NH_4-N previously found in chaparral.

Nitrate-nitrogen did not differ between our burned and unburned marsh sites through six months postfire. Wilbur and Christensen (1983) found increased NO_3-N levels the first growing season postfire but were uncertain whether this was due to ash addition, reduced plant uptake, or increased nitrification relative to denitrification caused by drying of the surface peat. We have observed increased NO_3-N levels occurring six months or more postfire in upland systems in this region (Schmalzer and Hinkle, 1989; Hinkle et al., 1990).

Nitrite-nitrogen was not reported by previous wetland fire studies. It was a minor component of nitrogen present in the system and showed no apparent response to fire.

Nitrogen transformations in flooded soils differ markedly from those in well-drained soils (Patrick, 1982; Reddy and Patrick, 1984). In flooded soils, a shallow surface layer usually remains aerobic, while the rest of the soil profile is anaerobic. Reactions requiring oxygen such as nitrification are restricted to the surface layer, while anaerobic reactions such as denitrification and methanogenesis occur in the anaerobic layer (Patrick, 1982; Reddy and Patrick, 1984). Ammonification can occur in either. Decomposition is slower under anaerobic conditions than aerobic, but the release of NH_4-N is greater than would be expected due to lower N requirements of anaerobic bacteria (Patrick, 1982). Primary nitrogen transformations in submerged soils involve NH_4-N diffusing from the anaerobic layer to the aerobic where nitrification converts it to NO_3-N; NO_3-N then diffuses into the anaerobic layer where denitrification converts it to nitrogen gas (N_2) or nitrous oxide (Patrick, 1982; Reddy and Patrick, 1984).

Nitrogen fixation occurs in wetlands (Patrick, 1982; Reddy and Patrick, 1984). Fire may stimulate nitrogen fixation (Jorgensen and Wells, 1971) in uplands, but effects in wetlands are unknown. Volatilization of ammonia-nitrogen (NH_3-N) occurs in wetlands but is considered important only where high NH_4-N levels occur along with high pH in the soil-water system (Reddy and Patrick, 1984). Given the low pH in our marshes, volatilization would not be expected to be an important nitrogen loss mechanism. Ammonia volatilization from postfire soils has not been studied, however.

Considering these processes in flooded soils, we suggest the following explanation of nitrogen responses in our burned sites. Fires in the marshes were of lower intensity than chaparral, scrub, or coastal strand fires and produced less NH_4-N directly. Ammonia-nitrogen was detected in the smoke plume (LeBel et al., 1988; this volume, Chapter 29). Some NH_4-N was added to the marsh surface but this occurred at a time of year when soil NH_4-N levels were increasing even in unburned marshes, making separation of the effects difficult. Postfire emissions of nitric oxide (NO) from the marshes (Levine et al., 1990) were less than from chaparral (Levine et al., 1988), consistent with smaller additions of NH_4-N. Most of the nitrogen deposited in the marshes was organic but high levels and spatial variability of TKN did not allow its detection in the soil. Added organic nitrogen was more readily mineralized than organic nitrogen in the soil already, but flooded conditions and lower winter temperatures slowed ammonification. By six months postburn, water levels declined below the soil surface; aerobic conditions and warmer temperatures led to greater NH_4-N levels in burned soils, while the unburned site was showing a seasonal NH_4-N decline.

No increase in NO_3-N occurred by six months postburn due to the lag in NH_4-N increase and the possible denitrification in the anaerobic soil layer of any excess NO_3-N produced.

Analysis of soil samples collected 9 and 12 months postburn may help clarify fire effects in these soils. Further work is needed to determine long-term gaseous fluxes from wetlands after fire in relation to changes in soil chemistry and microclimate. The significance of elevated trace gases emissions postfire cannot be determined without long-term data. Tracer studies or other techniques may be needed to determine the fate of nitrogen released by fire. Studies on the changes in microbial communities and activity in wetlands soils after fire are needed to determine mechanisms responsible for the observed changes in gaseous fluxes and soil chemistry.

Conclusions

1. Marshes in this study are representative of a variety of graminoid wetlands in the southeastern United States that burn periodically in natural or prescribed fire. Biomass and nutrient standing crops are comparable to similar marshes.

2. Increases in organic matter, pH, cations (K, Ca, Mg), conductivity, and phosphorus are generally comparable to other results from uplands and wetlands. The delayed increase in cations and phosphorus is probably due to time required for leaching or mineralization of material from ash and burned residue.

3. Nitrogen transformations in flooded soils differ from those in well-drained soils, and these differences affect postfire changes in nitrogen species. Most of the nitrogen in marsh soils is organic but variability is high, masking additions from fire. Some ammonium-nitrogen was released by the fire but seasonal changes in soil concentrations prevented its clear detection. Only at six months postburn were burned sites higher in NH_4-N than the unburned marsh. Nitrate-nitrogen peaks do not occur through six months postfire; if excess NO_3-N is produced while the marshes are flooded, it is probably lost to denitrification.

4. Seasonal variability of soil parameters and interactions with varying water levels indicate the need for long-term studies and for the careful selection of unburned reference sites for comparison to burned sites. Long-term measurements of gaseous fluxes after fire are needed to determine how these fluxes respond to changing postfire conditions. Changes in composition

and activity of soil microbial communities after fire in wetlands are needed to determine mechanisms responsible for the observed changes in gaseous fluxes and soil chemistry

Acknowledgments

This study was conducted under NASA contracts NAS10-10285 and NAS10-11624 in cooperation with the NASA Biospherics Research Program at Langley Research Center (LaRC). We thank Joel Levine and Wesley Cofer III from LaRC for their cooperation and encouragement in this study. We thank Steve Vehrs, Dorn Whitmore, Ray Farinetti, and other staff of the Merritt Island National Wildlife Refuge for conducting the controlled burn. Joseph Mailander, Patrice Hall, and Ray Wheeler provided field assistance. Teresa Englert, Steve Black, Chris Cantrell, Jill Caudle, and Lori Lai provided chemical analyses of soil and biomass samples.

IV

Biomass Burning: Laboratory Studies

Experimental Evaluation of Biomass Burning Emissions: Nitrogen and Carbon Containing Compounds

Jürgen M. Lobert, Dieter H. Scharffe, Wei-Min Hao, Thomas A. Kuhlbusch, Ralph Seuwen, Peter Warneck, and Paul J. Crutzen

Today biomass burning is accepted to be an important source of many trace gases affecting atmospheric chemistry (Crutzen et al., 1979; Cofer et al., 1988a; Radke et al., 1988; Crutzen et al., 1990). Despite its global significance and in contrast to fossil fuel use, where detailed investigations on global amounts and their distributions are available, still little quantitative information is known about the emissions of some of the compounds emitted and the global amounts consumed by biomass burning. Estimates of these global quantities are difficult to derive because of a very uncertain data base. The most recent values published by Crutzen et al. (1990) indicate that 1900 to 5000 teragrams (Tg) biomass carbon is released annually.

In order to obtain more reliable emission data, improvement of measurement techniques as well as methods for calculating emission data are required. A growing number of measurements on biomass burning emissions have been published since the first estimates (Crutzen et al., 1979), but some of the trace gas emissions are still very uncertain or even unknown as, for example, in the case of some nitrogen-containing species.

Most of the data on biomass burning emissions were derived from airborne measurements above large-scale fires. An experimental system, on the other hand, is advantageous in cases when information cannot be derived from open fires or when field measurements are inappropriate. In this chapter we describe results obtained by the latter technique. An important advantage of the apparatus described is the possibility of learning about burning behavior and burning stages, and thus clarifying relationships between combustion processes and emissions, finally leading to new methods for estimating emissions.

To extend our knowledge of the inventory of nitrogen and carbon in biomass burning, we took a close look at the most important emissions, trying to identify the major fractions which complete the mass balance. Such a balance cannot be achieved from natural fires due to the impossibility of exactly determining the absolute mass of burned and unburned plants.

Experimental Section

Burning Apparatus

The apparatus which was used for conducting open fires is shown in Figure 36.1. A more detailed description of this system is given by Lobert (1990a). The main parts of the apparatus include a hood in the form of an inverted funnel and a burning table of 60×60 centimeters (cm), which is placed on a high-resolution balance (E1210 and EB60; Mettler). The balance itself is placed on a frame which stands on the ground outside to avoid vibrations. A simple mechanism allows the raising of the burning table to an angle of 45° in order to simulate different wind directions, i.e., heading and backing fires. All the major parts are made of stainless steel and are mounted in a mobile trailer (Fladafi), which also contains most of the equipment required for carrying out the experiments.

A fanwheel-anemometer (Höntzsch) was used to measure the flow rate of stack gases; more recently, the device was replaced by a differential pressure sensor due to a high dirt sensitivity of the mechanical fan wheel. An electrical fan was built into the top of the chimney in order to provide a minimum draft during low-temperature combustion and to prevent the reaction gases from leaving the funnel from the bottom. Temperature sensors (Ni/Cr-Ni, TICON) were used to determine both stack gas and fuel temperatures, and two optical monitors were built in for the determination of CO and CO_2 concentrations. Several heated sampling tubes (stainless steel and teflon; Heraeus) allowed the exhaust gases to enter instruments in an adjacent laboratory to determine NO_x or for the on-line injection of the compounds mentioned above.

All the continuous measurements were recorded with a data logger (WES) at a frequency of 1 to 0.2 sec^{-1} and stored online on a personal computer for further calculations. The duration of an experiment was typically 15 to 20 minutes from ignition until the fuel bed cooled. Two switches connected to this data logger enabled us to send signals to the computer for synchronizing experiment and sampling procedures.

Figure 1

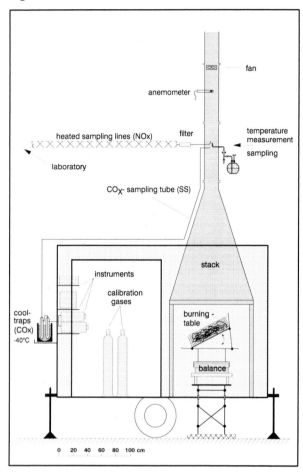

Figure 36.1 Burning apparatus for simulating open fires. The trailer is made of galvanized steel; all other parts are stainless steel or aluminum. Power and data transfer cables are connected to the adjacent laboratory.

A detailed description of sampling and analytical methods is given below.

Apparatus for N₂ Measurements

For the determination of molecular nitrogen emission we constructed a closed apparatus which enabled us to carry out burning experiments excluding atmospheric nitrogen (Kuhlbusch, 1990). This apparatus is shown in Figure 36.2 and consists of a stainless steel cylinder equipped with several sampling and supply valves in the top as well as two windows for observing the inside. The cylinder further contains a sample holder which is loaded with 0.5 to 1.0 gram (g) of biomass sample prior to the experiment and which is connected to the electrodes of a power supply. An additional port for evacuating the system is located at

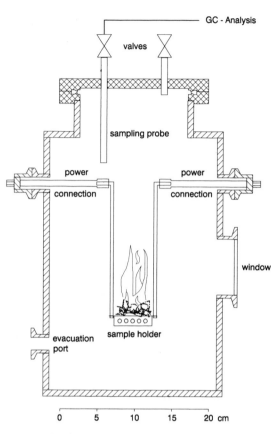

Figure 36.2 Apparatus for conducting uncontrolled burning experiments in an artificial atmosphere. The system is made of stainless steel and sealed with Small and ConFlat flanges.

the bottom of the apparatus. After reaching the final vacuum, which is determined with a Pirani pressure gauge ($1-10^{-4}$ hectopascals, or hPa) and a high-accuracy gauge (0–1400 hPa), the cylinder is flushed with a mixture of very high-purity helium or argon and oxygen (79% : 21%) and pumped out several times in order to remove traces of nitrogen which might be adsorbed on the (electropolished and very smooth) inside surfaces. Helium leak tests were carried out in order to ensure the tightness of the system.

Following several blank tests, the sample is ignited by resistance heating of the stainless steel or tantal sample holder. The duration of such an experiment (ignition and self-sustaining combustion) lasts for about one to two minutes. Afterwards, the combustion gases are allowed to spread and disperse homogeneously inside the cylinder for one hour, followed by direct injection GC-analysis using a valve with two sample loops.

Sampling Procedures

As shown in Figure 36.1, there are several ways of sampling the stack-gas continuously. For the analysis of CO and CO_2, gases are pumped through the analyzers using a stainless steel tube ($1/4'' \times 5$ m), connected to several flow controllers and passing three different filter units including a cold trap ($-40°C$), a glass fiber filter, and a stainless steel sinter-filter in order to remove water vapor, particles, and tar, which otherwise condense on the surface of the analyzer's optical parts. A dynamic on-line dilution of these gases is included due to the fixed concentration ranges of the analyzers. NO_x is determined by pumping the exhaust gases through the analyzers connected to a heated $1/4''$ Teflon tube of 8 m length passing a Teflon filter (0.5 μm) and a cold trap at $-20°C$ to remove big particles and water vapor.

The grab-sampling system which is used to determine most of the gases studied is shown in a cross-sectional view in Figure 36.3. Stack gas samples are taken from the middle of the chimney (see Figure 36.1) by sucking them into the evacuated 1 liter (L) stainless steel canisters. The canisters were electron beam welded and electropolished inside to ensure a smooth surface without diffusion-traps such as scratches and holes. They are equipped with two metal bellows valves (Nupro SS 4H). The biomass burning gases are pumped through the filter holder carrying glass fiber filters. After sampling, prior to the analysis, a dilution with high purity helium is performed in order to slow down possible reactions and to get a slight overpressure inside the canisters. Analysis starts about 10 minutes after taking the first of about 12 to 14 samples obtained during an experiment. After analysis, the canisters were evacuated and flushed several times at about 100°C for cleaning, and thereafter stored evacuated at about 10^{-2} hPa.

The sampling of ammonia and the aliphatic amines was carried out using several scrubber methods. While ammonia was sampled by bubbling the gases through a solution of 30 milliliter (ml) 0.1 N sulfuric acid in polyethylene bottles with a flow rate of 8.5 L/min, the amines and, in recent experiments, also ammonia were sampled using an advanced scrubber method, which is described by Cofer et al. (1985) and modified for our use by Hartmann et al. (1990). The efficiency for sampling ammonia and amines with these methods was determined to be 99%; there was no significant observable difference in the results using the different scrubber methods for ammonia. Teflon filters of 0.5 μm pore size in the front of the $1/4''$ Teflon sampling tube were used to remove particles from the emitted gases. Ammonia on the filters was ultrasonically desorbed into an H_2SO_4 solution before analysis.

Analytical Methods

Carbon monoxide and carbon dioxide were determined by using two BINOS nondispersive infrared analyzers (Heraeus), which operate at two different concentration ranges. The CO instrument contains one cell which is electronically split into ranges of 0 to 100 parts per million by volume (ppmv) and 0 to 2000 ppmv, while the CO_2 instrument contains two separate cells with ranges of 0 to 1500 ppmv and 0 to 6% by volume, respectively. The detection limits for these instruments are a few ppmv (noise level is 1 ppmv). Calibration of the instruments is carried out with commercial standard gas mixtures of CO and CO_2 (Steininger).

Methane analysis was carried out with a Shimadzu gas chromatograph (GC) Mini-3 equipped with a flame ionization detector (FID) and a stainless steel 3 m $\times$ $1/8''$ OD column packed with 60/80 mesh molecular sieve 10 Å and supplied with high-purity nitrogen carrier gas at a flow rate of 25 ml/min; the column temperature was maintained at 110°C isothermally. Two ml of gaseous sample were injected with a valve allowing a detection limit of about 9 ppbv (three

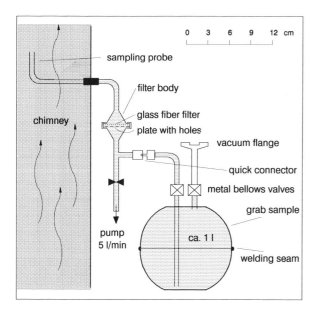

Figure 36.3 Cross-section of the grab sampling system. The 1 liter stainless steel samples are electropolished inside and electron beam welded. A pump connected to the filter holder provides a flow rate of 5 L/min.

times noise level) of methane, equal to 9 picograms (pg) C. For the calibration of methane measurements commercial standard gases of 1.5 to 1000 ppmv are used (Steininger).

To determine nonmethane hydrocarbons, a Siemens double oven GC L402 equipped with FID and WCOT fused silica capillary column 50 m × 0.32 mm ID coated with CP SIL 5 CB (1.19 μm) was used. Carrier gas flow rate (very high purity helium) was maintained at 1 bar inlet pressure equal to a linear gas velocity of 27 cm/sec. The method originally described by Matuska et al. (1986) was used for analyzing preconcentrated samples. An air sample of 0.4 ml is injected onto the first 10 cm of the column which is cooled to 77°K causing a cryofocussing of the sample, rendering sharp signals and enhancing detection limits. Injection of this sample is executed by removing the cryotrap followed by a temperature programmed heating (−40°C for 2 min; 15°C/min to +40°C; 10°C/min to +210°C; 10 min at 210°C). The detection limit (three times noise level) for this procedure is about 7 pg C or 18 ppbv for C_3 compounds. For compound identification, the absolute retention time method was used. Relative response factors (Dietz, 1967) were used for individual compounds and a response factor of 1.0 relative to CH_4 was used for the determination of $\Sigma(C_2-C_8)$. The absolute detector response was determined with a standard gas mixture of 10 ppmv propane in helium (AGA).

The sum of the nitrogen oxides NO and NO_2 (NO_x) is determined by using a Thermo Electron chemoluminescence NO_x analyzer type 14 A equipped with a stainless steel converter. The converter temperature was maintained at 600°C in order to minimize interference by other nitrogen-containing compounds. At this temperature there is no interference of, for example, acetonitrile, while at 800°C about 80% of an acetonitrile calibration gas is recorded with this detector. According to our experiments, fresh biomass burning NO_x emissions consist of about 90% NO and only 10% NO_2. For calibration of the instrument, standards of NO and NO_2 are used. The detection limit of this instrument for NO_x is around several ppbv in air.

Nitrous oxide was analyzed by taking 2 ml of gas sample from the canisters described above with gastight syringes and injecting them into a Dani GC equipped with Carlo Erba Electron Capture Detector (ECD). Gas samples first pass through a cold trap at −30°C to remove water, then through a precolumn (20 cm × ¼″) packed with NaOH coated asbestos to remove CO_2 followed by a 1.5 m × ⅛″ OD stainless steel column packed with 60/80 mesh Porapak N for separation. The flow rate of the carrier gas N_2 was maintained at 25 ml/min, adding 3 ml/min of makeup gas ($N_2/5\%CH_4$); oven temperature was held constant at 55°C. For calibration purposes gas mixtures containing from 310 ppbv to 360 ppmv were used. For further details see Hao et al. (1991a).

A new method for the sensitive determination of volatile C_1 to C_3 nitriles in one step was evaluated by Lobert (1990a, 1991b). A Delsi GC DI 200 equipped with a nitrogen specific Thermal Ionization Detector (TID) enables determining small amounts of nitrogen-containing compounds over a background of large quantities of hydrocarbons which are suppressed with a factor of 10^4 relative to the nitrogen substances. This detection method, described several times by other authors, was modified using a 30 m × 0.53 mm OD megabore column GS-Q of J&W (Porapak Q-like) and outlines for the first time a very good separation of cyanogen (NCCN), hydrogen cyanide (HCN), acetonitrile (CH_3CN), acrylonitrile (CH_2CHCN), and propionitrile (CH_3CH_2CN) in a single analysis. Even butyronitrile (C_3H_7CN) and higher compounds can be detected with this method, but were not determined during this work because they are only produced in small quantities. The parameters used were: 5 ml/min of helium-carrier gas, 3.5 ml/min of hydrogen, and 180 ml/min of synthetic air for the detector. Detector-bead temperature was maintained at 680° to 720°C with an electrode voltage of + 10V. An air sample of 0.5 ml was injected by a valve directly on column resulting in a detection limit (three times noise level) of about 2 pg N, which equals 10 ppbv of acetonitrile in air. Changing the detector parameters enables lowering the detection limit to a minimum of about 0.8 ppbv of acetonitrile in air. Calibration of the nitriles was realized by using gaseous cyanogen diluted by static dilution, solutions of KCN which are converted to HCN on an acidic precolumn, and solutions of the liquids in methanol (other nitriles). Gaseous standards were not used because they are very unstable and therefore require recalibration by the mentioned methods at least once a week.

The ammonia contents in the H_2SO_4 and HCl scrubber solutions were mostly determined by using an NH_3 sensitive electrode (Orion model 95-10) with a sensitivity of 10^{-5} mole/liter; in recent experiments ion chromatography was used. The sum of gaseous and particulate NH_3 are reported in this article. Detection limit for the electrode method is about 0.9 ppmv or 5.1 mg N for the integrated sample. The detection limit with ion chromatography is dependent on the sampling time at several pptv in air (Ernst, 1990).

For the determination of volatile aliphatic amines a method was developed by Seuwen (1990) using a nitrogen specific TID mounted on a Delsi GC DI 200. For separation of the amines, a glass column (2.3 m × 2 mm ID) packed with 60/80 mesh Carbopack B/4% Carbowax 20 M/0.8% KOH was used. A carrier gas flow rate of 20 ml/min He was maintained in addition to 4 ml/min of hydrogen for the detector. Several capillary columns proposed for the separation of amines had been tested and found to be unsuitable because of a very poor resolution for the low molecular amines. Six amines could be determined in biomass burning emissions and separated with the described method: methylamine, dimethylamine, ethylamine, trimethylamine, methylbutylamine and n-pentylamine, together referred to as "amines" in this paper. Calibration of the amines was carried out by using the amino hydrochlorides in water which had been converted on an alkaline precolumn (15 cm × ¼″ OD) packed with 0.8 to 1.6 mm NaOH on asbestos. Stable and certified gaseous amine standards are not available.

For the determination of N_2 emissions, combustion gases from the closed apparatus described above were directly injected into a Delsi GC DI 700 equipped with a Thermal Conductivity Detector (TCD) using an 8-port-valve with two 250 μl sample loops. The lower detection limit is about 10 ppmv of N_2 in air equal to 3.6 ng N in the sample loop. The 600 μl TCD was operated at 285 mA constant current (84% of maximum), 3.5 ml/min carrier gas, and 17 ml/min of makeup gas (both very high-purity helium), using a Chrompack 30 m × 0.32 mm OD 5Å/30 μm molecular sieve fused silica column at 30°C isothermally. Calibration of the N_2 measurement is realized by using a standard of 1000 ppmv N_2 in He followed by a series of static dilutions.

Elemental analysis of both biomass and ash samples was carried out using a Heraeus Rapid C/H/N analyzer (Dindorf, 1990). Samples were pulverized with a mill to an average particle size of about 60 μm, dried in an oven at 105°C for 24 hours, and allowed to cool in a desiccator prior to the analysis. Two to four single analyses were done on one ash or biomass sample, the detection limit is about 2 to 5 μg N which is equal to about 0.1% of nitrogen content in the sample.

Fuel

The fuels which were used for the open burning experiments were chosen to be as representative as possible of natural fires. Thus we burned two types of savanna grass as well as wood, hay, and straw representing agricultural wastes and forest-type fuels like pine needles and partially decomposed needle litter. Table 36.1 shows the fuel types, their carbon and nitrogen content as well as their molar N/C ratio and average moisture. Some of these fuels were also used for the N_2 measurements to compare open and closed burning. Also included in the table are some of the ash data, where all analyses are put together to give an average number which does not distinguish between different fuel types.

Table 36.1 Elemental composition of biomass burning fuels and ashes of our experiments

	C-content (%)	N-content (%)	Moisture (%)	N/C molar (%)
Savanna grass (Australian)	45.7 ± 0.1 (3)	0.15 ± 0.02 (3)	2.0 (1)	0.29 ± 0.04 (3)
Straw	45.0 ± 0.4 (15)	0.19 ± 0.08 (15)	4.1 ± 2.1 (5)	0.35 ± 0.15 (15)
Deciduous wood	48.3 ± 0.1 (3)	0.23 ± 0.02 (3)	3.4 (1)	0.47 ± 0.04 (3)
Savanna grass (Venezuelan)	44.1 ± 0.6 (24)	0.48 ± 0.06 (24)	5.2 ± 1.9 (7)	0.94 ± 0.12 (24)
Needle litter	48.1 ± 2.8 (9)	1.24 ± 0.46 (9)	8.7 ± 0.8 (2)	2.18 ± 0.69 (9)
Hay	42.0 ± 1.1 (18)	1.12 ± 0.14 (18)	6.7 ± 1.6 (4)	2.29 ± 0.29 (18)
Pine needles	49.5 ± 0.5 (24)	1.33 ± 0.22 (24)	8.7 ± 4.1 (3)	2.31 ± 0.38 (24)
Green grass	37.2 ± 0.1 (3)	2.13 ± 0.04 (3)	n.b.	4.92 ± 0.10 (3)
Tobacco	40.6 ± 0.0 (3)	2.37 ± 0.05 (3)	11.0 (1)	5.01 ± 0.09 (3)
Total average, fuels without tobacco	45.5 ± 3.3 (99)	0.86 ± 0.54 (99)	5.8 ± 2.7 (23)	1.63 ± 1.06 (99)
Total average, ashes	19.7 ± 13.8 (162)	0.58 ± 0.47 (162)	2.8 ± 1.4 (162)	2.73 ± 1.55 (162)

Note: Average ± standard deviation is followed by number of observations in parentheses.

Results and Discussion

Terminology

The term *volume mixing ratio* represents the volumetric fraction of a gaseous compound in the surrounding atmosphere.

Emission ratios are used for a relative comparison of different emissions and are defined as the above background mixing ratio of the compound studied, divided by the above background mixing ratio of a reference compound. In biomass burning calculations carbon dioxide is mostly taken to be the reference compound. The emission ratio of, e.g., nitrogen oxide to carbon dioxide is defined as

$$\Delta\mu_{NO}/ \Delta\mu_{CO_2}$$

where $\Delta\mu$ indicates the mixing ratio of the compound reduced by its ambient air level.

The calculation of the emission ratios of our experiments is based on a mass determination of the emissions. For that reason the mass flux of the compounds was determined during each experiment. Integration of this mass flux over one of the burning stages, or the whole fire, results in a total mass emitted during a fire, or during one of the burning stages, respectively, expressed in grams (g) carbon or g nitrogen. The ratio of the emitted mass of compound to the emitted mass of CO_2 gives the emission ratio of the compound relative to CO_2. All emission ratios presented here are expressed as *molar* ratios (mole compound per mole of CO_2).

Calculating the amounts relative to the fuel nitrogen or to the fuel carbon is an alternative way of expressing emission factors. This method is known as the carbon balance method, described by Radke et al. (1988). A balance of the nitrogen-containing compounds has not been presented previously. While Radke et al. use an approximate average percentage for the carbon content of 49.7%, we are able to use the elemental contents of both biomass and ash, which we measured and which, in the case of nitrogen, differ widely (Table 36.1).

Observations

Figure 36.4 shows mixing ratios of different compounds during the course of a typical biomass burning experiment. The abscissa shows the first 400 of about 900 to 1200 seconds of a complete fire, while the ordinate shows the volume mixing ratios of various gases. Curve 36.4g shows the temperature about 2 meters (m) above the fire, where gas samples are taken. The dotted, vertical line at 96 seconds repre-

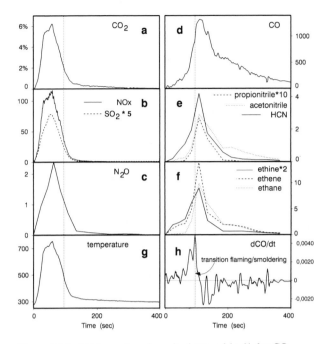

Figure 36.4 Mixing ratios above background in % for CO_2 and ppmv for the other gases, stack-gas temperature in °K, and the first derivative of the CO-curve during the first 400 seconds of a typical experiment. The maximum in dCO/dt is taken to define the transition between flaming and smoldering combustion.

sents the transition between flaming and smoldering combustion. These two burning stages are characterized by hot combustion with open flames, involving oxidation of pyrolysis products in the flame, and a colder, incomplete combustion without, or maybe with only small flames where the pyrolytic, oxygen-free fuel conversion predominates. In order to get a reproducible reference, the transition point was calculated from the first derivative of the CO mixing ratio versus time (Figure 36.4h). The maximum of the first derivative was defined to be the transition between flaming and smoldering stages. Indeed, this coincides closely with the cessation of the flames.

Some of the gaseous emissions are closely related to the temperature during a fire. The rapid increase in stack-gas temperature coincides with a similar increase in CO_2, NO_x, SO_2, and N_2O mixing ratios. These gases are shown in parts a to c of Figure 36.4. Parts d to f give examples of some gases with maximum mixing ratio during the smoldering stage of a burn. Substances matching this behaviour are defined to be *smoldering stage compounds* while those which are mostly emitted during the flaming stage are called *flaming stage compounds*. The latter are substances in a comparatively high oxidation state, while the former are less oxidized compounds, products of incomplete combustion such as carbon monoxide, hydrocarbons, ammonia, and the nitriles.

Gaseous emissions are basically a product of the pyrolytic cracking of the fuel. Additionally, depending on the presence or absence of flames, the pyrolysis gases are more or less oxidized. HCN, for example, with up to 90% of the fuel nitrogen, is one of the major products of pyrolytic (oxygen free) protein decomposition (Johnston et al., 1971; Lobert, 1990a), but during open fires it is emitted at a total of just several percent of the fuel nitrogen. This is a result of its oxidation to NO_x and CO_x in the flame. The flaming stage is supplied with plenty of oxygen by thermally induced *convection*. The smoldering stage produces much lower temperatures, which result in a low oxygen supply by *diffusion* into the fuel bed. If gases in this phase leave the fuel bed, they are no longer converted. Table 36.2 gives typical, maximum mixing ratios of the compounds during our experiments and the corresponding combustion phase.

Knowledge about the types of flaming stage and smoldering stage emissions allows correlation between compounds and gives insight into the use of different reference compounds for the calculation of emission ratios. Figure 36.5 shows graphs of the emissions of some compounds versus those of CO or CO_2. Part a shows a very good linear correlation between NO_x and CO_2 emissions, since both compounds are mostly produced in the flaming stage, while NO_x shows *no* correlation to CO (b). Acetonitrile, on the

other hand, shows a good correlation only with CO (d); both of them belong to the smoldering stage class of compounds. A linear regression with CO_2 (c) would be negative in this case. As CO_2 is used as a reference compound for biomass burning emissions, this finding shows that single measurements of smoldering stage compounds correlated to CO_2 are not useful for a comparison with other samples without information about the burning stage in which the sample was taken. If samples are taken from above large-scale, open fires, where a mixture of both burning stages exist, an emission ratio derived from regression analysis is only meaningful if both compounds belong to the same class.

For that reason we propose calculating emission ratios, depending on the type of compounds, relative to either CO_2 or CO, whichever gives a better linear correlation. There are only a few compounds which belong to the flaming stage class, such as CO_2, NO_x, N_2O, and SO_2. Most of the compounds emitted by biomass burning are produced mainly in the smoldering stage and therefore belong to the smoldering stage class. Taking carbon monoxide as a reference for emission ratios also has a practical advantage for aircraft measurements. The ambient air background mixing ratio of CO_2 is very high at about 350 ppmv, while an enhancement through biomass burning emissions seldom exceeds 50 ppmv and is often much lower. Because of the natural variability in CO_2 mixing ratios and measurement uncertainties, calculations of emission ratios to CO_2 may thus become very uncertain. For the smoldering stage compounds, on the other hand, which represent the majority of the emissions, it is much more reliable to use CO as the reference gas, since its background mixing ratio is only 50 to 150 ppbv and the fire plume excess can be up to several ppmv.

The procedure of taking both CO_2 and CO as reference compounds works well for most of the emissions, although some transition between flaming stage and smoldering stage type is observable. The C_2 hydrocarbons and the nitriles illustrate such a transition. Figure 36.4e shows the mixing ratios of three nitriles versus time. It can be seen that the amount which was produced in the flaming stage (area under the curve) increases from propionitrile over acetonitrile to hydrogen cyanide. The same effect is observable with CH_3CH_3, $CH_2 = CH_2$, and $HC \equiv CH$ (Figure 36.4f). It is interesting to note that, on average, more than 50% of both HCCH and NCCN are—probably because of their thermal stability—produced during the flaming phase (see Table 36.4) although, due to their rela-

Table 36.2 Typical maximum mixing ratios of several compounds during our experiments

Compound	Maximum mixing ratio during fire (ppmv)	Corresponding burning stage
CO_2	60,000–100,000	Flaming
CO	1500–3000	Smoldering
CH_4	43–350	Smoldering
$HC \equiv CH$	4–36	Flaming
$H_2C = CH_2$	11–46	Smoldering
$H_3C\text{-}CH_3$	5–36	Smoldering
NO_x	100–160	Flaming
N_2O	2–19	Flaming
NH_3	Avg ~5	Smoldering
HCN	4–30	Smoldering
CH_3CN	2–10	Smoldering
SO_2	20–30	Flaming

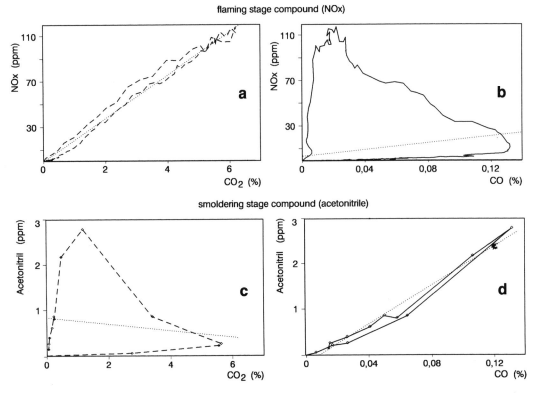

Figure 36.5 If compounds are produced in the same combustion stage, they show a good, linear correlation (a and d), otherwise a significant anti-correlation is observable (b and c). Emission ratios should be derived only from regression analysis, if a positive, linear correlation is given.

tively low oxidation state, one would expect them to be smoldering stage compounds.

Influences on Emissions

There are several factors influencing biomass burning emissions. The most important effect has been identified to be the kind and duration of the burning stage. The relative importance and distribution of the burning stages, on the other hand, is dependent on the type of fuel, the fuel moisture and density, and the wind direction during a burn. Since CO/CO_2 is an indicator of the quality of combustion, we took this ratio to show different influences on the emissions of CO during hay and savanna grass burns in Table 36.3. As can be seen, this ratio can vary between flaming and smoldering combustion by a factor of 20 to 30. The effect of moisture is demonstrated in the first two lines of this table. The much higher moisture of burn 2 (16%) causes an overall CO/CO_2 ratio which is twice as high as for the dry fire (5% moisture) due to the virtual absence of a flaming stage. On the other hand, there are no significant differences observable for a moisture content of under 10%. In order to be able to

study the different burning stages, we burned fuels with an average water content of 5%, which is typical of tropical savannas during the dry season.

The direction of the wind produces a further effect. We simulate different wind directions by raising the burning table to an angle of 30° to 45° and igniting the fuel either at the top or the bottom of the table. Since the wind is always moving into the chimney, ignition at the top forces the flames to burn against the air stream, thus simulating a *backing fire*. Ignition at the bottom of the table, on the other hand, results in a burn moving with the wind, a *heading fire*. Heading fires normally burn faster and less efficiently than backing fires, resulting in higher amounts of smoldering stage compounds. This can be confirmed from the last lines of Table 36.3, where the CO/CO_2 ratio increases by 40% from backing to heading fire. It is obvious that the effect of varying wind direction on the emissions is substantial, but much smaller than the differences between flaming and smoldering combustion, which are controlled by fuel moisture and fuel density.

One of the major influences on nitrogen emissions

Table 36.3 Examples of the influence of different burning factors on the emission ratios of compounds

Fuel	Moisture (%)	CO/CO₂ (%)			Wind direction
		Total	Flaming	Smoldering	
Hay	5%	7.1%	2.4%	11.9%	Backing
Hay	16	15.6	15.7	15.6	Backing
Savanna grass	4	5.0	0.7	20.0	Backing
Savanna grass	6	7.2	1.5	18.2	Heading

is the nitrogen content of the fuel. We found that most of the emission ratios of nitrogen-containing gases are significantly correlated to the N/C ratio of the fuel. Figure 36.6 shows emission ratios of CH_2CHCN, N_2O, and NO_x plotted versus the molar N/C ratio as well as the corresponding regression line. Due to this strong influence it is necessary to take into consideration the nitrogen content of biomass burning fuels when estimating global source strengths.

Owing to the design of our burning apparatus we can reduce variations in emissions to a minimum by using similar fuels and fire conditions during any experiment, and we can observe and separate the variations due to different burning stages with a high degree of control. Thus we are able to give emission ratios for any kind of fire behavior and any fuel variation. In this article, we present only overall mean values from our experiments, without distinguishing between different fuels and different fire types. Such values will be presented elsewhere (Lobert et al., 1991a).

Emission Ratios

Based on the above discussion, Table 36.4 gives an overview of the determined emission ratios of the carbon and nitrogen-containing compounds relative to CO or CO_2. Also included are the emission factors relative to the fuel nitrogen and carbon discussed below.

The table lists all mean values and the number of experiments for each compound, as well as the range of minimum and maximum values observed. The last columns give values for the distribution of the compounds to the burning stages. The bigger these numbers are, the more the compound belongs to the burning class indicated. The distribution values for RNO_x and NH_3 are results of one and two experiments only.

A maximum uncertainty of 20% for an emission ratio for a single fire is estimated for our measurements. As can be seen from our results, the variations

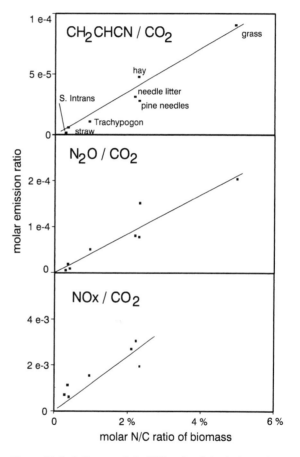

Figure 36.6 Influence of the N/C ratio of the fuels on the emissions of nitrogen containing compounds. The line represents the linear regression forced to zero, R^2 values are 0.95, 0.52, and 0.62, respectively. The scattering of the data shows clearly that the nitrogen content is just *one* important influence.

Table 36.4 Emission ratios of the described compounds relative to CO_2, CO, the carbon and the nitrogen content of the fuel

Compound	Formula	Molar emission ratio relative to CO$_2$, CO (%) or CO$_2$ (exponential)			Emission factor relative to the fuel C or N (%)			Fraction in burning stage (%)	
		Mean	Number of experiments	Observed range (min–max)	Mean	Number of experiments	Observed range (min–max)	Flaming	Smoldering
Carbon dioxide	CO_2	13.8	24		82.58	24	49.17–98.95	63	37
Carbon monoxide	CO	7.3E-02	12	3.2E-02–1.6E-01	5.73	24	2.83–11.19	16	84
Methane	CH_4	9.11	10	5.21–12.17	0.424	13	0.14–0.94	27	73
Ethane	CH_3CH_3	0.68	10	0.16–1.05	0.061	11	0.008–0.143	27	73
Ethene	$CH_2{=}CH_2$	1.21	10	0.40–2.08	0.123	11	0.021–0.420	43	57
Ethine	$CH{\equiv}CH$	4.2E-04	10	6.9E-05–1.7E-03	0.056	11	0.010–0.237	56	44
Propane	C_3H_8	0.15	10	0.021–0.319	0.019	11	0.002–0.042	22	78
Propene	C_3H_6	0.49	10	0.132–0.696	0.066	11	0.010–0.160	33	67
n-butane	C_4H_{10}	0.03	10	0.006–0.078	0.005	11	0.0006–0.010	21	79
2-butene (cis)	C_4H_8	0.02	10	0.003–0.032	0.004	11	0.0003–0.008	30	70
2-butene (trans)	C_4H_8	0.03	10	0.004–0.041	0.005	11	0.0005–0.011	28	72
i-butene, 1-butene	$C_4H_8 + C_4H_8$	0.19	10	0.025–0.398	0.033	11	0.003–0.082	28	72
1,3-butadiene	C_4H_6	0.11	10	0.039–0.185	0.021	11	0.004–0.056	31	69
n-pentane	C_5H_{12}	0.03	10	0.008–0.064	0.007	11	0.0012–0.024	44	56
Isoprene	C_5H_8	0.05	10	0.014–0.093	0.008	11	0.002–0.016	21	79
Benzene	C_6H_6	0.21	10	0.106–0.401	0.064	11	0.018–0.214	45	55
Toluene	C_7H_8	0.12	10	0.027–0.233	0.037	11	0.005–0.103	32	68
m-, p-xylene	C_8H_{10}	0.03	10	0.006–0.060	0.011	11	0.0012–0.025	29	71
o-xylene	C_8H_{10}	0.02	10	0.001–0.045	0.006	11	0.0003–0.015	27	73
Methyl chloride	CH_3Cl	0.16	9	0.018–0.437	0.010	10	0.0006–0.025	28	72
NMHC (as C)	C$_2$ to C$_8$	25.41	10	5.39–41.66	1.18	11	0.14–3.19	33	67
Ash (as C)					5.00	41	0.66–22.28		
Total sum C	C$_1$ to C$_8$				94.92 (including ash)				
Nitrogen oxides	NO$_2$O	2.0E-03	21	8.8E-04–7.7E-03	13.55	21	5.27–21.69	66	34
Ammonia	NH_3	1.55	15	0.118–4.89	4.15	15	1.04–11.74	15	85
Hydrogen cyanide	HCN	1.13	11	0.049–5.81	2.64	11	0.31–6.75	33	67
Acetonitrile	CH_3CN	0.25	10	0.013–0.53	1.00	10	0.079–2.323	14	86
Cyanogen	$NCCN$ (as N)	4.1E-06	9	2.0E-08–2.3E-05	0.023	10	0–0.133	56	44
Acrylonitrile	CH_2CHCN	0.05	10	0.008–0.135	0.135	10	0.030–0.343	32	68
Propionitrile	CH_3CH_2CN	0.02	9	0.006–0.055	0.071	10	0–0.188	17	83
Nitrous oxide	N_2O	9.2E-05	12	1.2E-05–2.8E-04	0.772	13	0.336–1.200	74	26
Methylamine	CH_3NH_2	0.08	5	0.040–0.131	0.047	5	0.025–0.076	27	73
Dimethylamine	$(CH_3)_2NH$	0.06	5	0.016–0.118	0.030	5	0.013–0.061	39	61
Ethylamine	$CH_3CH_2NH_2$	0.01	2	0.006–0.006	0.005	2	0.004–0.006	60	40
Trimethylamine	$(CH_3)_3N$	0.03	3	0.021–0.047	0.020	3	0.013–0.024	41	59
2-methyl-1-butylamine	$C_5H_{11}NH_2$	0.07	3	0.032–0.094	0.040	3	0.032–0.052	15	85
n-pentylamine	$n\text{-}C_5H_{11}NH_2$	0.22	4	0.132–0.329	0.137	4	0.039–0.238	32	68
Nitrates (70% HNO$_3$)	RNO_x	0.38	1		1.10	1	1.75–45.98	f	<s
Ash (as N)					9.94	41			
Total sum N (as N)					33.66 (including ash)				
Molecular nitrogen	N_2	4.7E-03	10	8.1E-04–9.9E-03	21.60	17	4.33–41.20	f	>s
Higher HC and particles					20				

Note: Presented are the mean values, number of experiments, and the range of minimum to maximum values ever observed as well as the distribution of the emissions to the burning stages. Emission ratios relative to CO are expressed in percent; those relative to CO$_2$ are given as exponential numbers. Emission factors are presented in percent of the fuel nitrogen or carbon.

between minimum and maximum emission ratios are fairly high, as observed by most other authors, and are to be explained by the high variability of the biomass burning process.

Mass Balance

On average, the carbon compound measurements account for nearly 95% of the carbon initially present in the fuel. CO_2 is the major emission (82.6%), followed by CO (5.7%) and CH_4 (0.42%). Nonmethane hydrocarbons contribute together 1.2%. Including the carbon found in the ash, the balance adds up to 94.9%, leaving only 5.1% unidentified. This is most likely hydrocarbons higher than C_8 emitted in gaseous or particulate form and also includes emissions of oxygenated or nitrogen-containing gases with low FID detector response. Table 36.4 also shows that the nitrogen-containing compounds determined in our work add up to only one third of the fuel nitrogen, including about 1% of nitrates (R-NO_x, mostly HNO_3) (Flocke, 1989). This leaves an unknown fraction of nearly 70% (Lobert et al., 1990b) and there are several candidates for this missing nitrogen.

First, the emission of molecular nitrogen contributes a significant amount to the nitrogen inventory. This emission is not detectable in natural fires due to the high nitrogen content of the atmosphere. Preliminary results obtained with the apparatus described above (Figure 36.2) show that during these experiments, on average, 22% of the fuel nitrogen was emitted as molecular nitrogen. Much larger N_2 emissions have been observed in our experiments during more intense combustion. Work is still in progress on this question. Detailed results are reported by

Kuhlbusch (1990) and Kuhlbusch et al. (1991). See also the discussion below.

Second, we did not measure the many high molecular weight compounds containing substantial amounts of nitrogen. To estimate this fraction, we considered the compounds found in tobacco smoke listed by Neurath (1969). We added all the reported compounds (as far as amounts had been given, excluding the compounds we measured) and found that, at most, 20% of the fuel nitrogen (tobacco) could be contained in these emissions. These are mainly aliphatic and aromatic amines and heterocyclic constituents with a high N/C ratio. Thus, tar and particulates may account for a substantial part of the nitrogen balance. This estimate, however, is uncertain until measurements on this subject are carried out.

Source Strengths

As described previously, the calculation of global emissions requires information on the amount of biomass that is burned worldwide. Unfortunately, such estimates are very uncertain because of a lack of information. The most recent estimate on global biomass burning emissions is given by Crutzen et al. (1990) and is used for our calculations. Table 36.5 summarizes the emissions from different types of ecosystems due to biomass burning.

Annual global source strengths of single compounds can be determined in several ways. The most common is by multiplying their emission ratios relative to CO_2 with the total CO_2 emissions from biomass burning per year. Again, this method is only reliable when the compound correlates with CO_2 (and there-

Table 36.5 Global amounts of biomass burned in different ecosystems expressed in terms of carbon and nitrogen

Source or activity	Carbon exposed (Tg C/yr)	Carbon released (Tg C/yr)	Molar (N/C ratio)	Nitrogen released (Tg N/yr)
Shifting agriculture	1000–2000	500–1000	1	5.0–10.0
Permanent deforestation	500–1400	200–700	1	2.0–7.0
Savanna fires	400–2000	300–1600	0.6	1.8–9.6
Firewood	300–600	300–600	0.5	1.5–3.0
Agricultural wastes	500–800	500–800	1–2	5.0–16.0
Prescribed + wild fires	150–300	120–250	1–2	1.2–5.0
Total	2900–7100	1900–5000	1.24[a]	17–51
			1.46[b]	28–72[c]
Quartile ranges		2700–4200		39–61[c]

Note: Amounts are expressed in terms of teragrams of carbon per year (10^{12} g C/yr). Corresponding nitrogen amounts are taken from Crutzen et al. (1990).
a. Average value of the table.
b. Average value of our 41 burning experiments.
c. Calculated from (1900–5000) $\times$ 1.455 and used for our estimates.

fore is nearly constant, not dependent on the burning stages), or when the emission factors are calculated on a mass basis for a complete fire, giving a true average, as in our measurements. If no correlation to CO_2 is given, global amounts should be calculated using emission ratios relative to CO multiplied by the corresponding ratio of CO/CO_2 and the global CO_2 amount. This method is proposed for use in field measurements of smoldering stage compounds, thus allowing comparison of emission ratios and source strengths from different fires.

In this article, we use a third way of estimating the global sources using our emission factors relative to the fuel carbon or nitrogen. The global sources which are presented in Table 36.6 are derived by dividing the emission factors relative to the fuel C or N by 0.95 or 0.90, which represent, respectively, the *emitted* fraction of C or N (total C or N minus ash content), and multiplying with the globally released carbon and an average N/C ratio of 1.455% (for the nitrogen compounds only) derived from our 46 experiments. Our N/C ratio is higher than the average value of 1.24% used by Crutzen et al. (1990). It must be noted that the global estimates for nitrogen compounds based on the emission factors relative to the fuel N show lower values than the same estimates based on the emission ratios relative to CO or CO_2, whereas the two computing methods give similar results in the case of carbon compounds. This might be due to a nonrepresentative N/C ratio used in our calculations, although higher sources would require an even higher N/C ratio.

For calculating the final range of global, annual emissions, we multiplied half the range of the estimated global carbon emissions with half the range of the emission factors. These *quartile ranges* were derived from the minimum and maximum values and the overall arithmetic mean:

Mean − ½(mean − minimum) or
Mean + ½(maximum − mean)

The range of possible CO emissions then is given by

2700 Tg C × 4.28% / 0.95 to 4200 Tg C × 8.46% / 0.95 = 121 to 373 Tg C / yr

This generates a narrower range for the global emission estimates and was used instead of the standard deviations which require a normally distributed data base. Due to the high variations in the emissions, we feel that this calculation provides more useful ranges.

Carbon Emissions

Our emission factors show that CO_2 is the major carbon compound which is emitted by biomass burning, at a rate of 1.9 to 4.0 Pg C/yr. It should be noted that these emissions represent the prompt release of CO_2 to the atmosphere; the net release is about one third to one half of the former (Crutzen et al., 1990) due to the uptake by plants covering the postburned areas. According to estimates of Andreae (1990) the amounts of carbon dioxide from biomass burning could contribute some 25% to the greenhouse effect produced by the global CO_2.

The range of our CO/CO_2 emission ratios covers most of the previously published results except for a few high values greater than 15% (Crutzen et al., 1979; Greenberg et al., 1984). Even such high values are observable in our experiments if we consider the pure smoldering stage values, which demonstrates clearly the importance of information about the burning stage when CO/CO_2 values are compared. Most authors present values between 4% and 15% (Andreae et al., 1988; Cofer et al., 1988a; Hegg et al., 1990), so our mean value of 7.3%, which was derived from complete fires, is a good working value. This CO/CO_2 ratio is close to the value presented by Andreae et al. (1988), who took integrated samples over a large volume and thus also considered an average of different burning stages. Our results show that carbon monoxide is the second but most important carbon emission with 121 to 373 Tg C/yr, which accounts for about 22% (16% to 34%) of the total CO source reported by Logan et al. (1981). Thus, pyrogenic CO is as important as fossil fuel emission, which, according to Logan, is in the order of 200 Tg C/yr.

For the hydrocarbons, which are discussed in more detail elsewhere (Hao et al., 1991b), the most important compound emitted from biomass burning is methane. CH_4/CO ratios average about 9.1%, in fairly good agreement with the values of 7.8% calculated from results of Greenberg et al. (1984) but substantially higher than the 3% determined by Hegg et al. (1990). The remaining hydrocarbons are also listed in Table 36.4. The highest emission ratios were measured for ethene (1.2%) and for C_3 and C_4 alkanes and alkenes as well as for benzene and toluene (0.1% to 0.5% each). Methane contributes strongly to the atmospheric greenhouse effect and also to global atmospheric chemistry. Our estimate results in a contribution to the global methane budget of 3% (2% to 4%) based on global source estimates by the WMO (1985).

Table 36.6 Global, annual source strengths of the compounds and the significance of biomass burning to their global budgets

Compound	Formula	Biomass burning source range (as C or N) (Giga-g/yr)			Total source range (as C or N) (Tg/yr)	Reference	Contribution to global source	
		Mean	Minimum	Maximum			Mean	Range
Carbon dioxide	CO_2	2,990,000	1,860,000	4,010,000				
Carbon monoxide	CO	207,000	120,000	370,000	610–1270	Logan, 1981	22%	16%–34%
Methane	CH_4	15,300	7,900	30,100	390–765	WMO, 1985	3%	2%–4%
Ethane	CH_3CH_3	2,190	972	4,495				
Ethene	$CH_2=CH_2$	4,460	2,030	12,000				
Ethine	$CH{\equiv}CH$	2,020	928	6,465	C_2–C_4			
Propane	C_3H_8	670	284	1,334	100	Greenberg, 1984	14%	6%–34%
Propene	C_3H_6	2,380	1,070	4,980				
n-butane	C_4H_{10}	169	75	327				
2-butene (cis)	C_4H_8	134	56	261				
2-butene (trans)	C_4H_8	171	73	338				
i-butene, 1-butene	$C_4H_8 + C_4H_8$	1,190	500	2,531				
1,3-butadiene	C_4H_6	754	352	1,707				
n-pentane	C_5H_{12}	266	121	701				
Isoprene	C_5H_8	302	139	548				
Benzene	C_6H_6	2,310	1,150	6,140				
Toluene	C_7H_8	1,340	594	3,098				
m-, p-xylene	C_8H_{10}	402	174	807				
o-xylene	C_8H_{10}	213	87	471				
Methyl chloride	CH_3Cl	357	147	780	0.55–1.2	WMO, 1985	41%	21%–75%
NMHC (as C)	C_2 to C_8	42,664	18,630	96,390				
Ash (as C)		180,866	79,770	602,070				
Total sum C	C_1 to C_8	3,432,000	2,004,000	4,506,000				
Nitrogen oxides	NO_x	7,520	4,070	11,930	25–99	Logan, 1983	12%	9%–15%
Ammonia	NH_3	2,310	1,120	5,380	20–80	Andreae, 1989	5%	3%–8%
Hydrogen cyanide	HCN	1,460	640	3,180				
Acetonitrile	CH_3CN	554	230	1,130				
Cyanogen	NCCN (as N)	13	5	53				
Acrylonitrile	CH_2CHCN	75	36	162				
Propionitrile	CH_3CH_2CN	39	15	88				
Nitrous oxide	N_2O	428	240	668	12–14	WMO, 1985	3%	2%–5%
Methylamine	CH_3NH_2	26	16	42				
Dimethylamine	$(CH_3)_2NH$	17	9	31				
Ethylamine	$CH_3CH_2NH_2$	3	2	4				
Trimethylamine	$(CH_3)_3N$	11	7	15				
2-methyl-1-butylamine	$C_5H_{11}NH^2$	22	15	31				
n-pentylamine	$n-C_5H_{11}NH_2$	76	38	127				
Nitrates (70% HNO_3)	RNO_x	610	476	745				
Ash (as N)		5,520	2,530	18,940				
Subtotal sum N		18,700	9,400	42,500				
Molecular nitrogen	N_2	11,990	5,610	21,270	100–170	Söderlund, 1976	9%	5%–14%

Note: The biomass burning source range is given in 10^9 g (gigagram) carbon or nitrogen per year, whereas the total source range is expressed in terms of 10^{12} g C/yr or g N/yr.

We also determined methyl chloride emissions and found an emission ratio of 1.64×10^{-3} relative to CO, which is much higher than values found by Crutzen et al. (1979) of 1.14×10^{-4}, but agrees reasonably well with the values by Tassios et al. (1985) which show a mean CH_3Cl/CO ratio of about 3×10^{-3} (assuming a CO/CO_2 ratio of 10%, not reported in that publication). The atmospheric input of methyl chloride has been reported to be 0.55 to 1.2 Tg C/yr (WMO, 1985). Adopting these amounts, biomass burning then could be the single most important source for CH_3Cl with 357×10^9 g C/yr, corresponding to a contribution of 41% (21% to 75%) to the global budget.

Nitrogen Emissions

Our values for NO_x/CO_2 lie well within the range of 0.3 to 3.5×10^{-3} published by other authors (Evans et al., 1974; Crutzen et al., 1979; Andreae et al., 1988). NO_x/CO_2 ratios do not change significantly during the different combustion stages, so *intra*fire variations are small and *inter*fire variations are due to varying fire and fuel properties. In nearly all of our experiments, NO_x is the most important reactive nitrogen emission. In a few fires, however, ammonia and hydrogen cyanide showed amounts comparable to NO_x. Calculation of the source strength yields a value of about 7.5 Tg N/yr, which can contribute 12% (9% to 15%) to the global budget of 25 to 99 Tg N/yr estimated by Logan (1983) and therefore represents an important source of atmospheric NO_x.

For N_2O, on the other hand, we estimate only minor emissions of about 0.4 Tg N/yr, contributing 3% (2% to 5%) to the global source of 12 to 14 Tg N/yr (WMO, 1985). The values for N_2O/CO_2 ratios presented in this paper are substantially lower than most other published data (Crutzen et al., 1979, 1985; Cofer et al., 1988a). Values of Crutzen et al. (1985) of 2×10^{-4} are much higher than our mean value of 9.2×10^{-5}. The reduction of published N_2O emission ratios is due to an artifact N_2O production in sample canisters of earlier measurements caused by an aqueous chemistry involving NO and SO_2, first determined by Muzio and Kramlich in 1988. We, too, observe this artifact production in our grab samples (Hao et al., 1991a), but since our SO_2 levels are fairly low (maximum mixing ratios of 20 to 30 ppmv during the flaming stage, see Table 36.2) and our samples were analyzed within 20 hours after sampling, we estimate, from repeated measurements on stored samples, a maximum artifact error in our values of +20%, so that our data on N_2O may still be a factor of 1.2 too high. Recent measurements of Griffith et al.

(1990) with an FTIR method (which avoids such an artifact) show values around 8×10^{-5} in good agreement with our data. In fact, there is only one publication reporting even lower values for N_2O/CO_2 of about (0.4 to 1.4) $\times 10^{-5}$ (Elkins et al., 1990). These values had been derived from burning wood, which has a very low nitrogen content; thus, these emission ratios are probably lower than in natural fires.

There are only a few published values for the biomass burning emission of ammonia. Our ammonia emissions are represented by a value of 1.55% relative to CO, which is slightly higher than the ratio derived by Andreae et al. (1988) for particulate NH_4^+ in aged plumes and the value given by Hegg et al. (1990) for seven forest fires in North America (both around 1%). Recent FTIR measurements over prescribed fires by Griffith et al. (1990), on the other hand, show average values of NH_3/CO which are a factor of 2 higher than ours in a range of (1.2 to 5.3) $\times 10^{-2}$. Their NH_3/NO_x ratio is also very high at 5:1, while our ratio is significantly lower at about 5:15. Thus, further measurements on ammonia emissions should be carried out. Information on the atmospheric budget of ammonia is very poor; the global source has been reported to be 20 to 80 Tg N/yr (Andreae et al., 1989), most of which is due to microbial and animal release. Our estimate shows that biomass burning, representing an amount of 2.3 Tg N/yr, on average accounts for about 5% (3% to 8%) of the total ammonia source.

Our emission ratios of total C_1 to C_5 amines are around 0.5% relative to CO. A comparison to other data is not possible, since they are published for the first time. Also there are no data on their atmospheric amounts. Cadle et al. (1980) determined the concentrations of amines in automobile exhausts and Seuwen (1990) estimated from this only negligible amounts of 1×10^6 g N/yr being emitted to the atmosphere. Nevertheless, when our emission factors from biomass burning are considered, the sum of these compounds may be in the order of 0.8 Tg N/yr; thus, further investigation might be worthwhile.

An interesting outcome of our experiments was the large amounts of nitriles, especially hydrogen cyanide and acetonitrile (Lobert, 1990a; Lobert et al., 1990b). Emission ratios for these compounds were determined for the first time and lie in a range comparable to ammonia emissions. Therefore, HCN and CH_3CN are two of the most important nitrogen-containing gases emitted by biomass burning, together releasing some 3.6% of the fuel nitrogen (Table 36.4). In some experiments, HCN emissions can be as high as NO_x emissions when incomplete combustion was predominant. The emission ratios for the remaining three nitriles—cyanogen, acrylonitrile, and propionitrile—lie a factor of 10 to 100 below those of HCN and CH_3CN. A few measurements over prescribed fires in Australia (Lobert, 1990a) show emission ratios of HCN/CO and CH_3CN/CO of 1.5×10^{-2} and 3.8×10^{-3}, respectively, which are in fairly good agreement with our experimental data of 1.1×10^{-2} and 2.5×10^{-3}. Recent measurements by Griffith et al. (1990) yield values for HCN which are ten times lower than our results. The reasons for these diverging data need to be investigated and may partially be explained by the low N/C ratio of only 1% in the burning fuels observed by Griffith et al.

Data on the global sources of nitriles are not available, but can be derived from their background mixing ratio and their lifetime against the reaction with OH radicals. The average mixing ratio in the free troposphere is about 170 parts per trillion by volume (pptv) for HCN (Cicerone et al., 1983) and about 80 pptv for CH_3CN (Hamm et al., 1990). Reaction with OH would remove 0.17 and 0.40 Tg N/yr of HCN and CH_3CN, respectively (Lobert, 1990a). The lifetimes were taken from Cicerone et al. (1983) and Harris et al. (1981). The computed amounts of 0.17 and 0.40 Tg N/yr are the required sources, which can explain the atmospheric background mixing ratio, assuming that the reaction with OH is the only important sink. In Table 36.6 we report values which are much higher than these estimates. This implies that other sinks for these gases must exist. Hamm et al. (1990) supposed the ocean to be a major sink for acetonitrile. However, it is conceivable that hydrogen cyanide can be consumed by plants (Lobert, 1990a; Selmar et al., 1988) and microorganisms (Postgate, 1982). Inherent in larger sinks, on the other hand, is a shorter atmospheric lifetime of these compounds, which causes a nonuniform distribution in the troposphere and therefore allows their use as tracer compounds for biomass burning emissions. Indeed, Hamm observed a strong acetonitrile increase in marine, equatorial latitudes during ship measurements in the dry season. Other important sources for HCN and CH_3CN are not known. Arijs et al. (1986) estimated the automotive and industrial exhausts for acetonitrile to be at most 0.02 Tg N/yr. A similar value of 0.04 Tg N/yr was derived from estimates by Lobert (1990a) for hydrogen cyanide. Thus biomass burning may well be the major source for the release of these gases into the atmosphere.

As stated above, molecular nitrogen could be an important emission of biomass burning, causing a *pyro-denitrification* of ecosystems (Lobert et al., 1990b), since N_2 is not deposited by dry or wet processes and is therefore removed from the soil irreversibly. The 17 experiments for determination of N_2 presently conducted yielded an average fraction of 22% of the fuel nitrogen. Most of these experiments showed a CO/CO_2 ratio of about 25%, which is an indicator for pure smoldering combustion or even pyrolysis. The most recent experiments were conducted to burn more efficiently with CO/CO_2 values of around 5%, emitting 41% to 46% of the fuel nitrogen in form of N_2. Thus, N_2 seems to belong to the flaming stage rather than to the smoldering stage. Based on our preliminary average of 22%, we calculate a minimum pyro-denitrification rate of 5.6 to 21.3 Tg N/yr, which equals 9% (5% to 14%) of the global, terrestrial nitrogen fixation rate of 100 to 170 Tg N/yr (Söderlund et al., 1976). Assuming N_2 emissions to be 50% of the nitrogen balance, which then is the upper estimate (Lobert et al., 1990b), it could equal 17% to 20% of the fixation rate. In either case, biomass burning has a significant effect upon the biogeochemical cycle and nutrient budget of tropical ecosystems. For detailed results see Kuhlbusch et al. (1991).

Conclusions

We presented data on the nitrogen and carbon emissions of biomass burning. The results of our experiments enable us to calculate new source strengths for many compounds, considering different burning stages and fire conditions on the one hand, and different fuel types and properties, on the other hand. We also presented a method for balancing elemental budgets of fires, which had already been described for carbon compounds by other authors but which is new for the nitrogen inventory. We feel that this is an important tool for estimating global amounts of nitrogen-containing gases. Measurements on quantifying the high molecular weight nitrogen compounds and particulate matter, which complete our balance, must be carried out in order to improve the rough estimate given in this chapter.

Based on our measurements we show that biomass burning contributes significantly to the global budgets of HCN, CH_3CN (possibly the major source), NO_x (12%), CO (22%), C_2 to C_4 hydrocarbons (14%), CH_3Cl (41%), and probably also to the global source of C_1-C_5 aliphatic amines. Further, pyrogenic CO_2 amounts are likely to represent a substantial contribution to the global greenhouse warming. An important result from our study is the identification of N_2 emissions, which causes a significant loss of fixed nitrogen (pyro-denitrification) in tropical ecosystems in the order of 5% to 20% of the global nitrogen fixation rate. Because of an interesting interplay between an enhanced postfire nitrogen fixation (Eisele et al., 1989) and an enhanced postfire N_2O emission (Levine et al., 1988), it is not yet known if losses due to pyro-denitrification are balanced by nitrogen fixation. Therefore, the potential ecological significance of N_2 emissions presents an interesting focus for further study.

Since our experiments had been conducted in an experimental system, we need to improve our data and analytical methods by measuring the same compounds over natural fires. In order to obtain more precise source strengths, on the other hand, an improvement of estimates or, even better, an experimental attempt in quantifying the amounts of globally burned biomass is of exceeding importance. Also, a reliable determination of average N/C ratios for the burned ecosystems is essential, since a lack of such information is one of the major problems connected to global source estimates.

Quantitative Assessment of Gaseous and Condensed Phase Emissions from Open Burning of Biomass in a Combustion Wind Tunnel

Bryan M. Jenkins, Scott Q. Turn, Robert B. Williams, Daniel P. Y. Chang, Otto G. Raabe, Jack Paskind, and Steve Teague

Agricultural practices and land use modification were estimated to produce 14% and 9%, respectively, of the total greenhouse gas emissions contributing to global warming in the decade preceding 1990 (Marshall, 1989). Carbon release rates from tropical forest conversion have been estimated at 0.9 to 2.5 gigatons per year (Gt/yr) (Myers, 1989). Specific determinations of emission factors for gaseous species and particulate matter from open fires in biomass fuels are often hampered by uncertainty in closure on mass balances relating quantity of emission released per unit of material burned. Uncertainty arises both from the inability to directly measure the amount of entrained air which dilutes the combustion products, and from the natural variability which occurs in fuel moisture, fuel loading, and wind speed in the open environment, factors which are known to influence the character of the fire and the resulting emissions.

The work undertaken in this project was specifically designed to quantify the effects of the fuel and environmental conditions on emission factors from open burning of agricultural and forestry wastes. To this end, a wind tunnel was constructed for conducting the burning and the emission sampling so as to reduce uncertainty associated with uncontrolled internal and external parameters of the fire. The combustion and emission formation mechanisms could be better described, and the results from the wind tunnel studies would be available for calibration of field studies and mathematical models developed to describe the fire under conditions outside the range of the tunnel.

The wind tunnel as constructed can accommodate both spreading and pile type fires. This tunnel is in some respects similar to other wind tunnels used elsewhere to study the fire propagation velocity in spreading type fires. Most of these tunnels had stationary floors and rather limited duration testing capabilities, which was seen to be undesirable from the standpoint of emissions monitoring because of the very small sample sizes involved. The tunnel described here includes a moving floor which can be used when conducting spreading fires. By translating the floor, the fire can be made to propagate indefinitely into the fuel bed, which is introduced into the upstream end of the tunnel. Emissions sampling can then occur for as long as necessary to capture representative samples or to integrate the emission load over the anticipated duration of an actual field fire. Primary emission sampling is done in a stack some 8 meters (m) above the fire. When testing spreading fires with an impressed wind, the gas temperature in the stack runs about 15°K above the temperature of ambient air entering the tunnel. The data collected therefore are representative of emission levels in the plume shortly after quenching and do not include any longer-term effects due to atmospheric reactions occurring far behind the fire. The atmospheric reactions may be understood to some degree from the primary emissions. The tunnel provides a means to control the factors influencing the fire and thus permits studies of how changes in the condition of the fire affect the emission levels observed—studies which are difficult to perform in the field.

Wind Tunnel Design

Design of the combustion wind tunnel was based on the need to determine and control variables of the fuel and the fire environment in order to understand the effects of these variables on the emission rates observed. The use of a wind tunnel was intended to enable well-prescribed conditions that could not be obtained under similar field experiments where the spatial and temporal variation of the major parameters influencing the fire are large, and the difficulties of sampling a moving emission stream in the case of a spreading fire are severe. The major variables to be controlled are freestream wind speed, wind velocity profile, turbulence intensity, fuel loading, and fuel moisture content.

The main concern in designing the combustion wind tunnel was that the fires produced in the tunnel and the resulting emissions be physically representative of those that occur in the field. In addition, a balance had to be found between blower capacity needed to generate model wind speed, tunnel cross-sectional area, and heat exchange between the fire and the wind tunnel structure. Consequently, important design considerations were identified—namely, the quantity of fuel to be burned during each test, the ability to reproduce the velocity profile and turbulence intensity of the wind, simulation of the fuel bed structure of the field, configuration of the tunnel to allow for the free expansion of the plume, and minimization of the effect of heat exchange between the physical tunnel structure and the fire. Unfortunately, compromises among these parameters reduced the size of fire that can be permitted in the tunnel, so that large pile fires cannot be directly tested. Because these types of fires can be accommodated with simpler structural considerations, however, the wind tunnel design was dominated by the potential to measure long-term emission rates in spreading type fires.

For the purposes of sampling the emissions from spreading fires, the use of stationary fuel beds requires extremely long combustion test sections to permit burns of sufficient duration. The concept employed here was instead to translate the fuel bed relative to the tunnel floor with a conveyor system. By moving the fuel bed at the fire propagation velocity, the flame could be held stationary in space. In principle, a fire could be sustained indefinitely with this approach. By fixing the position of the flame, sampling instruments could also be fixed without the need to follow the fire.

The major difficulty arising from the translation of the bed is the need to configure the bed so that the structure remains essentially intact during transport, and thus an accurate simulation of the fuel bed structure occurring in the field becomes more difficult. The structure used to convey the fuel also needs to provide minimum quench surface so as not to unduly influence the emission of volatile organics. The wind profile and turbulence characteristics are also problematical. In fact, complete simulation of the atmospheric boundary layer in a wind tunnel cannot be achieved. As discussed by Plate (1982), however, sufficient similarity in the lower part of the boundary layer can be obtained to warrant use of the wind tunnel in this kind of investigation. Because scaling of the fire itself appeared undesirable from the standpoint of emissions testing, the tunnel was designed with the intent of generating a sufficiently thick boundary layer above the fuel surface with turbulence

intensities of the same order as the field. As mentioned earlier, sampling was to be done from the immediate vicinity of the fire, without attempting to model plume processes.

Another design consideration was to allow for the free expansion of the plume behind the fire in the wind tunnel. The use of a closed channel wind tunnel leads to a reduced flame angle and recirculation behind the flame; neither of which appear to occur to any large extent in the field. An open channel would lead to deceleration of the jet-like flow at the exit of the flow development section (Fleeter et al., 1984). To deal with this problem, a semi-open channel was devised consisting of a large volume combustion test section incorporating a moveable ceiling, which was positioned at the leading edge of the fire in a manner similar to that used by Fleeter et al. (1984). The plume was then free to expand beyond the ceiling, and the problems of recirculation and turbulent mixing effects generated by closed and open channel tunnels were avoided. The adjustable ceiling need not be moved once the position of the flame has been established in the case of a spreading fire.

The width of the tunnel was selected to reduce the thermal dissipation at the walls to a reasonable fraction of the estimated energy release rate of the fire. Consideration was also given to the blower capacity required to reach the intended maximum wind velocity of 5 meters per second (m/s), the ability to maintain a uniform fire profile across the width of the tunnel in the case of a spreading fire, and the physical strength requirements of the conveyor elements supporting the fuel across the width of the tunnel. Using the results of prototype studies, the transverse thermal energy loss from the flame adjacent to the walls in a wind of 2.5 m/s and a loading rate similar to that of a rice straw fuel in the field was estimated to be approximately 17.5 kilowatts (kW). From the heating value of the fuel and the heating value of the residual char following the fire, the fireline intensity (energy released per unit width of fire) was estimated to be 130 kW/m. As the width of the tunnel increases, the fractional wall dissipation rate (the ratio of the wall loss to the energy release of the fire) declines in the manner shown in Figure 37.1. There is a rapid decline in the wall dissipation rate up to 1 m width. The incremental improvement is reduced past this point. An increase in width from 1.5 to 2.0 m, for example, results in a decline of only 2% in the wall dissipation rate but requires an increase of 33% in blower capacity to match the wind velocity. The final width was selected as 1.2 m and provided a wall thermal dissipation rate of approximately 10% under the assumptions made.

Wind Tunnel Description

The combustion wind tunnel is comprised of three major functional parts—the flow development section, the combustion test section, and the emissions sampling stack. An exterior design view of the wind tunnel is shown in Figure 37.2. The entire tunnel extends nearly 26 m in length and stands approximately 10 m high. The tunnel is a forced draft open circuit type, using a 45 kW centrifugal blower capable of generating an average 5 m/s wind speed approaching the fire in the combustion test section. A series of conveyors extend the length of the tunnel and are used to transport fuel to the fire at a speed matching that of the fire propagation rate so as to produce a flame standing stationary in space.

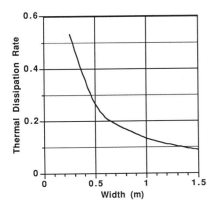

Figure 37.1 Wall thermal dissipation rate (dimensionless) as a function of wind tunnel width.

The flow development section consists of the blower, flow straightening elements, and the primary fuel feed conveyor used when testing spreading type fires. Fuel is loaded onto the primary conveyor behind the blower, and passes beneath the blower, diffuser, and flow straightening elements before entering the air stream 10 m upstream from the combustion test section. The primary fuel conveyor consists of a continuous flexible solid belt 36.5 m long. The belt travels along the inner floor of the flow development section on its delivery pass and returns on the outside below the tunnel. Fuel can be loaded continuously onto the belt as the experiment proceeds. The centrifugal blower is of fixed velocity but has variable dampers at the inlet and outlet to adjust the air flow rate. The blower is connected to the diffuser by a flexible coupling to reduce transmission of vibration from the blower to the tunnel structure. Flow straightening is done with five screens located downstream of the diffuser. Two 40 mesh stainless steel screens are followed by three 60 mesh stainless steel screens. Just following the last screen, the fuel enters the flow development section, providing a roughened lower surface for boundary layer development. At a distance of 2.4 m upstream of the combustion test section, a wire mesh fence was installed to trip the flow shortly ahead of the fire.

The combustion test section extends 7.4 m in length. The configuration of the conveyors, floor, and ash accumulation bin are shown in Figure 37.3. At the entrance to the combustion test section, the fuel is transferred from the primary conveyor to a stainless steel rod conveyor consisting of 12.7 mm diameter stainless steel tubes spaced every 100 mm along the tunnel and connected at the ends to continuous roller

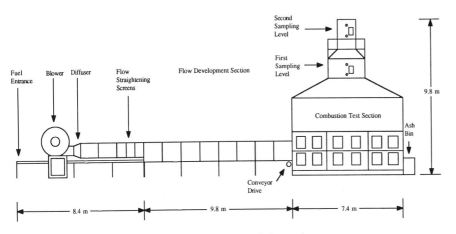

Figure 37.2 Exterior schematic of the combustion wind tunnel.

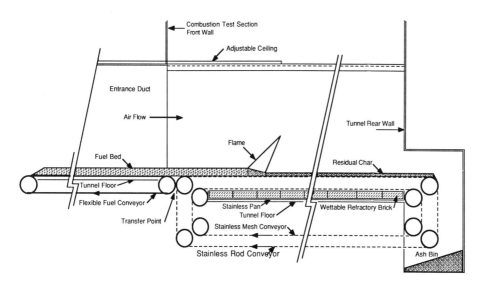

Figure 37.3 Schematic of the internal fuel conveyor system.

chains. The floor of the combustion test section consists of porous refractory brick which can be wetted to provide a moist surface below the fuel. The brick was placed in stainless steel pans connected to a manifold supplying water to saturate the brick. The brick has a specific gravity of 1.64 and is intended to remain wet in the presence of a flame, as in the case of a flame moving across a saturated soil surface.

Just above the brick and 150 mm below the rod conveyor is a third conveyor of 3 mm stainless steel mesh screen. This conveyor is used to convey fine fuel material, which drops through the rod conveyor downstream so that it will not serve as a source of ignition to incoming fuel. All three conveyors are driven at the same linear velocity by a variable speed direct current motor regulated to match the fire propagation velocity. The rod and mesh conveyors also return on the outside of the tunnel and are cooled to near ambient before they enter the tunnel. An ash collection bin is located at the discharge end of the conveyors to receive and accumulate residual char as a test proceeds. This bin is sealed so as to reduce air intrusion or loss of char. Twenty-five glass windows are located on the walls of the tunnel, 12 on each side and one in the end wall of the tunnel. These windows permit observation of the flame. Each window is 508 mm wide, 660 mm high, and 5 mm thick. During operation, the flame is held on the window surfaces to reduce catalytic reactions, rather than the sheet steel surfaces making up the remainder of the tunnel wall. Each window is held against a silicon gasket by a spring loaded frame to allow the windows to expand on heating, although the glass used is reported to have

zero thermal expansion at 800°C. The windows have not failed under thermal-mechanical stress generated by operating with the flame held on the glass. The lower six windows on each side of the tunnel are installed on three doors, hinged at the top, which may be opened for inspection purposes. The doors are also designed to be opened when conducting pile fires under no wind conditions when air should be drawn from around the fuel bed. An adjustable ceiling can be extended into the combustion test section at a height of 1.23 m above the rod conveyor. A water manifold is installed at the entrance to the combustion test section to extinguish any uncontrolled fires approaching the flow development section.

At a height of 3.7 m above the fuel bed the combustion test section begins to narrow into the stack sampling section. Sampling ports are located at two levels in the stack. The first level is in a section of duct 2.4 m long extending 1.2 m upwards from the initial transition. Above this is another transition leading into the second level sampling position situated in the vent stack of 1.2 m square cross-section. This second level is the principal sampling area and is situated nearly 8 m above the fuel bed. Windows are also provided at both levels to view the sampling instruments and the conditions of the exhaust. A ladder and catwalks provide operator access to the sampling ports.

Flow and Fire Characterization

Investigations of the velocity profiles and turbulence intensities generated at the end of the flow development section and in the stack at the second level sampling position under cold conditions were conducted by traversing the flow in these regions with a single element hot film anemometer. The anemometer output was connected to a high speed data acquisition system. The anemometer was sampled at 7 kilohertz (kHz). Long-term averages were collected over a period of at least one minute in each position of the traverse. Turbulence intensities at each point were computed from 15,000 data points according to:

$$I_1 = \frac{\overline{(u'u')}^{0.5}}{u} = \frac{\sigma_u}{u} \tag{37.1}$$

where I_1 is the turbulence intensity normalized to the mean velocity, u, in the streamwise direction; u' is the streamwise velocity fluctuation; and $\overline{(u'u')}^{0.5}$ is the root mean square of the fluctuation, i.e., the standard deviation, σ_u.

Velocity profiles and turbulence intensities for a mean velocity of 2.35 m/s are shown in Figures 37.4 through 37.7. The duct Reynolds number at this velocity was 1.9×10^5. Figures 37.5 and 37.6 show the velocity profiles through the center of the tunnel in the vertical and horizontal directions. These profiles were determined in the combustion test section with the ceiling extended and approximately 250 mm upstream from the edge of the ceiling. These data were collected without a fire in the tunnel, but with a layer of straw (600 g/m^2) extending the full length of the flow development section and into the combustion test section. Figures 37.7 and 37.8 are velocity profiles through the center of the stack at the second level sampling area in the transverse and longitudinal di-

rections. These were also taken without a fire in the tunnel. The vertical profile at the entrance to the fire (Figure 37.4) shows the boundary layer extending some 35 to 40 centimeters (cm) above the fuel surface as well as a boundary layer of some 10 to 15 cm thickness developed on the ceiling of the tunnel. A comparison of the boundary layer over the fuel to that computed for a natural grass surface estimated to be of similar roughness to the field shows similarity up to about 35 cm depth (Figure 37.8). The latter profile was computed from a logarithmic law of the wall (Sutton, 1960):

$$u = \frac{u^*}{k} \ln\left(\frac{z - d_0}{z_0}\right) \tag{37.2}$$

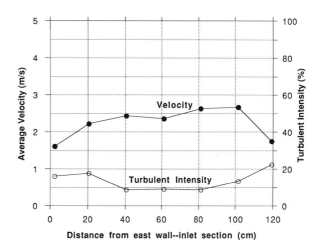

Figure 37.5 Horizontal velocity profile and turbulence intensity at the entrance to the combustion test section.

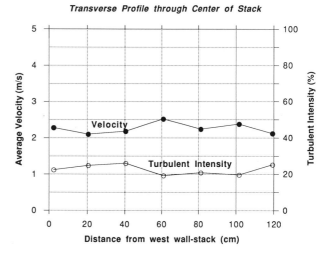

Figure 37.6 Transverse velocity profile and turbulence intensity across the sampling stack.

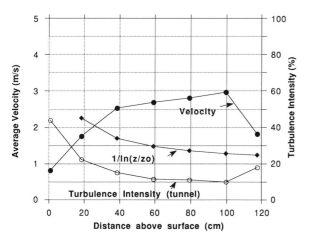

Figure 37.4 Vertical velocity profile and turbulence intensity at the entrance to the combustion test section. Shown for comparison is the computed turbulence intensity of the field estimated for a grass surface.

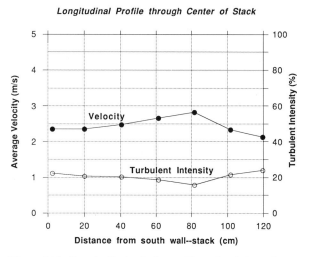

Figure 37.7 Longitudinal velocity profile and turbulence intensity across the sampling stack.

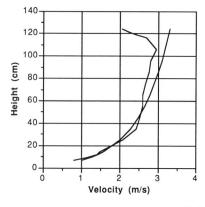

Figure 37.8 Comparison of the actual inlet velocity profile to a logarithmic law of the wall velocity profile based on a 5 cm high grass surface ($u^* = 0.32$, $z_0 = 0.02$, $d_0 = 0$).

where u is the wind velocity (m/s) at elevation z (m) normal to the surface; k is von Karman's constant, here taken numerically equal to 0.4; u^* is the friction velocity (m/s); d_0 is the zero plane displacement (m); and z_0 is the surface roughness parameter. Values used in developing Figure 37.8 were $u^* = 0.32$, $z_0 = 0.02$, and $d_0 = 0$. The 35 cm thickness corresponds roughly to the flame height of a field fire in rice straw propagating in opposition to the wind, as shown later. Plate (1982) has shown that equation (37.2) is valid with roughness height z_0 which depends only on the geometry of the roughness elements of the surface, and is independent of any Reynolds number for roughness Reynolds numbers $u^* z_0 / v > 5$ (where v is the kinematic viscosity, m^2/s), which is the situation of Figure 37.8.

The velocity profile across the horizontal direction is uniform (Figure 37.5) except for boundary layers of some 20 cm thickness at the wall. Such boundary layers do not appear in the stack, and the stack velocity profiles are uniform out to the walls (Figures 37.6 and 37.7). Conditions of uniformity for sampling the stack appear to be satisfactory.

Of primary importance is to attempt to match the gross velocity profile and the turbulence characteristics of the flow within the immersion depth of the flame. Wichman (1983) and Wichman et al. (1982) have demonstrated the dependence of the fire propagation characteristics on the velocity profile of the air entering the fire. In fact, Wichman (1983) has used this result to relate experiments conducted under different external flow conditions which yield similar flame spread rates. Comparing fires conducted under forced convection in wind tunnels with open flames in a free convection boundary, he finds the results to be identical based on velocity profile similarity over nearly an order of magnitude change in air velocity.

The vertical inlet velocity profile appears reasonably well matched to the field up to 35 to 40 cm above the fuel surface. The turbulence intensities are also of the proper order, although a comparison with atmospheric turbulence intensities suggests the wind tunnel intensities may be somewhat low. The logarithmic profile developed from Monin-Obukhov similarity theory and K theory (Lumley and Panofsky, 1964; Panofsky and Dutton, 1984; Pasquill, 1974; Nieuwstadt and van Dop, 1982; Plate, 1982; McBean, 1971) yields a streamwise turbulence intensity of:

$$\frac{\sigma_u}{u} = \frac{1}{\ln(z/z_0)} \tag{37.3}$$

For $z_0 = 0.02$ m (Figure 37.8), this yields estimated turbulence intensities of the magnitude shown in Figure 37.4. Estimates of intensities are about twice as large as those measured in the tunnel, but decrease upwards from the surface in the same manner. Turbulence intensities in the range of both the estimates of equation (37.3) and those measured in the wind tunnel (40% just above the fuel surface decreasing to about 10% at 60 cm height) have been reported in the literature. Plate (1982) summarizes field data for $z_0 = 0.05$ to 0.5 m which are nearly identical to those measured in the wind tunnel. Cionco (1972) measured horizontal turbulence intensities of 30% to 100% in the region immediately above a crop canopy. Maitani (1979) measured turbulence intensities over various surfaces and found values of 40% to 120% in the region just above plant canopies. Ohtou et al.

(1983) consistently measured turbulence intensities of about 50% at an elevation of 65 cm above a rice crop. Ohtaki (1980) measured turbulence intensities of about 40% at 80 cm above a rice crop.

Tests were also conducted to determine if the time-temperature relationships of fires in the field are adequately matched in the tunnel. Species emitted from a combustion process are directly dependent on the residence time of reactants at elevated temperatures. Field data were obtained for a fire spreading in opposition to the wind (backing fire) and compared against wind tunnel data.

Results of field and tunnel experiments in spread rice straw are shown in Figure 37.9. Temperature data were obtained by positioning 0.5 mm type K thermocouples a known distance apart in the fuel bed. The wind velocity in the field ranged from 1 to 2 m/s at 0.5 to 0.6 m elevation. The wind tunnel experiments were conducted at an average duct wind speed of 2.3 m/s.

Results from the field are shown in Figure 37.9a and for the tunnel in Figure 37.9b. The spreading velocity for the backing fires in the field was 0.95 m/min. The period of elevated temperature at the fuel surface lasted about 60 s. Measured peak temperatures were 800° to 900°C. The thermocouple time constant was too large to measure true peak flame temperatures, but the results are consistent from wind tunnel to field.

Comparable tests in the wind tunnel show similar trends in temperature and duration. Results of temperature measurements made at the fuel surface are included in Figure 37.9b. The peak temperature is about 800°C, and the duration is about 60 s. Spreading velocity was 0.94 m/min. This is in reasonable correspondence with the field data.

Emissions Testing

A number of experiments were conducted to determine if repeatable data could be obtained on emissions from fires conducted under similar conditions of wind speed, fuel type, fuel loading, fuel moisture, air temperature and relative humidity, and moisture status of the floor. Experiments on 1 September 1988 and 7 September 1988 were conducted using rice straw at a mean wind speed of 2.35 m/s and a loading rate to match that of the field at 7% moisture, or approximately 600 g/m^2. Each test was designed to accumulate information on fuel properties and fuel bed structure, residual solid phase composition, wind tunnel characteristics, and solid and gas phase emissions (particulate matter concentration and size distribution of CO_2, CO, NO_2, NO, SO_2, CH_4, C_2H_2, C_2H_4, C_2H_6, C_6H_6). Qualitative assessments of other organic compounds were made as well.

The sampling train used in conducting these experiments is illustrated in Figure 37.10. Samples were drawn from the second level sampling position through one of the ports in the sidewall of the stack. The sample was removed isokinetically with a stainless steel probe, 9.5 mm in diameter, located in the center of the stack. The end of the probe was bent and cut 90° from the probe axis, such that it faced into the exhaust stream. The probe flow rate was matched to the stack velocity measured by a hot film anemometer positioned near the probe. The probe was connected to a 25 mm inside diameter heated line of flexible stainless steel tubing lined with Teflon and wrapped with heating tape. The line was controlled to 100° ± 2°C by a temperature controller using an imbedded type K thermocouple. The sample line was fed to a heated aluminum manifold having a volume of 2.5 liters (L) from which various samples were drawn via stainless steel pipe connectors tapped into the wall. Filters and impactors were housed within an insulated heated chamber also maintained at 100° ± 2°C.

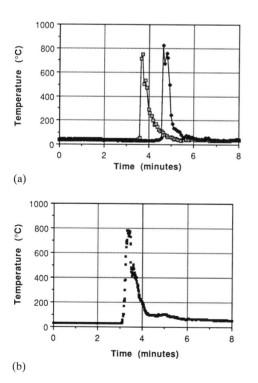

(a)

(b)

Figure 37.9 (a) Temperature at the fuel surface for a field backing fire in rice straw; (b) temperature at the fuel surface for a backing fire in rice straw conducted in the wind tunnel.

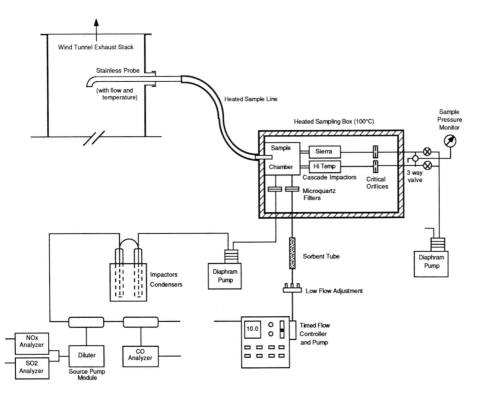

Figure 37.10 Sampling train used to collect particulate, volatile organic, and gas samples from the sampling stack.

A high temperature, high pressure (HTHP) impactor (Model 02-300, In-Tox Products, Albuquerque, N.M.) operated at 13.3 L/min average flow rate was used to determine particulate matter size distribution. This impactor was constructed of stainless steel and used wire gaskets made of brass or gold between stages to seal the impactor by crushing the wire. For the tests reported here, Gelman Micro-Quartz filters were used (Part #66089 47 mm, Gelman Sciences Inc., Ann Arbor, Mich.). The filters were hand cut to 37 mm diameter for all stages except the final filter stage, which was left at 47 mm diameter.

Two other samples were drawn from the manifold. One stream was used for collection of volatile organics on sorbent tubes. The other was used for gas analysis. Each line was filtered through a Micro-Quartz filter. Flow through each sorbent tube was pulled by a timed flow controller and pump at a rate of 100 ml/min for a period of ten minutes yielding a total flow of 1 L. The sorbent tubes used were 6 mm outside diameter packed with 50 mm of Tenax-TA 20×30 mesh and 38 mm of Ambersorb XE-340 (tubes supplied by T.R. Associates, Lewisburg, Pa.). The tube was placed immediately after the heated manifold and connected with a soft graphite ferrule on Teflon tubing. The gas sample line was connected via 6.4 mm Teflon tubing

to a diaphragm pump and an ice bath impinger train before descending to ground level. Samples were pulled from this line for each of the NO_x, SO_2, and CO analyzers located in an instrument trailer next to the tunnel. Sorbent tubes were analyzed by GC-MS.

The NO_x and SO_2 analyzers (Model 8840 and 8850, Monitor Labs, San Diego, Calif.) were supplied with sample from a source pump and dilution module (Monitor Labs Model 8730), which diluted the sample 20 to 1 with purified ("zero") air by means of a critical orifice. Sample dilution was specified on the basis of results from the prototype tunnel, which indicated levels within the source range of the instruments, which are designed as ambient monitors. Carbon monoxide was first monitored using an electrochemical polargraphic analyzer but was later monitored with an infrared gas analyzer (Anarad AR-500). Grab samples were collected in glass containers at regular intervals throughout each test. These were analyzed by GC/TCD for CO_2 and low molecular weight hydrocarbons. Ambient air background samples were also collected and analyzed.

Other instrumentation included various transducers for monitoring the conditions of the test, and the data acquisition system to read, process, and store the data. Inlet air temperature and relative humidity were monitored downstream of the blower by a thermistor and polymeric grid type humidity sensor (Campbell Scientific Model 207, Logan, Utah). Pressure differential across the five screens in the flow development section was monitored with a strain-gage type pressure transducer. Temperatures in the flame, in the stack, and at the outlet of the impingers were monitored with type K and type T thermocouples. Velocity in the stack was monitored using a hot film air velocity sensor (Kurz Instruments, Carmel Valley, Calif.) capable of withstanding the particulate load in the gas stream. All sensors except the velocity transducer were connected to an electronic data logger (Campbell Scientific Model CR21X, Logan, Utah), which was serially interfaced to a microcomputer (IBM PC-AT) running a custom data acquisition program to store the data on disk and output results to the screen in either text or graphics mode. Outputs from the NO_x, SO_2, and CO analyzers were also connected to the data logger and computer. Results were written to disk every three seconds throughout the test. Five 0.5 mm type K thermocouples were suspended from the leading edge of the adjustable ceiling into the flame. These were positioned at 44, 70, 127, 146, and 216 mm above the surface of the rod conveyor.

Results

Results of the fuel and ash analyses for both tests are included in Table 37.1. Fuel moisture was similar in both cases (6.5% and 7.3%, respectively, wet basis).

Material balances and characteristics of the fire appear in Table 37.2. The fire propagation velocities were 0.96 and 0.93 m/min. These values are consistent with values reported by Mobley (1976) for backing fires in grass fuels at similar moisture, and to the velocities obtained in the field tests described above. Total ash recovery is similar to what would be expected on the basis of the inorganic component of the fuel and ash. The amount recovered from the first test is somewhat higher than expected, and the amount from the second test is somewhat lower, but both are close to the anticipated value and are within the expected experimental error. There is some loss of ash to the floor of the tunnel which cannot be fully recovered and is lost during cleaning in preparation for the next experiment.

Inlet air conditions for both tests were 32° to 37°C and 15% relative humidity. Stack temperatures at the sampling point were 15° to 20°K above inlet air temperature. The flue gas heat loss is approximately 80 kW, or half of the energy released by the fire (160 kW) determined on the basis of the difference in heating values of the fuel and ash. Roughly 20 kW is transferred from the flame through the windows by radiation, conduction, and convection. The remainder is lost to the structure of the tunnel and by cooling of the ash after deposition in the ash collection bin.

Temperatures measured in the flame at 44 mm above the conveyor surface ranged from 400° to 800°C. At 216 mm above the surface, average temperature measured 100°C, with excursions to 400°C as flamelets burned past the thermocouple.

Table 37.1 Fuel and ash analysis for combustion trials of 1 September and 7 September 1988 in rice straw

	1 September		7 September	
	Fuel	Ash	Fuel	Ash
Proximate analysis				
Moisture (% wet basis)	6.5	2.8	7.3	3.3
Ash (% dry basis)	13.59	78.41	13.16	77.68
Volatiles (% dry basis)	69.44	11.11	69.94	10.34
Fixed carbon (% dry basis)	16.97	10.48	16.90	11.98
Higher heating value (MJ/kg, dry basis)	15.91	4.99	15.99	5.34
Ultimate analysis (% dry basis)				
Carbon	40.41	11.17	40.29	15.23
Hydrogen	5.42	0.49	5.39	0.56
Oxygen (by difference)	38.69	4.05	38.98	3.18
Nitrogen	0.53	0.24	0.69	0.31
Sulfur	0.05	0.09	0.08	0.05
Chlorine	0.20	0.06	0.27	—
Residual	14.7	83.9	14.3	80.7

Table 37.2 Material balances and fire characteristics for tests of 1 September and 7 September in rice straw

	1 September	7 September
Fire characteristics		
Effective fuel loading (g/m² wet basis)	584	577
Fuel consumption (g wet basis)	23,190	22,896
Fuel moisture (% wet basis)	6.5	7.3
Dry fuel consumption (g)	21,683	21,225
Burn duration (minutes)	34	35
Dry fuel consumption rate (g/s)	10.63	10.11
Fire propagation velocity (m/min)	0.96	0.93
Mean wind velocity (m/s)	2.35	2.35
Mean air temperature (°C)	37	32
Mean relative humidity (%)	15	15
Air density (kg/m³)	1.140	1.159
Air flow rate (g/s)	3,982	4,048
Overall air/fuel ratio (dry basis)	375	400
Stack gas mass flow rate (g/s)	3,991	4,057
Stack gas temperature (°C)	53	48
Stack gas flow rate at temperature (m³/s)	3.68	3.68
Ash recovery		
Total ash recovered (g)	3,993	3,538
Moisture content (% wet basis)	2.8	3.3
Dry ash recovery (g)	3,881	3,421
Fraction of dry fuel consumption (%)	17.9	16.1
Inorganic fraction of fuel (%)	14.7	14.3
Inorganic fraction of ash (%)	83.9	80.7
Expected ash recovery (% of fuel)	17.5	17.7
Nitrogen		
Fuel nitrogen (% dry basis)	0.53	0.69
Ash nitrogen (% dry basis)	0.24	0.31
NO concentration in stack (ppm)	3.0	2.8
NO_2 concentration in stack (ppm)	0.9	0.6
Fuel N (g)	115	146
Ash N (g)	9	11
Gas phase N from NO, NO_2 (g)	15	14
Excess fuel N (g)	91	121
Sulfur		
Fuel sulfur (% dry basis)	0.05	0.08
Ash sulfur (% dry basis)	0.09	0.05
SO_2 concentration in stack (ppm)	0.12	0.25
Fuel S (g)	11	17
Ash S (g)	3	2
Gas phase S from SO_2 (g)	1	2
Excess fuel S (g)	7	13
Carbon		
Fuel carbon (% dry basis)	40.41	40.29
Ash carbon (% dry basis)	11.17	15.23
Stack gas concentrations (ppm, excludes background):		
CO_2	2,000	2,000
CO	168	155
CH_4	8	4
C_2H_2	0.3	0.2
C_2H_4	1.7	1.7
C_2H_6	0.2	0.13
C_6H_6	0.137	0.082
Fuel C (g)	8,762	8,552
Ash C (g)	436	521
Gas phase C (g)	7,346	7,617
Particulate C (g)	ND	ND
Excess fuel C (g)	980	414
Closure (%)	89	95

Average concentrations of the gas phase emissions observed during each test are listed in Table 37.2. As shown by the material balance for nitrogen, the concentrations of N in the ash and in the gas phase emissions of NO and NO_2 account for only 20% of the fuel nitrogen.

SO_2 emissions were extremely difficult to measure, the signal from the analyzer being only slightly above the noise level of the instrument. The values reported are uncertain, but there does not appear to be a significant quantity of SO_2 in the gas. The discrepancy in the sulfur concentrations in the ash by ultimate analysis makes it difficult to compare the closure values on sulfur, but the gas phase emission of SO_2 is likely to be well below that computed on the difference of sulfur in the fuel and the ash as reported by Darley (1979). The low readings obtained on the SO_2 analyzer indicate that an ambient monitor should have been used because the actual concentrations are so much lower than those estimated from the fuel bound sulfur.

Carbon balances show reasonable closure levels, being 89% for the test of 1 September and 95% for the test of 7 September. The carbon balance is extremely sensitive to the concentration of CO_2 determined for the stack gas, and represents the critical problem in determination of emission factors from field studies where carbon balance is used to determine emitted fraction per unit mass of fuel.

The predominant hydrocarbon was methane at an average of 6 ppmv, although the concentration measured on 7 September was half that of the value on 1 September. Concentrations of other light hydrocarbons are comparable. Ethene was present at the next highest concentration after methane.

Benzene concentrations determined from the sorbent tubes were in the range of 100 ppbv. The analysis of benzene by trapping on the sorbent tubes was compared to grab sample analyses using tedlar bags by analyzing a calibration standard with a concentration of 112 ppbv. Chromatograms from the GC-MS exhibit a large number of peaks, some of which were tentatively identified using the library of the instrument and comparing boiling points and mass spectrograms. A sample chromatogram is given in Figure 37.11. Tentative qualitative identification has been made for toluene, ethyl benzene, ethenyl benzene, benzaldehyde (possible artifact from tube packing), phenol (possible artifact), and napthalene. Work is continuing to both qualitatively and quantitively determine volatile organics emitted from the fire. There is some concern that quench effects could be contributing to the concentrations of volatile organics observed. Further investigation is required in this respect as well.

Particle concentrations and size distributions measured by the HTHP cascade impactor are given in Table 37.3. The size distributions were matched to an equivalent mass median aerodynamic diameter using a log-normal distribution with the stage efficiency

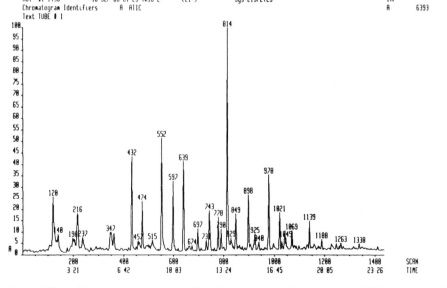

Figure 37.11 Sample chromatogram collected on sorbent tube during test of 7 September and analyzed by GC-MS.

Table 37.3 Particulate concentrations and size distributions for combustion trials of 1 September and 7 September 1988 in rice straw

Stage	Effective diameter (μm)	1 September Mass (mg)	1 September Cumulative fraction (%)	7 September Mass (mg)	7 September Cumulative fraction (%)
1	12.80	0.11	100.0%	0.12	100.0
2	7.64	0.18	97.1	0.17	96.7
3	4.31	0.16	92.3	0.17	92.1
4	1.97	0.15	88.1	0.17	87.5
5	1.30	0.15	84.1	0.27	82.9
6	0.75	0.35	80.1	0.43	75.5
7	0.46	1.16	70.9	0.91	63.8
8	Filter	1.52	40.2	1.42	38.9
Concentration (mg/m^3)			9.47		9.17
Aerodynamic diameter (μm)			0.47		0.51
σ_g			1.63		2.49

Table 37.4 Emission factors (kg/ton)[a] for preliminary combustion trials of 1 September and 7 September 1988 in rice straw (7% moisture), with comparison to Darley (1979) at 14.9% average moisture wet basis

	1 September	7 September	Average	Darley (1979)
Particulate matter[b]	3.28	3.34	3.31	1.06
NO	1.16	1.16	1.16	1.88
NO$_2$	0.54	0.38	0.46	0.65
NO$_x$ (as NO$_2$)	2.32	2.16	2.24	3.53
SO$_2$	0.10	0.22	0.16	1.00[c]
CH$_4$	1.66	0.88	1.27	0.82
C$_2$H$_2$	0.10	0.07	0.09	0.11
C$_2$H$_4$	0.62	0.66	0.64	0.41
C$_2$H$_6$	0.08	0.05	0.07	0.11
C$_6$H$_6$	0.14	0.09	0.12	0.05
CO	60.89	59.99	60.44	89.62[d]
CO$_2$	1,139	1,216	1,178	NR[e]

a. All values dry basis (zero moisture). Multiply by 2.0 to obtain lb/ton.
b. Based on results of HTHP impactor. Note that Darley (1977) reports 3.47 kg/ton particulate matter (dry basis) for a backing fire in 9.3% moisture rice straw on a 25° slope, and 4.64 kg/ton at 10.5% moisture and 15° burning slope.
c. Darley's value for SO$_2$ computed from the difference of sulfur in the fuel and ash.
d. Darley (1977). Corrected for average moisture of 9.9% wet basis. CO for rice straw is not reported by Darley (1979).
e. Not reported.

least squares method (Raabe, 1968). About 90% of the particles are below 4 μm in size, and 80% are less than 1 μm.

Emission factors computed on the basis of the concentrations obtained during the two experiments of 1 September and 7 September are listed in Table 37.4. These are preliminary factors only, due to the nature of the conditions of the tests conducted which are only partially representative of the field conditions pertaining to rice straw burning. The results of Darley (1979) are listed for comparison, as these values are in current regulatory use in California. There exist significant discrepancies in the particulate matter and SO$_2$ concentrations. The discrepancy in sulfur is due to the manner in which SO$_2$ concentrations were computed by Darley, who assumed the sulfur deficiency between fuel and ash could be accounted for by SO$_2$ emission. Even earlier work by Darley (1977) does not support the particulate matter emission rate reported by his later studies. This uncertainty suggests the need for more direct measurements of the type conducted here.

Summary and Conclusions

The combustion wind tunnel provides a means of conducting controlled experiments to directly quantify emissions from open burning of biomass materials. Emission data is therefore not subject to the uncertainty which accompanies field data determined by elemental balance methods. By permitting direct control over all fuel and oxidizer conditions, the wind tunnel experiments are also useful for understanding the combustion dynamics. The data collected are therefore suitable for calibrating model descriptions of the fire. The ability to translate the fuel bed relative to the floor allows long-term fires in spreading fuels to be conducted under the same controlled conditions, a capability not available earlier. The scale of the tunnel is not sufficient to conduct fires of the same size that occur in forest burning. It can, however, provide model predictions of how fire emission is influenced by fuel and environment conditions, and serve to validate larger-scale field measurements.

V

Biomass Burning and the Global Carbon Budget

Biomass Burning from the Perspective of the Global Carbon Cycle

Richard A. Houghton

The emissions of CO_2, CH_4, and CO to the atmosphere are of interest because they contribute either directly or indirectly to the heat balance of the earth. The accumulation of CO_2 in the atmosphere from fossil fuel and biotic emissions accounts for about half of the total increased radiative forcing from all gases combined, not counting water vapor (Ramanathan et al., 1987). The accumulation of CH_4 in the atmosphere contributes about 20% of the total radiative forcing currently. While CO is not radiatively important itself, it reacts chemically with OH radicals in the atmosphere and, thereby, affects the concentrations of CH_4.

Biomass burning is a source of CO_2, CH_4, and CO to the atmosphere. But despite the major importance of CO_2 in radiative forcing, its emission to the atmosphere from biomass burning is of less importance than the flux suggests. The paradox is explained by differences in the natural cycles of CO_2, as opposed to the other two gases. For CO_2, the emissions from biomass burning are largely balanced by the transfers of CO_2 back into biomass within months to years after burning. For example, the large emissions of CO_2 from burning of African savannas each year (Hao et al., 1990) are balanced by equally large accumulations of CO_2 in regrowing savanna grasses. The same balance is true for agricultural wastes and fuel wood. For CH_4 and CO, on the other hand, the emissions from biomass burning are largely net fluxes from land. Terrestrial sources of these gases are largely balanced by atmospheric sinks rather than terrestrial ones. Thus these other gases, unlike CO_2, affect the chemical properties of the atmosphere. The distinction between the cycling of CO_2 and that of other trace gases determines how the releases from biomass burning should be interpreted.

The differences between CO_2 and the other forms of gaseous carbon have led researchers to ask different questions, with the result that there seems little exchange between the two communities. The purpose of this chapter is (1) to compare estimates of the emissions of carbon from biomass burning with estimates of the net flux of carbon from terrestrial ecosystems and (2) to identify some areas for potential interchange between biomass burning and the global carbon cycle.

Estimates of the Annual Emissions of Carbon

Worldwide emission of carbon from biomass burning is estimated to range between 3.0 and 6.2 petagrams carbon (Pg C) annually (Crutzen and Andreae, 1990). The net flux of carbon to the atmosphere from changes in land use (deforestation and reforestation) is estimated to have been in the range of 1.0 to 2.6 Pg C in 1980 (Houghton et al., 1987). All but 0.1 Pg of the this net release was calculated to have come from tropical deforestation. Hence, the recent estimate for the tropics alone (Detwiler and Hall, 1988) extends the global range to 0.5 to 2.6 Pg C for 1980. In the 10 years since 1980, increases in the rates of tropical deforestation (Myers, 1989) suggest that the net annual flux from changes in land use is currently 1.1 to 3.6 Pg C (Houghton, 1991). This estimated land-use flux is, therefore, somewhat lower than, but overlaps, estimates of the emissions from biomass burning.

The global emissions of carbon from biomass burning and from changes in land use would appear to be approximately the same; a major contribution to each is burning of forests in the tropics. However, the dissimilarities are larger than this similarity. Biomass burning includes emissions not considered in the land-use flux, and the net flux from changes in land use includes emissions that are not the result of combustion.

Processes Included in Emissions from Biomass Burning and Not Included in the Land-use Flux

The emissions of carbon from biomass burning result from six processes (Crutzen and Andreae, 1990): (1) permanent deforestation, (2) shifting cultivation, (3) burning of savanna and bushland, (4) wildfires

Table 38.1 Estimates of the emissions of carbon from biomass burning (Tg C yr^{-1}), 1980 and 1850

	1980	1850
Permanent deforestation	400–1300	100–250
Shifting cultivation	500–700	300–600
Savannas, pastures with agricultural wastes	400–2400 1200–3200[a]	1000–2700[a]
Fires in temperate and boreal forests	100–165	200–300
Fuel wood	800	400–600
Agricultural wastes	800	0[a]
Total	3000–6200	2000–4500

Note: Estimates for 1980 from Crutzen and Andreae (1990). Estimates for 1850 described in text.
a. Agricultural wastes (burning of crop residues) included in burning of savannas and pastures.

and prescribed fires in temperate and boreal forests, (5) burning of industrial wood and fuel wood, and (6) burning of agricultural wastes.

Estimates of the emission of carbon from changes in land use include only the first two of these activities, or 0.9 to 2.0 Pg C in 1980 (Table 38.1) (Crutzen and Andreae, 1990). The range is almost identical to the range for tropical deforestation (0.9 to 2.5 Pg C) (Houghton et al., 1985), but the similarity is fortuitous. Only about 40% of the flux from changes in land use, or 0.4 to 1.0 Pg C, resulted from burning (see below).

Burning of savannas, fuel wood, and agricultural wastes, and fires in temperate and boreal forests are not included in analyses of the land-use flux. Such burning is ignored because most of the carbon released to the atmosphere from these fires is withdrawn from the atmosphere within the next years and accumulated again on land. Savanna is burned to replace the older, nutritionally poor grass with young, nutritious shoots. The carbon on land or in the atmosphere is not changed year to year as a result of this burning. Prescribed and wild fires in temperate and boreal forests also release gross amounts of carbon that are withdrawn from the atmosphere in subsequent years as a result of forest growth. In the long term, forests must be in balance with respect to carbon. Otherwise they would contain either much more or much less carbon than observed. In the short term, trends in the frequency of fires may store or release carbon, but such short-term trends have not been evaluated globally. The most conservative assumption is that natural systems are in steady state with

respect to carbon, neither releasing nor accumulating it.

Processes Included in the Land-use Flux and Not Included in Emissions from Biomass Burning

Analyses of the net flux of carbon from changes in land use provide an estimate of a net flux because they include both deforestation, which generally releases carbon to the atmosphere, and reforestation, which generally withdraws carbon from the atmosphere (Houghton et al., 1983, 1985, 1987; Detwiler and Hall, 1988). Moreover, the conversion of forests to agricultural lands represents a net release of carbon to the atmosphere because the carbon originally held in trees is not balanced by the carbon accumulated in crops or pasture grasses. Carbon is also lost from soils as a result of cultivation associated with agriculture.

Calculations of the net flux of carbon from changes in land use are based on two types of information: rates of land-use change and changes in the stocks of carbon in vegetation (biomass) and soils as a result of these changes in land use. Several different types of forests in each geographical region and different types of land-use changes have been considered in previous analyses (Houghton et al., 1983, 1985, 1987; Detwiler and Hall, 1988). The types of land-use change have generally included the conversion of natural ecosystems to permanent croplands, to shifting cultivation, and to pasture; the abandonment of croplands and pastures; the harvest of timber with subsequent regrowth; and the establishment of tree plantations with subsequent harvest. Analyses of Southeast Asia and Latin America have also included the conversion of forests to degraded lands (Palm et al., 1986; Houghton et al., 1991a, 1991b). The data and sources of information in these analyses have been documented in the studies referred to above.

Changes in the vegetation and soils as a result of changes in land use release carbon to the atmosphere from (1) burning of trees and ground vegetation; (2) decomposition of dead plant material, such as leaves, twigs, branches, and stumps, left on the soil surface during logging or clearing of a forest; (3) oxidation of wood products, such as paper, timber, and resins, harvested from the forest during logging; and (4) oxidation of soil carbon that often accompanies logging, and which almost always accompanies cultivation of the soil (Table 38.2). The net flux of carbon from changes in land use also includes (5) the accumulation of carbon on land as a result of the regrowth of vegetation and the redevelopment of soil organic matter.

Table 38.2 Net exchanges of carbon between terrestrial ecosystems and the atmosphere in 1980 (Pg C yr^{-1})

Net exchange of C (Pg C yr^{-1})	Percentage of net flux	Terrestrial ecosystem
0.7	39%	Burning of biomass on cleared and harvested sites
1.1	61%	Decay of biomass on cleared and harvested sites
0.4	22%	Oxidation of harvested products
0.3	17%	Oxidation of soil carbon
−0.7	39%	Regrowth of forests
1.8 (range 0.9–2.5)		Total net flux from deforestation and reforestation
3.0–6.2		Biomass burning (Crutzen and Andreae, 1990)
5.2		Fossil fuels (Marland and Rotty, 1984)

The contributions of these five components to the global net flux are shown in Table 38.2. The largest component to the net flux has been the decay of plant material left in the soil (roots) or on the forest floor (leaves, twigs, branches, stumps) at the time of harvest or clearing. Logging in the tropics is generally selective, but many of the trees not harvested are damaged and killed during harvest. The subsequent decay of these trees and of the plant material not burned with clearing is responsible for the large release of carbon from this source. Burning associated with clearing contributes about 40% of the total net flux (and only about 30% to the "gross" release). If harvested products are assumed to be oxidized through burning, rather than through decay, the percentage of the total flux that corresponds to biomass burning is about 60% of the net, and 44% of the gross, flux.

"Gross" is set off in quotation marks above because, as mentioned previously, the analyses ignore wildfires, fuel wood, agricultural wastes, and frequent burning of pastures, savannas, and bushlands. Furthermore, the analyses are often based on net changes in land use; that is, on data that include only net changes in the areas of forests, croplands, pastures, and other lands. These net changes are almost certainly composed of greater rates of clearing balanced, in part, with rates of abandonment. The difference between net and gross fluxes as a result of net changes in area, however, is thought to be small compared to the difference resulting from including, or

not including, gross emissions from biomass burning. The net contribution from soils is itself a net flux, composed of releases from decay and accumulations with recovery of soil organic matter. Such accumulations of soil carbon occur during the fallow period of shifting cultivation and, sometimes, following logging (Houghton et al., 1985; Detwiler and Hall, 1988).

The fact that burning accounts for only about 40% to 60% of the net flux from changes in land use (or 30% to 44% of the gross flux) raises the question of whether other trace gases are also emitted with the decay of organic matter. The emissions of CH_4 from land-fills and the releases of NO and N_2O from new pastures are examples of such emissions not directly related to burning. Are there other emissions of CH_4 or CO, or N_2O and NO_x, from the decay of plant material and soil organic matter? One would expect so, but few measurements have been made over disturbed landscapes. What are the emission ratios, for example, for the decay of organic matter?

Complementary Issues of Biomass Burning and Land Use

Elemental Carbon

There are at least two topics where analyses of biomass burning and of land-use change are not independent, but complementary. The first is the formation of elemental carbon, or charcoal, formed as a result of burning. Because this elemental carbon is only slowly oxidized, if at all, biomass burning removes carbon from the short-term carbon cycle and sequesters it in long-term storage (Seiler and Crutzen, 1980). The formation of elemental carbon represents a permanent sink for carbon. Between 0.3 and 0.7 Pg C are estimated to be converted annually to elemental carbon through fires (Crutzen and Andreae, 1990). Only about 0.1 Pg C is formed as a result of fires associated with shifting cultivation and deforestation and is already included in calculations of the net flux. Therefore, estimates of the net flux of carbon from changes in land use should be revised downward by 0.2 to 0.6 Pg C to account for the additional net sink resulting from gross burning. If elemental carbon does not exist indefinitely, but is in a steady state, then only variations in this steady state, or variations in global burning, represent a net flux. The production, fate, and half life of elemental carbon are poorly known, however, and the size of this sink is uncertain. The formation estimated by Crutzen and Andreae (1990) essentially cancels the net flux if the lower estimates of the land-use flux are correct. If the current net flux is near the high end of the range, the net flux is around 2 (± 1) Pg C yr^{-1}.

Historic Changes in Biomass Burning

It is generally accepted that the emissions from biomass burning have increased in recent decades, largely as a result of increasing rates of tropical deforestation. But quantitative estimates as to how much they have increased are difficult to obtain. Has the increase been 10% or 10X over the last century? The increase and the rate of that increase are important because they help constrain other sources and sinks in the global cycles of trace gases.

The different components to the global flux from biomass burning have probably changed at different rates. Emissions associated with permanent deforestation, for example, have probably increased in proportion to the net flux of carbon from changes in land use. That flux is estimated to have increased by a factor of 3 to 6 over the last 135 years (Figure 38.1), from about 0.5 Pg C yr^{-1} in 1850 to between 1.1 and 3.6 Pg C yr^{-1} currently (Houghton, 1991). Thus, the contribution of deforestation to biomass burning in 1850 was probably within the range of 100 to 250 Tg C yr^{-1} (Table 38.1).

The contribution from shifting cultivation is more difficult to determine because historical documentation of either the area in shifting cultivation or the number of people practicing it is difficult to obtain. Even current changes are controversial. In 1980, FAO/UNEP (1981; Lanly, 1982) reported increasing rates of shifting cultivation; Myers (1980, 1984) found

the area in shifting cultivation decreasing (Houghton et al., 1985). The reduction in area reported by Myers was a recent phenomenon, however, and between about 1945 and 1970 the area was probably increasing with population growth. In Latin America an increase of about 25% was estimated for this 25-year period (Houghton et al., 1991a). Most studies agree that the lengths of fallow periods have been decreasing recently, but the implications for biomass burning are unclear. Burning is more frequent, but the biomass burned per unit area is less. Overall, emissions of carbon from shifting cultivation have probably increased since 1850 (Table 38.1).

The burning of grasslands, agricultural lands, and savannas has certainly increased over the last century, because rarely burned ecosystems, such as forests, have been converted to frequently burned ecosystems, such as agricultural lands or grass/shrublands. In Latin America the area of grasslands, pastures, and croplands increased by about 50% between 1850 and 1985 (Houghton et al., 1991a). The area of natural grasslands actually decreased but was more than offset by increases in the area of pastures and croplands. Similarly, analyses of countries accounting for 75% of the area in South and Southeast Asia found a 50% increase in the area of temporary croplands and grass/shrublands between 1880 and 1980 (Flint and Richards, 1991, and personal communication). The area in temporary crops almost doubled; the area in grass and shrub increased by 15%.

The relative increases observed in Latin America and tropical Asia are probably larger than such changes in Africa, where large areas of savannas already existed before last century. Because almost half of all biomass burning is estimated to occur in the savannas of Africa (Hao et al., 1990), emissions from burning of savannas and agricultural lands, worldwide, have probably increased by 20% to 25%. These increased emissions from agricultural lands include the burning of agricultural wastes or crop residues (Table 38.1).

The net flux of carbon from conversion of forests to open lands is estimated to have released 90 to 120 Pg C over the 130-year period 1850 to 1980 (Houghton and Skole, 1990) (Figure 38.1). However, the emissions of carbon from the replacement of infrequently burned lands with frequently burned lands has released an additional 13 to 32 Pg C over this period (assuming a change of 200 to 500 Tg C yr^{-1} between 1850 and 1980, and a linear change over the period). If most of the change occurred after 1950, as seems likely, the additional release of carbon has been only 3

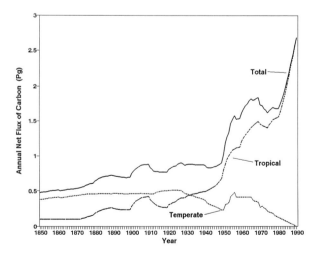

Figure 38.1 Annual net flux of carbon to the atmosphere between 1850 and 1990 from changes in land use.

to 8 Pg C. This additional release is a gross flux. It is largely unimportant for the carbon cycle (and for CO_2), but it represents an increasing source of other trace gases from biomass burning.

In temperate regions, the expansion of croplands and rangelands occurred earlier than it did in the tropics (Figure 38.1), and recent decades of fire suppression have probably reduced biomass burning in North America and Europe still further. In the United States, for example, the amount of burning has increased and decreased, perhaps several times over the last 150 years (Pyne, this volume, Chapter 61). Current emissions from biomass burning in such regions are, therefore, smaller than they were in the middle of last century. On the other hand, these reductions in burning since 1850 are probably small in comparison to recent increases in the tropics.

Changes in fuel wood over the last 130 years are difficult to determine. The increased number of people using wood fuels in the tropics has increased such burning, while the replacement of wood fuels with fossil fuels in the industrialized nations has reduced it. Overall, the number of people using wood for fuel has increased, and although the efficiency of its use has increased (Leach and Mearns, 1988), emissions have also probably increased over the last century.

These crude estimates suggest that biomass burning may have increased by about 50% since the middle of last century. Most of the increase seems to have come from ever increasing rates of deforestation. The other contributions, combined, have increased by only 15% to 40%, if these preliminary estimates are approximately correct. Nevertheless, the increased emissions of trace gases from biomass burning have contributed to the changes in atmospheric chemistry observed over this period.

Acknowledgments

Support for this work was provided by the National Science Foundation and by the U.S. Department of Energy (subcontract 19X-SB830C for Oak Ridge National Laboratory, Oak Ridge, Tenn. 37831-6335, operated by Martin-Marietta Energy Systems, Inc., for the U.S. Department of Energy under subcontract DE-AC05-84OR21400).

The Response of Atmospheric CO$_2$ to Changes in Land Use

Anthony W. King, William R. Emanuel, and Wilfred M. Post

The recent increase in the tropospheric concentration of CO$_2$ is well documented. Observations from monitoring stations at Mauna Loa, Hawaii, and elsewhere record an increase of approximately 11% over the past 30 years (Gammon et al., 1986; Keeling et al., 1989), and measurements from air trapped in old ice at Siple Station, Antarctica, indicate an increase of 26% since the middle of the 18th century (Neftel et al., 1985; Friedli et al., 1986).

The causes of this increase are less certain than the observed change, but it is agreed that the release of CO$_2$ to the atmosphere during the combustion of fossil fuels (i.e., the burning of fossil biomass) is a major contributor. The increase in tropospheric CO$_2$ is concurrent with the documented increase in global use of fossil fuels and the consequential increase in the release of CO$_2$ to the atmosphere (Keeling, 1973b; Marland and Rotty, 1984). The burning of biomass that accompanies deforestation and other changes in land use, including the rapid combustion of fuel wood and fires used in land clearing as well as the slow oxidation of decomposition, is also believed to release significant amounts of CO$_2$ to the atmosphere, but the relative contribution of this source to observed changes is even less certain than that of fossil fuels.

Part of the uncertainty has its origin in the observation that only a fraction of the CO$_2$ released in the combustion of fossil fuels is retained by the atmosphere. The 35 μL L^{-1} (or 75 GT C) increase in tropospheric CO$_2$ over the period of the Mauna Loa record (1959 to 1988) is equivalent to approximately 58% of the total CO$_2$ released from industrial sources during that same period (129 GT C, 96% of which is from fossil fuels) (Marland et al., 1989). If the observed CO$_2$ increase during this period were due solely to industrial sources, the biogeochemical dynamics of the earth's global carbon cycle would have had to redistribute approximately 42% of the industrial CO$_2$ emissions to oceanic and terrestrial reservoirs. Conventional models of CO$_2$ uptake by the ocean account for about 25% to 35% of fossil-fuel

CO$_2$ emissions from the period 1958 to 1980 (Peng, 1986), a shortfall of 7% to 17% that must be accounted for elsewhere. The terrestrial biosphere is a reasonable candidate, but the capacity of global terrestrial ecosystems to sequester the required amounts of CO$_2$ over that period of time is controversial. Recent work by Tans et al. (1990) suggests that oceanic uptake may be more on the order of 20% of industrial CO$_2$ emissions, requiring even greater sequestering by terrestrial ecosystems. It is difficult to account for any significant addition of CO$_2$ to the atmosphere from land-use change when current understanding of the global carbon cycle is unable to fully account for industrial CO$_2$ emissions alone. Thus, in the face of evidence for recent increases in the rates of tropical deforestation (Houghton, 1990) and presumably some nontrivial release of CO$_2$ to the atmosphere, the contribution of terrestrial ecosystems to observed changes in atmospheric CO$_2$ is uncertain and problematic. The uncertainty is compounded by difficulties inherent in reconstructing the history of CO$_2$ release that may have accompanied past changes in land use.

For example, the history of CO$_2$ release from land-use change can be inferred by deconvolution of data recording the history of changes in the global carbon cycle. In one approach, the equation representing changes in the mass of atmospheric CO$_2$

$$\frac{dc_a}{dt} = F_i + F_r - F_{as} \tag{39.1}$$

is solved for the unknown or residual flux F_r, where c_a is the observed mass of atmospheric CO$_2$, F_i is the known flux of CO$_2$ to the atmosphere from industrial sources, and F_{as} is the net CO$_2$ flux from the atmosphere to the ocean calculated by an appropriate carbon cycle model (Killough and Emanuel, 1981; Siegenthaler and Oeschger, 1987; Keeling et al., 1989). Similar formulations can be derived for changes in carbon isotope ratios, and other approaches or formalisms are possible (e.g., Oeschger and Heimann, 1983; Enting and Pearman, 1986; Enting and Mansbridge, 1987; Keeling et al., 1989).

By assuming that equation (39.1) completely describes the atmosphere's CO_2 balance, that all anthropogenic CO_2 emissions are distributed among atmosphere, ocean, and terrestrial biosphere reservoirs, and that F_{as} completely describes net oceanic uptake, the residual flux, F_r, can be interpreted as net flux between the atmosphere and the terrestrial biosphere. By assuming no net atmospheric CO_2 exchange with terrestrial ecosystems that have not been disturbed by land-use change, the residual flux can be interpreted as an estimate of the net terrestrial biospheric CO_2 exchange due to changes in land use.

Unfortunately, these interpretations are confounded by the potential for error in the carbon cycle models used to calculate F_{as} and potential error in the history of industrial CO_2 emissions used to define F_i (Killough and Emanuel, 1981; Siegenthaler and Oeschger, 1987). Error or uncertainty in either is reflected as error or uncertainty in the estimate of past terrestrial biospheric exchange. There is also the problem of error in the historical data being deconvolved, both atmospheric CO_2 concentrations and isotopic ratios. These same uncertainties impact estimates of terrestrial biospheric CO_2 release produced by deconvolution of changes in isotopic ratios recorded in tree rings (Peng et al., 1983; Emanuel et al., 1984; Stuiver et al., 1984). Tree-ring-based reconstructions are especially prone to limitations imposed by the intrinsic variability of tree-ring isotope records (Freyer, 1986).

Direct reconstruction of historical CO_2 release by explicit consideration of changes in land use and the response of carbon in soils and vegetation of affected ecosystems is preferable to inferential reconstructions by deconvolution, but this direct historical-ecological approach has its own limitations. There are no direct measurements of the CO_2 fluxes accompanying changes in land use at the spatial and temporal scales appropriate to large-scale land conversion (e.g., tropical deforestation) and the global carbon cycle. Thus, computer models are used to estimate CO_2 release to the atmosphere based on data that is available on rates of land-use change, the types of ecosystems affected, and carbon storage in vegetation and soils before and after the disturbance (Revelle and Munk, 1977; Moore et al., 1981; Houghton et al., 1983; Richards et al., 1983; Bogdonoff et al., 1985). However, uncertainty in these data contribute a great deal of uncertainty to the reconstruction of past land-use CO_2 release (Houghton, 1986). It is difficult enough

to estimate the amount of CO_2 released in a single year of the recent past for which relatively good data is available (Detwiler et al., 1985; Hall et al., 1985; Houghton et al., 1985, 1987). Reconstructing CO_2 release over the last 200 to 300 years, especially the timing of that release, is made even more difficult by the additional demands for historical data and the coincident increase in the uncertainty of those data. Consequently, there are very few direct historical-ecological reconstructions of CO_2 release from past changes in land use, especially those which resolve the timing of that release (Houghton, 1986).

Nevertheless, it is precisely the history of past CO_2 release to the atmosphere, both the amounts and the timing, that is crucial to understanding past changes in atmospheric CO_2, and consequently our ability to project future changes. As noted by Siegenthaler et al. (1978, p. 83), "If a large input pulse occurred only a few decades ago, it is still influencing the present CO_2 level by providing a decreasing atmospheric 'baseline,' since excess CO_2 is still being taken up by the ocean. The longer ago such an input occurred, the smaller is its influence on the present trend."

The purpose of this chapter is to examine how different histories of CO_2 release from past changes in land use influence the simulation of past and future changes in atmospheric CO_2. We first simulate past changes in atmospheric CO_2 using reconstructed histories of land-use CO_2 release from a historical-ecological model of land-use change and CO_2 release. We examine the impact of each history on the coincidence between simulated and observed atmospheric CO_2. We then compare these CO_2 release histories, and their contribution to coincidence or noncoincidence of simulation and observation, with histories reconstructed by deconvolution of the atmospheric CO_2 record. We conclude by exploring the implications of these deconvolved reconstructions for the simulation of future changes in atmospheric CO_2.

Methods

The Models

We used three models of global carbon cycling between ocean and atmosphere (Figure 39.1). The first is an implementation of the Oeschger et al. (1975) box-diffusion model and is referred to here as the *box-diffusion model* (Figure 39.1, model 1). The others are implementations of models described by Killough and Emanuel (1981). The first of these is a layered ocean model that considers only diffusive

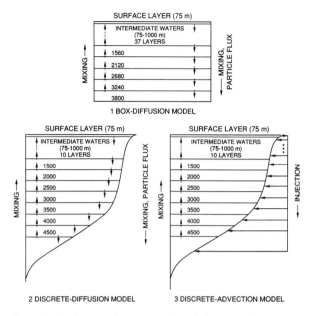

Figure 39.1 Structural representation of the ocean in the three carbon cycle models. Arrows indicate the flow of carbon.

150 m deep; the bottommost layer extends from 4500 m to the ocean floor at greater than 7000 m.

The exchange of CO_2 between the atmosphere and the surface mixed layer of the ocean is governed by the difference between the partial pressures of CO_2 in the atmosphere and in sea surface waters. The partial pressure of CO_2 in the atmosphere is proportional to the ratio of the masses of CO_2 and dry air, whereas the partial pressure of CO_2 in surface waters depends on the chemical equilibria between carbon compounds. We model this chemical equilibration with the dynamic carbonate chemistry of Takahashi et al. (1980). The result is a nonlinear function $p_s(c_s)$ that expresses the dependence of the partial pressure of CO_2 dissolved in surface waters on the total carbon content of those waters. This relationship buffers the increase in surface water CO_2 relative to the increase in atmospheric CO_2. Our implementations of the Killough and Emanuel (1981) models differ slightly from earlier implementations in our use of the Takahashi et al. (1980) formulation for sea surface carbonate chemistry rather than the Keeling (1973a) formulation. More importantly, our implementation of the box-diffusion model differs from the original Oeschger et al. (1975) implementation in that we use a dynamic nonlinear carbonate chemistry while Oeschger et al. (1975) represented sea surface buffering with a constant linear buffer factor.

Mixing and circulation remove carbon from the ocean's surface layer, transferring it to the deeper layers where it may be sequestered for several hundred years (Stuiver et al., 1983). The box-diffusion model represents these subsurface fluxes as a system of ordinary differential equations describing the layered finite-difference approximation of the partial differential equation representing subsurface mixing as continuous vertical eddy diffusion (Oeschger et al., 1975). The model's diffusion is best "regarded as a parameterized representation of the various processes of mixing and advection in the ocean, projected on a vertical axis" (Siegenthaler, 1983).

The discrete-diffusion model represents the transport of carbon by diffusive mixing and particle flux through successive layers, both upward and downward, as first-order linear kinetics between neighboring layers. The discrete-advection model represents upward transport in the same way, but downward transport is represented by first-order linear kinetics between the surface layer and individual subsurface layers (Figure 39.1). This downward pathway represents the sinking of cold dense polar waters and the accompanying advective transport of carbon from the

fluxes of carbon in the ocean (Figure 39.1, model 2); the second uses the same ocean structure but considers advective as well as diffusive fluxes (Figure 39.1, model 3). These models are referred to here as the *discrete-diffusion* and *discrete-advection* models, respectively. Structurally, the discrete-diffusion and discrete-advection models correspond to models 3b and 3a, respectively, of Killough and Emanuel (1981).

Each model represents ocean-atmosphere carbon dynamics with a system of ordinary differential equations describing changes in carbon mass of the various reservoirs or model compartments. The three models differ only in their structural representation of the deep ocean and their representation of carbon flux within that structure (Figure 39.1). The models each have a 75 meter (m) well-mixed surface layer. The box-diffusion model has 42 subsurface layers of constant horizontal cross-sectional area, 37 of 25 m depth down to 1000 m (intermediate waters) and five of 560 m depth to the ocean floor. The discrete-diffusion and discrete-advection models have 18 subsurface layers with horizontal areas that vary with depth according to the hypsography of Sverdrup et al. (1942) (Killough, 1980; Killough and Emanuel, 1981). The layer immediately below the surface layer is 50 m deep, the next is 75 m. The remaining eight layers of intermediate waters, to 1000 m, are 100 m in depth. The seven layers between 1000 m and 4500 m are surface to the deep ocean (Killough and Emanuel, 1981).

The Simulations

Initial Conditions We assume initial conditions of a preindustrial steady state carbon cycle. An initial atmospheric CO_2 concentration of 277 $\mu L\,L^{-1}$ in 1745 is indicated by the Siple ice core record (Neftel et al., 1985; Friedli et al., 1986). The assumption of steady state atmosphere-ocean CO_2 exchange determines an equivalent partial pressure of dissolved CO_2 in the surface ocean. The corresponding mass of surface ocean carbon is calculated by solving the surface water equations for chemical equilibria between carbon compounds with specified initial conditions. Initial deep ocean carbon inventories and ^{14}C activities are as given in Killough and Emanuel (1981). For the discrete-diffusion and discrete-advection models, transfer coefficients for the exchange of carbon between the atmosphere and surface ocean and the flux of carbon between ocean layers are determined from the steady state equations of carbon transfer and the assumed preindustrial carbon inventories and ^{14}C activities. Transfer coefficients for atmosphere-ocean CO_2 exchange and the diffusivity coefficient K of the box-diffusion model are determined from the steady state carbon inventories and fitting simulations of deep ocean ^{14}C to an idealized global preindustrial ^{14}C profile. This profile is derived from observations of ocean ^{14}C prior to the extensive release of ^{14}C in nuclear bomb testing (Oeschger et al., 1975; Broecker et al., 1985).

Anthropogenic Emissions The models are driven from their initial steady states by carbon added to the atmosphere from anthropogenic sources. For industrial CO_2 emissions (fossil fuels plus cement manufacturing) we used the history depicted in Figure 39.2. Emissions for 1950 to 1986 are those derived by Marland et al. (1989) from United Nations fuel production data and U.S. Department of Interior, Bureau of Mines data on cement manufacturing. T. A. Boden (Carbon Dioxide Information and Analysis Center, Oak Ridge National Laboratory, personal communication, 1988) provided the emissions estimate for 1987 based on these same data sources. Fossil fuel CO_2 emissions for 1860 to 1949 are from Keeling (1973b, Table 14). We estimated pre-1860 fossil fuel CO_2 emissions with Keeling's (1973b, equation 6.1) exponential extrapolation of the 1860 to 1949 data. Similarly, we estimated pre-1950 CO_2 emissions from cement manufacturing using the exponential extrapo-

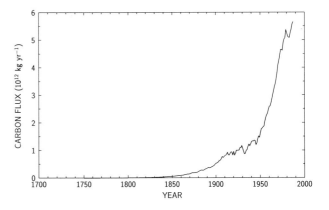

Figure 39.2 The history of annual CO_2 release to the atmosphere from fossil fuels, flaring of natural gas, and cement manufacturing (Keeling, 1973b; Marland et al., 1989). Fossil fuels contribute approximately 96.5% of the estimated annual industrial CO_2 emissions.

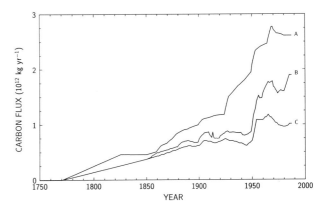

Figure 39.3 Three histories of global net carbon release to the atmosphere from past changes in land use. History A is the population-based estimate of Houghton et al. (1983). Histories B and C are from Houghton (1989). High estimates of tropical biomass carbon are assumed in history B, and low estimates are assumed in history C. History B approximates the land-use CO_2 release history presented in Houghton and Woodwell (1989, p. 39, graph d).

lation of cement emissions given by Keeling (1973b, equation 6.2).

We considered four different histories of CO_2 release to the atmosphere from the anthropogenic conversion of terrestrial ecosystems from their natural states, or at least their preindustrial states, to forestry, agricultural production, or other uses. In the first we simply assumed that there was in fact no net CO_2 release to the atmosphere from this biospheric source. The remaining histories (Figure 39.3) are from the work of Houghton et al. (1983) and Houghton (1989), using the Marine Biological Laboratory Terrestrial Carbon Model (Moore et al., 1981; Houghton et al., 1983; Houghton, 1986) and its derivative (Houghton, 1989). These models describe the net release of carbon to the atmosphere that accompanies changes in land use by tracking yearly changes in area, age, and carbon content (vegetation and soil) of terrestrial ecosystems that have been disturbed by various types of land-use change (e.g., conversion from forest to cultivated land). The areas and ecosystems involved are estimated from a variety of statistics on wood harvest and changes in agricultural land. Harvested material is distributed among decay pools with residence times based on different uses of the material, and the models calculate the subsequent annual loss of carbon from those pools. For example, fuel wood is accounted for in a decay pool with one-year residence time. Changes in the amount of carbon in vegetation and soils following the land-use disturbance are calculated from idealized response curves that describe temporal changes in carbon mass per unit land area. The curves are parameterized by ecosystem type and type of land-use change, and they account for factors affecting both loss and accumulation of carbon without actually modeling processes such as photosynthesis, growth, mortality, and decomposition (Moore et al., 1981; Houghton et al., 1983). For example, the impact of fires used in forest clearing is implicitly incorporated in response curve parameters defining the amount of carbon in vegetation before and after clearing. The net annual loss of carbon from the vegetation and soils of all ecosystem types and geographic regions plus the annual loss from the decay pools for harvested material is an estimate of annual global carbon release to the atmosphere from changes in land use. The models assume that there is no net exchange of carbon between the atmosphere and ecosystems not disturbed by land-use change.

History A (Figure 39.3) is the population-based estimate of Houghton et al. (1983). We estimated carbon release before 1825 by linear extrapolation from the 1825 value to zero release before 1770. We assumed constant release at the 1980 value for the period 1980 to 1987.

Histories B and C (Figure 39.3) are more recent estimates (Houghton, 1989) using revised estimates of land-use change, carbon stocks in tropical ecosystems, and the reduction in soil carbon following forest clearing (Houghton et al., 1985, 1987). History B is a high estimate based on medium to high estimates of tropical biomass; history C is the corresponding low estimate based on low estimates of tropical biomass.

We estimated pre-1860 carbon release by linear extrapolation to zero release before 1770, and we assumed constant release at the respective 1985 values for 1986 to 1987.

A fraction of the carbon added to the atmosphere from the anthropogenic sources penetrates the ocean surface as CO_2 and is distributed throughout the ocean according to the dynamics of each model. The mass of carbon remaining in the atmosphere is converted to CO_2 concentration using a C mass (GT C) to CO_2 concentration (μL L^{-1} or parts per million by volume, ppmv) conversion based on the molecular weights of carbon and air and the dry air mass of the atmosphere. The simulated history of changes in atmospheric CO_2 concentration is then compared with the observed history specified by CO_2 measurements from the Siple Station, Antarctica, ice core (Neftel et al., 1985; Friedli et al., 1986) and the monitoring station at Mauna Loa Observatory, Hawaii (Keeling, 1986). The Mauna Loa measurements approximate mean global concentrations within 0.2–0.3 μL L^{-1} (Gammon et al., 1986).

Results and Discussion

Simulation of Past Changes (1750 to 1988)

If we first assume that historically the terrestrial biosphere has made no net contribution to atmospheric CO_2 and simulate past changes in atmospheric CO_2 using only the history of industrial CO_2 emissions as atmospheric input, the models all underestimate observed concentrations during most of the Siple period (about 1750 to 1960) and throughout the Mauna Loa period (about 1960 to 1990) (Figure 39.4). The greatest deviations from the observations occur at about the middle of this century. The models overestimate the rate of change in atmospheric CO_2 during the Mauna Loa period. Consequently the differences between observations and simulations narrow during the latter years of the simulations (Figure 39.4). The simulations' overestimates of the change in atmospheric CO_2 during the Mauna Loa period are consistent with the problem of "missing" industrial CO_2 discussed in the introduction.

The box-diffusion model generates the smallest net uptake of CO_2 by the ocean and consequently yields the highest atmospheric concentrations, followed in turn by the discrete-diffusion model. The discrete-advection model with its more effective transfer of carbon to the deep ocean by advective flux has the largest net oceanic CO_2 uptake and yields the lowest

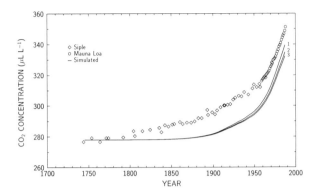

Figure 39.4 Comparison of model simulations (—) with the record of observed changes in atmospheric CO_2 from the Siple (◇) and Mauna Loa (○) records when the models' atmospheric CO_2 inputs are limited to the industrial source of Figure 39.2. Numbers refer to the models of Figure 39.1. The Siple record is the concentration of CO_2 in air bubbles trapped in old ice of an ice core extracted at Siple Station, Antarctica (Neftel et al., 1985; Friedli et al., 1986); the Mauna Loa record is the annual average atmospheric CO_2 concentration recorded at Mauna Loa Observatory, Hawaii (Keeling, 1986).

atmospheric concentrations. Although the differences among models increase gradually with time, the differences are always less than the differences between observations and simulations. For example, in 1987 the difference in CO_2 concentrations simulated by the box-diffusion and discrete-advection models is 7 μL L^{-1} while their respective deviations from the Mauna Loa record are 9 μL L^{-1} and 16 μL L^{-1}.

If we assume no net input of carbon from natural sources, an underestimation of past CO_2 concentrations when CO_2 input is limited to industrial sources suggests the presence of a significant contribution from past changes in land use. An accurate estimate of the size and timing of that input is indicated by coincidence between observed and modeled CO_2 in simulations that include histories of both land-use and industrial emissions.

Figure 39.5 shows the correspondence, or rather lack of correspondence, between the observed atmospheric CO_2 record and our simulations, combining the histories of land-use CO_2 release from Figure 39.3 with the industrial CO_2 emissions as atmospheric input. In each case the simulations overestimate concentrations during the Mauna Loa period. In the extreme, using the older estimate from Houghton et al. (1983) (history A, Figure 39.3), the models overestimate the 1987 concentration by 39 to 45 μL L^{-1} (Figure 39.5a). At best, using the low estimate of history C (Figure 39.3), the models overestimate the

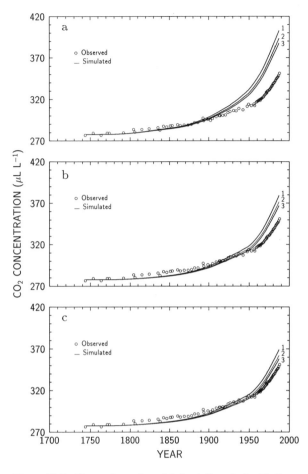

Figure 39.5 Comparison of model simulations (—) with the combined Siple–Mauna Loa record of observed changes in atmospheric CO_2 (○) when the histories of past land-use CO_2 release from Figure 39.3 are combined with the industrial CO_2 emissions of Figure 39.2 as model input. (a) Simulations with history A (Figure 39.3). (b) Simulations with history B (Figure 39.3). (c) Simulations with history C (Figure 39.3). Observed concentrations are as in Figure 39.4.

1987 concentration by 9 to 20 μL L^{-1} (Figure 39.5c). For comparison, the change in concentration over the Mauna Loa period is approximately 35 μL L^{-1}.

Coincidence between simulation and observation is more variable during the Siple period than the Mauna Loa period. Simulations using land-use emission history A do a reasonable job of matching the Siple record until around 1900, when they diverge radically from the observations (Figure 39.5a). Simulations with histories B and C tend to slightly underestimate CO_2 concentrations until about 1900, do reasonably well until about 1950, and then they too diverge from the observations, overestimating the concentration and rate of change for the rest of the simulation (Figures 39.5b and c).

The pattern among models seen in the simulations without CO_2 from land-use change (Figure 39.4) is repeated in these simulations (Figure 39.5). The box-diffusion model produces the largest atmospheric CO_2 concentrations, and in these comparisons the greatest deviation from the Mauna Loa observations. The discrete-advection model produces the lowest concentrations, and in these comparisons the closest agreement during the Mauna Loa period. The discrete-diffusion model is intermediate. As in the earlier simulations the models slowly diverge from one another with time. Note also that the differences among models and the differences attributable to land-use emission history overlap. Simulated atmospheric CO_2 from the discrete-advection model using emission history B approximates the simulation using the box-diffusion model and emission history C (Figures 39.5b and c). To a lesser extent this is also true for the discrete-advection model using history A and the box-diffusion model using history B (Figures 39.5a and b).

The improvement in coincidence between observation and simulation with the changes from land-use emission history A to history C is due principally to a change in the amount of CO_2 released from past changes in land use rather than the timing of that release. Except for nearly constant or slightly declining CO_2 releases from about 1920 to 1945 in histories B and C compared to increasing release for that period in history A, the pattern or timing of CO_2 release does not differ appreciably among the alternative emission histories (Figure 39.3). Furthermore, the difference between histories B and C and the accompanying improvement in correspondence between observation and simulation when history C is used (Figure 39.5c) are due to the lower estimates of tropical biomass used in producing history C.

Clearly, understanding of past global carbon cycle dynamics as summarized in the models and emission histories we use here is incomplete. Somewhere, either singly or in some combination, errors in the record of past CO_2 concentrations, the reconstruction of past industrial CO_2 emissions, the understanding and modeling of ocean-atmosphere carbon dynamics, or the understanding of terrestrial biospheric sources and sinks and the reconstruction of past CO_2 release from changes in land use are large enough to produce the level of disagreement between observation and simulation seen here and elsewhere (Peng et al., 1983; Enting and Pearman, 1986; Enting and Mansbridge, 1987; Keeling et al., 1989). For our present purpose

we will assume that errors in the record of past CO_2 concentrations and the reconstruction of industrial CO_2 emissions are negligible; indeed these are two of the better known components of the global carbon cycle, especially during the Mauna Loa period. Similarly, any error in modeling the net uptake of CO_2 by the ocean that might account for the disagreement between simulation and observation must exceed the uncertainty represented by differences among the three carbon cycle models we use since all three models fail to produce the desired agreement; we assume that error greater than this uncertainty is improbable and also negligible. Consequently, it remains that either the reconstructions of past land-use CO_2 release are in error, or there have been substantial historical exchanges between the atmosphere and those portions of the terrestrial biosphere not altered from preindustrial states by changes in land use.

In addition to the scientific interest in understanding the global biogeochemistry of carbon, resolution of the disagreement between simulation and observation of past changes in atmospheric CO_2 is important to the projection of future changes. First, if differences between observations and simulations were naively ignored and the simulations extended into the future with assumptions about future anthropogenic emissions, the initial trajectories of the projections would be biased by 9 to 54 $\mu L \, L^{-1}$, depending on the choice of model and history of CO_2 release from changes in land use (Figure 39.5). Similarly, the projections' initial rates of change, influenced by input from the recent past, would be slightly larger than the observed rate of change. These initial biases would influence projected concentrations, especially in the near future. Furthermore, conventional projections of future CO_2 concentrations assume that the global carbon cycle will continue to operate in the future as it has in the past. Consequently, confidence in future projections is proportional to the understanding of past changes. If some carbon cycle dynamic of the past has been overlooked, as the noncoincidence of simulation and observation in Figure 39.5 suggests is possible, confidence in the future behavior of the carbon cycle is compromised. Failure to correctly project that dynamic into the future could result in erroneous projections of future changes in atmospheric CO_2. In short, improved coincidence between simulations and observations of past changes in atmospheric CO_2 is needed for projection of future changes.

Reconstruction of Past Terrestrial Biospheric Exchange by Deconvolution

As noted in the introduction to this chapter, an estimate of past exchanges between the atmosphere and terrestrial biosphere can be obtained by constraining the ocean-atmosphere carbon model to reproduce the combined Siple–Mauna Loa record of atmospheric CO_2 and calculating the residual flux needed to achieve that fit (Siegenthaler and Oeschger, 1987; Keeling et al., 1989). Again, interpretation of the residual flux as terrestrial biospheric exchange with the atmosphere is confounded by uncertainties in the deconvolution, but the estimates that result are extremely useful as alternatives for comparison with the direct historical-ecological reconstructions. Furthermore they provide the additional benefit of allowing for the projection of future CO_2 changes from current concentrations and rates of change since they are produced by simulations that are constrained to reproduce these observations. Given the incompatibility of existing historical-ecological reconstructions of past land-use CO_2 release with conventional globally averaged one-dimensional models of ocean-atmosphere CO_2 exchange (Figure 39.5) (Enting and Pearman, 1986; Enting and Mansbridge, 1987), and with at least the first generation of regionalized two-dimensional tracer-based models (Peng, 1986; Enting and Mansbridge, 1987) and three-dimensional ocean general circulation models (Keeling et al. 1989), the constrained atmosphere approach to simulating past CO_2 changes is virtually indispensable for the projection of future changes.

In Figure 39.6 we compare our constrained simulation with the historical record of past changes in atmospheric CO_2. We used a modification of the approach to constrained atmosphere simulation used by Killough and Emanuel (1981), and we smoothed the simulation by cubic splines to just maintain a monotonic increase in simulated CO_2 concentration. Note that the simulated history of CO_2 is the same for each of the three models, but the residual flux required to achieve that simulation varies inversely with the oceanic uptake of the models (Figure 39.7). For example, the box-diffusion model simulates the smallest net oceanic uptake and consequently generates the largest residual flux in producing the simulation of Figure 39.6.

If the anthropogenic CO_2 emissions driving the model simulations are limited to industrial sources, the residual flux can be interpreted as the net exchange of carbon between the atmosphere and the entire terrestrial biosphere (Figure 39.7a). If CO_2

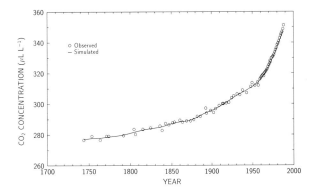

Figure 39.6 Simulation of past changes in atmospheric CO_2 (—) constrained to reproduce the observed changes of the combined Siple–Mauna Loa record (O). Observed concentrations are as in Figure 39.4.

release from land-use change is also included (e.g., the histories of Figure 39.3), the residual flux can be interpreted as the net exchange between the atmosphere and those portions of the terrestrial biosphere that have not been disturbed by land-use change (Figures 39.7b to d). In these latter simulations, the size and timing of the residual flux reflect the input from land-use change; the larger the estimate of CO_2 release from land-use change during the Mauna Loa period, the larger (more negative) the compensatory residual flux uptake must be to achieve the same observed change in atmospheric CO_2. Note the reduction in the residual flux from Figure 39.7b to Figure 39.7d with the change in land-use CO_2 history from the high estimate of Houghton et al. (1983) (history A, Figure 39.3) to the low estimate of Houghton (1989) (history C, Figure 39.3).

Interpretation of the residual flux as exchange with the terrestrial biosphere suggests either that net releases from past changes in land use have been appreciably lower than those estimated by the land-use change models or that there has been a significant amount of carbon uptake and sequestering by terrestrial ecosystems not experiencing land-use change. Indeed, with no CO_2 input from land-use change, the results from the box-diffusion and discrete-diffusion simulations suggest that the terrestrial biosphere as a whole, including both altered and unaltered systems, has been a small to moderate net sink for atmospheric CO_2 since about 1950 (curves 1 and 2 of Figure 39.7a). These results agree with those of Keeling et al. (1989, Figure 46) for their box-diffusion model calibrated with preindustrial ^{14}C and their version of the three-dimensional ocean model of Maier-Reimer and Has-

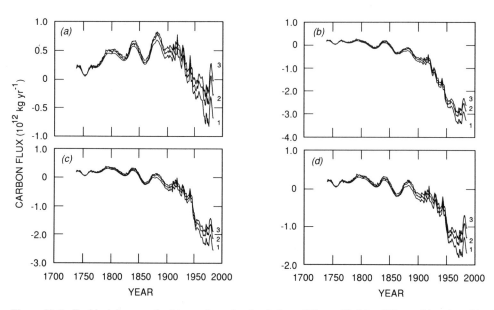

Figure 39.7 Residual flux required to produce the simulation of Figure 39.6 for different histories of land-use CO_2 release. Numbers refer to the models of Figure 39.1. (a) No land-use CO_2 release. (b) Simulations with history A (Figure 39.3). (c) Simulations with history B (Figure 39.3). (d) Simulations with history C (Figure 39.3).

selmann (1987). Our results from the discrete-advection model also suggest a recent global net biospheric sink, albeit much smaller and of shorter duration (curve 1, Figure 39.7a). This result is comparable to those of Siegenthaler and Oeschger (1987) and Keeling et al. (1989), based on atomic bomb ^{14}C calibrated box-diffusion models.

The negative residual flux values from the simulations using only industrial CO_2 inputs can be interpreted as global net CO_2 uptake by terrestrial ecosystems. This implies that terrestrial systems have in recent years assimilated and sequestered enough CO_2 to account for any CO_2 from industrial sources in excess of the increase in atmospheric CO_2 that is taken up by the ocean, and any additional CO_2 from changes in land use. Accordingly, the larger the estimate of CO_2 release from those portions of the terrestrial biosphere undergoing changes in land use and acting as a source of atmospheric CO_2, the larger the inferred uptake by the natural or unaltered portions of the terrestrial biosphere must be (Figures 39.7b to d) to achieve the same total net terrestrial biospheric exchange with the atmosphere (Figure 39.7a).

Interestingly, the estimates of terrestrial biospheric CO_2 uptake inferred from the residual flux results are comparable to the 2.0 to 3.4 GT yr^{-1} carbon sink required by Tans et al. (1990) to match the latitudinal gradient of atmospheric CO_2 and balance the carbon budget using an entirely different approach to model-

ing oceanic CO_2 uptake. Further analysis of the estimates of terrestrial biospheric flux and past behavior of the terrestrial biosphere (e.g., the capacity of unconverted ecosystems to sequester the required carbon and mechanisms explaining the apparent historical increase in that sequestering) is of considerable interest but beyond the scope and objectives of this chapter and will be treated elsewhere (see Houghton, 1986, and Keeling et al., 1989, for related discussion). For our present purpose we will next turn to the implications of these past terrestrial biospheric fluxes for projection of future changes in atmospheric CO_2 with simulations that have been constrained to reproduce past changes.

Implications for Future Changes

We begin by noting that future changes in atmospheric CO_2 tend to be dominated by the projected scenarios of change in industrial CO_2 emissions and uncertainty in what those emissions will be. Consequently we will limit ourselves to a single scenario of future industrial CO_2 emissions, one that approximates a continuation of current trends in the use of fossil fuels (Table 39.1), and consider the influence of projected terrestrial biospheric CO_2 exchange with the atmosphere on variability in projected atmospheric CO_2 concentrations, using that future scenario of industrial CO_2 emissions.

Figure 39.8 shows projected CO_2 concentrations if

Table 39.1 Projected industrial CO_2 emissions

Year	CO_2 emission (10^{12} kg C yr^{-1})
2000	6.7
2025	10.1
2050	13.8
2075	18.0
2100	22.0

Note: Industrial sources include commercial energy production and cement manufacturing (IPCC, 1989).

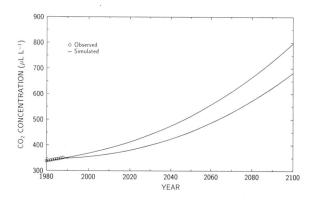

Figure 39.9 Envelope of uncertainty surrounding the discrete-diffusion model's simulation of future changes in atmospheric CO_2 in response to the industrial CO_2 emission scenario of Table 39.1 that results from uncertainty in past terrestrial biospheric CO_2 exchange with the atmosphere and the extrapolation of that exchange into the future. In these simulations we assume no future CO_2 input to the atmosphere from changes in land use. The upper bound of the envelope results from the assumption of zero future net CO_2 exchange with the terrestrial biosphere; the lower bound results from the assumption of a constant 3 GT C yr^{-1} uptake of CO_2 by the terrestrial biosphere.

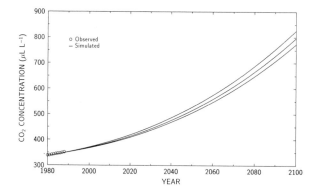

Figure 39.8 Simulations of future changes in atmospheric CO_2 assuming the industrial CO_2 emission scenario of Table 39.1 and no net exchange of CO_2 between the atmosphere and terrestrial biosphere. Simulations of past changes in atmospheric CO_2 are identical to the simulation of Figure 39.6. The upper curve is from the box-diffusion model (model 1, Figure 39.1), the middle curve is from the discrete-diffusion model (model 2, Figure 39.1), and the bottom curve is from the discrete-advection model (model 3, Figure 39.1).

we assume no net exchange between the atmosphere and terrestrial biosphere with future increases driven only by industrial CO_2 emissions. Despite identical constrained simulation of past changes (Figure 39.6), the model simulations slowly diverge as the simulations are continued into the next century. As before, the box-diffusion model yields the highest concentrations and the discrete-advection model the lowest with a difference in the year 2100 of 53 μL L^{-1}. Again, for comparison, the observed change over the Mauna Loa period is approximately 35 μL L^{-1}. Thus differences in projected atmospheric CO_2 attributable to differences in the carbon cycle models (i.e., their representation of ocean structure and carbon transport) are not trivial. However, as we will show below, differences attributable to uncertainty in terrestrial biospheric exchange can exceed the differences among models.

Based on our deconvolutions of the Siple–Mauna Loa record, recent net carbon uptake by those portions of the terrestrial biosphere not disturbed by land-use change could be quite small, nearly zero, or as large as 3 GT C yr^{-1} (Figure 39.7b to d). Assuming no net future terrestrial biospheric CO_2 release from land-use change or other sources, we can draw an envelope of uncertainty around the projected response to industrial CO_2 emissions that is a consequence of the uncertainty in past terrestrial biospheric exchange and the projection of that exchange into the future. One of many such possible envelopes is shown in Figure 39.9, with the upper bound formed by the discrete-diffusion model and zero future net biospheric exchange and the lower bound by the same model and a constant 3 GT yr^{-1} carbon uptake by the terrestrial biosphere. Projection of smaller terrestrial biospheric uptake would, of course, narrow this envelope, elevating the lower bound. Projection of the possibility that future biospheric uptake might increase with time, as the deconvolutions suggest may have occurred in the past, perhaps in response to changing climate or atmospheric CO_2, would expand the uncertainty envelope, depressing the lower bound.

Future changes in atmospheric CO_2 will, of course, be influenced by any significant release of CO_2 from

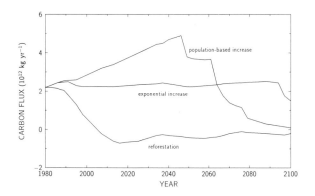

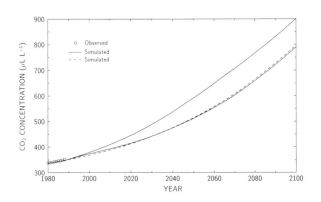

Figure 39.10 Projections of future CO₂ release to the atmosphere from future changes in land use (Richard A. Houghton, personal communication, 1988) from the Houghton (1989) implementation of the Houghton et al. (1983) model of land-use change and CO₂ release. In the exponential increase scenario, tropical deforestation was assumed to increase exponentially. In the population-based increase scenario, regional deforestation was assumed to increase at the same rate as the population growth of the same region. In the reforestation scenario, rates of deforestation were assumed to decrease to zero by 2025 with extensive reforestation through abandonment of areas of shifting cultivation, planted pastures of Latin America, and the establishment of tropical tree plantations.

Figure 39.11 Envelope of uncertainty surrounding the discrete-diffusion model's simulation of future changes in atmospheric CO₂ in response to the industrial CO₂ emission scenario of Table 39.1 that results from uncertainty in future CO₂ release from changes in land use. In these simulations we assume no future net CO₂ exchange with those portions of the terrestrial biosphere not altered by land-use change. The dashed curve (- - -) is the simulated response to the industrial CO₂ inputs alone, assuming no net CO₂ exchange with the terrestrial biosphere. The upper solid curve (—) is an upper bound produced using the population-based CO₂ release scenario of Figure 39.10; the lower solid curve (—) is a lower bound produced using the reforestation scenario of Figure 39.10.

future changes in land use. Projections of future land-use CO₂ release are scarce and uncertain. However, we can get some idea of their influence on future changes in atmospheric CO₂ by using projections of future land-use CO₂ release provided by R. A. Houghton (personal communication, 1988) (Figure 39.10) (see Houghton, 1990, for related projections). In Figure 39.11 we show the resulting envelope of uncertainty for the discrete-diffusion model assuming zero net exchange with terrestrial ecosystems not disturbed by land-use change. The upper bound is produced using the population-based scenario of Figure 39.10; the lower bound results from the reforestation scenario of Figure 39.10.

We want to emphasize that the uncertainty in projected CO₂ concentrations resulting from uncertainty in past terrestrial biospheric CO₂ exchange (Figure 39.9) is of the same magnitude as the uncertainty resulting from uncertainty in future land-use CO₂ release (Figure 39.11). Both exceed the uncertainty in ocean-atmosphere modeling (Figure 39.8) by about 50%. The combined uncertainty of ocean-atmosphere carbon cycle modeling and past and future terrestrial biospheric exchange is illustrated in Figure 39.12. The upper bound is from the box-diffusion model assuming no future net CO₂ exchange with the unaltered terrestrial ecosystems and the popula-

tion-based projection of future land-use CO₂ release. The lower bound is from the discrete-advection model assuming a constant 3 GT yr⁻¹ carbon uptake by unaltered terrestrial ecosystems and the reforestation scenario of future land-use change. Comparing Figures 39.11 and 39.12, uncertainty about terrestrial biospheric exchange with the atmosphere increases the range of projected CO₂ concentrations for the year 2100 from 774 to 827 μL L⁻¹ to 658 to 935 μL L⁻¹). The period in which a doubling of atmospheric CO₂ over preindustrial concentrations is achieved (i.e., a future concentration of approximately 555 μL L⁻¹) increases from 2056–2062 to 2041–2080.

Conclusions

Our results reaffirm earlier conclusions that the Houghton et al. (1983) population-based history of CO₂ release from past changes in land use, reconstructed by explicit consideration of carbon dynamics in terrestrial ecosystems disturbed by historical changes in land use, is incompatible with measurements of atmospheric CO₂ and conventional ocean-atmosphere carbon cycle models (Peng et al., 1983; Enting and Pearman, 1986; Enting and Mansbridge, 1987; Keeling et al., 1989; DOE Multi-Laboratory

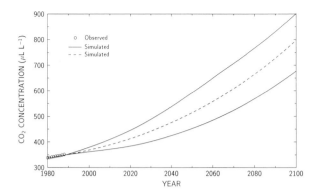

Figure 39.12 Envelope of uncertainty surrounding the discrete-diffusion model's simulation of future changes in atmospheric CO_2 in response to the industrial CO_2 emission scenario of Table 39.1 that results from the combined uncertainty in ocean-atmosphere carbon cycle modeling, uncertainty in past CO_2 exchange with the terrestrial biosphere and the extrapolation of that exchange into the future, and uncertainty in future CO_2 release from land-use change. The dashed curve (---) is the discrete-diffusion model's response to the industrial CO_2 inputs alone, assuming no net CO_2 exchange with the terrestrial biosphere. The upper solid curve (—) is an upper bound produced with the box-diffusion model assuming zero future net CO_2 exchange with the terrestrial biosphere not altered by land-use change and the population-based projection of future land-use CO_2 release from Figure 39.10. The lower solid curve (—) is a lower bound produced with the discrete-advection model assuming a constant 3 GT C yr^{-1} uptake of atmospheric CO_2 by that part of the terrestrial biosphere not disturbed by land-use change and the reforestation scenario of future land-use CO_2 "release" from Figure 39.10.

Climate Change Committee, 1990) (for related discussions see Broecker et al., 1980; Oeschger and Heimann, 1983; Elliot et al., 1985; and Enting and Pearman, 1986). We have also shown that more recent estimates of the history of land-use CO_2 release based on the same approach do not significantly resolve the incompatibility.

An understanding of past terrestrial biospheric behavior is important for projecting future changes in atmospheric CO_2. Disagreement between simulations and the historical record of atmospheric CO_2 can lead to initial bias in the projection of future changes, and the residual carbon flux needed to balance observation, model, and the history of CO_2 input to the atmosphere suggests unknown carbon cycle dynamics that can have an appreciable impact on the projection of future changes in atmospheric CO_2. Our results show that uncertainty surrounding the past source-sink behavior of the terrestrial biosphere can have as much influence on the projection of future atmo-

spheric CO_2 as uncertainty in future anthropogenic emissions.

The long-standing issue of the terrestrial biosphere's role as an historical source or sink for atmospheric CO_2 (e.g., Reiners, 1973; Broecker et al., 1980; Hobbie et al., 1984) has yet to be resolved. Broecker et al. (1980) concluded that understanding of global terrestrial biospheric dynamics was such that it was not possible to say whether the terrestrial biosphere of recent decades has been a source or sink for atmospheric CO_2, and consequently knowledge of the terrestrial biosphere could not be used to distinguish between alternative models of oceanic carbon uptake. A decade later, one can still reach that same conclusion.

There is need for continued efforts to reconstruct the role of the terrestrial biosphere in past changes in atmospheric CO_2, and to resolve incompatibilities between the deconvolution and historical-ecological approaches. The dilemma of needing ocean models with uncertain ocean-atmosphere exchange that cannot be resolved by terrestrial biospheric data to in turn define the uncertain behavior of the terrestrial biosphere by deconvolution may be with us for some time, but progress can be made in the direct and explicit reconstruction of historical CO_2 exchange between atmosphere and terrestrial biosphere based on knowledge of terrestrial ecosystems. These efforts must continue to consider the history of land-use change and the associated carbon dynamics of disturbed ecosystems (Moore et al., 1981; Houghton et al., 1983). The improved correspondence between observations and simulations using land-use CO_2 release history C (Figure 39.5) with its assumptions of low tropical biomass suggests that careful attention should be given to estimating the biomass of tropical forests before and after land-use change (e.g., Brown et al., 1989). Simultaneously, there must be further consideration of the history of carbon dynamics in ecosystems that have not been substantially altered from preindustrial states by land-use change (e.g., high-latitude boreal forests). From the perspective of this conference and these proceedings, improved estimates of fuel wood consumption, the use and impact of fire as an agent of land-use change, and the occurrence and impact of fire in natural ecosystems will all contribute to this effort.

Additional consideration should be given to the terrestrial biospheric dynamics of ^{13}C, including the influence of land-use disturbance. Keeling et al. (1989) have demonstrated how analysis of changes in atmospheric ^{13}C can aid in determining the terrestrial biosphere's contribution to changes in atmospheric CO_2 concentrations. Incorporation of ^{13}C into historical-ecological models of land-use change and carbon dynamics should prove useful in this analysis.

Finally, for ecosystems unaltered from preindustrial states as well as those disturbed by changes in land-use, additional consideration must be given to the past response of terrestrial carbon dynamics to rising concentrations of atmospheric CO_2 and other changes in the environment. Representations of terrestrial biospheric response to elevated CO_2 in global carbon cycle models that have attempted to model this dynamic (e.g., Bacastow and Keeling, 1973; Oeschger et al., 1975; Björkström, 1979; Goudriaan and Ketner, 1984; Kohlmaier et al., 1981, 1987; Esser, 1987) are often discounted for the lack of attention to physiological processes, ecological limitations, and the simple extrapolation of growth chamber results to the field and to global scales (e.g., Goudriaan and Ajtay, 1979; Strain and Armentano, 1982; Dahlman, 1985; Peterson and Melillo, 1985). If these representations are indeed too simplistic—for example, many do not include both climate and CO_2 response (but see Esser, 1987)—the appropriate level of ecological realism should be determined and incorporated into global carbon cycle models simulating historical changes in atmospheric CO_2 and tested against observed changes in ^{13}C and ^{14}C (Bolin, 1986).

Efforts to define current rates of tropical deforestation and the accompanying release of CO_2 should continue (e.g., Houghton et al., 1985, 1987; Detwiler and Hall, 1988). But it is equally important to resolve the terrestrial biospheric exchange of the past several decades (both release and uptake). Indeed the presence or absence of a peak or maximum in the input of CO_2 to the atmosphere from the terrestrial biosphere prior to the Mauna Loa observation period may be crucial to the coincidence of simulated and observed atmospheric CO_2 (Oeschger and Heimann, 1983; Enting and Mansbridge, 1987; Keeling et al., 1989). Many of the model deconvolutions of past terrestrial biospheric behavior point to the existence of such a maximum (Peng et al., 1983; Emanuel et al., 1984; Stuiver et al., 1984; Peng and Freyer, 1986; Siegenthaler and Oeschger, 1987; Keeling et al., 1989; Figure 39.7a, this chapter) (Stuiver et al.'s, 1984, deconvolution of the $\delta^{13}C$ record of Pacific coastal trees is an exception). With the exception of the Houghton et al. (1983) analysis based on FAO statistics on land use and an exercise assuming larger carbon stocks in forests cleared prior to 1900

(Houghton, 1986), this peak is absent from historical-ecological reconstructions. Resolution of this and other historical behaviors of the terrestrial biosphere will go a long way toward improving our confidence in past and future behavior of the terrestrial biosphere and its role in the global carbon cycle. In this way, knowledge of the terrestrial biosphere can better be used in the evaluation of ocean models, and our ability to project future changes in atmospheric CO_2 will be enhanced.

Acknowledgments

We thank Richard "Skee" Houghton for providing tabular data for the Houghton et al. (1983) estimate of past land-use CO_2 release and for providing us with his more recent estimates of past land-use CO_2 release and projections of future land-use CO_2 release. We also thank Tom Boden and the Carbon Dioxide Information and Analysis Center, Oak Ridge National Laboratory, for the data sets describing the industrial carbon emissions and the Mauna Loa CO_2 data used in this study.

Research was sponsored by the Carbon Dioxide Research Program, Atmospheric and Climate Research Division, Office of Health and Environmental Research, U.S. Department of Energy, under contract DE-AC05-84OR21400 with Martin Marietta Energy Systems, Inc., Publication No. 3698, Environmental Sciences Division, Oak Ridge National Laboratory.

The Contribution of Biomass Burning to the Carbon Budget of the Canadian Forest Sector: A Conceptual Model

Werner A. Kurz, Michael J. Apps, Timothy M. Webb, and Peter J. McNamee

The oxidation of organic matter releases energy, carbon dioxide, methane, and several other gases. From the perspective of the carbon cycle, different forms of oxidation such as wildfire, decomposition, and the use of biomass for the production of energy differ only in the rate of conversion of one form of carbon to others, the composition of the carbon types generated, and the use of the energy released in the process.

In most unmanaged Canadian forest ecosystems, the return frequency of wildfires determines the age-class distribution of the forest because wildfire is the predominant disturbance regime. Wildfires affect the amount of biomass carbon stored in a forest ecosystem and are an important and necessary disturbance to maintain forest productivity. Without such periodic disturbances, the productivity (though not necessarily total fixed carbon) of overmature temperate and boreal forest stands declines as an increasing proportion of the available nutrients becomes locked up in slowly decomposing organic matter (van Cleve and Viereck, 1981; van Cleve et al., 1983).

The recognition of the potential contribution of greenhouse gases such as CO_2 to global climate change has increased the interest in understanding and assessing the role of forest ecosystems in the global carbon cycle. Forest ecosystems are simultaneously taking up and releasing CO_2, with the net exchange of carbon between the forest and the atmosphere depending on factors such as tree species, site conditions, season, climate, and time since last disturbance. The relationship between net C uptake and time since last disturbance is difficult to quantify, particularly in the early stages of stand development. In the later stages of stand development, total ecosystem carbon generally increases, although the net carbon fixation rate decreases. In some cases, merchantable stem volume can decline because of tree mortality and the associated movement of carbon from the live biomass to the detrital and soil carbon pools. In these circumstances, total ecosystem carbon may continue to increase, although at a smaller rate than during immature and mature stand development stages. For particular ecosystem types, there appears to be a negative correlation between the amount of biomass carbon stored and additional net uptake of carbon. The absolute amounts of carbon as well as the time scales involved change in different ecosystems: Contrast, for example, the low-disturbance coastal forests of the Pacific Northwest with the fire-dominated aspen mixed woods of the southern boreal forest.

The role of biomass burning in the carbon budget is frequently assessed in complete isolation from its general role in the spatial and temporal dynamics of forest ecosystems. At a regional scale, forests are an assemblage of stands in various stages of stand development. The annual net exchange of carbon between the atmosphere and forest ecosystems must therefore be calculated by integrating the C-exchange of individual stands over the spatial and temporal scales of interest. Estimates obtained from studies of carbon release during wildfires represent only one-half of the carbon budget equation: the release rate. While these estimates of carbon released into the atmosphere may amount to impressive numbers, it is of far greater relevance to studies of global change and carbon cycles whether, at the spatial and temporal scales of interest, the balance of C-uptake and C-release is positive or negative.

This chapter provides an overview of a conceptual model of the carbon budget of the Canadian Forest Sector. It is currently being implemented in a computer model to enable the quantitative determination of carbon pool sizes and net-fluxes in forest biomass, forest soils, and the forest products sector. The computer models are designed specifically to address a series of policy questions about the possible role of the forest sector in sequestering and storing atmospheric carbon. Such policy questions could be aimed at reducing the rate of CO_2 increase in the global atmosphere through reforestation, altered fire protection strategies, substitution of fossil fuel–based energy through bioenergy, or increased recycling of forest products such as pulp and paper.

Carbon Budget Model Structure

Conceptual Model   Any analytical framework which attempts to address policy questions at a quantitative level must be comprehensive enough to account for all major aspects of the system under study. In the case of the carbon budget model, the goal is to quantify the role of forests and the influences of forest management on the carbon cycle. It is therefore necessary to include all major pools, sources, and sinks of carbon within the forest sector, including the use of fossil carbon for stand management and in the forest products sector. At present, the analytical framework has no economic component and does not consider functions of forest ecosystems other than carbon cycling. It is clearly recognized, however, that forest ecosystems provide many additional functions and that the results of any policy analysis cannot be considered without also assessing those economic, social, and ecological aspects which are currently external to the computer models.

The conceptual model consists of several carbon pools and fluxes between those pools (Figure 40.1). Carbon fluxes of both biomass and fossil fuel carbon are explicitly represented in recognition of their special role in the anthropogenic changes to the global carbon cycle. The three major biomass carbon pools are forest biomass, forest soils, and the forest products pool. One component of the forest products sector, biofuel production, is represented explicitly because of its potential contribution in substituting for fossil fuel carbon sources.

Carbon transported in aquatic systems is recognized as a potential long-term sink (Schlesinger and Melack, 1981), but this component of the carbon cycle is currently not implemented in the model, because of a lack of scientific understanding and of data at the regional forest level. Other carbon pools and fluxes are described in more detail below.

Disturbances   Several different events which redistribute C between carbon pools and which affect the rate of C accumulation following the event are treated conceptually in the same way and referred to as disturbances. Five disturbance types are currently recognized in the model: wildfires, insect-induced stand mortality, clear-cut logging without slash burning, clear-cut logging with slash burning, and partial cutting. The C fluxes between different biomass and soil C pools, the atmosphere (CO_2, CO, CH_4), and the forest products sector at the time of disturbance are quantified by disturbance matrices. These matrices reallocate the C in each pool among the other pools

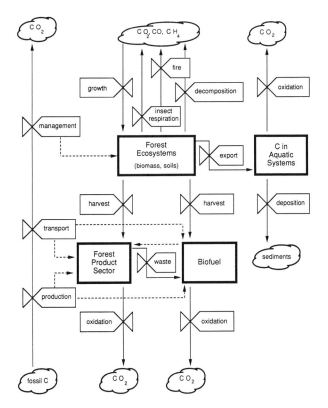

Figure 40.1   A diagram of the conceptual model of the carbon budget of the Canadian Forest Sector. Boxes represent pools (state variables), valve symbols represent transfer rates (rate variables), cloud symbols represent sources and sinks outside the system, solid arrows represent flow of carbon, and dashed lines represent flow of information.

according to the specific characteristics of the disturbance. Postdisturbance dynamics of the soil C pools are simulated as described in the section on soils below.

Provincial disturbance statistics are used to generate the annual disturbances which, based on a set of user-specified rules, are distributed across the forest ecosystem types in the inventory.

Biomass

Forest ecosystems are represented as two closely related carbon pools, biomass and soil, each of which contains several subcompartments. On a national scale, the available data as well as the scientific understanding about carbon cycling are far better developed for the biomass than for the soils component. Many of Canada's forests are naturally regenerated and have never been logged. Where growth and yield information is available, its emphasis is on merchantable volume and it does not generally provide any information about biomass of unmerchantable stand components. Forestry Canada's ENFOR (Energy from the Forest) program, which was started after the energy crisis of the early 1970s, has supported a national research program aimed at providing biomass data at a national scale.

Canada's Biomass Inventory (Bonnor, 1985) is the data source for the biomass data in this study. The national inventory data recognize nearly 50,000 spatial units, each representing on average 15 stands. This spatial resolution is too detailed for the purpose of this study. The data were therefore aggregated to produce 44 spatial units representing the 11 Ecoclimatic Provinces (Ecoregions Working Group, 1989) within Canada's 10 administrative provinces and two territories. Ecoclimatic Provinces represent broad regions of similar climatic conditions and vegetation types (Ecoregions Working Group, 1989). The same classification system has recently been used in an analysis of potential climate change impacts on the vegetation distribution in Canada (Zoltai, 1988; Rizzo and Wiken, 1990).

Within each of the 44 geographical units, six additional classifiers are recognized: land class, productivity class, stocking, forest type, site class, and maturity class (for definitions see Bonnor, 1985). Area and, to a lesser extent, biomass information is available for each spatial unit, stratified by these classifiers. The biomass data are area-weighted mean biomass per hectare values for each of eight biomass components and their total.

The biomass carbon pool is subdivided into eight additional pools, four each in the softwood and hardwood components of forest ecosystems. The four pools are: merchantable stems including bark, other merchantable wood and bark, foliage, and submerchantable biomass (see Bonnor, 1985, for definition of terms). Biomass dynamics are described in terms of net changes of biomass carbon in the sum of the eight biomass components. All net increases of biomass carbon are assumed to be taken up from the atmosphere; all decreases are assumed to be transferred to the soil carbon pools and are reflected in the corresponding soil C accumulation curves (see below). Direct inputs of biomass carbon to soil pools occur during disturbances (see below), at which time a series of simulations redistributes carbon between the various pools and the releases to the atmosphere.

Whenever stands were classified by maturity classes, net carbon accumulation curves as a function of stand age have been developed from the biomass data for each combination of the above classifiers. The underlying assumption is that the biomass data represent chronosequences from which biomass accumulation curves can be derived and that these curves will adequately describe current biomass accumulation rates. While on a national scale they represent the best data presently available, the limitations of these biomass accumulation curves are recognized. In particular, this chronosequence approach is not appropriate to predict future carbon accumulation curves as these may be affected by forest management and global change. In the next phase of this project, biomass accumulation curves will be derived using a more process-oriented modelling approach.

For the model execution, several look-up tables for carbon content and net annual change in biomass C are generated from the biomass accumulation curves. These look-up tables correspond to the six classifiers assigned to each spatial unit and contain data at five-year time steps. Biomass C accumulation curves are currently independent of the disturbance type which preceded stand establishment because insufficient data are available to indicate differences in net biomass accumulation rates following different types of disturbances.

Soils

Data describing soil carbon pools and their rates of change are extremely sparse. In many temperate and boreal forest ecosystems two-thirds or more of total carbon is in the soil component. Carbon uptake and carbon release are closely tied to vegetation dynamics and climatic conditions, but little is known, on a continental scale, about soil carbon dynamics following disturbances.

Four soil carbon pools are recognized in the model: fast, medium, and slow pools, as well as a "submerged" pool. The first three pools represent (1) fine litter, including foliage and branches with a retention time of zero to 15 years; (2) coarse litter, including woody materials greater than 10 centimeters (cm) in diameter with a retention time of 15 to 100 years; and (3) partially decomposed soil organic matter with retention times of over 100 years. The submerged pool is designed to account for forest ecosystems with substantial amounts of peat accumulation, characterized by extremely slow turnover rates due to waterlogged conditions. The proportion of gaseous C released as CH_4 is assumed to be greater in the submerged soil C pool. This structure of soil C pools will enable us to analyze the impacts of changes in the conditions which lead to the formation of peat, either through altered climatic conditions, flooding of soils, or through drainage of peatlands.

Soil C dynamics are simulated by a simple sub-model within the analytical framework. Soil carbon pool sizes are determined by the present average climatic conditions within a given ecoclimatic province, and the amount of above-ground biomass at a forest site type. The submodel calculates soil C content in each of the four C pools as a function of inputs from biomass C pools, decomposition, and movement of carbon from the fast and medium soil C pools to the slow soil C pool. Decomposition rates are linked to the ratio of current biomass C to maximum biomass C, with biomass estimates being derived from the biomass accumulation curves. Following disturbances, when the ratio of biomass C to maximum biomass C, is reduced, decomposition rates are increased to reflect the accelerated loss of soil C typically observed after disturbances.

Soil C content and annual net change in soil C are presented in look-up tables for all forest ecosystem types throughout Canada at five-year time steps. Although the soil C look-up tables are identical to the biomass C look-up tables in the structure of classifiers used, they have an additional level of stratification for the five disturbance types described below. This recognizes the differing impacts of disturbances on the size of the soil C pools at stand establishment and on the subsequent C dynamics.

The soil C look-up tables are generated in a two-pass process: in the first pass, soil C dynamics are simulated starting with the fast and medium soil C pools set to zero and the slow pool to a value determined from a regression model of soil C content as a function of temperature and precipitation (Birdsey, 1990). In the simulations, inputs into the soil C pools depend on the size of the biomass C pools as defined by the biomass look-up tables. At a user-specified maximum age, each of the five disturbance types is applied to the end conditions of the first pass of the simulation model, to generate the starting conditions for a second pass of the model. During this second pass, soil C dynamics are simulated for each of the five different disturbance types. The simulation results for the second pass provide the data for the soil C look-up tables.

Forest Products Sector
Harvesting forest ecosystems transfers biomass C from the biomass to the forest products C pools. Forest products retain C for different periods of time, depending on the type of product, its end-use, and the way by which it is disposed. In addition, the transport and production of forest products requires energy, which is frequently supplied from fossil C sources and which releases C into the atmosphere. In the analytical framework, historic harvest data and a set of assumptions about harvest efficiency, product conversion efficiency, and product types are used to build up the age-class structure of the forest products pool from which C is released.

Some forest products C pools retain C for several decades and may constitute an important component of the medium-term carbon cycle because the fate of this pool is easily affected by policy decisions. For example, emphasizing a buildup of forest products with slow C release rates may provide a carbon sink which contributes to the removal of C from the global atmosphere. The quantitative analysis of such policy options requires, however, that C release from energy utilization during production processes be accounted for. The analytical framework provides the necessary entry points for the energy analysis, although the energy data of the forest products sector are not yet included in the model.

Role of Biofuels in the Production of Energy
The release of fossil fuel carbon into the atmosphere is the major contributor to the problem of increasing atmospheric CO_2 concentrations. In global carbon cycles, there is no major pathway by which atmospheric C is returned to fossil C pools on biological time scales. Biomass burning for the generation of energy and the substitution of fossil energy sources can contribute to a reduction in the rate of the atmospheric C increase, provided that the burned biomass would otherwise have released its C content through oxidation by wildfires, slash burning, or decomposition. It must, of course, also be ensured that the harvested ecosystems are returned to a state in which net C accumulation in biomass occurs, i.e. that biomass carbon is accumulating at a rate faster than the C loss from the soil C pools.

The specific role in the forest sector of energy production from fossil C sources and its substitution potential with biomass C sources will be addressed quantitatively in the next phase of this study.

Model Integration
An integrating model has been developed to simulate the temporal and spatial dynamics of the carbon budget of the Canadian Forest Sector. Based on the C-content and net C exchange look-up tables for biomass and soil carbon pools, the rules about annual disturbances, the C content and energy use of the forest products sector, and the movement of biomass

C between the different C pools, the model integrates the net C exchange between the forest sector and the atmosphere. The model generates a comprehensive list of indicator variables, summarized at several spatial scales.

Application to Policy Questions

The model of the carbon budget of the Canadian Forest Sector provides the tool required for the assessment of the impact of biomass burning on the atmospheric C content. The framework places C emissions from wildfires and other disturbances in relation to the net C uptake of the previously disturbed forests. Fires create the growing space and release nutrients required for forest growth. The regrowing forest, in turn, sequesters carbon from the atmosphere, offsetting that released during fires. Forest harvesting disturbs the forest, creates growing space, and removes C from the forest ecosystems without releasing all of it into the atmosphere.

The ability of forests to sequester C is recognized in proposals to use large-scale reforestation projects to offset the atmospheric C increase (Marland, 1988; Sedjo, 1989a, 1989b). Wildfires and other disturbances will limit the amount of C stored in the forest ecosystem and the duration for which it can be removed from the atmosphere. Furthermore, considerable amounts of fossil fuel, biomass, and soil carbon may be released during site preparation, planting, stand tending, and fire-protection activities. The long-term net carbon uptake of proposed reforestation efforts can only be assessed quantitatively within an analytical framework of the type described here. Such analyses are currently in progress for a variety of policy questions relating to forest sector activities and their impacts on the atmospheric C budget.

Climatic change and its predicted impacts on fire frequencies and intensities, forest growth, and soil C dynamics will potentially affect the net C balance of the Canadian forest sector. The analytical framework has been developed such that after some additions to the current structure an analysis of climatic change impacts on the carbon budget can be conducted using a what-if scenario approach.

An example of the policy questions related to biomass burning that can be addressed within the analytical framework relates to fire protection and reforestation policies. Strictly from the question of carbon budgets, is it more effective to spend efforts on fire protection or on reforestation of burned areas? From a qualitative viewpoint, it is clear that the an-

swer to such a question depends on both the temporal and spatial scales at which this problem is addressed. In the short term, fire protection efforts will be more effective in retaining carbon in the ecosystems, but as the biomass C content builds up, the probability of fire increases as does the amount of C released at the time of disturbance. If climatic conditions change such that fire frequencies increase, the question arises as to when and if fire protection efforts should be reduced and replaced by more intensive reforestation programs. In a simplified policy scenario, forest harvesting could be used to transfer biomass carbon to the forest products sector before fire releases the C into the atmosphere. Clearly, ecosystem characteristics, forest access, economic criteria, and the retention of C in forest products must also be considered when exploring such policy questions. In a future publication we will report on the quantitative assessment of these questions using the model described here.

A second policy question which is receiving increasing attention relates to the utilization of forest biomass for the generation of energy and the substitution of fossil energy carriers. Many industries in the forest products sector—for example, the pulp and paper industry—are already using biomass-derived by-products of the production process for the generation of energy. There appears to be considerable potential for an increase in bioenergy production in the Canadian Forest Sector, but the quantitative implications for the carbon budget have not yet been analyzed. There are two primary considerations: (1) What would have been the fate of the biomass had it not been burned for the production of energy, and (2) does the energy supplied from biomass substitute for fossil energy? To assist the analysis of the second question, a study of the energy use in various parts of the Canadian Forest Sector is currently being prepared and these data will be used in a quantitative analysis with the carbon budget model.

Future Research

Preliminary analyses of the available data indicate that by far the greatest uncertainties are associated with the available information regarding soil C dynamics. Many of Canada's soils contain large quantities of C, but little is known about net C accumulation and the impacts of disturbances on soil C dynamics. Long-term C accumulation rates in Canada's peatlands (about 111 million hectares, ha) are estimated at 28 g m^{-2}, which amounts to 0.03 petagrams (Pg) annually (Gorham, 1988). This estimate is derived from measuring peat thickness and age, and reflects the average accumulation rate of 0.51 mm year^{-1} over the past 4100 years. Current and future climatic conditions could be affecting net C accumulation rates in peatlands and forest soils, and analyses based on historic data alone must be treated with great caution.

To illustrate this point, consider that approximately 440 million ha of land are classified as forest in Canada's forest biomass inventory (Bonnor, 1985). If net C accumulation rates in this forest land changed, on average, by 100 g m^{-2} year^{-1}, the global carbon budget would be altered by 0.440 Pg. This represents a substantial percentage of the C missing in global atmospheric C budgets (cf. Tans et al., 1990). Future research must address these uncertainties in net carbon accumulation rates in terrestrial ecosystems if global carbon budgets are to be better understood.

Detailed sensitivity analyses of these and other aspects of the carbon cycle of the Canadian forests will be undertaken in the next phase of this study. Three separate sensitivity analysis issues can be identified: global research issues as mentioned above, implications of data and algorithm uncertainties in the model, and the potential mitigative effects of alternate resource management and policy options.

With respect to scientific uncertainties, the area of greatest concern is in soil C dynamics. However, for the analysis of policy options, uncertainties in data and model formulation should not prevent a relative ranking of different options to assist in the decision-making process. Nevertheless, where resource management might directly affect soil C dynamics—for example, in short-rotation energy plantations—an improvement in the understanding of soil C dynamics should be gained to better assess the carbon cycle implications of management options. In lieu of data which may not be available in the near future, high and low estimates can be used to bound the possible range of data uncertainties.

The treatment of spatial entities within the analytical model will permit the analysis of carbon budgets at variable spatial scales such as a forest region, a forest management unit, a watershed, or a single stand. Although at the scale of individual stands, better representation by detailed ecosystem simulation models is possible, these models typically do not account for the fate of the harvested material and are not easily aggregated to the larger spatial scales where the carbon cycle impacts must be assessed. One reason for focusing the current model on the smaller spatial scales would be to facilitate validation activities and to allow direct comparison with the more detailed site-specific models.

Conclusions

Model structures of entire systems, such as the one described here, are required to quantitatively explore the current status quo, the consequences of various policy options, and future developments of regional carbon budgets in a meaningful way. Disturbances such as wildfires should not be evaluated only in terms of their carbon release rate, but should rather be analyzed in an integrated context which recognizes their role in the large-scale, long-term productivity of forest ecosystems. This is particularly true for fire-adapted boreal forests. Analyses of the impacts of policy options on the carbon budget can easily be misleading if they fail to account for all relevant carbon fluxes associated with the proposed management activity.

Acknowledgments

This project is funded by the Canadian Federal Panel on Energy Research and Development (PERD) through the ENFOR (Energy from the Forest) program of Forestry Canada. We thank the following experts from Forestry Canada, several universities, and the forest industry who gave freely of ideas and data at a three-day workshop sponsored by Forestry Canada: J. Balatinecz, M. Bonnor, L. Brace, J. Cihlar, J. Dobie, P. Hall, M. Harmon, O. Hendrickson, K. Lertzman, T. Lekstrum, J. Lowe, V. Mathur, J. Nyboer, D. Parkinson, D. Pollard, J. Richardson, B. Stocks, I. Simonsen, T. Trofymow, J. Volney, and S. Zoltai. We are also grateful for the assistance of J. Lowe, K. Power, and S. Gray, who supplied the data of the national biomass inventory.

Modeling Trace Gas Emissions from Biomass Burning

John A. Taylor and Patrick R. Zimmerman

Biomass burning is now recognized as a globally substantial source of a number of trace gases, including carbon dioxide, methane, carbon monoxide, nitrogen dioxide, nitrous oxide, and nonmethane hydrocarbons (Seiler and Crutzen, 1980; Crutzen et al., 1979; Fishman et al., 1986; Greenberg et al., 1984). Biomass burning also results in a substantial photochemical production of tropospheric ozone due to the release of ozone precursor compounds during biomass burning (Crutzen et al., 1979; Delany et al., 1985; Crutzen, 1987). Recent measurements of ozone in the tropical troposphere by Logan and Kirchhoff (1986) and the analysis of satellite total ozone data by Fishman et al. (1986) and Fishman (1988) support the hypothesis that biomass burning contributes significantly to tropospheric ozone formation.

Biomass burning has been recognized for some time to be a source of atmospheric methane (Crutzen et al., 1979) and nonmethane hydrocarbons (Greenberg et al., 1984). However, the spatial distribution, the seasonal pattern, and the magnitude of the emission of methane and other trace gases have been the subject of only a few studies (Seiler and Crutzen, 1980; Crutzen et al., 1985; Crutzen, 1987). Given the range of environmental factors which may affect the amount of biomass burned and the resulting trace gas emissions, only limited confidence can be placed in current estimates of total trace gas emissions from biomass burning.

Estimating the Emissions of Trace Gases from Biomass Burning

Estimating the emissions of trace gases from biomass burning can be achieved by measuring the flux of trace gases directly, by estimating the various contributing components to biomass burning such as the total land area, or by modeling the trace gas emissions. These methods all involve problems and uncertainties which are difficult to resolve.

In the case of direct measurement of the flux of trace gases during biomass burning it is not possible to sample all fires or to overcome the problems of spatial and temporal inhomogeneity. Direct sampling does provide important estimates of the flux of trace gases so that the relative strength of a source can be determined. Information on the patterns of variation of sources of trace gases can be gained using this approach. Improving the estimates of trace gas emissions by direct measurement can be achieved by increasing the spatial and temporal coverage of the sampling program.

Estimating trace gas fluxes by calculating and compiling statistical data on the individual components contributing to the total amount of biomass burned is achieved by estimating key factors such as the land area burned and organic matter available for combustion. Trace gas emissions are then computed as ratios of the estimated amount of CO_2 produced. This approach was adopted by Seiler and Crutzen (1980). They assumed that the total amount of biomass, M, burned annually in a biome is approximately given by the equation:

$$M = A \times B \times \alpha \times \beta$$
$$[M \text{ in units of g dry matter per year (g dm/yr)}] \quad (41.1)$$

where A is the total land area burned annually (m^2/yr), B is the average organic matter per unit area in the individual biomes (g dm/m^2), α is the fraction of the average above-ground biomass relative to the total average biomass B, and β is the burning efficiency of the above-ground biomass. Seiler and Crutzen (1980) note that the most uncertain parameters in equation (41.1) are the total land area burned annually and the burning efficiencies of the biomass in the individual ecosystems, whereas a better-known parameter, phytomass per unit/area, must be uncertain by maybe 20% to 40%.

We can evaluate the uncertainty in calculating M using the following standard expression:

$$\left(\frac{\Delta M}{M}\right)^2 = \left(\frac{\Delta A}{A}\right)^2 + \left(\frac{\Delta B}{B}\right)^2 + \left(\frac{\Delta \alpha}{\alpha}\right)^2 + \left(\frac{\Delta \beta}{\beta}\right)^2 \quad (41.2)$$

where ΔM, ΔA, ΔB, $\Delta \alpha$, and $\Delta \beta$ represent the uncertainty in the quantities M, A, B, α, and β, respectively. Equation (41.2) shows that if the uncertainty in any one component is large then a similar uncertainty in M will result. If all components have large uncertainties then very large uncertainties in M will result. Further uncertainty is introduced when the proportion of trace gas to biomass burned, M, is taken into account. Taking an average value of 20% to 40% for each of the terms in the right side of equation 41.2 and adding a further term ($\Delta f/f$), where f is the ratio of M to the release mass of trace gas of interest during biomass burning and assuming f is known to within 20% to 40%, leads to a range of uncertainty in M of between 43% to 89%.

This estimate of the uncertainty associated with M is consistent with the results obtained in other studies. For example, Lamb et al. (1987) estimated that problems associated with compiling inventories of biogenic hydrocarbon emissions lead to uncertainties in total emissions of the order of $\pm 210\%$, or a factor of three.

Modeling Approach

As illustrated above, much uncertainty remains in the estimation of the total global sources of trace gas emissions from biospheric sources, including biomass burning, and many anthropogenic sources, due to the difficulty of extrapolating trace gas flux measurements to the global scale. However, while the global total emissions from a particular source may be poorly quantified, the spatial distribution of sources and their seasonal variability may be much more accurately determined. Accordingly, a modeling approach is being developed to enable studies of the global total emissions of trace gases from individual sources such as biomass burning within chemical transport models (CTMs). The approach also allows the total emission of a trace gases to be adjusted as improved information becomes available. Should alternative hypotheses be developed they may readily be tested within the CTM.

In this chapter we use a generic approach to the problem of modeling trace gas emissions from biomass burning. This approach has been developed to enable the ready coupling of an emission model with CTMs. The approach allows selection of a range of total trace gas emissions from biomass burning to be used within a CTM, while still retaining the same spatial and temporal variation. Model predictions are compared with observed atmospheric concentration data. This approach facilitates the application of constrained inversion techniques where the relative spatial and temporal distribution of a trace gas acts as prior information to constrain the determination of the total global emission of trace gases from that source. The modeling approach allows the release of multiple trace gases to be treated in a consistent manner.

An important component of models which describe trace gas releases from the biosphere is the identification of the key climatic variables which drive the emissions. In the case of biomass burning we use precipitation to determine the seasonal variation in biomass burning. The advantages of this approach are that interannual variability based on changes in weather patterns may be computed which will allow feedbacks to the climate system to be studied. Clearly, if we are to anticipate the effects of climate change then feedbacks between climate and the biospheric processes governing trace gas emissions must be included in CTMs and climate models. However, if we are to avoid the problems of extreme computational expense, and problems of understanding arising from undue complexity of the model formulation, then simple models identifying key climatic variables must be developed.

Obviously, as our ability to quantify total trace gas releases improves, then further development of models of the spatial and temporal releases of trace gases will be required. Satellite observations of biomass burning are particularly well placed to assist here. However, the basic approach of determining the relative distribution of the flux of a trace gas from a biospheric source linked to important climatological processes such as precipitation should be retained.

Model Description

Henderson-Sellers et al. (1986) have examined global land-surface data sets for application in climate-related studies. Each land-surface data set was compiled using different methods. This variation in methodology is reflected in the wide range and number of vegetation types used to classify land cover and the differences in the resolution at which data sets were collected. In no case have the data sets examined by Henderson-Sellers et al. (1986) been prepared with the aim of estimating the flux of trace gases to the atmosphere.

In most cases it is difficult to evaluate the uncertainties associated with the underlying data sources due to the great variability in the origins, content, purpose of the data collector, resolution, and time of preparation (Matthews and Fung, 1987). For example, Matthews and Fung (1987) note that the evaluation of wetland areas is complicated, particularly in tropical regions, by the seasonal variation in inundated areas and whether the information recorded on maps reflects the minimum, mean, or maximum area inundated.

One source of systematic error associated with these land-cover data sets arises from the approach of reporting only the major ecosystem present with a grid cell (>50% land cover). This leads to a systematic underestimation of land cover for an ecosystem that is intermixed with another more dominant ecosystem and systematic overestimation if we assume that the dominant ecosystem represents 100% of the land cover when other ecosystems are present. It is hoped that, when taken over each region and the globe, these effects will average out. One solution to this problem is to increase the resolution of the data set. In this way land-cover data sets of 1° and 0.5° resolution are now common (Wilson and Henderson-Sellers, 1985; Olson et al., 1983). This approach to the problem of the resolution of ecosystem types is certainly a good compromise when faced with determining the actual percentage of land cover for each vegetation type within every grid cell.

Other approaches to the problem of properly representing mixed ecosystems include the introduction of additional categories which represent, for example, the intermixing of nonwooded vegetation types with wood or forest cover (Olson et al., 1983). Wilson and Henderson-Sellers (1985) included information on both a primary and a secondary vegetation category. The primary land-cover class was considered to represent the land cover that occupies ≥50% of a 1° × 1° grid area, while the secondary land cover represents cover types occupying 25% to 50% of the grid area.

In the modeling study reported here we have employed the data set developed by Wilson and Henderson-Sellers (1985). This data set was selected because it includes both primary and secondary classifications of land cover, and its land-cover classes could be easily related to methane source regions. Table 41.1 lists the Wilson and Henderson-Sellers (1985) land-cover classes.

In order to construct a spatial distribution for the release of trace gases from biomass burning, the vegetation types of Wilson and Henderson-Sellers (1985) where biomass burning occurs were identified. These regions are tropical savanna, tropical grassland plus shrub, open tropical woodland, tropical pasture,

Table 41.1 Land-cover classes

Code	Type
00	Open water
01	Inland water
02	Bog or marsh
03	Ice
04	Paddy rice
05	Mangrove
10	Dense needleleaf evergreen forest
11	Open needleleaf evergreen woodland
12	Dense mixed forest
13	Open mixed woodland
14	Evergreen broadleaf woodland
15	Evergreen broadleaf cropland
16	Evergreen broadleaf shrub
17	Open deciduous needleleaf woodland
18	Dense deciduous needleleaf forest
19	Dense evergreen broadleaf forest
20	Dense deciduous broadleaf forest
21	Open deciduous broadleaf woodland
22	Deciduous tree crop
23	Open tropical woodland
24	Woodland plus shrub
25	Dense drought deciduous forest
26	Open drought deciduous woodland
27	Deciduous shrub
28	Thorn shrub
30	Temperate meadow and permanent pasture
31	Temperate rough grazing
32	Tropical grassland plus shrub
33	Tropical pasture
34	Rough grazing plus shrub
35	Pasture plus tree
36	Semi-arid rough grazing
37	Tropical savanna
39	Pasture plus shrub
40	Arable cropland
41	Dry farm arable
42	Nursery and market gardening
43	Cane sugar
44	Maize
45	Cotton
46	Coffee
47	Vineyard
48	Irrigated cropland
49	Tea
50	Equatorial rainforest
51	Equatorial tree crop
52	Tropical broadleaf forest
61	Tundra
62	Dwarf shrub
70	Sand desert and barren land
71	Scrub desert and semidesert
73	Semidesert and scattered trees
80	Urban

Note: Wilson and Henderson-Sellers (1985). These classes and their associated land areas and locations were used in the present study to generate trace gas source functions.

thorn shrub, and cane sugar (codes 37, 32, 23, 33, 28, and 43 as listed in Table 41.1). To seasonally distribute the flux of a trace gas from biomass burning, a simple model based on precipitation was constructed. It was assumed that the flux of methane from biomass burning varied inversely with respect to precipitation. The results of Kirchhoff et al. (1989) clearly show that biomass burning in South America occurs during periods of low precipitation (from July to October).

Based on the above assumptions, the flux of methane attributed to biomass burning at each grid cell with longitude i, latitude j, and month t, was computed as follows:

$$f_{ijt}^{CH_4} = f_{ij}^{CH_4}[P_{ij}^{tot}/P_{ijt}]/\left[\sum_{t=1}^{12} P_{ij}^{tot}/P_{ijt}\right] \quad (41.3)$$

where P_{ij} is precipitation (mm), P_{ij}^{tot} is the annual precipitation in a grid square, and $f_{ij}^{CH_4}$ is the estimated annual total release of a trace gas to the atmosphere from biomass burning at that grid cell, which is the simple fraction of total biomass burning area represented by the grid cell multiplied by the annual total release of methane, derived from a study by Taylor et al. (1990).

Precipitation data are available as monthly means on a $2.5° \times 2.5°$ grid. Shea (1986) has prepared global maps of precipitation based on a compilation of data from a number of sources covering the period 1950 to 1979. Shea (1986) notes that precipitation data for the southern oceans are particularly unreliable. Fortunately we only wish to compute trace gas emissions from biomass burning over the land areas, for which the precipitation data are most reliable.

Model Results

Figure 41.1 shows the distribution of the flux of methane computed using equation (41.3) for the months of March, July, and October. Based on ozone measurements (Fishman, 1988), biomass burning over South America should be at a minimum in March, and July should be the commencement of the biomass burning season. The model results are in good agreement with ozone measurements (Fishman, 1988) and CO observations (Fishman et al., 1986). The October maximum, when biomass burning has been observed to be at its most extensive over the savannas of Africa and South America, is also clearly reproduced by the model. The model also predicts a significant contribution of methane emissions from biomass burning over Southeast Asia and the tropical north of Australia, as

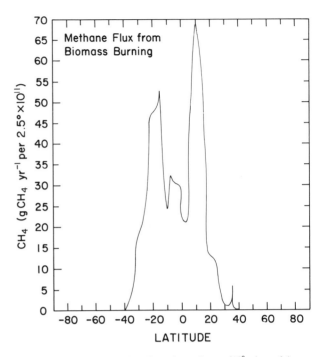

Figure 41.1 Model-predicted methane fluxes (10^8 g/month/$2.5°$ grid cell) for the months of March, July, and October.

would be expected (Seiler and Crutzen, 1980). It should also be noted that Fishman et al. (1986) found that the highest equatorial total ozone values are nearly always located near the Gulf of Guinea between July and November. The biomass burning model predicts high emissions over large areas of southern Africa, along the coast of the Gulf of Guinea in particular, during the period July to November, inclusive.

Figure 41.2 shows the computed annual average latitudinal distribution of the flux of methane to the atmosphere from biomass burning averaged over $2.5°$ latitudinal bands. The figure shows that methane released from biomass burning is restricted to the tropics. Clearly, biomass burning also occurs at higher latitudes; however, the frequency and extent of biomass burning at these latitudes is such that only a very small release of trace gases occurs as a consequence of biomass burning, when compared with the tropical releases. For example, Seiler and Crutzen (1980) estimated that the biomass burned annually in temporal and boreal forests represented ~4% of the total annually burned biomass.

The trace gas emissions model for biomass burning has been applied as part of a comprehensive study of the global budget and distribution of atmospheric methane using a global tropospheric tracer transport

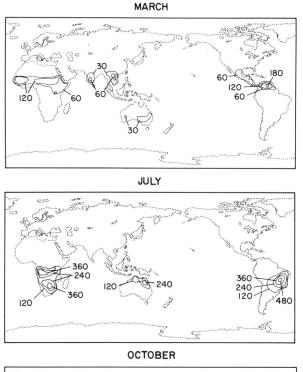

MARCH

JULY

OCTOBER

Figure 41.2 Model-predicted annual averaged latitudinal distribution of the flux of methane from biomass burning per 2.5° of latitude.

Table 41.2 Calculated global total emissions from biomass burning of selected trace gases assuming an annual global total methane emission of 63.4×10^{12} g and relative volume measurements of Crutzen et al.(1979) and Greenberg et al. (1984)

Trace gas	Ratio by volume to CH_4	Estimated annual emission (10^{12} g yr^{-1})
H_2	2.06[a]	16
CO	8.75[a]	971
	12.9[b]	1431
NO_x	0.29[a]	34 (as NO)
COS	9.88×10^{-4a}	0.13 (as S)
CH_3Cl	1.46×10^{-3a}	0.21 (as Cl)
NMHC	1.22[b]	77 (as CH_4)

a. Derived from Crutzen et al. (1979).
b. Derived from Greenberg et al. (1984).

CO, NO_x, COS, CH_3Cl, and total NMHCs. It should be noted that Crutzen et al. (1979) collected samples in stainless steel containers. An estimate of N_2O emissions does not appear in Table 41.2 due to problems associated with the storage of N_2O in stainless steel containers. Lyon et al. (1989) provide a discussion of this problem.

An estimate of the total carbon consumed during biomass burning can also be computed using the estimated annual methane emission. Based on the estimates of CH_4 to CO_2 volume ratios, Crutzen et al. (1979) and Greenberg et al. (1984) estimates of the total annual C from biomass burning are 4090×10^{12} g C yr^{-1} and 6835×10^{12} g C yr^{-1}, respectively.

Satellite Observations and Trace Gas Emissions from Biomass Burning

The success of the model of trace gas emissions from biomass burning, as described in this chapter, in reproducing the known qualitative information on the spatial and temporal variation of biomass burning indicates that further development of this approach is warranted. Aspects of modeling biomass burning which require further investigation are the interannual variability of trace gas emissions, the prediction of fuel load by estimating ecosystem productivity, and comparing trace gas concentrations predicted by global three-dimensional models with available observations with global coverage. If we are to determine the net impact of biomass burning on the emissions of trace gases to the atmosphere, then we must also estimate the fluxes of trace gases from the

model (Taylor et al., 1990). An annual average release of 63.4 Tg of CH_4 in combination with releases from other major sources of methane, as listed in Taylor et al. (1990), was found to be consistent with available atmospheric methane observations. Using the fluxes of trace gases relative to methane measured by Crutzen et al. (1979) and Greenberg et al. (1984), we can compute the global release of the trace gases, H_2, CO, NO_x, COS, CH_3Cl, and nonmethane hydrocarbons. Greenberg et al. (1984) also estimated the proportion of alkanes, alkenes, alkynes, aromatics, and oxygenated hydrocarbons included in their total nonmethane hydrocarbons. Table 41.2 lists the calculated global total emissions for the trace gases, H_2, natural unburned biosphere and from the biosphere after burning.

Satellite data can play an important role in conjunction with theoretical research programs whose objectives are to develop detailed models of the emissions of trace gases from biomass burning. Satellite data could be used in the following areas:

• To assist in determining the spatial distribution of biomass burning and the intensity of fires;

• To provide data on the interannual variability of biomass burning, even initially at a crude level, given the possible magnitude and importance of this variability;

• To provide leaf area index data or some other measure of total biomass or net primary productivity (NPP), particularly in grassland areas, so that estimates of fuel load and its spatial and temporal variability can be derived; and

• To provide trace gas observations with global coverage to be employed to calibrate models of flux of trace gases from biomass burning.

Limitations associated with satellite observations are that observed wavelengths become saturated at fairly low intensities, that small burned areas and fires will be below the instrument resolution, and that cloud cover will prevent collection of observations.

Conclusions

A high-resolution model describing the spatial and temporal variation of trace gas emissions from biomass burning has been developed. The trace gas emissions model has been applied within a global tropospheric tracer transport model. The model-predicted atmospheric concentrations have been found to be in good qualitative agreement with available observational data. Estimates of the total global emissions of selected trace gases have been derived. Finally, satellite observations important to the further development of mathematical models of the emissions of trace gases from biomass burning have been identified.

Acknowledgments

The authors would like to thank Jim Greenberg, Guy Brasseur, and Ralph Cicerone for useful discussions and comments on an earlier draft of the paper. The authors would like to thank Donna Sanerib for preparation of the typescript.

Impact of Biomass Burning on Tropospheric CO and OH: A Two-Dimensional Study

Jose M. Rodriguez, Malcolm K. W. Ko,
Nien Dak Sze, and Curtis W. Heisey

Removal of CO occurs in the troposphere through the reaction

$$CO + OH \rightarrow CO_2 + H \tag{42.1}$$

which is also the major pathway for converting OH to other HO_x species. The atmospheric concentration of the hydroxyl radical determines the removal rate of many important long-lived species in the troposphere, including many of the hydrochlorofluorocarbons (HCFCs) and hydrofluorocarbons (HFCs) which have been proposed as possible substitutes for the chlorofluorocarbons (CFCs). Changes in the OH concentration will alter the lifetimes, ozone depletion potentials, and greenhouse warming potentials of these species (AFEAS, 1990).

Oxidation of methane occurs (via multiple steps) through

$$CH_4 + OH \rightarrow \ldots \rightarrow CO + products \tag{42.2}$$

Methane is an important greenhouse gas in the troposphere. It also modulates the response of stratospheric O_3 to future increases in chlorine by regulating the partitioning between HCl and active chlorine species in the stratosphere. Reaction (42.2) could be a source or sink of HO_x in the troposphere, depending on the concentration of NO_x (see, e.g., Logan et al., 1981). We also note that reactions (42.1) and (42.2) introduce a strong coupling between CO, OH, and CH_4 in the troposphere (Sze, 1977), since (42.2) is also an important source of CO. Changes in the CO-OH-CH_4 system are modulated by the NO_x concentrations through the reactions

$$H + O_2 + M \rightarrow HO_2 + M \tag{42.3}$$

$$HO_2 + NO \rightarrow OH + NO_2 \tag{42.4}$$

which determine to first order the partitioning between OH and HO_2. Finally, reactions (42.1), (42.3), and (42.4), followed by

$$NO_2 + h\nu \rightarrow NO + O \tag{42.5}$$

$$O + O_2 + M \rightarrow O_3 + M \tag{42.6}$$

constitute an important source of tropospheric O_3.

Recent measurements of the emission of CO from biomass burning in the tropical regions (Crutzen et al., 1985; Andreae et al., 1988) have indicated that this process contributes about 30% of the global CO budget and is comparable to other known natural (oxidation of CH_4 and natural nonmethane hydrocarbons) and anthropogenic (commercial energy use) sources (Logan et al., 1981; WMO, 1985). On a regional basis, it may well be the primary source of CO in the tropical region during the burning season and contribute to the enhanced abundances of tropospheric O_3 deduced from satellite data during this season (Watson et al., this volume, Chapter 14). Changes in rates of biomass burning due to deforestation and agricultural practices could then have an important impact on future trends of tropospheric concentrations of CO, OH, CH_4, and O_3. There is thus a need for modeling studies and assessment of the impact of future changes in the above gases. Previous efforts include studies by Sze (1977), Thompson and Cicerone (1986), Levine et al. (1985), using one-dimensional (1-D) models; Crutzen and Gidel (1983), and Isaksen and Hov (1987), using two-dimensional (2-D) models; and Pinto et al. (1983), using a general circulation model.

We present here results from a 2-D model of the troposphere and stratosphere. We will first discuss briefly the modeling approach. The model was used to derive a budget for CO and to assess the contribution of different sources in maintaining the observed CO concentrations. We also present results of a time-dependent calculation designed to evaluate the impact of present-day biomass burning on CO, CH_4, and OH, and the time constants for the system to respond to changes in the biomass source.

Model Description

Utilization of a two-dimensional zonal-mean model for assessment of tropospheric changes implicitly assumes that we can realistically represent the basic chemical and transport mechanisms by zonal averages. In reality, one may expect large longitudinal asymmetries in sources and sinks of tropospheric source gases, differences in the chemistry of marine and continental environments, and variations in the meridional transport. In addition, important transport processes such as deep convection may play an important role in the redistribution of source gases over large regions. Nevertheless, previous two-dimensional studies (e.g., Crutzen and Gidel, 1983; Isaksen and Hov, 1987) have shown that these models provide a realistic tool for evaluating the global budgets and trends of tropospheric species.

Details of the AER 2-D model have been given elsewhere (Ko et al., 1985, 1989). Briefly, the model incorporates complete chemistry for the O_x, HO_x, Cl_x, NO_x (including PAN), and Br_x families, as well as complete oxidation schemes for CH_4 and C_2H_6. The radiation field for the present calculations assumes multiple scattering for clear sky conditions. A uniform washout rate is adopted, with an average removal time constant of about two weeks for soluble species.

Model resolution is 10° in latitude and 17 levels in the vertical dimension, from the surface to the stratopause. The present calculations have been carried out in constant pressure coordinates. The vertical grid size is about 3 kilometers (km); the number of levels in the troposphere ranges from 5 at the tropics to 3 at high latitudes. Since the midpoint of the lowest grid box is at an altitude of 1.5 to 2 km, we do not include a boundary layer in the present calculations. Boundary conditions and results at the lowest level must then be interpreted as pertaining to the free troposphere above the boundary layer.

Stratospheric transport in the model occurs through advection by the diabatic circulation and parameterized vertical and horizontal eddy coefficients. Derivation of the diabatic circulation from heating rates and validation of the stratospheric transport parameterization are described in Ko et al. (1985). Little attention has been paid to adjusting the tropospheric circulation in previous modeling work, since the interhemispheric exchange rate for the troposphere could be adequately modeled by appropriate choice of the horizontal mixing coefficient K_{yy}. The K_{yy} in the troposphere has a value of 1.5×10^{10} $cm^2 s^{-1}$. These values have been chosen to give agreement between calculations and the observed latitudinal gradients in CFC-11 from the ALE/GAGE data (Prinn et al., 1983; WMO, 1985). The model time constant for interhemispheric transport is about one year. A value of 10^5 $cm^2 s^{-1}$ is adopted for K_{zz} in the troposphere.

We adopt observed concentrations of CO and CH_4 as lower boundary conditions to our 2-D model. We then calculate global fields for OH, and lifetimes for CH_3CCl_3, $CHClF_2$ (HCFC-22), CH_4, and CO. Agreement between our calculated lifetimes and those derived from observations (Prinn et al., 1987) and three-dimensional (3-D) models (e.g., Prather, 1990) can serve as partial validation of the derived OH field. We then use the model to calculate sources of CO needed to maintain the observed mixing ratios which are imposed as boundary conditions. Consistency between our calculated budget for CO and existing budget studies (e.g., Logan et al., 1981; WMO, 1985) is taken as an additional criterion for the validity of our calculations.

Figure 42.1 shows the adopted mixing ratio boundary conditions for CO at the lowest grid point (about 2 km). These mixing ratios are taken from the climatology compiled by Spivakovsky et al. (1990). We have increased their mixing ratios at southern midlatitudes by 20% to 30%, in order to obtain positive CO fluxes in our model calculations for the latitude band centered around 28°S. Such corrections are within the uncertainty in the measurements from which the climatology was derived. Present-day inorganic chlorine and bromine concentrations (3.3 parts per billion by volume, ppbv, and 14 parts per trillion by volume,

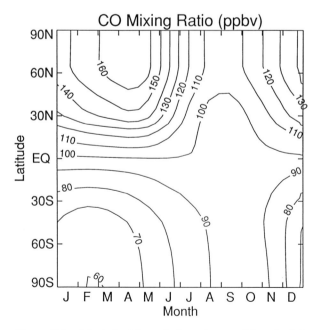

Figure 42.1 Adopted mixing ratio boundary conditions (ppbv) for CO at the lowest model grid point (about 2 km). Values are taken from climatology of Spivakovsky et al. (1990), with modifications described in text.

pptv, respectively, in the upper stratosphere) are assumed. Fixed mixing ratio boundary conditions at the lower boundary are adopted as follows: CH_4, 1.7 parts per million by volume (ppmv) in the Northern Hemisphere, decreasing to 1.6 ppmv in the Southern Hemisphere; N_2O, 310 ppbv; C_2H_6, 1.8 ppbv; O_3, 20 ppbv.

We have also considered two cases for NO_x (NO + NO_2 + $2xN_2O_5$ + HO_2NO_2) concentrations, in order to illustrate the sensitivity of our results to the adopted NO_x. The calculations denoted by "Low NO_x" adopt a fixed mixing ratio of 0.1 ppbv for total odd nitrogen (NO_x + HNO_3 + PAN) at the lowest grid point and a source from lightning in the tropics equal to 2 MT N per year (2 MT N/yr), distributed between 4 and 15 km (Ko et al., 1986). This case yields NO_x concentrations of about 6 to 12 pptv in the lower free troposphere, a factor of 2 to 3 smaller than those adopted in the Spivakovsky et al. climatology. They represent a lower bound for acceptable NO_x values. The case denoted by "High NO_x" adopts a flux of 20 MT N/yr at the surface, uniformly distributed with latitude and season. A lightning source of 4 MT N/yr in the tropics is also included. This assumption increases the calculated odd nitrogen in the troposphere by factors of 2 to 3, and yields NO_x concentrations between 20 and 30 pptv in the lower troposphere, consistent with the Spivakovsky et al. climatology.

Results

Steady-State Calculations

Model-Calculated OH and Lifetimes Figure 42.2 shows the calculated diurnally averaged OH density at the bottom grid point (about 2 km) for the "Low NO_x" (Figure 42.2a) and "High NO_x" (Figure 42.2b) cases, from a steady-state calculation. The seasonal and latitudinal pattern of the results are similar for both cases, but the absolute magnitude of the calculated concentrations are about 30% higher in the "High NO_x" case than in the "Low NO_x" results. This is a consequence of the partitioning of HO_2 and OH through reaction (42.4). Midlatitude OH values in the Southern Hemisphere are about 20% higher than in the Northern Hemisphere midlatitudes as a consequence of the interhemispheric differences in CO.

Table 42.1 summarizes results for the calculated global average tropospheric OH (averaged over volume) and lifetimes of selected long-lived trace species. The average OH concentrations for the "Low

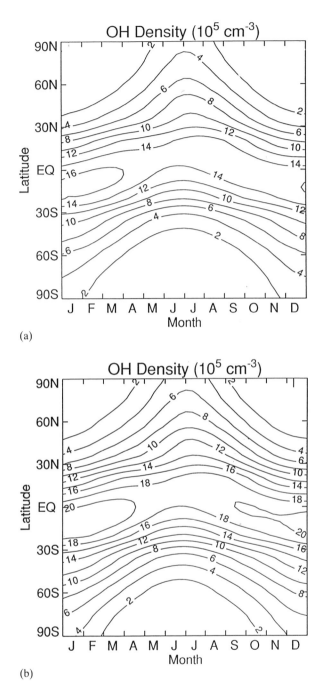

(a)

(b)

Figure 42.2 Calculated diurnally averaged number density of OH (10^5 cm^{-3}) near 2 km for (a) "Low NO_x" case and (b) "High NO_x" case.

Table 42.1 Calculated global average tropospheric OH and lifetimes of trace species

	Low NO$_x$	High NO$_x$
Global average tropospheric OH concentration (cm^{-3})	6.1×10^5	7.9×10^5
Lifetimes		
CH$_3$CCl$_3$	6.6 yrs	5.1 yrs
CHClF$_2$ (HCFC-22)	19.1 yrs	14.2 yrs
CH$_4$	12.0 yrs	8.9 yrs
CO[a]	3.0 mos	2.2 mos

a. Lifetime of CO is defined by removal due to reaction with OH.

NO$_x$" case are in better agreement with the 3-D results (6.5×10^5 cm^{-3}) of Spivakovsky et al. (1990) and Prather (1990) than the "High NO$_x$" results, which still fall within the estimated uncertainties in the 3-D analysis (about 30%). We note, however, that the latitudinal distribution of our calculated OH is somewhat different from that derived by Spivakovsky et al. (1990), due to differences in the details of the O$_3$, H$_2$O, and NO$_x$ distributions in the different models. Lifetimes for methyl chloroform bracket the 3-D results of Spivakovsky et al. (5.5 years), and are generally consistent with those derived from the ALE/GAGE data ($6.3^{+1.2}_{-0.9}$ years; Prinn et al., 1987). Lifetimes for HCFC-22 and CH$_4$ for the "High NO$_x$" case are the same as those calculated by Prather (1990). Given that the average OH calculated in the "High NO$_x$" case is larger than that of Spivakovsky et al. (1990) and Prather (1990), other factors in our model (e.g., OH distribution, temperature fields, transport) compensate for the differences in the average OH; the exact agreement in the HCFC-22 and CH$_4$ lifetimes is therefore fortuitous. Finally, the CO lifetime for the "High NO$_x$" case is consistent with the value of two months derived by Derwent and Volz-Thomas (1990).

The "Low NO$_x$" and "High NO$_x$" boundary values adopted in the calculations define the range of conditions which are consistent with existing data and analysis. Lowering the NO$_x$ further would decrease the tropospheric OH, but would yield abundances of NO$_x$ which are unacceptably small. On the other hand, concentrations of NO$_x$ higher than in the "High NO$_x$" case would increase tropospheric OH, and decrease the lifetime of CH$_3$CCl$_3$ below values consistent with the ALE/GAGE analysis.

CO Budgets The CO concentrations in the model simulations are sustained by *in situ* production initiated by oxidation of CH$_4$ (with a small contribution from C$_2$H$_6$ oxidation) and by surface sources from natural or anthropogenic processes. The CH$_4$ oxidation source yielded 602 MT CO/yr and 908 MT CO/yr for the "Low NO$_x$" and the "High NO$_x$" cases, respectively. This is to be compared with the estimate of 610 MT CO/yr from WMO (1985). The latitudinal distribution of the column-integrated production rate for the "High NO$_x$" case is shown in Figure 42.3.

The model is used to calculate surface fluxes of CO required to maintain the observed mixing ratios adopted as boundary conditions (Figure 42.1). From the continuity equation, we have for each of the bottom grid boxes:

$$\frac{dn(I,1)}{dt} = P(I,1) - L(I,1) - \frac{\partial \phi_v(I,1)}{\partial z} - \frac{\partial \phi_h(I,1)}{\partial y} \quad (42.7)$$

where n is the density (cm^{-3}) in the bottom grid box, P and L the photochemical production and loss (cm^{-3} s^{-1}), ϕ_v and ϕ_h the vertical and horizontal flux components (cm^{-2} s^{-1}), z and y vertical and horizontal (length) coordinates, and the indices $(I,1)$ denote the bottom grid box at the Ith latitude. The divergence of the vertical flux can be written in finite difference form as:

$$\frac{\partial \phi_v(I,1)}{\partial z} \simeq \frac{[\phi_v(I,1) - \phi_v(I,0)]}{\Delta z} \quad (42.8)$$

where $\phi_v(I,1)$ and $\phi_v(I,0)$ are the vertical fluxes through the top and bottom of the grid box. Equations (42.7) and (42.8) can be used to solve for $\phi_v(I,0)$, since all other terms can be determined from the imposed boundary conditions or the photochemical and transport calculations in the model. The flux through the bottom of the first grid box is given by:

$$\phi_v(I,0) = \phi_s(I) - v_d(I)n(I,1) \quad (42.9)$$

where ϕ_s is the surface flux of CO into the atmosphere due to natural or anthropogenic processes, v_d is the deposition velocity of CO at the surface, and $n(I,1)$ is the surface concentration. We adopt a value of 0.04 cm/s for the deposition velocity of CO over continents, and 0 cm/s over oceans (Logan et al., 1981), and obtain the effective deposition velocity in each latitude band by weighing these values by the relative fraction of land and ocean. Details of the latitudinal distribution of our calculated surface sources for CO are shown in Figure 42.3.

Our calculated budget for CO is compared in Table 42.2 to the budget analysis of Logan et al. (1981), as updated in WMO (1985). The magnitude of the surface sources in Logan et al. (1981) were estimated from economic and demographic data. Uncertainties in these data are about a factor of 2. Our calculated CO budget is consistent with the previous studies, both in magnitude and geographical distribution. Our calculated geographical distribution, however, is different from the recent analysis of Khalil and Rasmussen (1990), based on data from 16 stations and a mass-balance model. Our calculated total source of CO peaks in the tropical region, while Khalil and Rasmussen's source achieves a maximum near 40°N. This discrepancy may be due to differences in OH, K_{yy}, and distribution of CO mixing ratios in the two models, and points to the need for further assessment of the impact of existing uncertainties on the derived budgets.

We note that tropical sources contribute about 50%

of the total CO budget in our calculations. The fraction of these sources due to biomass burning has been estimated to be about 63% from the budget studies of Logan et al. (1981) and Crutzen et al. (1985). Given the uncertainties in these studies, however, larger fractions (close to 100%) are possible. We thus estimate that biomass burning in the tropics contributes 30% to 50% of the global CO source.

We performed additional calculations to determine the percentage of the observed CO concentrations at the bottom boundary (as shown in Figure 42.1), which can be accounted for by different sources of CO. In these calculations, the CH_4 and OH fields are fixed at the values obtained in the steady-state simulations. The change in CO is calculated when different sources are switched on and off. The results shown in Figure 42.4 have been taken from the "High NO_x" case; results for the "Low NO_x" case are similar.

The percentage of CO accounted for by methane oxidation is shown in Figure 42.4a. This source accounts for only 30% to 40% of the observed CO, and we need additional sources to explain the observations. Figure 42.4b includes both the methane oxidation and surface fluxes from natural sources. The percentage of the total source due to natural mechanisms has been taken from the study of Logan et al. (1981; Table 42.2). Methane oxidation and natural sources account for more than 70% of the observed CO in the Southern Hemisphere, but only for about 50% of the Northern Hemisphere data.

Results in Figures 42.4c and 42.4d illustrate the effects of the tropical sources. In Figure 42.4c we included all sources except those in the tropical regions ($\pm 22°$ of latitude). This case can account for most of the CO observed in the temperate zones, but only 50% to 60% of tropical CO. Figure 42.4d includes all sources, except 63% of the tropical source assumed to be due to biomass burning. In this case,

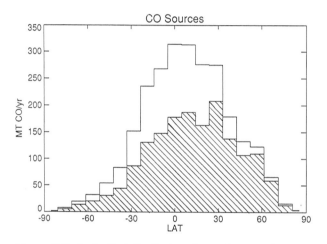

Figure 42.3 Latitudinal distribution of annual-averaged sources of CO (MT CO/yr, per 10° latitude band) calculated in the "High NO_x" case. The shaded area represents the surface fluxes. The white area denotes the *in situ* source due to methane and ethane oxidation. Both areas add up to the total source, represented by the histogram lines. The distribution pattern for the "Low NO_x" case is similar, but magnitudes are 20% to 30% lower.

Table 42.2 Surface sources of CO (MT CO/yr)

	Tropics	Northern temperate	Southern temperate	Global
Natural[a]	350 (36.5%)[b]	195 (26.2%)	110 (67.7%)	655 (35%)
Anthropogenic[a]	610 (63.5%)	550 (73.8%)	55 (32.3%)	1215 (65%)
Total[a]	960	745	165	1870
Calculated by model				
Low NO_X	664	532	135	1330
High NO_X	817	638	201	1660

Note: Biomass burning contribution to anthropogenic sources: 100% in tropics, 2.5% in temperate zones.
a. Logan et al. (1981); WMO (1985).
b. Numbers in parentheses denote percent contribution to regional total.

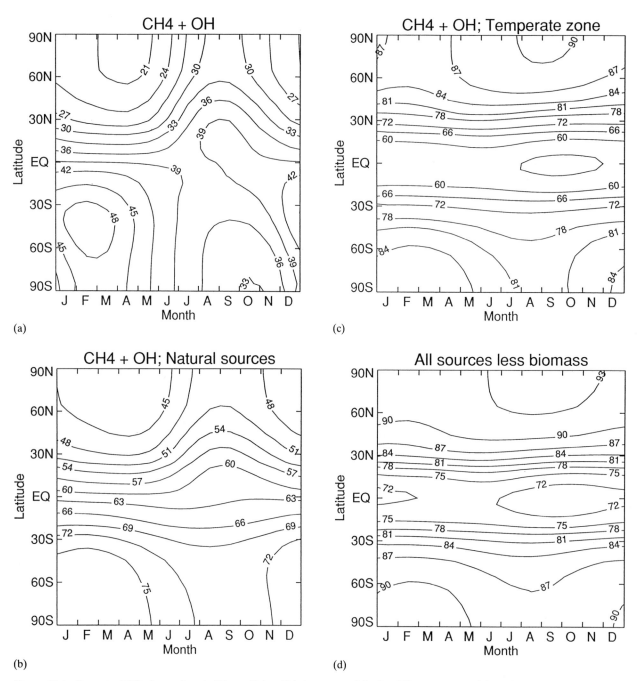

Figure 42.4 Percent of CO observations in Figure 42.1, which is accounted for by different sources: (a) methane oxidation only; (b) natural sources + methane oxidation; (c) all sources in temperate zones, plus methane oxidation; (d) all sources, except for estimated tropical biomass burning contribution. The percent of the total source due to different processes is estimated from Logan et al. (1981) and WMO/NASA (1986). See Table 42.2.

most of the observed CO in temperate zones and about 70% to 75% of the tropical observations are accounted for.

Time-Dependent Calculations

Previous assessments of time-dependent scenarios for tropospheric change are usually carried out by performing several steady-state "snapshot" calculations for conditions assumed for specific years (Thompson et al., 1989; Isaksen and Hov, 1987). However, one-dimensional studies of the CO-OH-CH_4 system by Sze (1977) indicated that, due to the feedbacks inherent in equations (42.1) to (42.4), the time constants with which the system adjusts to changes in the input fluxes could be much longer than even the CH_4 lifetime. The calculations presented in this section serve to illustrate the response of the 2-D model to a sudden change in input fluxes. Specifically, we address two questions: (1) What would be the concentrations of CO, OH, and CH_4 in the absence of tropical biomass burning, and (2) How long would it take for the atmosphere to adjust to a sudden (hypothetical) stop in biomass burning?

We have initialized the calculations with the steady-state results just described. The simulations were performed using flux boundary conditions for CO and CH_4. The surface fluxes for the temperate zones are set equal to the values calculated in the steady-state simulations, while the calculated tropical fluxes for CO and CH_4 are reduced by factors of 0.37 and 0.75, respectively, to simulate a sudden stop in tropical biomass burning. The biomass burning contribution to the tropical methane source has been estimated from the studies of Crutzen et al. (1985) and WMO (1985) and represents about 10% of the calculated global methane source [384 MT (CH_4)/yr for the "Low NO_x" case, 515 MT (CH_4)/yr for the "High NO_x" case]. Boundary conditions for other trace gases are the same as in the first part of the calculation. We then run the model forward in time for 30 years while maintaining the same boundary conditions.

Figure 42.5 illustrates the calculated time evolution of the percent change in CO (42.5a), OH (42.5b) and CH_4 (42.5c) concentrations in the first grid box (2 km) at the equator and midlatitudes, for the "Low NO_x" and "High NO_x" cases. The temporal behavior of CO and OH exhibits two components: a short, rapid response to the initial impulse during the first year, followed by a much slower approach to a new steady state. The initial rapid response is not apparent in the CH_4 calculations, due to the longer lifetime of this species. Examination of the curves in Figure 42.5

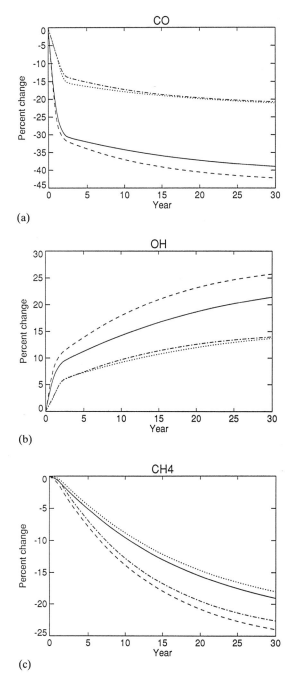

(a)

(b)

(c)

Figure 42.5 Percent change in CO, OH, and CH_4, as a function of year after sudden stop in biomass burning source. Results are shown for lowest grid box (2 km). The curves denote: "Low NO_x," case at equator (———); "Low NO_x," at midlatitude (. . . . . .); "High NO_x," at equator (- - - - -); "High NO_x," at midlatitude (———·———·———).

indicates that it takes about 30 years for the system to adjust to a new steady state.

The results in Figure 42.5 indicate that removal of the biomass burning source would decrease the CO concentrations in the lower troposphere by about 40% to 45% in the tropics and 15% at midlatitudes; increase the OH by about 20% to 25% in the tropics and 15% at midlatitudes; and decrease the CH_4 by about 20% to 25% in both tropics and midlatitudes. We note that the decrease in the methane is mostly due to the increase in its removal due to the calculated OH enhancement. Due to the long lifetime of CH_4, the percent change in CH_4 is smoothed out by transport, and thus exhibits less latitudinal dependence than the changes in CO and OH.

Table 42.3 illustrates the calculated changes in global average tropospheric OH and lifetimes of trace gases after 30 years. The average OH increases by about 15% for both "Low NO_x" and "High NO_x" conditions. As expected, the reduction in lifetimes for CH_3CCl_3, HCFC-22, and CH_4 scale roughly with the global OH increase. This scaling does not quite hold for CO, since, given the large seasonal and latitudinal variability of this species, the calculated lifetime gives more weight to the areas with larger concentrations of CO (Northern Hemisphere).

Discussion and Conclusions

We have used the model to derive a budget for CO from the observed CO concentrations near the surface. The calculations show that a source of between 1300 and 1600 MT CO/yr from the surface, in addition to the source from methane oxidation (600 to 900 MT CO/yr), is needed in order to maintain the observed CO concentrations. Calculated CO fluxes from the tropics contribute about 50% of the required surface source. The biomass burning source does not contribute significantly to CO concentrations at midlatitudes

but can account for between 25% and 50% of the CO abundance observed in the tropics.

Time-dependent calculations have also been presented to illustrate (1) the long time constants required by the CO-OH-CH_4 system to adjust to changes in forcing functions and (2) the contribution of the biomass burning source of CO and CH_4 to tropospheric concentrations CO, OH, and CH_4. Our calculations suggest a time constant of decades for the system to adjust to a sudden stop in the biomass burning source. The actual response to a particular scenario will depend on its specific assumptions. The calculations presented indicate that results from fully time-dependent calculations in the assessment of particular scenarios may be different from calculations which assume steady-state "snapshots" at different years.

Results from the time-dependent calculation indicate that a complete stop of the biomass burning source could decrease tropical CO and CH_4 by as 40% and 25%, respectively, and increase tropical OH by as much as 25% in 30 years. Although such a scenario is not realistic, the results suggest that biomass burning could be an important contributor to observed past trends in CO (2% to 3%/yr: Rinsland and Levine, 1985) and CH_4 (1%/yr: Rasmussen and Khalil, 1981; Rinsland et al., 1985; Blake and Rowland, 1988). Detailed analysis of past scenarios for biomass burning is needed to elucidate this impact. The calculated contribution to tropospheric OH indicates that biomass burning should be considered in any assessment of future changes in the tropospheric lifetime of long-lived trace species removed by the hydroxyl radical.

In our discussion, we have introduced several criteria which lend confidence to our approach in describing the large-scale budget and chemistry of the tropospheric gases in consideration. These criteria include: (1) testing our transport parameterization by

Table 42.3 Impact of removing tropical biomass burning sources of CO and CH_4 (30-year calculation)

	Low NO_x		High NO_x	
	Year 0	Year 30	Year 0	Year 30
Global average tropospheric OH concentration (cm^{-3})	6.1×10^5	7.0×10^5	7.9×10^5	9.2×10^5
Lifetimes				
CH_3CCl_3	6.6 yrs	5.8 yrs	5.1 yrs	4.3 yrs
$CHClF_2$ (HCFC-22)	19.1 yrs	16.5 yrs	14.2 yrs	11.9 yrs
CH_4	12.0 yrs	10.3 yrs	8.9 yrs	7.4 yrs
CO	3.0 mos	2.7 mos	2.2 mos	1.9 mos

comparison of model results to measurements of CFC-11 (WMO, 1985); (2) incorporation of boundary conditions for CO, CH_4, and NO_x consistent with observations; (3) consistency between our calculated average tropospheric OH and results from more detailed 3-D calculations; and (4) consistency between our calculated lifetimes for CH_3CCl_3, HCFC-22, CH_4, and CO, and those derived from ALE/GAGE data, or calculated by other multi-dimensional models.

In spite of the above consistency with other model results and observations, additional refinements should be implemented in order to achieve a completely validated model. Work in progress includes the following refinements to the model: improved resolution in the troposphere, particularly in the boundary layer, which will allow a more realistic treatment of the NO_x and O_3 budgets; incorporation of more detailed chemistry of nonmethane hydrocarbons; comparison of our transport parameterization with results from 3-D models and existing wind climatologies; assessment of the large-scale impact of deep convection and other regional processes; assessment of the effect of clouds on the radiation and chemistry; incorporation of a detailed data base from recent and future aircraft and ground-based campaigns to further constrain the model; and detailed comparison of our calculated OH field with 3-D results, and assessment of the impact of variations on the OH field on the calculated lifetimes.

Acknowledgments

The authors are grateful to Clarissa Spivakovsky for providing us with a copy of her manuscript before publication, to Jennifer Logan for valuable discussion of the CO budget, and to Debra Weisenstein and George Planansky for assistance in computing tasks. Work has been funded by the Alternative Fluorocarbon Environmental Acceptability Study (AFEAS) and by NASA's Upper Atmosphere Theory and Data Analysis Program.

VI

Biomass Burning and the Global Nitrogen Budget

The Global Impact of Biomass Burning
on Tropospheric Reactive Nitrogen

Hiram Levy II, Walter J. Moxim,
Prasad S. Kasibhatla, and Jennifer A. Logan

The two principal anthropogenic contributions to the reactive nitrogen budget of the atmosphere are fossil fuel combustion, occurring predominately in the industrial midlatitudes of the Northern Hemisphere, and biomass burning, occurring mainly in the tropics and subtropics. While dispersed over larger regions and generally more dilute, at least on an annual average basis, biomass burning emissions serve as a lesser developed nation's analogue to the industrial world's air pollution. In both cases the emissions strongly enhance the natural production of tropospheric ozone, thereby indirectly modifying the reactivity of the atmosphere, and add significantly to the natural levels of soluble nitrogen compounds in soils, fresh waters, and the ocean. In extreme cases this leads to photochemical smog and acid rain.

The natural sources of reactive nitrogen are injection of NO_y from the stratosphere, biogenic emissions from soil, and formation by lightning. Stratospheric NO_y, generally formed by excited atomic oxygen $[O(^1D)]$ attack on N_2O, is a small but significant source of NO_y in the free troposphere at mid- and high latitudes. Lightning is also thought to be an important source of NO_y in the free troposphere, particularly in the tropics and subtropics. Biogenic emissions, the only significant natural source of NO_y at the surface, appear to complement the biomass burning source. Current observations suggest that it is strongest in the same regions where biomass burning occurs, but that the emissions increase during the wet season and become much weaker during the dry/burning season.

In this chapter we will first review our current understanding of both the anthropogenic and natural sources of reactive nitrogen compounds in the troposphere. Then the available observations of both surface concentration and wet deposition will be summarized for regions with significant sources, for locations downwind of strong sources, and for remote sites. The obvious sparsity of the data leads to the next step: an attempt to develop a more complete global picture of surface concentrations and deposition of NO_y with the help of a global chemistry transport model (GCTM). The available source data are inserted into the GCTM and the resulting simulations compared with surface observations. The impact of anthropogenic sources, both downwind and at remote locations, will be discussed and the particular role of biomass burning will be identified.

Sources of NO_y

The current state of our knowledge regarding the two anthropogenic and three natural sources are summarized in Table 43.1. The estimates of the fossil fuel source shown in this table, as well as a recent compilation by Hameed and Dignon (1988) ($\sim$22 teragrams nitrogen per year, Tg N/yr) are all quite close and would suggest that this single largest source is well known. It should be noted that all these estimates have much of their raw data in common. Both Logan (1983) and Hameed and Dignon (1988) used similar United Nations fuel statistics and Levy and Moxim (1989) supplemented detailed emission data from North America and Western Europe with Hameed and Dignon's global data. Nonetheless, fossil fuel emissions are the single largest source and, as we can see from Figure 43.1a, are the dominant source of

Table 43.1 Reactive nitrogen sources

Source	Logan (1983) (Tg N/yr)	Current work (Tg N/yr)
Fossil fuel combustion	21 (14–28)	21.3[a]
Biomass burning	12 (4–24)	8.5[b]
Stratosphere	~0.5	0.64[c]
Soil emission	8 (4–16)	(5–14) (see Table 43.2)
Lightning	8 (3–20)	<10 (see text)

a. Levy and Moxim (1989).
b. Kasibhatla et al. (1990).
c. Kasibhatla et al. (1990).

NO_y in the northern midlatitudes. The three principal emission regions are the eastern United States and Canada, western and central Europe, and Japan and the more populated regions of China. The emissions are highly correlated with the more populated and industrial regions of the world. A more detailed discussion of this source is given in Levy and Moxim (1989).

Estimates of the biomass burning source strength in Table 43.1, as well as the recent estimate of 6 Tg N/yr by Hao et al. (1989) for the tropics and subtropics only, are in reasonable agreement though the range of uncertainty is greater. As shown in Figure 43.1b, this anthropogenic source is significantly weaker in the annual average and concentrated in the tropics and the subtropics. However, unlike fossil fuel emissions,

which normally vary by ~10% or less throughout the year, biomass burning is highly seasonal, and local emissions can be quite strong during the dry season. The NO_y source was constructed from a biomass burning CO source (Logan, private communication) and measured emission ratios of NO_x/CO_2 and CO/CO_2 (Andreae et al., 1988; Hao et al., 1989). For the simulations that will be discussed later, the source was activated only during the dry/burning seasons simulated by the model. Details are given in papers under preparation (Logan, private communication; Kasibhatla et al., 1990a).

Of the natural sources, stratospheric injection is the smallest, the most accurately known, and has a significant impact on the mid- and upper troposphere (Kasibhatla et al., 1990b). However, it has little im-

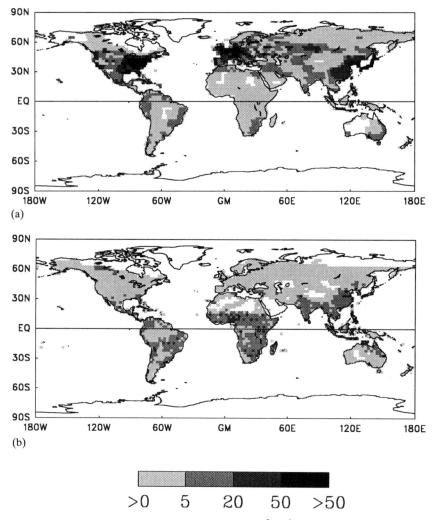

(a)

(b)

>0   5   20   50   >50

Figure 43.1 Anthropogenic sources of NO_y (m mol m^{-2} yr^{-1}): (a) annual source from fossil fuel combustion; (b) annual source from biomass burning.

pact on most surface mixing ratios or deposition (Logan, 1983; Savoie et al., 1989; Kasibhatla, 1990b), and will not be considered further in this chapter.

The estimates of biogenic emissions from soil are quite uncertain, although the emissions appear to be a significant natural source of NO_y in the tropical and subtropical grasslands, woodlands, and forests. Details of our most recent estimate of this source are given in Table 43.2. The principal difficulty is the extremely high variability in measured local emissions, which may fluctuate by a factor of 10 to 100 from season to season and location to location. The net flux is very dependent on soil moisture, nitrate level in the soil, and both soil and vegetation type. In Table 43.2 we present an estimate, based on a very sparse data set, of the range in yearly average fluxes for major biomes. The global flux calculation also required the estimate of total area of different vegetation types (Matthews, 1985). From Table 43.2, it is clear that biogenic emissions are predominately tropical and subtropical. They have a strong positive correlation with soil moisture, and the strongest fluxes have been measured in tropical woodland and grassland, which are also the strongest source regions for gases from biomass burning (Logan, private communication). Therefore, we expect the biogenic source to complement the biomass burning source with maximum emissions during the wet season rather than the dry/burning season. In the future, we hope to use the available observations of surface mixing ratios and wet deposition in the tropics to set an upper limit on the emission rate.

The final source of NO_y, lightning, is the least well known. However, it is possible to estimate a realistic upper bound with an extremely simple input-output model. Merely dividing the more generous estimates of 100 Tg N/yr by the surface area of the globe gives an average deposition of 14 m mol m^{-2} yr^{-1}. This greatly exceeds observed depositions in remote regions (see Table 43.3). In an earlier review, Logan (1983) chose an upper limit of 20 Tg N/yr. Our preliminary analysis of a recent simulation using a GCTM with a simple representation of the lightning source suggests that it should be no larger than 10 Tg N/yr. This would still make lightning an important, possibly the major, natural source and quite probably the dominant supplier of NO_y to the free troposphere of the tropics and subtropics.

As Figure 43.2 (Orville and Henderson, 1986) suggests, lightning appears to occur primarily over land, between 40°N and 40°S, with a maximum in the tropics. Other studies have found a less extreme separation between land and ocean, but all find that land-based lightning predominates. Studies also find that most lightning discharges are cloud to cloud. Therefore we expect that the lightning source will release NO_y into the mid- and upper troposphere. These conclusions regarding the distribution of the lightning source, while consistent with the available data, are based on a rather small sample. However, the fact that lightning releases most of its NO_y in the free troposphere where it is more easily dispersed, rather than in the boundary layer, makes the uncertainties regarding the distribution of the source less critical than they are for biogenic emissions from soil.

Observed NO_y Wet Deposition and Surface Mixing Ratios

Given the truth of the Newtonian adage that what goes up must come down, our first two questions should be: (1) Where is all the emitted NO_y depos-

Table 43.2 NO_y emissions from soil

Vegetation type	Net local flux (ng N/m²/sec)	Global flux (Tg N/yr)
Tropical rainforest	0–2[a]	0.0–0.7
Other forest	0.2–0.5[b]	0.2–0.4
Tropical woodland and grassland	10–30[c]	3.6–11
Other woodland and grassland	1–2[d]	0.8–1.6
Fertilized fields	<0.5% × 100 Tg N²	<0.5
Total		5–14

a. D. Jacob (private communication).
b. E. Williams (private communication).
c. Johansson and Sanhueza (1988).

Table 43.3 Surface deposition of NO_y (m mol m^{-2} yr^{-1})

Location number	Description	Range
1	Fossil fuel emissions	20–45[a]
2	South American biomass burning	5–10[b]
3	African biomass burning	15–30[c]
4	North Atlantic	3–7[d]
5	Remote sites	0.5–2[d]

a. See Levy and Moxim (1989).
b. Sanhueza (1988).
c. J. P. Lacaux (private communication).
d. J. N. Galloway (private communication).

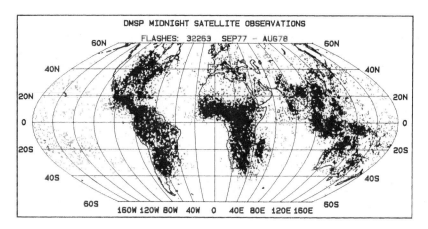

Figure 43.2 One year of midnight lightning locations for the period September 1977 to August 1978.

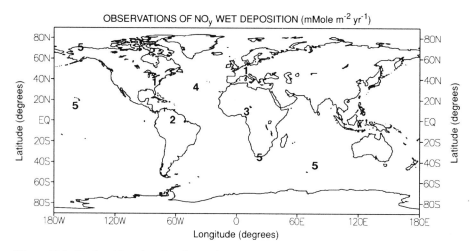

Figure 43.3 Regional locations for observations of NO$_y$ wet deposition.

ited, and (2) Is the amount deposited consistent with the amount emitted? A summary of available data for representative regions is presented in Table 43.3 and the companion Figure 43.3. While the values in a particular region may vary by as much as a factor of 2 to 4, a clear pattern emerges. The largest depositions occur in regions with large, generally anthropogenic, surface sources, while the minimum background deposition is a factor of 10 to 20 smaller and occurs over a large fraction of the earth's surface including most of the Southern Hemisphere. These data, while sparse, are already sufficient to set a reasonable upper limit on the least well-known source, lightning. The observations will be compared with GCTM simulations in the next section.

Representative ranges of surface observations of either NO$_y$ or, in the case of oceanic sites, the water soluble portion, are given in Table 43.4 for the regions

Table 43.4 Surface mixing ratio of NO$_y$ (ppbv)

Location number	Description	Range
1	Fossil fuel emissions	5–15[a]
2	South American biomass burning	<1[b]
3	North Atlantic	0.2–0.5[c]
4	North Pacific	0.1–0.2[d]
5	Remote sites	~0.05[d]

a. Parrish et al. (1990).
b. Andreae et al. (1988).
c. J. Prospero (private communication).
d. Savoie and Prospero (1989).

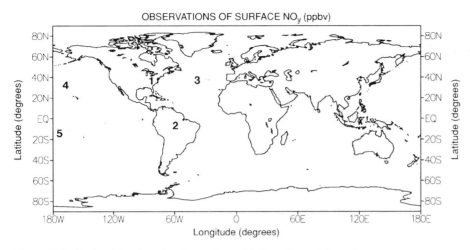

Figure 43.4 Regional locations for observations of NO$_y$ surface mixing ratio.

identified in companion Figure 43.4. Again the variability within a particular region is much less than the differences between source, downwind and remote locations. However, the gradients in surface mixing ratios, which depend strongly on the distance from a local surface source, are much greater than those for wet deposition, which removes NO$_y$ from the atmospheric column and is influenced both by local emissions and by NO$_y$ that has been transported through the free troposphere from distant sources. The surface levels of NO$_y$ in remote locations will help establish a limit on the magnitude of the less well known sources, particularly lightning.

GCTM Simulations and Comparisons with Observations

The GCTM has a horizontal resolution of ~265 km and 11 vertical levels at standard heights of 31.4, 22.3, 18.8, 15.5, 12.0 8.7, 5.5, 3.1, 1.5, 0.5, and 0.08 km. It is driven by six-hour time-averaged wind fields and a self-consistent precipitation field derived from a one-year integration of a parent general circulation (climate) model (Manabe et al., 1974; Manabe and Holloway, 1975). Details of the GCTM and its application to studies of emissions from fossil fuel and biomass burning are given in Levy and Moxim (1989) and in Kasibhatla et al. (1990a).

The simulated yearly wet depositions for fossil fuel and biomass burning are given in Figure 43.5. Deposition from fossil fuel emissions spreads in a broad midlatitude belt across the Northern Hemisphere, with the exception of the eastern Pacific. Biomass burning dominates in the tropical Atlantic, tropical Indian Ocean, and Africa, while deposition from both anthropogenic sources is important in South America, India, China, and Indonesia.

The simulated yearly averaged surface mixing ratios for both anthropogenic sources are given in Figure 43.6. The general patterns are quite similar to deposition, though the gradients away from source regions are steeper. In particular at midlatitudes, deposition patterns spread much farther from the source than do surface mixing ratios, due to the strong westerly winds in the free troposphere.

Detailed discussions and comparisons with observations are available in Levy and Moxim (1989) and Kasibhatla et al. (1990a). We will provide a brief summary. The simulation with fossil fuel emissions is in good agreement with observations in the midlatitudes of the Northern Hemisphere but makes a small contribution to the wet deposition and surface mixing ratios of NO$_y$ in remote regions of the Southern Hemisphere. Simulated surface mixing ratios from the sum of fossil fuel and biomass burning agree with the few observations in South America made during the burning season, and the simulated deposition can account for at least half of the observed yearly deposition. However, together these two anthropogenic sources account for less than half of the deposition observed in biomass burning regions of Africa, as well as far less than half of the NO$_y$ surface mixing ratios and wet deposition in the remote regions of the Southern Hemisphere.

This comparison suggests a possible major role for biogenic soil emissions and lightning throughout the tropics and most of the Southern Hemisphere. A preliminary simulation finds that lightning should account for most of the missing NO$_y$ in the remote Southern Hemisphere. However, the wet deposition

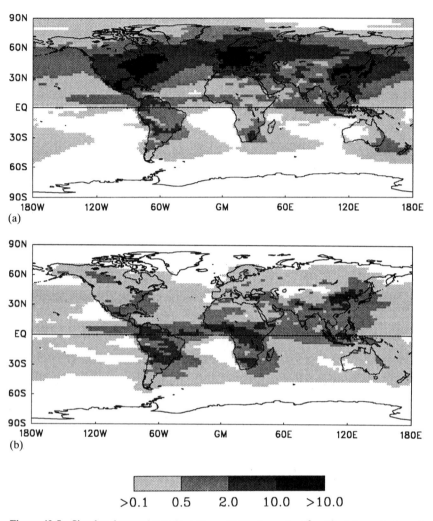

Figure 43.5 Simulated annual wet deposition of NO$_y$ (m mol m^{-2} yr^{-1}): (a) results for a fossil fuel combustion source; (b) results for a biomass burning source.

observed in Africa, which is quite high relative to that in South America, cannot be explained by our current estimates of biogenic emissions. This issue requires more observations and further analysis.

Summary

From this brief intercomparison of simulations and observations it is clear that biomass burning plays an important, but not necessarily dominant, role in the reactive nitrogen budget of the continental tropics and subtropics. It is equally clear that two natural sources, lightning and biogenic emissions from soil, also play an important role in the region. Moreover, the very limited NO$_y$ data on soil emission in South America, surface mixing ratios in the Amazon, and wet deposition in both Africa and South America do not appear to be consistent.

Acknowledgments

We wish to thank J. P. Lacaux, J. N. Galloway, and J. M. Prospero for providing data prior to publication.

This work was supported in part by funds from the National Science Foundation under grant ATM-8701289.

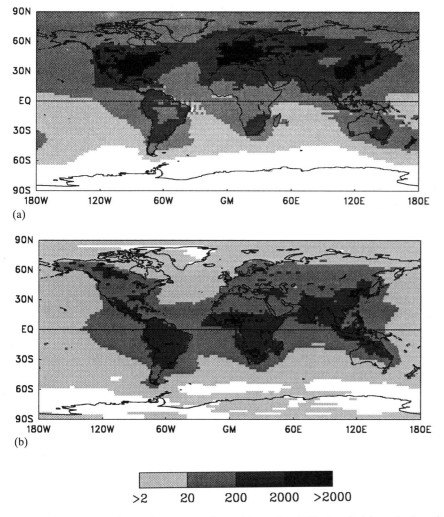

Figure 43.6 Simulated annual average surface mixing ratio of NO$_y$ (pptv): (a) results for a fossil fuel combustion source; (b) results for a biomass burning source.

Biomass Burning: A Source of Nitrogen Oxides in the Atmosphere

Jane Dignon and Joyce E. Penner

The importance of biomass burning in the atmospheric cycle of CO_2 and its role as a greenhouse gas have received considerable attention. Biomass burning is also a major source of nitrogen oxides ($NO_x = NO + NO_2$) in the atmosphere. NO_x is important in the atmosphere because it is a primary pollutant which forms nitric acid (HNO_3), a major contributor to acid wet and dry deposition, as well as a major limiting nutrient in some soil and ocean ecosystems. NO_x is also important because it acts to form ozone (O_3) in the troposphere. Tropospheric O_3 is important from both an atmospheric and a biological perspective. Ozone absorbs thermal radiation, thereby acting as a greenhouse gas and warming the earth's surface. It is also a chemical oxidant which causes injury to humans, plants, and animals at elevated concentrations. In the tropics, where the largest extent of biomass burning occurs, Browell et al. (1988) have measured O_3 concentrations exceeding 50 parts per billion by volume (ppbv) in areas downwind from sites of biomass burning, due in part to the elevated NO_x emitted from these fires.

In this chapter we present a spatially resolved (1 × 1 degree) inventory of the annual emissions of NO_x from biomass burning. Table 44.1 lists the major known sources of NO_x to the atmosphere. Here we estimate the release of up to 13 teragrams nitrogen (Tg N) annually from biomass burning. This is roughly 60% of the largest known source, ~22 Tg N from fossil fuels (Hameed and Dignon, 1988; Dignon, 1990). Thus, we show that biomass burning is a major source of NO_x globally and perhaps the most important source in many tropical regions. Results of this work will be used in a global chemical tracer model simulation to estimate the contribution of biomass burning on the cycle of reactive nitrogen in the tropics and globally (Penner et al., 1990a; Atherton et al., 1990).

Emission Factors

Clements and McMahon (1980) experimentally measured the amount of NO_x emitted from several types of vegetation. They found the emission factor for NO_x is linearly dependent on the nitrogen content of the fuel burned. Figure 44.1 shows the data from Clements and McMahon for the emission factors of NO_x plotted as a function of the nitrogen content in the biomass fuels. Regression analysis gives the linear expression

$$EF(NO_x) = -1.5 + 3.9n_f \qquad (44.1)$$

where $EF(NO_x)$ is the emission factor of NO_x in grams of nitrogen per kilogram of dry matter (g N/kg dm) burned and n_f is the percent of nitrogen bound within the plant matter. The relationship has a linear correlation coefficient of 0.95, which is statistically significant at better than the 99% confidence level. One may note that the least squares line does not go directly through the origin. Presumably the remaining nitrogen in the fuel is released in the form of other nitrogen compounds, e.g., NH_3, N_2O, HCN, and CH_3CN (Menaut et al., 1990a).

For our study the globe was divided into the nine major land-use categories described by Matthews (1982), which include (1) rainforest, (2) forest, (3) woodland, (4) scrub, (5) grassland, (6) tundra, (7) desert, (8) ice, and (9) agriculture. Of these nine categories all can undergo burning except desert and ice-covered lands. The nitrogen content in each of the seven remaining vegetation types is listed in Table 44.2 and was based on estimates from the ecological literature (Bate and Gunton, 1982; Buschbacher, 1987; Buschbacher et al., 1987; Duvigneaud and DeSmet, 1970; Gray and Schlesinger, 1981; Groves, 1983; Gulmon, 1983; Jordan, 1987; Medina, 1982; Nye, 1961; Ovington and Olsen, 1970; Perkins, 1978; Perkins et al., 1978; Read and Mitchell, 1983; Rundel, 1982, 1983; Russell, 1987; Saldarriaga, 1987; Scott, 1987; Shaver, 1981; and Wielgolaski et al., 1975). Where the data was available, we used a weighted average to determine the N content of the fuel burned for each biomass type. A higher weighting was given to the N content of leaves, underbrush, and branches, the plant parts most likely to be consumed, than was given to stems and boles. Please note

Table 44.1 Estimates for the global sources of reactive nitrogen

	Tg N yr^{-1}	Reference
Fossil fuel combustion	22	Hameed and Dignon (1988)
Biomass burning	5–13	This study
Lightning	3–10	Borucki and Chamiedes (1984); Penner et al. (1990a)
Soil microbial activity	1–11	Johansson (1984); Kaplan et al. (1988); Slemr and Seiler (1984); Williams et al. (1987)
Input from stratosphere	0.5–1	Jackman et al. (1980); Ellsaesser, (1981)

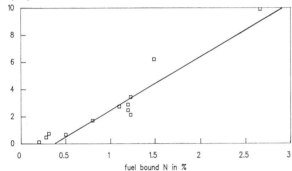

Figure 44.1 Emission factor of NO$_x$ plotted with respect to the percent of nitrogen bound in biomass fuels.

that the grassland and scrub categories have been combined in Table 44.2.

We used equation (44.1) together with these estimates of nitrogen content to determine the emission factor for each of the biomass types. The calculated emission factors are also shown Table 44.2. A list of measured emission factors for representative fuels from the different vegetation types is also presented. Comparison of the observed emission factors with those estimated using equation (44.1) shows that the equation provides a good method for estimating emission factors for all of the biomass types, although differences of up to a factor of 2 are noted for some cases (e.g., agricultural burning). These differences may, in part, be explained by different combustion conditions during the experiment. Because different plant components contain different amounts of nitrogen, these differences may also reflect differences in

the extent of burning for different components. Finally, these differences may be due to differences in the N content of the fuel itself. For example, in the case of agricultural burning, our estimate for N emissions assumes the N content estimated for live, healthy vegetation. Gerstle and Kemnitz (1967) measured the emissions from the burning of "landscape refuse." If this is dead leaves, it would have lower N content because plants use up most of their nutrients before dropping their leaves. We note that if most of the agricultural burning takes place after the crop is harvested and the plants are withered, our estimate for the emission factor in this biomass category would be too large. For this reason, in our figures below, we show estimates for N emissions that exclude the emissions from this category.

The calculated estimate of the emission factor for tundra listed in Table 44.2 is considerably higher than that for any other biomass category. This is due to the much higher nitrogen content of the fuel in this region (Wielgolaski et al., 1975). Unfortunately, there are no published measurements of N emissions from the burning of this biomass type. As we show below, however, tundra represents a very small contribution to the global total NO$_x$ emissions from biomass burning (i.e., less than 1%). Therefore, a large uncertainty in this emission factor does not affect our estimate of the total global flux of N from biomass burning.

Biomass

Biomass burning results primarily from anthropogenic activity and occurs globally for a variety of reasons, including deforestation for cultivation, clearing associated with shifting cultivation, annual bushfires (which are traditional in many savanna regions), burning of agricultural wastes, controlled burning, and simple carelessness. Natural wildfires are also minor contributors.

The amount of biomass burned, M, is given in Seiler and Crutzen (1980) as

$$M = A \times B \times \alpha \times \beta \qquad (44.2)$$

in kg dry matter per year (kg dm/yr). In equation (44.2), A is the total land area burned in m^2/yr, B is the average above-ground organic matter in grams of dry matter per unit area (g dm/m^2), α is the fraction of above-ground biomass relative to the total biomass B, and β is the burning efficiency of the above-ground biomass. There is considerable uncertainty about the magnitude of these parameters, and an extensive dis-

Table 44.2 Comparison of measured emission factors with those determined from fuel N content using equation (44.1) in g N/kg dm

Fuel type	% N content	Calculated emission factor	Measured emission factor	Reference
Rainforest	1.11	3.1	1.2	Buschbacher (1987)
			3.5	Russell (1987)
Forest	.85	2.1	1.3	Clements and McMahon (1980)
			.15–3.4	Rundel (1981)
			.24–2.2	Hegg et al. (1990)
Woodland	.77	1.8	1.1–2.3	Rundel (1983)
				DeBano and Conrad (1978)
Scrub & grassland	.82	2.0	2.1–2.2	Delmas (1982)
			1.5–5.3	Medina (1982)
			~.8	Crutzen et al. (1989)
			1.0–2.7	Hegg et al. (1990)
Tundra	2.0	6.6		
Agricultural	.67	1.42	.62	Gerstle and Kemnitz (1967)

Table 44.3 Estimates of global N emissions from biomass burning for each fuel type

Land use	A^a (Area 10^9 m²)	B^b (kg dm m^{-2})	α	β	M (10^9 kg dm)	Parameterized EF (Tg N)	Crutzen et al. EF (Tg N)
Rainforest	71	25	.81	.25	359	1.1	.29
Forest	131	22	.71	.25	442	.93	.35
Woodland	146	11	.81	.5	585	1.1	.47
Scrub	9	.6		.8^b	4	.01	.003
Grassland (shifting cultivation included)	7012	.6		.8^b	3366	6.7	2.7
Tundra	38	.2	.5	.9	3.4	.02	.003
Agriculture	2724	.6		.8^b	1308	1.9	1.1
						12.5	5.2

a. Most of the estimates for A and B are based on the study reported by Logan et al. (1990). The estimates of A for forest and woodland areas have been updated. The estimate for tundra is new.
b. Estimates from Logan et al. (1990); α and β combined.

cussion of the uncertainties can be found in Robinson (1989). The values used in equation (44.2) for this model study are listed in Table 43.3 and are based on the work of Logan et al. (1990), which will be reviewed briefly here.

Logan et al. (1990) consider the seven major vegetation types discussed above, which are from Matthews (1983). They also include an additional category called shifting agriculture (sometimes referred to as "swidden" agriculture). This form of food production has been practiced in tropical regions for centuries. It is perhaps the oldest method of agricultural production and allows for the natural regeneration of the soil fertility to be accomplished by the regrowth of forest or grassland by the practice of field-to-forest, or bush-to-fallow rotation.

In equation (44.2), A is the area burned in each biome annually. The most well-known area burned in these vegetation types is that of rainforest, forest, and woodland. Data for these types have been compiled for some countries or regions by various authors including Myers (1980), Lanly (1981a, 1981b, 1981c), Statistics Canada (1984), U.S. Forestry Service (1988), and Yamate (1975). For tropical nations where the largest amount of burning occurs, Logan et al. 1990 use the data of Lanly (1981a, 1981b, 1981c), who gives estimates for the amount of deforestation. Not all countries are included in Lanly (1981a, 1981b, 1981c). For the unlisted tropical countries, Lanly used the fraction of burned area in neighboring countries with similar vegetation types to scale the vegetation in the unlisted country. These values were then compared to the deforestation estimates of Myers (1980) where available. Logan et al. (1990) have also made estimates for the area of shifting agriculture for tropical rain forest, forest, and woodland regions. These are included in the figures for each biome in Table 44.3.

For countries outside of the tropics, the forest fire areas for the United States and Canada were given by individual state/province in Harrington (1982), Coe (1989), Alaska Fire Service (1988), Yamate (1975), and Statistics Canada (1982). Burn areas for some European and Mediterranean countries have been presented in LeHouerou (1974), Siren (1974), Ozyigit and Wilson (1976), Brandel et al. (1988), Valasco (1983), and Goldammer (1979). Some qualitative data for China has also been published in Fuchs (1988) and Salisbury (1989). In other midlatitude countries, Logan et al. (1990) estimate that 0.3% of the forested area is expected to burn each year based on an extrapolation of the Canadian data of Statistics Canada (1984) and estimates of Yamate (1974) for the area burned in a number of different continents.

The values of A for each of the other vegetation types are almost completely undocumented. Deschler (1974) reports that 40% to 60% of African savannas are burned each year. Hao et al. (1989) assume that one-half of the total savanna area is burned each year in Latin America, three-quarters of the area is burned each year in Africa, and two-thirds of the area is burned each year in Australia. Logan et al. (1980) assumed that one-half of the savanna is burned annually in these regions. We have used this estimate; however, Menaut et al. (1990b) have suggested that this burning frequency may be a factor of 2 to 3 too high. Shifting cultivation in the grassland areas is included in this estimate. In Australia, where the aborigines still practice burning, Walker and Gillison (1982) suggest that the very dry scrub regions are burned every 10 to 15 years, while the northern regions, which experience more rainfall, are burned more frequently. Thus, Logan et al. (1990) assume that one-quarter of the savanna area in Australia is burned each year.

Logan et al. (1990) do not include the burning of tundra in their calculation. Van Wagner (1989) estimates that burning in tundra regions cycles once every 100 to 200 years. We chose to use the more conservative quantity of 0.5% of the tundra area consumed annually.

The total area of cultivated land on the globe is taken from Matthews (1983). The area of agricultural burning in the United States and United Kingdom was taken from Yamate (1974) and Bullen (1974), respectively. Like most of the United States, Western Europe is assumed to keep open burning at a minimum due to air pollution restrictions. Its contribution to this source is considered negligible. In other regions of the world, the area of agricultural burning was assumed to be 29% of the total cultivated land area. Logan et al. (1990) derive this estimate by adopting the fraction of cultivated area that is harvested each year from Matthews (1983) (i.e., 60%) and assuming that 50% of this is burned annually.

The values for B, the average biomass per unit area, are also given in Table 44.3 for each land-use type. These were derived from Olson et al. (1983) and Seiler and Crutzen (1980). The fraction of aboveground biomass, α, and the burning efficiency, β, were taken from Seiler and Crutzen (1980) and Logan et al. (1990). We note that there are significant uncertainties in the parameters presented in Table 44.3, and these uncertainties will propagate through to our estimates of NO_x emissions.

NO_x Emissions

Table 44.3 compares the total emissions of NO_x, derived using the emission factors calculated from equation (44.1) based on the fuel nitrogen content, with the emissions of nitrogen, derived using the emission factor of ~0.8 g N/kg dm recommended by Crutzen et al. (1989). This latter emission factor was derived from laboratory burns of savanna fuels and is consistent with the field measurements from Brazil made by Andreae et al. (1988), which range from 0.3 to 1.4 g N/kg dm. Our value for savanna fuels, based on the relationship to fuel N content is a factor of 2 larger than the value derived by Crutzen et al. (1989) but compares well with the field measurements of northern African savanna burns by Delmas (1982). There may be considerable variability between the nitrogen content and burning conditions of the biomass fuels in Brazil and those of savannas in northern Africa. As a result of these differences, our estimate of the total emission of N from grasslands, the largest category listed in Table 44.3, is a factor of 2 larger than that derived by Hao et al. (1989), who used the emission factor measured by Crutzen et al. (1989). At the present time, it is not clear whether the emission factor we derived here is too high, or whether the emission factor derived by Crutzen et al. (1989) is too low. Other emission factors measured for scrub and grassland vegetation range from 1 to 5.3 g N/kg dm (see Table 44.2). Further experiments taking into account the variation in burning conditions actually experienced in the field are needed. We also note that our estimated emissions from the other fuel categories listed in Table 44.3 are larger than the emissions that would be estimated using the emission factor of Crutzen et al. (1989).

Figure 44.2 shows the distribution of N emissions from biomass burning around the world using the emissions derived from fuel nitrogen content, while Figure 44.3 shows the distributions of N using the emission factor of Crutzen et al. (1989). The results presented in Figures 44.2 and 44.3 are presented on the grid used by the NCAR Community Climate Model General Circulation Model. These figures were aggregated from 1×1 degree emissions model results. The 1×1 degree model results are available from the authors. The distribution of emissions shown in Figures 44.2 and 44.3 do not include the component from agricultural burning. Geographical regions of high emissions in the model correspond well with areas observed with remote sensing techniques which experience a large frequency of fires (Kendall and Justice, 1990; Menzel et al., 1990; Stocks, 1990). The contours in Figure 44.4 represent the ratio between the estimates shown in Figure 44.2 and Figure 44.3—i.e., the ratio between the parameterized method which determines emission factors from N fuel content versus the method of Crutzen et al. (1989). This figure shows that in most regions the parameterized method is generally a factor of 2 to 3 higher than the estimates made using the single emission factor

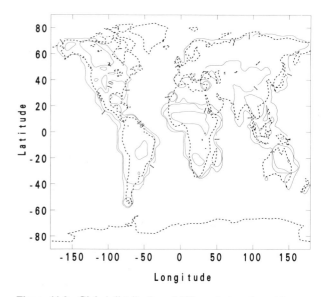

Figure 44.3 Global distribution of NO_x emissions from biomass burning estimated using the emission factor derived by Crutzen et al. (1989) presented on NCAR CCM grid (kg N km^{-2} yr^{-1}).

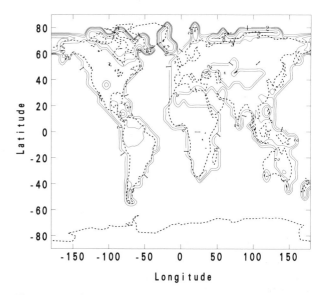

Figure 44.4 Comparison of NO_x inventories presented using emission factors derived from equation (44.1) and the emission factor derived by Crutzen et al. (1989). Figure shows the ratio of estimates presented in Figure 44.2 divided by the estimates presented in Figure 44.3 (contours shown are 1, 2, 3, 5, and 8).

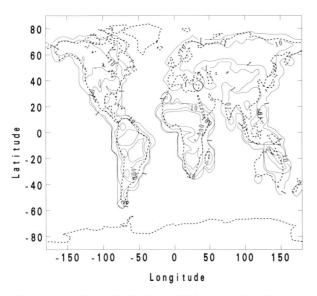

Figure 44.2 Global distribution of NO_x emission from biomass burning estimated using emission factors derived from equation (44.1) presented on NCAR CCM grid (kg N km^{-2} yr^{-1}).

method of Crutzen et al. (1989). In the forest areas of Indonesia and Central and South America, it is a factor of 3 higher. In the tundra regions there is a significant difference, up to a factor of ~8, but the source strength is very small.

The total amount of N emitted in biomass burning was estimated in several previous studies. Hao et al. (1989) estimated a mean value of ~6 Tg N yr^{-1} with a range of 3 to 9 Tg N yr^{-1}; however, their estimate only addresses tropical regions. Logan (1983) estimated a global flux of 12 Tg N yr^{-1} with a possible range of 4 to 24 Tg N yr^{-1}. A high estimate of 19 Tg N yr^{-1} is given by Hegg et al. (1990), but they suggest that roughly half of that may be the revolatization of NO$_x$ previously deposited on the vegetation and ground. Our estimate of 13 Tg N yr^{-1} using the parameterized emission factor method is compatible with these previous estimates, and is up to 60% of the estimate of ~22 Tg N yr^{-1} for the global flux of N from fossil fuel burning (Dignon and Hameed, 1988; Dignon, 1990).

Conclusions

Atherton et al. (1990) used the fuel inventory of Hao et al. (1989) and emission factors similar to those derived by Crutzen et al. (1989) to estimate the relative role of biomass burning in the reactive nitrogen cycle of Africa, South America, Australia, and Asia. In the case of Africa, they found that biomass burning (followed by soil emissions and fossil fuel burning) was the most important source of reactive nitrogen. For South America, the biomass burning source was second in importance after soil emissions. The total emission of nitrogen derived by Atherton et al. (1990) was 5.8 Tg N yr^{-1} (see also Penner et al., 1990b). The total emission derived here, using our parameterization of emissions based on fuel N content, is ~13 Tg N yr^{-1}, a factor of 2 higher than the estimates used by Atherton et al. (1990). If these new estimates are correct, these emissions would clearly dominate the budget of reactive nitrogen in the tropical areas of the world. Measurements of reactive nitrogen, including measurements of deposition of nitrate in precipitation, could help delineate the role of biomass burning in the budget for reactive nitrogen in these tropical areas.

Acknowledgments

This work was performed under the auspices of the U.S. Department of Energy by the Lawrence Livermore National Laboratory under Contract W-7405-Eng-48. Computer time was supplied by the Office of Health and Research, Department of Energy. We would also like to thank Jennifer Logan for permission to use the biomass data in Table 44.3.

Emissions of Nitrous Oxide from Biomass Burning

Edward L. Winstead, Wesley R. Cofer III,
and Joel S. Levine

Nitrous oxide (N_2O) is a key atmospheric trace gas typically found in the earth's troposphere at concentrations of about 320 parts per billion by volume (ppbv). N_2O is an extremely important trace gas since it is both a greenhouse gas (contributing to global warming) and the major contributor (after undergoing photolysis in the stratosphere to form NO) to stratospheric ozone destruction (Turco, 1985). N_2O is currently increasing in the atmosphere at about 0.2% per year (Weiss, 1981). Until recently, fossil fuel combustion was considered to be a major contributor of N_2O to the atmosphere (Hao et al., 1987; Weiss and Craig, 1976; Pierotti and Rasmussen, 1976). However, Muzio and Kramlich (1988) reported results that seriously questioned the measurements that the above assessments were based on. Their results indicated that chemical reactions in samples of combustion-produced emissions, when collected and stored in stainless steel grab sample containers, could actually generate N_2O. Since almost all measurements of fossil fuel combustion-produced N_2O have involved grab bottle collections with periods of storage, questions began to surface about the validity of the amount of N_2O produced by fossil fuel combustion. The results of Muzio and Kramlich have been confirmed by others (Linak et al., 1990; de Soete, 1989) and fossil fuel combustion-produced estimates have been revised significantly downward.

Estimates of N_2O produced during biomass burning have almost exclusively been based on grab sample collections in stainless steel canisters. The potential for similar reactions leading to serious errors is obvious. A study was therefore undertaken to compare our N_2O results determined over large prescribed fires or wildfires (Cofer et al., 1988a, 1989, 1990a, 1990b) in which grab sampling with storage had been used with N_2O measurements made in near real time. This chapter summarizes the results.

Experiment Procedure

Plume smoke samples from eight biomass fires (see Table 45.1) were collected using a Bell 204B helicop-ter. Aluminum sampling probes protruded about 76 centimeters (cm) forward of the helicopter nose through a modified nose access hatch. Flexible Tygon hoses were attached to the base of each probe and run about 4 meters to Hi-Vol air pumps used to fill 10 liter (L) Tedlar sampling bags; samples were then transferred (usually within 10 to 15 minutes) to stainless steel bottles before chemical analysis. All analyses were performed using aliquots from the stainless steel canisters.

Small laboratory test fires were conducted using loblolly pine branches and needles, small branches and twigs of red oak, marsh grasses, and mixtures of each, for fuels. Sampling and analysis was performed as in the helicopter measurements except that a sample of the smoke was analyzed for N_2O immediately (withdrawn from the Tedlar bag by syringe) after collection, before transfer to steel grab bottles for aging studies.

Analysis for N_2O was performed by electron capture gas chromatography using a 3 meter porapak Q column at 80°C and a Ni^{63} detector at 340°C. CO_2 analysis was performed using gas chromatography with thermal conductivity detection. Our collection, chemical analysis procedures, and a more comprehensive description of the prescribed fires and wildfires studied can be found in Cofer et al. (1988a, 1989, 1990a, 1990b).

Results

Mean CO_2-normalized emission ratios ($\Delta N_2O/\Delta CO_2$; volume/volume; differences are with respect to background concentrations) are presented in Figure 45.1 without respect to combustion phase (flaming or smoldering) for the eight fires studied. Fires 1 and 2 were Lodi Canyon chaparral fires in California, fires 3 and 4 were grassy wetlands fires at the Kennedy Space Center in Florida, and fires 5 to 8 were fires in the Canadian boreal forest system. It is immediately apparent that the mean emission ratio from fire 6 is distinctly different. Fire 6 was the Thomas prescribed fire, consisting of 375 ha of tramped forest, conducted

Table 45.1 Biomass fires sampled

Burn	Fire number	Date	Location	Vegetation
Lodi 1	1	12 Dec. 1986	California	Chaparral
Lodi 2	2	22 June 1987	California	Chaparral
Tyranite	5	29 Aug. 1987	Ontario, Canada	Boreal forest
KSC 1	3	9 Nov. 1987	Florida	Wetlands
Thomas	6	29 July 1988	Ontario, Canada	Boreal forest
Peterlong	7	22 Aug. 1988	Ontario, Canada	Boreal forest
KSC 2	4	7 Nov. 1988	Florida	Wetlands
Hill	8	10 Aug. 1989	Ontario, Canada	Boreal forest

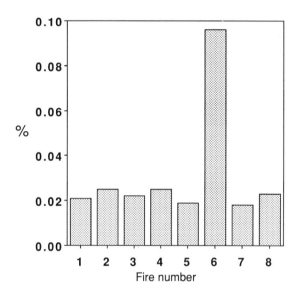

Figure 45.1 Mean emission ratios for nitrous oxide (N_2O) for fires in boreal forest, chaparral, and wetlands ecosystems.

on 29 July 1988. There was nothing particularly unique about this fire except that the chemical analysis of the trace gas emissions collected was not performed until about six weeks after collection (rather than the typical hours to days after a fire), translating into an exceptionally long period of storage in the stainless steel grab sampling bottles. Results for other trace gases (e.g., CO_2, CO, CH_4, etc.) for samples from this fire, however, were consistent with values from the other fires. A complete discussion of these samples appears in Cofer et al. (1990b). This anomaly, coupled with the findings of Muzio and Kramlick (1988) for fossil fuel samples, suggested that experiments were needed to determine if artifact N_2O was

being generated during storage of biomass burning samples.

Results from the laboratory test fires appear in Table 45.2, where concentrations of CO_2 and N_2O are shown for the aging studies. CO_2 concentrations did not change with time. N_2O concentrations in about half of the 27 samples were found to increase significantly after 10 days of storage. Almost all of these samples had much higher initial concentrations of CO_2 and N_2O in the grab bottles than for any of the large fires that were sampled. Thus, any other combustion products (e.g., SO_2, NO, etc.) might be presumed to have been at higher concentrations also. Most of the samples showing significant increases in both the amount and rate of growth of artifact-produced N_2O were at the higher initial mixing ratios. In fact, growth rates and amounts of artifact N_2O produced correlated well with initial concentrations. In several instances, N_2O concentrations almost doubled during the 10-day test period.

CO_2-normalized emission ratios (ERs) for initial (t_o) N_2O concentrations for the laboratory test fires and a similar and representative number of aircraft determinations from large biomass fires are shown in Figure 45.2. The initial CO_2-normalized ERs obtained from the laboratory fires are about an order of magnitude lower than those determined for the large-scale fires. Other researchers have recently reported similarly low N_2O ERs form laboratory fires (Lobert et al., 1990; Elkins, personal communication, 1990). The large difference in ERs between the laboratory and large field fires is yet to be resolved.

Discussion

Results from the small laboratory test fires and actual large-scale biomass fires have confirmed that a similar (or the same) N_2O-producing artifact as identified for fossil fuel combustion samples collected in steel grab bottles can occur with biomass burn samples. The effect for biomass burn–generated samples, however, appears to be much less predictable. Generally, growth rates of artifact N_2O and amounts produced tended to correlate with initial concentrations—that is, higher initial mixing ratios produced artifact N_2O faster and in larger quantity. However, some of the samples with lower initial concentration were also observed to rapidly generate artifact N_2O.

CO_2-normalized ERs, obtained initially (t_o) from the laboratory fires, are substantially lower (about an order of magnitude) than ERs obtained over the actual large-scale biomass fires. The overall consistency in ERs obtained for the large-scale fires, when coupled with the

Table 45.2 Summary of results from small laboratory fires of loblolly pine (P), marsh grass (G), and equal mixes of both (P/G)

Sample	CO_2 (ppmv)	t_o	½ hr	1 hr	6 hr	12 hr	1 da	3 da	5 da	10 da
						N_2O (ppmv)				
1 P	3200	0.402	0.400	0.403	0.403	0.402	0.401	0.403	0.403	0.402
2 P	17900	0.749	0.754	0.789	0.848	0.891	0.937	1.233	1.371	1.811
3 P	2700	0.381	0.379	0.380	0.381	0.382	0.383	0.382	0.383	0.383
4 P	6700	0.533	0.533	0.534	0.536	0.535	0.533	0.561	0.593	0.629
5 P	5200	0.441	0.441	0.443	0.442	0.442	0.443	0.443	0.445	0.445
6 P	4000	0.382	0.384	0.382	0.384	0.381	0.381	0.382	0.381	0.379
7 P	10800	0.677	0.677	0.679	0.684	0.688	0.711	1.011	1.103	1.251
8 P	17800	0.788	0.799	0.887	0.933	0.941	0.973	1.102	1.173	1.213
9 P	3040	0.396	0.396	0.395	0.397	0.397	0.396	0.395	0.395	0.395
10 G	9300	0.601	0.599	0.603	0.602	0.687	0.785	1.122	1.619	2.336
11 G	4700	0.461	0.461	0.463	0.462	0.463	0.463	0.493	0.508	0.549
12 G	7500	0.527	0.526	0.527	0.526	0.527	0.597	0.697	0.728	0.751
13 G	1900	0.348	0.347	0.347	0.347	0.347	0.346	0.351	0.351	0.351
14 G	1870	0.423	0.423	0.421	0.421	0.421	0.420	0.422	0.421	0.421
15 G	8100	0.556	0.554	0.555	0.556	0.557	0.575	0.661	0.725	0.833
16 G	4450	0.449	0.451	0.451	0.451	0.451	0.453	0.474	0.491	0.533
17 G	3700	0.381	0.381	0.379	0.381	0.380	0.379	0.378	0.377	0.381
18 G	2100	0.354	0.353	0.353	0.353	0.351	0.353	0.354	0.355	0.354
19 P/G	22800	0.834	0.856	0.911	1.012	1.044	1.092	1.199	1.213	1.358
20 P/G	4170	0.415	0.415	0.414	0.416	0.414	0.416	0.561	0.593	0.672
21 P/G	2100	0.348	0.348	0.349	0.351	0.349	0.348	0.349	0.350	0.348
22 P/G	10900	0.661	0.661	0.669	0.677	0.759	0.796	0.833	0.877	0.997
23 P/G	2400	0.389	0.388	0.388	0.388	0.385	0.386	0.389	0.395	0.456
24 P/G	1960	0.385	0.385	0.386	0.386	0.387	0.385	0.385	0.384	0.385
25 P/G	1600	0.340	0.340	0.341	0.339	0.341	0.339	0.339	0.339	0.339
26 P/G	1100	0.336	0.336	0.335	0.337	0.335	0.335	0.334	0.336	0.336
27 P/G	1850	0.422	0.423	0.423	0.422	0.423	0.427	0.498	0.537	0.577

Note: All samples collected during flaming combustion.

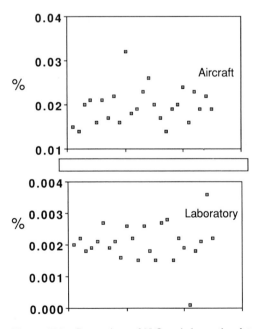

Figure 45.2 Comparison of N_2O emission ratios determined from aircraft collections and for near real time laboratory fires.

observation that the chemical analyses for the field determinations varied from about 2 to 48 hours after collection, suggest that the difference may be real. If the difference was largely the result of the N_2O-producing artifact, then more difference in ERs between individual large-scale fires would be expected.

Combustion itself may not be the only source of the N_2O in large fire smoke plumes. Physical, chemical, and biochemical processes in the soil (Levine et al., 1988) may be altered as a result of large biomass fires. For example, large releases of N_2O could result from the rapid heating of soils, and its escape into the atmosphere could contribute to the higher levels of N_2O measured over the large-scale fires. In any case, laboratory simulations of biomass burn–produced emissions of N_2O may not adequately represent the actual emissions from large-scale biomass fires. For a recent discussion on emissions of nitrous oxide from biomass burning, see Cofer et al. (1991).

VII

Biomass Burning and Particulates

The Particulate Matter from Biomass Burning: A Tutorial and Critical Review of Its Radiative Impact

Jacqueline Lenoble

The scattering and absorbing particles that constitute smoke modify the transfer of radiative energy in the atmosphere. The solar energy reaching the ground level is reduced, and Robock (1988) has shown that the surface temperatures were reduced between 1.5°C and 4°C during daytime over North America under the extended smoke plume from a large forest fire in British Columbia, Canada, at the end of July 1982. Modeling of this event (Westphal et al., 1989) has confirmed the surface cooling and shown a heating rate of about 2°C per day in the mid-troposphere due to the smoke absorption. The smoke particles also influence the longwave radiation, but at a much lower rate; the scattering is negligible because the particles are small compared to the wavelengths, and the absorption-emission processes occur at a temperature not very different from the ground temperature. In the above-mentioned event, no changes of the surface temperatures during nighttime were observed.

The smoke particles, acting as condensation nuclei, can modify the cloudiness. Imbedded in a cloud layer, the absorbing particles lead to a decrease of the cloud albedo and an increase of the cloud absorption (Chylek et al., 1984); this indirect radiative impact of the smoke particles can be as important as the direct impact in the clear atmosphere, but it is more difficult to analyze.

The modification of the upward radiance observed by satellite instruments allows the remote sensing of the smoke plumes (Kaufman et al., 1989); if the ratio of gas and particles is known, this provides a possibility for an indirect remote sensing of the gas emission of the fire.

Several experiments have been carried out, either on natural or prescribed forest fires or on laboratory fires, in order to observe the gas and particles emissions and to measure as many parameters as possible simultaneously (Patterson et al., 1986; Andreae et al., 1988; Cofer et al., 1988a,b; Radke et al., 1988). The influence of important injection of smoke into the atmosphere has been studied with climate models in the context of the so-called nuclear winter problem (SCOPE, 1986; Ramaswamy and Kiehl, 1985; Crutzen et al., 1984; Turco et al., 1983).

In the next section we review the particles' radiative characteristics, as they are linked to their physicochemical nature. The following section shows how the influence of smoke particles on solar radiation depends on a few major parameters, the most difficult to obtain being the single scattering albedo (final section).

Physicochemical and Radiative Characteristics of Smoke Particles

Biomass burning produces solid and liquid particles of various shapes and sizes and complex chemical structure. Liquid particles are more or less spherical, but solid particles (as, for example, carbon black crystals) are far from being spheres and can present complex shapes. For the purpose of computing their radiative characteristics, nonspherical particles are generally replaced by equivalent spheres, which can be defined either as equivalent in volume or in section (Janzen, 1980; Chylek and Ramaswamy, 1982); it seems that the best choice of equivalence depends on the type and size of the particles. The sizes of the particles range from particles smaller than one tenth of a micrometer (μm) (nucleation mode) to particles of several micrometers (coarse mode); the particles between approximately 0.1 μm and 2 μm (accumulation mode) contain most of the mass and are responsible for most of the radiative effects.

The chemical nature of a material reflects on its radiative characteristics only by its complex refractive index, the imaginary part being proportional to the material absorption coefficient. The smoke particles are made of a mixture of several different materials; from a radiative point of view, we can consider two types of components: (1) the organic and inorganic nonabsorbing or slightly absorbing components, with a real refractive index around 1.50 and an imaginary part smaller than 0.01; and (2) the strongly absorbing soot, or graphitic carbon, with a real refractive index around 1.9 to 2.0 and an imaginary part between 0.5 and 1.0 (Janzen, 1980). For a spherical particle of known radius r and refractive index $m = m_r - i m_i$, the radiative characteristics defined below can be computed from the Mie theory (Van de Hulst, 1957); for a given size distribution $n(r)$ of the same kind of particles, the contribution of each particle is weighted by $n(r)$. For an external mixture of two kind of particles (Figure 46.1), the radiative characteristics of the particles have to be computed for each component using its refractive index m_1 or m_2 and then weighted by the relative number of each kind of particle. For an internal volume mixture, the particle refractive index has to be computed from the two component refractive indices weighted by the volume concentration of each component; then the Mie theory is applied to this unique type of particle. Internal mixtures, either with an absorbing shell or an absorbing core, need to be handled by a more complex Mie code for nonhomogeneous particles.

The radiative characteristics of the smoke are defined in Figure 46.2. Locally the extinction coefficient, the single scattering albedo, and the phase function derive from the physicochemical characteristics by the Mie theory, as discussed above. The details of the phase function are, for most problems, of secondary importance, and the asymmetry factor g is defined as an average cosine of the scattering direction weighted by the phase function. The whole smoke layer is characterized by its optical depth δ, which is the extinction integrated over the layer, an average single scattering albedo $\bar{\omega}$, and an average

asymmetry factor g. The vertical profiles of extinction and absorption influence only the repartition of heating rate in the smoke layer. All these radiative parameters are wavelength dependent and should be averaged with the solar spectrum as a weighting function, to study the particle radiative impact.

The optical depth is easily measured by a ground-based sunphotometer (see Figure 46.2); from an aircraft flying in the smoke plume the optical depth above the flight level can be measured, leading to a vertical profile of the extinction coefficient. Measurements of the spectral variation of the optical depth may allow the retrieval of the particle size distribution by inversion of the Mie equation, if the refractive index is known. The optical depth is highly variable with the smoke density but is often between 0.1 and 2 in the visible (Pueschel et al., 1988). The asymmetry factor can be deduced only from measurements of the phase function, which are delicate, or it can be computed from the particle size distribution; however it varies only over a limited range of values from 0.5 for small particles to 0.8 for large particles, and, as we will see in the following section, its influence is not predominant. The single scattering albedo is a very important parameter, and in the final section we will

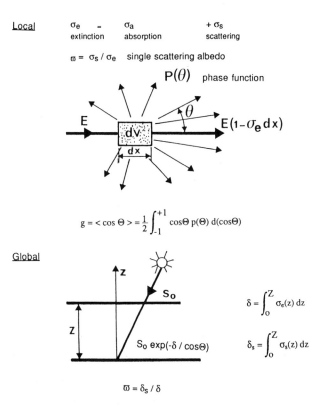

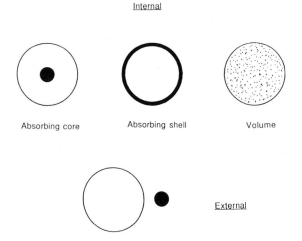

Figure 46.1 Mixtures of absorbing and nonabsorbing materials.

Figure 46.2 Definitions of the radiative characteristics.

Table 46.1 Some measured values of $\bar{\omega}$ for biomass burning

Conditions	$\bar{\omega}$	Authors
Laboratory fires		
Flaming	0.66	Patterson and McMahon (1984)
Smoldering	0.97	
Several prescribed or	0.85	Radke et al. (1988)
natural forest fires	0.89	
(averaged $\bar{\omega}$)	0.80	
	0.87	
	0.88	
	0.82	
Same fire		
Flaming	0.93	Radke et al. (1988)
Plume	0.78	
Several forest fires in	0.95	Kaufman et al. (1989)
Brazil (remote	0.98	
sensing)	0.90	
	0.92	
	0.94	
	0.93	

discuss its measurements in more detail. Table 46.1 presents some values of simple scattering albedo observed in forest fires; the dispersion of these values probably reflects a real variability, but also the measurement uncertainty.

Influence of a Smoke Layer on the Solar Radiation

As mentioned previously, the major direct climatic influence of a smoke layer is its effect on the solar radiation. The solar energy available at the ground level is reduced, whereas a part of the incoming radiation is reflected, leading to a possible modification of the radiation budget of the atmosphere above the smoke, and a part is absorbed in the smoke layer itself, which is heated.

Figure 46.3 illustrates the influence of the main parameters of the smoke in the simple case of a plane parallel layer above a black ground. The data are taken from Van de Hulst (1980) and are presented versus the single scattering albedo to underline the important effect of this parameter. Figure 46.3a presents the results for an overhead sun and particles with an asymmetry factor $g = 0.5$; the optical depths δ are 0.5, 1.0, and 2.0. As expected, the influence of the smoke increases with the optical depth (related to the total amount of particles). When the single scattering albedo $\bar{\omega} = 1$ (nonabsorbing particles), about 82% of the solar energy incoming on the top of the layer reaches the ground and 18% is reflected, for $\delta = 1$, which corresponds to an average smoke layer. For the

same $\delta = 1$, if $\bar{\omega}$ decreases to 0.8, only 68% of the solar energy reaches the ground, and 10% is reflected, whereas 22% is absorbed in the smoke layer itself; the vertical distribution of this absorbed radiative energy depends, of course, on the vertical structure of the smoke and commands the modification of the temperature profile. From Table 46.1, the range of likely values of $\bar{\omega}$ is between 0.8 and 1; however, much smaller values can be found when the black carbon component increases. Figure 46.3b illustrates the small influence of g; when g increases due to the presence of larger particles, the forward scattering dominates, leading to an increase of the transmittance and a decrease of the reflectance. Figure 46.3c shows the influence of the solar zenith angle; when it increases the transmittance decreases and the reflectance increases, whereas the longer path in the layer leads to an increased absorption; in the case $\delta = 1$ for the sun at 18° above the horizon, and $\bar{\omega} = 0.8$, the absorbed energy is as large as 40% of the incoming solar energy, the remaining 60% being approximately half transmitted to the ground and half reflected.

This study confirms that the radiative impact of an extended smoke layer strongly depends on the sun height, which is easily known; on the smoke optical depth, which can be measured from the ground; and on the single scattering albedo, which is much more difficult to obtain. The horizontal and vertical inhomogeneity of the layer may, of course, lead to strong local variations.

Measurement of the Single Scattering Albedo

The particle single scattering albedo characterizes the particle absorption and decreases from one for nonabsorbing particles to values which can be close to zero for strongly absorbing particles. It is therefore related to the absorption coefficient, or to the imaginary part m_i of the refractive index of the material which the particle is made of. A first category of method is based on laboratory measurements of the absorption (or of m_i) for the particle material, using particles collected in the smoke on filters. A lot of work has been done to improve this kind of method, but in all cases a further step is needed to compute $\bar{\omega}$ using m_i and the particle size distribution $n(r)$ in the Mie code; $n(r)$ has to be determined independently, either by sampling of the particles and microscopic observations, or by various in situ particle counters; a further possibility is by inversion of the optical depth $\delta(\lambda)$ measured at several wavelengths. The relation between $\bar{\omega}$ and m_i is not at all direct; Figure 46.4 illus-

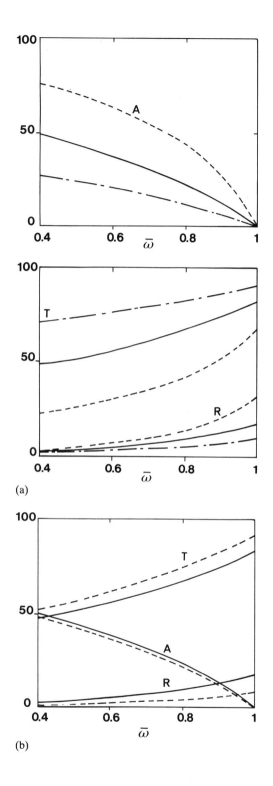

(a)

(b)

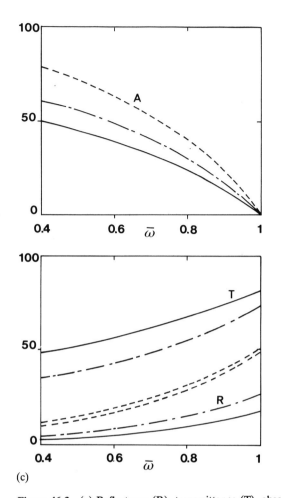

(c)

Figure 46.3 (a) Reflectance (R), transmittance (T), absorptance (A) of a plane-parallel homogeneous smoke layer over a black ground (from tables in Van de Hulst, 1980); solar zenith angle = 0°; g = 0.5; dash-dotted line δ = 0.5; solid line δ = 1; dotted line δ = 2. (b) Same as Figure 46.3a; δ = 1; solid line g = 0.5; dotted line g = 0.75. (c) Same as Figure 46.3a; δ = 1; g = 0.5; solar zenith angle = 0° (solid line), 45° (dash-dotted line), 72° (dotted line).

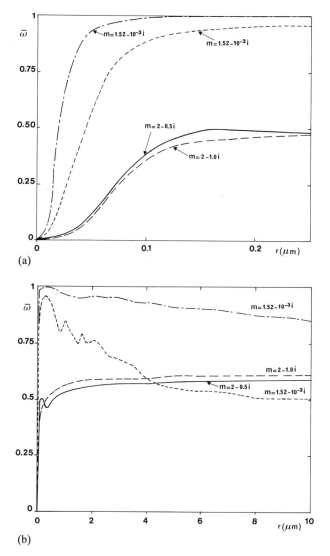

(a)

(b)

Figure 46.4 Single scattering albedo $\bar{\omega}$ versus particle radius for $\lambda = 0.63$ μm; $\bar{\omega}$ is unchanged when r/λ is kept constant. (a) Radius range 0–0.2 μm; (b) radius range 0–10 μm.

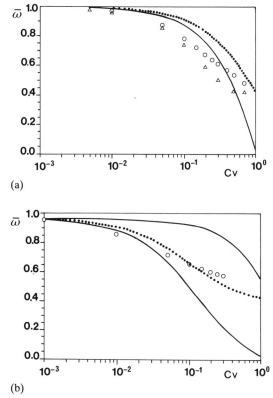

(a)

(b)

Figure 46.5 Single scattering $\bar{\omega}$ versus soot volume ratio C_v, from Ackerman and Toon (1981). (a) Curves are for accumulation mode sulfate particles externally mixed with nucleation mode (solid) and accumulation mode (dotted) soot. Symbols are for internal mixtures of sulfate core and soot shell (circles) and soot core and sulfate shell (triangles). (b) Curves are for coarse mode soil particles externally mixed with nucleation mode (lower solid), accumulation mode (dotted), and coarse mode (upper solid) soot. Circles are for an internal mixture of soil core and soot shell.

trates for monodisperse spherical particles how $\bar{\omega}$ varies with the radius r, for four different values of the refractive index; the two curves $m = 1.52 - 10^{-3}i$ and $m = 1.52 - 10^{-2}i$ correspond to slightly absorbing soil or organic particles; the curves $m = 2 - 0.5i$ and $m = 2 - 1.0i$ correspond to values expected for soot particles. For small particles, absorption always dominates scattering ($\bar{\omega} \approx 0$); for $m_i = 10^{-3}$, $\bar{\omega}$ rapidly tends to values close to 1 when the particle size increases; for large values of m_i, $\bar{\omega}$ reaches only values close to 0.5 in the size range 0.15 to 1 μm. For very large particles of radius r, the Mie solution can be approached by assuming that a fraction πr^2 of the

incident radiation is diffracted in a strong forward peak, and another fraction πr^2 intercepted by the particles is more or less strongly absorbed (depending on m_i). Even for moderate absorption, only a small part of this intercepted πr^2 energy is transmitted by very large particles, but the surface reflexion increases with m_r, leading to values slightly larger than 0.5 for $\bar{\omega}$. As mentioned before, the existence of different kinds of mixtures leads to more complexity. Figure 46.5 reproduces Figures 1 and 2 from Ackerman and Toon (1981) and shows how $\bar{\omega}$ varies with the soot volume ratio in different mixtures of the same components.

Direct measurements of $\bar{\omega}$ can be performed either locally inside the smoke, or globally for the whole smoke layer. The local measurements are based on simultaneous measurements of σ_e and either σ_s or σ_a;

Table 46.2 Single scattering albedo $\bar{\omega}$: Comparisons of measurements by various methods

Participant numbers	Soot	Soot and salt	
		50% soot	4% soot
1			
2			
3	0.180	0.569	0.882
4	0.115	0.610	
5			
6	0.180		0.970
7			
8			
9	0.146	0.590	0.949
10a	0.204	0.849	0.977
10b	0.180	0.568	0.933
11	0.150		
12	0.090	0.431	0.894
13			
14			
15	0.292	0.710	0.971
Average	0.171	0.618	0.939

From Gerber and Hindman (1982).

the extinction coefficient σ_e is relatively easy to measure; the scattering σ_s is generally measured by a integrating nephelometer and the absorption σ_a by an acoustic spectrophone; both the measures of σ_s and σ_a are rather delicate. On a global scale various indirect approaches have been proposed to obtain the single scattering albedo: flux divergence measurement in the layer, ratio of diffuse to direct flux at the ground level, remote sensing over surfaces of variable reflectance. Numerous references on instruments and methods for measuring $\bar{\omega}$ can be found in Gerber and Hindman (1982).

In 1980 a workshop on light absorption by aerosol particles took place at Colorado State University, Fort Collins, Colorado. During 10 days, 15 experimenters performed measurements with different kinds of instruments on the same artificial aerosols produced in a chamber. Table 46.2 presents the results of this comparison, for pure soot particles and for mixtures of soot and salt particles (Gerber and Hindman, 1982); the dispersion of the results underlines the difficulty of obtaining $\bar{\omega}$ with good accuracy. Although instruments are regularly improved, there has been no definitive breakthrough during the last 10 years in the measurement of the single scattering albedo.

Conclusion

The impact of a smoke plume on solar radiation, which may influence the weather and climate and which makes possible its remote sensing, depends mainly on two parameters: the average optical depth at solar wavelengths and an average single scattering albedo. Next parameters, in order of importance, are the asymmetry factor and the vertical profile of the extinction coefficient. A major effort remains necessary to obtain reliable measurements of the single scattering albedo.

The Role of Biomass Burning in the Budget and Cycle of Carbonaceous Soot Aerosols and Their Climate Impact

Joyce E. Penner, Steven J. Ghan, and John J. Walton

The large-scale burning of biomass for agricultural land clearing purposes in the tropics has been estimated to have a major impact on the atmospheric cycles and budgets of a suite of trace species (Seiler and Crutzen, 1980; Hao et al., 1989), with a variety of possible implications. In the case of the budget for NO_x, for example, it has been estimated that biomass burning can create levels of nitrate in precipitation that are similar to those caused by fossil fuel burning at more northern latitudes (Atherton et al., 1990; Dignon et al., 1990).

In addition to trace gas emissions, biomass burning creates particles. These may be important for climate through both direct and indirect radiative effects. The direct effect is caused by the scattering and absorption of sunlight by the particles themselves. For example, it has been estimated that for a global average model, the scattering of sunlight by particles with an optical depth of 0.125 and a single scattering albedo of 0.95 would lead to a cooling of the planet by 1.2°K (Charlock and Sellers, 1980). On the other hand, a more absorbing aerosol, with a single scattering albedo of 0.75 and the same optical depth, was predicted to lead to a net warming of 0.5°K.

The indirect effect of aerosols on climate is caused by their interaction with clouds and their effect on cloud albedo (Twomey, 1977). In this mechanism, particles act as cloud condensation nuclei (CCN). An increase in the number of CCN can lead to an increase in cloud droplet concentrations. Since cloud optical depth determines the cloud albedo and is proportional to cloud droplet number concentration to the one-third power, this increase can make clouds more reflective of solar radiation.

In this chapter, we estimate the climate impact from biomass burning due to both the direct radiative effects of smoke and the indirect effects on cloud albedo. We use the model of Walton et al. (1988) run in conjunction with the Lawrence Livermore National Laboratory version of the National Center for Atmospheric Research Community Climate Model 1 (NCAR CCM1). We describe the sources of aerosol particles from biomass burning as well as their soot content. After describing our model, we compare this latter quantity (or the predicted soot or black carbon concentration) to measured soot concentrations in the Southern Hemisphere. This provides a partial confirmation of our modeling procedures. Further confirmation of our predicted aerosol concentrations is also presented by comparing our predicted smoke optical depths with typical optical depths measured from satellites. Finally, we present the predicted change in outgoing solar radiation from biomass burning and summarize our estimate of the climate impact from biomass burning together with our conclusions.

Sources of Aerosol Particles and Soot in the Tropics

Besides biomass burning, the continental sources of aerosol include direct emissions from fossil fuel combustion, gas-to-particle conversion of the gas phase products from fossil fuel combustion, gas-to-particle conversion of organic emissions from vegetation, other biogenic sources of particles and low-volatility gases, and wind-blown dust (Prospero et al., 1986). While wind-blown dust has been estimated to provide the largest source of the atmospheric aerosol on a mass basis (e.g., Peterson and Junge, 1971), most of the particle mass in dust storms is in sizes that are greater than one micron. Therefore, we do not expect dust to dominate the number distribution of the submicron aerosol that makes up most of the particles that act as CCN. In tropical regions, the remaining continental sources of aerosol have also been estimated to be small relative to the source from biomass burning. For example, Cros et al. (1981) have shown that the dry season production rate of Aitken particles in the African savanna is increased by a factor of 100 over that in the wet season. In tropical forest areas, the production rate is also increased in the dry season. Because the fossil fuel sources should be relatively constant with season and the natural sources should be enhanced during the wet season, if these sources were dominant, one would expect greater production rates in the wet season. Therefore, the enhanced production during the dry season is likely due to biomass burning. We shall therefore assume that the number concentration of submicron aerosol particles is dominated by the source from biomass burning. Further, we assume that the aerosol particles produced by biomass burning are active as CCN. This is consistent with the reports of Hallett et al. (1989) and Rogers et al. (1990). They report that the fraction of particles that act as CCN (at 1% supersaturation) from the burning of forest slash and chaparral brush is between 80% and 100%. Radke et al. (1978) and Eagan et al. (1974) report a similar high production rate of CCN from burning vegetation. Below, we estimate the magnitude of the source of CCN from biomass burning by combining estimates of the amount of fuel burned with estimates of the emission factor for smoke from vegetation, assuming that all of the particles thus formed act as CCN.

Hao et al. (1989) and Logan et al. (1990) present estimates of the mass of vegetation burned during logging and land clearing practices in tropical areas. In addition, Logan et al. (1990) estimate the amount of wood burning throughout the world, as well as the amount of agricultural burning outside of the tropical regions of Central and South America, Africa, Indonesia, and Australia. By far, the largest amount of burning takes place in tropical regions and is primarily due to the burning of grassland areas for land clearing purposes. Dignon et al. (1990) compared the estimates of biomass burning from Hao et al. (1989) and Logan et al. (1990) and showed that although there could be a significant variation in the amount of fuel burned in any given 5° by 5° latitude/longitude zone, the amount of burning assumed over each continent was quite similar in the two studies, as was the gross spatial distribution (perhaps because of the similar methods and data sources used). In the following discussion, we shall adopt the estimates of Hao et al. (1989) for the amount of biomass burned in each 5° by 5° latitude/longitude zone and estimate the emission of particulates from this burning. We also separately estimate the amount of soot produced during this burning. Because soot is a unique indicator of a combustion source of particulates, our comparison of predicted soot concentrations with measured concentrations can provide some measure of confidence in our modeling procedures (see the section on results).

The emission of smoke and soot from biomass burning can vary considerably, depending on the conditions during the burn, including the moisture content of the fuel. Studies have shown that much more smoke is produced during smoldering conditions, while the fraction of smoke that is soot is much higher during flaming conditions (Patterson et al., 1986). Open burning conditions usually result in a combination of smoldering and flaming combustion. Radke et al. (1988) report emission factors for particulates less than 3.5 μm in diameter from 10 different vegetation fires that took place in the Pacific Northwest and California between 1986 and 1987. These fires involved a combination of wild and prescribed forest burns and one agricultural burn of wheat stubble. They derived an average emission factor of 16 grams (g) of smoke per kilogram (kg) of fuel burned. They also separately measured the specific absorption coefficient for smoke, deriving a value near 0.7 square meters per gram (m^2/g). We may use this to estimate a soot emission factor, by multiplying the measured smoke emission factor by the ratio of the measured absorption coefficient of the smoke and the absorption coefficient for pure soot (7 to 10 m^2/g). Table 47.1 summarizes the resulting smoke and soot emissions in tropical regions, combining the estimated biomass burned in each continent from Hao et al. (1989) with these smoke and soot emission factors. As shown there, a total of 6 teragrams per year (Tg/yr) of soot and 80 Tg/yr of smoke may be produced by biomass burning.

We may also estimate the smoke emissions by using the measured emission factor for smoke from the burning of rice straw with only 10% moisture content

Table 47.1 Smoke emissions from biomass burning

	Biomass burned[a] ($\times 10^{12}$ g/yr)	Soot carbon emissions[b] ($\times 10^{12}$ g/yr)	Fine particulate emissions[c] ($\times 10^{12}$ g/yr)
Tropical America	1356	1.6	22.0
Africa	2818	3.3	45.0
Asia	351	0.4	5.6
Australia	425	0.5	6.8
Total	4950	5.7	79.0

a. Hao et al. (1989).
b. Developed using the emission factors for smoke from Radke et al. (1988) together with an estimate of its soot content derived by multiplying the smoke emission factor by the ratio of the measured absorption coefficient of the smoke, and the absorption coefficient for pure soot (~10 m^2/g).
c. Estimated using the average emission factor for particles less than 3.5 μm in diameter reported by Radke et al. (1988) for vegetation fires.

and rice straw with 25% moisture content (Goss and Miller, 1973). These estimates result in a total of 25 to 80 Tg smoke/yr. The high estimate in this case (for 25% fuel moisture content) is about a factor of 2 smaller than the estimate obtained using the emission factor for wheat stubble derived by Radke et al. (1988), but is a factor of 3 larger than the emission factor for smoke measured by Andreae et al. (1988) (obtained from only two measurements). The emission factor for soot derived by these latter authors ranged from 8% to 22% of total smoke emissions and is in reasonable agreement with the relative emission factor for soot derived above from the data of Radke et al. (1988) (i.e., 7% to 10%). In the following discussion, we adopt estimates based on the emissions of smoke and soot derived from the work of Radke et al. (1988). The spatial distribution of the derived smoke source is shown in Figure 47.1. The approximate spatial distribution of the derived soot source may be obtained by scaling the contour lines shown in Figure 47.1 by 1/10th.

Model Description

For this study, we use the three-dimensional Lagrangian transport and deposition model of Walton et al. (1988) together with the LLNL version of the NCAR CCM1 model. A description of the current model treatment of mixing and deposition (al-

tered from that described in Walton et al., 1988) may be found in Penner et al. (1991), while the CCM1 model is described in Williamson et al. (1987). The LLNL version of the CCM1 has been altered to allow for the radiative effects of changes in droplet size distribution as described in Ghan et al. (1990). Removal of smoke takes place by both dry and wet deposition. In this version of the transport model, smoke is treated as having a deposition velocity of 0.1 centimeter per second (cm/s), which represents a reasonable average of the estimated deposition velocities for particles near 0.1 μm radius under a variety of atmospheric stability, wind speed, and surface roughness conditions (Sehmel, 1984). We note, however, that a range from 0.05 to 0.2 cm/s might be considered reasonable. The rate of precipitation scavenging of smoke in the model, R, is assumed to be proportional to the precipitation rate, dp/dt (in cm/hour). The proportionality constant, S, is derived based on Clarke's (1989) measurement of the washout ratio for soot. Clarke measured a ratio of the soot mixing ratio in Arctic snow to the soot mixing ratio in Arctic air equal to 100. Assuming the same washout ratio applies to biomass smoke and that the smoke mixing ratio in the lowest atmospheric layer does indeed result from precipitation scavenging in that layer, this results in a scavenging coefficient of $S = 1$ cm^{-1}. This may be compared to the coefficient for nitrate scavenging derived by Penner et al. (1991) of 2.4 cm^{-1}.

In the model, the effect of smoke on cloud albedos is calculated after establishing a steady state for the smoke distribution by running a perpetual July (the dry season) simulation in the transport model for a period of approximately six months. For the calculations shown below, we have used the annual average biomass emissions estimated from the data of Hao et al. (1989). After reaching a steady state, the transport model and climate model are run concurrently for 90 days, allowing the smoke to affect cloud droplet number distributions in the following manner. Two simulations are run. In the background simulation, droplet concentrations are assumed to be 50 cm^{-3} at all locations. In the model with smoke, the droplet concentration is assumed to be 50 cm^{-3} plus the predicted CCN concentration up to a maximum of 200 cm^{-3} (a typical droplet number concentration in continental air masses). The CCN concentration is derived from the predicted smoke mass concentration by assuming that all of the smoke would be present as 0.1 μm radius particles. This radius is close to the mean mode radius for the measured smoke volume

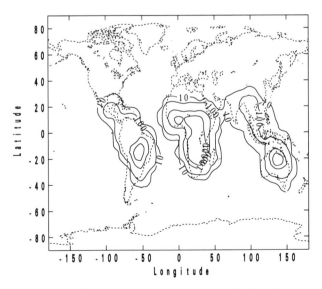

Figure 47.1 The annual average source of smoke from biomass burning. The total source strength is 80 Tg/yr. Contour intervals are 10, 100, 1000, and 2000 kg/km²/yr.

distributions typically present above vegetation fires (Radke et al., 1988).

Results

Table 47.2 shows a comparison of the predicted concentration of soot from the model for July with measured concentrations at rural and remote sites in the Southern Hemisphere. Our comparison is restricted to soot (as opposed to total smoke or aerosol) because soot provides a unique signature for combustion. In the comparisons, we expect biomass burning to provide a significant fraction of the total soot found to be present at Southern Hemisphere locations. Similarly, at Northern Hemisphere locations, we expect our predicted soot concentrations from biomass burning to be too low, reflecting the fact that we have not included the fossil fuel sources of soot (see Ghan and Penner, 1990). As expected, the predicted soot concentrations in the Northern Hemisphere are too small. However, as shown in Table 47.2, soot from biomass burning appears to explain roughly 25% to 30% of the rural concentrations observed in Ecuador and Brazil, while close to all of the observed concentrations in the eastern Pacific are explained, as are those at Mauna Loa and the South Pole. The implication is that fossil fuel sources may contribute a large fraction of the measured concentrations at rural sites in South America. If these additional fossil fuel sources were exported to the remote Pacific and the South Pole, the model would overpredict soot concentrations at these more remote sites. A substantial

overprediction would call into question our modeling procedures, especially our assumed rates of removal of soot by dry and wet deposition. Thus, in this respect, the present predictions for the climate effects from biomass burning could be upper limits. However, the measured concentrations in both Equador and Brazil or at the more remote sites could also be explained if they were poor representations of the climatologically averaged concentrations that the model simulations intend to represent. With only a very few measurements it is impossible to know whether the agreement with the measurements is a strong confirmation of our model or not. However, we are encouraged by the degree of agreement with the data, which span over 4 orders of magnitude in the variation of soot concentrations. Furthermore, the comparison shows that our biomass burning source is certainly a major source of soot in the Southern Hemisphere.

Figure 47.2a shows the predicted column abundance of smoke from biomass burning, while Figure 47.2b shows the inferred optical depths assuming specific scattering and absorption coefficients of 5 and 1 m^2/g, respectively. We note that the absorption coefficient adopted for dry particles is comparable to that measured by Radke et al. (1988), while the scattering coefficient is close to that derived by combining Radke et al.'s (1988) measured single scattering albedo with his measured specific absorption coefficient. The scattering optical depth would increase by a factor of 10 over that depicted in Figure 47.2b if the particles were swollen under the influence of high

Table 47.2 Measured and predicted soot concentrations in July (in ng/m^3)

	Measured concentration	Predicted for 5.7 Tg biomass burning	Latitude	Reference
Rural locations				
Florida	830.0	8.8	24°N–29°N	Andreae et al. (1984); Andreae (1983)
N. Carolina	520.0	3.0	33°N–38°N	Andreae et al. (1984); Andreae (1983)
Southwest U.S.	220.0	3.1	29°N–42°N	Andreae et al. (1984); Andreae (1983)
Ecuador	520.0	161.0	2°S–2°N	Andreae et al. (1984); Andreae (1983)
Brazil	620.0	250.0	2°S–2°N	Andreae et al. (1984); Andreae (1983)
Remote locations				
Arctic (NyAlesund)	3.0	0.1	78°N–82°N	Heintzenberg (1982)
Arctic (Barrow)	80.0	0.1	69°N–73°N	Rosen et al. (1984)
Central North Pacific	4.0	1.7	2°S–51°N	Clarke (1989)
Eastern Pacific	4.7	5.8	29°S–51°N	Clarke (1989)
Mauna Loa	2.5	2.8	20°N	Clarke et al. (1984)
South Pole	0.3	0.3	90°S	Hansen et al. (1988)

Note: The predicted concentrations are low in the Northern Hemisphere because we have not included the sources of soot from fossil fuel burning (see Ghan and Penner, 1990).

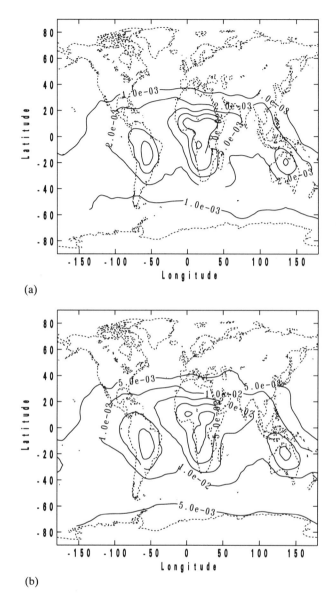

(a)

(b)

Figure 47.2 (a) The predicted column abundance of smoke in July. Contour intervals are 0.001, 0.002, 0.005, 0.01, 0.015, and 0.02 g/m^2. (b) The predicted optical depth for dry smoke (assumes $k_a = 1$ m^2/g and $k_s = 5$ m^2/g). Contour intervals are 0.005, 0.01, 0.02, 0.05, and 0.1 optical depths.

humidities to an equilibrium size of 0.2 μm radius. This latter radius is the predicted size of a fully soluble particle of dry radius 0.1 μm at 95% relative humidity. The resulting total optical depths bracket those derived by satellite measurements in the eastern United States which lie between 0.4 and 1.4 optical depths for clear and hazy days, respectively (Kaufman and Sendra, 1988). This comparison adds support to our assumption that biomass burning is a dominant source of aerosol and cloud condensation in these regions in the dry season, because CCN concentrations are measured to be quite similar in North America and Africa (Twomey and Wojciechowski, 1969).

Figure 47.3 shows the predicted number concentration of smoke particles at the surface (Figure 47.3a) together with the zonally averaged profile of smoke particle concentration (Figure 47.3b). For this calculation, we assumed a radius for the smoke particles of 0.1 μm and a material density of 1 g/cm^3. As shown there, concentrations at the surface reach values larger than 1000 cm^{-3} over large regions in South America, Africa, and Australia. For a larger extended region covering parts of the Atlantic and Pacific oceans, concentrations of over 100 cm^{-3} are predicted. The predicted concentrations decrease significantly with altitude, so that their numbers would be much smaller at the level of cloud formation. If all the particles represented in Figure 47.3 act as CCN during cloud formation, we expect an increase in droplet number concentration over the tropical regions associated with biomass burning. That such increases in cloud droplet concentrations occur is, of course, demonstrated by the increased concentrations of cloud droplets over continents relative to those seen over ocean regions. These increased concentrations are thought to be directly related to the larger CCN concentrations found over continents. Although not all of the predicted particles would necessarily form droplets, our model assumes a simplified treatment for the purpose of demonstrating the possible magnitude of the climate impact from these aerosols.

Figure 47.4a shows the predicted outgoing shortwave radiation at the top of the atmosphere over land as a function of latitude for the background case (which assumes cloud droplet number concentration equals 50 cm^{-3} at all locations) and for the case with biomass burning. In this calculation the optical properties for a dry smoke aerosol were assumed. Approximately half of the increase in outgoing shortwave radiation results from the induced change in cloud

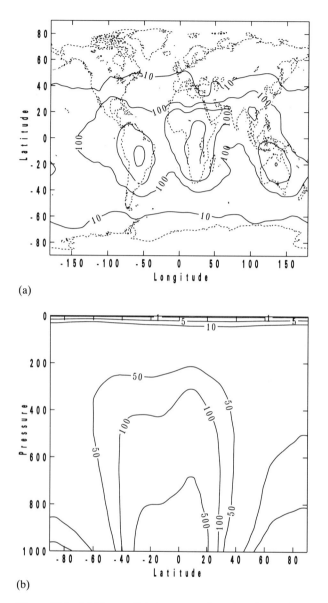

(a)

(b)

Figure 47.3 (a) The predicted concentration of smoke particles and cloud condensation nuclei at the surface in July. Contour intervals are 10, 100, 1000, and 2500 particles cm^{-3}. (b) The predicted zonal average concentration of smoke particles as a function of latitude and altitude. Contour intervals are 1, 5, 10, 50, 100, and 500 particles cm^{-3}.

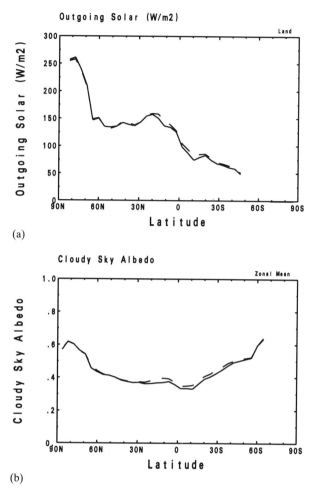

(a)

(b)

Figure 47.4 (a) The predicted zonal average of the outgoing solar radiation over land as a function of latitude in W/m^2 for the background case (solid line) and for the case with biomass burning (dashed line). (b) The predicted zonal average cloudy sky albedo as a function of latitude for the background case (solid line) and the case with biomass burning (dashed line).

albedos. As shown there, there is a significant increase in outgoing radiation for the latitudes between 30°N and 30°S. Figure 47.4b shows the longitude-averaged cloudy-sky albedo for the background case and for the case with biomass burning. Apparently, cloud albedos are significantly increased as far south as 60°S. This change in cloud albedo results in an increase in the total planetary albedo of 0.5% from 0.295 in the background case to 0.300 in the case with biomass burning.

Conclusions

The predicted increase in planetary albedo of 0.005 is equivalent to a net change in the earth's radiation balance of -1.8 Watts per square meter (W/m^2). This figure could be increased substantially if we also included the direct radiative effect of hazy smoke particles under the influence of high relative humidities. The predicted change in radiation balance is comparable in magnitude to the 1.7 W/m^2 expected from the increases in CO_2 and other greenhouse gases that have been experienced to date (Ramanathan et al., 1985). For this estimate, we scaled the predicted temperature change between 1880 and 1980 reported by Ramanathan et al. (1985) to that for doubled CO_2 from the same model (see Ramanathan et al., 1987) and assumed that the direct radiative effect of doubled CO_2 is approximately 4 W/m^2. (These model estimates neglect most feedbacks as well as the thermal inertia of the oceans.) According to these estimates, biomass burning has helped to mask the atmospheric temperature increase expected from increasing greenhouse gases to date. We expect that an even greater mask may be present due to the particulate and sulfur emissions associated with fossil fuel use, since the emissions of the latter were equal to between 63 and 80 Tg S/yr for 1980 (Hameed and Dignon, 1988; Moller, 1984) and are thus approximately a factor of 2 larger than the estimate of smoke emissions given here, if half of the sulfur is converted to $SO_4^=$. In fact, preliminary results from our global model suggest that the zonal average column abundance of sulfate from fossil fuel emissions of SO_2 is approximately twice that predicted here. Since the atmospheric lifetime of smoke particles and other aerosols is short, if the production of anthropogenic particles were to stop, we would expect to experience the full level of greenhouse gas forcing almost immediately. It is imperative that we carry out the research needed to identify and quantify the effects of anthropogenic sources of aerosol on clouds and the earth's radiation budget.

Acknowledgments

This work was performed under the auspices of the U.S. Department of Energy by the Lawrence Livermore National Laboratory under Contract W-7405-Eng-48 and supported in part by the D.O.E. Office of Energy Research's Climate Linkages Program.

Characteristics of Smoke Emissions from Biomass Fires
of the Amazon Region–BASE-A Experiment

Darold E. Ward, Alberto W. Setzer, Yoram J. Kaufman,
and Rei A. Rasmussen

Biomass burning is a major source of emissions of "greenhouse gases" and particulate matter to the atmosphere (Seiler and Crutzen, 1980; Crutzen et al., 1985; Mooney et al., 1987; Radke, 1989). The net effect on global climate is not well quantified (Mooney et al., 1987; Houghton and Woodwell, 1989), and there is a need for better source information regarding the total biomass consumed globally and the quantity and time of release of the important emissions. Measurements of particulate matter emissions are important due to their effect on cloud microphysics and reflectivity (Radke et al., 1988; Radke, 1989). Furthermore, fires and emitted particulate matter are the only components of biomass burning that can be monitored from satellites (Malingreau and Tucker, 1988; Holben et al., 1990; Setzer et al., 1990). Remote sensing of particulate matter and fires from Advanced Very High Resolution Radiometer (AVHRR) was used to estimate the emissions of trace gases and particulate matter from deforestation in the Amazon Basin (Kaufman et al., 1990). This estimate uses measured emission ratios of trace gases and particulate matter.

Global emissions of gases and particles from biomass burning are usually estimated based on extrapolations from work performed in North America (Ward and Hardy, 1984) or Europe (Hao et al., 1990). A few measurements of both gas concentrations and particulate matter concentrations were taken for fires in the Amazon region by Andreae et al. (1988). Artaxo et al. (1988) studied the inorganic fraction of aerosols collected from ground monitoring stations. Several have studied the gaseous composition of the atmosphere relative to biomass burning in Brazil (Crutzen et al., 1979; Greenberg et al., 1984; Crutzen et al., 1985; Andreae et al., 1988). The Biomass Burning Airborne and Spaceborne Experiment–Amazonia (BASE-A) was designed to study both aerosol and gaseous emissions from fires in the Amazon region of Brazil from an airborne sampling platform. There were many fires observed during the several days of the experiment. One of the objectives of BASE-A was to validate estimates of deforestation burning made using satellite imagery (Setzer et al., 1990).

Previous research by Kaufman et al. (1990) demonstrated the feasibility of sensing the concentration of atmospheric aerosol from space. The ratios of smoke aerosol concentration to the concentration of specific gases were needed to improve the measurements of concentrations of gases important from a greenhouse gas standpoint (Holben et al., 1990). This chapter describes the airborne measurements of PM2.5 and trace gases in the *cerrado* and tropical forested areas of Brazil and relates them to those made for fires in North America.

Algorithms for emissions of total particulate matter (PM), particulate matter containing particles less than 2.5 μm diameter (PM2.5), CO, CO_2, CH_4, and nonmethane hydrocarbons (NMHC) for logging residue fires burned in the Pacific Northwest were shown to be functions of combustion efficiency (Ward and Hardy, 1991). They used towers to support sample packages above prescribed fires and sampled for a wide range of combustion conditions. Combustion efficiency depends on fuel characteristics and fire conditions, which are often different in the tropical areas of the Amazon Basin from those in North America. As a result, it is important to measure the combustion products from burning of biomass in the tropics (e.g., Brazil). In addition to the emission measurements, the chemical composition of the forest biomass burned by one fire in Brazil is compared to the fuel chemical composition for biomass burned in North America.

Methods

Sample Collection
An airborne sampling system was employed to collect grab samples of smoke for analysis of both in-plume smoke characteristics and the ambient air. The sampling system was transported from the United States and installed on the Instituto de Pesquisas Espaciais

(INPE) aircraft while at San Jose dos Campos, Brazil (Figure 48.1). Two types of samples were needed that involved installing different systems for collecting samples of gases through a stainless steel tube from outside the aircraft and samples from high concentration plumes over a limited spatial region of the plume.

In-plume sampling was done using a system consisting of a plastic polyvinyl chloride (PVC) pickup tube, 38 millimeters (mm) in diameter, that extended from outside the aircraft to a large sample bag in the aircraft. The tube was designed to sample from under the fuselage approximately 25 cm from the fuselage and well in front of any possible influence from the engine exhaust or wings of the aircraft. A valving system was used for directing a sample of "ram air" from outside the aircraft through an exhaust port or into the large 1.5 cubic meters (m^3) polyethylene sampling bag. The bag was protected by a heavy cloth cover that was used to suspend the bag from the ceiling of the airplane. The bag required approximately 10 seconds to fill and integrated the sample over approximately a 500 meter (m) path length. After each sample, the large bag was flushed with ambient air and evacuated before repeating the sampling protocol. Cross-sectional penetrations of plumes were made and samples extracted from portions of the plume considered to contain high concentrations of smoke from the biomass fires. Similar systems have been used for airborne studies in the United States (Einfield et al., 1990; Radke et al., 1990).

Subsampling was accomplished onto pairs of 37 mm filters (one Teflon and one glass fiber filter) from the large bag through a cyclone (mean mass cutpoint diameter of 2.5 μm). These filters were used for determining the mass concentration of PM2.5 and the

content of the particles for inorganic and organic fractions. The aerosol samples were sampled at a rate of 2 liters per minute over 15 to 20 minute periods. A soap bubble calibrator was used at each altitude to sample the volume flow of the pumps, and these were corrected for temperature and pressure.

In addition, CO_2, CO, and CH_4 concentrations were determined from canister samples collected from the large bag. These canisters were pumped to 2 atmospheres pressure and returned to the United States for analysis using gas chromatography procedures.

Ambient gas samples were collected using a separate sampling system for collecting canister samples from outside the aircraft through the stainless steel sample tube that extended directly out of the cabin through a side port of the fuselage. A separate pump was used for sampling directly into canisters and these pumped to approximately 2 atmospheres pressure. The pressurized canisters were returned to the United States for gas analyses for CO_2, CO, and CH_4.

Calculations

Emission factors were calculated using the carbon mass balance (CMB) method to compute the mass of fuel consumed in producing the emissions. The technique was first used by Ward et al. (1979) and later perfected by Ward et al. (1982) with laboratory testing by Nelson (1982). The method is based on the stoichiometric partial oxidation of fuel ($C_6H_9O_4$) to CO_2 and incomplete combustion products. The carbon contained in the fuel is about 50% of the mass of the fuel (Byram, 1959). The measured carbon contained in the combustion products is multiplied by 2 to calculate the mass of fuel consumed in producing the combustion products. Emission factors for specific emission components are calculated by dividing the mass of the emission by the fuel consumed and are expressed in units of grams of emission released per kilogram of fuel consumed (g/kg).

Here *combustion efficiency* is defined as the mass of carbon released in the form of CO_2 divided by the total mass of carbon released. It is expressed as a ratio or percentage of total carbon. In practice, combustion efficiency can be approximated by dividing the sum of the molar concentrations of CO and CO_2 into the molar concentration of CO_2. Generally, the molar concentration of CH_4, and the methane equivalent molar concentration of the carbon contained in both the nonmethane hydrocarbon and particulate matter fractions, are added to that of CO and CO_2. The carbon released in the form of compounds and particulate matter other than CO and CO_2 makes up less

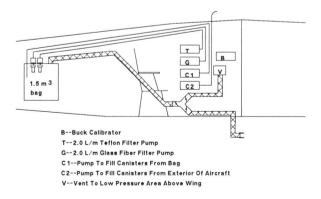

B--Buck Calibrator
T--2.0 L/m Teflon Filter Pump
G--2.0 L/m Glass Fiber Filter Pump
C1--Pump To Fill Canisters From Bag
C2--Pump To Fill Canisters From Exterior Of Aircraft
V--Vent To Low Pressure Area Above Wing

Figure 48.1 Schematic of sampling apparatus used on board the INPE aircraft for sampling emissions from biomass fires in Brazil.

than 5% of the total. Hence, the ratio of CO to the sum of CO plus CO_2 is an excellent first-order approximation of combustion efficiency.

Sample Collection and Concentrations

The INPE aircraft was flown from San Jose dos Campos to Brasillia and northward to Alta Florista (Figure 48.2). One sample was collected from a *cerrado* (savanna-like area) during this flight segment. The second and third plume penetrations of smoke from tropical deforestation burns were made during flight missions out of Alta Florista on 2 and 3 September 1989. After flying from Alta Florista to Manaus and while in flight to Curiabá, a fourth penetration of

a plume from a tropical deforestation fire was accomplished. In addition, many ambient air samples were collected, and these were examined as a function of time and altitude. Three spirals were flown to collect samples for profiling the atmosphere from near the surface to over 3600 m (see Figure 48.2 for locations on 4, 6, and 7 September). The profiles sampled on 4 and 6 September are discussed in this chapter.

In-plume Samples

The net concentrations of gases and particulate matter (after subtracting the background concentrations) are presented in Table 48.1. Background concentration data were averaged for a number of samples collected near the same altitude of the plume penetrations. Unlike the aged plumes sampled by Andreae et al. (1988) at concentrations near that of background, the particulate matter concentration of the ambient air was considered to be low relative to the in-plume concentrations.

On 2 September 1989 the airborne sampling system was first tested flying out of Brasillia for Alta Florista in an active clean-burning fire of mostly grass and low shrubs and trees (*cerrado* fuel type). A sample was collected at 1058 Local Standard Time (LST) using the large-bag sampling apparatus on the second penetration of the plume at about 500 m above terrain. There were no clouds, with surface winds from the southwest at no more than 2 meters per second (m/s). There was little turbulence.

A series of background samples were collected prior to the plume penetration and indicated an average background concentration of 362, 0.192, and 1.721 ppmv of CO_2, CO, and CH_4, respectively. The PM2.5 concentration in the plume was measured from an average of the Teflon and glass fiber filter mats of 399 $\mu g/m^3$.

Figure 48.2 General base map showing the time of sample collection and approximate relative locations of major events during the BASE-A experiment.

Table 48.1 Net concentration of gases and PM2.5 for one *cerrado* (low trees, brush, and grass fuel type) and three tropical deforestation fires sampled from the INPE aircraft using grab sampling techniques

			Concentrations			
Sample date	Time	Vegetation type	CO_2[a]	CO[a]	CH_4[a]	PM2.5[b]
2 Sept. 89	1058	*Cerrado*	91	1.96	0.08	339
3 Sept. 89	1256	Deforestation	129	1.88	1.04	
4 Sept. 89	1500	Deforestation	35	1.47	0.14	193
7 Sept. 89	1358	Deforestation	10	1.20	0.13	177

a. Parts per million by volume (ppmv).
b. Micrograms per cubic meter ($\mu g/m^3$).

On 3 September 1989 a second plume penetration was made of a fire near Alta Florista at 1256 UT for a deforestation burn. The meteorological conditions of the atmosphere were conditionally stable with a condensation level near 2000 m. The wind speed was light with scattered clouds. The gas concentrations inside the plume were higher than measured for the *cerrado* fire. The particulate matter sampling system became detached during the rough ride through the plume and no sample was collected.

On 4 September 1989 a third plume penetration was made at 1500 LST of a fire near Alta Florista. This was a planned deforestation fire that was carefully observed from the ground by two of our research team members. Surface winds were less than 2 m/s with air temperature of 31° to 36° C and an unstable atmospheric layer to at least 3000 m. A cumulus cloud formed along the top boundary of the smoke plume.

The gas concentrations inside the plume were low as compared to those needed for performing accurate carbon mass-balance calculations of the fuel consumed in releasing the emissions. The gas concentrations in the plume averaged 390, 1.972, and 1.901 ppmv with the background concentrations of 355, 0.498, and 1.764 ppmv for CO_2, CO, and CH_4, respectively. The particulate matter concentration averaged 193 $\mu g/m^3$.

In addition, nine samples of the more common fuels consumed by the fire were returned to the United States for C, H, and N analysis. (Oxygen was estimated by subtracting the percent C, H, and N from 100%. All results were corrected for water and mineral content.) The fuels consisted of stems, twigs, and leaves, which had an average elemental composition of 53.4 ± 1.8, 5.9 ± 0.1, 1.6 ± 0.7, and 39.1 ± 2.3% C, H, N, and O, respectively. The C and H composition is similar to fuels of North America with the N content being a factor of 2 higher than measurements for the boreal forest of Canada (Susott et al., 1990). The results suggest that the carbon mass balance technique described in the methods section of this report should be equally valid for the tropical forests of South America.

On 7 September 1989 a fourth plume penetration was made at 1358 UT of a plume from a deforestation fire near the state of Rondonia while flying from Porto Velho to Curiabá (Table 48.1). There were numerous towering cumulus clouds resulting from the heat generated from the higher intensity fires. The cloud condensation level was near 3000 m. It was not possible to estimate surface conditions at the time of sample collection. The gas concentrations inside the plume were

373, 1.840, and 1.887 ppmv with a background concentration of 363, 0.643, and 1.762 ppmv, respectively. The particulate matter concentration from the glass fiber filter was 177 $\mu g/m^3$.

Spiral Profiles

While ascending and descending through the atmosphere, profiles were made of the concentrations of CO_2, CO, and CH_4 through the inversion layer from within 150 m of terrain to over 3500 m altitude. In a companion paper (Holben et al., this volume, Chapter 49), the concentration data are compared with the sun photometer measurements taken from the aircraft.

On 4 September 1989 between 1100 and 1200 hours, a vertical profile of the atmosphere was sampled 50 km to the west of Alta Florista from 150 m above terrain to an altitude of 3600 m. Scattered cumulus clouds were present with a lower condensation level of 3000 m. The CO concentration near the surface to 2000 m averaged 350 ppbv and then decreased to 100 ppbv after passing through the top of the mixing layer located at an altitude of between 2000 and 2600 m. The CH_4 concentration paralleled that of CO, ranging from a high of 1773 ppbv near the surface to 1695 ppbv at 3600 m.

On 6 September 1989 between 0830 and 0930 hours, a vertical profile of the atmosphere was sampled again about 50 km west of Alta Florista from 50 m above terrain to 3600 m altitude. The atmosphere was reasonably stable with a noticeable smoke pall near the surface within 500 m of the surface. The CO concentrations near the surface close to Alta Florista were as high as 600 ppbv, and from 1000 m to 1500 m the concentration was near 300 ppbv. At approximately 2000 m, the concentration decreased to 100 ppbv and reached a low of 67 ppbv at 3300 m.

Results and Discussion

Emission Factors and Combustion Efficiency

The emission factors were calculated using the CMB technique for the smoke emissions released from the *cerrado* and tropical deforestation burns. The data are presented in Table 48.2. Emission factors for CO averaged 88 ± 30 g/kg for the three samples collected of smoke from fires for deforestation purposes. The emission factor for CO can be compared with the measurements made by Hegg et al. (1989) of 91 ± 21 g/kg, the calculated emission factors of 91.4 g/kg from data of Andreae et al. (1988), and those of Ward et al. (1989) from tower sampling of logging slash fires of

Table 48.2 Emission factors for PM2.5, CO₂, CO, and CH₄ for one *cerrado* and three deforestation fires in Brazil during September 1989

| Sample date | Time | Vegetation type | Emission factors (g/kg) | | | | Combustion efficiency (%) |
			CO₂	CO	CH₄	PM2.5	
2 Sept. 89	1058	*Cerrado*	1783	24.5	0.56	4.4	97.2%
3 Sept. 89	1256	Deforestation	1666	97.7	4.90	—	90.8
4 Sept. 89	1500	Deforestation	1741	46.6	2.48	5.3	94.9
7 Sept. 89	1358	Deforestation	1586	120.9	7.18	15.6	86.4

156 ± 37 g/kg for Douglas fir slash and 128 ± 22 g/kg for hardwood species slash. The samples of Hegg et al. (1989) were collected similarly to this study. The measurements by Ward et al. (1989) were for the flaming and smoldering combustion phases and weighted for the entire fire based on the fuel consumed during each of the combustion phases. Of major concern is the possibility of not collecting samples that are representative of the fire producing the emissions. Ward and Hardy (1991) discuss the fire buoyancy effect on lofting the emissions from the flaming phase and releasing these emissions in a stratum of the atmosphere higher than those emissions from the smoldering phase.

Emission factors for CH₄ averaged 4.8 ± 2.0 g/kg compared to the average by Hegg et al. (1989) of 2.9 ± 0.66 and that of Ward et al. (1989) of 5.5 ± 2.0 for Douglas fir logging slash.

Particulate matter emission factors require a sample with a weighable mass of particles from a sample space concurrent with the collection of a sample of the gases. The gas concentrations must be accurately measured to differentiate them from the background concentrations. The large-bag sampling system enhances the collection from portions of the smoke plumes where the concentrations are differentiable from background. PM2.5 emission factors ranged from 4.4 for the *cerrado* sample to 15.6 g/kg for one of the tropical deforestation fires.

The measured emission factors are plotted against the algorithms of Ward and Hardy (1991) as a function of combustion efficiency (Figure 48.3). The algorithms were developed from 38 test fires in the western United States and include numerous samples from both the flaming and smoldering combustion phases. The combustion efficiency for the *cerrado* fire was 97% and for the three tropical deforestation fires averaged 90.7% ± 4.3%. This compares to the measurements of Hegg et al. (1989) of 89.6% ± 2.0%, of Andreae et al. (1988) of 92% as calculated from the

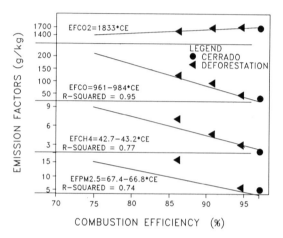

Figure 48.3 Emission factors for CO₂, CO, PM2.5, and CH₄ as functions of combustion efficiency. The data for the *cerrado* and tropical deforestation fires are superimposed. The similarity of emission factors for various emissions for different combustion efficiencies fit well on the data derived for logging slash fires in the United States (Ward and Hardy, 1991).

CO/CO₂ molar ratio, and to the measurements of Ward et al. (1989) of 84% to 89% dependent on fuel type and burning conditions. The measurements for emissions from fires in tropical fuel types agree well with the measurements of fires in North America if the combustion efficiency effect is factored into the analysis.

The ratios of emission factors as measured in the BASE-A experiment are summarized in Table 48.3 and compared with similar measurements in North America (Ward and Hardy, 1986) and Brazil (Andreae et al., 1988; Greenberg et al., 1984; Crutzen et al., 1985). It was found that although the fire combustion efficiency varied between the three observed fires, and as a result the ratio of CO/CO₂ varied (see the standard deviations in Table 48.3), the ratios of PM2.5 to CO or to CH₄ are markedly constant. The constant ratio between PM2.5 and trace gas emissions shows that remote sensing of smoke particles from

Table 48.3 Ratio of emission factors as measured in the BASE-A experiment

	BASE-A		Ward (1986)		Andreae et al. (1988)	Greenberg et al. (1984)	Crutzen et al. (1985)
	Cerrado	Forest	Flaming	Smoldering			
CO/CO$_2$	1.4%	5.4 ± 2.1%	4.3 ± 1.6%	18.1 ± 9.2%	5.4%	3 to 20%	9%
CH$_4$/CO$_2$	0.03%	0.3 ± 0.1%	0.1 ± 0.1%	0.5 ± 0.2%		0.2 to 0.8%	
PM2.5/CO$_2$	0.3%	0.6 ± 0.3%	0.4 ± 0.3%	0.8 ± 0.3%			
PM2.5/CO	18.0%	12.1 ± 0.8%	9.6 ± 3.5%	5.6 ± 3.3%			
PM2.5/CH$_4$	785.7%	215.5 ± 1.8%	325.5 ± 182.9%	176.6 ± 62.8%			

Note: Percent on a mass basis.

space, in addition to remote sensing of fires themselves, can be a useful tool for regional and global estimates of trace gas emissions from deforestation fires.

Andreae et al. (1988) discuss the difficulty of differentiating background concentrations of particulate matter and CO$_2$ from in-plume concentrations. None of their measurements show elevations of CO$_2$ of more than 2 ppmv. The average molar ratio of CO/CO$_2$ of 0.085 was determined from multiple samples. Individual fires produce ratios of emissions dependent on the burning conditions, fuel type, and method of ignition (Ward, 1991). Often, especially for airborne sampling work, the initial flaming phase burns at a high combustion efficiency with the combustion efficiency decreasing as the products of incomplete combustion from the smoldering phase dominate the emissions released from the fire. The molar ratio of CO/CO$_2$ can change dramatically from 0.25 to 0.023 over a range of combustion efficiencies of 75% to 95%.

Development of Gas-to-Particulate Matter Mass Ratios

To estimate the emissions of other trace gases from measurements of particulate matter loading on a regional basis requires reliable measurements of the ratio of trace gas to particulate matter. We measured both PM2.5 concentrations and the concentrations of trace gases in three plumes during the BASE-A experiment. The multiplier to estimate PM2.5 concentration in the background air was derived from the plume penetration data for the three tropical deforestation fires. Samples of PM2.5 and gases were collected from only two of the deforestation fires. The ratio of PM2.5 to gas concentration for the *cerrado* fire was 0.2032 mg of PM2.5 per m³ per ppmv of CO. Figure 48.4 shows a good fit with the ratio data for the

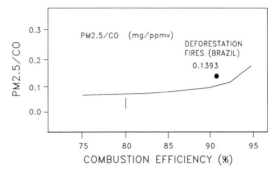

Figure 48.4 The ratio of mass of PM2.5 to CO for fires in the United States and the average ratio for the tropical deforestation fires as a function of combustion efficiency.

smoke emissions sampled for prescribed fires in the United States when the effect of combustion efficiency is factored into the comparison. The two deforestation fires averaged 0.1393 mg of PM2.5 per m³ per ppmv CO (average combustion efficiency of 90.6%). This also is comparable to measurements for fires in the United States of similar combustion efficiency. The removal of smoke particles from the atmosphere due to scavenging through cloud processing was not factored into these calculations. In fact, the selective removal of particles could have easily exceeded 50% of the mass released for the fires that induced precipitation (Radke, personal communication, 1990).

Calculation of Atmospheric Loading of PM2.5

The PM2.5 to CO concentration ratios were used to compute the mass of PM2.5 from the surface to 3600 m. This was done by collecting canister samples of the gases at several altitudes and applying the ratio of 0.1393 mg of PM2.5 per m³ per ppmv CO for each of the CO concentrations at the different altitudes of measurement. For integration purposes, the concentration was considered constant from midpoint to midpoint between the sample altitudes. The lowest CO concentration measured at 3600 m was assumed to be above the mixing height for the particles from the deforestation burns, and the CO concentration at this altitude was subtracted from the other measurements before multiplying by the ratio of PM2.5 to CO concentration.

Figure 48.5 shows the CO concentration data for 4 and 6 September and the calculated concentration of PM2.5 as a function of altitude. The integrated total PM2.5 for each height interval is shown as well by the bar diagram superimposed on the line drawings. The total atmospheric PM2.5 loading from ground level to 4000 m was determined from this analysis to be 92.8 and 67.5 mg/m^2 on 4 and 6 September, respectively.

Inorganic and Organic Content of PM2.5

The lightly loaded filters may have contributed to some of the differences between measurements of trace elements for samples collected in the *cerrado*

and tropical deforestation fires and those for North America. Generally, the filter samples collected in the United States contain upward of 100 µg of sample per filter and often as much as 1 mg. The filter samples for Brazil contained 13 and 14 µg for the two deforestation fires and 29 µg for the *cerrado* fire. The filter weighing was done meticulously with full use of check weights and control filters.

The inorganic content of the PM2.5 was determined using X-ray fluorescence techniques by NEA Laboratory in Beaverton, Oregon. The concentrations of a few of the trace elements that are most interesting are shown in Figure 48.6. Of special interest is the absence of lead, generally considered to be a deposition product residual from leaded gasoline. Mercury was found for the tropical deforestation samples collected in the Amazon tropical forested areas but not the *cerrado*. The presence of mercury with the smoke particles may be a result of the widespread use of mercury to remove gold from ore in this region of Brazil. Of further interest is the high content of Si, Ca, Cl, and K for the *cerrado* area. The S and Cl are in low concentration for the tropical deforestation smoke relative to the smoke from the *cerrado* fires. Future analyses of the elemental composition of the biomass fuels should help explain these differences between the inorganic elemental composition of the fuel types in Brazil as well as the differences between typical smoke samples collected in other parts of the world.

The carbon content of PM2.5 was determined using a volatilization/combustion technique first used by Johnson and Huntzicker (1981) and now used commercially by Sunset Laboratory of Beaverton, Oregon. The ratio of graphitic to organic carbon measured with the technique was considered to be a valid measurement even though the absolute values of the mass of carbon lost from the sample was higher

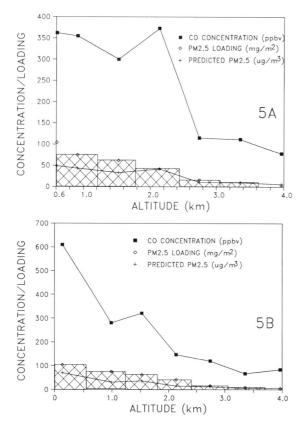

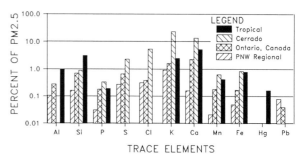

Figure 48.5 Vertical profile of CO was used with ratios shown in Figure 48.4 to compute the mass concentration of PM2.5 and the optical thickness of the atmospheric loading of PM2.5 for two profiles: (a) 4 September 1989 and (b) 6 September 1989.

Figure 48.6 Inorganic content of PM2.5 for the *cerrado* and tropical deforestation fires as compared to a fire in Canada and the average of 38 test fires in the United States.

Table 48.4 Apportionment of carbon released in the form of particles for the *cerrado* and two deforestation fires sampled in Brazil during September 1989

Vegetation type	Date	Net weight per filter		Net wt PM2.5 (μg)	Ratio of carbon		Percentage carbon	
		Graphitic (μg)	Organic (μg)		Graphitic	Organic	Graphitic (%)	Organic (%)
Cerrado	2 Sept.	0.97	16.78	29	0.054	0.946	3.3%	56.7%
Deforestation	4 Sept.	1.29	6.49	13	0.166	0.834	9.9	
Deforestation	7 Sept.	0.40	3.35	14	0.107	0.893		49.9

Note: The filter loadings were not sufficient to provide true mass loadings of each of the types of carbon. However, the relative proportion of each is thought to be within ±20% of the percentage values listed in the table.

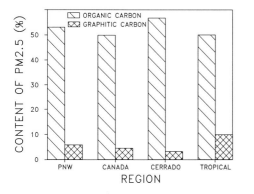

Figure 48.7 Comparison of the organic carbon content of the emissions of PM2.5 for fires of the *cerrado* and tropical areas as compared to a fire in Canada and the average of 38 fires in the United States.

than previous samples of smoke from biomass burning. The total carbon contained with the PM2.5 was assumed to be similar to that for samples of smoke collected in the United States (60% carbon) and the proportion of graphitic to organic used to apportion the carbon between graphitic and organic fractions (Table 48.4). The ratio of graphitic to organic was lower for the fire in the *cerrado* fuel type than for the tropical deforestation fires by a factor of two. The comparison with other fuels in the United States and Canada is shown in comparison to the *cerrado* and tropical deforestation fires of Brazil (Figure 48.7).

Conclusions

Emission factors for combustion products from four fires suggest that the proportion of carbon released in the form of CO_2 is higher than for fires of logging slash burned in the western United States (higher combustion efficiency). The combustion efficiency was 97% for the *cerrado* test fire and 86% to 94% for the deforestation fires.

Although ratios of PM2.5 to CO were measured in the fire plumes, the actual ratios in the ambient air may be lower. The removal of PM2.5 due to cloud processing needs to be evaluated and could be a dominant factor in the tropics. Systematic measurements of the ratios of particles to CO_2 concentrations close to the fire and in the aged mixed air could be used to show the removal of smoke particles from the air.

Source characterization needs to account for the mass of biomass consumed relative to the emissions released because wild-land biomass fires have high combustion efficiencies during the flaming phase. (Carbon released in the form of CO_2 exceeds 90% of total carbon released during the flaming phase and typically ranges from 70% to 85% during the smoldering phase.)

The inorganic content of particles from fires in biomass fuels in the tropics may be different from fires of the western United States. This needs further investigation.

This limited data set suggests that tropical biomass combustion may be markedly different than for temperate forest fuels. Combustion efficiencies are higher, as are emission factors for CO_2. These results are consistent with our visual observations of the fuel beds where the small-sized fuels dry quickly and are fully consumed during the flaming combustion process, but the large-diameter fuels dry more slowly and are only partially consumed during the early part of their dry-down. The heavy large-diameter fuels, when wet, are quickly extinguished following the consumption of the fine fuels.

We found that although the combustion efficiency varied between the three observed fires, and as a result the ratio of CO/CO_2 varied from fire to fire, the ratio of PM2.5 to CO or to CH_4 remained markedly constant. The smoke particle concentration was measured instantaneously and only in two fires; therefore, these conclusions have to be further substantiated by additional tests. Better measurements are needed closer to the source to establish the concentrations of trace gases and particles and measurements of the well-mixed aged smoke. The constant ratio between PM2.5 and trace gas emissions shows that remote sensing of smoke particles from space, in addition to remote sensing of the fires themselves, can be a useful tool for regional and global estimates of trace gas emission from deforestation fires.

Note: The use of trade or firm names in this chapter is for reader information and does not imply endorsement by the U.S. Department of Agriculture of any product or service.

Optical Properties of Aerosol Emissions from Biomass Burning in the Tropics, BASE-A

Brent N. Holben, Yoram J. Kaufman, Alberto W. Setzer,
Didre D. Tanre, and Darold E. Ward

The optical properties of smoke aerosol are discussed in the general context of biomass burning. Biomass burning in the tropics, a large source of trace gases, has expanded drastically in the last decade due to increases in the controlled and uncontrolled deforestation in South America (Tucker et al., 1984; Malingreau and Tucker, 1988; Setzer et al., 1988; Setzer and Pereira, 1989) and due to increases in cultivated areas associated with expansion of population in Africa and South America (Seiler and Crutzen, 1980). Trace gases and particulates are emitted to the atmosphere from the burning process. The biomass burning process is linked to the need to understand the sources and sinks of the atmospheric trace gases and their atmospheric variations (GTC, 1986). The importance of these trace gases (in addition to CO_2) is in their greenhouse effect, which is comparable to that of CO_2 (Ramanathan et al., 1985), and in their effect on tropospheric chemistry (Crutzen, 1988).

Knowledge of the optical characteristics of smoke particles is important in regard to the following two aspects. The first aspect deals with the relation between particles and gases as biomass burning products. Smoke particles are one of the emission products in the process of biomass burning. Since they are formed due to incomplete combustion, as is the case for trace gases emitted from biomass burning, their relative abundance is proportional to the abundance of trace gases (Kaufman et al., 1990a). As a result, remote sensing of smoke particles (Kaufman et al., 1990b; Ferrare et al., 1990) can be used to asses the magnitude of the emission of trace gases from biomass burning associated with deforestation in the tropics (Kaufman et al., 1990a). These remote sensing methods are based on optical characteristics of the smoke particles derived from ground-based and airborne measurements (Puschel et al., 1988; Radke et al., 1988; Kaufman et al., 1990b) as well as on analysis of satellite imagery (Kaufman et al., 1990b; Ferrare et al., 1990). The derivation of the trace gases is based on measurements of the ratios of trace gases

to particles emission rates (Ward and Hardy, 1984; Ward, 1986). To the extend of our knowledge, except for lidar measurements of the backscattering ratio (Browell et al., 1990), there are no measurements of the optical characteristics of smoke particles emitted in the tropics. In order to use smoke particles as a tracer of the emission of trace gases, the particle mean size and the resulting relation between the smoke particles' mass and their effect on the outgoing radiation have to be established.

The second aspect deals with the effect of particles on atmospheric composition and energy balance. Particles emitted from biomass burning may affect the radiation budget and boundary layer meteorology by reflecting sunlight to space and absorbing solar radiation (Coakley et al., 1983). Smoke particles are also a major source of cloud condensation nuclei (CCNs) (Squires and Twomey, 1960; Hobbs and Radke, 1969; Radke, 1989). Therefore, an increase in the aerosol concentration may cause an increase in the reflectance of thin to moderate clouds (Twomey, 1977; Coakley et al., 1987) and a decrease in the reflectance of thick clouds (Twomey, 1977). The aerosol effect for most clouds is dominated through an increase in cloud albedo. Twomey et al. (1984) suggested that the net cooling effect from increases in aerosol concentrations can be as strong as the heating effect from the increase of global CO_2 and other trace gases and thus acts in the opposite direction (Coakley et al., 1987). The net effect of increased concentration of aerosol and trace gases on the earth's energy budget is therefore not obvious. Since the number of potential CCNs generated from biomass burning is inversely proportional to the cube of the particle radius, the mean mass radius of smoke particles has to be established.

Biomass burning in the tropics is of particular interest, because of the large extent of forest clearing and agricultural burning. Estimates based on satellite imagery are of ~200,000 square kilometers (km^2) in Brazil alone (Setzer et al., 1988; Setzer and Pereira, 1989). The high solar irradiation in the tropics enhances atmospheric chemical reactions (due to ultraviolet radiation) and the climatic effects of aerosols (more solar radiation reflected to space).

Sunphotometry has been a useful tool to measure optical properties of aerosols for many years; however, until this past decade when tropical biomass burning became a significant issue in the research community, few researchers studied the optical properties of smoke with sunphotometry. Kaufman et al. (1990b) measured aged smoke near Washington, D.C., generated from Canadian forest fires and traveling across 4000 km (Ferrare et al., 1990). In order to assess the smoke particle size from remote measurements, Kaufman et al. (1990b) measured by a sunphotometer from the ground a wavelength dependence of the aerosol optical thickness (τ_a) of $\alpha = 1.65 \pm 0.15$ (where $\alpha = d\ln \tau_a/d\ln \lambda$), and an aerosol optical thickness of $\tau_a \sim 1$ at 500 nm. This value of α indicates a mean mass radius of the particles around 0.2 μm. Pueschel et al. (1988) studied several fires in California using the Ames airborne autotracking sunphotometer. They found that the wavelength dependence of the smoke optical thickness varied with the residence time and the source of the fuel and that forest fire smoke showed a wide range of wavelength dependencies, indicating a wide range of particle size.

In this chapter, ground-based and airborne measurements of the smoke particles' optical properties are reported. The method of measurements is summarized, followed by a summary of the results of ground-based and airborne measurements and a discussion of the smoke aerosol optical model. Aerosol absorption is discussed, and measurements are reported of the effect of smoke particles on the outgoing radiation.

Method

Measuring the optical properties of smoke emissions from biomass burning in Brazil required a multitemporal and multielevational approach to characterize smoke aerosol background conditions, fresh smoke from the flaming phase, aged smoke from flaming, fresh smoke from the smoldering phase, and a mixture of all. Sunphotometers were used at ground level and in the Institute for Space Research (INPE) aircraft on 3, 4, and 6 September 1989, during prescribed fire events near Alta Floresta, Matto Grosso, in Brazil. Ground observations of mixed smoke representing background conditions were made in the mornings before the fires were started. Plume observations were made from flaming and smoldering conditions and in the case of the flaming source,

observations from 10 m, 500 m, 1 km, and 10 km from the source allowed evaluation of the effects of short-term aging on the aerosols.

The airborne sunphotometer was operated through an open window of the twin-engine INPE aircraft. Profiles were made at heights from 300 m to 4000 m from the surface in approximately 600 m intervals flying in 5 km ovals such that a series of readings could be made while the sunphotometers faced the sun at a constant elevation; then the climb to the next measurement elevation could be made on the turns. We assumed constant optical depth during each 5 km transect. Profiles were made in late morning before fires were started (background mixed smoke) and mid-afternoon at the peak of burning (flaming smoke). A profile was also made in transect to Alta Floresta in the semiarid *cerrado* to characterize the aerosols from that substrate.

All sunphotometers were cross-calibrated immediately before, during, and after the BASE-A experiment. It was found that the aerosol optical thickness can be measured to a precision of ± 0.02 between the three instruments and an accuracy of ± 0.05. The absolute calibration used was obtained by cross-calibration with a spectrometer calibrated by viewing the sun from a tethered balloon at an altitude of 30 km. The precision of calibration between the spectral bands is estimated to be ± 0.02. These relatively large uncertainties resulted from instabilities detected in one or more of the sunphotometers. In the section on airborne observation the aerosol optical thickness of the mixing layer alone is obtained from the differences between two levels, 500 m and 3500 m, which removes the absolute calibration uncertainty. Nevertheless, spatial inhomogeneities and temporal evolution of the mixing layer lead to an estimate of the precision of ± 0.02 as well. These uncertainties in the precision lead to uncertainties in the derived particle size of ± 0.02 μm for the mixing layer, or the whole atmosphere. The estimate of the particle size of the optically thin layer above 3.5 km is more uncertain, and we can only roughly estimate the particle size.

The airborne instrument had, in addition to 3 "aerosol" channels (at 450, 650, and 850 nm), a broad and narrow band centered in the water vapor absorption region, 0.935 to 0.950 μm. This allowed water vapor retrieval. The retrievals were compared to measured radiosondes at Dulles Airport near Washington, D.C., and found to agree within ± 0.4 centimeters (cm) for total column water vapor, for a range of 1 to 5 cm of water.

The aerosol optical thickness in each spectral band is computed from the Bouguer-Lambert-Beer law—

$$V_\lambda = r^2 \times V_{0\lambda}\exp(-\tau_\lambda/\mu_o) \qquad (49.1)$$

where V_λ is the output voltage of the sunphotometer for a bandwidth λ; $V_{o\lambda}$ is the zero air mass voltage intercept at that bandpass for a mean normalized Earth-sun distance (r) of 1.00; τ_λ is the wavelength-dependent vertical optical depth above the sunphotometer; and μ_o is the cosine of the solar zenith angle at the observation site.

The wavelength dependence of the aerosol optical thickness may be computed in terms of the Ångstrom exponent α:

$$\alpha = 1n(\tau_2/\tau_1)/1n(\lambda_1/\lambda_2) \qquad (49.2)$$

where λ_1 and λ_2 can be any two aerosol spectral channels.

α was modeled for spherical aerosols assuming an index of refraction of $1.43 - 0.0035i$ using Dave's Mie scattering code. Assuming that the smoke aerosol size distribution is a log normal function, a look-up table of α was computed as a function of the size distribution parameters (mean geometric radius, r_g, and the standard deviation of the natural logarithm of the radius, σ). r_g was converted to the mean mass radius and the effective radius, r_m and r_{eff}, respectively, according to

$$r_{eff} = r_g\exp(2.5\sigma^2) \text{ and } r_m = r_g\exp(1.5\sigma^2) \qquad (49.3)$$

The effective radius r_{eff} is defined as the ratio of the volume and the geometrical cross-section

$$r_{eff} = \int r^3 n\,(r)\,dr/\int r^2 n\,(r)\,dr \qquad (49.4)$$

It is an appropriate parameter for radiative computations and it is directly related to the mass radius r_m by $r_{eff} = r_m\exp(\sigma^2)$ for a log normal size distribution. We express the results also in terms of r_m, since r_m describes the median mass distribution. The derived value of r_{eff}, in contrast with r_g, does not have a strong dependence on σ. The analysis of the data showed that the spectral range of the present optical thickness observations (at wavelengths of 450 to 850 nm) is not wide enough to measure σ (see Figure 49.1). Therefore, in the following the results for r_{eff} and r_m will be given for a range of σ between 0.3 and 0.9. This range was chosen based on the size distributions measured from aircraft by Radke et al. (1978) and Stith et al. (1981). A best fit to the volume distributions in Figure 2 of Radke et al. (1978) and Figure 2 of Stith et al. (1981) results in σ in the range 0.6 to 0.8.

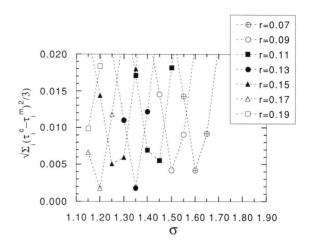

Figure 49.1 Sum of the differences between the retrieved, τ^c, and measured, τ^m, aerosol optical thicknesses for three bands (450, 650, and 850 nm) are plotted as a function of the standard deviation, σ, for several mean geometric radii r_g.

Three spectral aerosol optical thicknesses were measured allowing two independent data sets of α to be computed and were modeled for a range of effective sizes ($r_m = 0.01$ to 2.0 μm and $\sigma = 0.10$ to 2.00). The results from the ground observations are presented in terms of the size distribution under the constraints of our ability to evaluate directly the in-plume smoke particles. The aircraft observations were also interpreted for the size distribution of the mixing layer and of the atmosphere above it. Because the aircraft measured profiles of the optical thicknesses, the optical thicknesses data were converted into the extinction coefficient (km^{-1}) allowing comparisons to *in situ* aircraft measurements of trace gases and particulates within a layer. The water vapor profile is also presented.

Ground-Based Observations

Background Observations
Background observations of mixed smoke are made away from the plume and were taken on all observation days at various times. These measurements represent mixed smoke conditions, and can be used to extract the smoke properties of a particular plume from the total measurements of the optical thickness. The background optical depths increased from morning through the afternoon at all wavelengths owing to the increase of emission products introduced to the mixing layer from biomass burning. The spectral dependence of the optical thickness, α, increased during

the day from 1.5 to 2.2 indicating a decrease in the effective particle radius from 0.45 μm to 0.22 μm from the aged morning smoke to the fresher afternoon smoke (Figure 49.2). In the next section we shall show that fresh smoke had smaller particle sizes. We noted an increase in optical depth of the background conditions from 3 to 6 September; however, the trend was significant with the limited data available and did not affect α or the size distribution. We therefore used the smoothed time-dependent background values for all dates to correct our individual plume observations. This was accomplished by subtracting the background aerosol optical thicknesses from the total aerosol optical thicknesses, to result in the net plume optical thickness.

A second background correction for the measured plume optical thickness was made by subtracting the optical thicknesses measured by the aircraft above 4 km. These high atmospheric optical thicknesses were

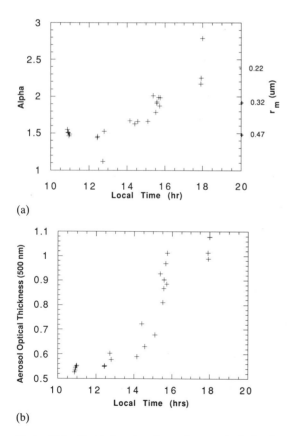

(a)

(b)

Figure 49.2 Sunphotometer observations made during 3 and 4 September 1989 were used as background conditions for computation of alpha and the aerosol optical thickness. (a) Alpha increased from morning to afternoon and the mass radius decreased by about one-half during the same time period. (b) The aerosol optical thickness approximately doubled from late morning to late afternoon.

shown to be relatively stable through time and date from the plume observations. This resultant optical thickness characterizes a combination of the plume and mixing layer. We therefore analyzed observations for smoke from active fires in three conditions, termed "whole atmosphere," "mixing layer," and "plume." These were analyzed for flaming, smoldering, and background conditions.

Fresh Smoke

Observations of fresh smoke (approximately 5 to 10 seconds old) from the flaming phase were highly variable due to the lack of homogeneity of the plume during the 40 seconds required to make a spectral observation series. A running average of 20 observations out of the 80 observations collected continuously from a specific fire decreased the high frequency variations in the wavelength dependence of the optical thickness. No further increase in stability was noted by averaging on a larger number of observations. The results are summarized in Table 49.1. The mean mass radius r_m of the fresh smoke in the plume is $r_m = 0.15$ to 0.24 μm (for $\sigma = 0.5$ and corresponding to $\alpha = 2.1$ to 2.7); the mixing layer results in $r_m = 0.13$ to 0.22 for the plume plus boundary layer smoke and $r_m = 0.20$ to 0.27 for the whole atmosphere.

Two other data sets of slightly older smoke (~1 minute and ~5 minutes) were analyzed as the previous data set. The wavelength exponent for the plume increased with aging time from 2.8 to 3.2, with a corresponding decrease in the value of r_m for the net plume to 0.16 μm and 0.14 μm, respectively (Table 49.1). Observations of a one-hour-aged plume, approximately 10 km downwind of the source, showed a decrease in the wavelength exponent to 2.85, with a corresponding smaller particle size $r_m = 0.11$.

Smoldering Smoke

Observations of smoldering/flaming phase smoke were made of relatively fresh smoke less than one minute old. The wavelength exponent of the net plume was 3.25, resulting in no solution for r_m under the specified conditions. Decreasing σ to 0.3 yields an r_m of 0.09. Note that these data were made late in the day when no background data was measured. Therefore it is unlikely our background correction made by extrapolation is applicable.

Discussion of the Ground-Based Observations

This investigation shows that background smoke characteristics are different from fresh smoke characteristics. Therefore, background smoke observations

Table 49.1 Summary of ground-based measurements

Smoke type	Plume values		r_m		
	τ_a (0.50 μm)	α (0.50/0.875)	Whole atmosphere	Mixed layer	Plume
Fresh smoke (10 sec)	1.0–1.7	2.4 ± 0.3	0.20–0.27	0.13–0.22	0.14–0.24
Less fresh smoke (1 min)	1.0–2.1	2.8 ± 0.2	0.19–0.20	0.12–0.15	0.10–0.15
Less fresh smoke (5 min)	0.45–0.75	3.15 ± 0.05	0.20–0.22	0.11–0.15	0.07–0.09
Aged smoke (1 hour)	1.0–1.1	2.85 ± 0.03	0.19–0.20	0.12–0.15	0.11–0.12
Smoldering	1.7–1.9	3.25 ± 0.10	0.18–0.19	0.11	0.07 and lower
Background smoke	0.5–1.0	1.5–2.2	0.22–0.45[a]		

Note: τ_a (0.50 μm) = the aerosol optical thickness; α(.50/.875) = the wavelength dependence of the optical thickness; r_m = the mean mass radius of the particles.
a. Observations taken over a range of days and burning conditions.

are critical for evaluating the optical properties of the net smoke generated from biomass burning in Brazil. Accounting for the background contribution to the smoke optical properties reduced the mean mass radius by 0.06 to 0.08 μm, or 30% to 50%. Aging of the smoke in the first hour decreased the mean mass radius for the flaming phase from 0.17 to 0.13 μm. The smoldering phase smoke indicated the smallest size distribution. However, the background correction for the smoldering phase is questionable. The decrease of the particle size within the first hour can be attributed to the deposition of larger ash particles generated in the dynamic flaming phase but not in the smoldering phase. The increase in the particle size in the next several hours (from the fresh smoke to the background smoke measured the next morning) can be due to coagulation of the liquid organic aerosol particles (Radke, private communication, 1988) and interaction with water vapor. The smoke aerosol model, which is applicable for remote sensing of fresh biomass burning, is therefore defined by a particle mean mass radius between 0.10 and 0.20 μm. The background aged smoke model would be represented by a mean mass particle radius of 0.2 to 0.4 μm. For comparison, aircraft measurements of volume size distribution made by Radke et al. (1978) and Stith et al. (1981) showed mean mass radii between 0.12 and 0.22 μm, which is similar to the present measurements.

Airborne Observations

Water Vapor Profiles

Three water vapor profiles from 3, 4, and 6 September indicated almost identical results (Figure 49.3), indicating that approximately 50% of the total water vapor is within the first km. Above 3 km the total

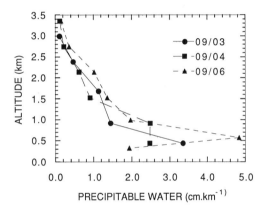

Figure 49.3 The vertical profile of the precipitable water was measured for three conditions on three dates: flaming phase (3 September), smoldering phase (4 September), and background (6 September).

precipitable water vapor is less than 0.5 cm. Because the mixing layer was below 3 km throughout the experiment, virtually all of the water in the atmosphere is available for interaction with the emission products. Note that on 6 September the high concentration of water vapor at 0.6 km is due to the presence of thin cirrus clouds.

Aerosol Profiles

Three aerosol profiles were measured corresponding to three times of the day (Figure 49.4): a 10:00 flight which we term *background condition* (6 September); a 12:00 noon flight in which the mixing layer is becoming active but with only aged (>1 day) smoke and smoke from smoldering, termed *smoldering* (4 September); and a 15:00 profile in which many flaming fires are contributing to the aerosol loading, termed *flaming* (3 September). The background profile resulted in large extinction coefficients (>0.5 km⁻¹) for all wavelengths below 500 m above ground level

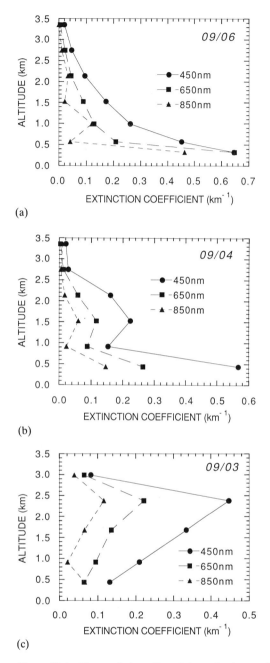

(a)

(b)

(c)

Figure 49.4 The vertical profiles of the extinction coefficient measured at 450, 650, and 850 nm are plotted for (a) the background conditions, (b) the smoldering phase, and (c) the flaming phase.

(AGL). The extinction coefficients decreased exponentially with elevation to less than 0.1 km^{-1} above 2.0 km AGL (Figure 49.4a).

The smoldering phase profile indicated high extinction coefficients below 0.5 km and between 1.5 and 2.0 km with a much stronger wavelength dependence than the background conditions. Above 2.5 km the results were very similar to the background (Figure 49.4b). The flaming phase profile showed emission from of the flaming smoke directly to a layer at 2 to 3 km AGL. Thin high smoke layers were already observed in Brazil during the burning season (Andreae et al., 1988). In contrast to the other profiles, lowest extinction coefficients occurred near the ground and, like the other profiles, almost no wavelength dependence was observed for the highest layer (>3 km, Figure 49.4c).

The mean optical thickness above 3.5 km for each phase was comparable in the three days of observations: 0.13 ± 0.02 at $\lambda = 650$ nm (Table 49.2). It was less wavelength dependent in background conditions than in smoldering or flaming conditions; the corresponding effective radius r_{eff} decreased from more than 1 μm for background conditions to 0.5 μm for flaming conditions (Table 49.2).

Within the mixing layer, the optical thickness at 650 nm was 0.36 for the flaming phase, 0.26 for the smoldering phase, and 0.31 for the background smoke (see Table 49.2). The effective radius was smallest in the flaming phase (0.16) and increased to 0.20 and 0.22 for smoldering and background conditions. These results confirm the smaller particle size for fresh smoke as deduced from ground measurements.

Comparison to CO Profiles

CO samples were made during the smoldering and background profiles. The emissions of particulates and CO are highly correlated at ground levels; however, because of different dispersion rates and atmospheric lifetimes, owing to very different mass and physical properties, it is not known how well they correlate in a large convective laboratory such as the Amazon Basin. In Figure 49.5a the background extinction coefficient profile at 450 nm was plotted against the CO profile. Both demonstrate an exponential decrease in value with elevation over nearly an order of magnitude of change for each variable (Figure 49.5a). The smoldering profile also demonstrated a high degree of similarity although the correspondence was not as good as the background profile. These data suggest that remote sensing of the optical properties of the aerosols is potentially a good estimator of CO emissions under these two conditions. With

Table 49.2 Smoke aerosol optical thickness, τ; wavelength dependence, α; and particle mean mass radius, r_m for the mixing layer and the layer above 3.5 km, from aircraft measurements

Date	Smoke type	τ (0.45 μm)	τ (0.65 μm)	τ (0.85 μm)	r_{eff} (μm)
Layer above 3.5 km					
3 September 89	Flaming	0.17	0.12	0.10	≈0.5
4 September 89	Smoldering	0.16	0.15	0.14	≈1.0
6 September 89	Background	0.13	0.12	0.11	≥1.0
Mixing layer					
3 September 89	Flaming	0.76	0.36	0.18	0.16 ± 0.03
4 September 89	Smoldering	0.55	0.26	0.12	0.20 ± 0.03
6 September 89	Background	0.57	0.31	0.16	0.22 ± 0.03

sufficient CO-to-particulate ratios the particulate matter could be computed and compared to estimates made from optical thickness observations.

Summary of the Smoke Optical Model

Spectral measurements from the ground and from aircraft of the smoke aerosol optical thickness indicate large Ångstrom coefficients of α = 2.1 to 2.5 except for the upper troposphere (above 3.5 km), as measured from the aircraft and during morning conditions where the measured smoke was emitted at least one day earlier (α = 0 to 1.5). These results are in agreement with measurements conducted in North America from aircraft (Pueschel et al., 1988) and lidar backscattering in Brazil (Browell et al., 1990). Assuming a log-normal distribution of the smoke particles, with σ in the range 0.3 to 0.9 the effective radius r_{eff} for the plume is in the range 0.10 to 0.18 μm for fresh smoke, 0.10 to 0.13 μm for smoldering, and 0.2 to 0.4 μm for background aged smoke. For the whole atmospheric mixed layer, far away from the plume, the effective radius is, respectively, 0.16, 0.20, and 0.22 μm.

Single Scattering Albedo and Graphitic Carbon

The single scattering albedo measures the effectiveness of the aerosol absorption, and it depends on the refractive index and on the size distribution of the particles. Smoke aerosol is a mixture of graphitic carbon and organic particulates. The characteristics of each component are

	Organic particulates	Graphitic carbon
Refractive index	1.43–0.0035i	2.00–0.66i
Mass radius (μm)	0.15	0.03–0.10

The graphitic carbon component was taken from Ackerman and Toon (1981) and organic particulates characteristics are derived from our results.

During the BASE-A experiment, mass measurements were made only through the plume of smoke generated in the flaming stage, and 10% of the total mass was found to be graphitic carbon (Ward et al., 1990). Accounting for liquid water, 70% relative humidity consists of 50% of the particles volume (Fraser et al., 1984; Kaufman et al., 1990b), and 5% of the total particles mass is graphitic carbon. For simplicity, the volume ratio was assumed identical to mass ratio. For three intermediate mass radii within the previous range—0.03, 0.05, 0.10 μm—we computed the single scattering albedo from the Mie theory by assuming an external mixing. The results are within 0.89 to 0.91 which is more absorbing than indicated from previous remote sensing studies (Kaufman et al., 1990). Since the mass ratio of graphitic carbon was measured through the plume only, it seems reasonable to assume that it would be smaller far away from the fire.

Application to Radiance Measurements

In this section the optical model is used to predict the difference between the upward radiance measured below the smoke layer and at the top of the smoke layer. The radiances were measured above a uniform undisturbed forest on 6 September 1989, during simultaneous measurements from the aircraft of the aerosol spectral optical thickness. The analyzed radiances were measured at 500 and 3600 m above the ground at 0.44 μm and 0.65 μm bands of a four-channel radiometer. The radiometer was calibrated in a NASA/GSFC calibration facility using an integrating sphere. Table 49.3 and Figure 49.6 summarize the radiances and optical thickness measured from the aircraft on 6 September. The radiances are normalized to reflectance units.

Since the low-level radiance was measured at 500 m

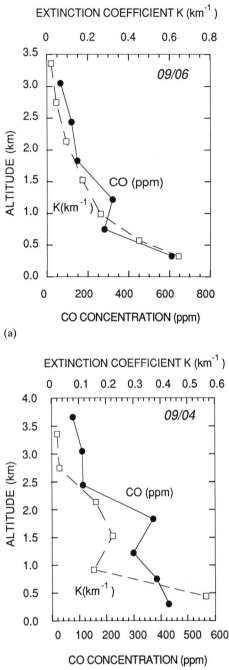

(a)

(b)

Figure 49.5 The vertical profiles of the extinction coefficient in the blue band (450 nm) and the CO concentration (ppm) are plotted for (a) the background conditions and (b) the smoldering phase.

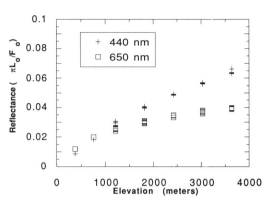

Figure 49.6 A nadir-view aircraft profile of the spectral reflectance from a closed forest canopy near Alta Floresta demonstrates an increase in atmospheric effect with altitude. Coincident sunphotometer observations allowed comparisons of the wavelength dependence of the reflectances to that of the aerosol optical thickness.

rather then at the ground, the difference in the measured optical thickness at 500 m and 3500 m (also shown in Table 49.3 for two wavelengths) was used to compute r_{eff} for this smoke layer for three values of σ. For each of the σ values in Table 49.4, the optimum value of r_{eff} was found and the predicted difference in the upward radiance was computed. As expected, for each value of σ a value of r_{eff} was found to fit the measured optical thickness with residual errors much smaller then the measurement accuracy. Surprisingly, though, the measured values of the radiance difference were not reproduced for $\sigma = 0.5$, but rather for $\sigma = 0.2$. This may be an indication that the wavelength dependence of the radiance is much more sensitive to the width of the size distribution than the wavelength dependence of the optical thickness, and that the width of the size distribution of particles emitted from biomass burning in the tropics is much smaller then the width of the distribution measured from fires in North America. This part of the investigation requires additional measurements and sensitivity analyses to establish the value of σ and its effect on the radiance.

Conclusions

Measurements of the tropical smoke aerosol optical characteristics were reported for several biomass fires and smoke conditions, as a function of time and height in the atmosphere. It was found that the wavelength dependence of the optical thickness can be explained by a log-normal size distribution with particle's effective radius varying between 0.1 and 0.2 μm. Analysis

Table 49.3 Upward radiance (in reflectance units) and corresponding aerosol optical thicknesses measured from the aircraft on 6 September 1989 for two flight altitudes

Wavelength thickness	Measured radiance		Measured aerosol optical thickness	
	500 m	3500 m	(500 m)	(3500 m)
0.44 μm	0.015 ± 0.004	0.067 ± 0.005	0.62	0.13
0.65 μm	0.018 ± 0.005	0.039 ± 0.005	0.35	0.12

Table 49.4 Comparison between the measured radiance difference between the two flights and the computed radiance difference for the measured optical thickness, and several values of the effective radius and the width of the particles size distribution σ that fit the measured wavelength dependence of the optical thickness

r_{eff}	r_m	σ	Difference in the upward radiance			
			Measured		Computed	
			450 nm	650 nm	450 nm	650 nm
0.17	0.15	0.26	0.047	0.018	0.047	0.019
0.17	0.17	0.18	0.047	0.018	0.047	0.017
0.11	0.09	0.51	0.047	0.018	0.058	0.027

of the upward radiance indicated that perhaps the width of the smoke particles size distribution is much narrower then reported from fires in North America. Surprisingly, strong sensitivity of the wavelength dependence of the upward radiance on the width (σ) of the particle size distribution generated a poor fit for σ = 0.5 but a good fit for σ = 0.2. Further measurements of the relation between the upward radiance and the smoke optical thickness are necessary to verify this finding. A strong correlation was found between the aerosol particles profile and the CO profile. This high correlation indicates that smoke particulates can be a good tracer for emission of trace gases from biomass burning in the tropics. Further investigations are also required to characterize the plume aerosols, taking better account of background conditions. Measurements of the solar aureole are required to invert into a full-size distribution. A more complete trace gas and particulate profile is required for specific burning conditions to compare with sunphotometer data profiles. Without this, remote sensing of biomass emissions will remain only an educated guess.

Effects of Fire Behavior on Prescribed Fire Smoke Characteristics: A Case Study

Wayne Einfeld, Darold E. Ward, and Colin Hardy

Biomass burning on a global scale injects a substantial quantity of gaseous and particulate matter emissions into the troposphere. Some of these combustion products are known to accumulate in the atmosphere and are implicated in observed changes in tropospheric composition and chemistry. The practice of open burning of biomass has come under close examination as a result of its pollution potential. In the United States, most biomass burning is managed in order to limit pollutant release. Recently, it has been inferred that biomass burning may produce a net increase in the concentration of "greenhouse gases" and particles in the atmosphere. Estimates of the quantity of biomass consumed globally per year vary widely, but generally Seiler and Crutzen's (1980) estimate of 5×10^{15} g yr^{-1} (grams per year) is regarded as accurate to within a factor of 2. The degree of consumption of total available biomass is affected by the quantity of biomass present, size distribution of the fuel, precipitation history for the site, condition of the vegetation, local meteorology, and a host of other factors. Combustion characteristics vary widely from vegetation types ranging from tropical grasslands to subarctic forest. As a result, the total emissions released are a complex function of many variables (Ward, 1990). Factors contributing to the total release of a particular pollutant species from a biomass fire can be quantitatively described by the following expression:

$$M_x = M_f \times A \times f_c \times EF_x$$

where M_x is the mass yield of a particular gaseous or particulate species, M_f is the fuel load in mass per unit area, A is the area burned, f_c is the fractional consumption of the total fuel load, and EF_x is the emission factor for the species of interest. Here, *emission factor* is defined as the mass release of a particular species per unit mass of fuel consumed. Ward and Hardy (1984) showed that emission factors for particulate matter and gaseous products are functions of the phase of combustion (flaming or smoldering). The work described here further demonstrates the impor-

tance of combustion phase on the release of incomplete products of combustion. More specifically, in this chapter we report results from a study that was designed to measure all of the terms noted in the above expression in an effort to derive an estimate of the total release of important pollutant species from a well-characterized fire. Ground and aircraft measurements were coordinated to yield a relatively complete picture of fire behavior and accompanying smoke production. Results from these measurements are then integrated over the lifetime of the fire and compared to less rigorous methods of estimating pollutant yield.

Experimental Methods

Fire Site Characteristics and Measurements

A 0.43 square kilometer (km^2) site in northwestern Montana, shown in Figure 50.1, was selected for this particular study. The fuel consisted of logging slash resulting from the clear-cut harvesting of true fir and western larch commercial timber from the site. The material left on the site was not of commercial value and provided much of the biomass fuel consumed by the fire. Prefire site characterization included a fuel inventory accomplished by taking a random transect across the site and tallying all fuel material within the transect into fuel size categories. Extrapolations were then made to the total area to be burned. Fuel moisture was also determined from a number of wood sections by specimen weighing, oven drying, and final weighing. Estimates of fuel consumption were obtained by fastening wires around the circumference of selected fuel in each size category. Following the fire, the wires were pulled tight, and the excess wire length used as a measure of fuel mass consumption in each fuel category. The fire was lit with hand-carried torches from the center of the plot outward in concentric circles. An observer was positioned in a suitable location to map the fire progression on all nine subunits throughout the duration of the fire. Fire characteristics such as the ignition time, duration and extent

Table 50.1 Sampling and analysis methods used on the instrumented aircraft

Analyte or parameter	Analytical method
Particle size and count (0.01–0.5 μm)	Grab sample—mobility analyzer
Particle size and count (0.1–3 μm)	Wing-mounted laser optical spectrometer
Particle size and count (2–32 μm)	Wing-mounted laser optical spectrometer
Light scattering coefficient @ 475 nm	Integrating nephelometer
Fine particle (<3 μm) mass concentration	Collect particulate on Teflon filter—gravimetric analysis
Fine particle nonvolatile and volatile carbon concentration	Collect particulate on quartz filter—combustion analysis
Carbon dioxide and carbon monoxide	Continuous analyzers—gas filter infrared absorption

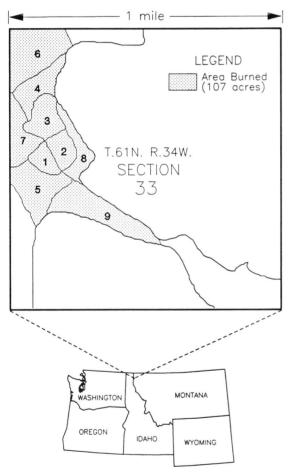

Figure 50.1 Map of burn location with inset showing the nine subunits.

of the flaming phase, and the onset of the smoldering phase were recorded for each of the nine subunits.

Instrumented Aircraft Measurements

The Sandia National Laboratories instrumented De-Havilland Twin Otter aircraft was used as a sampling platform during the fire. The aircraft was equipped with a number of sampling systems as outlined in Table 50.1. Particle size distributions over the range of 0.01 to 30 micrometer diameter were measured using mobility and wing-mounted laser optical particle counters. Gas sampling was carried out for carbon dioxide and carbon monoxide with continuous monitors using gas filter correlation infrared absorption techniques. Particulate material was collected through an external probe that extended forward of the aircraft windshield and terminated in a 1 m³ capacity bag positioned in the rear compartment of the aircraft, as schematically shown in Figure 50.2. The bag was filled to capacity in about 5 seconds while the aircraft was flying through smoke. Immediately following sampling, the grab sample was pumped through a series of cyclones with a 3 micrometer aerodynamic diameter cut point. The sample was then routed through a flash lamp nephelometer and finally through quartz and Teflon filters. Sample volumes were determined by time integrated measurement of

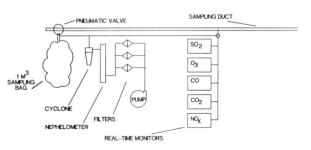

Figure 50.2 Schematic diagram of the sampling configuration on the instrumented aircraft.

air flow through each of the filters using mass flow meters output continuously logged by a data acquisition system. Postsampling analysis on the filters included gravimetric analysis, X-ray fluorescence elemental analysis, and particulate carbon analysis by a two-step combustion procedure as given by Johnson et al. (1981). Eight smoke samples were collected that covered all phases of the fire from the initial lighting phase to the final smoldering phase.

Emission factors for the gases and particulate categories of interest were calculated from samples collected with the aircraft using a chemical element balance method (Ward et al. 1982; Radke et al., 1988) whereby total combustion-derived carbon in the plume is used as a tracer for total mass consumption at the ground. The concentration ratio of a particular species to the total combustion-derived carbon concentration in the same parcel of plume air sampled with the aircraft is corrected to account for the carbon fraction of the fuel consumed. The resulting expression is given by

$$EF_x = \frac{[X] \times f_c}{[C_{total}]}$$

where EF_x is the emission factor for a given plume component x, $[X]$ is the measured plume concentration of the component, $[C_{total}]$ is the total combustion-derived carbon concentration in the same plume sample, and f_c is the dry fuel carbon fraction, which in this case is taken to be 0.5 according to Byram (1959).

Results

Fuel Loading and Consumption

Results from the pre- and postfire biomass inventories provided a measurement of the fuel consumption listed in Table 50.2. The total solid wood fuel loading was estimated at 13.5 kg m^{-2}. About 70% of the mass of the large diameter fuel was contained in the 8 to 23 and 23 to 50 cm diameter categories. The thick duff layer consisting of partially decomposed needles, cones, and twigs covering the forest floor increased the total fuel loading at the site to 24.1 kg m^{-2}. Nearly all of the fuel was consumed in fuel-size categories smaller than 23 cm. Duff consumption was estimated at about 60% of the total duff on the site. Total fuel consumption (woody fuels, litter, and duff) was 17.6 kg m^{-2}, which was 73% of the total biomass fuel on the site. The percent consumption by fuel categories and phase of combustion was as follows: 58.5% and 3.4% for the woody fuel for flaming and smoldering

Table 50.2 Measured fuel loading and consumption

Fuel diameter (cm)	Mass loading (kg m^{-2})	Mass consumption
0–0.5	0.87	100
0.5–2.5	1.46	100
2.5–7.5	1.25	100
7.5–23	6.81	87
23–50	2.44	50
50+	0.52	30
Total woody fuel	13.35	81
Duff layer	10.71	62
Total woody + duff	24.06	73

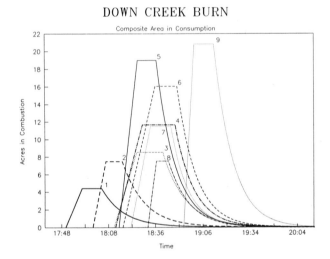

Figure 50.3 A plot of combustion history for each of the nine subunits. The lighting phase is indicated by a steep linear increase, the flaming phase by a horizontal segment, and the smoldering phase by an exponential decay.

phases, respectively; and 12.5% and 25.6% for the duff fuel for the flaming and smoldering combustion phases, respectively. The high proportion of total fuel consumed by this fire was attributed to the low moisture content of the fuels (ranging from 13% to 28%) at the time of the fire.

Time-Resolved Fire Behavior

Figure 50.3 is a plot of the progress of ignition of the nine subunits of the plot shown in Figure 50.1. A ground-based observer periodically drew in isopleth lines showing the location of the fire front on each of the subunits. The time of ignition is indicated by the steep positive-sloped lines in Figure 50.3 for each of the subunits. The duration of the flaming phase for

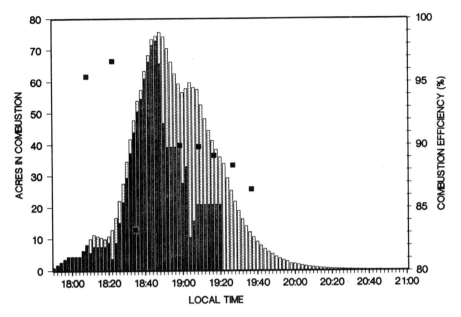

Figure 50.4 A plot of combustion history with all subunits combined. Dark shading represents flaming combustion and light shading represents smoldering combustion. Combustion efficiency for each of the eight aircraft smoke samples is shown by black squares.

each subunit is indicated by the flat portion of the curve, and the smoldering period is shown by the exponential decay curves for each of the subunits. An empirically derived half-life of 9.5 minutes (Ward and Hardy, 1984) is used to define the rate of change for the fuel consumption during the smoldering combustion phase. A composite merging of the flaming and smoldering phase of combustion activity from all nine subunits is shown in Figure 50.4. The period of extensive flaming phase combustion during the first 60 minutes of the fire is of particular significance and is well correlated with the characteristics of the emissions produced during that period.

Combustion Efficiency

A term called *combustion efficiency* has been used by Ward and Hardy (1984) to characterize the completeness of the oxidation of released carbon during the combustion of biomass fuels. Combustion efficiency is a particularly sensitive parameter that represents the degree of contribution from flaming and smoldering combustion in a volume of smoke. It is quantitatively given by the following:

$$\text{Comb eff (\%)} = \frac{[C_{CO_2}]}{[C_{total}]} \times 100$$

where C_{CO_2} is the background corrected CO_2 carbon as measured in the plume sample, and C_{total} is the sum of all forms of background corrected carbon in the

same plume sample. C_{total} in this case includes carbon from both gaseous (CO_2, CO, methane, and non-methane hydrocarbons) and particulate species (volatile and nonvolatile carbon) in the plume. The combustion efficiencies for the eight samples collected throughout the burn are shown in Figure 50.4 superimposed on the time history of the fire to illustrate the changes in combustion efficiency throughout the fire period. The combustion efficiencies range, with the exception of one outlier, from about 96% at the onset of the fire when flaming conditions were predominant, to about 85% during smoldering conditions during the later stages of the fire. An observed 83% combustion efficiency for the third sample, although collected early in the fire, suggests that this particular plume sample originates from smoldering combustion from a smoke region outside the major smoke column that was normally sampled with the aircraft. The range of combustion efficiencies observed during this fire are similar to measurements made in fires in Canada (Susott et al., 1991), Brazil (Ward et al., 1991), and the western United States (Ward and Hardy, 1990).

Particle and Gas Emission Factors

A plot of fine particulate and volatile particulate carbon emission factors is given as a function of sample combustion efficiency in Figure 50.5. Fine particle emission factors in the range of 2 to 4 g kg^{-1} were measured during the flaming periods. Increases on

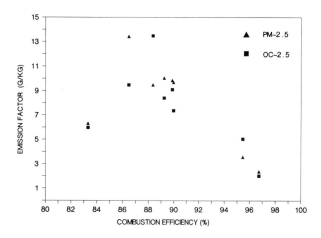

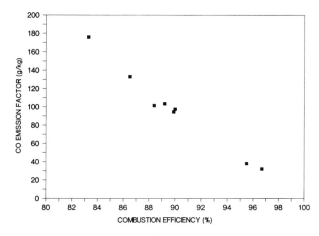

Figure 50.5 PM2.5 and nonvolatile carbon particulate (OC2.5) emission factors as a function of combustion efficiency for all samples collected with the instrumented aircraft.

Figure 50.6 Carbon monoxide emission factors as a function of combustion efficiency for all samples collected with the instrumented aircraft.

the order of threefold are observed for samples collected during periods when smoldering combustion was dominant. An opposite although more variable trend was observed for nonvolatile or elemental carbon measurements where emission factors were high during flaming combustion and decreased during smoldering combustion. During flaming combustion, the fine particles contained about 30% elemental carbon by weight. In the smoldering phase the elemental carbon content of the fine particles dropped to about 4%. These measurements clearly suggest that changes in the combustion efficiency of smoke samples are accompanied by changes in the composition of the particulate matter. During the predominantly flaming phase, the content of the particulate matter is low in organic carbon and high in elemental carbon. During smoldering combustion the opposite relation appears valid. These observations are consistent with measurements made by Ward and Hardy (1991) on the particulate matter content of smoke collected from both flaming and smoldering combustion using tower-based sampling systems positioned over large area fires.

Elements such as potassium, sulfur, iron, and lead were detected by X-ray fluorescence on most of the filter samples. Potassium emission factors were relatively constant throughout the duration of the fire with an average emission factor of 0.107 g kg^{-1}. The ratio of fine particle mass to fine particle potassium ranges from about 90 for the flaming periods to 180 for the smoldering periods. Particulate sulfur emissions showed a clear correlation with fine particle emission factors and ranged from low levels of about 0.03 g kg^{-1} during flaming phase to about 0.07 g kg^{-1}

during the smoldering phase. Lead emissions were relatively uniform over the entire fire period and at a low level of about 2 mg kg^{-1}.

Carbon monoxide emission factors calculated for each of the eight plume samples are plotted as a function of sample combustion efficiency and are shown in Figure 50.6. The CO yield can be observed to increase from about 40 g kg^{-1} to about 120 g kg^{-1} as the fire progresses into the smoldering phase. A least squares regression analysis with the CO emission factor as the dependent variable and combustion efficiency as the independent variable reveals a linear relationship with a slope of -10.52 (0.58), an intercept of 1043 (7) with a coefficient of determination (R^2) of 0.982.

A summary of differences between smoke emission factors for selected pollutant species from flaming and smoldering phase combustion is given in Table 50.3. An arithmetic mean of the emission factors for the first two samples was chosen to represent the flaming phase of the fire. Similarly, the smoldering category is represented by an average of the final two aircraft samples collected during the latter stages of the fire. A comparison of the overall emission factor averages from all eight aircraft samples with the combustion phase specific results shows that the overall averages do not fall at the midpoint between the flaming and smoldering phases but are closer to the smoldering values. Even though an analysis of the fuel consumption shows that approximately 71% of the fuel was consumed during the flaming phase, the average emission factors more closely represent the mix of emissions that would be expected for the smoldering phase. Since the rate of fuel consumption during

Table 50.3 Measured smoke emission factors for flaming and smoldering combustion phases and overall averages

Species or parameter	Emission factor (g kg^{-1})		
	Flaming	Smoldering	Overall
Fine particulate (<3 micron)	3.0	11.5	8.10 (3.47)
Elemental carbon (<3 micron)	0.8	0.5	0.55 (0.27)
Organic carbon (<3 micron)	3.6	11.5	7.61 (3.19)
Carbon dioxide	1733	1576	1621 (74)
Carbon monoxide	35.0	117.2	96.9 (43.7)
Single scatter albedo	0.75	0.97	0.90 (0.10)
Specific scattering	5.8	9.1	8.00 (1.80)
Coefficient @ 480 nm (m^2 g^{-1})			

Note: The uncertainty (one standard deviation) for the overall average is given in parentheses.

flaming combustion is high, the time duration of the flaming phase was short relative to the much longer smoldering phase, during which the remaining 29% percent of the fuel was consumed. As a result, emission factor averages from the eight samples collected over the duration of the fire are skewed toward the smoldering condition.

Particle Size Distributions

Particle size data collected during the first two and final two aircraft passes were examined in detail in order to observe effects of fire behavior on particle size distributions. The first two aircraft samples were collected during a flaming phase as illustrated by calculated combustion efficiencies of 0.95 and 0.96. The last two samples were collected during smoldering conditions as evidenced by combustion efficiencies of 0.88 and 0.87. Number distributions for these four passes are shown in Figure 50.7. Although significant differences between the flaming and smoldering samples are not observed, some subtle differences can be noted. The samples from flaming combustion show an elevated condensation nuclei mode in the 0.02 to 0.04 micron particle diameter range when compared to the smoldering phase distributions. An upward shift in the number mode diameter can also be observed in the smoldering samples, although it is, at best, only about 0.1 micrometer in magnitude. A plot of cumulative volume less than 3 micrometer particle diameter is shown in Figure 50.8 for the same four samples. Here again a slight upward shift of about 0.05 micron in the median volume diameter is observed in the smoldering phase smoke samples.

Carbon Distribution

The apportionment of carbon among the major products of combustion is shown in Table 50.4 for the

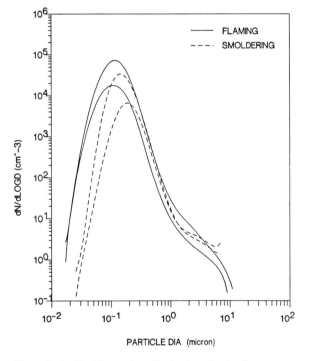

Figure 50.7 Particle number distributions for the first two (flaming) and final two (smoldering) smoke plume penetrations with the instrumented aircraft.

flaming and smoldering phases of combustion. Gaseous volatile organic carbon compounds were not quantified during this experiment. We have used a model of Ward et al. (1990) to predict emission factors for methane and nonmethane hydrocarbons. This model is derived from measurements of combustion efficiency as well as methane and nonmethane hydrocarbon emissions from a number of prescribed burns in the Pacific Northwest. During both flaming and smoldering conditions, in excess of 96% of the carbon is released as CO_2 and CO. During flaming conditions the volume ratio of CO to CO_2 is 3.1×10^{-2}. This ratio increases to 1.2×10^{-1} during smoldering conditions. Volatile and nonvolatile carbon particulate material accounts for 0.85% of the total released carbon in the flaming phase and nearly 2.4% during the

smoldering phase. The first three categories—namely, CO_2, CO, and volatile organic particulate carbon—account for in excess of 99% of the carbon released in both flaming and smoldering phases.

Estimates of Total Pollutant Release

We calculated the total release of various gas and particle species during both flaming and smoldering periods by breaking down the total fire interval into flaming and smoldering periods and applying phase-specific emission factors (Table 50.3, columns 2 and 3) and measured fuel consumption for each phase. These estimates are given in units of kg m^{-2} in Table 50.5. The total from flaming and smoldering phases is compared to estimates of total release using average emission factors (Table 50.3, column 4) from all aircraft samples collected and total fuel consumption. Using the average CO yield results in a 64% overestimate of the total CO emissions when compared to the

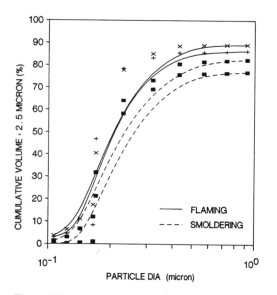

Figure 50.8 Cumulative volume (<2.5 micron diameter) distribution for the first two and final two smoke samples.

Table 50.4 Species contribution to total mass release of carbon from flaming and smoldering phases

Species	Carbon contribution (% by weight)	
	Flaming	Smoldering
Carbon dioxide	95.8%	86.6%
Carbon monoxide	3.0	10.1
Organic carbon particulate	0.7	2.3
Elemental carbon particulate	0.2	0.1
Methane[a]	0.2	0.5
Nonmethane hydrocarbons[a]	0.1	0.4

a. Estimated from earlier work by Ward (1991).

Table 50.5 Total mass release of significant pollutants

Species	Mass release (kg m^{-2})				
	Flaming	Smoldering	Total[a]	Average[b]	Ratio[c]
CO_2	21.5	8.0	29.4	28.3	0.96
CO	0.43	0.60	1.03	1.69	1.64
O	0.062	0.055	0.117	0.133	1.13
EC	0.010	0.003	0.012	0.010	0.77
CH_4	0.035	0.045	0.080	0.102	1.22
TNMHC[d]	0.031	0.031	0.062	0.076	1.22
PM2.5	0.037	0.058	0.095	0.141	1.48

a. Total is the sum of mass release from the flaming and smoldering categories.
b. Average is determined by use of average emission factors as measured over the duration of the fire.
c. Ratio is calculated with "Average" as the numerator and "Total" as the denominator.
d. TNMHC refers to total nonmethane hydrocarbons.

sum of CO yields from flaming and smoldering periods. Similarly, the PM2.5 emissions determined by an average emission factor are about 50% higher than the sum of the phase-specific emission yields. The agreement for other carbonaceous species such as volatile and nonvolatile carbon is somewhat better and is likely a result of the fact that the difference between flaming and smoldering emission factors is less pronounced for these species.

Conclusions

Measurements of smoke emissions were made of the 0.43 km^2 Down Creek fire in northwestern Montana using an instrumented aircraft and smoke sampling system flown periodically through the smoke column. Results from these tests demonstrated an expected change in the characteristics of the emissions as the fire progressed from flaming to smoldering conditions over a period of about two hours. Combustion efficiencies ranged from 95% conversion of released fuel carbon to CO_2 during the flaming phase to 85% conversion during the smoldering phase. The overall fuel mass consumption was about 70%, with consumption of 81% of the woody fuel and 62% of the duff layer.

The median particle diameter became larger for smoke derived from smoldering combustion sampled toward the latter stages of the fire. Although these particle size changes are not likely to be significant in terms of the mass release of particulate matter and its atmospheric transport, they are accompanied by appreciable changes in smoke optical properties. An enhanced scattering deficiency is observed in smoldering samples and is consistent with predictions based on Mie theory for an upward shift in the particle median volume diameter. This observation becomes important in the context of radiative transfer through the smoke plume.

A comparison of measured emission factors from flaming and smoldering periods indicates about a threefold increase in CO, PM2.5, and volatile organic particulate carbon (for particles <2.5 micrometers in diameter) for emissions from smoldering combustion. These emission factor increases are at least partially offset by a reduced fuel consumption during the smoldering phase. Total fuel consumption for the Down Creek prescribed fire was 12.4 kg m^{-2} for the flaming phase and 5.0 kg m^{-2} for the smoldering phase. An approximate threefold increase in emission factors is then offset by about a 2.5-fold reduction in fuel consumption in the smoldering phase. Taken together, the net result is an approximate 50% increase in the total release of such pollutant species as PM2.5 and CO during smoldering conditions as compared to flaming periods.

Results suggest that knowledge of fuel consumption by phase of combustion (flaming vs. smoldering) is important in making accurate estimates of the characteristics of smoke emissions from individual fires. Contributing factors such as fuel type, fuel loading, and meteorological history vary significantly by region and should be taken into account when compiling estimates of fuel consumption rates during both flaming and smoldering fire conditions.

Acknowledgments

We recognize the assistance of Janet Hall and her documentation of the prescribed fire history. We thank the Champion Paper Company for their generosity in allowing this particular fire to be used for detailed study. The assistance of the Libby Ranger District of the U.S. Forest Service in site preparation is also noted. We also recognize the Defense Nuclear Agency, Global Effects Program, for their financial support of this work.

A Numerical Simulation of the Aerosol-Cloud Interactions and Atmospheric Dynamics of the Hardiman Township, Ontario, Prescribed Burn

Joyce E. Penner, Michael M. Bradley, Catherine C. Chuang,
Leslie L. Edwards, and Lawrence F. Radke

The interaction of aerosols with clouds is important for climate for two major reasons. First, the abundance of aerosols in the atmosphere has an important direct radiative effect that may be important for climate (Charlock and Sellers, 1980), and the scavenging of aerosols by cloud droplets and hydrometeors, in part, controls their abundance by determining their removal rate and lifetime. Second, the interaction of aerosols with clouds affects the cloud droplet number and size distribution. This may have a profound impact on climate by changing the scattering of solar radiation by clouds (Twomey, 1977; Charlson et al., 1987).

The scavenging of aerosols by cloud droplets, their subsequent removal by precipitation, and their effect on droplet spectra is determined by both the aerosol properties and abundance and the atmospheric dynamics of the cloud system. In this chapter we focus on the effects of nucleation scavenging on the removal of aerosols and the determination of droplet spectra. The aerosol properties of importance here (size distribution, chemical composition, and total number concentration) determine the ease with which aerosols are scavenged by nucleation. The dynamics of the cloud system determine the updraft velocities and mixing processes (which affect total water and aerosol abundance) in the cloud.

Our goal is to develop a method for treating these processes accurately in a dynamical cloud model so that these interactions can eventually be treated in a climate model. This initial study focuses on the nucleation of smoke above the August 1987 Hardiman Township, Ontario, prescribed burn. This situation presents a particularly good opportunity for checking our understanding of the microphysical and dynamical interactions because the forcing which gives rise to the cloud and the aerosol source can be relatively well quantified. Fuel inventories before and after the burn were used to determine the average burn rate and surface heat flux for the fire (Stocks, private communication, 1988). In addition, an instrumented aircraft determined the local background moisture, wind, and temperature profiles as well as the smoke emission factor and size distribution (Radke et al., 1988). In this chapter, we use these measured quantities as input to drive both a dynamical cloud model and a microphysical model of droplet formation. We briefly describe the models used in this study together with the choice of aerosol properties used in this work and the development of a parameterization for aerosol scavenging by nucleation. The nucleation parameterization is used to determine the amount of aerosol scavenged by nucleation for the Hardiman fire. As a check on the inferred aerosol nucleation properties and the results of the dynamical model, the predicted drop size spectra is compared to the measured spectra. Our conclusions appear in the final section.

Description of Models

The dynamical model used in this study has been developed from the Klemp and Wilhelmson (KW) cloud model (Klemp and Wilhelmson, 1978). The KW model is a compressible, nonhydrostatic cloud model with open (wave-permeable) lateral boundaries. It uses a time-splitting numerical solution technique with a small time step to integrate the acoustic terms in the momentum equations and a larger time step to integrate the advection, gravity wave, and turbulence and microphysical parameterization terms. This dynamical framework has been tested by application of the model to a variety of situations.

The dynamical prognostic variables represented by the original KW model include the three velocity components—pressure, potential temperature, and the subgrid-scale kinetic energy (which is used to compute a first-order closure turbulence parameterization for the effects of eddy mixing). In addition, a Kessler-type bulk microphysical parameterization is used that includes prognostic equations for water vapor, cloud water, and rain (Kessler, 1969). To these basic equations, we have added prognostic equations for the bulk aerosol mixing ratio, the mixing ratio of aerosol particles in cloud water, and the mixing ratio of aerosol in rain. The current version of the model at Lawrence Livermore National Laboratory also treats the ice phase (Bradley and Molenkamp, 1990; Molenkamp and Bradley, 1990), but for the study reported here, ice was not present and these features were not utilized.

The microphysical model used in this study is described in Penner and Edwards (1986), Edwards and Penner (1988), and Chuang et al. (1990). It treats the nucleation of aerosol particles in a rising air parcel. The full aerosol size distribution is described as a function of dry aerosol size, r, the mass fraction of soluble material in the aerosol, ϵ, and time, t. The amount of water condensed on the aerosol is calculated as a function of these variables as well. When the wet radius of the aerosol particles becomes larger than the critical radius, the aerosols are "activated" to drops. Numerically, this may be treated either by leaving the activated particles in the aerosol array and allowing them to grow and separate from the smaller unactivated aerosols by the condensation of water associated with the dry radius of the aerosol or by moving them to a separate drop array, whose independent variable is thereafter treated as the wet radius. In this last procedure, details of the original aerosol size distribution can be lost because several aerosol sizes may combine to form a single drop size due to a narrowing of the drop size distribution under the process of condensation. However, for this study, we found little difference in the predicted scavenging using these two techniques. In addition to condensation and activation, the model treats coagulation of aerosols and coalescence of drops. The capture of aerosols by drops due to phoretic, electrical, or impaction processes may also be treated in the model provided the appropriate kernals are specified; however, these processes were not included in this study and appear to be unimportant for the derived scavenging processes and drop size distribution.

A Parameterization for Nucleation Scavenging of Smoke

Figure 51.1 shows the critical supersaturation of aerosol particles as a function of dry radius, r, and mass fraction of soluble material, ϵ. For the purposes of this figure and the following analysis, we have assumed that both the soluble and insoluble material in the smoke have densities of 1 gram per cubic centimeter (g/cm^3) and that the soluble fraction may otherwise be treated as ammonium sulfate $(NH_4)_2SO_4$. Our choices for other model parameters are described in

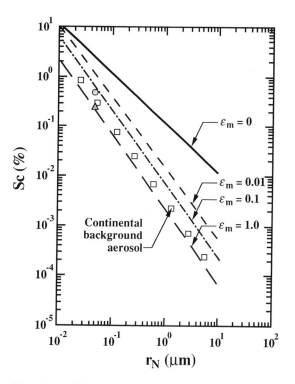

Figure 51.1 Critical supersaturation as a function of dry particle radius. The mass fraction of soluble material in the aerosol is indicated by ϵ. Open squares show the critical supersaturation for the background continental aerosol, while the circle and triangle show two estimates of the critical supersaturation of smoke particles from burning forest material.

Edwards and Penner (1988). Also shown in the figure is the critical supersaturation as a function of dry radius for background continental aerosol (Pruppacher and Klett, 1978).

Very little is known about the nature of the soluble material in smoke although we do know that smoke from burning vegetation is easily activated (Hallett et al., 1989). Hallett et al. (1989) report that the measured mass fraction of soluble ions from a northern California forest fire smoke was between 4% and 5%, but this is only a lower limit because the analysis was restricted to only nine ionic species. Andreae et al. (1988) report emission factors for burning vegetation from Amazonia that are consistent with about 35% of the particle mass composition being in the form of the ionic species $SO_4^=$, NO_3^-, and NH_4^+, if it is assumed that half of their reported emission factor for NO_x + aerosol NO_3^- + gaseous HNO_3 is due to NO_3^-, and half of their reported emission factor for SO_2 + aerosol $SO_4^=$ + MSA is due to $SO_4^=$. If all of the reported nitrogen and sulfur emissions are NO_3^- and $SO_4^=$, respectively, the three measured ionic species

(NH$_4^+$, SO$_4^=$, and NO$_3^-$) would comprise 60% of the total aerosol mass. These estimates of the average soluble mass fraction can be misleading, however, because the soluble ionic species may be distributed differently in different size fractions of the aerosol. Although our detailed model is able to distinguish different aerosol compositions as a function of size, this feature was not implemented for this study due to lack of detailed information. Instead, we made a rough estimate of the average fraction of soluble material in the smoke using measurements reported by Radke et al. (1978), Eagan et al. (1974), and Hobbs et al. (1978), in the following manner. Radke et al. (1978) reported the measured size distribution for smoke generated in several prescribed burns of forest slash, while Eagan et al. (1974) reported the number of particles active as CCN at 0.5% supersaturation (albeit for a different prescribed fire). Equating the number of particles that act as CCN with the larger particles in the size distribution yields a critical aerosol size of 0.05 μm radius for the smallest aerosol particle active at 0.5% supersaturation. This data point is shown by the circular symbol in Figure 51.1. A similar analysis using the measured aerosol size distribution and the measured CCN concentration at 0.2% supersaturation for the prescribed burns reported by Hobbs et al. (1978) yields the data point shown by the triangular symbol in Figure 51.1. As shown in the figure, smoke from the burning of forest slash appears to behave as if it has a soluble mass fraction of $0.1 < \epsilon < 1.0$. In the development of the parameterization for nucleation scavenging of smoke that follows, we have chosen a mass fraction parameter of $\epsilon = 0.3$. In the section on simulated and measured drop size distributions, we test the sensitivity of the predicted droplet spectra and fraction of smoke nucleated to the assumed soluble mass fraction.

As shown in previous work (Edwards and Penner, 1988), the fraction of smoke nucleated depends on both the quantity of smoke present at cloud base and the updraft velocity. To develop a parameterization for nucleation scavenging for use in large-scale models, we used our microphysical parcel model of nucleation and condensation with a fixed and specified updraft velocity. We varied the updraft speed through the range from 2 to 20 m/s while the mass concentration of smoke was varied from 5.0×10^{-12} to 1.0×10^{-8} g/cm^3. In this analysis the assumed smoke size distribution was that given by Radke et al. (1988) for the Hardiman fire. Thus, by varying the mass concentration, the number concentration is also varied. Figure 51.2 shows the resulting fraction of

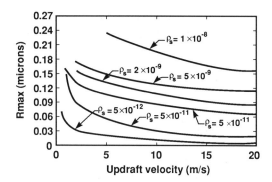

Figure 51.2 Fraction of aerosol mass incorporated into cloud droplets as a function of updraft velocity and smoke mass concentration at cloud base. The smoke mass density (in g/cm^3) is indicated by ρ_s.

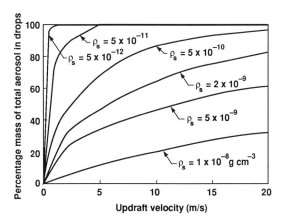

Figure 51.3 Critical radius of activated smoke particles as a function of updraft velocity and smoke mass concentration at cloud base. The smoke mass density (in g/cm^3) is indicated by ρ_s.

aerosol mass that was activated as a function of velocity and smoke mass concentration. Figure 51.3 shows the critical radius (smallest particle activated) for the same range of conditions. For a given smoke mass, as the velocity is increased, a higher fraction of smoke is activated and the critical radius is decreased. This is caused by a faster cooling of the rising air parcel (due to expansion), creating a higher saturation ratio. Thus, because the total number of aerosol particles is constant at any given altitude, the rate of condensation cannot increase quickly enough to compensate for the increased cooling rate, leading to a higher supersaturation. The figures also demonstrate that as the smoke mass concentration (which is proportional to number concentration) is decreased, a higher fraction of smoke is activated to drops. In this case, the decreased number of aerosol particles available for nucleation causes a decrease in the rate of condensation, thereby leading to an increase in the maximum saturation ratio that is experienced by the parcel. We note that although we assumed a particular background atmosphere [a dry adiabatic standard atmosphere with a relative a humidity of 61% at 900 millibars (mb)], we verified that the results are not very sensitive to the assumed pressure and temperature at which saturated conditions are reached. The results in Figure 51.2 were used in table form and interpolated for the parameterization of smoke scavenging in the dynamical model.

A Numerical Simulation of the Hardiman Fire

The Hardiman Township fire was estimated to burn about 3.25 kilograms per square meter (kg/m^2) of fuel. Since most of the burn took place over a period of three hours, we estimated a resulting average surface heat flux of 6.0 kilowatts per square meter (kW/m^2). The smoke emission factor was taken as 10.5 grams per kilogram (g/kg) fuel, as determined for this fire from the measurements of Radke et al. (1988). This results in a smoke injection rate of 0.003 $g/m^2/s$. The background moisture, temperature, and wind profiles for the simulation were those measured by the University of Washington instrumented aircraft. These were extrapolated to the surface using radiosonde data from the nearest available stations. The actual fire geometry was used in the simulation, giving a total area burned of 325 hectares. The model resolution was 250 m in the horizontal and 100 m in the vertical. Separate simulations (not reported here) showed that this resolution was adequate for representing the vertical structure above the fire.

Figure 51.4a shows a wire diagram of the predicted smoke mass mixing ratio concentrations above the fire after 40 minutes of simulation. At this point in time, the simulation had reached a near steady-state, and a small cloud had formed. The smoke captured in cloud water is shown in Figure 51.4b. Both the smoke and water clouds from the simulation were similar in visual appearance to portions of the real-time video records of the fire. In particular, the gravity-wave feature seen at the top of the smoke plume in Figure 51.4a was evident in pictures of the actual plume, as was the sinking smoke field shown just below and to the right of the central updraft region. The predicted height of the smoke plume agreed well with the observed height, as did the predicted water content of the small cloud that formed. Although no precipitation was observed on the ground, the water cloud did

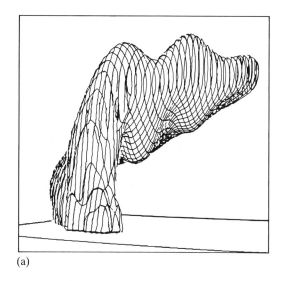

(a)

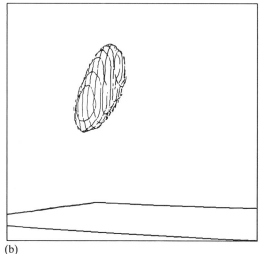

(b)

Figure 51.4 (a) Wire diagram of the predicted smoke mass mixing ratio above the fire after 40 minutes of simulation. The contour interval for this diagram is 10^{-3} g/kg. (b) Wire diagram of the predicted smoke mass mixing ratio in cloud water after 40 minutes of simulation. The contour interval is 10^{-3} g/kg.

develop drops that were large enough to have a significant fall velocity. The simulation, as well, produced a liquid water content in the cloud of somewhat more than 1 g/m^3, which would normally have produced a slight amount of precipitation in the model (the autoconversion mechanism for precipitation formation was deactivated in the simulation reported here, however).

The parameterization of nucleation has been implemented in the dynamical model to recognize two processes which result in the nucleation of smoke. In the first, smoke is nucleated to form droplets as new

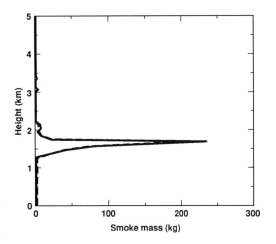

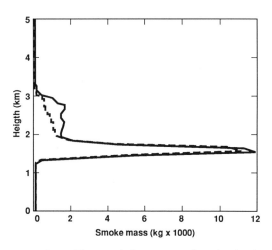

Figure 51.5 The cumulative amount of smoke exposed to new cloud formation above the fire (solid line) after 40 minutes of simulation. The amount of smoke activated during this exposure is also shown (broken line) but is hardly visible due to the scale of this figure. This process operates primarily early in the simulation.

Figure 51.6 The cumulative amount of smoke that has entered an area of cloudy air (solid line) and the amount of smoke that is activated under these conditions (broken line) after 40 minutes of simulation. The overall efficiency for this process is about 47%.

cloud water forms. Thus, if a previously dry region reaches saturated conditions, a fraction of the smoke present becomes smoke in cloud water depending on the local smoke concentration and updraft velocity. Figure 51.5 shows both the total amount of smoke exposed to new cloud in this manner and the amount that has activated after 40 minutes of simulation. At this time, roughly 250 kg of smoke has been exposed to new cloud formation, while 210 kg was actually activated. Figure 51.5 shows that this process operates primarily at the cloud base altitude of 1.5 km, while further analysis shows that much of the exposure to new cloud formation conditions and activation occurred early in the simulation.

In the second nucleation process simulated in the model, smoke may pass from a region with no cloud into a region of pre-existing cloud. The smoke is nucleated according to the local conditions of smoke and water concentration as before. The total amount of smoke exposed to this second process and the amount activated after 40 minutes are shown in Figure 51.6. As shown by comparison with Figure 51.5, this is by far the dominant process. Approximately 12,000 kg of smoke are exposed to these conditions and 11,000 kg are activated. Again, most of the activation occurs near cloud base, but, in this simulation, a smaller amount of smoke is activated above 2 km. This high-level activation occurs because smoke enters the cloud from the side.

The overall efficiency of nucleation scavenging for

this fire was about 47%. Thus, of the 23,400 kg of smoke emitted into the domain in the first 40 minutes, 11,000 kg were activated to form cloud droplets.

A Comparison of the Simulated and Measured Drop Size Distributions

As an integral check on the dynamical model simulation, the microphysical model, and the choice of nucleation characteristics for the smoke, we used the microphysical model together with the predicted updraft velocities at cloud base from the dynamical model to predict the drop size distribution. The parcel model simulation was stopped when the liquid water content of the predicted cloud in the parcel was equal to the measured liquid water content of the actual cloud during the aircraft sample of droplet spectra. Figure 51.7a shows the simulated aerosol and drop size distributions for $\epsilon = 0.3$. Although the simulated distributions do not show a clear separation of the droplet and aerosol size distributions, the activated aerosols are clearly visible as the particles above the notch in the spectra. Had new cloud water been allowed to continue to condense, the droplets would have eventually grown much larger than the unactivated aerosol particles.

The measured droplet spectra is also shown in Figure 51.7a. It should be noted that no measurements were obtained below about 2.5 μm radius. Furthermore, the droplets that are simulated with radius greater than about 200 μm have formed on a

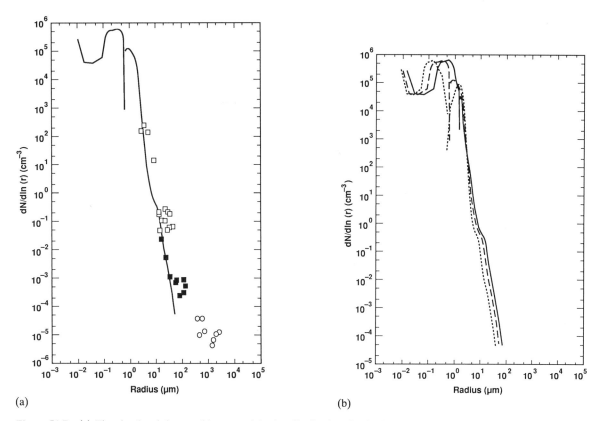

Figure 51.7 (a) The simulated drop and haze particle size distributions in the detailed microphysical model are compared to the measured drop size distribution. The three symbols indicate measurements taken by an FSSP (open squares), by a cloud probe (solid squares), and by a precipitation probe (open circles) (see Radke et al., 1988, and references therein for details). No droplet measurements were available for sizes less than 2 μm radius. Also, the measured drops beyond 50 μm formed on a part of the aerosol distribution that was not measured. (b) The simulated drop and haze particle size distribution in the detailed microphysical model for an assumed mass fraction of soluble material in the aerosol of ϵ = 1.0 (solid line), ϵ = 0.3 (long-dashed line), and ϵ = 0.05 (short dashed line).

part of the aerosol spectra that was not measured. (We simply extrapolated the measured aerosol size distribution out to these sizes.) Comparison of the observed spectra in this size range with the simulated spectra indicates that either the actual size distribution of aerosol had a greater number of large particles than we assumed or that some degree of coalescence took place among the cloud droplets. Either possibility may explain the presence of large drops in the measurements. Thus, Radke et al. (1983) have observed supermicron particle size distributions of smoke from a variety of fires that support a size distribution with the majority of the mass present in sizes greater than 45 μm. Similarly, simulations with the coalescence mechanism activated in the model indicate that the coalescence of cloud droplets might also explain the data.

Figure 51.7b shows the simulated aerosol and drop size distributions for the cases ϵ = 0.05 (short dashed

line), 0.3 (long dashed line), and 1.0 (solid line). Comparison of these simulated spectra with the data shown in Figure 51.7a demonstrates that the choice of ϵ = 0.3 provides an adequate measure of the nucleation tendency of the smoke aerosol, but is not unique in this regard. More highly resolved measurements that included the aerosol portion of the size distribution might be able to distinguish between these three cases.

Summary and Conclusion

We have used a microphysical model of condensation and nucleation to develop a parameterization that relates the mass of smoke scavenged by nucleation to the updraft velocity and total smoke concentration at cloud base. The parameterization was developed for a typical smoke size distribution and used a description of smoke nucleation properties that was consistent with measurements of the fraction of CCN from vegetation fires (Hallett et al., 1989) as well as measured CCN concentrations and smoke size distributions (Hobbs et al., 1978; Eagan et al., 1974; Radke et al., 1978).

We used a comparison between the measured droplet spectra and the predicted spectra from our microphysical model to test whether the updraft velocities and smoke mass concentrations predicted at cloud base by the dynamical model were adequate. Then, the dynamical model was used to predict the efficiency of nucleation scavenging of smoke above the fire. We found that more than 47% of the smoke on a mass basis was nucleated to drops above the fire. The parameterization of nucleation scavenging developed here may be extended in an obvious way to predict the cloud droplet number concentration and spectra resulting from the nucleation of aerosols. In future work, we plan to extend our analysis to enable us to parameterize the effects of CCN on droplet spectra for use in our global model simulations of the sulfur cycle (Erickson et al., 1990) and the effects of smoke on climate (Penner et al., 1990).

Acknowledgments

This work was supported in part by Dr. David Auton and Major Richard Hartley, Defense Nuclear Agency. The Lawrence Livermore Laboratory operates under the auspices of the U.S. Department of Energy under contract No. W-7405-Eng-48.

Impact of Carbonaceous Fuel Burning on Arctic Aerosols: Silicon as an Indicator of Long-Range Transport of Coal Smoke

Jozef M. Pacyna, John W. Winchester, and Shao-Meng Li

It has been known for some years that the arctic troposphere contains suspended particulate matter of pollution as well as natural origin, at levels regulated by the rates of inflow from primarily lower latitude sources and the rates of removal by wet or dry deposition. In late winter the transport of pollutants northward into the Arctic may be especially efficient, and deposition fluxes to the surface may be minimal owing to generally stable air conditions with little precipitation. High levels of visible haze in the Arctic during late winter have been attributed to these balancing effects. However, an accurate apportionment of contributions from pollution and natural sources has only recently become possible as carefully designed field measurement programs have been carried out for this purpose, principally north of western Europe (e.g., Ottar et al., 1986) and North America (e.g., Schnell, 1984; Barrie, 1985; Herbert et al., 1989; Shaw, 1988).

In this chapter some of the apportionment results are presented, with emphasis on contributions of aerosol from carbonaceous fuel combustion in populated areas south of the Arctic. We emphasize here the contributions from fossil fuel combustion, since these are now rather well documented, but we also call for future research to assess contributions from the burning of living biomass to the arctic atmosphere, with consequences for air quality and global environmental and climate effects.

Concentrations of silicon in the arctic aerosol are used as an indicator of long-range transport of coal smoke from various emission source regions in lower latitudes to the Arctic.

Contribution of Carbonaceous Fuel Combustion Emissions to the Arctic Haze

The measurements of the arctic aerosol at ground level and during aircraft flights have indicated clearly that very long-range transport of air masses, over several thousand kilometers, can carry mixtures of pollutants and natural substances and affect arctic air quality during both seasons. Polluted air masses, car-

rying mixtures of anthropogenic substances from a variety of sources in different geographical areas, have been identified in the Arctic up to 5 kilometers (km) altitude. The layers of polluted air below 2.5 km can be traced to episodic transport from Eurasian and North American sources (e.g., Pacyna and Ottar, 1988).

Various techniques have been applied to explain the origin of air pollutants in the Arctic, particularly in the lower 2 km of the troposphere. Receptor modeling of chemical composition measurement data has been performed by chemical mass balance and principal component analysis (e.g., Lowenthal and Rahn, 1985). Dispersion modeling, mainly by a Lagrangian trajectory model, has also been performed (e.g., Pacyna et al., 1985). The results showed that carbonaceous fuel combustion emissions were one of the most important source categories contributing to arctic air pollution at lower altitudes. The emissions include trace metals, e.g., V, Ni, As, Cd, Se, and Sb, from fossil fuels, measured in highest quantities when air masses had passed over major anthropogenic source regions in Eurasia and were transported to the Norwegian Arctic and Alaska. The enrichment factors of these metals relative to common soil mineral constituents were also high, as shown in Table 52.1. Average regional apportionments of trace metals in the winter and summer aerosol from the Norwegian Arctic are presented in Table 52.2 after Maenhaut et al. (1989).

A quantitative assessment of the contribution of carbonaceous fuel combustion emissions to Norwegian arctic haze layers has been made using results of absolute principal component analysis of aerosol composition measurement data from Table 52.2 and a survey of emissions from different nations in Eurasia prepared by Pacyna (1984) and Pacyna et al. (1985). The assessment showed that 10% to 50% of various air pollutants, measured during winter and summer, were from fuel burning, although sulfur could not be quantitatively assessed owing to uncertainties in the magnitude of natural marine biogenic emissions and

Table 52.1 Crustal enrichment factors relative to Ti and average crustal rock for carbonaceous fuel combustion trace metals measured in the Norwegian Arctic during winter episodes of long-range transport of air pollution

Trace metal	Enrichment factors (Ti, crustal rock)	
	Range	Average
As	700–3000	1300
Cd	500–6000	1500
Mo	60–150	95
Ni	30–80	60
Pb	2500–4000	3500
Sb	1000–3200	1700
Se	3700–36000	12100
V	35–50	45

Table 52.2 Average regional apportionments (in percent) of trace metals in the winter and summer aerosol from the Norwegian Arctic

Element	Winter		Summer	
	Europe	Central USSR	Europe	Central USSR
As	29	71	71	29
Sb	46	54	84	16
Se	44	56	83	17
V[a]	46	54	84	16
Mn[a]	59	41	90	10
Pb	55	45		

a. Noncrustal.

their fluxes to the Arctic. There is an indication, however, that over 50% of the non–sea salt sulfate present in the Norwegian Arctic (Maenhaut et al., 1989) is from fuel burning, and a similarly high percentage has been estimated for Barrow, Alaska, as well (Li and Winchester, 1989).

The presence of emissions from burning fossil fuels, and perhaps other carbonaceous fuels, in the Arctic at lower altitudes can be illustrated by elements considered up to now to be mainly constituents of soil dust from crustal erosion and weathering, e.g., Si, K, and Fe. These elements can exhibit bimodal particle size distributions with minimum abundances around 1 or 2 μm diameter. Generally, the mass size distribution of the carbonaceous fuel combustion component of the arctic aerosol is presented in Figure 52.1 for all elements in the component and for all

elements except S. It appears that the latter, with abundant Si, is skewed strongly to ultrafine particle sizes. Additional sulfur is most evident in the 0.5 to 1.0 μm diameter range, the accumulation mode often found to contain considerable sulfur. The average percentage distribution of the Si mass in the arctic aerosol during winter is presented in Figure 52.2. The explanation of the high concentrations of Si, K, and Fe in <1 μm diameter particles is somewhat complicated. Some earlier investigators reasoned that these elements must be of crustal origin and suggested that natural soil erosion could generate fine as well as coarse particles (e.g., Heintzenberg et al., 1981; Pacyna et al., 1984). However, it is now known that another abundant crustal element, Al, was often not present when Si, K, and Fe were at high concentrations. This finding is most extensively documented by late winter measurements at Barrow, Alaska, during periods of air flow across the Arctic Ocean from northern and eastern Europe.

Finally, the contribution of fuel burning emissions can be compared with contributions of sea salt, soil dust, and particles from conversion of biogenic gases to arctic haze. Windblown dust is most important in higher-altitude haze layers, and dust from as far away as Asian deserts can be transported to the Norwegian Arctic at 2 to 5 km altitude (Pacyna and Ottar, 1989). In addition, during summer, up to 50% of non–sea salt sulfate at Spitzbergen may originate from natural marine biogenic sources (e.g., Maenhaut et al., 1989). From measurements of both sulfate and methanesulfonate, a natural marine biogenic sulfur gas reaction product, at Barrow during late winter, it was estimated that about 20% of sulfate was from this source (Li and Winchester, 1989), with the major part of sulfur being combustion-derived, as already discussed.

Silicon as an Indicator of Long-Range Transport of Coal Smoke to the Arctic

Coal contains many different minerals composed of more than 60 chemical elements. The major elements are Si, Al, Fe, Ca, Mg, and S. Their relative amounts in the coal ash mineral mixture do not depend on the ash content of the coal. However, Si is usually the most abundant, often twice as abundant as Al, and present as the oxide silica (SiO_2, e.g., quartz), as well as in aluminosilicates and other minerals. During combustion of coal, a part of the SiO_2 from the coal ash is reduced by carbon to form a volative compound SiO. Once outside the coal particle boundary layer,

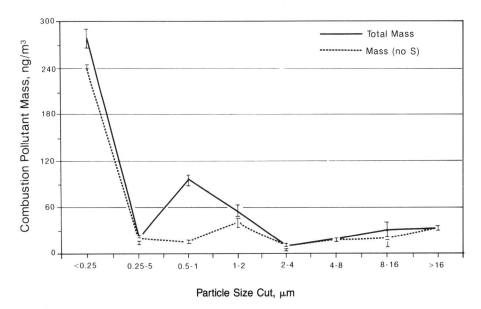

Figure 52.1 Carbonaceous fuel combustion aerosol mass size distribution. The error bars stand for one standard deviation.

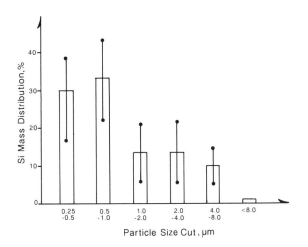

Figure 52.2 Size distribution of silicon mass as measured in winter arctic aerosol during the episodes of long-range transport of air pollutants. The error bars present a range of percentage contributions measured in given size fractions.

the reoxidation of SiO to SiO_2 is very rapid and highly exothermic. According to Flagan and Taylor (1981), the high condensation rate results in the formation of large numbers of very small nuclei, formed when the SiO vapor diffuses into the oxidizing atmosphere away from the surface of the burning coal particle.

Vaporization of SiO_2, which leads to the formation of SiO_2 nuclei, seems to occur preferentially for silica rather than aluminosilicates and moreover to be affected by the presence of other oxides, such as MgO and CaO. When coal ash contains substantial amounts of MgO, e.g., from 7% to 10%, this oxide rather than SiO_2 can comprise most of the submicron aerosol. Neville et al. (1981) studied the vaporization and condensation of mineral matter during combustion of Montana lignites, rich in MgO and CaO, and Illinois bituminous coals with low MgO contents. They concluded that for low-rank coals (the Montana lignites), MgO is the dominant oxide in the fume produced on combustion, while SiO_2 predominates for bituminous coal, even though SiO_2 still predominates in the original inorganic matter of the tested coals. One explanation is that alkaline earth metals inhibit SiO vaporization by reacting with silica to form nonvolatile silicates. Of course, the suppression is quite limited when there are only small amounts of alkaline earth metals available, so that SiO vaporization is not decreased and SiO_2 nuclei can be a major part of submicrometer particles. Thus, fine aerosol silicon can result from the combustion of some, but not all, coals, raising the possibility that it can serve as a tracer for certain types of coal.

In either case, Al bound in aluminosilicates is quite nonvolatile during coal combustion. Although Al_2O_3 can be between 15% and 20% of coal ash, and SiO_2 between 25% and 50%, there may be only between 0.1% and 1.5% Al_2O_3 in submicrometer coal combustion aerosol, compared with between 1% and 40% SiO_2, an enrichment of Si over Al by an order of magnitude. Measurements of submicrometer aerosol composition from the combustion of Illinois bituminous coal showed a ratio SiO_2/Al_2O_3 of 50. Thus, we believe that the presence of Si and absence of Al in very fine aerosol that is rich in S can be a good tracer for emissions from the burning of at least some kinds of coal.

Once in the atmosphere on fine particles (smaller than 1 or 2 μm diameter), silicon can be subject to long-range transport within air masses in the same way as can Cd, As, Sb, Mo, and other trace metals. Owing to their small particles size, the SiO_2 particles do not readily undergo dry or wet deposition during this transport, and chemical reactivity is also limited, so that combustion-derived fine SiO_2 particles could remain suspended in air masses for a long time and be transported perhaps for several thousand kilometers.

It should be noted that Si can be a good tracer of coal combustion, but not for all coals. The element should be the best tracer for high-rank coals with a high Si content. Such coals are burned in eastern Europe in larger quantities than elsewhere in Europe (e.g., Pacyna, 1984).

Final Remarks

Our studies of the arctic aerosol have emphasized fossil fuel combustion because of its probably much greater impact than from the burning of living biomass. However, the latter needs an independent assessment. Moreover, some of the trace metal emissions and our finding of unique tracer compositions, especially Si in relation to Al and to particle size distribution, which may be indicative of fossil fuel burning effects in the Arctic, may be of value in assessing burning of any biomass, fossil or living, in other areas, since fossil fuels are derived from biomass that once was living. And the strategies for aerosol composition measurement and analysis of data using statistical receptor modeling procedures and interpretation with the aid of meteorological information may be applied with profit in other areas of biomass or other carbonaceous fuel burning. Their

impacts can be of importance to the global troposphere, and the relative interplay of different forms of fuel burning in different geographic regions should be studied carefully to understand these impacts more fully.

Acknowledgments

This study was supported in part by U.S. National Science Foundation Grant ATM-86-01967 to Florida State University and by the Norwegian Institute for Air Research (NILU). A portion of the work was carried out at the National Center for Atmospheric Research (NCAR) with additional support by a graduate assistantship (S.M.L.), a visiting scientist appointment (J.M.P.), and an appointment for special assignment (J.W.W.). NCAR is sponsored by the National Science Foundation.

Cloud Condensation Nuclei from Biomass Burning

C. Fred Rogers, James G. Hudson, Barbara Zielinska,
Roger L. Tanner, John Hallett, and John G. Watson

Twomey et al. (1987) have shown the importance of variations in cloud properties, through their influence on the reflectivity of solar radiation. Relatively small changes in the extent or droplet concentration or size of low-altitude (e.g., stratocumulus) clouds could radiatively offset the effect of a doubling of CO_2 (Slingo, 1990). Parameterization of cloud cover and its associated variability and uncertainty constitutes a major impediment to the further development of global climate models (e.g., Cess et al., 1989). Aerosol particles which are active as cloud condensation nuclei (CCN) are hypothesized to exert a strong influence on the concentrations and sizes of cloud droplets (Squires, 1958; Twomey et al., 1984); variations in CCN may cause the most marked effects when superimposed on relatively clean air, such as is the case when anthropogenic emissions impact marine stratocumulus (cf. Hudson, 1983; Hudson and Rogers, 1986; Albrecht, 1989; Hudson, 1990). The identification and evaluation of globally significant CCN sources is therefore an ongoing and important activity which will benefit future global climate models.

The purpose of this chapter is to discuss the likely generation of CCN by tropical biomass burning. We examine two issues: (1) Are tropical biomass smoke particles active as CCN, and (2) if these particles are active as CCN, what is their estimated source strength and eventual fate, i.e., can their influence on global cloud reflectivity be deduced from existing data? The first issue can be resolved, but answering the second involves difficulties that have been addressed earlier with respect to CCN derived from sulfur compounds (Schwartz, 1988; Henderson-Sellers and McGuffie, 1989; Charlson et al., 1989; Gavin et al., 1989; Ghan et al., 1989; Wigley, 1989).

CCN are usually atmospheric aerosol particles of 1 μm or less in size. Their ability to serve as nuclei for water condensation at typical atmospheric supersaturations of 1% or less usually results from their chemical composition. As little as 10^{-16} grams (g) of a soluble electrolyte in a single particle (constituting less than 1% by mass for smoke particles a few tenths of a micrometer in diameter) is sufficient to cause activity as a CCN at a water supersaturation of 1% (Pruppacher and Klett, 1978). Ammonium sulfate, for example, is often cited as the most common CCN constituent, at least in Northern Hemisphere measurements (Twomey, 1971).

The process of deforestation and agricultural clearing in tropical regions including the Amazon Basin and the African savanna involves large-scale dry-season burning, which is an important source of both gaseous and particulate emissions (Seiler and Crutzen, 1980; Robinson, 1989). The emissions of CO_2 may be comparable to those of fossil fuel combustion (Seiler and Crutzen, 1980). There are very few direct measurements of the activity of biomass smoke particles as CCN. Warner and Twomey (1967) observed elevated concentrations of CCN in the plumes of sugar cane fires in Australia and in laboratory burns of sugar cane leaves; elevated concentrations of cloud droplets associated with these plumes were also observed in marine cumuli off the west coast of Australia. Their data did not include measurements of the total number concentrations of all aerosol particles, so the percentage of the smoke particles active as CCN cannot be calculated. Data were also taken during brushfire events in the Ivory Coast region of West Africa (Desalmand et al., 1985a, 1985b). These measurements were limited to two events on two different days and showed that from 25% to 100% of the submicron particles are active as CCN. Radke et al. (1978, 1988) have observed high levels of CCN activity in field observations of Northern Hemisphere biomass fires.

There are some chemical data available for tropical biomass smoke particles: Measurements were taken during the burning season in the Amazon Boundary Layer Experiment (ABLE) (July and August 1985) in extensive haze layers (Andreae et al., 1988). These measurements show enrichment of particulate-phase water-soluble components such as potassium, chloride, sulfate, and nitrate ions. As discussed below, polar organic compounds, such as methoxylated phenols which result from the pyrolysis of wood lignin (Hawthorne et al., 1988, 1989), may also contribute to the CCN activity of biomass smoke particles.

In this work, we have analyzed biomass and crude oil smoke samples for ionic and organic species. The CCN activities of the smoke particles are discussed in terms of the measured chemical compositions of the smoke samples. The implications of biomass burning to global climatic change are discussed.

Experimental Procedure

Particle Generation and Collection
This study includes both laboratory and field-scale burns of grass, wheat, and barley stubble, chaparral brush species, eucalyptus twigs and leaves, and coniferous trees. Laboratory studies were performed on freshly generated smoke samples. Fuel masses of 20 to 30 g were burned (in both flaming and smoldering phases) for each experiment, and the smoke was stored in a 6 cubic meter (m^3) electrically conducting plastic (Velostat, 3M Company, St. Paul, Minn.) bag as described previously (Hallett et al., 1989). For comparison purposes, crude oil smoke samples were also generated, by dripping 3 milliliter (ml) quantities onto a hot quartz surface; more detail is given in Rogers et al. (1990). For chemical analyses, smoke samples were collected from the 6 m^3 bag or from ambient air at flow rates of 20 to 113 liters per minute (L/min), on 47 millimeter (mm) ringed Teflon filters (Membrana-Ghia, 2 μm pore size). Smoke samples were also collected on 0.2 μm pore size Nuclepore filters (Nuclepore Corp., Pleasanton, Calif.) for morphological analysis by scanning electron microscopy (SEM). Field CCN data were taken in aircraft studies of Oregon and northern California forest fires described in detail by Hallett et al. (1989) and Hudson et al. (1990).

CCN Measurements
CCN spectral data were obtained from bag or ambient samples using a spectrometer designed and built at Desert Research Institute (Hudson, 1989), which provides in real time the concentrations of particles active as CCN over a range of water supersaturations. The concentrations of all particles in the samples were determined by taking simultaneous total particle or "condensation nucleus" (CN) measurements, using a commercial CN counter (TSI Model 3020, TSI, Inc., St. Paul, Minn.). The ratio of the concentration of CCN active at a supersaturation of 1% (approximately the upper limit of supersaturations in typical clouds) to the CN concentration is designated as the "CCN/CN ratio," and provides a comparative measure of the ability of different samples to be active as CCN.

Chemical Analyses of Ionic Species
The Teflon filters for ion analyses were extracted using 10 ml of distilled, deionized water to which 200 μl of ethanol were added as a wetting agent. The anion analyses were performed by ion chromatography (IC) on Dionex 2020i and 4000i systems (Dionex Corporation, Sunnyvale, Calif.), with results reported for Cl^-, NO_3^-, $SO_4^=$, acetate, and formate. Water-soluble potassium and sodium analyses were performed on a Perkin-Elmer 2380 atomic absorption spectrophotometer (Perkin-Elmer Corporation, Norwalk, Conn.). Ammonium analyses were performed on a Technicon Traacs-800 automated colorimetric analyzer (Bran-Leubbe Analyzing Technologies, Elmsford, N.Y.). Appropriate standards and quality control were applied for each analysis type.

Chemical Analyses of Organic Compounds
Filters loaded with wood smoke and oil smoke particles were Soxhlet-extracted separately for six hours in methylene chloride (CH_2Cl_2). The extracts were concentrated by rotary evaporation under vacuum to ~2 ml, filtered through 0.45 μm Acrodiscs (Gelman Scientific), and evaporated to dryness under a stream of nitrogen. Table 53.1 lists the weights of particles and extracts and the extractable mass fractions for oil and wood smoke samples.

Methylene chloride (4 ml) was added to the residues and the samples were analyzed with a Hewlett-Packard 5890 Gas Chromatograph (GC) interfaced to a 5970B Mass Selective Detector (MSD), operated in a full scan mode. A 12.5 m × 0.25 mm i.d. HP-1 (Hewlett-Packard) capillary column was used, with injections made in the splitless mode. Temperature

Table 53.1 Particle and extract weights and extractable mass fractions for oil and wood smoke samples

Sample	Particle weight (mg)	Extract weight (mg)	Extractable mass fraction (%)
Oil smoke	14.2	4.0	28%
Wood smoke	8.7	8.2	94

programming was 80° to 300°C at 4°C/min and held at 300°C for 15 minutes.

Results

Data from the ion analyses are shown in Table 53.2, which indicates the mass percentages of several common water-soluble inorganic and organic ions, and the corresponding CCN/CN ratios as percentages when available. For comparison purposes, data are also given for high- and low-sulfur crude oil smokes. Sulfate appears to be in stoichiometric ratios with ammonium for the forest fire, chaparral, and wheat/barley smoke samples. Potassium and chloride are both elevated and are nearly in stoichiometric ratios for the chaparral, wheat/barley, and eucalyptus cases. Nitrate is probably underestimated, as no special precautions were taken to quantitatively collect and preserve this labile species. Additional ionic constituents could have been present but were not analyzed.

Additional information concerning organic constituents of the crude oil and eucalyptus smoke were obtained by gas chromatograph/mass spectrometer (GC/MS) analysis. Figure 53.1 shows the GC/MS traces for crude oil (Figure 53.1a) and eucalyptus smoke (Figure 53.1b) unfractionated particulate ex-

tracts. Organic compounds identified or tentatively identified in these mixtures are listed in Table 53.3.

The identification of polycyclic aromatic hydrocarbons (PAH) was based on the comparison of their mass spectra and retention orders with those recently published (Atkinson et al., 1988). Aliphatic hydrocarbons were identified based on their mass spectra and no attempt was made to identify individual isomers. Identification of all compounds eluting in the eucalyptus wood smoke extract is tentative due to the lack of standards; it is based on the comparison of the mass spectra of these compounds with those from a mass spectral library (NIST Database 49K.1) and from published data (Hawthorne et al., 1988, 1989).

The morphologies of the smoke particles as determined by scanning electron microscopy (SEM) were markedly different for the biomass and crude oil cases. In laboratory and field studies (Hallett et al., 1989; Rogers et al., 1990) of coniferous forest and chaparral smoke properties, the predominant morphology of particles examined by SEM was that of single, nearly spherical globules narrowly distributed about a mode diameter of about 0.5 μm. Radke et al. (1988) discuss a similar finding for Canadian and Pacific Northwest forest fire smokes, although their mode diameters were closer to 0.3 μm, and on some

Table 53.2 Water soluble compounds in biomass smoke samples (percentages by mass as determined by ion chromatography)

Fuel type	Cl⁻	NO₃⁻	SO₄⁼	CH₃COO⁻	HCOO⁻	Na⁺	NH₄⁺	K⁺	Percentage of particles active as CCN[a]
Northern California forest fire, average of two samples	0.06 (0.02)[b]	1.80 (0.29)	1.59 (0.17)	<0.2 (<0.03)	<0.2 (<0.04)	0.50 (0.22)	0.80 (0.44)	0.63 (0.16)	100%
Southern California[c] chaparral brush mixture, average of two samples	4.15 (1.17)	1.34 (0.22)	2.20 (0.23)	2.08 (0.35)	<0.2 (<0.04)	0.33 (0.14)	0.82 (0.45)	4.86 (1.24)	80%
California wheat and barley stubble	13.4 (3.78)	0.47 (0.08)	2.58 (0.27)	n/a —	n/a —	0.14 (0.06)	1.17 (0.65)	15.51 (3.97)	n/a
Oregon rye grass	1.17 (0.33)	0.11 (0.02)	1.00 (0.10)	n/a —	n/a —	0.24 (0.10)	0.16 (0.09)	0.89 (0.23)	n/a
California[c] eucalyptus, average of two samples	4.33 (1.22)	0.17 (0.03)	0.99 (0.10)	n/a —	n/a —	n/a —	n/a —	4.9 (1.25)	100%
Low sulfur[c] Alberta sweet crude oil	0.25 (0.07)	0.98 (0.16)	0.30 (0.03)	1.92 (0.32)	1.48 (0.33)	n/a —	0.43 (0.24)	0.10 (0.03)	5%
High sulfur[b] Arabian light crude oil	0.23 (0.06)	0.88 (0.14)	0.50 (0.05)	1.54 (0.26)	1.49 (0.33)	0.12 (0.05)	0.43 (0.24)	0.12 (0.03)	9%

a. Typical uncertainty ±20%; data are for unaged samples.
b. Mmoles of specified ion/100 g sample, in parentheses.
c. Laboratory burn.
Note: n/a = Data not available.

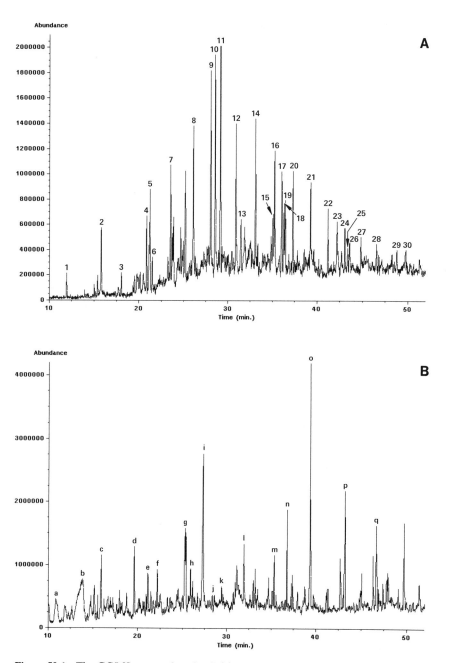

Figure 53.1 The GC/MS traces of crude oil (a) and eucalyptus smoke (b) unfractionated methylene chloride extracts. See Table 53.3 for peak identification. The *y*-axis represents total ion current.

Table 53.3 Identities of chromatographic peaks

Figure 53.1a		Figure 53.1b	
Peak id[a]	Oil smoke (mw)[b]	Peak id[a]	Eucalyptus smoke[c] (mw)[b]
1	$C_{10}H_{20}$[c] (140)	a	Benzenetriol (126)
2	$C_{12}H_{24}$[c] (168)	b	Levoglucosan (162)
3	$C_{16}H_{34}$ (226)	c	Acetonylguaiacol (180)
4	$C_{17}H_{36}$ (240)	d	Propenylsyringol (194)
5	Phenanthrene (178)	e	Acetonylsyringol (210)
6	Anthracene (178)	f	Tetradecanoic acid (228)
7	$C_{18}H_{38}$ (254)	g	Syringol derivative (252)
8	$C_{19}H_{40}$ (268)	h	Syringol derivative (252)
9	Fluoranthene (202)	i	Hexadecanoic acid (256)
10	Acephenanthrylene (202) + $C_{20}H_{42}$ (282)	j	Fluoranthene (202)
11	Pyrene (202)	k	Pyrene (202)
12	$C_{21}H_{44}$ (296)	l	Octadecanoic acid (284)
13	Benzofluorenes [a,b + c] (216)	m	$C_{23}H_{48}$ (324)
14	$C_{22}H_{46}$ (310)	n	Dioctyl adipate (370)
15	Benzo (ghi) fluoranthene (226)	o	Dioctyl phthalate (390) (226)
16	$C_{23}H_{48}$ (324)	p	$C_{27}H_{56}$ (380)
17	Cyclopenta (c,d) pyrene (226)	q	$C_{29}H_{60}$ (408)
18	Benz(a)anthracene (228)		
19	Chrysene/triphenylene (228)		
20	$C_{24}H_{50}$ (338)		
21	$C_{25}H_{52}$ (352) + phthalate ester		
22	$C_{26}H_{54}$ (366)		
23	Benzofluoranthenes [b,j + k] (252)		
24	$C_{27}H_{56}$ (380)		
25	Benzo(e)pyrene (252)		
26	Benzo(a)pyrene (252)		
27	$C_{28}H_{58}$ (394)		
28	$C_{29}H_{60}$ (408)		
29	Indeno (1,2,3-cd) pyrene (276)		
30	Benzo(ghi)perylene (276)		

Note: Shown in Figure 53.1a (oil smoke) and 53.1b (eucalyptus smoke).
a. Numbers and letters refer to chromatograms in Figures 53.1a and 53.1b, respectively.
b. mw = molecular weight; numbers in parentheses.
c. Tentative identification.

occasions a smaller-size mode, less than 0.1 μm, was observed. The crude oil smoke particles were always observed to be chain aggregates (Rogers et al., 1990).

The percentage of fresh forest fire, chaparral, and eucalyptus smoke particles active as CCN is shown in Table 53.2 and ranges from 80% to 100%. Aging of laboratory-generated pine wood smoke particles has shown that the percentage active as CCN only increases with time; no aging mechanism causing passivation with respect to CCN activity has ever been observed. The link between CCN activity and chemical constituents can be established because, at least for these three cases, the water-soluble constituents appear to be present as an approximately uniform internal mixture. That is, the narrow size distributions of the dry smoke particles translate into narrow, single-mode distributions of activated CCN when sampled by the spectrometer. Figure 53.2 shows an example of this behavior for the fresh eucalyptus

smoke sample. Figure 53.2 plots dN/dlog (Sc) versus Sc, where N is the concentration (cm^{-3}) of activated CCN, and Sc is the critical water supersaturation required to activate them. The integrated concentration of CCN over the indicated supersaturation range (0.01% to 1.0%) equals the total particle concentration in this case; i.e., 100% of the particles are active as CCN. The log normal mode value of Sc, about 0.1%, is in good agreement with theoretical predictions for a 0.5 μm diameter particle containing several percent ammonium sulfate or some other water-soluble compound as discussed by Pruppacher and Klett (1978).

Table 53.2 shows crude oil smoke data as a contrast to the biomass smoke results. Ion analyses indicated totals of about 5% by mass of the indicated soluble species, similar to the forest fire smoke case. The CCN/CN ratios, however, are lower than any of the biomass smoke cases (by a factor of 10).

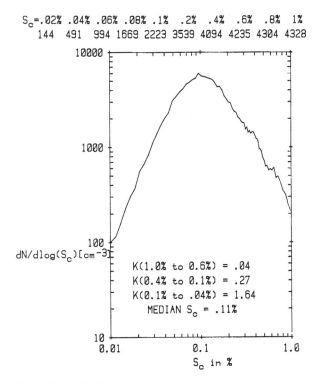

S_c= .02% .04% .06% .08% .1% .2% .4% .6% .8% 1%
144 491 994 1669 2223 3539 4094 4235 4304 4328

K(1.0% to 0.6%) = .04
K(0.4% to 0.1%) = .27
K(0.1% to .04%) = 1.64
MEDIAN S_c = .11%

Figure 53.2 Cloud condensation nuclei (CCN) spectrum for eucalyptus smoke sample (N = number concentration of activated CCN; Sc = critical supersaturation; K = slope of spectrum).

Discussion

The comparison of GC/MS profiles for oil and wood smoke particulate extracts (Figure 53.1) reveals important differences between both samples. The main organic constituents of crude oil smoke are nonpolar aliphatic and aromatic hydrocarbons (see Table 53.3). In contrast, the preliminary data on the organic constituents of eucalyptus wood smoke indicate the presence of polar oxygenated phenol derivatives together with fatty acids and esters of some dicarboxylic acids. The oxygenated phenol compounds are tentatively identified as derivatives of guaiacol (2-methoxyphenol) and syringol (2,6-dimetoxyphenol), which result from the pyrolysis of wood lignin (Hawthorne et al., 1988, 1989). Thus, these data show that biomass smoke is enriched and crude oil smoke depleted in polar organic compounds.

The CCN percentages are very much less for crude oil smoke than for biomass smoke, as can be seen from Table 53.2. The ion chromatography data identify the ionic species which, if present as internal mixtures in both the biomass and oil smoke samples we have studied, are more than adequate to explain the observed CCN percentages. Since the ionic content is not always much lower for the oil smokes sampled than for biomass smokes (see the forest fire and low sulfur crude data in Table 53.2), this difference may be due to one or both of the following reasons:

1. Polar organic compounds present in biomass smoke elevate the CCN/CN ratio. The GC/MS data reveal the important differences between the nature of organic constituents of oil and wood smoke. The presence of highly polar organic compounds in eucalyptus smoke, as opposed to hydrophobic nonpolar hydrocarbons in crude oil smoke, may help to explain the higher CCN activity of biomass smoke.

2. The available water-soluble ionic species are not uniformly internally mixed in the crude oil smoke particles; i.e., a large percentage of the particles do not contain enough water-soluble material to become active as CCN at a water supersaturation of 1%.

The findings reported above were obtained in laboratory and Northern American field experiments but may apply to a wide range of biomass species. The ion data are similar to data taken in haze layers associated with dry-season fires in the Amazon Basin by Andreae et al. (1988), in the sense that ions including sulfate, nitrate, potassium, chloride, formate, and acetate were enriched in the Amazon smoke samples. Sulfate and potassium alone accounted for more than 10% by mass of the aerosol composition. Andreae et al. (1988) did not have a CCN measurement capability. However, in measurements in a well-aged dry-season plume in the Guinean savanna, Desalmand et al. (1985b) observed CCN/CN ratios of 1.0 and thought that biomass smoke was responsible. These widespread observations, including the CCN data of Warner and Twomey (1967), support the following reasoning:

1. Water-soluble compounds, including a variety of inorganic and organic species, are generated as particulate-phase constituents in the smoke from every biomass fuel analyzed to date; in contrast to crude oil, eucalyptus smoke particles are enriched in polar organic compounds.

2. In limited observations on Northern Hemisphere forest fire, chaparral, and eucalyptus smoke particles, the compounds appear to be approximately uniformly internally mixed.

3. In separate observations, similar water-soluble ionic species have been observed in Amazon Basin smoke plumes, and CCN/CN ratios of unity have

been observed in the African savanna. These observations support the hypothesis that the high CCN activities observed for the Northern Hemisphere biomass smoke samples will also be found when more extensive CCN measurements are made in tropical biomass smoke plumes.

Extrapolation to Global Climatic Implications

The climatic importance of increases in CCN has been discussed at length (e.g., Twomey et al., 1987; Wigley, 1989); briefly, for all but the thickest clouds, increasing CCN implies increased reflectivity of incoming solar radiation. This is especially true for the climatically important types such as marine stratus. The decrease in downward radiative flux is proportional to the fractional increase in the CCN concentration (C) on which the cloud forms, $\Delta C/C$. For example, Wigley (1989) recently noted that a decrease of 2.2 watts per meter squared (Wm^{-2}) would result if Northern Hemisphere CCN concentrations increased by about 50%, i.e., if $\Delta C/C$ in the Northern Hemisphere were about 0.5; this decrease represents a cooling term equal to the estimated warming term due to anthropogenic greenhouse gases accumulated in the past few centuries.

Published estimates of the yearly global production of tropical biomass smoke particles are, like similar estimates for fossil fuel and other emissions, subject to large uncertainties, because usually these estimates are derived by multiplication of several terms, the uncertainty in each being several tens of percent (Robinson, 1989). However, the best available information indicates that tropical biomass burning accounts for more than half of the total global biomass burning emissions (Robinson, 1989), and generates a submicron particle mass flux estimated to be in the range of 10 to 30 teragrams per year (Tg/yr) (1 Tg = 10^{12} g) (Andreae et al., 1988; Cachier et al., 1985). For comparison, the global mass flux of carbonaceous particles resulting from fossil fuel combustion has been estimated to be about 16 Tg/yr (Duce, 1978). The total global emission of sulfur compounds is estimated to be about 200 Tg/yr (Schwartz, 1988), but not all of this sulfur is expected to be converted to additional numbers of sulfate CCN particles (Ghan et al., 1989). It seems to be a conservative estimate that the submicron particle mass flux from tropical biomass burning is about 10% of the global submicron sulfate particle mass flux (and about 50% of the global sulfur mass flux attributed to marine phytoplankton sources (Schwartz, 1988). As previously argued, there is

strong support for the hypothesis that the tropical biomass particle flux is mostly composed of particles active as CCN.

These estimates of the global fluxes of biomass smoke and sulfate particles are on a mass basis, and at best are an imperfect indicator of CCN concentrations. Our data and that of Radke (1988) do indicate that biomass smoke particles occur mostly in the submicron size region, which may be comparable to the typical sizes of sulfate particles, but even if the size distributions of the two types of particles were identical, transforming mass emissions data into CCN concentrations would involve large uncertainties. At this time the necessary data are not available to allow an estimate of the seasonal number concentration flux of CCN due to tropical biomass burning emissions. The available mass flux data suggest that biomass smoke particle concentrations are significant (i.e., on the order of 10%) with respect to those of sulfate CCN, but rigorous quantification is not possible until the actual number and mass distribution emissions data can be compared.

The correct procedure and controls necessary to evaluate cloud reflectivity changes due to CCN, based on satellite radiometer or other data, may be easier to derive for the limited scale of the tropical biomass fires than has been the case for the widespread CCN emissions from fossil fuel combustion (Schwartz, 1988). Recently Dr. P. A. Durkee (personal communication) has analyzed satellite radiometer data taken in April 1983 near Borneo. Regions of enhanced maritime cloud reflectivity were found associated with plumes of smoke from large burns. More generally, tropical biomass smoke CCN are injected in dry-season periods, near the Inter-Tropical Convergence Zone (ITCZ), and may be advected to either hemisphere before encountering precipitating clouds or other processes which may then remove them from the atmosphere. The smoke generated in the dry season (August) in the Amazon Basin would, according to prevailing surface wind maps, circulate in a general counterclockwise direction (corresponding to Southern Hemisphere anticyclones), and may affect low-level clouds over the southern Atlantic Ocean. The smoke generated in western Africa in the dry season (January) would be transported out to sea by easterly winds if injected north of the ITCZ; Savoie et al. (1989) have detected the likely chemical signature (high nitrate/sulfate ratios) of West African emissions advected to their sampling site at Barbados.

Conclusions

We have observed that smoke particles from the burning of Northern Hemisphere biomass fuels are enriched in water-soluble ions and, in the samples analyzed, with polar organic species compared to crude oil smoke particles. These ionic species are present in sufficient concentrations to allow the smoke particles to be active as CCN. The presence of polar organic compounds in the eucalyptus smoke particles, in contrast with only nonpolar organics found in crude oil smoke, is consistent with the enhanced CCN/CN ratios observed for the biomass smoke particles. Available mass emissions data suggest analogously that tropical biomass burning contributes a CCN flux which is of global significance. The enhanced reflectivity of tropical clouds, apparently due to the input of smoke CCN, has been observed in at least one case, consistent with the inference that tropical biomass smoke CCN must be taken into account in estimating global cloud albedo.

Acknowledgments

Part of this work was supported by Lawrence Livermore National Laboratory and the Defense Nuclear Agency. The authors gratefully acknowledge discussions with Dr. P. A. Durkee, Naval Postgraduate School, Monterey, California, and with Dr. Melanie Wetzel, Mr. Lyle Pritchett, and Dr. Kelly Redmond of Desert Research Institute. Dr. J. Chow of Desert Research Institute provided the filter analyses.

VIII

Biomass Burning and Climate

The Contribution of Biomass Burning to Global Warming: An Integrated Assessment

Daniel A. Lashof

Global Emissions from Biomass Burning

Biomass burning is a major contributor to the buildup of greenhouse gases in the atmosphere, which is predicted to lead to dramatic changes in climate during the coming decades if "business-as-usual" emission patterns continue (Houghton et al., 1990). Although emissions from agriculture and extensive land-use practices are notoriously difficult to understand and quantify, their impact on greenhouse forcing is probably only surpassed by carbon dioxide (CO_2) emissions from fossil fuel combustion and release of chlorofluorocarbons (CFCs) from various industrial processes (Burke and Lashof, 1990; Lashof and Tirpak, 1989). Biomass burning, in particular, plays a central role in emissions from both forested and agricultural ecosystems. Biomass burning contributes to the net flux of CO_2 to the atmosphere from terrestrial ecosystems because burning is one important means by which land is cleared and converted from one use to another (e.g., primary forest to pasture). But even when there are no net CO_2 emissions because the biomass is renewed, the trace gases released during combustion add significantly to the greenhouse effect. In particular, the impact of carbon monoxide (CO), methane (CH_4), nonmethane hydrocarbons (NMHC), nitrogen oxides (NO_x), and nitrous oxide (N_2O) emissions will be considered here.

Global emissions from biomass burning were examined by Hao et al. (1989). They estimate total biomass burned, gross emissions of CO_2, and net emissions of CO_2 released promptly from combustion. Total emissions of CO, CH_4, NMHC, NO_x, and N_2O were estimated based on the average emission rate of each species compared to gross CO_2 emissions. The low, middle, and high estimates based on this analysis are given in Table 54.1. To account for emissions in mid- and high latitudes the high values for CO, CH_4, and NMHC have been adjusted upward by 15% to 25% from the range given by Hao et al. for the tropics. The best estimate from an independent assessment by Andreae (this volume, Chapter 1) is shown for comparison. These values all fall within the range from

Hao et al. except for net CO_2 emissions, which presumably reflects a different definition that is not restricted to prompt combustion emissions. The $N_2O:CO_2$ ratio given by Andreae is also substantially lower than that given by Hao et al., reflecting recent evidence that some emission measurements were influenced by N_2O produced within the sample container. A recent report by Lobert et al. (1990) on measurements not affected by this sampling problem also gives lower N_2O emission estimates. On the other hand, these N_2O emission estimates do not include excess emissions from soil observed to continue for an extended period following biomass burning (Anderson et al., 1988; Levine, personal communications, 1990). Also, if lower emission values for fossil fuel and biomass combustion are accepted, then currently identified N_2O sources are too small to balance the global budget.

Biomass burning is a significant contributor to global emissions of both the carbon and nitrogen compounds listed in Table 54.1. The estimates of total anthropogenic emissions of each compound used here are included in the last column. For CO and NMHC the upper bound of the biomass burning source is equal to the estimate for total emissions, indicating that total emissions would have to be higher if the high biomass burning estimate is accepted (this adjustment has not been made here, so the high estimate for the percentage of warming potential caused by biomass burning given below must be somewhat too high).

Global Warming Potential

The global warming potential (GWP) of greenhouse gas emissions as described by Lashof and Ahuja (1990) can be used to obtain a first-order estimate of the overall impact of biomass burning compared to other anthropogenic sources of greenhouse gases integrated over time. Because biomass burning contributes most to emissions of short-lived gases, its relative impact will be greatest when future radiative forcing

Table 54.1 Emissions from biomass burning (Tg C or N)

Gas	Low	Middle	High	Andreae	Total
CO_2 gross	1500	2600	4200	3500	—
CO_2 net	100	200	300	1800	6000
CO	120	260	500	350	500
CH_4	12	31	60	38	250
NMHC	15	34	60	24	60
NO_x	3	6	9	8.5	30
N_2O	0.5	1.4	2.2	0.8	3.3

Note: The low, middle, and high estimates used here are based on Hao et al. (1989). Values from Andreae (this volume, Chapter 1) are given for comparison. The column labeled "Total" gives the assumed 1985 emissions from all anthropogenic sources (based on Lashof and Tirpak, 1989) used to calculate biomass burning's share of total greenhouse forcing.

Table 54.2 Global warming potentials of greenhouse gas emissions (degree-years / Pg of C or N)

Gas	Low	Middle	High
CO_2 gross	0.0	0.0	0.0
CO_2 net	1.6	0.30	0.19
CO	0.64	0.62	0.61
CH_4	4.3	3.8	3.4
NMHC	2.7	2.7	2.7
NO_x	22	22	22
N_2O	120	46	29

Note: Global warming potentials (GWPs) for CO_2, CO, CH_4, and N_2O are based on Lashof and Ahuja (1990) with a discount rate of 0.0, 0.01, and 0.02 for the low, middle, and high cases, respectively. The GWPs for NMHC and NO_x are from Derwent (1990).

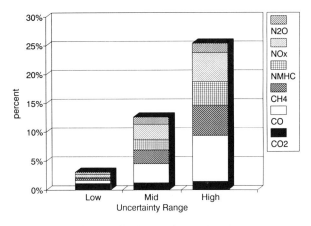

Figure 54.1 The contribution of biomass burning to the total global warming potential of anthropogenic greenhouse gas emissions circa 1985. Calculated by combining the assumptions in Table 54.1 and Table 54.2.

is given less weight than current radiative forcing (or equivalently when a short time horizon is adopted). I adopt equal weighting over time (discount rate $r = 0$) for the low estimate of biomass burning's contribution, a discount rate of 0.01 per year (approximately equivalent to a time horizon of 100 years) for the midcase, and a discount rate of 0.02 per year (approximately equivalent to a time horizon of 50 years) for the high case. The resulting GWPs, expressed in degree-years per petagram (Pg) of emissions, are given in Table 54.2. The GWP of CO_2 is most sensitive to the time weighting factor because its atmospheric retention is described by multiple time constants (Lashof and Ahuja, 1990). The GWPs of NMHCs and NO_x were not estimated by Lashof and Ahuja, and were based on Derwent (1990). This estimate of the GWP from NO_x does not appear to take into account its effect on OH and therefore may be an overestimate (Rotmans et al., 1990).

The results of applying these GWPs to the biomass

burning emission estimates discussed above are shown in Figure 54.1. Results are expressed as the percent of the total global warming potential of anthropogenic greenhouse gas emissions from all sources. For central estimates, biomass burning is responsible for 13% of the GWP of all sources. The largest impact of biomass burning is through the indirect effects of CO emissions, followed by the impact of NO_x and CH_4 emissions. In the low case, net CO_2 emissions have the largest impact because the GWP of the other gases relative to CO_2 is lowest under these assumptions.

Assessment Model

As an alternative to GWPs for calculating biomass burning's contribution to global warming, I used an assessment model for atmospheric composition devel-

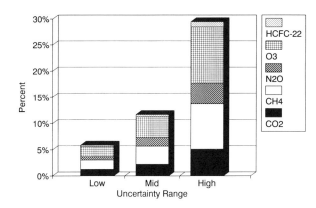

Figure 54.2 The contribution of biomass burning to total radiative forcing as calculated by the Assessment Model for Atmospheric Composition. Individual gas shares are based on concentration changes rather than emissions. HCFC-22 is effected through the influence of biomass burning on OH concentrations (see Figure 54.3).

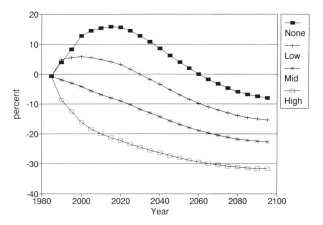

Figure 54.3 The impact of biomass burning on OH concentrations as calculated by the Assessment Model for Atmospheric Composition. Concentrations are given as the percentage change relative to concentrations under reference conditions. Biomass burning emissions are scaled from their midcase values to their assumed value for each case between 1985 and 2000.

oped for the Environmental Protection Agency (Prather, 1989). This parameterized model of atmospheric chemistry calculates changes in the concentration of radiatively active gases as a function of emissions using first- and sometimes second-order relationships. The model was run from 1985 to 2100 with and without constant biomass burning emissions after 2000. The difference in radiative forcing at a given date is then approximately equivalent to the integrated radiative forcing to that date from an emission pulse (the latter being the basis of the GWP definition used by Lashof and Ahuja; see Rotmans et al., 1990). Radiative forcing differences for the year 2100 were used in the low and midcases because this was the endpoint of the model calculations. This has the effect of increasing the relative impact of biomass burning in the low case compared with the GWP calculation which was based on indefinite integration. In the high case, the radiative forcing difference for 2050 was used in correspondence to the discount rate of 0.02 used in the GWP calculation.

The results based on these model calculations are shown in Figure 54.2. Unlike Figure 54.1, the impact of CO, NO_x, and NMHC emissions is not shown explicitly. Rather, the impact of these emissions is expressed through the radiative forcing from changes in CH_4, O_3, and HCFC-22 concentrations. In particular, HCFC-22 concentrations are influenced by biomass burning through its effect on OH concentrations (see Figure 54.3). Based on these model results, ozone formation from biomass burning is the largest climate forcing impact, followed by increases in methane. Unfortunately, the ozone forcing is highly uncer-

tain because it depends on the vertical distribution of ozone, which was not explicitly modeled.

Discussion

The two alternative methods of calculating the impacts of biomass burning on climate forcing give broadly consistent results. Both produce a central estimate between 10% and 15% of the forcing from all sources, with a range of about 5% to 30%. Most of this range can be explained by the range of emission estimates, although the chosen range of time horizons (or discount rates) is also a significant factor. The two approaches also give complementary and broadly consistent information about how this impact breaks down by gas. Although the model calculation gives results on the basis of concentration changes rather than emissions, both methods suggest that CO emissions, through their impact on O_3 and OH, make the largest contribution to radiative forcing from biomass burning. Methane emissions appear to be a close second.

The direct net CO_2 emissions from biomass burning make a relatively small contribution, except in the low case with an indefinite integration period. If the indirect CO_2 emissions from decomposition associated with biomass burning were included, the relative share of CO_2, as well as the overall share of biomass burning, would increase considerably. Whether or not such emissions should be attributed to biomass burning is unclear. The CO_2 emissions from deforestation fundamentally result from the change of land use, rather than the method of clearing. On the other hand, biomass burning is currently an integral part of the clearing process.

The close agreement between the results of the GWP and model calculations must be at least partially accidental. The uncertainties, particularly regarding the impact of ozone changes, are large. While the fact that the two methods agree reasonably well is encouraging, there is no guarantee that both are not significantly in error. Because atmospheric chemistry and radiative forcing are nonlinear, the impact of biomass burning also depends on assumptions about emissions from other sources. The GWP calculation assumes implicitly and the model calculation assumes explicitly that emissions from other sources will be roughly stabilized at current levels. If this is not the case, the integrated impact, even of past biomass burning, could change significantly.

Conclusion

The results presented here suggest that while biomass burning is less significant than fossil fuel combustion on a global basis, it is a major contributor to the greenhouse gas buildup, responsible for perhaps 10% to 15% of the total forcing from current emissions. Uncertainties about emissions and the relative impact of different gases are large, yielding a range of 5% to 30%. Nonetheless, biomass burning is probably the dominant source of greenhouse gases in some regions. A comprehensive policy to limit global climate change must, therefore, address biomass burning.

Policy Options for Managing Biomass Burning to Mitigate Global Climate Change

Kenneth J. Andrasko, Dilip R. Ahuja, Steven M. Winnett,
and Dennis A. Tirpak

Biomass Burning: The Policy Context

The summer of 1988 may have marked a turning point for policy consideration of biomass burning. Congressional testimony on a possible relationship between the U.S drought and the greenhouse effect shared headlines with photographs and reportage of new remote sensing studies of vast fires in Amazonia. Recognition of the link between biomass burning in the tropics and its role as a major source of greenhouse gas emissions moved tropical deforestation beyond its former status largely as a regional and biodiversity issue. Forest conversion emerged as a truly global concern directly affecting all countries, suddenly figuring prominently in the machinations of foreign policy by heads of state and policy makers.

The United Nations–sponsored Intergovernmental Panel on Climate Change (IPCC) convened in late 1988 by WMO and UNEP and has stimulated attempts at quantification of biomass burning as a major source of greenhouse gas emissions. The IPCC process has also spawned policy analyses of potential reductions in emissions in anticipation of the commencement of climate convention negotiations in early 1991. Potential policy options under review include reducing emissions from biomass burning during forest conversion to other land uses, increasing carbon storage in forest and agricultural systems (especially in the tropics), as well as increasing reliance on and improving the efficiency of bioenergy systems.

Most recently, debate has begun on the merits of negotiating a Global Forestry Agreement or convention, following support for the concept in the declaration from the G7 Houston economic summit in June 1990. Such an agreement might address biomass burning by providing disincentives for land conversion using fire, by establishing a system to track greenhouse gas emissions from land conversion, and by offering "credits" for countries that reduce emissions by implementing appropriate biomass management policies and practices.

Little work has been done to systematically identify and assess the set of potential biomass burning policy response options for implementing silvicultural and grassland fire-management practices, improved stove designs, and integrated bioenergy production and utilization systems. Nor have the relative costs and benefits and the environmental implications been evaluated. The discussion below takes an initial step in this direction.

Biomass Burning from Land-Use Management

Fire has long been a tool relied on to magnify human abilities. The term *biomass burning* refers to all (mostly anthropogenic) combustion of biomass, including burning forest for clearing, savanna for stimulating regeneration of more nutritious grass fodder for livestock, fuel wood and charcoal for energy, and agricultural residues left after harvest in the field as fuel.

Most anthropogenic biomass burning occurs in the tropics. The area of savanna burned each year may reach 750 million hectares (ha) (Hao et al., 1990), about half of it in Africa, an expanse about 40 times larger than the area of tropical forest converted in the tropics by fire and axe. Shifting cultivation—in which forest is cleared, used for agricultural cropping for 2 to 5 years, and then left fallow for 7 to 12 years—is practiced by 200 million people worldwide, on 300 to 500 million ha (Seiler and Crutzen, 1980).

Of the 7.5 million ha total primary moist forest cleared in the tropics in 1980 (Lanly, 1982), 25% to 45% is due to shifting cultivation. Burning efficiency is quite low on these lands, often only 10% to 30% for the first burn, leaving most of the carbon standing in large trunks, which are often fired again when dry conditions occur (Fearnside, 1985, 1989). Other estimates indicate that 55% to 90% of virgin moist forest cleared is lost permanently to agriculture or development. Burning on this land is likely to be far more intensive and efficient, and repeated at regular intervals to eliminate tough, low nutrition grasses or crop residues (Andreae, 1990; Lanly, 1982; Hao et al., 1990).

Total land clearing releases in the tropics are on the

order of 1.8 ± 0.8 petagrams of carbon per year (Pg C/Yr) or 25% of the 5.2 Pg C emitted from fossil fuel burning annually. Adding in savanna burning, charcoal production and combustion, and crop residues, fully 87% of all biomass burning takes place in the tropics (Andreae, 1990).

All of these estimates of emissions and area burned remain uncertain and subject to frequent revision. Biomass burning may contribute as much as 50% of gross CO_2 emissions from combustion from all sources (including fossil fuels), 10% to 25% of net CO_2 emissions from deforestation (taking carbon sequestration from regeneration of vegetation into account), 10% to 40% of total global CO releases, 5% to 15% of CH_4, 5% to 15% of N_2O, and 5% to 20% of NO_x (Andreae, 1990; Lashof, 1990). Of the anthropogenic part alone, the percentages are: CO_2–15%, CH_4–15%, N_2O–30%, NO_x–35% and CO–60% (USEPA, 1990). Information presented in this volume seems to indicate that direct emissions of N_2O from biomass burning may be lower, but offset by emissions from soil postcombustion.

Burning in Household Biofuel Cooking Energy Systems

Biomass in one form or another continues to be the predominant source of primary energy for at least half of the world's population. In some developing countries, over 90% of total energy used comes from biomass (Goldemberg, 1988). Although on a global basis it accounts for only one-seventh of all energy consumed (Smith, 1987), for over 2 billion people, it is close to being the only source of energy.

The most common use of biomass as energy is for cooking. Making an assumption that biofuel use in stoves is 1 kilogram per capita per day (kg/cap/day) (reasonable for areas with scarcity of biofuels), the annual global consumption would be 730 million tons (MM tons). Kohlmaier et al. (1987), deriving their numbers from Smith (1987), estimate the consumption of traditional fuels in developing countries to be close to 900 MM tons/yr. Fuel wood accounts for more than half of this—460 MM tons/yr; crop residues are 340 MM tons/yr, and animal dung about 100 MM tons/yr. The use of biofuels in stoves can also be assumed to be in the same proportion as the global use of these fuels.

The burning of all of crop residues and animal dung in stoves and other applications is more or less neutral with respect to carbon emissions, i.e., the amount of carbon emitted into the atmosphere during burning is subsequently fixed as plant biomass during regeneration of the burned area. A slight difference may arise

from the long-term sequestration of unburned charcoal. Contrary to earlier beliefs, most of the fuel wood that is burned in stoves also seems to be recycled and does not cause widespread land-use changes responsible for exacerbating climate change. In some locations where fuel wood has become increasingly scarce, however, its use in inefficient stoves contributes to the scarcity and a decline in the local areal biomass (and carbon) densities.

In addition to being a source of carbon dioxide, most of which is merely recycled, biomass burning in stoves (as elsewhere) is a significant source of emissions of carbon monoxide, methane, nonmethane hydrocarbons, and nitrogen oxides, including some nitrous oxide. Methane and N_2O are potent greenhouse gases, CH_4 and CO are oxidized to CO_2 in the atmosphere, and CO, NMHCs, CH_4, and NO_x are important in the formation of tropospheric ozone, also a greenhouse gas. As Lashof estimates (this volume, Chapter 54), most of the contribution to global radiative forcing from biomass burning comes from gases other than CO_2.

Our estimates of trace gas emissions from biomass burning are uncertain because of our limited knowledge of the distribution of types of biomass burned and of the emissions from different sources. This is an important first step in remedying this state of affairs.

Except for a few measurements of CO and suspended particulates (Ahuja et al., 1987; Butcher et al., 1984; Joshi et al., 1989), very little is known about the emissions of trace gases from stoves burning solid, unprocessed biofuels. For these two pollutants, however, it is known that emission factors can vary easily by factors of up to 3 or 4, depending on the combustion conditions, even for the same species of wood or biofuel.

Potential Response Options to Reduce Biomass Burning

Two major categories of strategies can be identified (Figure 55.1) to reduce biomass burning:

1. Reduce the frequency, area, and amount of biomass burned in natural and intentional fires in various ecosystems.

• Minimize conversion of forest and savanna to other land uses, especially agriculture and pasture, through use of fire.

• Reduce use and improve efficiency of fire as a land management tool on forest, grasslands, and agricultural lands.

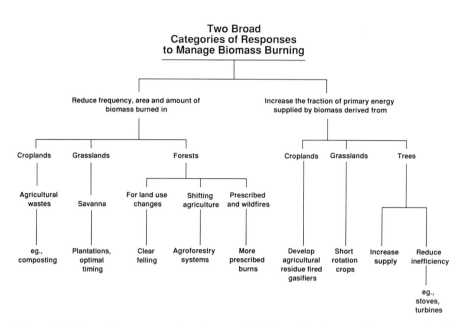

Figure 55.1 Potential response options for managing biomass burning are shown, categorized by major approaches (2d line) and by land-use system.

• Substitute sustainable land management practices and systems with minimal use of fire for fire-dependent or fire-proclivity systems.

2. Increase the fraction of primary energy supplied by biomass derived from natural systems, and the efficiency of its utilization (Figure 55.2).

• Replace unsustainably harvested biomass with sustainably managed (plantation) biomass as a fuel stock.

• Increase energy extraction efficiencies for biomass conversion.

Three major classes of land uses are targets for response strategies for reducing the frequency, area, and amount of biomass burned: crop lands, grasslands, and forests.

This set of options could be tailored to a specific land use, unit of analysis (e.g., rural development project, conservation or production forest, or political unit), and set of biophysical and socioeconomic factors (for a fuller set of options, see Andrasko, 1990a, 1990b). Biophysical, social, economic, and technological constraints to each of these options abound and are only addressed in passing here.

Cropland Strategies

Option 1: Incorporate crop residues into soils instead of burning. Emissions from burning agricultural residues could be reduced by incorporating residues

into soils, or composting them, instead of burning. Both alternatives increase soil fertility, the object of burning, although both are more labor-intensive than burning. In some cases, the thickness of the vegetative stalk may hinder organic breakdown.

Option 2: Burn residues as fuel in household energy systems. The burning of field wastes may also be reduced by increasing their use in stoves and in high-efficiency combustion systems such as gasifiers. Efficient stove designs and gasifiers would need to be developed, and extension services would have to be provided to make these strategies possible. In addition, increasing the productivity of existing croplands through biotechnologies or improved management practices could increase the amount of field wastes for energy uses and reduce the conversion and burning of other lands for energy purposes.

Option 3: Replace annual or seasonal crops with tree crops. Burning of some crop lands could be eliminated by replacing some annual or seasonal crops with tree crops (fruits, nuts, fodder, poles) that are not burned on a cyclic basis. Land tenure issues might need to be resolved for agriculturalists to encourage investments in time and capital required to grow trees to maturity. Food crop plantations could be phased in to allow for sufficient food from field crops until trees began producing. Markets would need to be developed and extension services provided to help the farmers make the transition to tree-crop farming.

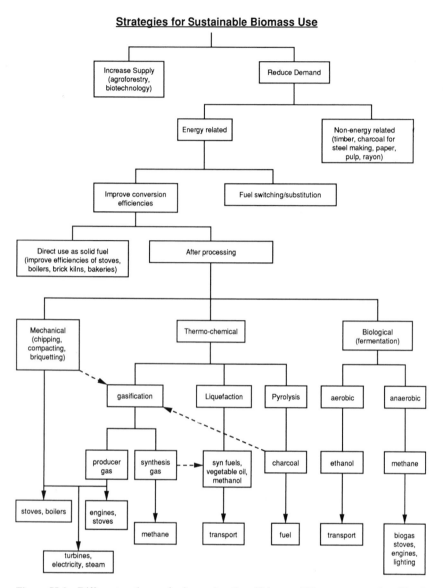

Figure 55.2 Different pathways for improving the efficiency of biomass use and making it sustainable.

Grassland Strategies

Emissions from savannas and pastures are usually the result of the naturally occurring ignition of dry biomass by lightning or of fires intentionally set by pastoralists who want to improve the quality of the forage for their livestock. The primary goal is to reduce fires intentionally set. It may be possible to reduce emissions of natural fires through prescribed burning at appropriate times.

Option 1: Increase grassland management to reduce fire frequency and area. The frequency, area, and amount of grasses burned might be reduced through intensified management of some grasslands and savannas. Data needed to evaluate the viability of can-

didate practices include the moisture content of biomass, timing of burns in relation to macro- and microclimate (humidity, temperature, and precipitation), and the ecology and chemistry of vegetation. One major constraint is the difficulty of changing culturally entrenched management practices. Instituting improved practices would require improved data on current practices, their greenhouse gas emissions by ecosystem and practice, and remote sensing data on land-use patterns to allow scaling up of site-level estimates of benefits.

Option 2: Substitute game ranching for domestic livestock systems. Domestic livestock are not naturally adapted to the native grasses and often require young-shoot forage with a higher nutritional value that arises after burning. Game ranching native ungulates and other families adapted to naturally occurring vegetation could replace domestic livestock and increase yield per hectare, reduce costs (Hudon et al., 1989), and reduce methane emissions (AFOS, 1990). Another alternative is to replace grazing of domestic livestock by feeding in stalls, although CH_4 emissions could increase if feedlot emissions are not managed. Constraints on these strategies include their labor-intensive nature, technology transfer difficulties, and potential disruption of existing lifestyles of pastoralists.

Option 3: Introduce silvipastoral systems to provide fodder. A strategy complementary to stall feeding is establishment of plantations to produce animal fodder in place of pasture regularly burned to support domestic livestock. Establishment of fodder trees might be phased in to provide livestock with feed until trees reach sufficient maturity to sustainably produce fodder.

Forest Strategies

Option 1: Intensify forest management to reduce wildfires. Biomass burned in wildfires might be reduced by increasing the frequency of prescribed burns to eliminate excess understory and litter fuel, and by introducing fire-management, detection, and suppression systems like those that have evolved in Australia, Canada, and the United States. A recent integrated forest management project in Guatemala designed to offset the CO_2 emissions of a power plant in Connecticut includes a fire-management program and estimates its CO_2 emissions-reduction potential (Trexler et al., 1989). Disturbed moist tropical forests are subject to wildfire to varying extent, and dry and open forests are often both fire dependent and fire prone. However, the potential for reducing greenhouse gas emissions from wildfires may be low. Studies of the forest ecology of candidate ecosystems are needed to develop fire management strategies and extension programs.

Option 2: Lengthen rotation times and improve productivity of shifting agriculture. Increasing the length of the fallow period between burns in shifting agricultural systems would allow abandoned land to reforest and reduce average emissions per annum, if cropping practices with improved yields can be introduced and sustained (Sanchez, 1989). Harvesting of nontimber, noncrop products from forest could be expanded as a source of significant supplemental revenue available to buy food (Peters et al., 1989), if markets are developed and demand intensified.

Option 3: Increase productivity of existing agricultural lands. Shifting agriculture could be replaced by more intensive, rather than extensive, and alternative (e.g., agroforestry) cropping systems on existing croplands, to relieve conversion pressure on adjacent forests from population pressures and declining yields per hectare. New or genetically improved species, biotechnologies, improved silvicultural treatments which shorten rotations, alternative crop rotations, and extension education are needed to increase productivity.

Option 4: Incorporate charcoal into the soil after burning. Plowing under charcoal would keep it from being consumed in subsequent fires, although this may affect soil characteristics. Cost-effective, low-labor practices to increase the carbon component in soils are needed, as are data on changes in greenhouse gas emissions from soils. Soil management practices are likely to be labor-intensive.

Option 5: Clear-fell forest before or instead of burning. Clear-felling promotes use of merchantable wood for durable products, which both reduces biomass ground fuel available and increases the storage of carbon in those products. Clear-felling and marketing may not be feasible where timber supply greatly exceeds demand and the capacity of market and transport systems.

Methods

Policy Considerations in Designing Biomass Burning Response Options

Steps necessary to identify, evaluate, and compare competing biomass burning reduction practices, technologies, and policies are outlined below (Table 55.1). These steps have been followed in detail for several forestry sector response options to climate change, at the forestry project level in the tropics (Trexler et al., 1989) and at the national level in the United States (Andrasko, 1990b).

Of the set of policy options potentially available, we review two representative ones here—reducing tropical forest conversion and improving stove efficiency—and attempt a first-order assessment of their contribution to reducing greenhouse gas emissions from biomass burning.

Table 55.1 Sequence for designing and evaluating response options

1. Identify land-use changes and activities involving burning.
2. Determine area affected.
3. Perform biomass (dry matter) accounting per hectare.
4. Estimate emissions by gas and by type of biomass burned.
5. Estimate net gas emissions by current practice and by recommended policy response to determine net greenhouse balance.
6. Perform global warming potential (GWP) calculations.
7. Formulate response options:
   • Field practices
   • Technologies
   • Policies
8. Calculate cost effectiveness per unit of biomass or warming avoided for each option.
9. Perform feasibility and implementation assessment.
Goal: Tailoring responses to local needs and conditions.

Model Scenario Descriptions

The U.S. Environmental Protection Agency (EPA), in its report to Congress *Policy Options to Stabilize Global Climate* (USEPA, 1990), constructed six scenarios of future emissions of greenhouse gases. Three scenarios explore how the world may evolve assuming unimpeded growth of greenhouse gas emissions. The rapidly changing world (RCW) scenario assumes rapid economic growth and technical change, while the slowly changing world (SCW) scenario assumes more gradual growth and change. The rapidly changing world with accelerated emissions (RCWA) scenario assumes that nations pursue perverse policy choices that directly conflict with concerns about global warming.

The fourth and fifth scenarios incorporate the same economic and population change assumptions as SCW and RCW but also assume that policies to contend with global climate change are also adopted. These scenarios are labeled the slowly (and rapidly) changing worlds with stabilizing policies—SCWP (and RCWP, respectively). Finally, a variant of the RCWP case assumes more aggressive policies are implemented to address the threat of climate change than those assumed in the RCWP scenario, called the rapidly changing world with rapid emissions reductions (RCWR).

In the three scenarios with unimpeded emissions growth, the population engaged in traditional agriculture and in shifting cultivation continues to increase, as do demands for fuel wood and speculative land clearing. These factors lead to accelerated deforestation until tropical forests are almost eliminated during the second half of the next century. The other set optimistically assumes a global effort to reverse de-forestation that transforms the biosphere from a source to a sink for carbon.

The single most important determinant of future greenhouse gas emissions is the level of energy demand and the combination of sources that is used to supply that energy. While policies affecting demand will have the largest impact on near-term emissions, changes in the supply mix are critical in the long term. Biomass energy technologies include exploitation of fuel wood, forest residues, agricultural residues, municipal solid waste, animal wastes, and energy plantations.

The policy set also assumes that research and development into nonfossil energy supply options, chiefly biomass-derived fuels (rather than coal-based synfuels) and biomass combustion turbines, makes them competitive after 2000. Starting around 2010, biomass production and conversion is assumed to become competitive with (imported) oil and gas in many developing regions. Biomass, like coal, can be converted to a liquid or a gaseous fuel at some cost and some energy loss. The efficiency of energy conversion is assumed to be 75% after 2010.

Land available worldwide for biomass development projects is variously defined and estimated, so, following the U.S. Department of Energy, EPA optimistically assumed that 10% of total lands classified as forest lands, wood lands, and crop lands would be available (i.e., 4.3% of global land area). Modest productivity increases were also assumed, in response to widespread use of improved genetic varieties and more intensive management techniques.

Bioenergy plays a major role in meeting a substantial fraction of the total future primary energy demand in all three policy scenarios (Figure 55.3). Biomass contributes 250 EJ by 2100 in SCWP (48% of total) and 275 EJ in RCWP (32%). In the RCWR case, since lower conversion costs were assumed, the biomass contribution is the highest—470 EJ (or over 60%). In the SCW, RCW, and the RCWA cases, both the lack of research and development into bioenergy and an unwillingness or inability to commit land for biomass development prevents biomass from competing better with other fuels, until later periods.

Policy Option Example 1: Reduce Tropical Deforestation as a Source of Biomass Burning

Constructing Scenarios of Biomass Burning from Forest Conversion   A computer model built to track the emission of carbon from changes in land use on three tropical continents (Asia, Africa, and Latin America)

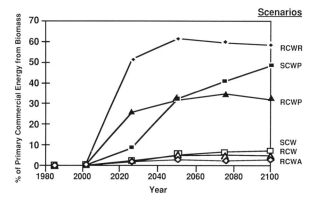

Figure 55.3 Percentage of primary commercial energy derived from biomass in the twenty-first century for different scenarios analyzed by the EPA in its draft report to Congress (USEPA, 1990).

Table 55.2 Forecast dates of total depletion of forests in the three tropical regions

Forest type	Africa	Asia	Latin America
Primary closed	2045–2050	2020	2065–2070
Primary open	2035–2040	2045–2050	2015
Logged	2050–2055	2050–2055	—

simulated the effect of several potential mitigation strategies on emissions. (Note that the six new scenarios reported below depict only land conversion and are different from the six multisector mitigation scenarios from USEPA [1990] described above.) The fraction of land clearing by burning is high in most regions but has not been incorporated into this or other models of deforestation, so forecasts given are for total clearing, a working proxy for clearing by burning, as yet. Preliminary model results are reported here.

The model, building directly on work by R. Houghton for EPA (USEPA, 1990) and currently being expanded to provide country-level scenarios (Braatz and Pepper, 1990), accounts for emissions resulting from clearing forest land, including primary, open, fallow, logged, unproductive, and plantation forests. Cleared lands are used for permanent crop and pasture land, and shifting agriculture.

The demand for crop and pasture land drives land clearing in most areas. Agricultural product demand is a function of population growth, consumption per capita, and yields per unit area of land. Assumptions about future population are based on a World Bank study (Zachariah and Vu, 1988) that is optimistic about the success of efforts to reduce future population growth. Assumptions about future changes in yield and food consumption per capita follow an analysis by FAO (1990), and those about recent past forest conversion reflect work by Myers (1989) (the best available until the 1990 FAO Forest Assessment is available).

Plantation establishment projections are driven by population estimates from Zachariah and Vu (1988).

Wood is assumed to be grown for fuel wood and industrial wood. Consequently, 50% of the biomass harvested goes into a long-term carbon pool and the remainder is burned for fuel or is treated as slash. Plantations are grown for 40 years and then selectively harvested on 10-year rotations. The harvest leaves the equivalent of a 30-year-old plantation.

The reference case developed shows the emission of carbon from 1955 to 2100, assuming no mitigating measures are taken, and land conversion is completely driven by agricultural demand. In the reference case, emissions begin to level off around 1995 and reach their culmination in approximately 2010. Emissions drop as population-growth rates slow and agricultural productivity increases. The exhaustion of the forests through conversion in large part drives changes in the rate of decline of carbon emissions over time. Table 55.2 gives the approximate dates of the final depletion of the various regions' forests. As nearly all forests are depleted, emissions from land conversion cease and net accumulation of carbon begins to take place.

Figure 55.4 shows a comparison among Houghton's population-based-growth and exponential-growth cases in his original model (in USEPA, 1990); Grainger's (1990) high and low cases; and finally both the EPA reference case and the combined sensitivity case incorporating response options (see discussion below). Strict comparison is not possible, given varying time frames, and assumptions about past conversion rates, future rates of growth in population, land productivity, and consumption. Figure 55.4 does illustrate the relative ranking of the models with regard to the total net carbon emissions, the shape of the emissions curves, and their rate of change over time.

Scenarios were developed to depict the effect of three forest practice response options potentially available to reduce biomass burning resulting from land-conversion activities in tropical regions. Scenario 1 (Figure 55.5) replaced burning with increased use of clear-felling to produce merchantable timber, a practice feasible in some conditions. Beginning in

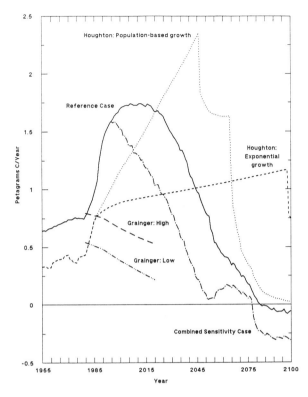

Figure 55.4 Comparison of global carbon emission projections among the models of Houghton (USEPA, 1990), Grainger (1990), and the EPA model's reference case and combined-sensitivity case.

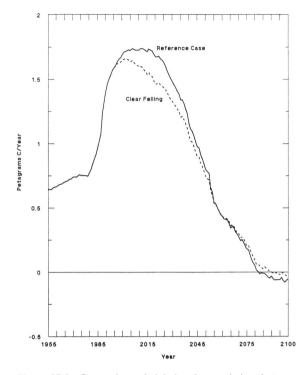

Figure 55.5 Comparison of global carbon emissions between the EPA model's reference case and a scenario in which clear-felling and the production of durable wood products is increased.

1995, 20% of the cleared biomass goes into durable products (a 50- to 100-year carbon sink) and the biomass burned is reduced proportionately. The amount of land so treated increases yearly so that by 2020, 75% of land cleared for permanent agriculture is cleared in this manner, an admittedly optimistic assumption.

Emissions drop immediately and are lower than the reference case until around 2055, when the carbon stored in durable products begins to decay, and emissions increase above the level in the reference case. The clear-felling strategy reduces net carbon emissions by 6% over the reference case and increases the slight net accumulation in the final 20 years of the century.

A second scenario was run which simulated the effect of increasing the rate of establishment of plantations by four times the rate used in the reference case. Simulation results (Figure 55.6) show a reduction of emissions which grows with time until approximately 2055. Net accumulation begins in approximately 2055 and increases until approximately 2085. The strategy decreases net carbon emissions by

38% and increases the net accumulation by more than 1900%.

Scenario 3 modeled the effect of replacing shifting agriculture with sustainable agriculture (Figure 55.7). Beginning in 1995, a fraction of the land in shifting agriculture is converted to sustainable agriculture yearly, resulting in a complete elimination of land in shifting cultivation by 2050. Carbon stocks of sustainable agriculture land are assumed to be midway between those of shifting and permanent agriculture.

As the shifting cycle is eliminated, demand for additional land decreases due to the greater productivity of "sustainable" land. Farmed-stage carbon stocks increase, further decreasing net emissions. Reduced demand for additional land allows carbon stocks on fallow-stage land to increase above stocking of normal, mature fallow-stage lands. Carbon emissions decline until all shifting land has been eliminated.

When all the farmed-stage land in the shifting cycle has been converted to higher-yield, sustainable condition, new demand must be relieved by cutting into mature fallow, which has greater carbon stocks than it would have under the reference case. When it is cut,

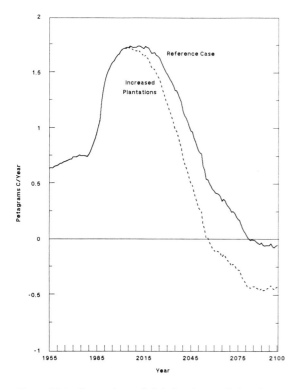

Figure 55.6 Comparison of global carbon emissions between the EPA model's reference case and a scenario in which the establishment of plantations is increased.

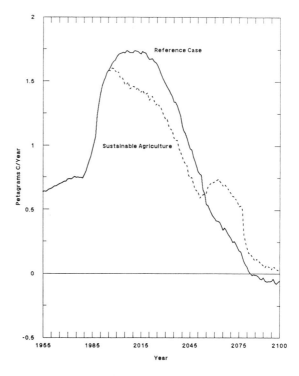

Figure 55.7 Comparison of global carbon emissions between the EPA model's reference case and a scenario in which shifting agriculture is replaced with sustainable agriculture.

emissions rise dramatically to a point higher than the reference case. Finally, when all the mature fallow land has been cut, the carbon dynamics of the system begin to resemble that of the reference case and emissions begin to drop.

Emissions remain above the reference case because the scenario's 55-year reprieve from cutting primary forest extended the existence of the forest beyond the point at which forests had been completely exhausted in the reference case. The strategy decreases net carbon emissions by 4% over the reference case, but sacrifices net accumulation of carbon which would have otherwise begun in the last 20 years of the century.

A final scenario combined the three forest practices previously described (see Figure 55.3 above, combined sensitivity case). Emissions decrease immediately but increase beginning in approximately 2055, following the dynamics of the scenario in which shifting agriculture was replaced by sustainable agriculture. Net accumulation begins in approximately 2075 and levels out around 2090. The scenario decreased net emissions by 48% and increased net carbon accumulation by 800% over the reference case.

Policy Option Example 2: Replace Current Traditional Stoves with More Efficient Designs

Improved stoves frequently have been seen as a low-cost solution to a host of problems in developing countries. Low fuel efficiencies have been observed in traditional stoves, and demonstrated improvements achieved under laboratory settings by stove designers. Despite problems encountered in the acceptance of these designs, the promise of carbon emission savings has led to the recommendation of disseminating high-efficiency stoves as a cost-effective option to mitigate the risk of global climate change (Gadgil and Modera, 1990).

If higher efficiencies are achieved in practice under field conditions and the current biomass is derived from unsustainable harvesting of fuel wood, then the climate problem is mitigated by improving the efficiency of stoves. However, a possible trade-off exists between the emissions of CO_2 and those of other greenhouse gases, such as CO, CH_4, NMHCs, etc., resulting from incomplete combustion (Smith, 1990).

More efficient designs have tended to improve overall stove efficiency by increasing heat transfer efficiency, but often at the expense of reducing combustion efficiency. For some improved designs, the increase in emission factors is compensated by the decrease in fuel requirements, but for other designs this compensation is not possible and the total emissions (on a task or a heat-delivered basis) can increase by a factor of 2 (Joshi et al., 1989). Therefore, it is critical that net emission balance be evaluated before widespread implementation of this response option justified on climate change grounds.

Gadgil and Modera (1990) have shown that the contribution of the household biofuel cooking energy system in developing countries to the anthropogenic emissions of greenhouse gases is significant, and the potential exists to halve this contribution cost effectively. They have calculated the costs of conserving carbon by mounting cookstove efficiency improvement programs. Making the reasonable assumptions about averages—that wood is 45% carbon, that fuelwood consumption is 1 kilogram per capita per day (kg cap^{-1} day^{-1}), that one stove is used to cook for five persons, that a new stove will cost $10 for manufacture and installation, will last two years, will double the present efficiency, and will depreciate linearly over its lifetime—the cost of conserved carbon is estimated to be 1.2 cent kg^{-1}.

In comparison, the cost of conserving carbon by electricity end-use efficiency programs where the electricity is produced by coal-fired power plants is about 7.6 cent kg^{-1}. This calculation assumes that the current global average carbon content of coal is 70.7%, the higher heating value is 8.1 kWh kg^{-1}, power plant efficiency is 33%, and the cost of conserved electricity equals 2 cents per kilowatt hour (kWh^{-1}).

Similarly, the cost of plantation and operation for proposed U.S. biomass plantation programs has been estimated to be $10 to $20 per ton of carbon captured (Andrasko, 1990a). This works out to be 1 to 2 cents kg^{-1} of carbon conserved. This estimate does not take into account costs of acquiring land and is certainly expected to be higher as plantations move to poorer quality lands.

Directions for Future Analyses

This provisional survey of response options for limiting biomass burning illustrates analytic approaches, data requirements, and a wide array of biophysical, technical, and social constraints (Table 55.3). Future directions may include:

1. Net greenhouse gas balance analyses of specific practices in specific ecosystems and energy-use environments;

2. Country-level analyses of mixes of responses within or across sectors (e.g., comparing costs and greenhouse benefits of mixes of responses in the energy and land-use sectors);

3. Design and evaluation of integrated land-use and energy responses (e.g., fire management and suppression, fuel-wood plantation establishment, and improved stoves); and

4. Development of deforestation models specific to biomass burning.

Policy Implications of Greenhouse Gas Emissions from Biomass Burning

A number of policy implications of efforts underway to quantify biomass burning have emerged.

1. Biomass burning estimates are presently sufficient to devise policy responses. Despite considerable scientific attention to the uncertainty in biomass burning estimates, the range of values for CO_2 emissions has not significantly narrowed or changed (from a policy perspective) over the past decade (Hall, 1989). This relative stasis suggests two key policy points: (1) The magnitude of carbon emissions identifies biomass burning as a major global source of greenhouse gases, and the primary source in many countries; and (2) uncertainties about these estimates are not likely to be resolved soon or to change significantly (in a policy sense). Therefore, the development of policy options to slow biomass burning, and exploration of mechanisms for implementing them, need not await refinement of CO_2 estimates, although other gas estimates remain uncertain.

2. Net gas balance analysis of response options is necessary to devise policy interventions that reduce emissions on a greenhouse equivalent basis. Biomass burning emits a wide range of greenhouse gases besides CO_2, which has been the subject of most analytic efforts at fabricating response options thus far, including CH_4, N_2O, NO_x, and CO. Burning of savanna and grasslands are often considered greenhouse-neutral in terms of CO_2, since carbon uptake during subsequent regeneration of grasses roughly equals the carbon emitted during burning.

Most estimates of biomass burning emissions present values of gross emissions, since gross emissions are more readily estimated than net fluxes, which require a second set of estimates for carbon uptake during regeneration of biomass in burned vegetation communities. However, potential response options need to be assessed on a *net* greenhouse gas balance basis, by discriminating between gross and net emissions. Both emissions and sinks of the full range of

Table 55.3 Overview of constraints, research, design, and development needs associated with response options to reduce the contribution of biomass burning to global warming

Response	R&D needs	Constraints

Forests: shifting agriculture
  1. Increase fallow periods.
  2. Promote sedentary and sustainable agroforestry systems.
	Remote sensing of fires
	Extension services; determine sustainability
		Increasing population pressure
		Terraced farming could result in soil loss on steep slopes

Forests: land-use changes
  1. Encourage clear-felling instead of burning for land conversions.
  2. Incorporate charcoal into soils, prevent consumption in subsequent fires.
  3. Increase productivity of existing lands, minimize need for conversions.
	Monitoring of remote fires; estimates of prompt, net, and gross emissions; regulatory and enforcement mechanisms
	Modification of soil characteristics
	Net greenhouse impacts of fertilizers, irrigation, crop switching
		Termites may increase net emissions; access to product markets
		Labor intensive
		Capital costs, land tenure, pro-deforestation policies

Forests: prescribed and wildfires 1. Increase frequency of small, managed, prescribed fires to reduce amounts burned in wildfires.	Determine optimal sizes and frequencies; net dry matter accounting by forest type	Potential for reducing contribution is low; may not apply to moist tropical forests

Crop lands: agricultural residues burning
  1. Reduce field burning; increase composting and soil incorporation.
  2. Increase use in stoves, high-efficiency applications.
  3. Increase biomass productivity of existing croplands.
	Measure net changes in emissions; develop field management practices
	Design and develop stoves and gasifiers
	Cultivars and practices that increase biomass without reducing yields
		Lack of alternative use, stalk thickness, labor-intensive
		Decentralized; but agricultural extension services available
		Use as fuel for increased biomass

Grasslands: savanna and pastures burning
  1. Reduce frequency, area and amount burned by optimal timing, etc.
  2. Replace grazing by stall feeding, game ranching.
  3. Replace grasslands by plantations (orchards, energy crops, etc.).
	Remote sensing of fires, emissions from current practices
	Establish net greenhouse emissions benefit
	Matching site characteristics with species
		Decentralized; culturally entrenched, difficult to change
		Labor intensive, acceptance
		Absence of existing extension services

Primary energy options: grow more biomass 1. Increase supply of biomass from new plantations.	Matching site characteristics with species	Competition for land, water by crops; sapling protection; soil infertility, fencing increase costs

Increase energy extraction efficiencies
  1. Disseminate high efficiency stoves, etc.
  2. Biomass cogeneration systems and gasifiers.
	Net greenhouse impacts; designs of low emission stoves
	Demonstration, feasibility projects
		Performance replication, acceptance difficult
		Decentralized fuel sources

gases need to be quantified, then weighted using a global warming potential (GWP) approach that reflects relative gas lifetime and greenhouse potency (Lashof and Ahuja, 1990). For example, grassland burning for livestock management may be relatively neutral in terms of CO_2 flux, but it releases major quantities of other gases with higher radiative forcings than CO_2 and may increase emissions of CH_4 from introduced livestock.

3. The time frame and method of GWP analysis used determine the importance of biomass burning in projected global warming. Thus the GWP method chosen influences the degree of policy attention. GWP methods that use shorter time frames (e.g., 20 to 50 years) will tend to produce relatively higher estimates of the relative importance of biomass burning, compared to those using longer time frames (over 100 years), since CO_2 has a long lifetime (e.g., 100 to 200 years) and low radiative forcing (1) compared to short-lifetime gases like CH_4 (lifetime of 10 years and radiative absorptance 21 times that of CO_2 on a molecule-for-molecule basis). Similarly, GWP methods assigning relatively higher values to gases other than CO_2 will tend to generate higher estimates of biomass burning importance, since this will tend to decrease the relative importance of the carbon uptake and sink function performed by grasslands and forests converting atmospheric carbon into biomass during photosynthesis while regenerating after burning.

4. Biomass burning response options are possible, and offer nongreenhouse benefits. Fire management has been practiced on limited scales for decades in all of the ecosystems of interest here, and stove efficiency has been increased in numerous demonstration projects. Improved fire management practices and alternative agricultural, livestock, and forestry systems with less reliance on fire can be identified and implemented by land managers at the forest stand, tract, or regional level.

In Brazil, for example, the federal environmental agency IBAMA launched a vigorous burning-management program during the dry season in 1989. The federal space agency INPE identified areas being burned through remote sensing, and then dispatched environmental police to the site to ascertain if permits had been obtained and to levy fines. IBAMA maintains that this enforcement program, along with the unusually long wet season, contributed to a major reduction in number of fires and area burned in 1989, although the program is being challenged in court at present, and few fines have been collected as yet (Setzer, 1989).

5. Our ability to intervene in burning practices is limited. Alternative field management practices and stove technologies need to be evaluated in terms of net gas emissions, costs, nongreenhouse benefits, and potential for technological transfer and adoption. Regulatory and incentive policy options need to be assessed to encourage implementation of promising options.

6. A suite of feasible response options to biomass burning needs to be developed for target conditions and locations. Fire-management field practices and policies have been utilized in some regions for decades, especially fire suppression and prescribed burning as a silvicultural treatment in temperate and boreal forests and in the semiarid tropics. Improved stove design and dissemination programs have been funded over the past two decades throughout the tropics, with varying degrees of success.

Reducing biomass burning to achieve climate benefits is a new concern, however. Work has begun forecasting potential effects of climate change on fire regimes, and management options (e.g., Mabbutt, 1989, for the semi-arid tropics; Andrasko, 1990c, reviews work thus far). Further work is needed on comparing the streams of benefits from a set of detailed, pragmatic responses feasible at site, development project, and regional or national levels.

The kinds of response practices and policies—and incentives to encourage their adoption—most likely to reduce emissions from biomass burning may well be ones already recognized for the stream of benefits they offer irrespective of greenhouse concerns. Several grassland, forest, and woodlot biomass management systems presently exist in field demonstrations or full-scale programs that offer many of the right qualities, from a greenhouse perspective: enhanced carbon storage, minimized use of fire as a management tool, minimal emissions of gases other than CO_2, low cost, reasonable economic returns, low or moderate technological complexity, and social acceptability.

Most of the existing and proposed practices and systems are mixed blessings. They contain trade-offs in terms of their ecological, economic, productivity, efficiency, social, and public policy attributes, and their greenhouse gas emissions profiles. One next step for analysts is to compare systematically and contrast options from both a greenhouse and a social development perspective and assemble an array of options for each situation likely to be funded and embraced.

Acknowledgments

The authors appreciate the major contribution of Barbara Braatz, William Pepper, and Parvadha Suntharalingam of ICF, Inc., Vienna, Virginia, for modeling support under their contract with the U.S. Environmental Protection Agency, Washington, D.C.

Amazonia: Burning and Global Climate Impacts

Luiz Carlos Baldicero Molion

Tropical forests are dynamical systems which have been disturbed naturally throughout the ages by climate fluctuations. At the end of last ice age, about 15,000 years ago, most of the humid tropics was colder and drier than at present, and rainforests on several continents shrank to an area much smaller than their current range (Dickinson and Virji, 1987). On the other hand, during the so called "climatic optimal" about 6,000 years ago, warmer and wetter climate conditions allowed the tropical forests to expand towards higher latitudes. The Amazon forest, for example, may have had extended as far south as 25°S and eastward to the Atlantic coast. Tropical forests are also affected by short-term natural disturbances, such as interannual variability of climate and volcanic eruptions. The Kalimantan forest (Indonesian Borneo) had 3 million hectares (ha) destroyed by fires during the severe drought resulting from the 1982 to 1983 El Niño–Southern Oscillation event (Malingreau et al., 1985).

In recent years, humans have been playing a major role in reducing the natural forest cover in the tropics through different forms of slash and burn. The most serious destruction, it is said, is occurring in the Amazon, which is the largest expanse of tropical forest remaining on the planet. This chapter reviews briefly the causes and the extent of Amazonian deforestation and focuses on its global and local climate impacts.

Causes, Extent, and Rates of Deforestation

The deforestation process in the Brazilian Amazon is highly complex; its history varies in different parts of the region. Its primary driving forces seem to have been geopolitical and economical—the decision to occupy the region and the increase of the regional population. The first has led to opening of highways—especially the Belem-Brasilia (BR-010) and the Cuiaba-Porto Velho (BR-364), where the largest deforestation nuclei are found—and fiscal incentives policies, such as tax reductions and negative interest loans, which large enterprises and landholders have benefitted from. The regional population growth was mainly the consequence of migration of small farmers, due to changes in agriculture patterns in southern parts of the country and to real estate speculation to a lesser extent. When compared to the above motives, mining and prospecting, subsistence agriculture, and commercial logging are relatively minor contributions, although the latter may increase in the near future as the Asian and African forests are reduced. Detailed discussions on causes of deforestation are found, for example, in Fearnside (1983a, 1990), Hecht et al. (1988), and Mahar (1979, 1988).

Although there is a general consensus about the causes of deforestation, the same is not true regarding the total area deforested. The estimates are 400,000 square kilometers (km^2) (Fearnside, 1990), 600,000 km^2 (Mahar, 1988), and 660,000 km^2 (Myers, 1989), respectively 10.9%, 16.4%, and 18.0% of the Brazilian Amazon forest taken equal to 3.66 million km^2 according to IBGE (1985). The disagreements emerge from the fact that these estimates are based on exponential projections of "desk surveyed" numbers. Brazil's Institute for Space Research published an inventory of alteration of forest cover evaluated from Landsat-TM satellite imagery. The results are shown in Table 56.1, together with estimates of 1978 and 1988, reported by Fearnside et al. (1990) using the same methodology. The altered area by deforestation was about 299,000 km^2, to which should be added 98,000 km^2 of deforestation prior to 1970s, resulting in total deforested area of 397,000 km^2, or 10.8% of the Brazilian Amazon forest.

In order to assess the impact of deforestation on the global climate, it is necessary to know the present rates of destruction of tropical forests. Myers (1989) made a survey of deforestation rates which includes 34 countries and together comprise 97.3% of the existent tropical forests of the globe. According to this author's appraisal, during 1989 tropical forests lost 142,000 km^2 of their expanse; together Brazil, Indonesia, and Zaire account for 46% of this total. He estimated that the current amount of deforestation

Table 56.1 Deforested area in the Brazilian legal Amazon from comprehensive Landsat Multispectral Scanner and Thematic Mapper surveys

State	Area of state (sq. km)	January 1978	April 1988	August 1989
Acre	153,698	2,464 (1.6%)	7,292 (4.7%)	8,836 (5.7%)
Amapa	142,359	167 (0.1%)	781 (0.5%)	1,016 (0.7%)
Amazon	1,567,954	1,725 (0.1%)	18,565 (1.2%)	21,584 (1.4%)
Maranhao	260,233	6,076 (2.3%)	24,451 (9.4%)	30,840 (11.9%)
(In addition to "old" deforestation: 57,824 sq. km or 22.2%)				
Mato Grosso	802,403	20,005 (2.5%)	71,414 (8.9%)	79,594 (9.9)
Para	1,246,833	16,525 (1.3%)	88,535 (7.1%)	99,786 (8.0%)
(In addition to "old" deforestation: 39,819 sq. km or 3.2%)				
Rondonia	238,379	4,242 (1.8%)	29,678 (12.4%)	31,476 (13.2%)
Roraima	225,017	132 (0.1%)	2,745 (1.2%)	3,621 (1.6%)
Tocantins	269,911	3,166 (1.2%)	20,959 (7.8%)	22,327 (8.3%)
Legal Amazon	4,906,784	54,502 (1.1%)	264,421 (5.4%)	299,079 (6.1%)
(In addition to "old" deforestation: 97,643 sq. km or 2.0%)				

Source: Fearnside et al. (1990). Deforested area in square kilometer (percentage of area of state).

for the Amazonian countries (Bolivia, Brazil, Colombia, Ecuador, Peru, and Venezuela) to be 66,000 km^2 per year and that Brazil is the overall leader, with a rate of 50,000 km^2 per year. As it would be expected, evaluations differ considerably from each other. Myers's numbers for the Brazilian Amazon forest are about 138% higher than INPE's recent rate of 21,000 km^2 per year. The deforestation annual rates, averaged for the period 1978 to 1989, vary from 0.57% to 1.37%, the lower rate being the one evaluated using satellite imagery.

Global Impacts

Due to its large extension, the Amazon forest may be important for the stability of the global climate. This section attempts to demonstrate this possibility, based on the present knowledge of both physical and chemical forest-atmosphere interactions. Two hypotheses will be developed; first, the forest as controller of the greenhouse effect, and second, the region as an important heat source to maintain climate stability.

Amazonia and the Greenhouse Effect

The role that tropical forests have on the chemical composition of the atmosphere is not known quantitatively yet. It is said they are important sources of methane (CH_4) and carbon monoxide (CO); the CH_4 is produced by the decomposition of organic matter in lakes, marshes, and floodplains, and the CO mainly by biomass burning. Both CO and CH_4 are oxidized through different catalytic processes which involve nitrogen oxides (NO_x) also produced by biomass burning. With high concentrations of NO_x, ozone

is formed, and with low concentrations of NO_x, ozone is destroyed. Crutzen (1987) showed that the vertical distribution of tropospheric ozone were higher over the *cerrados* than over the Amazon during the dry season. He attributed this fact to the higher number of fires in the *cerrados* with consequent photochemical production of ozone, which can be transported to the high troposphere by the tall cumulus nimbus (Cbs) clouds and then to other parts of the world through atmospheric circulation of the type Hadley-Walker Cell.

To study the influence that the Amazon forest exerts on the chemical composition of the atmosphere, two campaigns of the Global Tropospheric Experiment (GTE/ABLE) were made, one during the dry season of July and August 1985 (Harriss et al., 1988) and another during the wet season of April and May 1987 (Harriss et al., 1990). Preliminary results show that the forest is a sink of tropospheric ozone and the region as a whole is a source of methane. A surprising result was that the forest was a net CO_2 sink, i.e., photosynthesis minus respiration = 0.25 kg of carbon per hectare per hour (Song Miao et al., 1990). Just to have a feeling for the magnitude of this number, if it could be generalized for the whole of Amazonia, the forest is incorporating 1.2 billion tons of carbon per year, that is, the equivalent to 25% of the amount that human activities release into the atmosphere through fossil fuel burning, estimated to be 5 billion metric tons of carbon per year. Although these latter measurements were carefully made and analyzed, they encompass only 12 dry, sunny days during the wet season, when trees are well supplied with water and can transpire freely. Therefore, they may be biased toward optimal conditions and high carbon net intake rates. It is possible that this absorption rate may not be valid for the entire year because some areas in Amazonia present a marked dry season when plants reduce photosynthesis activity due to water stress. During that period, the balance may even be negative, so that the annual average carbon intake rate may possibly be lower than these measurements suggest, as pointed out by the authors themselves. Nevertheless, the numbers are, by far, much lower than the values obtained outside of the tropics during the growing season, and they are the only measured data for tropical forests so far.

Some scientists criticize this hypothesis under the argument that the forest is in a climax and therefore could not have a net absorption of carbon. This latter hypothesis is of more difficult verification than the previous one because, even though the forest apparently is not increasing its geographic domain, the present knowledge does not allow one to affirm that the forest is not becoming denser. It could be in fact that the forest does not retain much of the carbon itself but acts as a transfer mechanism from the atmosphere to other reservoirs—namely, the soil and the aquatic ecosystems—and rivers, in turn, carry the carbon to the ocean. The carbon in the ocean and in the soil has a higher residence time than in the forest. Richey (1989) estimated that annually the Amazon River, at Obidos (longitude 55°W), transports to the ocean a carbon load of 32 million metric tons. This rate, however, may represent only part of the carbon that is being sequestered by the system as a whole. For example, Nepstad et al. (1990) found that the root mass in Eastern Amazon forest was 13.4% of the above-ground biomass, estimated to be 264 metric tons per hectare, and only 9% of the root mass was renewed every year. Long et al. (1989) reported results of an experiment in Amazonian floodplains, where a grass *Echinochloa polystachya* presented a net primary productivity (npp) of 100 metric tons of dry matter per hectare per year. It is estimated that natural flooded grasslands occupy about 0.3 million km^2 in Amazonia. Again, just for the sake of example, if it is assumed that all grasses in this area have the same npp, the annual absorption in the floodplains alone would be 1.5 billion metric tons of carbon. Unfortunately, there are no estimates of how much carbon is incorporated in the aquatic ecosystems annually. Bringel et al. (1990) studied the sediments composition of 29 Central Amazonian lakes; sediments at the bottom of tropical lakes are, most of the time, under anoxic conditions and, thus, reduced organic matter decomposition. Among these lakes,

there are 15 which are located in the Rio Negro Basin, and their top layer sediments presented 4% of organic matter of which 60% were carbon. Walker (1990) estimated that the litter production of *igapo* (flooded forest) is 5 to 7 metric tons per hectare per year and that, during the flooding period, the weight of sediments, including fine sand and clay varies between 4.3 to 8.2 metric tons per hectare per year. Another evidence of incorporation of carbon into the system is the presence of peat in Amazonia with an estimated volume of 20 billion metric tons, about 85% of the Brazilian deposits (Suszczynski, 1982). Evidence of carbon incorporation have been reported for other tropical forests (Lugo, 1990). In summary, it is possible that the carbon transfer from the atmosphere to the Amazonian ecosystems may not be negligible when compared to the total carbon released annually by fossil fuel burning.

The burning of the forest, on the other hand, would contribute to increasing the CO_2 and other greenhouse gases concentrations. During the dry season of 1987, Setzer et al. (1988), using NOAA satellite imagery, estimated that 20 million hectares of land were burned in the Legal Amazon, of which at least 40% were natural forests; according to these authors, those fires produced about 600 million metric tons of carbon, which is about 10% of the total carbon released globally every year. Currently, it is estimated that biomass burning in the tropics contributes with 1,200 million metric tons per annum of carbon equivalent to increase CO_2 concentrations, and that the Brazilian share would be 540 million metric tons, that is 45% of the total (WRI, 1990). These estimates may be higher than the actual rates because of the uncertainties related to the parameters involved in the calculations.

The release of carbon by burning depends on the product of four parameters—namely, the percentage of carbon in the biomass, the annual rate of deforestation, the biomass density, and the efficiency of carbon release in the burning process. The percentage of carbon in the biomass is taken equal to 50% as a consensus. A survey of literature, however, shows that deforestation rates vary from 2.1 to 5.0 million hectares per year; that biomass density spans from 180 to 730 metric tons per hectare; and, finally, that the efficiency coefficient range is 20% to 100%. With these numbers, one obtains carbon release rates that vary from a minimum of 38 to a maximum of 1,825 million metric tons per year, a factor of 48 between extremes. The Amazonian contribution can be anywhere within these limits.

The above discussion may be merely academic because probably the contribution of Amazonian burnings to increasing CO_2 atmospheric concentration may be relatively insignificant. According to Fearnside et al. (1990), the deforested area of the Brazilian Amazonia is 40 million hectares. Assuming that the biomass density is 300 metric tons per hectare (Brown, 1990), that all carbon stored has been released, and, in the worst possible case, that there is no biomass regrowth (that is, no carbon has been sequestered), Amazonia burnings have contributed with 3 billion metric tons to increase atmospheric CO_2—that is, 2.1% of the global release in the past 150 years or 0.4% of the total CO_2 present in the atmosphere. If the remaining 550 million hectares, including all Amazonian countries, were burned and the 82 billion metric tons of stored carbon discharged in the atmosphere at once, without carbon uptake by regrowth, the CO_2 level would increase by 20 ppm or less than 6% of the present global concentration estimated to be 350 ppm, a modest contribution to doubling CO_2.

The hypothesis is, therefore, that deforestation of Amazonia may contribute to the enhancement of the greenhouse effect in two ways: first, by biomass burning, which is not too extensive when compared to fossil fuel burning; second, by destroying the trees, which fix carbon through photosynthesis. In addition, rising of the sea levels, as a result of warmer climate, would have a damming effect on the Amazon River discharge, which will tend to flood permanently part of the *varzeas* and increase the methane release rates.

Amazonia: A Heat Source for the Atmosphere
The solar energy reaching the surface is primarily used for evaporating water (latent heat) and heating the air (sensible heat). In Central Amazonia, micrometeorological studies (e.g., Molion, 1987; Shuttleworth, 1988) have shown that 80% to 90% of the available energy is used for evapotranspiration (evaporation + plant transpiration); the rest warms the air. Over terra firme forest, the water vapor flux is basically constituted of 70% transpiration of plants and 30% evaporation of rainfall water intercepted by the forest canopy and the litter layer. Direct soil evaporation was found to be negligible. In the annual mean, evaporation in Amazonia is about 50% of the total rainfall; in other words, considering the climate stable, in the long range half of Amazonia's rainfall comes from the local evaporation and the other half from the Atlantic Ocean (Molion, 1976; Salati, 1987). This local contribution is considered high compared to what occurs in temperate latitudes where it is esti-

mated that local evapotranspiration makes up about 10% of local precipitation. When the water vapor condenses, forming clouds and rain, it releases the heat (latent heat) that was used in the evapotranspiration process.

Over a tropical continent, the warm and moist air rises (convection), and it is replaced in the lower levels by air coming from the oceans (convergence); in the high troposphere, the air is transported away (divergence) from the continent and sinks over regions near 30° latitude, thus closing a circulation cell. In the tropics, there are three regions where ascending motions predominate: the Maritime Continent (Indonesia, North of Australia), the Congo River Basin, and the Amazon River Basin. The latter two are of truly continental origin and therefore depend on the surface cover. The first one is of different nature; it is a result of heat transfer from the ocean to the atmosphere as the Pacific waters in that region have surface temperatures of 28°C or higher. The latent heat released by these sources is transported away from the tropics to temperate and polar regions by the circulation cells, part of the general circulation of the atmosphere (GCA). Due to this transport, the global climate remains stable but presents year-to-year variations which may be tied up to the fluctuating power of the sources.

The hypothesis is, then, that Amazonia is an important heat source for the GCA and that a large-scale deforestation may reduce the power of this source. As mentioned earlier, in the average, about 50% of Amazonian rainfall comes from water evaporated locally. During GTE/ABLE-2 wet season campaign, Nobre et al. (1988) concluded that 58% of the rainfall, in fact, came from local evapotranspiration. Deforestation is known to reduce evapotranspiration, therefore reducing precipitation and latent heat release. Experiments of large-scale deforestation of Amazonia were performed using global circulation models (e.g., Dickinson and Henderson-Sellers, 1988; Lean and Warrilow, 1989; Shukla et al., 1990). The results showed a 20% to 30% reduction in rainfall (latent heat) over the basin, which corresponds to 5% to 8% of the total heat transported poleward by the atmosphere across latitudes 10°N and 10°S, based on data published in Hastenrath (1985).

Using a climate feedback equation proposed by Dickinson (1986), where the heat variation is equal to the temperature change times a feedback parameter ($\lambda = 1.5$ wm^{-2}C^{-1}), i.e., $\Delta Q = \lambda \Delta T$, and assuming that only regions poleward of 30° latitude would be affected, a reduction of 5% to 8% in atmospheric heat transport to the extratropics would reduce the mean temperature of these regions by 0.7 to 1.1°C. One of the possible consequences of increasing the equator-to-pole temperature gradient is the equatorward displacement of the subtropical anticyclones (Flohn, 1981) and of the storm tracks (Paegle, 1987), conditions which are sufficient to change the present climate. Cooling, also, may shorten the growing season, thus decreasing grain production.

Although the two hypotheses—namely, a greenhouse effect "controller" and the heat source—and the related arguments are physically sound, the deforestation effects on the global climate are controversial issues, mainly because the numerical models used for testing such hypotheses are yet in their beginning, they do not represent the atmospheric physical processes, especially feedbacks, adequately enough and, therefore, their results must be interpreted and used with care.

Conclusion

The main causes of deforestation appear to be the geopolitical decisions for occupying Amazonia, which have led to the opening of roads and growth of population, mainly through migration to the region. Secondary causes, such as real estate speculation and fiscal incentives, contribute to the current deforestation rates. The principal land use after the removal of the forest has been pastures for cattle raising, which is the worst possible type of use for that environment (Hecht et al., 1988).

The evaluation of land use transformation in the Brazilian Amazon, using Landsat-TM recent images, resulted in a deforested area of about 397,000 km², which represents 10.8% of the area covered with forests, and a deforestation rate of 21,000 km² per year.

The main objective of this work was to bring to debate the two hypotheses of climate change related to removal of the Amazonian forest—namely, the forest as a controller of the "greenhouse effect" and the forest as a main "source of heat" for the extratropics. By burning Amazonia, in principle, the greenhouse effect would be accelerated. On the other hand, deforestation may reduce the power of the heat source, resulting in an equatorward displacement of subtropical anticyclones and storm tracks, and cooling of the extratropics. The two hypotheses are not antagonistic since one deals with the additional energy which would be trapped in the Earth-atmosphere system, and the other deals with the latitudinal distribution of energy. The two unresolved questions

should be addressed immediately, if global climate changes are to be understood.

Other consequences of deforestation should at least be mentioned briefly. Deforestation will cause local climate changes. Soil surface temperatures may increase by 2° to 5°C and air temperatures by 1° to 3°C; evapotranspiration may decrease by 20% to 50% and local rainfall may be reduced by 20% to 30%. Another important component of the hydrologic cycle, the surface runoff, may decrease by 10% to 20% (Dickinson and Henderson-Sellers, 1988; Lean and Warrilow, 1989; Shukla et al., 1990). The general circulation models (GCMs) results also suggest that deforestation may change both the temporal and spatial distribution of hydrologic variables and increase the length of the dry season. Another problem linked to the variation of the climatic elements and the removal of forests is the soil degradation and consequent erosion due to exposure of fragile soils to high rainfall rates. Jansson (1982) reviewed the existent literature on tropical soil erosion and found rates up to 334 metric tons per hectare per annum. Erosion will silt the river channels and reservoirs, changing the water quality and the aquatic life.

The great biodiversity of the region would be lost completely in a few years if present deforestation rates persist. It is said (see, e.g., Mori and Prance, 1987) that the rainforest may contain 30% to 50% of the species of vegetation, animals, and insects existent in the world. Although the forest is rich, 90% of its soil is said to be of old formation, leached and poor in nutrients. Closed nutrient cycle plays important role in maintaining the rainforest. Deforestation exposes the soil and the top 10 to 20 centimeter (cm) layer of organic matter is removed by weathering very quickly; soils cannot sustain agriculture or pasture for more than two to three years and, in the fallow land, the forest takes a long time to recover (Uhl et al., 1989). Changing the water quality and aquatic life by increasing sediment load and mercury concentration due to gold prospecting and mining, as well as increasing fertilizers and pesticides concentrations tied up with the expansion of agriculture frontiers, are among the problems that threaten the aquatic system which contains 20% of the unfrozen fresh water of the planet.

The region also guards the remnants of Indian cultures and societies which are simply disappearing or being transformed by Western civilization. The *caboclos*—people of mixed European, Indian, and African ancestry, who live mainly on the river banks and constitute an estimated population of 2 million people in Amazonia—are a vanishing human species. One reason for the extinction is that the *caboclos* have neither the legal protection given to Indians nor the rights of the white population; also, they used to live off extraction of rubber, nuts, and other forest products, an economical subsistence practice which does not generate an adequate income in the modern society.

If a conservative population growing rate of 2% per annum is considered, one finds that, by the middle of next century, the world may have 12 billion inhabitants. It is obvious that a tropical region such as Amazonia—where in principle there are no climatic limitations for food production, even considering future scenarios that are predicted, for example, in case of an enhanced greenhouse effect—should not remain marginal to this process. The development of the region should be rational and careful in view of the arguments presented here. The question that follows is, What is the most appropriate land use for Amazonia? Several authors (e.g., Goodland, 1980; Fearnside, 1983b) have dealt with this question. Molion (1986) also proposed some solutions for land use in Amazonia. In conclusion, Molion wrote that the best solution seems to be a balance of natural forests, agricultural, and pasture fields, with a higher proportion of forests and smaller proportion of pastures.

Acknowledgments

This work was done during a granted fellowship at the Wissenschaftskolleg zu Berlin (WKB). Both the financial and staff supports are acknowledged in the preparation of this paper, especially to Dr. W. Lepenies, Rector of WKB, and Dr. H. G. Lindenberg. The support of the International Geosphere Biosphere Program (IGBP/ICSU), through its executive director, Dr. Thomas Rosswall, and his deputy director, Dr. Hassan Virji, is also acknowledged.

Surface Cooling Due to Smoke from Biomass Burning

Alan Robock

Biomass burning is known to dramatically affect the chemistry of the atmosphere, as the other chapters in this volume discuss in great detail. Smoke is another atmospheric input, and although in normal circumstances its lifetime is on the order of one week, large smoke clouds can last longer in the atmosphere and, in the extreme case, "nuclear winter" can persist for years (Turco et al., 1990). In all cases of biomass burning, smoke is generated, and in continuous burning cases, the smoke can persist for as long as the burning continues, which can be for weeks or months during agricultural or deforestation burning.

In this chapter, results from five different forest fire cases are presented which illustrate the surface temperature effects from this type of smoke. In all cases, daytime cooling and no nighttime effects were found. These results correspond to theoretical estimates of the effects of smoke, and they serve as observational confirmation of a portion of the nuclear winter theory. This also implies that smoke can have a daytime cooling effect of a few degrees over seasonal time scales. In order to properly simulate the present climate with a numerical climate model in regions of regular burning, it may be necessary to include this smoke effect.

Wexler (1950) presented anecdotal evidence of daytime cooling in Washington, D.C., at the surface of 2° to 5°C from "the Great Smoke Pall" caused by extensive Canadian forest fires in 1950. Veltishchev et al. (1988) also presented anecdotal observations of daytime surface cooling caused by extensive Siberian forest fires in 1915. These works suggest that forest fire smoke can produce net cooling of surface air temperature but do not provide an objective measure of the effect. In this chapter, comparisons to numerical model forecasts and to climatology provide an objective measure of this cooling effect. In addition, a case of burning in the Amazon, where synoptic-scale advection provides the dominant control on surface air temperatures and where there is not a measurable effect from smoke, is presented. However, in this case, better data may show a smoke effect.

As will be shown later, smoke clouds were easy to detect with visible wavelength imagery but were invisible in infrared imagery on satellite photos. Submicron smoke particles strongly interact with the incoming short-wave solar radiation, scattering and absorbing it, producing visible images of smoke. The long-wave outgoing terrestrial radiation ($\lambda > 10\ \mu$m) can easily pass through the smoke particles, allowing the radiation to be detected by the satellite but preventing smoke detection in these wavelengths. Thus, based on only the satellite images, we would expect the smoke to have a net cooling effect on the surface, with less solar radiation reaching the surface to warm it, but long-wave radiation able to leave the surface and cool it normally. The downward long-wave radiation from the smoke cloud would not be expected to compensate for the loss of solar radiation (Vogelmann et al., 1988; Turco et al., 1990). This is in contrast to the effects of the volcanic dust particles from Mount St. Helens, which were larger and, while also cooling during the day, caused compensating warming at night (Robock and Mass, 1982; Mass and Robock, 1982).

All the cases discussed here have applications not only to normal biomass burning but also to nuclear winter. Crutzen and Birks (1982) first suggested that smoke from forest, urban, and industrial fires ignited by nuclear weapons would be extensive enough to block out significant amounts of sunlight. Subsequent works, such as Turco et al. (1983), National Research Council (1985), and Pittock et al. (1986), have pointed out that the smoke from urban and industrial fires (especially oil refineries) would probably be much more effective at preventing solar radiation from reaching the earth's surface than forest fire smoke after a large-scale nuclear war. With both urban and rural targets, not only would more urban smoke be generated, but its optical properties would make it more effective at blocking sunlight. However, the optical properties and surface temperature effects of forest fire smoke are important parts of the study of nuclear winter. Much forest fire smoke would still be generated in many nuclear war scenarios, especially those that include only nonurban military targets. In addition, it is useful to have some actual observations of the effects of smoke to compare to theoretical models of nuclear winter. In fact, the 1982 case presented here has been modeled by Westphal and Toon (1991), and comparisons of the results are discussed later.

Fortunately, extensive urban and industrial smoke plumes are not readily available for study. Each year, however, forest fires generated by lightning and by human activities produce extensive smoke plumes. In this chapter, several of these cases are discussed.

British Columbia Fires of 1981 and 1982

The location of smoke clouds from extensive Canadian forest fires was determined from satellite imagery for cases in the summers of 1981 and 1982. The smoke clouds were easy to detect with visible wavelength imagery (Figures 57.1 to 57.3), but were invisible in infrared imagery. In the 1981 case (Figure 57.1), the smoke appeared to be in streaks, with patches of less dense smoke in between. Within three days of being produced, the smoke plumes in the 1982 case grew into an extensive shield covering about 10^6 km^2. There was some patchiness evident (Figures 57.2 and 57.3), but the smoke veil maintained a coherent structure for several days. The total area covered was about the same in both cases. Robock (1988b) presented a preliminary summary of these cases.

As these smoke clouds passed over the midwestern United States, surface air temperature effects were determined by comparing actual temperatures with those forecast using model output statistics (MOS) by the United States National Weather Service (Glahn and Lowry, 1972). This MOS error technique has been used successfully before, to determine surface air temperature effects of the Mount St. Helens volcanic eruption of 1980 (Robock and Mass, 1982; Mass and Robock, 1982). The analysis was done in regions of high pressure where synoptic disturbances were not affecting the temperature. The errors are attributed to the presence of aerosols in the atmosphere, since the aerosol content is not a MOS predictor. The locations of all the MOS stations used in this analysis are shown in Figure 57.4.

1981 Case

During the second week of August 1981, numerous forest fires burned in western Canada (Schneider et al., 1986). Figure 57.5 shows MOS forecast errors for

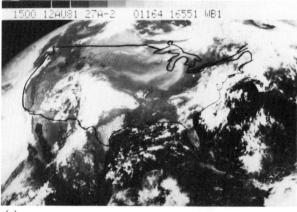

(a)

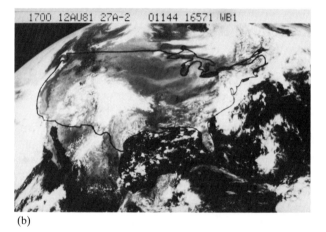

(b)

(c)

Figure 57.1 GOES satellite image in visible wavelengths, 12 August 1981, for (a) 1500, (b) 1700, and (c) 2100 GMT.

(a)

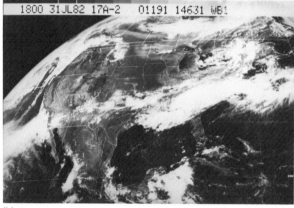

(b)

(c)

Figure 57.2 GOES satellite image in visible wavelengths, 31 July 1982, for (a) 1501, (b) 1800, and (c) 2100 GMT.

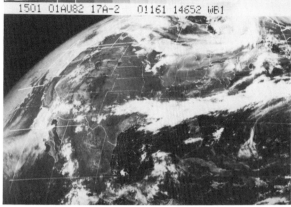

(a)

(b)

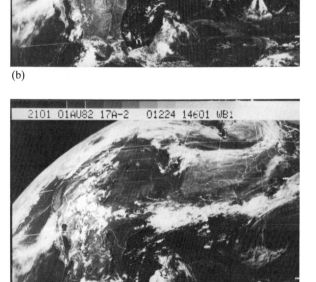

(c)

Figure 57.3 GOES satellite image in visible wavelengths, 1 August 1982, for (a) 1501, (b) 1801, and (c) 2101 GMT.

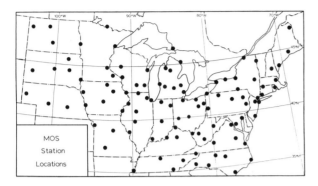

Figure 57.4 MOS station locations for analyses shown in Figures 57.5 to 57.7.

1500, 1800 and 2100 GMT (11:00 A.M., 1:00 P.M., and 4:00 P.M. CDT local time) for the 12 August 1981 case, corresponding to the images in Figure 57.1. Daytime cooling of $-1.5°$ to $-3°C$ is seen under the smoke.

Forecast errors of the same or greater amplitude are also evident in other locations in the figures. On close inspection, each can be attributed to the presence or absence of water clouds or mesoscale features that were not well forecast by the limited-area fine mesh (LFM) numerical forecasting model, which provides the predictors for the MOS forecasts. In clear areas, however, the only large errors are under the smoke cloud.

For this case and the 1982 case, the results presented are from forecasts made from LFM runs 12 to 24 hours before the forecast time. Comparisons were made with forecasts made with earlier and later model runs, and the results were virtually the same. The errors (cooling) found were much larger than the run-to-run differences.

1982 Case

More information is available for the intense forest fires which burned in British Columbia, Canada, on 29 July 1982 (Cowell, 1983). Smoke from these fires was transported by the prevailing winds and crossed the U.S. border in North Dakota on 31 July. It proceeded across the Midwest and then east over the Mid-Atlantic states, and was reported over Western Europe on 5 and 6 August. Such was also the case for the Great Smoke Pall (Wexler, 1950); this demonstrates the ability of the atmosphere to transport smoke over long distances before it is removed.

Figures 57.6 and 57.7 show MOS surface temperature forecast errors for 31 July and 1 August 1982 at 1500, 1800, and 2100 GMT (11:00 A.M., 1:00 P.M.,

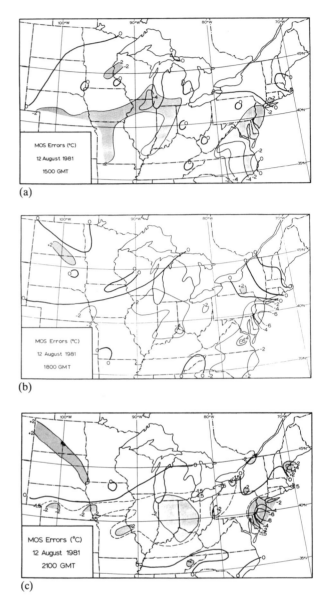

(a)

(b)

(c)

Figure 57.5 MOS surface-air temperature errors, 12 August 1981, for (a) 1500, (b) 1800, and (c) 2100 GMT.

and 4:00 P.M., CDT) each day corresponding to the satellite images in Figures 57.2 and 57.3. Smoke plumes appear gray in these images, and can easily be detected over the Midwest. On the original images, the smoke plume that was headed for Europe can be seen on 1 August over the Atlantic Ocean. Figure 57.8 shows MOS forecast errors for 1 August 1982 at 0600 GMT (1:00 CDT).

Forecast errors of $-2°$ to $-4°C$ are evident under the smoke plumes. Forecast errors are not evident under the smoke plumes at other times of the day. Since the errors are only evident at times of maximum

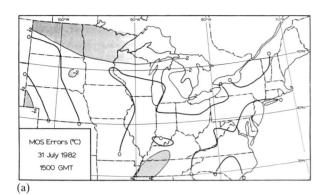

(a)

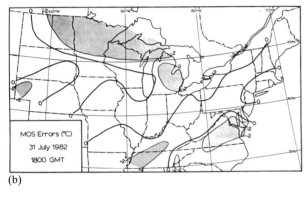

(b)

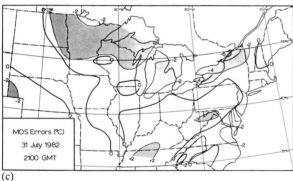

(c)

Figure 57.6 MOS surface-air temperature errors, 31 July 1982, for (a) 1500, (b) 1800, and (c) 2100 GMT.

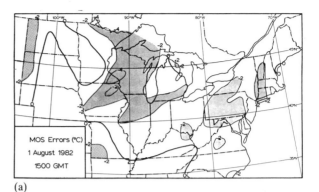

(a)

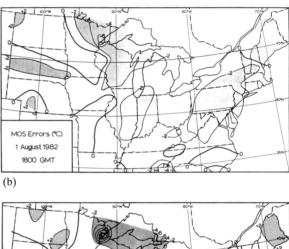

(b)

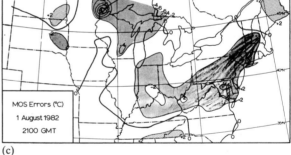

(c)

Figure 57.7 MOS surface-air temperature errors, 1 August 1982, for (a) 1500, (b) 1800, and (c) 2100 GMT.

daily insolation, their predominant effect on short-wave radiation is demonstrated.

Again, forecast errors of the same or greater amplitude are also evident in other locations in the figures, and on close inspection, each can be attributed to the presence or absence of water clouds or mesoscale features that were not well forecast by the LFM. In clear areas, however, the only large errors are under the smoke cloud. For example, at 1500 GMT on 31 July (Figure 57.6a), there is a negative area in eastern Michigan that can be attributed to an unforecast mesoscale region of cloudiness. A positive error re-

gion in western Tennessee and Kentucky is associated with a mesoscale clear area in an otherwise overcast region.

In Figure 57.6b for 1800 GMT, 31 July, the same error regions are present as at 1500 GMT, except that an additional negative area appears in Virginia under a heavy water cloud bank, and small negative areas appear east and west of Lake Ontario. These latter areas become more extensive and of higher amplitude three hours later (Figure 57.6c). They are associated with strong cold advection behind a rapidly developing mesoscale low, the center of which is located on

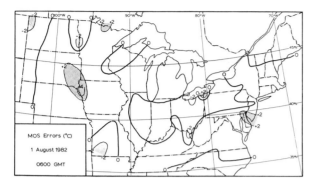

Figure 57.8 MOS surface-air temperature errors, 0600 GMT, 1 August 1982. The smoke cloud location cannot be determined from the satellite image. It is intermediate between those in Figure 57.2c and 57.3a.

the Vermont-Canadian border. This is easily seen on the synoptic weather maps (not presented here) and presumably was not well forecast by the LFM, which does not handle small-scale weather systems well. In Figures 57.6b and 57.6c the moderating effects of Lakes Erie and Ontario are evident, as the maximum cooling associated with this cold advection is over the land areas between the lakes. The only other negative error area, and by far the largest one, is under the forest fire smoke veil. Areas of fair weather cumulus over Illinois, Indiana, and Ohio do not produce large MOS errors.

In Figure 57.7, on 1 August during the daytime, the only large error regions (both negative) shown are associated with a small region of water clouds over Wisconsin and the forest fire smoke veil. The region around the Great Lakes under the smoke veil has errors between $-1°$ and $-2°C$, evidence again of a moderating effect of the lakes. The thickest patch of smoke appears over Baltimore, Maryland, at 2100 GMT, and has an error of $-4.1°C$ associated with it (Figures 57.3c and 57.7c).

Discussion

Attempts were made to estimate the height of the smoke as it passed over the Midwest. For the 1981 case, pilot reports over the Midwest on 12 and 13 August reported smoke from 2.5 to 7.5 kilometers (km) and smoke motions for the thickest smoke corresponded to 400 mb (7.5 km) winds (Schneider et al., 1986). For 1982, winds were examined at different levels during this time and compared to the motion of the smoke. The winds corresponding most closely to the motion were at 700 mb (3.2 km). The jet of smoke headed to Europe on 1 August corresponded to winds between the 700 mb and 500 mb levels (3 km to 5.5 km).

A vertical profile study of lower-level haze was conducted by chance on the next day over the eastern shore of Chesapeake Bay, due east of Baltimore at 1700 GMT (Eloranta, personal communication, 1987). The portion of the smoke veil that was over Kentucky on 1 August appears from satellite images to have been over Maryland at that time. Eloranta flew up into the base of the forest fire smoke plume at an altitude of approximately 3.5 km, and at 4.5 km (the highest altitude reached), he was still in the smoke.

A cooling of 1.5° to 4°C is found in the daytime under forest fire smoke plumes in two cases in 1981 and 1982. No effect is found at night. This corresponds to theoretical estimates of the effects of an elevated smoke plume (Veltishchev et al., 1988) with optical depth of approximately 2. The optical depth of the 1982 smoke cloud, based on multispectral satellite measurements, has been reported to be 2 to 3 (Ferrare et al., 1990).

The 1982 case was investigated by Westphal and Toon (1991) by use of a high-resolution numerical model which simulated the transport and radiative effects of the smoke. They found virtually identical cooling at the surface to that found here with the MOS error technique. When they did the same simulation, but changed the optical properties of the smoke to be more like that from a burning city or industrial area (more absorbing and less scattering), they found surface cooling of 8°–10°C.

China and Siberia Fires of 1987

In May 1987 extensive forest fires burned in the People's Republic of China and in Siberia, USSR. In less than one month, the fires burned more than 7,500,000 hectares (ha) in China and 20,500,000 to 37,500,000 ha in Siberia (Salisbury, 1989). Smoke from these fires was seen in polar orbiting satellite images over Alaska for several days starting on 10 May. This demonstrates the potential for long-range transport of smoke in the atmosphere (> 4000 km in this case).

The same technique of using errors of the MOS surface air temperature forecasts in regions of little synoptic disturbance was used to search for surface temperature effects of the smoke in Alaska. Surface temperature forecasts are produced by MOS for 26 locations in Alaska (Figure 57.9) verifying at six-hour intervals (Maglaras, 1983). Since Alaska is between 130° and 170° west of Greenwich, forecasts verifying at 0000 GMT are for the approximate time of day of maximum temperatures, the 1800 and 0600 GMT

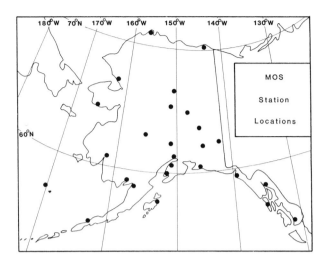

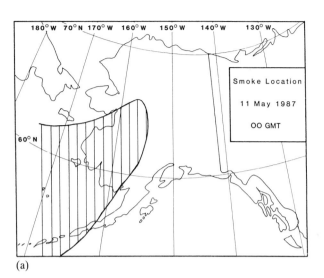

Figure 57.9 Locations of stations in Alaska where MOS forecasts of surface-air temperature are made every six hours.

forecasts are for early morning and early evening, and the 1200 GMT forecasts are for the middle of the night.

The locations of the smoke plumes from the fires are shown in Figure 57.10 for 0000 GMT for 11 to 13 May as determined from satellite images at the NOAA/Navy Joint Ice Center. The poor quality of the images prevented their reproduction for this chapter. It can be seen in Figure 57.11 that the smoke locations correspond to MOS errors representing surface cooling of 2° to 6°C and there are no large areas of positive MOS errors. For all other times, such patterns do not appear, and the MOS errors are much smaller in amplitude and scale (see, for example, Figure 57.12).

Yellowstone Fires of 1988

On 7 September 1988 forest fires already burning in Yellowstone National Park in northwestern Wyoming erupted into a massive conflagration, pumping a tremendous smoke plume into the atmosphere (Figure 57.13a). The next day, the fires died down, but the smoke generated on 7 September had spread to cover the midwestern United States (Figure 57.13b). Three days later, on 11 September, as the attention of weather watchers was riveted on the approach of Hurricane Gilbert, the smoke from the fires of 7 September was clearly seen as it passed over New York City on its way into the Atlantic Ocean (Figure 57.13c). Again, the smoke clouds were easy to detect with visible wavelength imagery, but were invisible in infrared imagery (Figure 57.13d).

For 8 and 9 of September MOS errors were calcu-

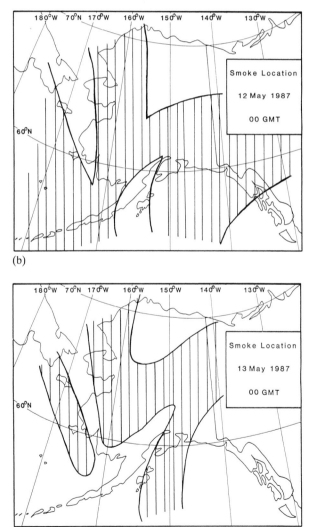

Figure 57.10 Location of smoke as determined from visible satellite imagery at 0000 GMT, for (a) 11 May 1987, (b) 12 May 1987, and (c) 13 May 1987.

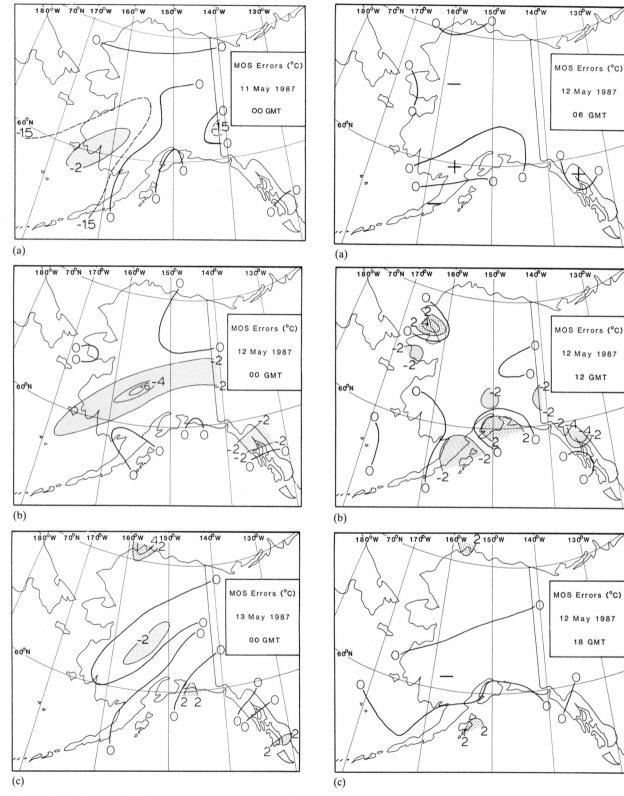

Figure 57.11 Errors of MOS surface temperature forecasts for for 0000 GMT, for (a) 11 May 1987, (b) 12 May 1987, and (c) 13 May 1987. Contours are every 2°C. Errors less than −2°C are shaded. Errors greater than 2°C are shaded with grainier shading.

Figure 57.12 Errors of MOS surface temperature forecasts for (a) 0600, (b) 1200, and (c) 1800 GMT, 12 May 1987, showing smaller amplitude and scale of errors as compared to 0000 GMT error patterns, as shown in Figure 57.11.

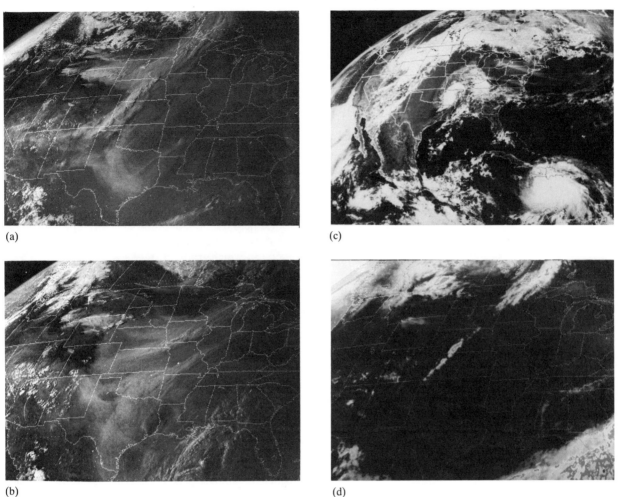

(a)

(c)

(b)

(d)

Figure 57.13 (a) Geostationary satellite image in visible wavelengths (Channel 1) for 7 September 1988 at 2331 GMT (1731 local Mountain Daylight Time, MDT) showing smoke from the Yellowstone forest fires. (b) As in (a) for 8 September 1988, at the same time. Note the smoke clouds covering most of the Great Plains. In Colorado the western half of the state is black, indicating no smoke, since the Front Range of the Rocky Mountains, which runs north-south through the center of the state, blocks the smoke. (c) As in (a) for 11 September 1988 at 1931 GMT (15:31 local Eastern Daylight Time). Note smoke seen as gray streak from Michigan over northern Pennsylvania and New Jersey out into the Atlantic Ocean. Remnants of Tropical Storm Florence are seen over Arkansas, and Hurricane Gilbert is at bottom, south of Hispaniola. (d) Geostationary satellite image in infrared wavelengths (Channel 4) for 7 September 1988 at 2301 GMT (17:01 MDT), 30 minutes before the visible image in (a). Note that the only portion of the smoke cloud visible in this image is the very thickest smoke directly over the fire in northwest Wyoming and water clouds associated with the cold front running from southwest to northeast in western Kansas, southern Nebraska, and northern Minnesota.

Figure 57.14 Location of MOS stations used for analysis shown in Figure 57.15.

lated every six hours based on the most recent MOS forecasts made at 0000 GMT. The MOS station locations are shown in Figure 57.14, and the error patterns for three of the times are shown in Figure 57.15. It can be seen that during the daytime, there is a large negative MOS error under the smoke, up to 7°C, indicating a cooling effect of the smoke. At night, there is no net effect.

To test the robustness of the above results, the MOS errors for the same times, but based on MOS forecasts from four different LFM runs, were examined. For example, for the 0000 GMT 9 September errors, MOS forecasts were used, based on LFM runs at 0000 GMT on 7 September, 12 GMT on the 7th, 0000 GMT on the 8th, and 1200 GMT on the 8th. In all cases, the error patterns were virtually identical. Therefore, the error patterns shown are not a quirk of a particular numerical forecast. Since the errors are close to 0 for the areas not under the smoke, and only isolated stations have errors with absolute value greater than 2°, with no large-scale biases, the error patterns shown are representative of the radiative effect of the smoke.

The cooling effect of forest fire smoke shown here is larger than that found previously for smoke which had traveled several thousand kilometers already. This is to be expected, as the optical depth of the smoke is larger in the present case. The only case where larger surface temperature effects were found from smoke was exactly one year earlier in the Klamath River

Canyon of California where smoke was trapped by an inversion in the valley and the smoke cloud became much thicker than the one here (see next section).

California Fires of 1987

On 30 August 1987, orographic thunderstorms in northern California and southern Oregon ignited severe forest fires that burned for more than one month. In one month, 203 km^2 of forest burned in northern California. (This is less than 0.1% of the area that might burn in a nuclear holocaust.) For the first 2 weeks of the fires, with the exception of a weak cold frontal passage on 2 September, a high pressure system covered the area. The result was a subsidence inversion which trapped smoke in the mountain valleys, particularly in the Klamath River Canyon running from Happy Camp at the north to Orleans at the south (Figure 57.16). A detailed description of this case is given by Robock (1988a).

Each day, more smoke accumulated beneath the inversion, with the surface cooling produced by the blockage of sunlight strengthening the inversion, hence trapping more smoke (personal communication, C. Fontana, D. Willson, and D. Gettman, 1988). The smoke has a higher albedo than the wooded surface (Figure 57.17). This results in a net cooling of the entire atmosphere-surface system. Virtually all the sunlight that is not reflected by the smoke is absorbed before it reaches the ground, thereby strongly cooling the ground while slightly heating the air, although not enough to destabilize it with respect to the synoptic scale inversion. This positive feedback effect of the smoke enhanced both the amplitude and duration of the cooling.

Surface-air temperature data were examined for the region between 39°N and 45°N and west of 120°W to the Pacific Ocean, in northern California and southwestern Oregon (Figures 57.16). For this region data were available from 96 National Weather Service stations for which 30-year normals have been computed, another 65 National Weather Service climatological observing stations, and 85 Forest Service stations, for a total of 242 stations (Figure 57.16). Not all of these stations made observations each day during this period, but for September, it was possible to use more than 70 stations in the region to calculate deviations from normal and more than 200 stations to calculate daily maximum, minimum, and range of temperatures.

Since there are only a few MOS stations in this area, and several of them are on the coast, where errors would be dominated by local circulations, the MOS

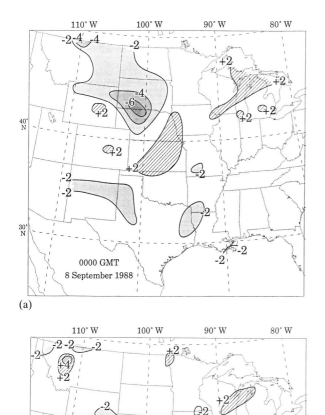

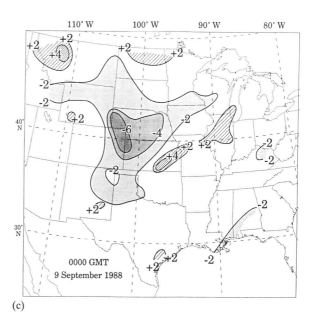

(a)

(b)

(c)

Figure 57.15 (a) MOS errors (°C, observations minus MOS forecasts) for 0000 GMT 8 September 1988 (18:00 local time: MDT, 7 September), only 29 minutes after the image shown in Figure 57.13a, based on the MOS forecasts made 0000 GMT on the 7th. Contours are every 2°C, with the 0°C line left out. Errors less than −2°C are shaded, and errors greater than +2°C are striped. Note the negative MOS error under the smoke-covered region, indicating the cooling effect of the smoke. (b) As in (a) for 1200 GMT 8 September 1988 (06:00 local time), based on the MOS forecasts made 0000 GMT on the 8th. For this early morning period, there are no large scale errors, indicating no net radiative effect of the smoke. (c) As in (a) for 0000 GMT 9 September 1988 (18:00 local time, 8 September), only 29 minutes after the image shown in Figure 57.13b, based on the MOS forecasts made 0000 GMT on the 8th. Again, note the negative MOS error under the smoke-covered region, indicating the cooling effect of the smoke during the daytime.

error technique was not applicable in this case. Since the temperature signature was so large, however, the anomalies with respect to climatology served to isolate it.

On 7 September 1987, which was typical of all the days from 4 September through 12 September, a region of large negative anomalies of maximum temperature can be seen in northwest California (Figure 57.18). This is the Klamath River Canyon. At the same time, no minimum temperature anomalies are evident. Anomalies in this region began on 2 September and lasted through 22 September, with only a

brief respite on 14 and 15 September due to a strong cold frontal passage.

Maximum temperature anomalies were analyzed for Happy Camp and Orleans, California, which were in the smoke-filled canyon, and Medford, Oregon, nearby to the north, which was not (Figure 57.19). Orleans, near the mouth of the canyon, did not cool quite as much as Happy Camp because it experienced some ventilation from the ocean to the west. The August average temperatures should also be compared to those in Figures 57.18 and 57.19. The maximum temperature in Happy Camp on 30 August, the day of

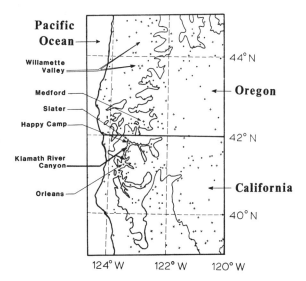

Figure 57.16 Station locations in California for maximum and minimum temperature observations of September 1987 forest fire effects.

Figure 57.17 Satellite image for 2115 GMT (local time—2:15 P.M. PDT). The smoke can be seen at the northern border of California.

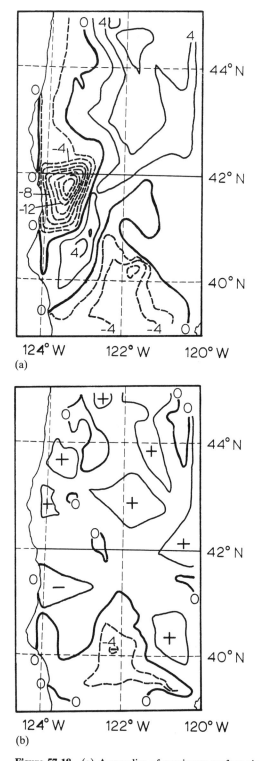

(a)

(b)

Figure 57.18 (a) Anomalies of maximum surface-air temperature (maximum temperature *minus* normal maximum temperature) for 7 September 1987. Contours are every 2°C. The 0°C contour is thick. Negative contours are dashed. (b) Anomalies of minimum surface-air temperature for 7 September 1987.

the lightning that started the fires, was 42.2°C, so the cooling shown in the figures was even larger than the difference from the normal maximum temperature, although some of this cooling may have been due to synoptic variation.

Northerly winds on 16 September blew the smoke out of the Klamath Canyon into a long plume moving southward off, and parallel to, the California coast. This ended the most dramatic effect of the smoke in the Klamath River Canyon. For the next week cooling persisted in the Klamath Canyon but was not as large as for the earlier period.

The effects of the smoke trapped in the Klamath

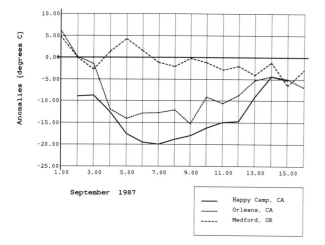

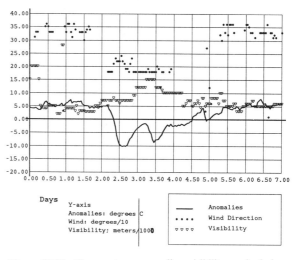

Figure 57.19 Maximum temperature anomalies for Happy Camp, Calif., Orleans, Calif., and Medford, Ore., for 1–16 September 1987. See Figure 57.16 for station locations.

Figure 57.20 Temperature anomalies, visibility, and wind direction for Cuiabá, Brazil, 9–15 September 1987.

River Canyon were dramatic for those living there. By the end of the first week, more than 400 persons per day were being treated for respiratory problems. Tomato plants in gardens in Happy Camp died and produced no fruit.

A mechanism has been identified which enhances the surface cooling effect of forest fire smoke. Smoke was trapped in a valley by an inversion which was strengthened by the surface cooling. This strengthening trapped more smoke, which produced more cooling, creating a positive feedback. Only a strong synoptic-scale front was able to finally destroy this amplifying cycle. It had been previously suggested (Turco et al., 1983; Veltishchev et al., 1988) that while elevated aerosol layers would produce cooling, layers at the surface would have a net warming effect. In this case, because of the high albedo of the smoke, not only did an aerosol layer at the surface produce cooling, but it enhanced the cooling.

Amazon Fires of 1987

During August and September 1987, just after the Great Black Dragon Fire and during the northern California fires, large smoke clouds covered the Rondonia region of the Amazon forests of Brazil due to deforestation burning. Carlos Nobre provided hourly surface data for the stations of Cuiabá and Vilhena for these two months. Since MOS forecasts and the climatologies for these stations were not available, means were calculated for each hour for this 61-day period, and then anomalies with respect to these means were calculated (Stutzer, 1990).

Figure 57.20 shows one of several cases of dramatic cooling that occurred during this period. It can be seen that, in contrast to the cases examined above, the cooling occurred during periods of high visibility, while the warmer period had low visibility, implying that it was warmer when smoke was present than when it was not. On further examination of this figure, it can be seen that winds were southerly during the clear periods, implying a cold front with clear cold-air advection behind the front producing both the cooling and the clearing. This was confirmed by examining weather maps for this period. The same pattern was found for other coolings during this period.

Thus, synoptic-scale advection provides a dominant control on surface temperature during this smoky period. No radiative effect of the smoke is seen. This does not prove that smoke does not cause cooling in the tropics. The question is still open. It could be that the temperatures for the entire period were depressed due to the smoke, and that anomalies with respect to the mean for this period could therefore not show a smoke effect. Further analysis is needed.

Conclusions

The results reported here correspond to theoretical estimates of the effects of smoke and serve as observational confirmation of a portion of the nuclear winter theory. If this smoke had been more absorbing, characteristic of smoke from burning petroleum or plastics such as would result from a burning city or industrial facility, the surface temperature effects would have been even larger (Westphal and Toon, 1990; Turco et al., 1990). This reminder of the validity of the nuclear winter theory should serve as a warning to society. Although the Cold War appears to be over, the nuclear arms race is not. Even proposals to reduce strategic arsenals by 50% will not rid us of the threat of global starvation should nuclear weapons be used in warfare. As described by Robock (1989), policy implications of nuclear winter must be taken seriously by society, which must work toward the elimination of nuclear weapons. The only ultimate solution is to learn to routinely resolve conflicts in nonviolent ways.

These results also imply that smoke from biomass burning can have a daytime cooling effect of a few degrees over seasonal time scales. In order to properly simulate the present climate with a numerical climate model in regions of regular burning, it may be necessary to include this smoke effect.

Acknowledgments

I thank Rich Ferrare for valuable discussions of the 1982 case and for providing preliminary satellite photos, Mike Matson and Will Gould for information on availability of satellite images of smoke, Russell Schneider for providing information on and a satellite loop of the 1981 case, Harrison Salisbury and George Golitsyn for valuable discussions of the China/Siberia case, the NOAA/Navy Joint Ice Center for access to satellite images of Alaska, Paul Dallavalle for MOS forecasts and surface temperature observations, Ed Eloranta for providing in situ smoke cloud data taken 2 August 1982 over Maryland, Gil Grodzinsky for analyzing the Yellowstone data, Carlos Nobre for providing data for the Amazon case, David Stutzer for analyzing the Amazon data, Sunny Bae and Qing Xiao for plotting the maps, Marco Rodriguez and Nono Kusuma for drafting the figures, and Sherri West for proofreading the manuscript. Data analysis, plotting, and word processing were done on the Department of Meteorology Apollo computer system. This work was partially supported by NOAA grants NA87AA-D-CP003 and NA84-AA-H-00026, which were funded by the Defense Nuclear Agency and by the University of Maryland Undergraduate Apprenticeship in Research and Scholarship Program.

A Study of Climate Change Related to Deforestation in the Xishuangbanna Area, Yunnan, China

Chungcheng Li and Cong Lai

Xishuangbanna (21°8′ to 22°36′N, 99°56′ to 101°50′N) is a Dai Nationality Autonomous Prefecture consisting of the three counties of Jinghong, Menghai, and Mengla, bordered by Burma and Laos in southwestern China. Out of its total 1,920,000 hectare (ha) area, 94% is mountain area and 6% is basin. Located south of the Tropic of Cancer, this region has a warm and humid climate: The annual sunlight time is between 1900 and 2200 hours; the annual mean temperature is between 15° and 23°C; annual relative humidity is 78% to 88%; there are 180 to 250 windless days and 90 to 180 thick fog days yearly, of which 75% to 90% appear during the dry season (December through April); and most of its total 1000 to 2500 millimeter (mm) precipitation is concentrated in the wet season (May through October). Because of the favorable living conditions provided by the long hours of daylight and abundant rainfall, together with the varied topography, this region is the most important and largest tropical ecosystem in China's mainland. Its huge virgin jungles are rich in plants and animals: There are more than 4000 species of plants, along with more than 400 species of birds and 540 species of land animals, which are one-third and one-fourth, respectively, of the total number in China. Therefore, this area is known as the "Green Treasure on the Tropic of Cancer" and the "Kingdom of Plants and Animals of China."

With economic development and the increase in population, the situation has been changed greatly in recent decades. The tropical rainforest in this region is acutely endangered as are similar rainforest areas elsewhere on the planet. This chapter is one of the research series on the deforestation in Yunnan Province, China, and its possible impact on climate change. We begin with a brief review of deforestation, then evaluate the meteorological data in detail.

Deforestation

Xishuangbanna is historically a region fully covered by rainforest. Until the early 1950s, 62.5% of the total

area was still covered by forest. Unfortunately, this relatively large figure has been decreased rapidly in the last 30 years. Table 58.1 shows the statistics of the forest decrease in this area, which shows the forest destruction rate. We can also see this change in Figure 58.1.

Our investigation shows that the situation is getting serious because of the increasing population and expanding economy. According to a recent statistic, in the late 1980s, the total wood consumption in this prefecture is 3.2 million cubic meters, all of which comes from deforestation. Figure 58.2 gives the distribution of the consumed wood, from which we can see that only 19.69% of the total consumed wood is used for timber purposes. We can also find out the causes of deforestation of this prefecture from Figure 58.2. Similar to the other rainforest areas on the earth, the causes of deforestation can be divided into the following categories: (1) deforestation for cultivable land to meet the requirement of economic development and population increase, (2) deforestation for fuel wood to meet the local people's fuel demand, (3) destruction by forest fires, and (4) deforestation for timber harvesting.

The most important factors for forest destruction in this area is the demand of cultivatable land. First, this area is one of a few areas where tropical economic plants like rubber, coffee, cocoa, shellac, cinchona, etc., can grow in China. Second, the population of this region has increased more than three times in the last 30 years (from 200,000 in the early 1950s to 650,000 in the 1980s). To meet the demand of the economic and population increase, 200,000 ha of forest have been destroyed in the past 30 years. The second factor of the forest destruction is forest fire. Slash and burn is the traditional cultivation method of the local people. Although it has been gradually changed with the changing of their life-style, burning forest to form farmland still happens frequently, especially in the deep forest area. In recent years, it has become one of the major causes of forest fires. In just the years of 1973 to 1979, there were 1,254 forest fires,

Table 58.1 The forest area change in the Xishuangbanna area

	1950s	1960s	1970s	1980	1982
Forest area ($\times 10^6$ ha)	1.2	0.84	0.61	0.55	0.47
Forest covering ratio (%)	62.5	43.8	31.8	28.6	24.7

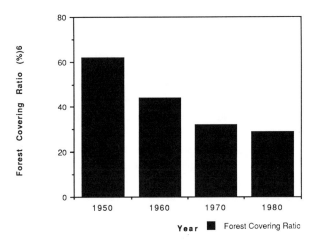

Figure 58.1 The change of forest covering ratio of Xishuangbanna.

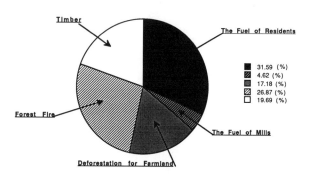

Figure 58.2 The annual consumed wood in the Xishuangbanna area.

which destroyed 46,000 ha of forest. The third factor is energy demand. Wood is the traditional fuel of the local people. As the result of population and economic increase, this demand increased rapidly in recent years: In the 1980s, this prefecture needed 870,000 and 280,000 cubic meters (m^3) of wood as fuel wood for cooking and industry separately each year. To meet this huge demand, 6,700 ha of forest are being felled yearly.

Climate Change

In order to understand the possible impact of deforestation to climate change, we collected the last 30

Table 58.2 Some climate factors of the Xishuangbanna area

	1950s	1960s	1970s	1980s
Precipitation (mm)				
Jinghong	1206 mm	1209 mm	1188 mm	1161 mm
Menghai	1518	1376	1366	1327
Mengla	1635	1485	1579	1563
Average	1453 mm	1357 mm	1378 mm	1350 mm
Temperature (°C)				
Jinghong	21.4°C	21.8°C	21.9°C	22.4°C
Menghai	17.6	18.3	18.1	18.5
Mengla	20.6	20.9	21.1	21.5
Average	19.9°C	20.3°C	20.4°C	20.8°C
Relative humidity (%)				
Jinghong	84%	83%	82%	79%
Menghai	84	82	82	80
Mengla	86	85	86	83
Average	85%	83%	83%	81%
Number of fog occurrence days (yr^{-1})				
Jinghong	166	129	118	120
Menghai	141	139	138	120
Mengla	130	151	163	144
Average	146	140	139	117

years of meteorological data from the local meteorological stations of the three counties. The historical data go back to the early 1950s when the stations were established. After smoothing these data and setting the data of the 1950s as a reference, we reached some meaningful results. Table 58.2 presents some of these results, it shows clearly that the climate of this region has been changed with the forest destruction.

Temperature

Figure 58.3 is the temperature increase of this region in the past 30 years. Compared with 1950s, we can see a general increase. In the early 1950s, the annual mean temperature of this region was 19.9°C; it kept going up in the 1960s and 1970s; in the first five years of the 1980s, it reached 20.3°C. Compared with 1950s, the temperature of the 1980s was increased by 0.9°C. We also noticed that this tendency is not affected by the season, although there are seasonal differences in the temperature increase. The increase is 1.20°C in the dry season and 0.8°C in the wet season. From Figure 58.4, which is a typical temperature increase curve of Jinghong County, we can see how the temperature was increasing in that region.

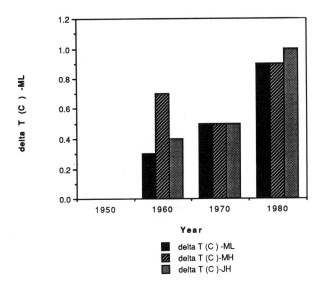

Figure 58.3 Temperature variation of Xishuangbanna area.

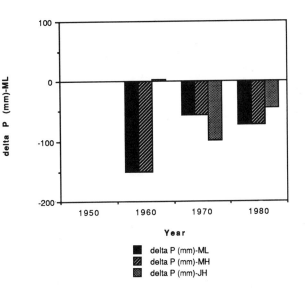

Figure 58.5 Precipitation variation of Xishuangbanna area.

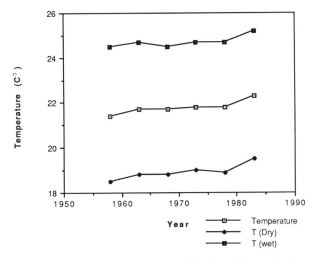

Figure 58.4 Five-year temperature of Jinghong County during 1958 to 1983.

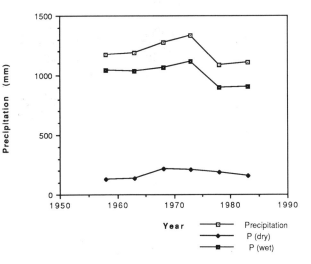

Figure 58.6 Five-year precipitation of Jinghong County during 1958 to 1983.

Precipitation and Relative Humidity

Analysis of the meteorological data also indicated that the precipitation and relative humidity of this region have changed. The data shows that during the past 30 years, both precipitation and relative humidity went down: In the 1950s, the annual mean precipitation was 1453 mm; it became 1357 and 1378 mm in the 1960s and 1970s; and in the first five years of 1980s, the precipitation was 1350 mm, which is 88 mm less than the 1950s. The trend of precipitation decrease in this region is shown in the Figure 58.5. An interesting fact we noticed is that despite the annual mean precipitation decrease, it still showed an increase in the dry season. Figure 58.6 shows how the seasonal precipitation changes with the annual value. Compared with the 1950s, the precipitation in the dry season of early 1980s increased by 20.8% but decreased by 6.8% in wet season. The largest difference is found in Jinghong county: The precipitation increase in the dry season is 45.3% and the decrease in the wet season is 9.4%. However, we did not observe seasonal differences in the change of relative humidity—it has kept going down at both dry and wet season during the past 30 years. The tendency can be seen in the Figure 58.7. For the whole region, the change of relative humidity can be seen in Figure 58.8.

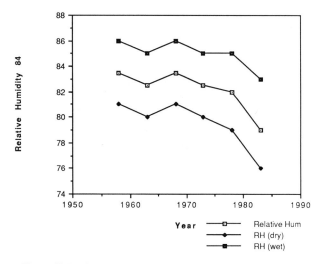

Figure 58.7 Annual relative humidity of Jinghong County during 1958 to 1983.

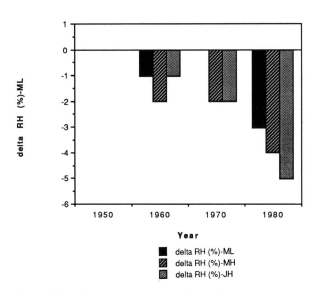

Figure 58.8 Relative humidity variation of Xishuangbanna area.

Fog Formation

Thick fog is a characteristic of the regional weather. According to observations, fog appears almost half the year. However, with the forest decrease, fog has become less and less frequent. The average number of fog days was 146 in the early 1950s—in the 1980s, it became 117 days. The other obvious change is its duration. Table 58.3 gives the description of these changes. We also find the changes in the fog amount: In the 1950s, the surface fog was usually as thick as 1000 m with the maximum amount at 0.4 to 0.5 mm,

Table 58.3 Fog occurrence variation of Jinghong country (January)

	1950s	1960s	1970s	1980s
Forming time (hr:min)	00:30	03:00	04:10	05:00
Dissolve time (hr:min)	08:39	09:22	10:00	09:12
Duration (hr)	8.1	6.7	5.8	4.5

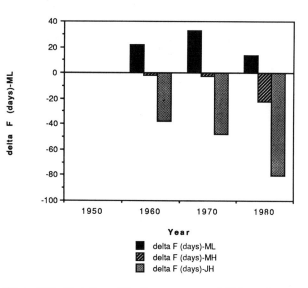

Figure 58.9 Variation of the fog occurrence of Xishuangbanna area.

Table 58.4 Temperature variation of Xishuangbanna area

	1950s	1960s	1970s
Annual temperature variation (°C)			
Jinghong	0.5°C	0.7°C	1.2°C
Menghai	—	0.8	0.9
Mengla	0.1	0.7	1.0
Extreme lower temperatures (°C)			
Jinghong	4.2°C	4.4°C	2.7°C
Menghai	—	−2.7	−5.4
Mengla	3.6	3.2	0.5
Occurrence of temperatures lower than 5°C (decade^{-1})			
Jinghong	1	3	5
Menghai	—	1	1
Mengla	1	2	5

Table 58.5 The results of t-test to the climate factors of Jinghong county

	CK	A	B	C	D	E	Diff. Ave.	t_{cal}
Temperature (°C)								
Annual	21.4	21.7	21.7	21.8	21.8	22.3	+ 0.46	4.11[a]
Dry season	18.5	18.8	18.8	19.0	18.9	19.5	+ 0.5	3.84[a]
Wet season	24.5	24.7	24.5	24.7	24.7	25.2	+ 0.26	2.23[b]
Precipitation (mm)								
Annual	1173	1190	1275	1332	1085	1108	+ 26	0.53
Dry	132	140	221	208	190	159	+ 51	3.45[a]
Wet	1043	1038	1066	1119	897	907	− 38	0.86
Relative humidity (%)								
Annual	84	83	83	82	82	79	− 2	2.99[a]
Dry	81	80	81	80	79	76	− 2	2.09[b]
Wet	86	85	86	85	85	83	− 1	2.45[b]
Numbers of fog occurrence days								
Annual	168	140	124	123	119	88	− 49	5.79[c]
Dry	125	103	94	98	86	80	− 33	7.84[c]
Wet	45	37	27	28	30	28	− 15	8.35[c]

Note: CK—Reference group (1954–1958), A—1959–1963, B—1964–1968, C—1969–1973, D—1974–1978, E—1979–1983.
a. Confidence = 0.05.
b. Confidence = 0.1.
c. Confidence = 0.01.

but in the 1980s, the thickest fog is only 700 to 800 m, and there is little fog condensation. The exception is the county of Mengla which is located farthest south. The reason is not clear. Figure 58.9 shows the change of fog occurrence in the whole region.

Temperature Variance

Another factor we want to point out is the change of the stability of temperature. Despite the increasing annual mean temperature, the frequency of low temperatures increased. If we consider a temperature lower than 5°C, which can cause damage to the local plantation, as a threshold of cold damage, we can find from Table 58.4 that this kind of damage is increasing rapidly with the increase of the temperature variance (from 0.5°C in the 1950s to 1.2°C in the 1970s).

To further understand the climate change of this region, we chose the county of Jinghong, which has the most complete meteorological data and the largest increases in deforestation, population, and industry as an example for the further analysis. Table 58.5 is the results of the t-test to the local meteorological data. By this test, we find that compared with the reference data group (data of 1954 to 1958), during the past 30 years, the annual mean temperature increased by 0.46°C, annual relative humidity decreased by 2%, and the number of fog occurrence days decreased by 29%. We also find there are seasonal differences among these changes: In the wet season, the temperature average increased by 0.26°C, the relative humidity decreased by 1% and the fog days decreased by 38.6%; in the dry season, the temperature increase is 0.5°C, precipitation increase is 38.6%, relative humidity decrease is 2%, and the decrease of the fog occurrence days is 26.4%. Therefore, the changes in the precipitation of annual and wet/dry season values are different.

Conclusion

The analysis of the results of deforestation and the meteorological data of the Xishuangbanna region shows that there are possible relations between the deforestation and climate change. With the forest area decreased by 33% during the past 30 years, the climate of this region has also been changed. The annual mean temperature has been increased by 0.7°C, of which the increase is 0.97°C in the dry season and 0.53°C in the wet season. Together with the annual temperature increase the temperature variations have also been increased, which has resulted in more frequent low temperature damage to the local plantation agriculture. The relative humidity decreased by 3% annually; and the annual precipitation also decreased, with a decrease in the wet season of 6.8% and an increase in the dry season of 20.8%.

Acknowledgments

We thank Dr. Michael R. Rampino for reading and correcting the manuscript. We also thank Dr. Martin I. Hoffert and the members of the Earth System Group of the Department of Applied Science, New York University, for helpful discussions. A special thanks is due to Dr. Joel S. Levine of the NASA Langley Research Center. This research is partially supported by NASA grant number NAGW-1 697.

IX

Biomass Burning: Historic and Prehistoric Perspectives

Major Wildfires at the Cretaceous-Tertiary Boundary

Edward Anders, Wendy S. Wolbach, and Iain Gilmour

The first hint of a fire at the Cretaceous-Tertiary (K/T) boundary was observed by Tschudy et al. (1984) at a Raton Basin site in Colorado. They found "large amounts of fusinite"[1] in the basal coal layer containing the Ir anomaly and attributed it to "periods of fire consuming the vestigial or dead organic matter." Evidence for the worldwide nature of these fires was obtained by Wolbach et al. (1985), who found soot at K/T sites in Denmark, Spain, and New Zealand, and attributed it to global fires of biomass (and perhaps fossil carbon) triggered by the impact. We review the current status, updating our most recent paper on this subject (Wolbach et al., 1990b).

Evidence for a Global Fire

Measurement of Soot

The "insoluble carbon" remaining after acid dissolution of a sediment is usually dominated by kerogen, which prevents recognition, let alone measurement, of soot (Figure 59.1a and 59.1b). A perfect separation is impossible, since kerogen and elemental C form a structural continuum, with aromatic ring systems growing ever larger while H and other foreign atoms are progressively eliminated. The original extraction procedure improvised by Wolbach et al. (1985) was too gentle, recovering all of the soot and other elemental C but leaving variable amounts of kerogen undissolved. Wolbach et al. (1988a) used a harsher but more effective procedure involving dichromate oxidation under controlled conditions (Wolbach and Anders, 1989). It destroys all the reactive kerogen and much of the resistant kerogen, while etching off only a small, easily correctable, fraction of the elemental C. The etched sample gives the abundance and $\delta^{13}C$ of elemental C, whereas the unetched sample gives analogous data for kerogen (by difference). The mass fractions of soot and coarse carbon are determined by planimetric analysis on the SEM.

The "coarse carbon" has platy or pitted morphology (Figure 59.1c). It can include charcoal, detrital sedimentary carbon, and resistant (terrigenous) kerogen. Detrital carbon is hard to distinguish from charcoal, but usually is only a minor component.[2] Resistant kerogen can be largely or entirely destroyed by a long, 600-hour (hr) etch, but this has been done only for a few samples, mainly boundary clays (Wolbach et al., 1990a). Thus our elemental C values, except for the boundary clay, may be too high by variable factors.

Carbon Profiles across the K/T Boundary

Figure 59.2 shows C and Ir profiles for the exceptionally well-preserved Woodside Creek site, which has a very thin (0.6 centimeter, cm) boundary clay layer with a very sharp Ir spike (Brooks et al., 1984; Wolbach et al., 1988a). Ir, elemental C, and especially soot rise steeply at the boundary, by factors of 1400, 210, and 3600. (The rise may be steeper than indicated in the graph, since both the -4.5 to 0 and -1.5 to 0 cm samples abut the boundary clay and may have been contaminated by slight mixing, imperfect sampling, or Ir diffusion.)

Similar enrichments of Ir, C, and soot are found at Chancet Rocks and even at the severely burrowed and sheared Stevns Klint site, which shows a ≥ 340-fold increase in soot[3] (Table 59.1). Wolbach et al. (1985) had found only a four-fold increase in total C at the latter site (mainly due to incomplete removal of kerogen; see Wolbach et al., 1988a, 1990a) but argued that the actual difference was as large as $\sim 10^3$ if one took into account the very short deposition time—≤ 1 year (yr)—of the boundary clay. Opponents of the impact theory balk at accepting such short deposition times, apparently failing to realize that Stokes' Law, which allows volcanic ash to settle in weeks or months (Ledbetter and Sparks, 1979), also applies—with majestic impartiality—to impact ejecta, ensuring equally prompt fallout for a given particle size and water depth. Moreover, a detailed calculation based on the actual size distribution of quartz grains in boundary clay from Elendgraben (Austria) gives an *e*-folding time of 0.58 to 0.75 yr for primary K/T

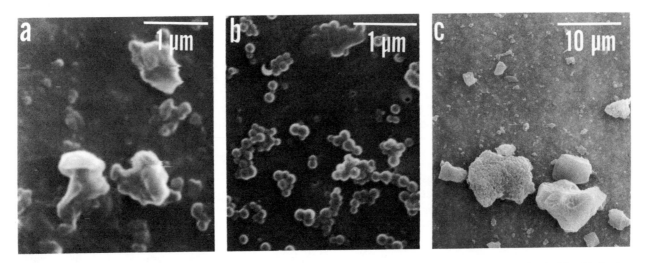

Figure 59.1 SEM photographs of demineralized carbonaceous fraction from K/T boundary clay at Woodside Creek, New Zealand. (a) Soot particles are barely discernable through the kerogen film. (b) After removal of kerogen by a 60 hr dichromate etch, soot spherules, with their characteristic "chained cluster" morphology, become clearly visible. (c) Coarse carbon, of platy or pitted morphology. It is mainly charcoal in this sample, with minor amounts of resistant (terrigenous?) kerogen and detrital, metamorphic carbon. Some detrital minerals—mainly rutile—are also present.

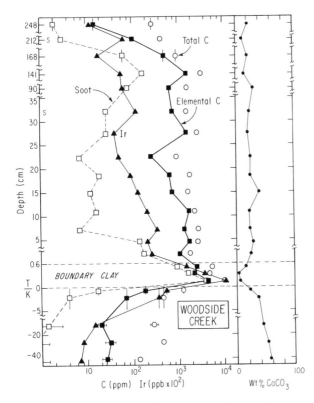

Figure 59.2 Abundance profiles across the K/T boundary at Woodside Creek. Note expanded scales from −5 to +5 and especially 0 to 0.6 cm. Ir, soot, and elemental C all rise at the boundary by 2 to 3 orders of magnitude, and then slowly revert to Cretaceous levels. Soot appears in the basal 0.3 cm of the boundary clay, showing that the fires were in progress before the primary fallout had settled. (From Wolbach et al., 1988a.)

Table 59.1 Iridium, carbon, and soot enrichments at the K/T boundary

	Increase at K–T boundary over cretaceous values		
Site	Ir	C	Soot
Woodside Creek, NZ	1400	210	3600
Chancet Rocks, NZ	290	≥790[a]	≥660[a]
Stevns Klint, DK	2000	≥1200[b]	≥340[b]
Agost, E	—	37	—
Sumbar, SU	55	10	≥610

a. As the Cretaceous values were smaller than their errors, we used 2σ upper limits.
b. Cretaceous marl from Nye Kløv, 190 km away, reanalyzed by Wolbach et al. (1988a, 1990). The residue of 77 ppm C remaining after 60 hr oxidation with $Cr_2O_7^=$ disappeared after 132 hr, and thus must have been resistant kerogen rather than elemental C, which has $t_{1/2}$ ~600 hr. Some of the elemental C and soot values cited by Wolbach et al. (1988a) for Stevns Klint or Nye Kløv are in error, as the soot values were based on the mass of the 0.08 to 3 μm size fraction rather than on planimetric measurement of particles with soot morphology. The correct values, in ppm, are (soot in parentheses): Cretaceous ≤2 (≤2), boundary clay 2500 (680), Tertiary ≤430 (17).

fallout (Eder and Preisinger, 1987). But even without this additional factor, the new carbon and soot values show striking enrichments at the boundary that cannot be explained by condensation of the sediment or by other enrichment processes (Anders et al., 1986).

In the Tertiary, Ir, elemental[4] C, and soot all remain elevated to at least ~2 meters (m), reverting to Cretaceous levels only above 213 cm (Figure 59.2). A similar trend for Ir alone has been seen at other sites (Alvarez et al., 1980; Kastner et al., 1984; Preisinger et al., 1986; Strong et al.,1987) and is usually interpreted in terms of two kinds of fallout: *primary*, represented by the boundary clay or even only its basal layer (Kyte et al., 1985), and *secondary*, represented by the remaining material of elevated Ir content (Preisinger et al., 1986). This secondary fallout apparently was eroded from elevated sea-floor sites and redeposited in topographic lows. Such lows have the best chance of retaining their primary fallout, but at the price of becoming sinks for secondary fallout. For this reason, carbon and Ir values should be integrated only for the boundary clay.

Coagulation of Soot with Ejecta

At Woodside Creek soot correlates not only with Ir but also with As, Sb, and Zn throughout the boundary clay and for the next 1.4 cm in the Tertiary (Figure 59.3). This correlation is very surprising in view of their separate origins: Ir from the meteorite, instantaneously; soot from fires, gradually over the next few weeks; As, Sb, and Zn, mainly adsorbed from seawater during final descent, perhaps with some contribution from the crater (Gilmour and Anders, 1989; Hildebrand and Wolbach, 1989).

The most likely explanation is that soot and Ir-bearing ejecta coagulated in the stratosphere, and then swept out As, Sb, and Zn from ocean water, which had turned anoxic due to the mass death of plankton (Gilmour and Anders, 1989; see this reference for alternative origins of As, Sb, and Zn). The initial coagulation may have been aided by the opposite electric charges of the soot and rock dust.

The soot/Ir ratio remains nearly constant for the first 6 cm of the Tertiary but then fluctuates widely about the mean (Figure 59.2; see also Wolbach et al., 1988b). Perhaps soot and Ir gradually became decoupled during lateral transport and redeposition, while the impact glass weathered to clay (Kastner et al., 1984).

This correlation needs to be checked at other undisturbed sites. The only check made thus far (Ir-soot at Sumbar; Wolbach et al., 1990a), showed some divergence, with Ir and soot peaking at 0 to 1 cm and 7 cm, respectively. No data were available for As, Sb, and Zn. However, two resistant kerogen components are present at Sumbar, which interfered in the soot measurement and perhaps even partly decoupled soot and ejecta during settling through the water column.

Global Distribution of Carbon at the K/T Boundary

Table 59.2 summarizes carbon data for boundary clays at 11 sites. All show substantial amounts of elemental C of fairly constant δ^{13}C and percentage of soot. Seven sites have 11% to 19% soot, but the three New Zealand sites are higher (55% to 69%) and Stevns Klint is intermediate (27%).

The surface abundances of K/T carbon and soot show remarkably little variation among sites (Figures 59.4a and 59.4b). If we exclude Elendgraben, where part of the C was dissolved by an overly harsh procedure, the range for soot and C is only ~10^1, compared to ~10^2 for Ir. Part of the variation may reflect local variations in deposition, preservation, and sampling. Moreover, although these two elements apparently became associated at one time by coagulation of soot and ejecta (Figure 59.3), they did not move in lockstep from beginning to end. Ir was injected into the atmosphere all at once, whereas carbon was injected gradually, and with different lateral and vertical distributions.

The variations in soot/charcoal ratio can also be rationalized. Only smoke plumes from very intense

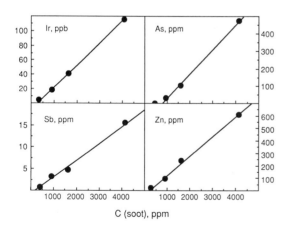

Figure 59.3 In the boundary clay and the next 3 cm at Woodside Creek, soot from biomass correlates tightly with Ir from the meteorite and As, Sb, Zn from terrestrial sources, despite their different origins. Presumably soot coagulated electrostatically with Ir-bearing impact ejecta in the stratosphere, and picked up As, Sb, Zn from vaporized rock, ocean water, or both. (From Gilmour and Anders, 1989.)

Table 59.2 Carbon and iridium at the K/T boundary

Site	Carbon abund.* (mg/cm²)		Soot (%)	$\delta^{13}C_{PDB}$ (%)	Iridium abund. (ng/cm²)	Ref.† C	Ir
Woodside Creek, NZ	**4.8**	± 0.5	69	−25.23	91	1	1
Chancet Rocks, NZ	**35**[a]	+5/−20	68	−25.42	32[a]	1	1
Stevns Klint, DK	11[be]	+11/−1	27	−25.81	250[e]	2	5
Caravaca, E	10[bceh]	+10/−1	14	−25.00	38[e]	2	6
Gubbio, I	13[bceh]	+13/−1	17	−25.48	15[e]	2	7
Agost, E	≥**3.6**[g]	+3.8/−0	11	−26.04[i]	≤24.5	2	8
Raton, USA	**4.0**[ag]	+6.5	13	−26.70[i]	7.2	2	9
El Kef, TN	15[begh]	+15	19	−26.57[ij]	49[e]	2	10
Flaxbourne River, NZ	≤**1.3**[d]	+0/−0.6	55	(−26.08)[d]	134	3	11
Elendgraben, A	>**0.8**[f]	+8/−0	12	(−23.37)[i]	125	2	12
Sumbar, SU	11[g]	±3	15	−25.96[i]	580	4	13
Mean	11	±3		−25.80±0.58			

*Total carbon remaining after dichromate etch (60 h except 600 h for Agost, Raton, El Kef, and Sumbar), corrected to 0 h on the basis of actual or assumed (600 h) half-life. No "background" corrections were made, as the *concentrations* of elemental C in the uppermost Cretaceous are typically 2 to 3 orders of magnitude lower than at the K/T (Wolbach et al., 1990; Wolbach, 1990), and the *fluxes* are much lower still, because of the shorter deposition time of the boundary clay. Some resistant kerogen (generally ≤10% to 20%) may have survived, especially in samples etched for only 60 h. Values in **boldface** are for complete boundary clay layer. C and Ir were integrated only for the boundary clay (or part thereof), excluding secondary fallout, and therefore differ from published values in some cases. Complications and qualifications are indicated by the following code.

a. transition from primary to secondary fallout poorly determined.
b. sample was only a portion of boundary clay layer; error covers missing part.
c. sample thickness was not known.
d. not etched, hence includes kerogen.
e. carbon and Ir analyses were done on different samples, of different thicknesses and sampling location.
f. carbon values low and $\delta^{13}C$ probably fractionated, due to use of H_2O_2 procedure.
g. etched for 600 h, to remove resistant kerogen.
h. only basal layer was measured.
i. sample still brown after 600 h.
j. H_2O_2 treatment gave −25.66§t.

†References:
  (1) Wolbach et al. (1988a)
  (2) Wolbach (1990)
  (3) I. Gilmour, unpublished data
  (4) Wolbach et al. (1990)
  (5) Hansen et al. (1986)
  (6) Smit and Hertogen (1980)
  (7) Alvarez et al. (1980)

  (8) J. Smit, private communication
  (9) A. Hildebrand, private communication
  (10) Kuslys and Krähenbühl (1983)
  (11) Strong et al. (1987)
  (12) Preisinger et al. (1986)
  (13) Alekseev et al. (1988)

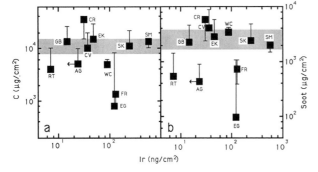

Figure 59.4 Black carbon and soot have fairly constant abundances at 11 K/T boundary sites, averaging 11 ± 3 mg/cm² and 2.2 ± 0.7 mg/cm². Ir shows much larger variations, perhaps reflecting patchier distribution and greater mobility.

fires extend into the stratosphere (Pittock et al., 1986, p. 110; hereafter SCOPE), and other things being equal, small soot particles should rise higher than large charcoal particles. This may explain the high soot percentage at all three New Zealand sites. New Zealand apparently was far away from the K/T impact site, as judged from the small size and scarcity of shocked quartz grains (Bohor et al., 1987), and because it also was far from the main biomass areas, there was ample opportunity for size sorting of smoke and ejecta during transit. The soot enrichment at the New Zealand sites thus may actually signify a charcoal deficiency. Similarly, the soot-Ir correlation at Woodside Creek (Figure 59.3) may represent a winnowing of coarser or low-altitude particles from the soot-ejecta plume, leaving mainly coagulated soot-ejecta particles from the stratosphere.

In summary, it appears that there exists at the K/T boundary a global layer of soot and charcoal, rather uniform in isotopic composition, surface abundance, and soot content. Second-order variations exist, but they are smaller than those for the global Ir layer, and can be explained by variations in deposition, preservation, and sampling.

Is the K/T Soot Layer Unique?

The absence of similar soot horizons elsewhere in the stratigraphic column can never be proved rigorously. However, only 9 of the 50 non-K/T samples examined by Wolbach (1990) have shown detectable amounts of soot, and in at least 37 cases, the detection limit was $\leq 0.01 \times$ of the median value of 505 ppm soot for the K/T boundary.

More comprehensive evidence comes from the work of Herring (1985). He analyzed 186 deep-sea core samples covering the last 70 Ma for charcoal and found only a smooth decline with age, without any spikes indicating major fires. Such spikes should have been readily observable, because a K/T-like carbon layer of 11 milligrams per square centimeter (mg/cm^2) would have contributed 250 mg in one-half of a 3-inch core sample. But no such values appear even among his 7 cores on continental margins, which have the highest charcoal abundances. Among 87 samples covering the last 10^7 yr, each typically representing an interval of about 5×10^4 a, the highest value is only 36 mg and the mean and median are 9.2 ± 7.0 mg and 7.7 mg, respectively—far below the 250 mg expected for a K/T-like soot layer.

Thus no charcoal layers approaching 11 mg/cm^2 have been found anywhere, let alone layers that are (1) global, (2) isochronous, and (3) enriched in soot. The K/T soot layer, coinciding with a very sharp, worldwide marker, appears to be unique.

Sources of Charcoal and Soot

An asteroid or comet is not a plausible source. The total amount of C may be barely adequate, but it is present in combined, not elemental form. Conversion to soot in the fireball is at best inefficient and perhaps impossible, because the gas composition is fairly oxidizing (Wolbach et al., 1985), and any soot formed would combust as the fireball expands into the atmosphere. Moreover, $\delta^{13}C$ of K/T carbon differs from that of meteoritic or cometary carbon, -25 or -16 vs

$+380$§t (HCN in Comet Halley; Wyckoff et al., 1989), but is in the range of photosynthetic C, -20 to -30§t.

Biomass or Fossil Carbon?

The $\delta^{13}C$ values in Table 59.2 are remarkably constant and thus imply a single global source, either intrinsically uniform or homogenized by mixing. The mean of -25.8 ± 0.6§t is suggestive of a biomass source, since it resembles values for natural charcoal from forest fires (43% of the charcoal samples fall between -25 and -26§t; Deines, 1980) and for carbon aerosols from biomass fires, -26.5 ± 2§t (Cachier, 1989), both of which are slightly fractionated relative to the parent plants. However, this is a weak argument, since at least some samples of fossil carbon fall in the same range (oil 4%, coal 18%).

A stronger argument comes from the polynuclear aromatic hydrocarbon (PAH) *retene* (1-methyl-7-isopropyl phenanthrene), which is present in boundary clay at Woodside Creek (Gilmour and Guenther, 1988), Stevns Klint, and Gubbio (Venkatesan and Dahl, 1989), and at DSDP site 605, off the east coast of the United States (Simoneit and Beller, 1987). This compound appears to be diagnostic of resinous wood fires, and probably forms by low-T pyrolysis of abietic acid, a common constituent of plant resins in conifers and some angiosperms (Ramdahl, 1983; Gilmour and Guenther, 1988). Apparently such trees provided at least part of the fuel for the K/T fire.

The boundary clay also is enriched up to 10^3-fold in various PAH with three to six rings (Simoneit and Beller, 1987; Gilmour and Guenther, 1988; Venkatesan and Dahl, 1989). Diagenesis can account for only part of the distribution; the remainder apparently formed by combustion. Lacking alkyl side chains (in contrast to retene), these compounds evidently formed by high-T synthesis from C_2 units rather than by gentle degradation of precursor molecules. Although not diagnostic of any particular fuel, these PAH further strengthen the case for a major fire. The presence of the six-ring PAH coronene (Gilmour and Guenther, 1988; Venkatesan and Dahl, 1989) implies conditions favoring growth of large ring systems, of which soot is the limiting case (a sheet of hexagons, curving onto itself due to the presence of ~ 12 pentagons; Kroto, 1988).

It is not yet possible to estimate the fraction of resinous wood in the K/T fire. The retene/PAH ratio at Woodside Creek (~ 0.01 to 0.02) is smaller than that in air from Elverum, Norway (0.07), where half the total fuel is pine wood (Ramdahl, 1983), and since resinous trees accounted for only a moderate fraction of the Maastrichtian biomass, these numbers are superficially consistent with a dominant biomass source. However, the retene/PAH ratio reflects not only the mass fraction but also the resin content of resinous wood, and—more important—the conditions of combustion. Retene requires low temperatures and limited access of oxygen, and lacking such data for modern fires, let alone the K/T fire, we cannot draw any quantitative conclusions. We merely note that a major contribution of fossil C is not supported by the $\delta^{13}C$ values or the absence of cenospheres.

Required Amount of Fuel

At first sight, biomass carbon would seem to be an ample source, as the amount of elemental C at the K/T, 11 mg/cm^2 (Table 59.2) is much smaller than the present above-ground biomass carbon of 0.2 g/cm^2 (Seiler and Crutzen, 1980), let alone the maximum Cretaceous value of 0.6 g/cm^2 (estimated on the assumption that a land area equal to the present one had the same biomass density as present-day forests, 2 g/cm^2). However, this comparison is misleading; what matters is not the total biomass carbon A but the fraction F that was converted to elemental C and dispersed as smoke ("soot yield"). This fraction is

$$F = A f_b \epsilon f_c$$

where f_b = fraction of biomass carbon burned, ϵ = the fraction of f_b converted to smoke, and f_c = the fraction of elemental C in smoke.

The "baseline" parameters assumed for post-nuclear forest fires and their estimated uncertainties (National Research Council, 1985; hereafter NRC) are: f_b = 0.2 (2 to 3×), ϵ = 0.03 (2×), and f_c = 0.2 (3×). The uncertainties in the latter two parameters are 2× even in a single, controlled test fire (SCOPE, p. 48). Using these values and A = 0.6 g/cm^2, we obtain F = 0.72 mg/cm^2, which is 15× less than the K/T value. Coincidentally, the combined uncertainties of the above parameters are 12 to 18×, but since these uncertainties are not independent, this does not eliminate the discrepancy. Either the soot yield was much higher in the K/T fire, or fossil carbon rather than biomass was the major fuel source.

Actually, there is good reason to believe that the soot yield was much higher in the K/T fire. Seiler and Crutzen (1980) have pointed out that about one-third of the "unburned" C in forest fires is converted to charcoal, i.e., 160 mg/cm^2 in the above example. Less than 10% of this would need to be lofted to account

for the 11 mg/cm^2 of K/T carbon. Such coarse (10 to 20 μm) charcoal—which actually accounts for most of the carbon at the K/T (Table 59.2 and Figure 59.1)—was missed in test fires where f_c was determined from light absorption (SCOPE, p. 51), but it has been detected in copious amounts by particle collections (NRC, p. 72).

Moreover, there are several profound differences between the test fires on which the NRC or SCOPE parameters are based and the K/T fire. (1) Test fires typically have dimensions of 10^{-3} to 10^{-1} km, in contrast to 10^3 to 10^4 km for the K/T fire. It is known that ϵ and f_c increase for larger fires, as the air supply decreases and is more strongly preheated (NRC, pp. 59, 62; SCOPE, pp. 49, 52). (2) The impact and subsequent winds would strip the forest canopy, shatter heavy timbers, knock them down into the burning zone, and desiccate the fuel, thus raising fuel density, f_b, ϵ, and f_c (NRC, pp. 49, 56, 59). (3) Smoke plumes and impact dust clouds would backscatter thermal radiation, likewise raising the above parameters (NRC, p. 56). (4) The resulting larger and hotter fires would loft a larger proportion of the charcoal.

Given the very large potential reservoir of charcoal (some of which would reach the sea as river-borne runoff rather than atmospheric fallout), it appears that biomass alone may be an adequate source of the K/T carbon. However, fossil carbon remains a potential additional source. The amount of carbon excavated by the crater would not be large; a 150 km crater in *average* terrain with 2300 g/cm^2 of carbon would eject 4×10^{17} g of C, or 0.08 g/cm^2, of which only an unknown, probably small, part would be converted to soot/charcoal in the relatively oxidizing medium of the fireball (Wolbach et al., 1985). Larger amounts might be produced by an impact into carbon-rich terrain, and by post-impact fires of oil seeps, exposed coal deposits, or carbonaceous shales.

Ignition Problem

Delayed Fires

Living trees do not burn readily, and several authors have therefore proposed that the trees were first killed and freeze-dried, and ignited by lightning only after the sky had cleared and thunderstorms had resumed (P. J. Crutzen, private communication, 1985; Argyle, 1986; S. H. Schneider, unpublished manuscript, 1986). A prediction of this model is that carbon should overlie the Ir layer, but the data show no such trend: carbon appears even in the lowermost 0.3 cm of the boundary clay at Woodside Creek (Figure 59.2) and Sumbar (Wolbach et al., 1990a). Evidently the fires began well before the ejecta had settled.

High Atmospheric O_2

Another alternative is that the O_2 content of the atmosphere was higher: at 24 vol. % O_2, living trees with ~20% moisture have an ignition probability of ~15%, compared to <1% at 21% O_2 (McMenamin and McMenamin, 1987; Watson et al., 1978). For a more typical moisture content of 10%, these probabilities are 99% and 20%. However, there are three reasons why a higher O_2 content is not the answer.

First, fires would have started more easily if the O_2 content were higher; forests would burn down as fast as they grew. Indeed, Watson et al. (1978) contend that O_2 levels of 25% to 35% *"would be incompatible with land-based vegetation."* Second, at present O_2 levels only about (2 to 4) $\times$ 10^{-3} of land biomass burns down each year (Seiler and Crutzen, 1980). This results in an aeolian carbon deposition rate of 3.9 $\times$ 10^{-6} g cm^{-2} yr^{-1} (Cachier et al., 1985), which in turn should yield mean carbon contents of 200 to 400 parts per million (ppm) in sediments of deposition rates of 10 to 5 cm/10^3 yr (which are typical for Cretaceous shelf sediments). At higher O_2 contents a greater fraction of the land biomass would burn down each year, leading to correspondingly higher sedimentary carbon contents—provided soot yields remain the same, which is not certain. But the observed elemental carbon contents in uppermost Cretaceous sediments generally are much *lower* than 200 to 400 ppm, even though oxidative loss is ruled out by the presence of kerogen: ≤20 ppm (Woodside Creek), ≤2 (Chancet Rocks), ≤2 (Nye Kløv), and ≤2 (Agost). Third, the report of 30% O_2 in Cretaceous air trapped in amber (Berner and Landis, 1988) has been questioned on a variety of grounds (Horibe and Craig, 1988).

Ignition Mechanisms

A very effective mechanism of global scope is thermal radiation from reentering ejecta (Öpik, 1958; Melosh et al., 1990). The condensing vapor plume from a 10 km impactor would rise above the atmosphere, expanding to blanket the globe and depositing dust grains that are heated to incandescence as they reenter the atmosphere. The thermal radiation levels during the next few hours greatly surpass the solar input (50–150 vs 1.37 kW m^{-2}) and exceed the threshold for ignition of dry or moist wood over most of the Earth,

except in densely clouded areas. Meanwhile, trees will have been killed and dried by side effects of the impact, e.g., prompt heating of the atmosphere, the "extraordinarily powerful wind, capable of flattening forests out to a distance of 500–1000 km" which lasted for ~1 hr and was followed by a very strong return wind (Emiliani et al., 1981).

Those trees that were not ignited by the impact and its immediate aftereffects might be burned down by a second mechanism: charge separation during settling of ejecta through the atmosphere, which, by analogy to volcanic ash falls (e.g., Perret, 1935), should lead to extensive lightning activity, but on a global rather than local scale. With multiple ignition points ("a thousand points of light"), fires would cover the globe.

Effects of Fire

A fire producing 7×10^{16} g of soot would aggravate most of the environmental stresses of an impact, making a bad situation worse (Alvarez et al., 1980; Alvarez, 1986; Wolbach et al., 1985, 1988a).

Darkness and *cold* would last longer, since soot absorbs sunlight more effectively than does rock dust (optical depth of 11 mg/cm^2 of soot is 1600). The cooling would be no greater, though, as it is already at the maximum.

Poisons such as NO and NO_2 (Lewis et al., 1982; Prinn and Fegley, 1987) would be accompanied by ~100 ppm CO and by a variety of organic pyrotoxins such as dioxins and PAH.

Mutagens

Many pyrotoxins—especially PAH—also are mutagens. They may have caused the delayed extinctions of some Cretaceous "survivors" and may have speeded up evolution of others.

Greenhouse Effect

The ~900 ppm of CO_2 expected to accompany the observed amount of soot (SCOPE, p. 217) would increase the greenhouse effect due to water vapor ~8°C, (Emiliani et al., 1981); by another 5° to 10°C (Wolbach et al., 1986; Crutzen, 1987), but it would last decades rather than months—depending on the rate of CO_2 absorption by the oceans. The net temperature change of course also depends on other factors such as cloud cover and albedo (Rampino and Volk, 1988) as well as persistence of the last remnant of dust/soot in the stratosphere.

Table 59.3 Environmental stresses caused by K/T impact

Stress	Time scale	Reference
Darkness	Months	ab
Cold	Months	abc
Winds (500 km/h)	Hours	d
Tsunamis	Hours	de
H_2O-greenhouse	Months	d
CO_2-greenhouse	Decades	fg
Fires	Months	hi
Pyrotoxins	Years	hijk
Acid rain	Years	l
Destruction of ozone layer	Decades	lm
Impact-triggered volcanism	Millennia?	n
Mutagens	Millennia	j

a. Alvarez et al. (1980)   h. Wolbach et al. (1985)
b. Toon et al. (1982)   i. Wolbach et al. (1988a)
c. Thompson (1988)   j. Gilmour and Guenther (1988)
d. Emiliani et al. (1981)   k. Venkatesan and Dahl (1989)
e. Bourgeois et al. (1988) l. Lewis et al. (1982)
f. Wolbach et al. (1986)   m. Prinn and Fegley (1987)
g. Crutzen (1987)   n. Rampino and Stothers (1988)

Selectivity of Extinction Patterns

Some authors (Hickey, 1981; Officer et al., 1987) have argued that the observed selectivity of the extinction patterns is inconsistent with the simple darkness-and-cold scenario of Alvarez et al. (1980). However, this scenario has been greatly extended in subsequent years by inclusion of other factors, exceeding the seven biblical plagues in number if not severity (Table 59.3). Virtually all of these 12 stresses are selective, inasmuch as they affect different species to different degrees. Some stresses pervade the entire Earth, whereas others vary with latitude, terrain, elevation, etc. And some taxa have hardy dormancy forms (spores, seeds, roots) or high reproduction ratios even for small breeding populations, whereas others do not. In any case, the selectivity problem may have been overstated; some authors contend that a cold spell alone could cause the observed pattern of plant extinctions (Wolfe and Upchurch, 1986).

A first-order scientific task awaiting an imaginative ecologist is to rationalize the observed extinction patterns in terms of the above stresses. Some steps in this direction have already been taken, explaining the extinction of reptiles relative to mammals (Cowles, 1939) or ammonites relative to nautiloids (Emiliani et al., 1981). Constructive work on this problem would be far more valuable than specious attacks on the impact theory.

Acknowledgments

This work was supported in part by NSF grant EAR-8609218 and NASA fellowship NGT-50015.

Notes

1. *Fusinite* is carbonized woody material (i.e., natural *charcoal*) that is found in coal. It is $\sim 10^1$ to 10^2 μm in size and sometimes shows the cellular structure of wood. *Soot* in the broad sense is the elemental carbon component of smoke, consisting of up to 4 sub-components (Medalia and Rivin, 1982). (1) Irregularly shaped, *charcoal* or coke, particles of charred fuel lofted into the smoke plume. (2) *Aciniform* (grape-bunch-like) carbon, consisting of 10^{-2} to 10^{-1} μm spherules of amorphous C that are welded into characteristic clusters and chains (Figure 59.1b); it is formed not by solid-phase charring of fuel but by gas-phase polymerization of C_2 radials to icospirals (Kroto, 1988). (3) *Carbonaceous microgel*, i.e. spheroidal carbon particles embedded in a carbonaceous matrix. (4) *Carbon cenospheres:* 10 to 100 μm spheres formed by carbonization of liquid drops. We use the term "*elemental C*" for the first three components (cenospheres are absent in the K/T), serving *soot* in the narrow sense for the spheroidal components (2) and (3), which are distinctive in morphology and origin (having formed by polymerization of C_2 units in the gas phase rather than by charring of a solid).

Kerogen is insoluble organic matter derived from plants. Depending on its source (i.e., *terrigenous* or *marine*) and thermal history, it can range from largely aliphatic to largely aromatic, with a parallel decrease in reactivity and H/C, O/C ratios. With increasing metamorphism, it grades continuously into *elemental C*, ranging from *amorphous C* to crystalline *graphite*. We shall use the generic term *carbon* or *elemental C* for the insoluble carbon remaining after removal of soluble minerals and reactive kerogen. It consists mainly of charcoal and soot, but with variable, usually small, amounts of resistant kerogen and detrial carbon from metamorphosed sediments.

2. Biomass burning produces about 1 to 3 $\times$ 10^{-4} g cm^{-2} yr^{-1} elemental C (Seiler and Crutzen, 1980), compared to $\sim 2 \times 10^{-5}$ g cm^{-2} yr^{-1} from weathering of sediments (Holland, 1978). Moreover, the major part of the latter is oxidized to CO_2 rather than redeposited, as the ^{14}C ages of modern sediments show no evidence of such "dead" carbon (Holland, 1978). For this reason, essentially all elemental C in marine sediments is assumed to be charcoal (Goldberg, 1985; Herring, 1985).

3. Hansen et al. (1987) have reported that carbon black at Stevns Klint first appears 3.5 m below the KTB, and suggest that it was produced by volcanism rather than by impact-triggered fires. This suggestion has been cited approvingly by Officer and Drake (1989). However, both the data and the conclusions are faulty, as shown by Wolbach et al. (1990a). The carbon black actually was kerogen, and the argument for volcanic carbon is a complete misrepresentation.

4. The "elemental carbon" was taken to be the carbon remaining after a 60-hr oxidation with dichromate. This is a fairly good assumption for the boundary clay but not for the Tertiary samples, which, like other near-shore samples (Sackett and others, 1974), contain resistant, "recycled" kerogen from land. Spot checks on 4 Tertiary samples showed that only one (+ 141 cm) survived a 600-hr etch and thus contained elemental C (60 ppm after 600 hr, or 120 ppm after correction for carbon loss during etching, compared to 1570 ppm after 60 hr). Three other samples (+ 11, 18, and 27 cm) dissolved more or less completely, leaving 0% to 3% residues after 72 to 446 hrs (Wolbach, 1990). Presumably the elemental C curve in Figure 59.2 in reality runs close to the "soot" curve.

Impact Winter in the Global K/T Extinctions: No Definitive Evidences

Dewey M. McLean

About 66 million years ago, Earth experienced a global extinction event so profound that it marks the boundary between the Mesozoic and Cenozoic eras and, on a finer scale, the Cretaceous (K) and Tertiary (T) periods. Long the topic of scientific inquiry and debate, the Cretaceous-Tertiary, or K/T, extinctions are cited as one of the top 10 to 20 unsolved mysteries in science. In the past decade, the debate has attracted scientists from many disciplines, and has expanded into one of the truly great debates in the history of science.

The K/T debate is a classic example of conceptual polarities, or antitheses, emerging from a common data base. The most fundamental polarities involve (1) extra-terrestrial asteroid/comet impact versus terrestrial volcanism as causative factors in the extinctions, (2) sudden impact-induced catastrophe spanning a few months to years versus gradual volcanic-induced bioevolutionary turnover spanning several hundreds of thousands of years, and (3) "impact winter" via impact dust blasted into the stratosphere blocking out sunlight, and plunging Earth into darkness and refrigeration versus "greenhouse" warming via volcanic CO_2 release into the atmosphere that triggered climatic warming, marine chemistry changes, and ecosystem collapses of a CO_2-induced "greenhouse." The K/T debate can only be resolved by eliminating polarities.

This chapter addresses the latter polarity, and suggests that if a K/T boundary impact winter occurred, it was too transitory, or feeble, to be recorded in the geological record, and not of sufficient magnitude to trigger global biological catastrophe. On the other hand, a major long-duration K/T transition carbon cycle perturbation associated with coeval climatic warming is indicated in the record. Throughout, I will discriminate between (1) short-duration K/T boundary impact-type phenomena that begin at the K/T contact and persist for a few months to a few years, and (2) long-duration K/T transition volcanic-type phenomena that begin in the Upper Cretaceous and persist into Early Tertiary, spanning several hundreds of thousands of years.

K/T Boundary Climatic Cooling/Warming Background

Climatic cooling as a factor in the K/T extinctions, and especially those of the dinosaurs, has long been espoused. Through the 1970s, Dale Russell related K/T climatic cooling to a supernova explosion in the vicinity of the solar system (Russell, 1977, 1979a, 1979b). According to Russell (1977), evidences for a supernova explosion in the K/T extinctions were: (1) the extinctions themselves and (2) evidences of an enormous ring of interstellar neutral hydrogen (Lindblad's ring), some 18,000 billion miles in diameter near the solar system, whose speed of expansion indicates an explosion relatively close to the sun about 65 million years ago. Supposedly, climatic effects are due to increase in visible light from the supernova, and to changes in ionizing radiation which would disturb the radiation of the upper atmosphere. Russell (1979a) suggests that such atmospheric perturbations would have triggered climatic cooling.

The modern K/T cooling/warming antithesis dates from the late 1970s when I proposed CO_2-induced greenhouse warming as a factor in the K/T extinctions (McLean, 1978). Initially, I evoked the failure of the dominant planktonic marine algae, the coccolithophorids, as the cause of the greenhouse. Those algae served as a major sink for the uptake and storage of CO_2. Prior to the K/T extinctions, the coccolithophorids had been so abundant that their photosynthesis and $CaCO_3$ shell production had drawn down atmospheric pO_2, causing antigreenhouse climatic cooling. Their K/T failure would have triggered a carbon cycle perturbation. Later, I evoked mantle CO_2 degassing perturbation of the carbon cycle via the Deccan Traps volcanism in India as the cause of coccolithophorid failure, and the combination of volcanism and coccolithophorid failure for a major trans-K/T greenhouse (McLean, 1981, 1982a, 1982b; see also McLean, 1985a, 1985b, 1985c). Recently, I subsumed the Deccan Traps volcanism, and extinctions, into a broader theory involving thermal evolution of Earth (McLean, 1988).

K/T boundary climatic cooling was reinforced by Alvarez et al. (1980) who proposed that 66 million years ago a giant asteroid 10 ± 4 km diameter struck Earth, blasting dust into the stratosphere and blocking sunlight for several years until the dust settled to Earth, plunging Earth into darkness and refrigeration. Loss of sunlight suppressed photosynthesis, collapsed most food chains, and triggered global extinctions. Theoretically, the asteroid was rich in the element iridium, and the impact dust settled out globally as a thin, iridium-rich clay layer that marks the K/T boundary. The K/T boundary clay was taken as primary evidence of a sunlight-blocking dust shroud. A K/T impact winter was reinforced by discovery of elemental carbon, mostly in the form of soot, in the K/T boundary clay (Wolbach et al., 1985). Theoretically, asteroid impact ignited global wildfires that consumed most of Earth's forest biomass, injecting soot into the atmosphere simultaneously with the impact dust, enhancing the dark and cold of the K/T impact winter. The soot can also be taken as primary evidence of a K/T impact winter.

The notion of a K/T boundary impact winter was further reinforced by the proposals of Wolfe and Upchurch (1986, 1987) that plant physiognomy-climate relationships in the Rocky Mountain paleobotanical record supports pronounced, short-duration K/T boundary climatic cooling. Assuming that physiognomy-climate relationships are valid as pertains to the K/T paleofloral record, a postulated short-duration K/T boundary cooling can also be taken as primary evidence of a K/T impact winter.

In this paper, I will examine validity of the primary evidences supporting the concept of a K/T boundary impact winter against the geobiological record. These involve: (1) genesis of the iridium-rich K/T boundary clay, (2) genesis of K/T boundary clay soot enhancement that provided the basis of K/T wildfires, and (3) the K/T paleofloral record that is claimed to indicate sharp, pronounced, K/T cooling supportive of an impact-induced "impact winter."

K/T Boundary Iridium Clay Implication in Impact Winter

K/T Boundary Interval Clays (Western North America)

The picture presented by K/T impact-oriented literature is that dust blasted into the stratosphere by the Alvarez asteroid settled out as a simple dust layer that marks the global K/T boundary. In fact, the K/T

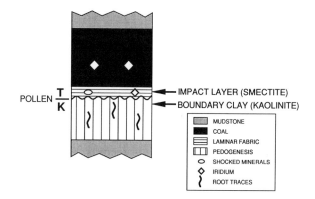

Figure 60.1 Composite Rocky Mountain K/T boundary interval showing the boundary clay couplet composed of the K/T Boundary Clay and the K/T Impact Layer. These units are separated by a depositional hiatus. The pollen-based K/T contact occurs at the top of the Boundary Clay.

boundary sedimentary record is too complex to be explained so simply.

The American Rocky Mountain region contains the most complete, and examined, terrestrial K/T sections known. Figure 60.1 (adapted from Izett, 1987) shows that the K/T boundary interval is not a simple unit, but instead, a complex couplet composed of a lower K/T Boundary Claystone (mostly kaolinite), and an overlying K/T Boundary Impact Layer (mostly smectite). Those two units are separated by a hiatus (Izett, 1987; Fastovsky et al., 1989). The Impact Layer is itself directly overlain by the Boundary Coal Bed.

K/T Boundary Claystone (abstracted from Izett, 1987)

This 1 to 2 cm thick, largely kaolinitic unit, displays no evidence of an impact origin. It lacks both iridium and shock-metamorphosed mineral grains. Its lithology indicates slow accumulation by normal sedimentary processes rather than rapid accumulation by fallout of impact material. Horizontal vitrinite laminae, blebs of amber, degraded plant material, fossil twigs, fossil roots, angular and rounded claystone fragments, rounded fragments of woody material, and carbonaceous structures that flare out at the contact with the overlying Impact Layer indicate a complex origin. It lacks the angular shard shapes and broken crystals typical of impact fallout material; it is not altered tektite glass. Authigenic spherules of kaolinite, goyazite, and jarosite, resembling objects in marine K/T boundary clays described as impact melt droplets, are likely not of impact origin. Its interpretation as a widespread continuous layer may have been misinterpreted as it has been found only at about

25 localities in low-energy coal-swamp environments in north-trending Rocky Mountain coal basins. It has not been found outside of these coal basins. It underwent pedogenesis; its upper surface is a depositional hiatus.

K/T Boundary Impact Layer (abstracted from Izett, 1987) This 4.0 to 8.0 mm thick, largely smectitic unit, contains the maximum iridium abundance in the Rocky Mountain K/T boundary interval, and also shock-metamorphosed grains; the latter constitute no more than 2.0% by volume and are not concentrated at the base of the unit as would be primary air-fall materials. Its microlaminated structure contrasts with the underlying Boundary Clay. Microlaminations are warped over and under clastic mineral grains reflecting soft sediment compaction. Microlaminae surfaces display impressions of macerated plant material, suggesting slow accumulation during normal depositional processes. Formed mostly of locally derived sediment, and not from altered volcanic or impact glass, the unit reflects slow accumulation of clay particles. Ubiquitous pellets of white cryptocrystalline kaolinite are common. That it does not contain a large component of asteroidal materials is attested by low nickel and cobalt contents. It is overlain at most localities by the Boundary Coal which, itself, ranges from 4 to 16 cm in thickness.

Izett (1987) states that shocked minerals and iridium of the Impact Layer settled out on a Boundary Clay paleosurface that was intermittently covered with water and local floras. Impact upset the hydrologic regime via climate change, which triggered a new but amazingly gentle erosional-depositional cycle during which shocked minerals were reworked gently into a slowly accumulating deposit of clay and decaying plant material. The presence of only one K/T boundary shocked mineral spike indicates but one impact event and not multiple impacts as suggested by some authors. Izett (1987) notes that the K/T boundary impact event was considerably smaller than that proposed by Alvarez et al. (1980).

Comment The Impact Layer, which reflects slow accumulation by normal processes of locally derived materials, contains a miniscule amount of shocked mineral grains. They are distributed throughout the Impact Layer and not at the base as they would be if they represented simple air-fall of impact debris. Microlamination indicates lack of bioturbation; thus, it is difficult to see how the shocked mineral grains mixed throughout the Impact Layer were "reworked" into it. Most grains illustrated in publications have

rounded edges, suggesting transport into the coal basins via fluvial processes. Grains settling directly from an impact related air fall would be sharply angular. Most plausibly, the shocked mineral grains were transported into topographically low coal basins by natural sedimentary processes together with other sediments that make up the Impact Layer.

Iridium generally peaks in the Impact Layer but also occurs in carbonaceous-rich layers either above or below it (Izett, 1987). Izett assumes that the iridium was originally concentrated in the Impact Layer and was later mobilized and transported to adjacent carbonaceous sediments. The Impact Layer is directly overlain by the Boundary Coal. The latter represents a return to coal-producing conditions after a long coal-barren interval and seems to have developed as a part of the flooding of the region during the K/T transition (Fastovsky et al., 1989). Iridium existing about on the landscape as a result of either impact or volcanism (see section on K/T Boundary Iridium and Deccan Traps Volcanism below) would have been mobilized by abundant humic acids and transported into, and concentrated in low-lying swamp areas (McLean, 1985b), along with the shocked minerals. The Impact Layer could be underclay-equivalent base of the Boundary Coal.

Conclusion The terrestrial Rocky Mountain K/T boundary clays do not definitively support a K/T boundary blockage-of-sunlight-induced impact winter.

K/T Boundary Clay (Marine)

Impact Origin Alvarez et al. (1980) proposed that the K/T boundary clay represents impact dust that settled out as a global iridium-rich clay unit. Kastner et al. (1984), noting that smectite, the dominant clay mineral in the K/T boundary at Stevns Klint, Denmark, is rich in magnesium, proposed that the K/T boundary clay is unique and dissimilar from other smectites in the rock record. They concluded that the unusual Mg-smectite originated via alteration of impact ejecta.

Terrestrial Origin Most authors propose an Earthly source for the marine K/T boundary clay. Because of the high smectite content, Christensen et al. (1973) suggested a volcanic origin for the K/T boundary clay. Elliott et al. (1989) contend that the Mg-smectite is not unique to the K/T boundary and that it was not derived by a bolide impact but by alteration of volcanic ash. Hansen et al. (1988) note that smectite represents devitrification of volcanic materials; their study of chromite chemistry of the Danish boundary clays indicate a terrestrial ultrabasic origin. Rampino and Reynolds (1983) indicated that the K/T boundary clay is neither mineralogically exotic, nor distinct from locally derived clays above and below it. The Schmitz (1985) study of metal enrichment in the K/T boundary clay in Denmark indicated that the most likely source for the metals is dissolved detritus in the underlying chalks which may be volcanic debris with some inputs of terriginous and meteoritic material. Schmitz (1988) concludes that the "impact scenario of Alvarez et al. (1980) in which a world-encompassing dust cloud led to suppression of photosynthesis and extinction, is not supported by chemical and mineralogical characteristic of the marine K/T boundary clays."

K/T Marine Clay Deposition Time The long duration of deposition of the K/T marine boundary clay argues against rapid air-fall from an impact event. Smit (1982) notes that the K/T boundary clay represents 5,000 to 15,000 years deposition. Joseph Hazel (personal communication, 1989) indicates from sophisticated time-rock relationships that the K/T boundary clay represents about 4,000 years deposition.

Conclusion Studies of the K/T boundary clay layer from various parts of the world indicate that it represents altered volcanic materials and not a short-duration air-fall of impact debris. The marine K/T boundary clay does not support a K/T boundary dust-shroud-induced impact winter.

K/T Boundary Refrigeration via Impact Dust
A K/T boundary refrigeration of magnitude to trigger global catastrophic extinctions would be recorded in the oxygen isotope record as a sharp positive excursion precisely at the K/T boundary. In fact, a negative excursion begins below the boundary and continues well into the Early Tertiary coeval with the Deccan Traps volcanism in India.

Global Blackout via Impact Dust
There are no definitive evidences of an impact-induced K/T boundary global blackout. A test of a global blackout exists via the marine algal record. Microscopic, planktonic algae such as the diatoms, dinoflagellates, and coccolithophorids, depend on light for photosynthesis. A six-month to one-year blackout would have triggered massive extinctions of many species precisely at the level of the K/T boundary iridium clay.

Diatoms The K/T event did not much affect the diatoms. Harwood (1988), based on studies from Seymour Island, eastern Antarctic Peninsula, the first to record siliceous microfossil assemblages across a K/T boundary sequence, notes that diatom survivorship across the K/T boundary was above 90%. Resting spores increase from 7% below to 35% across the K/T boundary.

Dinoflagellates Dinoflagellates also were little affected by the K/T event (Bujak and Williams, 1979). Brinkhuis and Zachariasse (1988) record no accelerated rates of extinction across the K/T boundary in Tunisia. Nor does Hultberg (1986) in Scandinavia. Danish dinoflagellates responded more by appearance of new species than by extinctions (Hansen, 1977), as did Seymour Island assemblages (Askin, 1988).

Coccolithophorids Marine calcareous microplankton, the coccolithophorids and planktonic foraminifera (discussed below), were hit hardest of all by the K/T event. Thierstein (1981) proposes that the coccolithophorids extinctions were the most severe plankton extinction event in geologic history; via a "deconvolution" process, Thierstein (1981, 1982) reduced a Cretaceous-Tertiary "transition," in which Cretaceous assemblages were replaced by "new" Tertiary taxa, to an instantaneous catastrophe.

Comment Perch-Nielsen et al. (1982) note that the "catastrophic event" at the K/T boundary did not result in geologically instant extinction of the calcareous nannoplankton and that most Cretaceous species survived the event. At DSDP Site 524, a sample above the K/T boundary contains 90% Cretaceous species. Isotopic analyses indicated that the Cretaceous species were not reworked specimens but represented survivors of the K/T event that continued to reproduce in the earliest Tertiary oceans. The relict species became extinct some tens of thousands of years after K/T boundary time, probably via environmental stresses.

Antia and Cheng's (1970) work on survival times of phytoplankton species in complete darkness indicate that one month of complete blackout would produce 13% extinction, three months 68% extinction, and six months 81%. Thus, the six month to one-year global blackout predicted by Wolbach et al. (1985) would have obliterated diatoms, dinoflagellate, and coccolithophorids precisely at the K/T boundary. A blackout event is not reflected in the algal record. The calcareous coccolithophorids and foraminifera were likely affected by pH change of the marine mixed layer via CO_2 mantle degassing by the Deccan Traps volcanism.

Conclusion The marine algal record does not support a K/T boundary global blackout.

Foraminifera
Studies at El Kef, Tunisia, and Brazos River, Texas (Keller, 1988a, 1988b, 1989a, 1989b; Keller and Lindinger, 1989), indicate that (1) an extended period of species extinctions spans the K/T boundary from about 300,000 years below to 200,000 to 300,000 years above the boundary in shallow continental shelf sections; (2) apart from the K/T boundary, there are two distinct extinction episodes beginning at about 300,000 years below the boundary, and at about 50,000 years above the boundary; (3) at El Kef, the K/T event affects only 26% of the planktic foraminiferal species; 29% disappear earlier, 11% disappear 15 cm above the K/T boundary, 18% disappear near Zone P0/P1a about three m above the K/T boundary, and the remaining Cretaceous survivors disappear gradually 150,000 to 300,000 years after the K/T boundary; (4) at Brazos River, there are no species extinctions or measurable faunal changes associated with the K/T boundary. There are two extinction events. The first, with 46% of the species extinct, occurs 27 to 35 cm (about 310,000 years) below the K/T contact. The second, with 45% of the species extinct, occurs 25 cm (about 50,000 years) above the K/T contact, at the Zone P0/P1a boundary. No species extinctions or major faunal assemblage changes occur at the Brazos River K/T boundary.

Keller and Lindinger (1989) note that primary productivity in the oceans was drastically reduced for at least 230,000 years. $CaCO_3$%, benthic, and fine fraction ^{13}C values reach near pre-K/T levels about 300,000 to 400,000 years after the K/T boundary. This period was coeval with "Strangelove" ocean conditions that were coeval with the Deccan Traps volcanism.

Conclusion Foraminifera, the organisms most affected by K/T event, do not reflect a single catastrophe precisely at the K/T boundary but, instead, a gradual process that affected them over a long K/T transition interval of time. The Deccan Traps volcanism, that began erupting below the K/T boundary, and whose main activity was in Early Tertiary, would have produced long-duration K/T transition pH change of the mixed layer, and the Strangelove oceans implicated in low Early Tertiary marine productivity.

Reptilian Extinctions and Impact Winter
Carpenter (1983) notes that the assumption that all latest Cretaceous dinosaurian taxa persisted to the end of the Cretaceous before becoming extinct permeates the popular and scientific literature; therefore it is no wonder that single-cause catastrophic hypotheses abound.

Russell (1975, 1979a, 1979b, 1982, 1984), who contends that there was no decline in reptilian diversity toward the end of the Cretaceous, and that over half of Earth's species became extinct in a terminal Mesozoic crisis, has advanced the idea of a K/T asteroid impact. He visualizes 18 families of dinosaurs—three of pterosaurs, one of ichthyosaurs, three of plesiosaurs, and one of mosasaurs—becoming extinct at the K/T boundary.

Sullivan (1987), who has developed the first global survey of reptilian species-level diversity spanning the K/T transition, contradicts Russell. Sullivan notes that of the 44 reptilian families existing before the K/T boundary, 13 died out before the boundary, 9 died out at the boundary, and 22 survived it. The extinction rate at the boundary is thus about 20%. For the 19 dinosaurian families existing in the last 20 million years of the Cretaceous, most disappeared before K/T boundary time, and 8 at the K/T boundary. The 8 families are represented by 12 species in the final 3 million years of the Cretaceous, and most of the species by only a few specimens (2 to 10, average 5.7).

One turtle family died out prior to the K/T, and six survived the K/T boundary event. Twelve lizard and snake families survived the K/T event; one, the mosasaurs, had died out at the beginning of the Maastrichtian. The four crocidilian families survived seemingly unaffected. Of two pterosaur families, one died prior to the K/T, and one at the K/T boundary. The single families of ichthyosaurs and plesiosaurs died out in the Campanian long before K/T boundary time.

Sloan et al. (1986) noted that the dinosaurian extinctions were a gradual process beginning 7 million years before the end of the Cretaceous and accelerating rapidly in the final 0.3 million years. They propose that in Montana, the last dinosaurs occur stratigraphically above (about 40,000 years after) the K/T iridium anomaly. Of the 30 dinosaur genera living in the area 8 million years before the end of the Cretaceous, 12 genera were still living just before the K/T boundary event, and between 7 to 11 survived into the Paleocene. Only 1 to 3 genera became extinct at the K/T boundary. They propose that the dinosaurs survived well into the early Paleocene in tropical India, the Pyrenees, Peru, and New Mexico.

Conclusion Evidence exists that the dinosaurs could tolerate periods of darkness and cold. Rich et al. (1988) indicate from work in the high latitudes of southeastern Australia that cold and darkness may not have been a prime factor in the extinction of dinosaurs at the K/T boundary unless the duration exceeded three to five months. Dinosaurs living near the poles was an enduring event in which they had coped with high latitudes for at least 65 million years. Brouwers et al. (1987) note for the high latitudes of the Alaskan North Slope 66 to 76 million years ago, that dinosaurs remained at high latitudes the year-round, challenging the idea that short-term periods of darkness and cold via bolide impact triggered the dinosaurian extinctions. The reptilian record does not support a K/T boundary impact winter.

K/T Boundary Soot Implication via Wildfire-Induced Impact Winter

K/T Boundary Wildfires and Soot

Wolbach et al. (1985) and Wolbach et al. (1988) reported the presence of graphitic carbon, primarily in the form of soot, at several K/T boundary sites from various localities: Stevns Klint and Nye Klov, Denmark; Woodside Creek and Chancet Rocks, New Zealand; and Caravaca, Spain. They attribute the soot to impact-ignited global wildfires that burned down most of Earth's forests. Theoretically, radiation from the impact fireball and cloud of rock vapor can ignite wildfires more than 1,000 km away; once started, fires can spread over entire continents, distributing soot worldwide; trees dried by the impact via heat and winds exacerbated ignition and combustion. The authors note that because soot enrichment begins in the first 0.3 cm of the K/T boundary clay, wildfires started before the impact rock dust had settled. Because soot absorbs sunlight more effectively and settles more slowly than does rock dust, it would have caused darkness and cold to last longer than would rock dust alone.

Fusinite as a Test for K/T Wildfires Charcoal (structured fusinite of petrographic usage, after Cope and Chaloner, 1985) provides evidences of wildfire. K/T global wildfires of the magnitude envisioned by Wolbach et al. (1985) and Wolbach et al. (1988) would have produced a global layer of fusinite (charcoal) precisely at the K/T boundary in the Rocky Mountain region and, because of continental runoff, at the K/T boundary in marine sections as well. Fusinite is not

recorded in the Rocky Mountain K/T boundary sediments in sufficient quantities to support the idea of K/T wildfires. Abundant vitrinite (coalified plant tissue) is recorded.

Marine Algae as a Test for K/T Soot-Induced Blackout Same as for the K/T boundary impact rock dust blackout discussed above.

Terrestrial Origin of K/T Soot Hansen et al. (1987) note that carbon black occurs frequently from Devonian to present and is generally interpreted as soot from forest fires or burning of grasslands. In Danish sections, carbon black enrichments begin 3.5 m below the boundary clay, or about 50,000 years earlier than the iridium anomaly. The carbon black is consistent with an earthly origin from reduced carbon or carbonados; carbonados are amorphous lumps of pure carbon associated with kimberlites. The Danish carbon black reflects pulsed or multistage deposition lasting for about 50,000 years.

Hansen et al. (1988) note that elemental carbon in the Danish K/T boundary clays reflects terrestrial origin, unrelated to meteoritic carbon. Isotopic evidence, and the long duration of carbon enrichment, indicate that it cannot originate from world forest fires and that ultrabasic volcanism is the likely origin of the carbon black.

K/T Clay Sedimentation Rates For the duration time of deposition of the boundary clay, Wolbach et al. (1988) assume that the clay represents impact ejecta that could not have stayed aloft for more than six months. Thus, they propose that the soot "enrichment" in the K/T clay resulted from short-duration global wildfires. In fact, Smit (1982) indicates a 5000- to 15,000-year deposition time for the K/T boundary clay. Joseph Hazel (personal communication, 1989) indicates on the basis of sophisticated time-rock relationships a 4,000-year deposition time.

Figure 2a of Wolbach et al. (1988) shows iridium and soot peaking in abundance in the K/T boundary clay; iridium and elemental carbon concentrations rise sharply at the boundary by factors of 1500 and 210 compared with the Cretaceous values, and soot by 3600. However, both iridium and soot begin increasing in abundance below the K/T clay, and remain above Cretaceous values well into the Early Tertiary. At the time of K/T clay deposition, $CaCO_3$ deposition had nearly ceased. I have argued (e.g., McLean, 1985b) that K/T cessation of $CaCO_3$ sedimentation would allow clay particles that are ordinarily admixed in $CaCO_3$ to become concentrated as a clay layer at the K/T boundary. Slow accumulation of the K/T clay would allow iridium and soot to become concentrated in it. By this mechanism, soot from natural forest fires becoming concentrated via slow accumulation of the K/T clay would create the illusion of rapid fallout of soot over a short period.

Conclusion The K/T boundary soot enrichment represents concentration via normal Earthly sedimentary processes and is not supportive of K/T impact winter blackout and refrigeration.

K/T Boundary Paleobotanical-Climate Implications in Impact Winter

Physiognomy-Climate Relationships
According to Wolfe (1971, 1981), foliar physiognomy provides one of the firmest bases for evaluating the paleoclimatic significance of fossil plants. Wolfe and Upchurch (1986, 1987) propose that because vegetation and climate can be directly inferred from the physiognomy of leaves, and because leaf species typically represent low taxonomic categories, leaf floras of western North America can be used to infer K/T climatology. They propose that those floras support a K/T boundary impact winter with a one- to two-month mean temperature near 0°C. They note that sharpness and magnitude of vegetational and floral changes at the K/T boundary are dramatic, and attribute mass-kill, ecological disruption, climatic change, and the K/T iridium to bolide impact.

Wolfe and Upchurch follow the physiognomy-climate methodology of Bailey and Sinnott (1915, 1916). The latter reported that in mesic, tropical environments, the percentage of dicotyledonous species with entire (untoothed) margins is high, but low in mesic, temperate environments. Wolfe and Upchurch (1987) note that in mesic east Asia environments, the percentage of dicotyledonous species with entire margins closely parallels the mean annual temperature. Reportedly, an increase of 3% of entire-margined species correlates with an increase of 1°C in mean annual temperature. Eastern Asia is their standard by which all other foliar physiognomic analyses should be judged.

However, the physiognomic-climate relationships utilized by Wolfe and Upchurch are more complex than they present, open to other interpretations, and fraught with flaws. According to Dolph (1990), Wolfe's work contains fatal flaws that bring his climate inferences based on foliar physiognomy into question. Dolph notes that several crucial questions must be answered before leaf margin analysis can be used to precisely estimate paleoclimate; most basic is whether or not the foliar physiognomy of eastern Asia should be used at a standard by which all other physiognomic analyses are judged. Dolph (1990) notes that analysis of Wolfe's data on the foliar physiognomy of eastern Asia shows that that area is unacceptable as a standard for foliar physiognomic studies. He points out fatal flaws both in Wolfe's choice of eastern Asia as the standard, and in methodological errors. Other readings on problems associated with physiognomy-climate interpretations may be found in Dolph and Dilcher (1979) and Dolph (1984, 1987).

Comments Wolfe and Upchurch (1987; after Wolfe, 1971) base their physiognomy-climate analyses on the assumption that their assemblages represent nonsuccessional, climax vegetation. Wolfe believes that more than one million years must pass before the vegetation takes on its climax characteristics; vegetation of disturbed habitats can give anomalously low temperature estimates because of the high proportion of toothed-leaf species in disturbed versus "climax" vegetations. This assumption bears upon their interpretation of a K/T impact winter. During the K/T transition, flooding in the Rocky Mountain region would have caused continual migration of plant biotas, impeding development of climax vegetational patterns, and producing temperature estimates based on physiognomy-climate relationships lower than they actually were.

Palynological K/T Boundary Claims that Rocky Mountain paleofloras reflect an impact winter must be based on evidence that the American K/T boundary is isochronous with European K/T boundary stratotypes. Such isochroneity has not been convincingly established. Use of a K/T iridium spike is not definitive in that some localities display multiple iridium spikes. For example, Lattengebirge, Bavarian Alps, has three iridium anomalies, below, at, and above, the K/T boundary (Graup and Spettel, 1989). The oldest anomaly antedates the K/T boundary by 14,000 to 9000 years. To which of these anomalies might the Rocky Mountain "K/T boundary" iridium spike be isochronous, if any?

The pollen-based K/T boundary currently used in the Rocky Mountain region was arbitrarily picked at the top of the Boundary Clay (Izett, 1987). Izett notes that the K/T boundary, placed by Tschudy et al. (1984) and Tschudy and Tschudy (1986) at the top of the Boundary Clay, was an arbitrary choice because only rare palynomorphs have been recovered from the Boundary Clay, and these might have been re-worked from underlying Cretaceous sediments.

K/T Fern Spike An abrupt increase in the ratio of fern spores to angiosperm pollen at this stratigraphic level has been attributed to recolonization of a devastated landscape following asteroid impact (Tschudy et al., 1984; Tschudy and Tschudy, 1986). However, Sweet (1988) notes that absence of the fern spore spike in most western Canadian localities may denote rapid flooding rather than recolonization after a catastrophic destruction of vegetation. Fastovsky et al. (1989) note that in eastern Montana, the earliest Paleogene sediments are mainly ponded water deposits, and that the K/T transition was concommitant with extensive flooding of the landscape, indicated by coal deposition and the ponding. I have previously discussed problems relating to chroneity of Tschudy's pollen-based Rocky Mountain K/T boundary relative to European K/T stratotypes (McLean, 1985c).

Canadian Floras at K/T Sweet and Jerzykiewicz (1985) note that many species in western Canada range across the K/T boundary and provide a strong sense of continuity; evolutionary trends also continue uninterrupted across the boundary. Those authors note that they cannot personally add support to Dr. Alvarez's hypothesis of an extraterrestrial catastrophic event. Lerbekmo et al. (1987) note that western Canada iridium anomalies and palynological floral events do not support a simple, North American-wide catastrophic destruction of vegetation resulting in extinctions coincident with an iridium event at the time of initiation of swamp condition in western Canada.

Seymour Island Terrestrial Floras The high latitudes of Seymour Island, Antarctic Peninsula, supported humid temperate coniferous forests during the Maastrichtian and Paleocene (Askin, 1988). Over that time interval, plant associations remain essentially unchanged except for slight differences in relative abundances which may be facies related. The only changes in terrestrial palynomorphs species are among angiosperms. Of 51 species recorded, only eight species disappear, and not simultaneously but over 20 m. Less than 10% of the total preserved terrestrial species disappeared near the K/T boundary, and these were supplanted by new species. The apparent lack of disturbance in the Seymour terrestrial vegetation, similar to K/T successions throughout much of the world, is consistent with a gradual K/T transition. Askin notes that a K/T event that caused the demise of many marine and nonmarine

taxa at the end of the Cretaceous had little effect on the terrestrial floras of the Seymour Island region. The nonmarine palynomorph succession of Seymour Island is more compatible with gradual climate change. A K/T impact winter blackout would likely have had major impact on those high-latitude floras.

Conclusion The Wolfe and Upchurch proposal that Rocky Mountain K/T paleofloras indicate an impact winter supportive of bolide impact seemingly requires reinterpretation. More plausibly, floral change are a consequence of flooding during the K/T transition (Fastovsky et al., 1989; Sweet, 1988). In fact, Wolfe and Upchurch note that a transgression of the North America epeiric sea had reached the Western Interior by the early Paleocene. They further note that the period of instability encompassed by the mid-Maastrichtian regression and the Paleocene transgression might have contributed to climatic and extinction events associated with the boundary. In light of documented flooding in the Rocky Mountain region during the K/T transition, it seems unnecessary to evoke bolide impact as cause of floral perturbations.

K/T floras in the Rocky Mountain region do not support a K/T impact winter.

K/T Boundary Iridium and Deccan Traps Volcanism

Whether K/T iridium represents impact or volcanism has been controversial. Alvarez et al. (1980) attributed it to an "abnormal influx of extraterrestrial material." Early in the debate, I attributed it to the Deccan Traps volcanism in India at the 1981 Toronto AAAS national meeting, the 1981 Ottawa K/TEC II, and the Snowbird I Conference. Later, Zoller et al. (1983) discovered iridium in the gas phase of the modern Kilauea volcano, and suggested that the Deccan Traps was of sufficient magnitude to have produced the K/T iridium. Recently, Toutain and Meyer (1989) reported that the hotspot volcano that produced the Deccan Traps is yet liberating iridium.

The Deccan Traps, whose lavas yet cover much of northern and western India, is part of a hotspot track produced by northward drift of the Indian lithosphere plate over a deep-origin mantle plume. Courtillot et al. (1986) note that the hotspot now located under Reunion Island was beneath the Deccan Traps when they erupted. Southward from India, the hotspot track system later generated the Chagos-Laccadive ridge, the Mascarene plateau, and then the Mauritus and Reunion islands (Toutain and Meyer, 1989).

The Deccan Traps volcanism was the major K/T boundary event capable of perturbing Earth's outer physiochemical spheres sufficiently to trigger biological overturn in the biosphere. Morgan (1971, 1972a, 1972b) originally related the Deccan Traps to the Reunion hotspot, and estimated that duration of the Deccan Traps was within 1 million years (1981). Subbarao and Sukheswala (1981) cite it as possibly the greatest volume of continental basalt on Earth's surface. Alexander (1981) notes that of the major episodes of flood basalt volcanism, the Deccan Traps has the greatest volume of all and was erupted over the shortest time span. Courtillot et al. (1986) and Courtillot and Cisowski (1987) cite it as one of the greatest episodes of flood basalt volcanism in Earth's history, and Courtillot et al. (1988) as the largest volcanic eruption during the Mesozoic and Cenozoic eras. Duncan and Pyle (1988) cite it as one of the most remarkable volcanic provinces on Earth in sheer extent and volume, and that it can be correlated in time with events at the K/T boundary.

Today, after extensive erosion, Deccan Traps lavas cover about 5×10^5 km^2 of western and central India to a thickness of about 2000 m in western India to 100 to 200 m in central India (Bose, 1972). Pascoe (1964) suggested that the original lava coverage and related volcanics was greater than 2.6×10^6 km^2. Eruption was extremely rapid, perhaps spanning 0.5 million years (Courtillot and Cisowski, 1987). Eruptive rates for the Deccan Traps of 1.5 km^3 were far greater than for the later Reunion hotspot track of about 0.04 km^3 (Richards et al., 1989). Vogt (1972) had visualized the Deccan Traps as part of globally synchronized mantle plume activity in the K/T transition.

The hotspot volcano Piton de la Fournaise on Reunion Island that produced the Deccan Traps is still producing iridium. Toutain and Meyer (1989) reported that sublimates deposited at 250° to 450°C contain 7.5 particles per billion (ppb) iridium, and the lavas 0.25 ± 0.03 ppb. They noted that because Piton de la Fournaise is related to the hotspot track that generated the Deccan Traps that volcanic gaseous iridium can be related to the K/T iridium anomaly. Rocchia et al. (1988) calculated that iridium amounts three times less than in lavas of Piton de la Fournaise could have produced the K/T iridium anomaly.

Multiple iridium spikes at and about the K/T boundary suggest volcanic source. An almost complete K/T section at Lattengebirge, Bavarian Alps, has three iridium-bearing events over an extended period from latest Maastrichtian into early Danian (Graup and Spettel, 1989). The oldest spike predates the K/T boundary by 14,000 to 9,000 years. Geochemically those spikes display the same signature as the K/T boundary layer and should have a common source. The Brazos River, Texas, locality has two iridium spikes, one at the K/T boundary, and one below (Ganapathy et al., 1981). At Gubbio, Italy, both the iridium and shocked minerals do not occur as a sharp K/T spike; enrichment begins two m below the boundary and extends two m above it for a total of about 400,000 years (Crockett et al, 1988; Rocchia et al., 1989). Koeberl (1989) reported iridium concentrations up to 7.5 ppb in volcanic dust bands in blue ice fields from Antarctica.

Comment Asteroid impact has been suggested as initiator of the Deccan Traps volcanism. However, White (1989) notes that even gigantic impacts are not capable of generating large quantities of basaltic melt by removing the crust and allowing the underlying mantle to decompress. The kinetic energy of a bolide could generate some melt but, for realistic bolide sizes, is inadequate to explain observed flood basalt volumes. Morgan (1979) noted that the Reunion hotspot was beneath the northern part of India 80 million years ago. If an impact event triggered eruption of the Deccan Traps, then an extraterrestrial-origin iridium spike would have to be 80 million old and not 66 million years.

K/T iridium was the basis for the Alvarez et al. (1980) impact theory. However, a suitable crater of the right age and magnitude to trigger the K/T extinctions has not been located. The Manson crater in Iowa has been suggested as the K/T impact site (French, 1984; Izett and Pillmore, 1985; Hartung et al., 1986). Kunk et al. (1989) argue that its age, 65.7 ± 1.0 million years, is indistinguishable from the K/T boundary, estimated to be 66.0 million years. However, Cisowski (1988; personal communication, 1989) notes that the Manson structure is in a normal magnetic polarity chron, whereas the K/T boundary is in a reversed polarity chron (R29). The Manson structure—if it is an impact event—was too small to have triggered an impact winter.

The Kara and Ust-Kara impact structures in the USSR are too old for a 66 million year K/T boundary event (Koeberl et al., 1990). Proposed impact sites in Cuba (Bohor and Seitz, 1990), and the Colombian Basin of the Caribbean Sea (Hildebrand and Boynton, 1990) are too tentative, and open to terrestrial tectonic and volcanic interpretations at this point, to be currently considered at the K/T impact site.

Conclusion The Deccan Traps volcanism was the major documentable biospheric perturbing event of the K/T transition, and is of the correct age. The Reunion hotspot volcano that produced the Deccan Traps is yet producing iridium after 66 million years. Originating near earth's core-mantle boundary, it could have liberated siderophiles in cosmical proportions onto Earth's surface. India's near-equatorial location during the K/T, and the fact that thermal plumes above flood basalt type volcanos can penetrate to the stratosphere, could have carried iridium bearing dust particles into both the Northern and Southern hemispheres.

Having a documentable terrestrial source of iridium negates having to evoke an exotic exterrestrial source for the K/T boundary iridium, and further weakens the concept of a K/T impact winter.

Stable Isotope Record

Climatic cooling associated with a K/T impact winter would be reflected as a positive excursion of the oxygen isotope record in the K/T clay on a global scale. In fact, a general pattern of a negative excursion beginning below the K/T boundary, and extending well above it into the Early Tertiary, is emerging. This general signal of K/T transition warming is coeval with both the Deccan Traps volcanism and a major carbon cycle perturbation. Scholle and Arthur (1980) and Perch-Nielsen et al. (1982) note that a major negative shift in carbon isotopes has been detected at every K/T boundary sequence examined.

Mount et al. (1986) note that several oxygen and carbon isotopic excursions occurred before and during the K/T boundary event. Each involved synchronous negative shifts in ^{18}O and ^{13}C values of about 2 per mil. They note that whereas some workers have related those excursions to an impact event, they are not significantly different from several others that precede it. They further suggest that the K/T boundary excursion is not unique; the excursions are interpreted as episodic warming events. Extinctions of some marine invertebrates were associated with surface-water warming events and decreases in productivity beginning before the K/T boundary event.

Conclusion Globally, the ^{13}C and ^{18}O negative excursions begin below the K/T boundary, and extend above it into the Early Tertiary, coevally with the duration of the Deccan Traps volcanism. Because mantle CO_2 is depleted in ^{13}C, I have proposed that the trans-K/T ^{13}C excursion is reflective, at least in part, of mantle CO_2 degassing associated with the

Deccan Traps volcanism (McLean, 1985a, 1985b, 1985c). The ^{18}O excursion is reflective of trans-K/T climatic warming. The combination of major carbon cycle perturbation and climatic warming would seem to indicate a trans-K/T greenhouse. Stressed conditions in the oceanic mixed layer that persisted for about 500,000 years into the Early Tertiary, referred to as "Strangelove" oceans (Hsu and McKenzie, 1985), were coeval with the Deccan Traps mantle CO_2 degassing.

The isotope record does not support a K/T boundary impact winter.

K/T Shock-Metamorphosed Minerals

Shock-metamorphosed quartz grains displaying multiple intersecting sets of planar lamellae discovered at the K/T boundary by Bohor et al. (1984) were proposed to be definitive evidences of a K/T boundary impact event. Supposedly, such a high degree of shock metamorphism can only be produced by the hypervelocity impact events. Shocked minerals remain the most compelling evidences of a K/T boundary impact event. However, Carter et al. (1986) have described microstructures in the Toba Tuff of Sumatra, proposing the shock metamorphism can be produced by explosive volcanism. The origin of the K/T shocked mineral is currently the topic of lively debate.

Shocked mineral grains are most abundant at western North American sites, but have been reported from several sites in Europe, a core from the north-central Pacific Ocean, and a site in New Zealand (Bohor et al., 1987). Izett (1987) notes that the grains outside North America are exceedingly rare.

Bohor et al. (1987) note that the ubiquitous presence in the K/T boundary layer of shocked quartz, feldspar, and composite siliceous grains argues against an oceanic asteroid impact and, instead, support the idea of a large asteroid impacting into a continental crustal terrain. In this light, an impact crater estimated to range in diameter up to 200 km (Alvarez et al, 1980) ought to be plainly visible somewhere on the North American continent.

Conclusion Shocked mineral grains at the K/T boundary remain the most convincing evidence of a possible K/T impact event.

Conclusions

Little definitive evidence supports the concept of a bolide-induced K/T boundary impact winter. Thus, impact advocates must seek another killing mechanism capable of triggering global extinctions. A K/T transition greenhouse, that is supported by the record, is a possibility.

The recent Williamsburg "Chapman Conference on Global Biomass Burning: Atmospheric, Climatic, and Biospheric Implications" provided stimulating thinking on the possibility of a combined impact-volcano unification of the K/T record via a CO_2-induced greenhouse. A small impact event, by starting fires that injected greenhouse gases into the atmosphere, could have reinforced an already existing volcano-induced greenhouse, pushing the biosphere "over the edge" at the K/T boundary.

A combination model would have a short-duration K/T *boundary* impact-induced greenhouse superimposed upon, and intensifying, a long-duration volcano-induced K/T *transition* greenhouse. Such unification would accommodate the K/T boundary shocked minerals, and intensification of ecological stresses, within the long-duration K/T transition carbon cycle and bioevolutionary perturbations that are preserved in the record. This unification, which accords with the actual record, offers a step forward in isolating the cause of the extinctions while other details are being sorted out down through the years. Theory coupling greenhouse warming to embryogenesis dysfunction in global extinctions has been developed (McLean, 1988). In fact, some impactors have advocated greenhouse conditions as a K/T killing mechanism in their models (Emiliani et al., 1981; O'Keefe and Ahrens, 1988, 1989; Hsu et al., 1982).

Acknowledgments

I thank Dr. Joel Levine, senior scientist, Atmospheric Sciences Division of NASA Langley Research Center, Hampton, Va., for the invitation to participate in the timely and educational "Chapman Conference on Global Biomass Burning: Atmospheric, Climatic, and Biospheric Implications" held at Williamsburg, Va., March 1990, and to write this chapter.

Sky of Ash, Earth of Ash: A Brief History of Fire in the United States

Stephen J. Pyne

Thus sped the days—fearful days—but they brought no relief. The sky was brass. The earth was ashes.

—Frank Tilton, *Sketch of the Great Fires in Wisconsin* (1871)

We are uniquely fire creatures on a uniquely fire planet, and through fire the destiny of humans has bound itself to the destiny of the planet. The capture of fire by early hominids, perhaps as long as 1.5 million years ago, is part of our biological heritage as a species. It is a capacity we will never allow another species to acquire. But the torch was only one tool among many, and fire, one environmental influence among a multitude of manipulations that reworked fire regimes. Whatever humans did, directly or indirectly, to affect fuels affected fires. Together they reconstituted the geography and history of fire on Earth. Together, too, they reconstituted human society.

As humans spread, they propagated like a slowly expanding ring of fire. Within that ring lived humanity; outside it, the wild. Much as cooking rendered edible foodstuffs that were otherwise poisonous or unpalatable, so broadcast fire converted landscapes into forms more readily accessible to humans. Hominids set about slowly cooking the planet. Societies advanced across continents like a fire drive, reshaping habitats and flushing out valuable species. In much the same way fire allows us to extract new meaning from the historic record.

From Asia to America

Certainly fire is an indelible part of our national history. Lightning fires were abundant, requiring only the right proportion of wet to dry, the one to create fuels and the other to ready them for burning, a two-cycle engine for which lightning supplies the spark. Still, the number of ignitions does not, by itself, measure the biotic power of fire: A single point source on the High Plains could propagate over millions of acres, and fire does not have to occur annually to be influential.

Besides, by the time the last glaciers were in retreat, human firebrands supplemented lightning as a source of ignition. The fire regimes that evolved reflected *both* influences in ways that are difficult, perhaps impossible, to disentangle. With their torch American Indians inscribed a new matrix of fuels—some burned annually as hunting grounds, sites for foraging, or corridors of travel; others, as drought, windfall, or the vagaries of human history favored them. Accidental fires proliferated. Long a variable in the equations that governed free-burning fire, ignition now became a near constant; the most heavily used environments were the ones most heavily fired. The pattern was, I suspect, very similar to that witnessed today in Subsahara Africa. It was within this reconstituted geography that natural ignitions had to operate. The torch became a flaming lever that, suitably positioned, allowed otherwise technologically primitive peoples to move a continent.

In broadcast forms, fire was applied to the cultivation and harvest of natural grasses, forbs, berries, tubers, and nuts—wild rice and dry grains, sunflowers, blueberries, camas and bracken, acorn and mesquite beans. Broadcast fire was used to drive off flies and mosquitoes, to keep open major thoroughfares of travel, to encourage rain, and to attack enemies. Where agriculture was practiced in forests, fire was essential to slash-and-burn cultivation. Escape fires from cooking and signal fires were the camp followers of transient hunters and war parties. Writing from Jamestown, Virginia, Captain John Smith encountered an "aboundance of fires all over the woods"; "they cannot travell but where the woods are burnt," he added. When he asked a Manahoac Indian what lay beyond the mountains, he was told "the Sunne," that nothing more was known "because the woods were not burnt" (Babour, 1986).

But of all Indian uses, the most widespread was probably the most ancient: fire in the service of hunting. The strategy of surrounding or driving the principal grazers of a region by fire was universal. In the East it was used for deer; in the Everglades, for alligators; on the tallgrass prairies, for buffalo; in California and along the Colorado River, for rabbits and wood rats; in Utah and the Cordillera, for deer and antelope; in Alaska, for muskrats and moose; in the Great Basin, for grasshoppers. Once horses were acquired, tribes burned for pasturage as well. It is in fact a remarkable coincidence that virtually all fauna prized as game come from environments that are sustained by periodic fire—that are, in brief, susceptible to anthropogenic burning. Additionally, tribes could control the large-scale movement of herds by alternately firing and greening up sites. And only with fire could they interrupt the reinvasion by trees of traditional hunting grounds.

Thus Cabeza de Vaca reported in 1530 how Texas tribes "further inland . . . go about with a firebrand, setting fire to the plains and timber so as to drive off the mosquitoes, and also to get lizards and similar things which they eat, to come out of the soil. In the same manner they kill deer, encircling them with fires, and they do it also to deprive the animals of pasture, compelling them to go for food where the Indians want." In New England Thomas Morton wrote in 1637 that "the Savages are acustomed to set fire of the Country in all places where they come; and to burne it, twize a year, vixe at the Spring, and the fall of the leafe. The reason that mooves them to doe so, is because it would other wise be a coppice wood, and the people would not be able in any wise to passe through the Country out of a beaten path." In Virginia (again) John Smith reported how tribes—"commonly two or three hundred together"—staged annual fire drives, some of which sought to harrass deer into water where they they could more easily be slaughtered from canoes. A century later William Byrd reported that the Indians regularly burned the Virginia forests and barrens. Later, Thomas Jefferson informed John Adams that, indeed, hunting in circles has been known among the Virginia Indians. "It has been practiced by them all; and is to this day, by those still remote from the settlements of whites. But their numbers not enabling them, like Genghis Khan's seven hundred thousand, to form themselves into circles of an hundred miles diameter, they make their circle by firing the leaves fallen on the ground. . . . This is called fire hunting, and has been practiced within this State within my time, by the white inhabitants." The practice, he continued, is "the most probable cause of the origin and extension of the vast prairies in the western Country" (Pyne, 1988). He was not far off.

The American Indian was not unique in these fire practices. The Seminole used fire to hunt alligators identically to the way Indonesians hunted crocodiles; the Iroquois and Cherokees practiced swidden agriculture in much the same way as did pre-Columbian Brazilians or hill cultivators in Thailand; pasturage fires for wildlife in America are indistinguishable from those set on the steppes of central Asia, on the veldt of southern Africa, or on the kangaroo ranges of Australia. The great division of lands in pre-Columbian America was between those sites that were chronically dry and those that were perennially wet; the dry sites were burned routinely. The patterns of Indian burning did not, moreover, merely recapitulate natural fire. The gross geography of fuels and ignitions, the relative proportion of fires by season, the frequency of successive burns—all changed, and with them changed the spangled fire regimes of North America.

The power to start fires, however, is only a kind of vandalism without the power to also control them. The American Indian did practice fire prevention and fire control adequate to the needs of his society. Encampments were kept clear of fuel by grazing or were burned off prior to occupation. Wildfires that threatened a camp would be met by backfires, blankets, and hides. Buffalo robes sometimes served as protective shelters, and threatened parties could kindle escape fires. But the fundamental solution was nomadism or the segregation of important settlements from the scene of broadcast burning. A tribe could simply move on, following the herds or the seasonal cycles of hunting and foraging. Without a fixed land base, the standards for fire control were flexible and the opportunities for fire use extensive. The friction between a mobile people and a tindered land kindled fire everywhere.

At first blush it seems impossible that relatively small numbers of seminomadic hunters, foragers, and swidden cultivators could have exerted much influence over their environment. But fire, properly used, has a multiplicative effect. It propagates. It compounds and magnifies the effects of other processes with which it is associated. It can create, but even more powerfully, it can sustain. Intelligent beings armed with fire can apply it at critical times for maximum spread and effect. The consequences, of course, varied with environments and tribal histories, but by and large the effects were to replace woody vegetation with grassy vegetation, to keep forests (especially pine and oak) in a seral stage, and to reduce understories.

Anthropogenic fire coincided with the climatic warming that ended the last glaciations—together they shaped the biota of the continent. By the time the English made landfall, they often discovered a landscape that resembled the parks and "champion fields" of the Old World. In some places, like the Great Plains—where climate, topography, and vegetation allowed fire to run over vast distances—the outcome could be dramatic. Without anthropogenic fire the tallgrass prairie could not survive. When Indian fire was removed, grasslands spontaneously reverted to brush and trees. The mutual consequences of fire use and withdrawal were particularly evident in the seasonally dry Far West.

In his celebrated 1878 *Arid Lands* report, John Wesley Powell, pioneer conservationist then on his way to the directorship of the U.S. Geological Survey and the Bureau of American Ethnology, declared that the "protection of the forests of the entire Arid Region of the United States is reduced to one single problem—Can these forests be saved from fire?" Fires annually destroyed "larger or smaller districts of timber, now here, now there, and this destruction is on a scale so vast that the amount taken from the lands for industrial purposes sinks by comparison into insignificance." Powell had personally witnessed two fires in Colorado and three in Utah, any one of which "destroyed more timber than that taken by the people of the territory since its occupation." The chief condition for the fires was, admittedly, the seasonally arid climate, but the chief source of ignition was local tribes of Indians (Powell, 1878).

These perceptions have, in fact, been borne out. Throughout the public lands of the West, the past century has witnessed an astonishing efflorescence of woody vegetation. Grasslands have succumbed to succulent desert, hardy scrublands, and colonizing conifers; forests have thickened and aged, in places so densely that they have become impassable, stunted, little more than swards of woody tinder. Study almost any volume of comparative photographs, one set taken in the late 19th or early 20th century and the other in the past 20 years, and the record is almost everywhere the same—in high desert grasslands, across sagebrush plains, through pine savanna, or amid mixed conifer forests (cf. Progulske, 1974; Humphrey, 1987; Rogers, 1982; Gruell, 1983, nd; Hastings and Turner, 1965). Naturally, over such diverse landscapes, the contributing causes are several, but all share a proximate source in the temporary exclusion of fire. In some instances, this reflects the expulsion of fire-prone frontier economies; in some,

the onset of massive overgrazing, which swept the landscape free of grassy fuels; and in some, the aggressive actions of firefighting institutions.

Old World, New World

Europeans and Africans brought additional fire practices to the New World. Some easily merged with existing patterns—fire herding readily replaced fire hunting, for example. Others were novel, or represented such an acceleration of usage that they might as well have been exotic. More significantly, these fire practices accompanied a new ecological invasion and a new economic order. The Europeans brought a swarm of Eurasian flora and fauna, including deadly microbes and avaricious livestock that attacked the old sources of ignition and fuels, so that even when the newcomers exploited similar fires, they did so within a new context that yielded new outcomes. But many of the introduced fire practices intended not to warmly massage the indigenous biota but to replace it violently with exotic plants and animals. The old fire regimes crumbled under the shock. When they were reconstituted, they were rebuilt out of different pieces and they obeyed new, often troubled dynamics. Pioneers, too, found themselves threatened by fires loosed from their former confinement.

There had been plenty of fire in Europe's past. In the ancient world, it had been customary for new colonies to carry with them a torch from the sacred fire of the mother polis. That, in a sense, is what the Europeans did in the New World. Wherever they came from, the new immigrants transported that fire lore—fire practices, fire sources, fire ceremonies and beliefs. The process dated from the *landnam* of neolithic times. It accelerated, for central Europe, with the agricultural crusades directed at the heathen wastelands that bordered medieval Christendom. Even the intensification of land use known as the agricultural revolution of the 18th and 19th centuries exploited fire to rework wastelands; pastoralism expanded into Scottish moors behind a screen of fire, slash-and-burning pioneered the opening of remote sites, and paring-and-burning brought the techniques of swidden to peat and moor. And of course fire was endemic throughout the Mediterranean Basin, an indelible brand of transhumant pastoralism. But gradually cultivation became so intensive and pyrophobic flora so extensive that they restricted open burning to debris piles and hearths. Instead, free-burning fire persisted only along the margins, and especially in overseas colonies stocked with a metastable compound of native fire and transplanted burning.

American pioneers adopted fire practices from both the American Indian and the European. The relative proportion varied by time and place. But it may be said that fire was everywhere used and, in most places, controlled relative to needs. Revealingly, the earliest fire codes attempted to regulate both the transplanted and the indigenous, to redefine the seasons by which farmers might burn fields and surrounding woods and to restrict such adopted techniques as fire hunting. Extensive broadcast burning was itself an indispensable instrument of fire control; in many regions protective burning was mandated by custom and occasionally by statute.

An official with the survey crew on the Plumas Forest Reserve in northern California—this is 1904—summarized the scene well:

> The people of the region regard forest fires with careless indifference. . . . To the casual observer, and even to shrewd men . . . the fires seem to do little damage. The Indians were accustomed to burning the forest over long before the white man came, the object being to improve the hunting by keeping down the undergrowth, which would otherwise shelter the game. The white man has come to think that fire is a part of the forest, and a beneficial part at that. All classes share in this view, and all set fires, sheepmen and cattlemen on the open range, miners, lumbermen, ranchmen, sportsmen, and campers. Only when other property is likely to be endangered does the resident of or the visitor to the mountains become careful about fires, and seldom even then (Pyne, 1988).

Over many landscapes of active settlement smoke hung like a pall during the late summer and fall. Droughts brought the infamous Dark Days. Thus in the 1860s the *Oregonian* declared that "much of the sickness which prevails among us at present is attributed to the heated state of the atmosphere and the immense volumes of smoke created by the vast fires"—fires attributed to landclearing and transhumant pastoralists. "We read about Egyptian darkness, but it is smoke, Josephine smoke—smoke in the morning, at noon, and at night. Meet a neighbor, it is smoke; parting from one, it is smoke. Hogs running around are smoked through and through—live, running bacon . . . we live in the days of smoke. It is smoke, smoke, smoke!" (Pyne, 1988).

Yet pioneers demanded access to fire, without which many lands were biologically worthless to their frontier economies of hunting, shifting cultivation, and pastoralism. They burned to create new environments and to sustain those landscapes. It was as-

sumed in laissez-faire fashion that settlement itself was the principal form of fire control. With reclamation, wildfire would go the way of wild land and wild animals. Wildfire would recede before a muted rural fire, and rural fire, before urban fire. The process of settlement would, of itself, strangle free-burning fire from the landscape.

In some places this did occur, though residual fire practices could endure, as they did in the South. There woods-burning persisted, a protest as much ecological as social. (Early in the 20th century one outraged forester estimated that 105% of Florida burned in one year, the unlikely figure resulting from the double burning of many sites in both spring and fall.) In many other places, however, this evolution never came—never would come. Such was the case of public lands reserved as parks and forests from which settlement was excluded. Other lands had barely encountered the vanguard of frontier reclamation before industrialization set into motion a counterreclamation that rolled back subsistence agriculture.

Everywhere the critical time was the period of transition. Fire practices of Indians or frontier hunters mixed unhappily with those of farmers, and agricultural fire practices could, in turn, interact with industrial settlement in strikingly violent ways. Examples include the early settlement of the tallgrass prairies, where grass fires raged amid planted fields; the opening of the North Woods, catalyzed by the railroads, where farming, logging, and chronic ignitions compounded into deadly holocausts; the abandonment of northeastern farmland, reclothed with resinous conifers and relogged with fiery axes.

Counterreclamation

The Industrial Revolution demanded new conceptions of nature, new valuations of natural resources, a new geography of land usage and settlement—all of which required, and were reflected in, new fire practices. For the most part, the advocates of reform condemned as unacceptable that fiery amalgam then thriving on the frontier. Whatever new fire regimes might evolve, they required first that the old practices be shattered. There was no excuse for the American fire scene, scowled Bernhard Fernow, Prussian-trained forester and emigré; it was all the result of "bad habits and loose morals" (Pyne, 1988). But even in those days a fire-addicted folk found it hard to "just say no."

Clearly American wildlands would not, unassisted, evolve into a fire-free zone like the heartland of European husbandry in which every stalk of domesticated grass was eaten, thatched into roofs, or burned in ovens, or for which woodlots were scraped clean of needles for bedding and branches were harvested for the hearth or field. Some wildlands would, by law, remain wild. So began the movement for organized fire protection that, over the course of the 20th century, determined suitable fire practices and oversaw the creation of suitable fire regimes. Excluding settlement, of course, excluded most anthropogenic fire, but not all of it, and it did nothing about lightning; instead, it excluded the traditional means of fire control. No one set fires, but no one put them out either.

Thus it was that fire control became itself a frontier institution, and vast expanses of public wildland were effectively settled in the name of wildland fire protection—this in the face of enormous pressure from public opinion that considered fires inevitable, fire control impossible, and fire exclusion unwise. Gifford Pinchot recalled

very well indeed how, in the early days of forest fires, they were considered simply and solely as acts of God, against which opposition was hopeless and any attempt to control them not merely hopeless but childish. It was assumed that they came in the natural order of things, as inevitably as the seasons or the rising and setting of the sun. . . . Forest fires were allowed to burn long after the people had means to stop them. The idea that men were helpless in the face of them held long after the time had passed when the means of control were fully within our reach. It was the old story that "as a man thinketh, so is he"; we came to see that we could stop forest fires, and we found that the means had long been at hand (Pinchot, 1910).

Among important experiments were the Adirondacks and Catskills preserves established in New York in 1885, the General Land Office (then responsible for the administration of the forest reserves), private timber protective associations particularly in the Pacific Northwest, and the U.S. Army in the national parks. The cavalry took over in Yellowstone in 1886 and were greeted by fires when they rode into the park, and suppressed 60 fires that first summer—the origins of fire protection by the federal government. All these institutions, however, lacked a genuine basis in theory or scholarship; they were instruments for applying techniques—the common techniques of the day for the control of fire. For real power those skills needed a larger rationale, and they needed to tap a reservoir of moral energy. This only came with the introduction of forestry.

The U.S. Forest Service, in particular, organized fire programs into an integrated, national *system* of fire management. It gave fire protection a federal infrastructure, and to it professional foresters brought the intellectual order of the academy, the rigor of natural science, and the crusading fervor of a reformist era. Its charismatic chief, Gifford Pinchot, thundered that "like the question of slavery, the question of forest fires may be shelved for a time, at enormous cost in the end, but sooner or later it must be faced" (Pyne, 1988). Beginning with the Transfer Act of 1905, which placed the forest reserves under its jurisdiction, the Forest Service did just that.

The Trying Fire

Then, in 1910, the whole pot boiled over. Taft had just succeeded Theodore Roosevelt to the presidency, fired Pinchot for insubordination, criticism of Forest Service mounted—and then the fires came. In the western national forests some 5 million acres burned, more than 3 million concentrated in the Northern Rockies—the fabled Big Blowup. That same month, in California, a long-smoldering controversy over appropriate fire policy flared into public view as critics demanded that the routine light-burning of surface fuels supplement, or even replace, programs aimed at outright fire suppression. Shortly afterwards Congress passed the Weeks Act (1911), which allowed for the expansion of national forests into the East, set up the machinery for federal-state cooperative programs in fire control, and even made possible interstate compacts for fire protection.

The challenges were staggering. The 1910 fires questioned the technical ability of the Forest Service to fight fire, the light-burning controversy, its capacity to formulate adequate policy, and the Weeks Act, its competence to transform a national forest policy into a national fire policy. Almost immediately the management of new forests in Arkansas and Florida was overwhelmed by folk burning. But it was the 1910 fires—millenial fires, the fires of reference for a generation—that set the agenda for the next half century. For the first time in American experience, a serious campaign was waged against wildfires. Results were mixed, naturally. But they climaxed with the incredible saga of Ranger Edward Pulaski, holding his fire crew in their mineshaft shelter at gunpoint while outside a firestorm raged, a modern Horatio at the bridge of American conservation.

It was all high drama. But against it light-burners argued for a fundamentally different strategy of fire protection. The controversy itself was not unique to America; nearly identical versions were played out wherever European forestry moved beyond the ecological and social confines of Western Europe. In the U.S. light-burning brought together an unlikely coalition of frontiersmen and home-grown intellectuals like sentimentalist poet Joaquin Miller and novelist Stewart Edward White, the Secretary of the Interior and the state engineer of California, all of whom argued for routine surface burning out of a conviction that fuel reduction was the only logical and economical strategy for protecting wildlands from wildfire. Besides, light-burning was "the Indian way" of benign forest management. Against such folkways, professional foresters countered with the perceived exemplars of Western Europe in which intellectuals, almost to a man, condemned any use of fire.

All this occurred nearly overnight; amidst the trauma of 79 dead firefighters and an agency debt of over $1 million (1910) dollars; within a society whose political leader, Teddy Roosevelt, advocated the life of "strenuous endeavor," whose greatest philosopher, William James, had that same August 1910 argued for a "moral equivalent of war" to be waged against the threatening forces of nature, whose most popular author, Jack London, wrote novels and stories about struggle in the wilderness, and whose eastern establishment was enamored of the western experience as a place to test manhood. The Yale foresters who went west to fight fires shared the same cultural milieu as the Harvard swells who rode to war with the Rough Riders. The 1910 holocaust was the right event at the right time: It decided the issue of whether, in the United States, fire management would be based on fire control or on fire use. It had to start someplace, and it began with control. By 1923 a panel of foresters officially condemned light-burning as heretical. Instead the U.S. adopted the self-evident strategy that all fires could be stopped if they were detected and attacked soon enough. The U.S. Forest Service established hegemony over American fire policy, equipment, manpower, and research.

From Strategy to System

The subsequent history of an American system of fire protection can be thought of in terms of roughly 20-year cycles, each dominated by a particular kind of problem fire and each subject to the distortions wrought by particular kinds of surpluses made available to fire agencies. The first era addressed the problem of frontier fire, what to do with the specter of widespread folk burning. It attacked—and extinguished—light-burning on front-country lands of un-

questioned value, and it responded to an enormous landbase and virtually unlimited funds available for fire suppression. Foresters brought system to what had been folklore; they even applied the industrial engineering of the day—Taylorism—to the problem of fire suppression.

By the early 1930s the question of backcountry fires dominated the national agenda—lands remote in space and time, forests deep in the public reserves, lands cut over and burned over for which fire protection and reforestation were investments in a distant future. Thanks to the Roosevelt administration, the means at hand swelled enormously, and the Forest Service adopted an otherwise improbable end, the 10 A.M. Policy, an "experiment on a continent scale," the Chief Forester called it, which stipulated as an agency objective that fire officers should plan to control *every* fire by 10 A.M. the morning following its report. The means at hand had so expanded that they made otherwise improbable ends possible.

The great transition came with the Second World War. This was a fire war, culminating in a terrifying new fire weapon, the atomic bomb—nationalized fire consciousness, galvanized fire research, and, after the Korean War, liberated surplus military equipment in such quantities that fire protection mechanized virtually overnight. The war equated firefighting with military defense, and while it raged, Walt Disney Studios released *Bambi,* which became a cultural icon in American consciousness of fire that quickly metamorphosed into a national fire prevention campaign. The problem of the big fire, the mass fire, dominated attention; and in response, fire protection expanded into the final frontiers like the Alaskan interior. Everywhere it intensified. All fire was hostile fire. The United States embarked on what appeared to be a cold war on fire and assumed a self-appointed role as global fireman.

Then came the breakup. Fire control reached a point of diminishing returns, a new ecology doubted the legitimacy of the biological premises behind fire suppression, and the wilderness movement restored moral fervor—now in articulate opposition—to the debate over fire practices. Fire suppression was not cheap, and it was not an ecologically neutral act; the removal of fire could affect ecosystems as powerfully as its introduction. In particular, the conundrum of how to manage fire in wilderness preserves stimulated a debate that, by 1970, forced a reexamination of policy, fragmented Forest Service hegemony, and installed a pluralism of fire practices. Between 1968 and 1978 a policy of fire by prescription replaced the moribund 10 A.M. policy among federal agencies. Fire managers struggled to *restore* fire, in some form, wherever possible. With almost metronomic regularity, this cycle ended in the Yellowstone Gotterdammerung of 1988.

The contemporary era has apparently selected the exurban fire—the problem of fire in mixed wildland and suburban usage—as its informing concern. But issues of biomass burning may challenge that choice. When measured against planetary combustion and potential climatic warming, how to protect amenities communities from themselves seems parochial, even dilettantish.

Global Fire: Means and Ends

A history, even one as sketchy as this, nonetheless questions the choice of a baseline for measuring environmental change. There is less open burning in the United States now than a century ago, less even than 50 years ago; almost certainly there is less than in pre-Columbian times. What has changed is the character of combustion, the geography of carbon sinks and sources, and the biological timing of burning. Industrial combustion has superseded free-burning fires; combustion now proceeds with almost savage indifference to timing, burning winter and summer, day and night; the chemistry of this new combustion has created products not readily scavenged and reabsorbed by the biosphere; its fuels are not part of biospheric cycles, large and small, but geospheric rhythms, exhumed from ancient sediments, long outside the ecological mechanics of free-burning fire.

It is not self-evident, however, that the overall amount of combustion is greater now than in the past. In some respects it is clearly less. Nor is it obvious that trees, as carbon sinks, are significantly fewer. The real biotic losers in American settlement are grasslands and wetlands, not woodlands. The 20th century makes a much compromised calibration point. Just when soaring industrial combustion compensated for plummeting acreages subjected to open fires—if it has so replaced it—is not obvious.

Even so, it is quixotic to speak of fire exclusion because fire is too fundamental to human technology and too elemental in its planetary ecology. Humans cannot deny themselves fire and remain human; fire control and fire use are complementary practices; where humans practice one, they must also practice the other. Likewise, few American ecosystems can thrive without fire at some point, to some extent, recharging their dynamics. The only way to eliminate

wildland fire is to eliminate wildlands, and that typically shifts the burden to some other form of combustion. The downside to pyromanticism is both economic and ecological: We cannot, without huge economic and environmental costs, convert forests into biotic vaults to store carbon bullion. In North America there is a fire tithe levied against wildlands. It can be paid voluntarily, or it can be extracted forcibly.

All this is to affirm that fire burns within in a cultural no less than a natural environment. Humans and fire have coevolved, like the double strands of a DNA molecule. For millenia the geography of fire and the geography of humans have increasingly intertwined to the point that they have become virtually coextensive. But while we are genetically programmed to handle fire, we do not come equipped knowing how to use it. From the earliest days of *Homo erectus,* that is, fire entered into a moral universe.

Questions about appropriate fire practices far transcend scientific inquiries into fire ecology and technical issues of fuel loads, aerosols, and air tankers. They interact with all those concerns that influence human behavior. They involve politics, economics, religion, literature, art, and philosophy; they interpenetrate with questions about how people have behaved in the past and how they should behave in the future, about the legitimacy of societies, about identity, about values. Fire has always forced these issues, even as it forces stubborn biomes to yield nutrients necessary for human sustenance. It can meld ideas and values that otherwise have little in common, and it can, by a kind of intellectual pyrolysis, dissolve bonds that have traditionally held the complex values by which we evaluate our behavior as fire creatures.

As old fire practices enter new landscapes or as new forms of combustion absorb old landscapes, they are defining a new calculus of values. The scene can be confusing. The need to incorporate fire in some ecosystems coexists with the need to reduce it in others. Some fires are good, some bad, some are sometimes good and sometimes bad. Ancient fire practices may be condemned in developing countries and applauded in developed nations. Not only are the criteria changing by which we determine good fires from bad, but the procedures, both scientific and political, by which those determinations are made are also in flux. Fire is a multiple, subtle, nuanced phenomenon; human fire practices and policies must reflect that variety.

In a preindustrial Earth, broadcasting burning on a continental scale could be acceptable, part of a venerable practice in which, bit by bit, humanity has cooked the biosphere. In an industrializing Earth,

simmering with fossil fuel combustion, it may help
convert the planet into a crock pot. By midcentury,
inspired by the doctrine of conservation, American
intellectuals convinced a skeptical public that fire was
unnecessary and undesirable. More recently, en-
flamed by an ideology of wilderness, the belief propagated that free-burning fire was natural and healthy, a
vestal flame for America's virgin lands; the American
fire community has spent the last 20 years vigorously
exploring ways to put fire back into the landscape;
through prescribed fire the goodness of the wilderness could be distributed universally. In this way it
was possible to applaud a million acres of conifer
forest burned in Yellowstone at a cost of $130 million.
Within the context of potential global warming, however, that spectacle can look different, more like a
leveraged biotic buyout, an exotic self-indulgence
made possible by an ill-disciplined American afflu-
ence. In the emerging environmental calculus of
global change, all fires have their price. The new
physical geography of fire, that is, must be reconciled
with its new moral geography.

For this end, too, fire is a competent means—the
fire not only of forest, field, and furnace, but the
trying fire, the fire of judgment. The issue is not only
how we apply and withhold fire but how, through fire,
we evaluate ourselves. Thus, in this—one of the great
eras of fire on Earth—we are witnessing new declen-
sions of an eternal flame. To the hearth fire as symbol
of the family, to the vestal fire as symbol of society, we
are adding a yet-unnamed global fire as symbol of a
planet.

A Five-Century Sedimentary Geochronology of Biomass Burning in Nicaragua and Central America

Daniel O. Suman

Charcoal, or black carbon, is produced during incomplete combustion of biomass. Small carbonized plant particles are mobilized by winds and rivers, carried to the sea, and deposited in near-shore sediments. During these processes, the carbonized materials remain relatively inert and maintain their identifying structures. Thus, marine and lacustrine sediments may hold historical records of burning activities (Griffin and Goldberg, 1975, 1979, 1981, 1983; Suman, 1986).

Biomass burning in Central America has been amply documented (Bartlett, 1956; Batchelder, 1967; Batchelder and Hirt, 1966; Bennett, 1968; Cook, 1909; Daubenmire, 1972; Johannessen, 1963; Myers, 1981; Parsons, 1976; Wagner, 1958). Fire plays an important role in agriculture in this region. It is used during the dry season (December to April) to clear the savanna of dead growth, stimulate new growth of grasses with the first rains, maintain the savanna free of shrubs and woody material, and provide clean fields for cattle. Farmers also practice traditional slash-and-burn agriculture by igniting the cleared and dried woody biomass on their small plots at the end of the dry season. With the onset of the first rains, they plant maize and beans in their cleared plots which the ash has "fertilized."

In spite of the extensive use of fire as an agricultural agent in Central America today, little is known of its history of biomass burning or agriculture. As an indicator of the burning practices on the adjacent land, a sedimentary record of carbonized particles sheds light on the trends in frequency and areal extent of biomass burning. This research focuses on a sediment core recovered from an anoxic site in the Pacific Ocean adjacent to the Central American Isthmus and reports a five-century record of charcoal deposition.

Methodologies

Sampling

During Leg 73/16 of the R/V *Knorr,* I collected a 45 cm box core (Core 2) adjacent to the Nicaraguan coast at location 11°28.5′N, 87°16.4′W (Figure 62.1).

This core site was 84 kilometers (km) from the Central American Isthmus and located on the continental slope at 586 meter (m) water depth.

On board, I divided the top 21 cm of the core into 2 cm sections, except for the top 1 cm section. Section thickness from 21 cm depth to the bottom of the core was 4 cm.

Dating Techniques

Much of the lead-210 (Pb-210) produced by the decay of radium-226 in seawater and added to the ocean from the atmosphere is scavenged and reaches the sediments before decaying with a half-life of 22.3 years. The activities of this excess Pb-210 are useful for determination of sedimentation rates and have successfully been employed in lacustrine and near-shore sediments (Faure, 1986). In cores with a constant sedimentation rate, the logarithm of the excess Pb-210 activity is a linear function of core depth, permitting calculation of the sedimentation rate.

I introduced time frames into the sediment core through Pb-210 geochronologies using the analytical techniques of Koide et al. (1973) and Koide and Bruland (1975). The Pb-210 was leached from the sediments with HCl and HNO_3 and separated from the solution along with the Pb carrier by anion exchange chromatography. The Pb was precipitated as the sulphate and its activity determined by beta emission of its short-lived bismuth-210 daughter.

Quantitative Determination of Sedimentary Charcoal

Prior to calculation of the fluxes of charcoal to the sediments, the sediment samples were chemically digested in order to isolate the charcoal fraction. This involved treatment with HCl and HF to dissolve carbonates and silicates and, subsequently, mild oxidation with alkaline H_2O_2 to destroy organic materials. Charcoal is inert to this chemical treatment (Griffin and Goldberg, 1975).

I determined the charcoal concentrations in the residues using the infrared spectrometric technique of

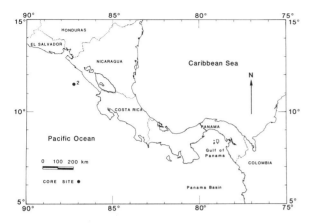

Figure 62.1 Map of sediment core location and the Central American region.

Smith and Griffin (1975). Grinding of the samples in air produced surface oxidation of the charcoal. In the infrared spectra these carbonyl and anhydride bonds give rise to a series of absorption bands which are linearly related to the mass of charcoal in the sample.

Morphological Analysis of Sedimentary Charcoal

I studied the morphologies of charcoal particles greater than 38 μm in diameter using optical and scanning electron microscopy and developed a morphological classification scheme from observation of standard combustion materials (Suman, 1983). Charcoal produced from grasses (*Gramineae*) and woods can be distinguished on the basis of the structures of their >38 μm charcoal particles. Grass charcoal particles originating from epidermal tissues are easily identifiable by their flat and serrated characters and stoma-like structures (Figure 62.2). Woods produce characteristic carbonized xylem particles which show horizontal medullary rays and unique banding patterns (Figure 62.3). A useful concept for expressing the relative contribution of grass and wood particles in a sample is the "Gramineae Index," which is the percentage of particles of grass origin divided by the total number of identifiable carbonized grass and wood particles.

Results

Geochronology

The excess Pb-210 activities (Figure 62.4, Table 62.1) from the sediment core sections suggest that the recovered sediments were undisturbed and preserve a high-quality record of sedimentary deposition. The Pb-210 activities of the surface section of the core (0 to

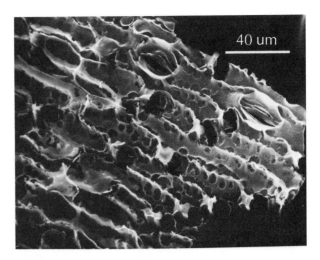

Figure 62.2 Electron micrograph of a carbonized grass particle. These epidermal particles are generally flat and serrated and may contain stoma-like structures.

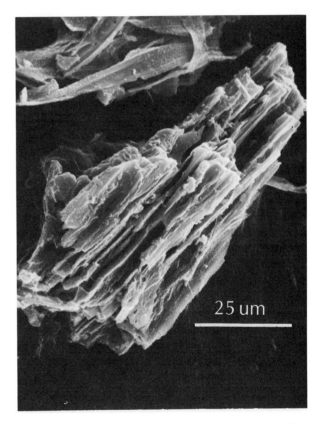

Figure 62.3 Electron micrograph of a carbonized wood particle. These xylem particles show longitudinal banding and sharp breakage.

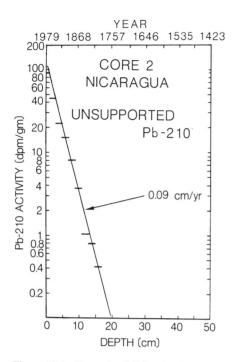

Figure 62.4 Excess lead-210 activity as a function of core depth and deposition date.

Table 62.1 Results of Pb-210 analyses of Core 2 sediments

Depth (cm)	Total Pb-210 activity (dpm g^{-1})	Background activity (dpm g^{-1})	Unsupported Pb-210 activity (dpm g^{-1})
0–1	92.94	1.85	91.14
1–3	46.52		44.72
3–5	23.93		22.13
5–7	16.91		15.11
7–9	9.95		8.15
9–11	5.51		3.71
11–13	2.83		1.03
13–15	2.60		0.80
15–17	2.22		0.42
17–19	3.24		1.44
19–21	2.07		0.22
21–25	1.77[a]		0
25–29	1.68[a]		0
29–33	1.83[a]		0
33–37	2.57[a]		0
37–41	1.46[a]		0
41–45	1.78[a]		0

a. Background.

1 cm) reach a high value of 92 dpm g^{-1}, indicating the absence of Pb-210 remobilization from deeper strata. The linear relationship between the logarithm of the Pb-210 activity and core depth is also consistent with the absence of bioturbation.

Additional evidence (high values of sedimentary organic carbon between 14% to 19%, observed hydrogen sulfide smell during sampling, and high values of alkalinity [>12 meq/L] in the pore waters) suggests that the sedimentary conditions at the core site are anoxic and that the overlying waters have low oxygen concentrations (Suman, 1983). Such anoxic conditions minimize bioturbation, which is a cause of sediment remobilization. At 586 m water depth beneath the intense upwelling zone located westward of Central America, the oxygen minimum in the water column probably impinges on these sediments. Murray et al. (1983) noted that in the nearby Guatemala Basin the oxygen minimum zone extends from 75 to 1000 m in the water column with oxygen concentrations less than 10 μmol kg^{-1}. The R/V *Knorr* 73/16 hydrographic data at a location several kilometers from the site of the core (11°30.7′N, 87°15.6′W) also substantiate anoxic conditions (M. Bacon, personal communication). Oxygen concentrations in the water column below 366 m to the sediment-water interface at 628 m were less than 5 μmol kg^{-1}.

The Pb-210 activities reached background levels at 21 cm in Core 2 sediments. The excess Pb-210 activities were calculated by subtracting the average activity in the six deepest sections (1.85 dpm g^{-1}) from the total Pb-210 activities. Graphical calculation yielded a sedimentation rate of 0.09 cm yr^{-1}. I have assumed that the sedimentation rate was constant at this site and extrapolated deposition dates for the entire core. Based on this model, the 45 centimeters of sediments represent approximately 500 years of deposition history.

Charcoal Concentrations and Fluxes to the Sediments

The charcoal concentrations in the sediments of Core 2 range between 358 and 1061 ppm dry sediment (Table 62.2). Charcoal fluxes to these sediments are the product of the Pb-210 determined sedimentation rate, measured sediment densities, and the charcoal concentration determined by infrared spectrometric analysis. Charcoal fluxes to the sediment site vary from 44 to 135 μgC cm^{-2} yr^{-1} (Table 62.2, Figure 62.5).

The core displays a decrease in the charcoal flux from the 15th century toward the end of the 19th

Table 62.2 Charcoal fluxes to sediments

Core depth (cm)	Sedimentation rate (cm yr^{-1})	Deposition date (yr)	Charcoal concentration (ppm)	Charcoal flux (μgC cm^{-2} yr^{-1})
0–1	0.09	1968–1979	607	64
3–5		1923–1946	614	73
7–9		1879–1901	699	84
11–13		1835–1857	368	45
13–15		1812–1835	422	52
15–17		1790–1812	358	44
17–19		1768–1790	505	63
21–25		1701–1746	861	107
25–29		1657–1701	736	91
29–33		1612–1657	834	104
33–37		1568–1612	710	88
37–41		1523–1568	1061	133
41–45		1479–1523	1056	135

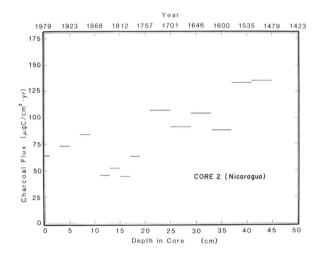

Figure 62.5 Charcoal flux as a function of core depth and deposition date.

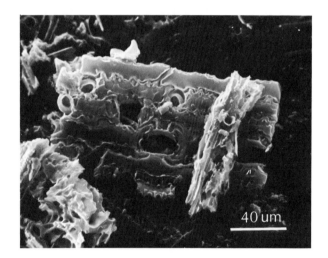

Figure 62.6 Carbonized grass particle from the 41 to 45 cm section of Core 2, corresponding to A.D. 1479 to 1523.

century. The 20th century strata show a modest increase in charcoal flux from the low 19th century values, but they do not reach the high fluxes of the 15th, 16th, 17th, and 18th centuries. The deepest sections of the core, which may date to the pre-Columbian period, exhibit the core's highest charcoal fluxes.

Charcoal Particle Morphologies
In spite of the combustion, transport, and sedimentation processes, the large charcoal particles produced by biomass burning were well preserved and retained

morphological characteristics of their original plant structure (Figure 62.6). I determined the "Gramineae index" for >38 μm charcoal particles in eight sections throughout the length of the core (Table 62.3). The Gramineae Index was consistently high in all sections and ranged between 86% to 97%. The top two sections of the core (0 to 1 cm and 9 to 11 cm, corresponding respectively to deposition periods 1968 to 1979 and 1857 to 1879) contained the lowest Gramineae indices (86% and 87%, respectively). Deeper core sections displayed abundant and well-preserved carbonized grass particles with Gramineae indices over 90%. The maximum value was 97% at 29 to 33

Table 62.3 Gramineae index of charcoal particles

Depth (cm)	Deposition date (yr)	Gramineae index (%)
0–1	1968–1979	86%
9–11	1857–1879	87
11–13	1835–1857	95
15–17	1790–1812	90
21–25	1701–1746	93
29–33	1612–1657	97
33–37	1568–1612	95
41–45	1479–1523	92

Table 62.4 Comparison of charcoal fluxes to marine sediments

Site	Flux (μgC cm^{-2} yr^{-1})
Nicaragua (this work)	44–135
Gulf of Panama (Suman, 1986)	87–469
Santa Barbara Basin, California (Griffin and Goldberg, 1975)	9–37
Saanich Inlet, British Columbia (Griffin and Goldberg, 1975)	42–100
Central Equatorial Pacific Ocean (Herring, 1977)	<0.011

cm (corresponding to deposition period 1612 to 1657).

Discussion

Comparison of Present-Day Charcoal Fluxes

The charcoal fluxes to the continental slope core site adjacent to Nicaragua range between 44–135 μgC cm^{-2} yr^{-1} and are the same order of magnitude as charcoal fluxes to other near-shore sites, such as three sites in the Gulf of Panama (Suman, 1986), the Santa Barbara Basin (Griffin and Goldberg, 1975), and the Saanich Inlet (Griffin and Goldberg, 1975). All these near-shore charcoal fluxes are several orders of magnitude greater than fluxes of charcoal to deep-sea sediments of the equatorial Pacific Ocean (Herring, 1977). Table 62.4 compares the charcoal fluxes to these sites.

The present-day charcoal fluxes to the uppermost section of the Nicaraguan core (64 μgC cm^{-2} yr^{-1}) are notably lower than the corresponding fluxes to the three sites in the Gulf of Panama (127, 165, and 353 μgC cm^{-2} yr^{-1}). While these fluxes might suggest that biomass burning is more extensive in Panama than further north in Central America, perhaps their differences are more a function of the distance between the core sites and the shore. The Nicaraguan site was 84 km offshore, while the three Panama cores mentioned above were recovered at 36, 29, and 17 km offshore, respectively. Given the simplistic assumption that the charcoal source function is relatively constant throughout the Pacific watershed of Central America, charcoal fluxes should be greater when the core location is closer to the burning area. Muller (1959) demonstrated this principle in the Orinoco Delta where pollen concentration in the sediments generally decreased with increasing distance from the coast.

Extent of Biomass Burning during the Last Five Centuries

An underlying assumption of this discussion is that the charcoal fluxes to the sediments are proportional to the biomass burned in the region upwind or upcurrent of the core site. Whether the charcoal particles are transported from their source by runoff or winds, or a combination of the two, their likely origin is the Pacific watershed of Nicaragua and northwestern Costa Rica. The strong surface drift currents in these coastal waters range between 22 to 46 cm sec^{-1} and flow to the west and west-northwest, or offshore, during the dry and rainy seasons (Hubbs and Roden, 1964). Tchernia (1980) indicates variable longshore currents in this area of the Pacific Ocean. During the dry season, the prevailing winds blow from the east and northeast, again pointing to Nicaragua as the source of the eolian charcoal (Forsbergh, 1969; Hubbs and Roden, 1964; Tchernia, 1980). The major rivers flowing into the Gulf of Fonseca, which is located between Nicaragua, Honduras, and El Salvador, may also be important charcoal carriers to these waters.

The variation of the charcoal flux to the Nicaragua core site (Figure 62.5) suggests that, when averaged over decades, the biomass burned in Central America during dry seasons has generally decreased since the Spanish Conquest, which occurred five centuries ago. Moreover, the interval with the highest charcoal flux (41 to 45 cm) and, therefore, the most biomass burning, represents the four decades *prior* to the arrival of the Spanish in 1522. Although this conclusion may seem inconsistent with the popular notion that biomass burning in Central America has never been more extensive than it is today, biogeographers offer supportive evidence and hypotheses.

The pre-Conquest inhabitants of Central America had a tremendous impact on the land due to activities such as vegetation clearing and burning before planting (Bartlett and Barghoorn, 1973; Tsukada and Deevey, 1967; Turner, 1978). Many researchers suggest that the Central American savannas probably are a cultural artifact (Taylor, 1963). Johannessen (1963) estimated that in 1522 when the Spanish occupied Honduras, savannas covered most of the valley floors of that nation's interior. Similarly, Sauer (1969) and Bennett (1968) reported that the first Spanish explorers in central Panama found savanna without forest. Native peoples must have maintained the savannas through extensive burning.

After the Conquest, the Indian population throughout Central America rapidly collapsed as a result of disease, military campaigns, and the disruption of their food supplies. Based on colonial documents, Newson (1986) estimated that within two decades of the Conquest, the Indian population in Honduras shrank from 600,000 to 32,000, or to as little as 5% of its pre-Conquest numbers. During the same period, the population in Nicaragua was reduced to 5% of its pre-Conquest levels or from 600,000 to 30,000 people (Denevan, 1961). In Panama, the population experienced a 90% decrease soon after the Conquest (Jaén Suárez, 1979).

The Spanish also introduced cattle to Central America, and the Iberian cattle culture soon began to thrive on the savannas or former croplands of the region which were left empty as a result of the Indian population decrease. Chroniclers (Cibdad Real, 1975) report many cattle ranches by the end of the 16th century in Honduras, Nicaragua, and Costa Rica. By 1800, over 200,000 cattle grazed in Honduras, while the human population was less than half this number (Johannessen, 1963). Panama conformed to this pattern (Jaén Suárez, 1980). Johannessen (1963) suggested that the absence of livestock management in Central America resulted in overgrazing and deterioration of the savannas. Heavy cattle grazing may have led to an actual invasion of fire-resistant shrubs and trees in areas that once were pure grasslands, as well as to a decrease in biomass on these degraded savannas. According to Johannessen (1963), "the intensity and probably the frequency of fires on the savannas are greatly reduced from the former conditions as a result of the heavy grazing that removes the fuel over much of the surface." Both overgrazing and the Indian population collapse could account for the decreased charcoal source function since the Conquest. However, because the charcoal flux remained high during the interval 1523 to 1568 (37 to 41 cm), the decrease in fire frequency may not have been contemporaneous with the population decrease or the introduction of cattle.

The pollen analyses of lake sediments in Guatemala and El Salvador also support the hypothesis that biomass burning and human disturbance of vegetation significantly declined after the Spanish occupation of the region (Tsukada and Deevey, 1967). The abundant presence of carbon fragments, probably of grass origin, in these lake cores (measured in number per gram of sediment) is generally lower during the post-Conquest period A.D. 1523 to the present) than during the post-Classic period (A.D. 900 to 1523). Sediments of the Classic Mayan period (A.D. 200 to 900) contain the highest incidence of carbon fragments.

In the late 19th and 20th centuries, the rural population in Central America increased dramatically (International Institute for Environment and Development, 1987). This population rise may be responsible for an increase in the use of traditional slash-and-burn agricultural practices and a modest increase in the charcoal source function and charcoal flux from the core's low values during the period from 1770 to 1860.

Additional factors in the 20th century tend to mitigate the effect of population increase on the charcoal source function. Production of export crops, such as coffee, cotton, and bananas, is widespread in the Pacific watershed of Central America (Parker, 1964) and does not involve regular agricultural burning. Moreover, the modern livestock management techniques which employ introduced grasses and irrigated plots do not depend substantially on burning.

Although biomass burning is widely practiced today in Central America's Pacific watershed, the burning frequency and the biomass burned have decreased when compared to pre-Conquest levels.

Type of Vegetation Burned

Throughout the sediment core, carbonized grass particles were much more abundant than carbonized wood particles. This suggests that (1) the Pacific watershed of Central America was deforested and converted into savanna by native peoples prior to the 15th century or (2) the region's natural vegetation due to climatic and/or edaphic conditions was grassland savanna. The first hypothesis is favored by biogeographers who note that in most of Central America, savanna is at a disclimax; the region's rainfall, temperature, and soil types would support tropical forests if the vegetation were not disturbed (Porter, 1973; Taylor, 1963). On the other hand, in *certain* areas, such as the Nicoya Peninsula in northwestern Costa Rica, the savanna is probably favored by the volcanic dust which covers the plains (Wagner, 1964).

Early Spanish chroniclers observed extensive savannas in the Central American Isthmus. However, the post-Conquest land abandonment may have stimulated a partial resurgence of forests in areas where grasslands were once predominant. Although today's deforestation in Central America is occurring in the Atlantic watershed, most of the remaining Pacific forests have also been destroyed and replaced by grasslands and export crops during the 20th century (Parsons, 1965; Taylor, 1963; Terán and Barquero, 1964; Wagner, 1958). Today, grasslands with scattered trees cover most of the Pacific coastal plain from Guatemala to Panama.

The results from this sediment core contain little evidence for this recent forest destruction in the Pacific watershed. Rather, carbonized grass particles predominate throughout the core's five-century period. Whichever theory regarding savanna formation in Central America is correct, the high fluxes of carbonized grass particles to the nearshore sediments strongly suggest that the Pacific savannas of Central America were well-developed in pre-Columbian times and that grasses constitute most of the biomass burned during the last five centuries.

Conclusion

The research illustrates that biomass burning has been an important ecological factor in the Pacific watershed of Central America at least during the past five centuries. Fluxes of charcoal have generally decreased toward the present suggesting a reduction in the charcoal source function. Perhaps, five centuries ago, the frequency of biomass burning was greater than it is today, larger areas were burned, or biomass per unit area of burned grassland was greater. The major type of biomass burned throughout this five-century period has been grass, as opposed to woods, indicating that any major deforestation of the Pacific watershed of Central America occurred prior to the Conquest.

The Great Lakes Forest Fires of 8–10 October 1871

Steven W. Leavitt

Wilson (1978) suggested a "pioneer agricultural revolution" took place worldwide from 1850 to 1900 as evidenced by an abrupt change in $^{13}C/^{12}C$ ratios of a time series from bristlecone pine tree rings. This "revolution" took place with the conversion of large tracts of forest to agriculture in North America, New Zealand, Australia, South Africa, and East Europe. Although that carbon-isotope time series has not been reproduced, indeed, lumbering in Michigan, Minnesota, and Wisconsin was taking place at a frantic pace in this time period. This deforestation activity was a contributing factor to a number of large forest fires in the American Midwest, including the largest historic fire in North America on 8 to 10 October 1871.

This October 1871 fire was actually a collection of fires (Figure 63.1) which burned in Wisconsin, the Upper and Lower Peninsulas of Michigan, and Illinois. That in Illinois is best known as the Great Chicago Fire, although in terms of loss of life, the forest fire centered at Peshtigo, Wisconsin, was much more devastating. Historic accounts yield many clues as to the nature and characteristics of the fire, and provide insight into the importance of this event in the context of the global carbon cycle.

Contributing Factors

Three prevalent beliefs seemed to have guided the forestry activities of the 1800s (Davis, 1959; Holbrook, 1944). The first was that there existed an inexhaustible supply of virgin timber. The second was that cutting of timber was a necessary prelude to taking advantage of the rich agricultural land beneath the forest. Third, fire in second-growth forest was not of major concern because it would help clear the land faster. Forest fires seemed to pose a problem only when they threatened settlements, railroads, or valuable timber.

Logging in the area began in Michigan in the 1840s and spread westward. By the 1860s, tremendous activity in the Saginaw area of Michigan elevated that state to the top producer in the nation. By 1871,

however, there were still immense quantities of timber to be harvested. At Peshtigo, Wisconsin, for example, it was noted that "in spite of logging for more than a decade, the immense forest began at town's edge and ran west and north; broken by several clearings in a hardwood area a few miles from Peshtigo" (Holbrook, 1944). In the upper Midwest, in general, Holbrook (1944) states that "by 1871, the vast forests of white pine were being cut at a great rate . . . some of the pines were one hundred-twenty feet tall and three hundred years old . . . [but] cutting methods wasted one-fourth of each tree." The latter statement suggests the potential abundance of fuel even after the forests were logged over.

Several climatic conditions preceded the fire (Haines and Sando, 1969), the most important of which was a period of prolonged moisture deficiency beginning as early as May in Lansing and Grand Rapids, June at Sturgeon Bay and July at Chicago, and extending into October. Soil moisture was also low for several months prior to the fire as seen in the Lansing and Sturgeon Bay records where soil moisture was below wilting point for two months. Relative humidity in August and September also tended to be quite low, often 20% to 30% in the 10 days preceding the fires. Temperature actually tended to be somewhat below normal over the drought period and in the immediate days prior to the fire.

Several practices contributed to the fires (Holbrook, 1944), which would eventually flare up to the devastating burn of 8 October. These activities included burning of the forest waste (slash) by loggers and burning of slash and second-growth vegetation by settlers to open up the land. The latter type of fire is believed responsible for the Williamsonville fire of the Door Peninsula in Wisconsin. Land-clearing operations of the railroad were another source of fire, and believed to be the primary source of fire in the Peshtigo area.

During the summer and fall months it was not uncommon to have such fires burning unchecked in the area unless settlements were threatened. Up at

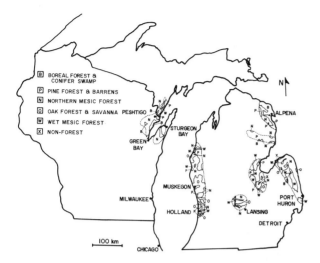

Figure 63.1 Map showing the locations of the 8 to 10 October 1871, forest fires (after Haines and Sando, 1969), and the nature of the presettlement vegetation in these locations (after Stearns and Guntenspergen, 1987).

Peshtigo, such fires in September contributed "dense, suffocating smoke . . . over hundreds of square miles" (Tilton, 1871), resulting in the necessity of steamers in Green Bay to navigate by compass and to blow foghorns because the shores were not visible. Such preliminary fires in 1871 also destroyed railroad camps near Peshtigo and eliminated 20 miles of telegraph line (Tilton, 1871). On 7 October the *Marinette and Peshtigo Eagle* reported "fires are still lurking in the woods . . . ready to pounce upon any portion of the village in the event of a favorable wind" (Holbrook, 1944).

With abundant fuels and fires in place, only the mechanism to drastically intensify the fires was lacking. An intensifying low-pressure system tracking from South Dakota into southwestern Minnesota on 8 October, together with a corresponding high-pressure system over West Virginia, drove southwesterly winds of 25 miles per hour or greater into northern Illinois and Wisconsin (Haines and Kuehnast, 1970). The ultimate passage of the associated front brought some rain to put out fires: 0.2 to 0.5 inches at Peshtigo and ≤0.25 inches in western Lower Michigan on 9 October, although none was reported in eastern Lower Michigan through October 10 (Haines and Kuehnast, 1970).

Nature of the Fire

The October 1871 fires, especially those at Peshtigo and Williamsonville, were characterized by their extreme intensity. Kegs of iron nails were heated until the metal ran "like water" (Tilton, 1871). A revealing account (published in 1874) by an eyewitness, Father Pernin (1971), states:

> The iron tracks of the railroad had been twisted into all sorts of shapes, whilst the wood which had supported them no longer existed. The trunks of mighty trees had been reduced to mere cinders, the blackened hearts alone remaining. All around these trunks, I perceived a number of holes running downwards deep in the earth. They were sockets where the roots had lately been. I plunged my cane into one of them, thinking what must the violence of that fire have been, which ravaged not only the surface of the earth, but penetrated so deeply into its bosom.

Tilton (1871) reported "boulders of rock in the middle of large clearings cracked to pieces by the heat, and the stumps so completely burned out that the roots, even to the small fibers, are gone, and holes left in the ground where the stumps had been."

Some unfortunate souls caught by the intense, rapidly moving fires, were reduced to piles of ash, sometimes only with some metal items (e.g., coins or jewelry) remaining as a clue to their human provenance. A member of the relief party marveled that the bodies of large lumberjacks were reduced to "nothing . . . [more] than a mere streak of ashes that would scarce fill a thimble" (Holbrook, 1944). Suffocation and asphyxiation were also important causes of death as evidenced by the deaths of many who sought refuge from the rising heat in wells and root cellars (Pyne, 1982). The majority of survivors at Peshtigo were those who went into the Peshtigo River, keeping their heads continuously wetted to prevent combustion. Water was not always a place of safety, exemplified in the macabre case at Williamsonville where two people who sought refuge in a water tank were boiled alive (Holbrook, 1944).

The intense nature of the fires was enhanced by the associated self-generated winds. Eyewitness reports of "tornado of fire," "hurricane of fire," "waterspout of fire," "cyclone of flames," and fire "balloons" (Pyne, 1981) attest to the devastating winds. Like accounts of meteorological tornadoes, the noise from these fire tornadoes was described as being akin to the roaring sound of freight trains or to a rumbling even greater than Niagara Falls (Holbrook, 1944). Out in Green Bay, falling coals fell onto a lumber schooner 2 miles (3 km) offshore, and a flaming board fell on the deck of a steamer 7 miles (12 km) offshore (Holbrook, 1944). The intense heat of the fires producing rapidly rising convective plumes coupled with the strong horizontal regional wind components of 15 to 25 mi h^{-1} (25 to 40 km h^{-1}) were necessary ingredients to induce formation of cyclonic and anticyclonic vortices with wind speeds in excess of 50 mi h^{-1} (80$^+$ km h^{-1}) (Moran and Stieglitz, 1983). These vortices helped to rapidly propagate the fire into the area in advance of the main burning front. Pyne (1982) notes that the origin of the term "firestorm" dates to contemporary descriptions of the 1871 fire, and it was independently "discovered" and applied to the firebombing of German cities during World War II.

Consequences

Approximately 1.28 × 10^6 acres (0.5 × 10^6 ha) were burned in the Peshtigo-Williamsonville fire of Wisconsin and the Upper Peninsula of Michigan, and another 2.5 × 10^6 acres (1.0 × 10^6 ha) were burned in Lower Michigan. Estimates of deaths range from about 1200 to 2000, the uncertainty arising from the remoteness of some of the communities and the itinerant population of the region. This loss of life far overshadowed the 200 deaths in the concurrent Great Chicago Fire.

The effect on the forests was indeed devastating. In the Peshtigo area, a logger reported seeing not a single green the entire winter (Holbrook, 1944). Pernin (1971) observed that in the Peshtigo area trees lay "heaped up one over the other in all imaginable positions, their branches reduced to cinders, and their trunks calcined and blackened." Holbrook (1944) noted that "only a fraction of the damaged softwoods could be salvaged before the trees were infested by wood-boring insects." Over on the Lower Michigan side the fire in the "thumb" region (Port Huron) an observer states, "pine and hemlock stands are all destroyed . . . [and] whole townships of timber are burned up by the roots and have fallen in every conceivable shape" (Holbrook, 1944). A man from Port Huron reportedly "walked for a mile on the trunks of fallen trees without touching the ground" (Pyne, 1982). These are evidences of wholesale and extreme destruction, but the apparently "miraculous" survival of people and some areas of trees could be summed up in Pernin's (1971) words, "the tempest did not rage in all parts with equal fury."

The amount of gaseous carbon released from these fires may be estimated with some elementary calculations. The presettlement vegetation depicted in Figure 63.1 indicates these areas were heavily forested. Nearly 50% of the area was occupied by northern mesic forest (hard maple, birch, hemlock, white pine), 18% by the pine forests and pine barrens (white, red, and jack pine), and 15% by wet mesic forests (soft maple, elm, cottonwood). Oak forest and savanna (white, black, red and burr oak) comprised about 11% of the burned area, and boreal forests and conifer swamp (white spruce, fir, tamarack, black spruce) another 7%. The average amount of carbon in living vegetation in this area should be in the range Olson, Watts and Allison (1988) estimate of 4 to 25 kg C m^{-2} for cool hardwood-conifer forest with minor pastures. Wilson (1978) used a value of 30 kg C m^{-2} for standing forests. It is difficult to select a single average value to use: On one hand, lumbering had already removed timber from the areas and settlement had begun, but on the other, much slash was probably left from the cutting as dead biomass. On top of this, detritus density for temperate evergreen forests is reported at 4 kg C m^{-2} by Bolin et al. (1979). Given the area burned (1.5 × 10^{10} m^2), the carbon emissions were calculated for the range of 4 to 25 kg C m^{-2} in living biomass plus the detritus density of 4 kg C m^{-2}. Frequently, a fraction representing the portion of above-ground living biomass is entered in such calculations, but because of the evidence cited for below-ground burning of biomass and because emissions are calculated over a wide range of possible average living biomass density, such a factor is not included herein. Finally, a factor representing efficiency of combustion is needed. Seiler and Crutzen (1980) have suggested a value of 20% for temperate forests, whereas Turco et al. (1984) indicate 35% is realistic for "intense natural forest fires." For the 1871 fires, the descriptions indicate that in local areas the burning efficiency was likely much higher than these values, but 20% to 35% is probably a good range for the average efficiency over the whole fire area.

Table 63.1 provides the results for various combinations of these terms. The carbon gas emissions range from 1.2 × 10^{12} gC for the situation in which only living biomass with density of 4 kg C m^{-2} was burned at a 20% efficiency, to 1.5 × 10^{14} gC emitted from burning of living biomass with density of 25 kg C m^{-2} plus detritus density of 4 kg C m^{-2} at a 35% efficiency. These numbers may be compared to estimates by Seiler and Crutzen (1980) of current annual release of carbon gas from temperate forests of 0.6 to 1 × 10^{14} g. At the upper end of the 1871 fire emission estimates, the carbon release is similar to the Seiler and Crutzen estimates, but the 1871 fires released the quantity effectively in two and three days rather than over a whole year.

The quantity of smoke generated by these fires must likewise have been immense. On 9 October,

Table 63.1 Estimates of gaseous carbon release from 1871 Great Lakes fires

Living biomass density[a]	Detritus density[b]	Burning emission[c]	Gaseous emission
4 kg C m^{-2}	0 kg C m^{-2}	0.2	1.2×10^{12} gC
25	0	0.2	7.5×10^{13}
4	0	0.35	2.1×10^{13}
25	0	0.35	1.3×10^{14}
25	4	0.2	8.7×10^{13}
25	4	0.35	1.5×10^{14}

Note: Emission = area × carbon density × burning efficiency, where the burned area is 1.5×10^{10} m^2.
a. For cool hardwood-confier forest and minor pastures, Olson et al. (1983) estimate 4 to 25 kg C m^{-2}.
b. For temperate (evergreen) forests, Bolin et al. (1979) give a value of 4 kg C m^{-2} in soil.
c. Seiler and Crutzen (1980) employ efficiency of 20% for temperate forests; Turco et al. (1984) indicate 35% for "intense natural forest fires."

smoke obscured the sun totally for a distance of 200 miles near Peshtigo, and relief boats in Green Bay had to be guided by the sound of planks being slapped against the water at the shoreline (Holbrook, 1944). On 10 October, the weather observer at Thunder Bay reported "smoke so dense and heavy . . . that the house had to be lighted up the same as at night. Could not see to read any newspapers . . . the chickens went to their roosts" (Haines and Kuehnast, 1970).

Conclusions

The catastrophic Great Lakes fires of 8 to 10 October 1871 were a consequence of logging and settling practices combined with favorable climate and weather conditions. Fire was a common and accepted element of this pioneer environment, but the intensity of these fires was beyond the limits of routine imagination. Although the destruction was not uniform, descriptions indicate that combustion was highly efficient and nearly complete in places where "firestorms" were active. The estimated carbon (gas) emission from this two- to three-day event rival the current annual emissions from global temperate forest fires.

References

Abbadie, L., A. Mariotti, and J. C. Menaut. 1990. Independence of Savannas Grasses from Soil Organic Matter for Their Nitrogen Supply. *Ecology.*

Abelson, P. H. 1986. Greenhouse Role of Trace Gases. *Science,* 231, 1233.

Ackerman, T. P., and O. B. Toon. 1981. Absorption of Visible Radiation in Atmosphere Containing Mixtures of Absorbing and Nonabsorbing Particles. *Appl. Optics,* 20, 3661–3668.

Adrian, F. W., R. C. Lee, Jr., and J. E. Sasser. 1983. Upland Management Plan for Merritt Island National Wildlife Refuge, USFWS/MINWR, Titusville, Fla.

AFEAS Report. 1990. *Appendix in World Meteorological Organization Global Ozone Research and Monitoring Project—Report No. 2. Scientific Assessment of Stratospheric Ozone: 1989.* Geneva.

AFOS (Agriculture, Forestry and Other Subgroup), Intergovernmental Panel on Climate Change, Response Strategies Work Group. 1990a. Draft Report and Draft Executive Summary. Geneva: IPCC Secretariat.

Ahuja, D. R. 1990. Anthropogenic Emissions of Greenhouse Gases: Estimation of Regional Fluxes. *Indian Geosphere-Biosphere,* edited by T. N. Khoshoo and M. Sharma. New Delhi: Vikas Publishing House.

Ahuja, D. R., V. Joshi, K. R. Smith, and C. Venkataraman. 1987. Thermal Performance and Emission Characteristics of Unvented Biomass Burning Cookstoves: A Proposed Standard Method for Evaluation. *Biomass,* 12(4), 247–270.

Ajaiyeoba, D. B. 1983. *Wildfire Damage to the Pine Plantations in Nigeria: An Economist's Viewpoint.* Paper presented at the Workshop on Forest Fires: Ecology and Environment, Forestry Research Institute of Nigeria. Ibadan. Mimeograph.

Akeredolu, F. A. 1989a. Atmospheric Environment Problems in Nigeria—An Overview. *Atmosph. Environ.* 23(4), 783–792.

Akeredolu, F. A. 1989b. Seasonal Variations of Deposition Rates, Concentrations and Chemical Composition of Air Particulate Matter in Ile-Ife. *Ife Technol. J.,* 1(2), 7–17.

Alaska Fire Service. 1988. *Fire Season Statistic 1987.* Washington, D.C.: U.S. Dept. of Interior, Bureau of Land Management.

Albini, F. A. 1984. Wildland Fires. *American Scientist,* 72, 590–597.

Albrecht, B. A. 1989. Aerosols, Cloud Microphysics, and Fractional Cloudiness. *Science,* 245, 1227–1230.

Alekseev, A. S., M. A. Nazarov, L. D. Barsukova, G. M. Kolesov, I. V. Nizhegorodova, and K. N. Amanniyazov. 1988. The K-T Boundary in Southern Turkmenia and Its Geochemical Characteristics. *Bull. Moscow Soc. Study of Nature, Geol. Sect.,* 63(2), 55–69.

Alexander, P. O. 1981. Age and Duration of Deccan Volcanism. In *Deccan Volcanism and Related Basalt Provinces in Other Parts of the World,* edited by K. V. Subbarao and R. N. Sukheswala, Geological Society of India, Memoir 3, 244–258. Bangalore: Rashtrothana Press.

Alvarez, L. W., W. Alvarez, F. Asaro, and H. V. Michel. 1980. Extraterrestrial Cause for the Cretaceous-Tertiary Extinction. *Science,* 208, 1095–1108.

Alvarez, W. 1986. Toward a Theory of Impact Crises. *Eos* 67, 649–658.

Alviar, G. 1983. *Report on Forest Industries in Nigeria,* UMDP/NIR/FAO, Nigeria Development of Forest Management Capability, FDF, Lagos, Nigeria.

American Society of Photogrammetry. 1983. *Manual of Remote Sensing, 2d ed.,* Falls Church, Va.: ASP, 94–98.

Ampadu-Agyei, O. 1988. Bushfires and Management Policies in Ghana. *Environmentalist,* 8(3), 221–228.

Anders, E., W. S. Wolbach, and R. S. Lewis. 1986. Cretaceous Extinctions and Wildfires. *Science,* 234, 261–264.

Anderson, D., and R. Fishwick. 1987. *Fuelwood Consumption and Deforestation in African Countries.* World Bank Staff Working Papers, 704. Washington, D.C.: World Bank.

Anderson, I. C., and J. S. Levine. 1986. Relative Rates of Nitric Oxide and Nitrous Oxide Production by Nitrifiers, Denitrifiers, and Nitrate Respirers. *Appl. Environ. Microbiol.,* 51, 938–945.

Anderson, I. C., and J. S. Levine. 1987. Simultaneous Field Measurements of Biogenic Emissions of Nitric Oxide and Nitrous Oxide. *J. Geophys. Res.,* 92, 965–976.

Anderson I. C., J. S. Levine, M. A. Poth, and P. J. Riggan. 1988. Enhanced Biogenic Emissions of Nitric Oxide and Nitrous Oxide Following Surface Biomass Burning. *J. Geophys. Res.,* 93, 3893–3898.

Anderson, I. C., and M. A. Poth. 1989. Semiannual Losses of Nitrogen as NO and NO from Unburned and Burned Chaparral. *Global Biogeochem. Cycles,* 3, 121–135.

Andrasko, K. J. 1989, 1990a. Forestry Chapter. In *Policy Options to Stabilize Global Climate,* Report to Congress by Office of Policy Analysis, U.S. Environmental Protection Agency, January 1989 and February 1990 (revised), pp. V 175–237.

Andrasko, K. J. 1990b. Designing National and Local Scale Plans to Offset and Sequester Carbon Dioxide Emissions from the Atmosphere. In *Proceedings of the Conference on Tropical Forestry Response Options to Climate Change,* Report of workshop January 9–11, São Paulo, Brazil. IPCC Secretariat, Geneva, and U.S. Environmental Protection Agency, Washington, pp. 180–205.

Andrasko, K. J. 1990c. Climate Change and Global Forests: Current Knowledge of Potential Effects, Adaptation, and Mitigation Options. Rome: Forestry Dept., FAO.

Andreae, M. O. 1983. Soot Carbon and Excess Fine Potassium: Long-Range Transport of Combustion-Derived Aerosols. *Science,* 220, 1148–1151.

Andreae, M., 1990. Biomass Burning in the Tropics: Impact on Environmental Quality and Global Climate. Paper presented at the Chapman Conference on Global Biomass Burning: Atmospheric, Climatic, and Biospheric Implications, Williamsburg, Va., 19–23 March (this volume, Chapter 1).

Andreae, M. O. 1991. The Global Biogeochemical Sulfur Cycle: A Review. *Proceedings, Global Change Institute,* Snowmass, Colo., August 1988, Office of Earth and Interdisciplinary Studies, National Center for Atmospheric Research, Boulder, Colo.

Andreae, M. O., B. E. Anderson, J. E. Collins, G. L. Gregory, G. W. Sachse, J. D. Bradshaw, D. D. Davis, and V. W. J. H. Kirchhoff. 1990a. Influence of plumes from biomass burning on atmospheric chemistry over the equatorial Atlantic during CITE-3. *Eos Transactions, American Geophysical Union,* 71, 1255.

Andreae, M. O., and T. W. Andreae. 1988. The Cycle of Biogenic Sulfur Compounds over the Amazon Basin in Dry Season. *J. Geophys. Res.,* 93, 1487–1497.

Andreae, M. O., T. W. Andreae, R. J. Ferek, and H. Raemdonck. 1984. Long-Range Transport of Soot Carbon in the Marine Atmosphere. *Science of the Total Environ.,* 3, 73–80.

Andreae, M. O., E. Browell, M. Garstang, G. L. Gregory, R. C. Harris, G. F. Hill, D. J. Jacob, M. C. Pereira, G. Sachse, A. Setzer, P. L. Silva Dias, R. T. Talbot, A. L. Torres, and S. C. Wofsy. 1988. Biomass-Burning Emissions and Associated Haze Layers over Amazonia. *J. Geophys. Res.,* 93, 1509–1527.

Andreae, M. O., A. Chapuis, B. Cros, J. Fontan, G. Helas, C. Justice, Y. J. Kaufman, A. Minga, D. Nganga. 1991. Ozone and Aitken Nuclei over Equatorial Africa: Airborne Observations during DECAFE 88. *J. Geophys. Res., Special issue on DECAFE 88,* in press.

Andreae, M. O., A. C. Delany, S. Liu, J. Logan, L. P. Steele, H. Westberg, and R. Zika. 1989. Key Aspects of Species Related to Global Biogeochemical Cycles. In *Global Tropospheric Chemistry: Chemical Fluxes in the Global Atmosphere,* edited by D. H. Lenschow and B. B. Hicks, 9–29. Boulder, Colo.: NCAR.

Andreae, M. O., and D. S. Schimel, eds. 1989. *Exchange of Trace Gases between Terrestrial Ecosystems and the Atmosphere.* Chichester, England: J. Wiley.

Andreae, M. O., R. W. Talbot, T. W. Andreae, and R. C. Harriss. 1988b. Formic and Acetic Acid over the Central Amazon Region—Brazil in Dry Season. *J. Geophys. Res.,* 93, 1616–1624.

Andreae, M. O., R. W. Talbot, H. Berresheim, and K. M. Beecher. 1990b. Precipitation Chemistry in Central Amazonia. *J. Geophys. Res.,* 95, 16,987–16,999.

Antia, N. J., and J. Y. Cheng. 1970. The Survival of Axenic Cultures of Marine Planktonic Algae from Prolonged Exposure to Darkness at 20°C. *Physologia,* 9, 179–183.

Argyle, E. 1986. Cretaceous Extinctions and Wildfires. *Science,* 234 and 261.

Arijs, E., and G. Brasseur. 1986. Acetonitrile in the Stratosphere and Implications for Positive Ion Composition. *J. Geophys. Res.,* 91D, 4003–4016.

Arkin, H., and R. R. Colton. 1953. *Statistical Methods,* New York: Barnes and Noble, 126–127.

Arrhenius, E. A., and T. W. Waltz. 1990. *The Greenhouse Effect: Implications for Economic Development.* World Bank Discussion Paper No. 78. Washington, D.C.: International Bank for Reconstruction and Development.

Artaxo, P. 1990. Trace Element Concentration and Size Distribution of Aerosols from the Amazon Basin in Burning Plumes and in the Dry and Wet Seasons. Paper presented at the Chapman Conference on Global Biomass Burning: Atmospheric, Climatic and Biospheric Implications, Williamsburg, Va., 19–23 March.

Artaxo, P., H. Storms, F. Storms, F. Bruynseels, R. Van Grieken, and W. Maenhaut. 1988. Composition and Sources of Aerosols from the Amazon Basin. *J. Geophys. Res.,* 93, 1509–1527.

Aselmann, I., and P. J. Crutzen. 1989. Global Distribution of Natural Freshwater Wetlands and Rice Paddies, Their Net Primary Productivity, Seasonality and Possible Methane Emissions. *J. Atmos. Chem.,* 8, 307–358.

Ashton, P. S. 1969. Speciation among Tropical Forest Trees: Some Deductions in the Light of Recent Evidence. *Biol. J. Linn. Soc.,* 1, 155.

Ashton, P. S. 1988. Dipterocarp Biology as a Window to the Understanding of Tropical Forest Structure. *Annual Rev. Ecol. Systems.* 19, 347.

Askin, R. M. 1988. The Palynological Record across the Cretaceous/Tertiary Transition on Seymour Island, Antarctica. In *Geology and Paleontology of Seymour Island, Antarctic Peninsula,* edited by R. M. Feldman and M. O. Woodburne, 155–162.

Aspliden, C. 1976. A Classification of the Structure of the Tropical Atmosphere and Related Energy Fluxes. *J. Apl. Meteor.,* 15, 692–697.

Atherton, C. S., J. E. Penner, J. J. Alton, and S. Hameed. 1991. Wet and Dry Nitrogen Deposition: Results from a Global, Three-Dimensional Chemistry-Transport-Deposition Model, submitted to *Atmo. Environ.,* also Lawrence Livermore National Laboratory report UCRL-JC-103403.

Atjay, G. L., P. Ketner, and P. Duvigneaud. 1987. Terrestrial Primary Production and Phytomass. In *The Global Carbon Cycle,* SCOPE 13, edited by B. Bolin, E. T. Degens, S. Kempe, and P. Ketner, 129–181. Chichester: John Wiley.

Atkinson, R., J. Arey, A. M. Winer, and B. Zielinska. 1988. *A Survey of Ambient Concentrations of Selected Polycyclic Aromatic Hydrocarbons (PAH) at Various Locations in California.* Final Report, prepared under Contract No. A5-185-32 for California Air Resources Board, May.

Ayers, G. P., and R. W. Gillett. 1988. Acidification in Australia. In *Acidification in Tropical Countries,* SCOPE 36, edited by H. Rodhe and R. Herrera, 347–400. Chichester: John Wiley.

Ayres, R. U., L. W. Ayres, and J. A. Tarr. 1987. *An Historical Reconstruction of Major Atmospheric Gaseous Emissions in the U.S., 1880–1980.* Final report prepared for Resources of the Future, Inc., by Variflex Corp., Pittsburgh, Pa.

Bacastow, R. B., and C. D. Keeling. 1973. Atmospheric Carbon Dioxide and Radiocarbon in the Natural Carbon Cycle: II. Changes from A.D. 1700 to 2070 as Deduced from a Geochemical Model. In *Carbon and the Biosphere,* AEC Symposium series 30, CONF-720510, edited by G. M. Woodwell and E. V. Pecan, 86–134. Washington, D.C.: U.S. Atomic Energy Commission.

Bailey, I. W., and E. W. Sinnott. 1915. A Botanical Index of Cretaceous and Tertiary Climates. *Science*, 41, 831–834.

Bailey, I. W., and E. W. Sinnott. 1916. The Climatic Distribution of Certain Types of Angiosperm Leaves. *American J. of Botany*, 3, 24–39.

Baker, H. G., K. S. Bawa, G. W. Frankie, and P. A. Opler. 1983. Reproductive Biology of Plants in Tropical Forests. In *Ecosystems of the World. 14A: Tropical Rain Forest Ecosystems*, edited by F. B. Golley, 183–215. Amsterdam: Elsevier.

Bandeira, A. G., and M. F. P. Torres. 1985. Abundancia e distribuiao de invertebrados do solo em ecossistemas da Amazonia Oriental. O papel ecologico dos cupins. Boletim do Museu Paraense Emlio Goeldl, *Zoologia*, 2(1), 13–38.

Barbour, Philip L., ed. 1986. *The Complete Works of Captain John Smith (1580–1631)*. 3 vols. Chapel Hill: University of North Carolina.

Barnes, A. C. 1974. *The Sugar Cane*. New York: John Wiley.

Barnes, R. A., A. C. Holland, and V. W. J. H. Kirchhoff. 1987. Equatorial Ozone Profiles from the Ground to 52 km during the Southern Hemisphere Autumn. *J. Geophys. Res.*, 92, 5573–5583.

Barnola, J.-M., D. Raynaud, and Y. S. Korotkevich. 1989. CO_2-Climate Relationship as Deduced from the Vostok Ice Core: A Reexamination Based on New Measurements and on Reevaluation of the Air Dating, *Proceeding of the Third International Conference on Analysis and Evaluation of Atmospheric CO_2 Data, Present and Past*, Federal Republic of Germany: Hinterzarten, October (in press).

Barnola, J.-M., D. Raynaud, Y. S. Korotkevitch, and C. Lorius. 1987. Vostok Ice Core Provides 160,000–Year Record of Atmospheric CO_2. *Nature*, 329, 408.

Barrie, L. A. 1985. Arctic Air Pollution: An Overview of Current Knowledge. *International Conference on Atmospheric Sciences and Application to Air Quality*. Seoul, Korea, May.

Barrie, L. A., and J. M. Hales. 1984. The Spatial Distributions of Precipitation Acidity and Major Ion Wet Deposition in North America during 1980. *Tellus*, 36B, 333–355.

Bartlett, H. H. 1956. Fire, Primitive Agriculture, and Grazing in the Tropics. In *Man's Role in Changing the Face of the Earth*, edited by W. Thomas, 692–720. Chicago: University of Chicago Press.

Bartlett, H. H. 1961. *Fire in Relation to Primitive Agriculture and Grazing in the Tropics: Annotated Bibliography*. vols. 1–3. Mimeo. Ann Arbor: University of Michigan Botanical Gardens.

Bartlett, A. S., and E. S. Barghoorn. 1973. Phytogeographic History of the Isthmus of Panama during the Past 12,000 Years (A History of Vegetation, Climate, and Sea-Level Change). In *Vegetation and Vegetational History of Northern Latin America*, edited by A. Graham, 203–299. Amsterdam: Elsevier Scientific Publishing Co.

Batchelder, R. B. 1967. Spatial and Temporal Patterns of Fire in the Tropical World. *Proceedings of the Tall Timbers Fire Ecological Conference*, 6, 171–190.

Batchelder, R. B., and H. F. Hirt. 1966. *Fire in Tropical Forests and Grasslands*, U.S. Army Natick Laboratories technical report ES-23.

Bate, G. C., and C. Gunton. 1982. Nitrogen in the Burkea Savanna. *Ecology of Tropical Savanna, Ecological Studies*, 42, 498–534. New York: Springer-Verlag.

Baudet, G. J. R., J. P. Lacaux, J. J. Bertrand, F. Desalmand, J. Servant, and V. Yoboue. 1990. Presence of an Atmospheric Oxalate Source in the Intertropical Zone: Its Potential Action in the Atmosphere. *Atmos. Res.*, 25, 236–248.

Bauer, E. 1986. *Nuclear Winter: Smoke Generation Deposition and Removal*. Report M-4 IDA, Alexandria, Va.: Institute for Defense Analysis.

Beavington, F., and P. A. Cawse. 1978. Comparative Studies of Trace Elements in Air Particulate in Northern Nigeria. *Sci. Tot. Environment*, 10,239–10,244.

Benchimol, S. 1989. *Amazonia: Planetarizacao e Moritoria Ecologica*. São Paulo: Centro de Recursos Educacionais (CERED).

Bennett, C. F. 1968. *Human Influences on the Zoogeography of Panama*. Berkeley: University of California Press.

Bennet, J. J., S. Krishnaswami, K. K. Turekian, M. G. Melson, and C. A. Hopson. 1982. The Uranium and Thorium Decay Series Nuclides in Mount St. Helens Effusives. *Earth Planet. Sci. Let.*, 60, 60–69.

Berger, K. C., W. H. Erhardt, and C. W. Francis. 1965. Po-210 Analyses of Vegetables, Cured and Uncured Tobacco, and Associated Soils. *Science*, 150, 1738–1739.

Berner, R. A., and G. P. Landis. 1988. Gas Bubbles in Fossil Amber as Possible Indicators of the Major Gas Composition of Ancient Air. *Science*, 239, 1406–1409.

Beven, S., E. F. Connor, and K. Beven. 1984. Avian Biogeography in the Amazon Basin and the Biological Model of Diversification. *J. Biogeogr.*, 11, 383.

Bingemer, H. G., M. O. Andreae, T. W. Andreae, P. Artaxo, G. Helas, N. Mihalopoulos, and B. C. Nguyen. 1991. Sulfur Gases and Aerosols in and above the Equatorial African Rainforest. *J. Geophys. Res.* (in press).

Birdsey, R. A. 1990. *Estimation of Regional Carbon Yields for Forest Types in the United States*. U.S.D.A. Forest Service, Washington D.C., unpublished manuscript.

Bjorkstrom, A. 1979. A Model of CO_2 Interaction between the Atmosphere, Oceans and Land Biota. In *The Global Carbon Cycle*, SCOPE 13, edited by B. Bolin, E. T. Degens, S. Kempe, and P. Ketner, 403–445. Chichester: John Wiley.

Blake, D. R., and F. S. Rowland. 1988. Continuing Worldwide Increase in Tropospheric Methane, 1978 to 1987. *Science*, 239, 1129–1131.

Bogdonoff, P., R. P. Detwiler, and C. A. S. Hall. 1985. Land Use Change and Carbon Exchange in the Tropics: III. Structure, Basic Equations, and Sensitivity Analysis of the Model, *Environmental Mgmt.*, 9, 345–354.

Bohor, B. F., E. E. Foord, P. J. Modreski, and D. M. Triplehorn. 1984. Mineralogic Evidence for an Impact Event at the Cretaceous-Tertiary Boundary. *Science*, 224, 867–868.

Bohor, B. F., P. J. Modreski, and E. E. Foord. 1987. Shocked Quartz in the Cretaceous-Tertiary Boundary Clays: Evidence for a Global Distribution. *Science*, 236, 705–709.

Bohor, B. F., and R. Seitz. 1990. Cuban K/T Catastrophe, *Nature*, 344, 593.

Bolin, B. 1986. How Much CO_2 Will Remain in the Atmosphere: The Carbon Cycle and Projections for the Future. In *The Greenhouse Effect, Climatic Change, and Ecosystem*, SCOPE 29, edited by B. Bolin, B. R. Doos, J. Jager, and R. A. Warrick, 93–155. Chichester: John Wiley.

Bolin, B., B. R. Doos, J. Jager, and R. A. Warrick, eds. 1986. *The Greenhouse Effect, Climatic Change, and Ecosystems.* SCOPE 29, Chichester: John Wiley.

Bolin, R., E. T. Degens, P. Duvigneaud, and S. Kempe. 1979. The Global Biogeochemical Carbon Cycle. In *The Global Carbon Cycle,* edited by B. Bolin, E. T. Degens, P. Duvigneaud, S. Kempe, 1–56. New York: John Wiley.

Bolle, H.-J., W. Seiler, and B. Bolin. 1986. Other Greenhouse Gases and Aerosols: Assessing Their Role for Atmospheric Radiative Transfer. In *The Greenhouse Effect, Climatic Change, and Ecosystems,* SCOPE 29, edited by B. Bolin, B. R. Doos, J. Jager, and R. A. Warrick, 157–203. Chichester: John Wiley.

Bolton, D. 1980. The Computation of Equivalent Potential Temperature. *Mon. Wea. Rev.,* 108, 1046–1053.

Bonnor, G. M. 1985. *Inventory of Forest Biomass in Canada,* Chalk River, Ont.: Canadian Forestry Service, Petawawa National Forestry Institute.

Bonsang B., M. Kanakidou, G. Lambert, and P. Monfray. 1988. The Marine Source of C2–C6 Aliphatic Hydrocarbons. *J. Atmos. Chem.,* 6, 3–20.

Bonsang B., M. Kanakidou, G. Lambert, and J. Rudolph. 1988. *Vertical Profiles and Fluxes of NMHC in the Equatorial Forest of Congo,* AGU Fall Meeting, San Francisco, 5–9 December.

Bonsang, B., G. Lambert, and C. Boissard. 1990. *Light Hydrocarbons Emissions from African Savannah Burnings.* Paper presented at the Chapman Conference on Global Biomass Burning: Atmospheric, Climatic, and Biospheric Implications, Williamsburg, Va., 19–23 March (this volume, Chapter 20).

Booth, W. 1989. Monitoring the Fate of the Forests from Space. *Science,* 243, 1428–1429.

Bormann, F. H. 1985. Air Pollution and Forests: An Ecosystem Perspective. *Bioscience,* 35, 434–441.

Borucki, W. J., and W. L. Chameides. 1984. Lightning: Estimates of the Rates of Energy Dissipation and Nitrogen Fixation. *Rev. Geophys.,* 22, 363–372.

Bose, M. K. 1972. Deccan Basalts. *Lithos,* 5, 131–145.

Bose, R. K., P. Malhotra, C. Puri, and V. Joshi. 1990. *Rural Energy Consumption Patterns in Two Agroclimatic Zones—A Case Study.* Paper presented at the International Symposium: Application and Management of Energy in Agriculture—The Role of Biomass Fuels, New Delhi, 21–23 May.

Bourgeois, J., T. A. Hansen, P. L. Wiberg, and E. G. Kauffman. 1988. A Tsunami Deposit at the Cretaceous-Tertiary Boundary in Texas. *Science,* 241, 567–570.

Bowden, William B. 1986. Gaseous Nitrogen Emissions from Undisturbed Terrestrial Ecosystems: An Assessment of Their Impacts on Local and Global Nitrogen Budgets. *Biogeochemistry,* 2, 249–279.

Bowen, H. J. M. 1966. *Trace Elements in Biochemistry.* New York: Academic Press.

Bowen, H. J. M. 1979. *Environmental Chemistry of the Elements.* London: Academic Press.

Braatz, B. V., and W. J. Pepper. 1990. Collected memoranda to USEPA, Climate Change Division, ICF, Inc., Vienna, Va.

Bradle, M. M., and C. R. Molenkamp. 1990. Numerical Simulation of Aerosol Scavenging by Ice-Bearing Convective Clouds. *Proceedings of the American Meteorological Society Conference on Cloud Physics,* San Francisco, California, 23–27 July 1990; also Lawrence Livermore National Laboratory Report UCRL-101997, Livermore, Calif.

Braga, P. I. S. 1979. Subdivisao fitogeografica, tipos de vegetago, conservao e inventario floristico da floresta amazonica. *Acta Amazonica,* 9(4) suplemento, 53–80.

Brain, C. K., and A. Sillen. 1988. Evidence from the Swartkrans Cave for the Earliest Use of Fire. *Nature,* 336, 464.

Braman, R. S., T. J. Shelly, and W. A. McClenny. 1982. Tungstic Acid for Preconcentration and Determination of Gaseous and Particulate Ammonia and Nitric Acid in Ambient Air. *Anal. Chem.,* 54, 358–364.

Brandel, K., M. Rogers, and G. Reinhart. 1988. Fire Management in Israel. *Fire Mgmt. Notes,* 43, 34–37.

Brazil, Instituto de Pesquisas Espaciais. 1989a. *Avaliasao da cobertura florestal na Amazonia Legal utilizando sensoriamento remoto orbital.* São Jose dos Campos, São Paulo: INPE.

Brazil, Instituto de Pesquisas Espaciais. 1989b. *Avaliasao da cobertura florestal na Amazonia al utilizando sensoriamento remoto orbital,* 2d ed. São Jose dos Campos, São Paulo: INPE.

Brazil, Ministerio da Agricultura, Instituto Brasileiro de Desenvolvimento Florestal (IBDF). 1982. Alterasao da cobertura vegetal natural do Estado de Mato Grosso: Relatorio tecnico. Brasilia: IBDF.

Brazil, Ministerio da Agricultura, Instituto Brasileiro de Desenvolvimento Florestal (IBDF). 1983a. Desenvolvimento 28 Florestal no Brasil. PNUD/FAO/BRA-82-008. Folha Informativa No. 5. Brasilia: IBDF.

Brazil, Ministerio da Agricultura, Instituto Brasileiro de Desenvolvimento Florestal (IBDF). 1983b. Alterao da Cobertura Vegetal Natural do Territorio de Roraima: Anexo Relatorio Tecnico. Brasilia: IBDF.

Brazil, Ministerio da Agricultura, Instituto Brasileiro de Desenvolvimento Florestal (IBDF). 1985. Monitoramento da alterao da cobertura vegetal natural da area do programa polonoroeste nos Estados de Rondonia e Mato Grosso: Relatorio tecnico. Brasilia: IBDF.

Brazil, Ministerio da Agricultura, Instituto Brasileiro de Desenvolvimento Florestal (IBDF). 1988. Alterasao da cobertura vegetal do Estado de Acre: Relatorio tecnico. Brasilia: IBDF.

Brazil, Ministerio da Agricultura, Instituto Brasileiro de Desenvolvimento Florestal (IBDF). 1989. Alterasao da cobertura vegetal do Estado de Rondonia: Relatorio tecnico. Brasilia: IBDF.

Brazil, Ministerio das Minas e Energia, Departamento Nacional de Produo Mineral (DNPM). 1973–1983. Projeto RADAMBRASIL.

Brazil, Ministerio do Interior Superintendencia do Desenvolvimento da Amazonia (SUDAM) and Instituto Brasileiro de Desenvolvimento Florestal (IBDF). 1988. Levantamento da alterao da cobertura vegetal primitiva do Estado do Para. Brasilia: SUDAM/IBDF.

Brazil, Secretaria de Planeamento (SEPLAN), Programa Grande Caraas (PGC), Companhla de Desenvolvimento de Barcarena (CODEBAR) and Ministerio do Interior, Superintendencia do Desenvolvimento da Amazonia (SUDAM). 1986. Problematica do carvao vegetal na rea do Programa Grande Caraas. Belem.: CODEBAR/SUDAM.

Bremner, J. M. 1986. Inhibition of Nitrogen Transformations in Soil. In *Perspectives in Microbial Ecology,* edited by F. Megushar and M. Gantar, 337, 342.

Bremond, M. P., H. Cachier, and P. Buat-Menard. 1989. Particulate Carbon in the Paris Region Atmosphere. *Environ. Techn. Lett.,* 10, 339–346.

Brinkhuis, H., and W. J. Zachariasse. 1988. Dinoflagellate Cysts, Sea Level Changes and Planktonic Foraminifers Across the Cretaceous-Tertiary Boundary at El Haria, Northwest Tunisia. *Marine Micropaleontology,* 13, 153–199.

Broecker, W. S., T.-H. Peng, and R. Engh. 1980. Modelling the Carbon System, *Radiocarbon,* 22, 565–598.

Broecker, W. S., T.-H. Peng, G. Ostlund, and M. Stuiver. 1985. The Distribution of Bomb Radiocarbon in the Ocean, *J. Geophys. Res.,* 90, 6953–6970.

Bromfield, A. R. 1974. The Deposition of Sulphur in the Rain Water in Northern Nigeria. *Tellus,* 26(3), 408–411.

Brooks, R. R., et al. 1984. Elemental Anomalies at the Cretaceous-Tertiary Boundary, Woodside Creek, New Zealand. *Science,* 236, 539–542.

Browell, E. V., C. F. Butler, P. Robinette, S. A. Kooi, M. A. Fenn, and S. Ismail. 1990. *Airborne Lidar Observations of Aerosols and Ozone in Plumes from Biomass Burning over the Amazon Basin and Over Alaska.* Paper presented at the Chapman Conference on Biomass Burning, Williamsburg, Va., 19–23 March.

Browell, E. V., G. L. Gregory, R. C. Harriss, and V. W. J. H. Kirchhoff. 1988. Tropospheric Ozone and Aerosol Distributions across the Amazon Basin. *J. Geophys. Res.,* 3, 1431–1451.

Browers, E. M., W. A. Clemens, R. A. Spicer, T. A. Ager, L. D. Carter, and W. V. Sliter. 1987. Dinosaurs on the North Slope, Alaska; High Latitude, Latest Cretaceous Environments. *Science,* 237, 1608–1610.

Brown, A. P., and K. P. Davis. 1973. *Forest Fire, Control and Use.* New York: McGraw-Hill.

Brown, D. W., W. E. Kenner, J. W. Crooks, and J. B. Foster. 1962. Water Resources of Brevard County, Florida, Rep. Invest. No. 28. Tallahassee: Florida Geological Survey.

Brown, I. F. 1988. *Bacias Hidrograficas.* Presentation at the 2nd Semana do Ambiente, 5–9 December 1988. São Paulo: Piracicaba.

Brown, I. F. 1990. *Uncertainties in Tropical Forest Biomass: An Example from Rondonia, Brazil.* Paper presented at the Chapman Conference on Global Biomass Burning: Atmospheric, Climatic and Biospheric Implications, Williamsburg, Va., 19–23 March.

Brown, S., A. J. R. Gillespie, and A. E. Lugo. 1989. Biomass Estimation Methods for Tropical Forests with Applications to Forest Inventory Data. *Forest Science,* 35(4), 881–902.

Brown, S., and A. E. Lugo. 1982. The Storage and Production of Organic Matter in Tropical Forests and Their Role in the Global Carbon Cycle. *Biotropica,* 14(3), 161–187.

Brown, S., and A. E. Lugo. 1984. Biomass of Tropical Forests: A New Estimate Based on Forest Volumes. *Science,* 223, 1290–1293.

Brummer, B. 1978. Mass and Energy Budgets of a 1 km High Atmospheric Box over the GATE C-Scale Triangle during Undisturbed and Disturbed Weather Conditions. *J. Atmos. Sci.,* 35, 997–1011.

Bryam, G. M., and K. P. Davis, eds. 1959. *Forest Fire Control and Use.* New York: McGraw-Hill.

Buat-Menard, P., and M. Arnold. 1978. The Heavy Chemistry of Atmospheric Particulate Matter Emitted by Mount Etna Volcano. *Geophys. Res. Lett.,* 5, 245–248.

Bujak, J. P., and G. L. Williams. 1979. Dinoflagellate Diversity through Time. *Marine Micropaleontology,* 4, 1–12.

Bullen, E. R. 1974. *Burning Cereal Crop Residues in England.* Proceedings of the Tall Timber Fire Ecology Conference, 13, 223–235.

Burke, L. M., and D. A. Lashof. 1990. Greenhouse Gas Emissions Related to Agriculture and Land-Use Practices. *Impact of Carbon Dioxide, Trace Gases, and Climate Change on Global Agriculture.* American Society of Agronomy Special Publication. 53, 27–43.

Buschbacher, R. J. 1987. Deforestation for Sovereignty over Remote Frontiers in Amazon Rain Forests. *Amazon Rain Forests: Ecological Studies,* 60, 46–57. New York: Springer-Verlag.

Busse, W. 1908. Die Periodischen Grasbrande in Afrika, Einfluss Auf Die Vegetation und Ihre Bedeutung fur Die Landeskultur. *Mitt. Deutscher Schutzgebiete,* 21, 113–139.

Butcher, S. S., U. Rao, K. R. Smith, J. F. Osborn, P. Azuma, and H. Fields. 1984. Emission Factors and Efficiencies for Small Scale Open Biomass Combustion: Toward Standard Measurement Techniques. Meeting of the American Chemical Society, Washington, D.C.

Butler, J. H., J. W. Elkins, T. Thompson, and K. B. Egan. 1989. Tropospheric and Dissolved N_2O of the West Pacific and East Indian Oceans during the El Niño-Southern Oscillation Event of 1987. *J. Geophys. Res.,* 94, 14,865–14,877.

Byram, G. M. 1959. Combustion of Forest Fuels. In *Forest Fire: Control and Use,* edited by K. P. Davis. New York: McGraw-Hill.

CACGP. 1989. *The International Global Atmospheric Chemistry (IGAC) Programme.* Paris: CACGP.

Cachier, H. 1990. Isotopic Characterization of Carbonaceous Aerosols. *Aerosol Sci. Tech.*

Cachier, H., J. Ducret, M. P. Bremond, V. Yoboue, J. P. Lacaux, A. Gaudichet, and J. Baudet. 1990. Biomass Burning Aerosols in a Savannah Region of the Ivory Coast. Paper presented at the Chapman Conference on Global Biomass Burning: Atmospheric, Climatic, and Biospheric Implications, Williamsburg, Va., 19–23 March (this volume, Chapter 23).

Cachier, H., M. P. Bremond, and P. Buat-Menard. 1989. Carbonaceous Aerosols from Different Tropical Biomass Burning Sources. *Nature,* 340, 371–373.

Cachier, H., M. P. Bremond, and P. Buat-Menard. 1989. Determination of Atmospheric Soot Carbon with a Simple Thermal Method. *Tellus,* 41B, 379–390.

Cachier, H., P. Buat-Menard, M. Fontugne, and J. Rancher. 1985. Source Terms and Source Strengths of the Carbonaceous Aerosol in the Tropics. *J. Atmos. Chem.,* 3, 469–489.

Cadle, R. D., A. L. Lazrus, B. J. Huebert, L. E. Heidt, W. I. Rose, Jr., D. C. Woods, R. L. Chuan, R. E. Stoiber, D. B. Smith, and R. A. Zielinski. 1979. Atmospheric Implication Studies of Central American Volcanic Eruption Clouds. *J. Geophys. Res.,* 84, 6961–6968.

Cadle, S. H., and P. A. Mulawa. 1980. Low Molecular Weight Aliphatic Amines in Exhaust from Catalyst-Equipped Cars. *Environ. Sci. Technol.* 14, 718–723.

Cahoon, D. R., Jr., J. S. Levine, W. R. Cofer III, J. E. Miller, G. M. Tennille, T. W. Yip, and B. J. Stocks. 1990. *The Great Chinese Fire of 1987: A View from Space.* Paper presented at the Chapman Conference on Global Biomass Burning: Atmospheric, Climatic, and Biospheric Implications, Williamsburg, Va., 19–23 March (this volume).

Campistron, B., G. Despaux, and J. P. Lacaux. 1987. A Microcomputer Data-Acquisition System for Real-Time Processing of Raindrop Size Distribution Measured with the RD69 Distrometer. *J. Atmos. and Ocean. Technol.*, 4, 536–540.

Canby, T. Y. 1984. El Niño's Ill Wind. *National Geogr.*, 165, 144–184.

Cardenas, J. D. R., F. L. Kahn, and J. L. Guillaumet. 1982. Estimativa da fitomassa do reservatorio da UHE de Tucurul, 1–11 In *Brazil, Ministerio das Minas e Energia,* Centrais Eletricas do Norte S.A. (ELETRONORTE) and Brazil, Instituto Nacional de Pesquisas da Amazonia (INPA). Projeto Tucurul, Relatorio Semestral Jan.–June 1982. Vol. 2: Limnologia, Macrofitas, Fitmassa, Degradaao de Fitomassa. Doenas Endemicasr Solos. Manaus: INPA.

Carney, J. 1989. Last Gasp for the Everglades. *Time,* 25 September, 26–27.

Carpenter, K. 1983. Evidence Suggesting Gradual Extinction of Latest Cretaceous Dinosaurs. *Naturwissenschaften,* 70, 611.

Carter, N. L., C. B. Officer, C. A. Chesner, and W. I. Rose. 1986. Dynamic Deformation of Volcanic Ejecta from the Toba Caldera—Possible Relevance to the Cretaceous-Tertiary Boundary Phenomena. *Geology,* 14, 380–383.

Carter, N. L., and B. Breithaupt. 1986. Latest Cretaceous Occurrence of Nodosaurid Ankylosaurs (Dinosauria, Ornithischia) in Western North America and the Gradual Extinction of the Dinosaurs. *Journal of Vertebrate Paleontology,* 6, 251–257.

Cess, R. D., G. L. Potter, J. P. Blanchet, G. J. Boer, S. J. Ghan, J. T. Kiehl, H. LeTreut, Z.-X. Li, X.-Z. Liang, J. F. B. Mitchell, J.-J. Morcrette, D. A. Randall, M. R. Riches, E. Roeckner, U. Schlese, A. Slingo, K. E. Taylor, W. M. Washington, R. T. Wetherald, and I. Yagai. 1989. Interpretation of Cloud-Climate Feedback as Produced by 14 Atmospheric General Circulation Models. *Science,* 245, 513–516.

Chandler, C., P. Cheney, P. Thomas, L. Trabaud, and D. Williams. 1983. *Fire in Forestry.* Volume 1: *Forest Fire Behavior and Effects.* New York: John Wiley.

Charlock, T. P., and T. D. Sellers. 1980. Aerosol Effects on Climate Calculations with Time-Dependent and Steady-State Radiative-Convective Models. *J. Atmo. Sci.*, 37,1327–37,1341.

Charlson, R. J., J. E. Lovelock, M. O. Andreae, and S. G. Warren. 1987. Oceanic Phytoplankton, Atmospheric Sulphur, Cloud Albedo and Climate. *Nature,* 32, 655–661.

Charlson, R. J., J. E. Lovelock, M. O. Andreae, and S. G. Warren. 1989. Sulphate Aerosols and Climate: Scientific Correspondence. *Nature,* 340, 437–438.

Chatfield, R., and H. Harrison. 1977. Tropospheric Ozone, 2: Variations along a Meridional Band. *J. Geophys. Res.*, 82, 5965–5976.

Chatfield, R. B., and A. C. Delany. 1990. Cloud-Vented Tropical Smog Can Produce High Ozone, but Ill-Simulated Dilution Leads Models to O_3 Over-Prediction. *J. Geophys. Res.*, 95, 18,473–18,488.

Chaturvedi, A. N. 1990. Personal communication.

Ching, J., and A. Alkezweeny. 1987. Tracer Study of Vertical Exchange by Cumulus Clouds. *J. Appl. Meteor. and Clim.*, 25, 1702–1711.

Christensen, L., S. Fregerslev, and J. Thiede. 1973. Sedimentology and Depositional Environment of Lower Danian Fish Clay from Stevns Klint, Denmark. *Bulletin of the Geological Society of Denmark,* 22, 193–212.

Christensen, N. L. 1973. Fire and the Nitrogen Cycle in California Chaparral. *Science,* 181, 66–69.

Christensen, N. L. 1977. Fire and Soil-Plant Relations in a Pine-Wiregrass Savanna on the Coastal Plain of North Carolina. *Oecologia,* (Berl.), 31, 27–44.

Christensen, N. L., and C. H. Muller. 1975. Effects of Fire on Factors Controlling Plant Growth in Adenosoma Chaparral. *Ecol. Monog.,* 45, 29–55.

Chuan, R. L. 1976. Rapid Measurement of Particulate Size Distribution in the Atmosphere. In *Fine Particles, Aerosol Generation, Measurement, Sampling, and Analysis,* edited by B. Y. H. Liu, 763–775. New York: Academic Press.

Chuan, R. L., and D. C. Woods. 1984. The Appearance of Carbon Aerosol Particles in the Stratosphere. *Geophys. Res. Letts.,* 11, 553–556.

Chuang, C. C., J. E. Penner, L. L. Edwards, and M. M. Bradley. 1990. The Effects of Entrainment on Nucleation Scavenging. *Proceedings of the American Meteorological Society Conference on Cloud Physics,* San Francisco, California, July 23–27, 1990; also, Lawrence Livermore National Laboratory Report UCRL-JC-103959, Livermore, Calif.

Chung, Y. S., and Le, H. V. 1984. Detection of Forest-Fire Smoke Plumes by Satellite Imagery. *Atmos. Environ.,* 18(10), 2143–2151.

Chylek, P., and V. Ramaswamy. 1982. Lower and Upper Bounds on Extinction Cross Sections of Arbitrarily Shaped Strongly Absorbing or Strongly Reflecting Nonspherical Particles. *Appl. Opt.,* 21, 4339–4344.

Chylek, P., V. Ramaswamy, and R. J. Cheng. 1984. Effect of Graphitic Carbon on the Albedo of Clouds. *J. Atmos. Sci.,* 41, 3076–3084.

Chynoweth, L. A. 1975. *Net Primary Production of Spartina and Species Diversity of Associated Macroinvertebrates of a Semi-Impounded Salt Marsh.* Tech. Rep. No. 1, Grant No. NGR10-019-009, NASA/Kennedy Space Cen., Fla., 147 pp.

Cibdad Real, A. 1975. Relacion Breve y Verdadera de Algunas Cosas de las Muchas que Sucedieron al Padre Fray Alonso Ponce en las Provincias de la Nueva Espana. *Siendo Comisario General de Aquellas Partes,* 1873, reprinted, Papelera Industrial de Nicaragua, Managua.

Cicerone, R. J. 1989. Analysis of Sources and Sinks of Atmospheric Nitrous Oxide (N_2O). *J. Geophys. Res.,* 4, 18,265–18, 271.

Cicerone, R. J., and R. S. Oremland. 1988. Biogeochemical Aspects of Atmospheric Methane. *Global Biogeochem. Cycles,* 2, 299–327.

Cicerone, R. J., and J. D. Shetter. 1981. Sources of Atmospheric Methane: Measurements in Rice Paddies and a Discussion. *J. Geophys. Res.,* 86, 7203–7209.

Cicerone, R. J., and R. Zellner. 1983. The Atmospheric Chemistry of Hydrogen Cyanide. *J. Geophys. Res.,* 88C, 10689–10696.

Cionco, R. M. 1972. Intensity of Turbulence within Canopies with Simple and Complex Roughness Elements. *Boundary-Layer Meteorology,* 2, 453–465.

Cisse, A. M., and H. Breman. 1982. La Phytoecologie du Sahel. In

La Productivite des Turages Sahliens, edited by F. W. T. Penning de Vries and M. A. Djiteye, 71–83. Wageningen: PUDOC.

Clarke, A. D. 1989. Aerosol Light Absorption by Soot in Remote Environments. *Aerosol Science and Technol.,* 10, 161–171.

Clarke, D., R. E. Teiss, and R. J. Charlson. 1984. Elemental Carbon Aerosols in the Urban, Rural and Remote-Marine Troposphere and in the Stratosphere: Inferences from Light Absorption Data and Consequences Regarding Radiative Transfer. *Science of the Total Environ.,* 3, 97–102.

Clements, F. G. 1916. *Plant Succession,* Carnegie Inst. Publ. 242, Washington, D.C.

Clements, H. B., and C. K. McIahon. 1980. Nitrogen Oxides from Burning Forest Fuels Examined by Thermogravimetry and Evolved Gas Analysis. *Thermochemica Acta,* 35, 133–139.

Coakley, J. A., Jr., R. L. Borstein, and P. A. Durkee. 1987. Effect of Ship Stack Effluents on Cloud Reflectance. *Science,* 237, 953–1084.

Coakley, J. A., Jr., R. D. Cess, and F. B. Yurevich. 1983. The Effect of Tropospheric Aerosol on the Earth's Radiation Budget: A Parameterization for Climate Models. *J. Atmos. Sci.,* 40, 116.

Coe, K. 1989. Personal communication.

Cofer W. R. III, V. G. Collins, and R. W. Talbot. 1985. Improved Aqueous Scrubber for Collection of Soluble Atmospheric Trace Gases. *Environ. Sci. Technol.,* 19, 557–560.

Cofer, W. R. III, J. S. Levine, P. J. Riggan, D. I. Sebacher, E. L. Winstead, E. F. Shaw, Jr., J. A. Brass, and V. G. Ambrosia. 1988a. Trace Gas Emissions from a Mid-Latitude Prescribed Chaparral Fire. *J. Geophys. Res.,* 93, 1653–1658.

Cofer, W. R. III, J. S. Levine, D. I. Sebacher, E. L. Winstead, P. J. Riggin, J. A. Brass, and V. G. Ambrosia. 1988b. Particulate Emission from a Mid-Latitude Prescribed Chaparral Fire. *J. Geophys. Res.,* 93, 5207–5212.

Cofer, W. R. III, J. S. Levine, D. I. Sebacher, E. L. Winstead, P. J. Riggin, B. J. Stocks, J. A. Brass, V. G. Ambrosia, and P. J. Boston. 1989. Trace Gas Emissions from Chaparral and Boreal Forest Fires. *J. Geophys. Res.,* 94, 2255–2259.

Cofer, W. R. III, J. S. Levine, E. L. Winstead. 1991. Chemistry of Trace Gas and Particulate Emissions from Mediterranean, Boreal, and Wetlands Fires. Paper presented at the Chapman Conference on Global Biomass Burning: Atmospheric, Climatic, and Biospheric Implications, Williamsburg, Va., 19–23 March (this volume, Chapter 27).

Cofer, W. R. III, J. S. Levine, E. L. Winstead, and B. J. Stocks. 1991. New Estimates of Nitrous Oxide Emissions from Biomass Burning. *Nature,* 349, 689–691.

Cofer, W. R. III, J. S. Levine, E. L. Winstead, P. J. LeBel, A. M. Koller, Jr., and C. R. Hinkle. 1990a. Trace Gas Emissions from Burning Florida Wetlands. *J. Geophys. Res.,* 95(D2), 1865–1870.

Cofer, W. R. III, J. S. Levine, E. L. Winstead, and B. J. Stocks. 1990b. Gaseous Emissions from Canadian Boreal Forest Fires. *Atmos. Environ.,* 24A, 1653–1659.

Cofer, W. R. III, and G. C. Purgold. 1981. An Automated Analyzer for Aircraft Measurements of Atmospheric Methane and Total Hydrocarbons. *Review of Scientific Instruments,* 52, 1560–1564.

Cohen, A. D. 1974. Evidence of Fires in the Ancient Everglades and Coastal Swamps of Southern Florida. In *Environments of*

South Florida Past and Present, edited by P. J. Gleason. Miami, Fla.: Miami Geological Society, 213–218.

Cole, M. M. 1986. *The Savannas: Biogeography and Geobotany,* London: Academic Press.

Colinvaux, P. A. 1989. The Past and Future Amazon. *Sci. Am.,* 260, 102.

Collins, N. M. 1977. Vegetation and Litter Production in Southern Guinea Savanna, Nigeria. *Oecologia,* (Berl.) 28, 1163–1175.

Collins, N. M., and T. G. Wood. 1984. Termites and Atmospheric Gas Production. *Science,* 224:84–85.

The Comprehensive Survey Report of the Xishuangbanna Nature Preserve. 1983. Yunnan Science Press, Yunnan, People's Republic of China.

Connell, J. H. 1978. Diversity in Tropical Rain Forests and Coral Reefs. *Science,* 199, 1302.

Connor, E. F. 1986. The Role of Pleistocene Forest Refugia in the Evolution and Biogeography of Tropical Biotas. *Tree,* 1, 165.

Conrad, R., and W. Seiler. 1986. Exchange of CO and H_2 between Ocean and Atmosphere. In *The Role of Air-Sea Exchange in Geochemical Cycling,* edited by P. Buat-Menard, 269–282, Dordrecht: Reidel.

Cook, O. F. 1909. *Vegetation Affected by Agriculture in Central America,* U.S. Department of Agriculture, Bureau of Plant Industry, Bulletin No. 145, Washington, D.C.: U.S. Government Printing Office.

Cooper, J. A. 1980. Environmental Impact of Residential Wood Combustion; Emission and Its Implications. *J. Air Poll. Control Ass.,* 30, 856–861.

Cope, M. J., and W. G. Chaloner. 1985. Wildfire: An Interaction of Biological and Physical Processes. In *Geological Factors and the Evolution of Plants,* edited by B. H. Tiffney, 257–277. New Haven: Yale University Press.

Cotton, W. R. 1985. Atmospheric Convection and Nuclear Winter. *Amer. Sci.,* 73, 275–280.

Countryman, C. M. 1964. *Mass Fires and Fire Behavior.* Res. Paper PSW-19 Berkeley, Calif.: U.S. Department of Agriculture, Forest Service, Pacific Southwest Forest and Range Experiment Station.

Courtillot, V. E., J. Besse, D. Vandamme, R. Montigny, J. Jaeger, and H. Cappeta. 1986. Deccan Flood Basalts at the Cretaceous/Tertiary Boundary? *Earth and Planetary Science Letters,* 80, 361–374.

Courtillot, V. E., and S. Cisowski. 1987. The Cretaceous-Tertiary Events: External or Internal Causes? *Eos,* 68, 193, 200.

Courtillot, V. E., G. Feraud, H. Maluski, D. Vandamme, M. G. Moreau, and J. Besse. 1988. Deccan Flood Basalts and the Cretaceous/Tertiary Boundary. *Nature,* 333, 843–846.

Courtney, W. J., J. W. Tesch, R. K. Stevens, and T. G. Dzubay. 1980. Paper presented at the 73rd Annual Meeting Air Pollution Control Association, Montreal.

Cowell, D. 1983. The Roar of the Fire. *Forestalk,* 7, no. 2, 14–20.

Cowles, R. B. 1939. Possible Implications of Reptilian Thermal Tolerance. *Science,* 90, 465–466.

Craig, H., and C. C. Chou. 1982. Methane: The Record in Polar Ice Cores. *Geophys. Res. Lett.,* 9, 1221–1224.

Crill, P. M., K. B. Bartlett, R. C. Harriss; E. Gorham, E. S. Verry, D.

I. Sebacher, L. Madzar, and W. Sanner. 1988. Methane Flux from Minnesota Peatlands. *Global Biogeochem. Cycles,* 2, 371–384.

Critchfield, W. B., and E. L. Little. 1966. *Geographic Distribution of the Pines of the World,* USDA For. Serv. Misc. Publ. 991, Washington, D.C.

Crockett, J. H., C. B. Officer, F. C. Wezel, and G. D. Johnson. 1988. Distribution of Noble Metals across the Cretaceous/Tertiary Boundary at Buggio, Italy: Iridium Variation as a Constraint on the Duration and Nature of Cretaceous/Tertiary Events. *Geology,* 16, 77–80.

Croft, T. A. 1977. *Nocturnal Images of the Earth from Space.* Final Report, SRI Project 5593, Stanford Research Institute, Menlo Park, Calif.

Cros, B., A. Lope, and J. Fontan. 1980. Estimation of the Aitken Particle Source Intensity in Rural Equatorial Africa. *Atmos. Environ.,* 15, 83–90.

Cros, B., R. Delmas, J. Clairac, J. Loemba-Ndembi, and J. Fontan. 1987. Survey of Ozone Concentrations in an Equatorial Region during the Rainy Season. *J. Geophys. Res.,* 92, 9772–9778.

Cros, B., R. Delmas, D. Nganga, B. Clairac, and J. Fontan. 1988. Seasonal Trends of Ozone in Equatorial Africa: Experimental Evidence of Photochemical Formation. *J. Geophys. Res.,* 93, 8355–8366.

Crozat, G., J. L. Domergue, J. Baudet, and V. Bogui. 1978. Influence des Feux de Brousse sur la Composition Chimique en Afrique de l'Ouest. *Atmos. Environ.,* 12, 1917–1920.

Crozat, G. 1978. *L'Aerosol Atmospherique en Milieu Naturel; Etude des Differentes Sources de Potassium en Afrique de l'Ouest (Cote d'Ivoire).* Ph.D. thesis, Université de Toulouse (France).

Crozat, G. 1979. Sur l'emission d'un aerosol riche en potassium par la forest tropicale. *Tellus,* 31, 52–57.

Crutzen, P. J. 1987. Acid Rain at the K/T Boundary. *Nature,* 330, 108–109.

Crutzen, P. J. 1990. *Biomass Burning: A Large Factor in the Photochemistry and Ecology of the Tropics.* Paper presented at the Chapman Conference on Global Biomass Burning: Atmospheric, Climatic, and Biospheric Implications, Williamsburg, Va., 19–23 March.

Crutzen, P. J. 1987. Role of the Tropics in Atmospheric Chemistry. In *The Geophysiology of Amazonia,* edited by R. E. Dickinson, 107–130. New York: John Wiley.

Crutzen P. J. 1988. Tropospheric Ozone: An Overview. In *Tropospheric Ozone: Regional and Global Scale Interactions,* edited by I. S. Isaksen, 3–32, Dordrecht: Reidel.

Crutzen, P. J., and M. O. Andreae. 1990. Biomass Burning in the Tropics: Impact on Atmospheric Chemistry and Biogeochemical Cycles. *Science,* 250, 1669–1678.

Crutzen, P. J., I. Aselmann, and W. Seiler. 1986. Methane Production by Domestic Animals, Wild Ruminants, Other Herbivorous Fauna, and Humans. *Tellus,* 38B, 271.

Crutzen, P. J., and J. W. Birks. 1982. The Atmosphere after a Nuclear War: Twilight at Noon. *Ambio,* 11, 115–125.

Crutzen P. J., A. C. Delany, J. Greenberg, P. Haagenson, L. Heidt, R. Lueb, W. Pollock, W. Seiler, A. Wartburg, and P. Zimmerman. 1985. Tropospheric Chemical Composition Measurements in Brazil during the Dry Season. *J. Atmos. Chem.,* 2, 233–256.

Crutzen, P., I. Galbally, and C. Bruhl. 1984. Atmospheric Effects from Post Nuclear Fires. *Climatic Change,* 6, 323–364.

Crutzen, P. J., and L. T. Gidel. 1983. A Two-Dimensional Photochemical Model of the Atmosphere: 2, The Tropospheric Budgets of the Anthropogenic Chlorocarbons, CO, CH, CH_3C and the Effect of Various NO_x Sources on Tropospheric Ozone. *J. Geophys. Res.,* 88, 6641–6661.

Crutzen, P. J., W. M. Hao, M. H. Liu, J. M. Lobert, and D. Scharffe. 1990. Emissions of CO_2 and Other Trace Gases to the Atmosphere from Fires in the Tropics, Our Changing Atmosphere. In *Proceedings of the 28th Liege International Astrophysical Colloquium.* University of Liege, Cointe-Ougree, Belgium, 449–472. Edited by P. J. Crutzen, J. C. Gerard, and R. Zander.

Crutzen, P. J., L. E. Heidt, J. P. Krasnec, W. H. Pollock, and W. Seiler. 1979. Biomass Burning as a Source of Atmospheric Gases CO, H_2, N_2O, NO, CH_3Cl, and COS. *Nature,* 282, 253–256.

Cunningham, R. H. 1963. The Effect of Clearing a Tropical Forest Soil. *J. Soil Science,* 14, 334–344.

DNES. 1988–89. Annual Report, Department of Non-Conventional Energy Sources (New Delhi).

DOE. 1990. *Multi-Laboratory Climate Change Committee, Energy and Climate Change.* Report of the DOE Multi-Laboratory Climate Change Committee. Chelsea, Mich.: Lewis Publishers.

Dahlman, R. C. 1985. Modeling Needs for Predicting Responses to CO_2 Enrichment: Plants, Communities, and Ecosystems. *Ecol. Modelling,* 9, 77–106.

Darley, E. F., F. R. Burleson, E. H. Mateer, J. T. Middleton, and V. P. Osterli. 1966. Contribution of Burning of Agricultural Wastes to Photochemical Air Pollution. *JAPCA,* 11, 685–690.

Darley, E. F. 1979. *Hydrocarbon Characterization of Agricultural Waste Burning.* Final Report, CAL/ARB Project A7-068-30, Statewide Air Pollution Research Center, University of California, Riverside, Calif.

Darley, E. F. 1977. *Emission Factors from Burning Agricultural Wastes Collected in California.* Final Report, CAIARB Project 4-011, Statewide Air Pollution Research Center, University of California, Riverside, Calif.

Dash, J. M. 1982. Particulate and Gaseous Emissions from Wood-Burning Fireplaces. *Environ. Sci. and Technol.,* 16, 639–645.

Daubenmire, R. 1972. Ecology of Hyparrhenia Rufa (Nees) in Derived Savanna in Northwestern Costa Rica. *J. Applied Ecol.,* 9, 11–23.

Davis, K. P. 1959. *Forest Fire Control and Use.* New York: McGraw-Hill.

De la Cruz, A. A. 1974. Primary Productivity of Coastal Marshes in Mississippi. *Gulf Res. Rep.,* 4, 351–356.

De Soete, G. G. 1989. *Formation of Nitrous Oxide from NO and SO_2 during Solid Fuel Combustion.* Paper presented at 1989 Joint Symposium on Stationary Combustion NO_x Control, Rep. EPA 60019-89-062b, Environmental Protection Agency, Research Triangle Park, N.C.

DeAngelis, D. G., D. S. Ruffin, J. A. Peters, and R. B. Reznik. 1980. *Source Assessment: Residential Combustion of Wood.* EPA-600/2-80-042b. Washington, D.C.: U.S. Environmental Protection Agency.

DeBano, L. F. 1974. Chaparral Soils. *Symposium on Living with the Chaparral,* 19–26, Riverside, Calif., 30–31 March.

DeBano, L. F., and Conrad, C. E. 1978. The Effect of Fire on Nutrients in a Chaparral Ecosystem. *Ecolo.*, 59, 489–497.

DeBano, L. F., G. E. Eberlein, and P. H. Dunn. 1979. Effects of Burning on Chaparral Soils. I. Soil Nitrogen. *Soil Sci. Soc. Amer. J.*, 43, 504–509.

DeBell, D. S., and C. W. Ralston. 1970. Release of Nitrogen by Burning Light Forest Fuels. *Soil Sci. Soc. Amer. Proc.*, 34, 936–938.

Deines, P. 1980. The Isotopic Composition of Reduced Carbon in the Terrestrial Environment. In *Handbook of Environmental Isotope Geochemistry*, edited by P. Fritz and J. Ch. Fortes, 329–407. Amsterdam: Elsevier.

Delany, A. C., P. Haagensen, S. Walters, A. F. Wartburg, and P. J. Crutzen. 1985. Photochemically Produced Ozone in the Emission from Large-Scale Tropical Vegetation Fires. *J. Geophys. Res.*, 90, 2425–2429.

Delmas, R. 1982. On the Emission of Carbon, Nitrogen and Sulphur in the Atmosphere during Bushfires in Intertropical Savanna Zones. *Geophys. Res. Lett.*, 9, 761–764.

Delmas, R., P. Loudjani, and A. Podaire. 1990. Biomass Burning in Africa: An Assessment of Annually Burnt Biomass. Paper presented at the Chapman Conference on Global Biomass Burning: Atmospheric, Climatic, and Biospheric Implications, Williamsburg, Va., 19–23 March (this volume, Chapter 16).

Delmas, R. A. 1990. Research Activities in Africa Related to the IGAC Programme. *IGAC Newsletter*, 2.

Delmas, R. A., A. Marenco, J. P. Tathy, and B. Cros. 1990. *Methane Emission from Combustions in Equatorial Africa.* Paper presented at the Chapman Conference on Global Biomass Burning: Atmospheric, Climatic, and Biospheric Implications, Williamsburg, Va., 19–23 March.

Denevan, W. M. 1961. The Upland Pine Forests of Nicaragua: A Study in Cultural Plant Geography. *University of California Publications in Geography*, 12(4), 251–320.

Derwent, R. G., and A. Volz-Thomas. 1990. The Tropospheric Lifetimes of Halocarbons and Their Reactions with OH Radicals: An Assessment Based on the Concentration of 1 CO. *AFEAS Report: Appendix in World Meteorological Organization Global Ozone Research and Monitoring Project—Report No, 2. Scientific Assessment of Stratospheric Ozone*, Geneva.

Desalmand, F. 1987. Observations of CCN Concentrations South of the Sahara during a Dust Haze. *Atmos. Res.*, 21, 13–28.

Desalmand, F., J. Baudet, and R. Serpolay. 1982. Influence of Rainfall on the Seasonal Variations of Cloud Condensation Nuclei Concentrations in a Subequatorial Climate. *J. Atmos. Sciences*, 39, 2076–2082.

Desalmand, F., J. Podzimek, and R. Serpolay. 1985b. A Continental Well-Aged Aerosol in the Guinean Savannah at the Level of a Trough along the ITCZ. *J. Aerosol Sci.*, 16, 19–28.

Desalmand, F., R. Serpolay, and J. Podzimek. 1985a. Some Specific Features of the Aerosol Particle Concentrations during the Dry Season and during a Brushfire Event in West Africa. *Atmos. Environ.*, 19, 1535–1543.

Desbois, M., G. Seze, and G. Szejwach. 1982. Automatic Classification of Clouds on METEOSAT Imagery: Application to High Clouds. *J. Appl. Met.*, 21, 401–412.

Deschler, U. 1974. *An Examination of the Extent of Grass Fires in the Savanna of Africa along the Southern Side of the Sahara.* Paper presented at the Ninth International Symposium on Remote Sensing of the Environment, Environmental Research Institute Mich., Ann Arbor, Mich., April.

Detwiler, R. P., and C. A. S. Hall. 1988. Tropical Forests and the Global Carbon Cycle. *Science*, 239, 42–47.

Detwiler, R. P., C. A. S. Hall, and P. Bogdonoff. 1985. Land Use Change and Carbon Exchange in the Tropics: II. Estimates for the Entire Region. *Environ. Mgmt.*, 9, 335–344.

Devol, A. H., J. E. Richey, W. A. Clark, S. L. King, and L. A. Martinelli. 1988. Methane Emissions to the Troposphere from the Amazon Floodplain. *J. Geophys. Res.*, 93, 1583–1592.

Dianov-Klokov, V., and L. N. Yurganov. 1989. Spectroscopic Measurements of Atmospheric Carbon Monoxide and Methane 2: Seasonal Variations and Long-Term Trends. *J. Atmos. Chem.*, 8, 153–164.

Dickerson, R., G. Huffman, W. Luke, L. Nunnermacker, K. Pickering, A. Leslie, C. Lindsey, W. Slinn, T. Kelly, P. Daum, A. Delany, J. Greenberg, P. Zimmerman, J. Boatman, J. Ray, and D. Stedman. 1987. Thunderstorms: An Important Mechanism in the Transport of Air Pollutants. *Science*, 235, 460–464.

Dickinson, R. E. 1986. Impact of Human Activities on Climate—A Framework. In *Sustainable Development for the Biosphere*, edited by W. C. Clark and R. E. Munn, 252–291, IIASA. Cambridge: Cambridge University Press.

Dickinson, R. E., and R. J. Cicerone. 1986. Future Global Warming from Atmospheric Trace Gases. *Nature*, 319, 109–115.

Dickinson, R. E., and A. Henderson-Sellers. 1988. Modeling Tropical Deforestation: A Study of GCM Land-Surface Parameterization. *Quart. J. Royal Meteorol. Society*, 114, 439–462.

Dickinson, R. E., and H. Virji. 1987. Climate Changes in the Humid Tropics, Especially Amazonia, Over the Last Twenty Thousand Years. In *The Geophysiology of Amazonia*, edited by R. E. Dickinson, 91–101. New York: John Wiley.

Dietz, W. A. 1967. Response Factors for Gas Chromatographic Analyses. *J. Gas Chromatogr.*, 2, 68–71.

Dignon, J. 1991. NO_x and SO_x Emissions from Fossil Fuels: A Global Distribution. *Atmos. Environ.*, in press.

Dignon, J., S. Hameed, J. E. Penner, C. S. Atherton, and J. J. Walton. 1990. Biomass Burning and Its Contribution to Nitrate Deposition in the Tropics. Paper presented at the Chapman Conference on Global Biomass Burning: Atmospheric, Climatic, and Biospheric Implications, Williamsburg, Va., 19–23 March (this volume, Chapter 44).

Dimitriades, B., and M. Dodge (eds.). 1983. *Proceedings of the Empirical Kinetic Modeling Approach (EKMA) Validation Workshop.* EPA Report No. 60019-83-014, August.

Dindorf, W. et al. 1990. Private communication, Johannes Gutenberg Universitat Mainz, Organic Chemistry Department.

Dinh, P. V., and E. Pique. 1988. Le Collecteur d'aerosols par impaction CRA-88. *Rapport du Laboratoire d'Aerologie*.

Dolph, G. E. 1990. A Critique of the Theoretical Basis of Leaf Margin Analysis. *Proceedings of the Indiana Academy of Science*.

Dolph, G. E. 1984. Leaf Formation of the Woody Plants of Indiana to Related Environment. In *Being Alive on Land: Tasks for Vegetation Science*, 13, edited by N. S. Margaris, M. Arianoustou-Farragitaki, and W. C. Oechel, 51–61. The Hague: Dr. Junk Publishers.

Dolph, G. E. 1987. The Variation of Four Bioclimatic Indexes in Indiana. *Proceedings of the Indiana Academy of Science,* 96, 99–111.

Dolph, G. E., and D. L. Dilcher. 1979. Foliar Physiognomy as an Aid in Determining Paleoclimate. *Palaeontographica,* 170, 151–172.

Donnelly, R. E., and J. B. Harrington. 1978. *Forest Fire History Maps of Ontario.* Report FF-Y-6, Forest Fire Research Institute, Canadian Forest Service, Department of the Environment, Ottawa, Ont. (6 sheets at 1:500,000).

Dozier J. 1981. A Method for Satellite Identification of Surface Temperature Fields of Subpixel Resolution. *Remote Sensing of Environ.,* 11, 221–229.

Duce, R. A. 1978. Speculations on the Budget of Particulate and Vapor Phase Nonmethane Organic Carbon in the Global Troposphere. *Pure Appl. Geophys.,* 116, 244–273.

Duncan, R. A., and D. G. Pyle. 1988. Rapid Eruption of the Deccan Flood Basalts at the Cretaceous/Tertiary Boundary. *Nature,* 333, 841–846.

Dunn, P. H., L. F. DeBano, and G. E. Eberlein. 1979. Effects of Burning on Chaparral Soils: II. Soil Microbes and Nitrogen Mineralization, *Soil Sci. Soc. Amer. J.,* 43, 509–514.

Duvigneaud, P., and S. Denaeyer-DeSmet. 1970. Biological Cycling of Minerals in Temperate Deciduous Forests. *Analysis of Temperate Forest Ecosystems,* Ecological Studies 1, 199–225. New York: Springer-Verlag.

Duxbury, J. M., D. R. Bouldin, R. E. Terry, and R. L. Tate. 1982. Emissions of Nitrous Oxide from Soils, *Nature,* 298, 462–464.

EDSG. 1986. *Report of the Energy Demand Screening Group.* New Delhi: Planning Commission, Government of India.

Eagan, R. C., P. V. Hobbs, and L. F. Radke. 1974a. Measurements of Cloud Condensation Nuclei and Cloud Droplet Size Distributions in the Vicinity of Forest Fires. *J. Appl. Meteorol.,* 13, 553–557.

Eagan, R. C., P. V. Hobbs, and L. F. Radke. 1974b. Particle Emissions from a Large Kraft Paper Mill and Their Effects on the Microstructure of Warm Clouds. *J. Appl. Meteorol.,* 13, 535–552.

Ecoregions Working Group. 1989. *Ecoclimatic Regions of Canada, First Approximation.* Ecoregions Working Group of Canada Committee on Ecological Land Classification. Ecological Land Classification Series, No. 23, Sustainable Development Branch, Canadian Wildlife Service, Conservation and Protection, Environment Canada, Ottawa, Ontario.

Eder, G., and Preisinger, A. 1987. Zeitstruktur Globaler Ereignise, Veranschaulicht an der Kreide-Tertiar-Grenze. *Naturwissenschaften,* 74, 35–37.

Edwards, L. L., and J. E. Penner. 1988. Potential Nucleation Scavenging of Smoke Particles over Large Fires: A Parametric Study. In *Aerosol and Climate,* edited by P. V. Hobbs and M. P. McCormick, 423–434. Hampton, Va.: Deepak Publishing.

Egunjobi, J. K. 1971. Savanna Burning, Soil Fertility and Herbage Productivity in the Derived Savanna Zone of Nigeria. *Wildlife Conservation in West Africa,* 52–58. Geneva, Switzerland: IUCN.

Egunjobi, J. K. 1974. Dry Matter, Nitrogen and Mineral Element Distribution in an Unburnt Savanna during the Year. *Oecologia Planetarium,* 9, 1–10.

Ehhalt, D., and J. W. Drummond. 1982. The Tropospheric Cycle of NO$_x$. In *Chemistry of the Unpolluted and Polluted Troposphere,*

edited by H. W. Georgi and W. Jaeschke, 219–251. Dordrecht: Reidel.

Ehhalt, D. H., J. Rudolph, and U. Schmidt. 1986. On the Importance of Light Hydrocarbons in Multiphase Atmospheric Systems. In *Chemistry of Multiphase Atmospheric Systems,* edited by W. Jaeschke, 322–350. Berlin-Heidelberg, FRG: Springer-Verlag.

Eimer, P., and C. Ndamana. 1987. Carbonisation, les ratios de transformation. *Rapport du Ministère de l'Agriculture du Burundi.* Paris: BDPA.

Einfeld, W., B. Mokler, D. Morrison, and B. Zak. 1989. Particle and Trace Element Production from Fires in the Chaparral Fuel Type. *Twenty-fifth Session of the Air and Waste Management Association.* Available from D. E. Ward, U.S.D.A. Forest Service, Intermountain Fire Science Laboratory, Missoula, Mont. 59807.

Einfeld, W., D. Ward, and C. Hardy. 1990. Effects of Fire Behavior on Prescribed Fire Smoke Characteristics. Paper presented at the Chapman Conference on Global Biomass Burning: Atmospheric, Climatic, and Biospheric Implications, Williamsburg, Va., 19–23 March (this volume, Chapter 50).

Eisele, K. A., D. S. Schimel, L. A. Kapustka, and W. J. Parton. 1989. Effects of Available P and N:P Ratios on Non-symbiotic Dinitrogen Fixation in Tallgrass Prairie Soils. *Oecologia,* 79, 471–474.

Eiten, G. 1972. The *Cerrado* Vegetation of Brazil. *Bot. Rev.,* 38, 201–341.

Eleuterius, L. N. 1972. The Marshes of Mississippi. *Castanea,* 37, 153–168.

Eleuterius, L. N. 1976. The Distribution of Juncus Roemerianus in the Salt Marshes of North America. *Chesapeake Sci.,* 17, 289–296.

Eleuterius, L. N. 1984. Autecology of the Black Needlerush, Juncus Roemerianus. *Gulf Res. Rep.,* 7, 339–350.

Eleuterius, L. N., and F. C. Lanning. 1987. Silica in Relation to Leaf Decomposition of Juncus Roemerianus. *J. Coastal Res.,* 3, 531–534.

Elkins, J. W., B. D. Hall, and J. H. Butler. 1990. *Laboratory and Field Investigations of the Emissions of Nitrous Oxide from Biomass Burning.* Paper presented at the Chapman Conference on Global Biomass Burning: Atmospheric, Climatic, and Biospheric Implications, Williamsburg, Va., 19–23 March.

Elkins, J. W., S. C. Wofsy, M. B. McElroy, C. E. Kolb, and W. A. Kaplan. 1978. Aquatic Sources and Sinks for Nitrous Oxide. *Nature,* 275, 602–606.

Elliott, W. C., J. L. Aronson, H. T. Millard, Jr., and E. Gierlowski-Kordesch. 1989. The Origin of the Clay Minerals at the Cretaceous/Tertiary Boundary in Denmark. *Geological Society of America Bulletin,* 101, 702–710.

Elliott, W. P., L. Machta, and C. D. Keeling. 1985. An Estimate of the Biotic Contribution to the Atmospheric CO$_2$ Increase Based on Direct Measurements at Mauna Loa Observatory. *J. Geophys. Res.,* 90, 3741–3746.

Ellis, R. C. 1985. The Relationships among Eucalypt Forest, Grassland and Rainforest in a Highland Area in North-Eastern Tasmania. *Aus. J. Ecol.,* 10, 297.

Ellsesser, H. 1981. Comment on Production of Odd Nitrogen in the Stratosphere and Mesosphere: An Intercomparison of Source Strengths by C. H. Jackman et al. *J. Geophys. Res.,* 6, 12157–12158.

Emanuel, W. R., G. G. Killough, W. M. Post, and H. H. Shugart.

1984. Modeling Terrestrial Ecosystems in the Global Carbon Cycle with Shifts in Carbon Storage Capacity by Land-Use Change. *Ecology,* C5, 970–983.

Emiliani, C., E. B. Kraus, and E. M. Shoemaker. 1981. Sudden Death at the End of the Mesozoic. *Earth and Planetary Science Letters,* 55, 317–334.

Enting, I. G., and J. V. Mansbridge. 1987. The Incompatibility of Ice-Core CO_2 Data with Reconstructions of Biotic CO_2 Sources. *Tellus,* 3B, 318–325.

Enting, I. G., and G. I. Pearman. 1986. The Use of Observations in Calibrating and Validating Carbon Cycle Models. In *The Changing Carbon Cycle: A Global Analysis,* edited by J. R. Trabalka and D. E. Reichle, 425–458. New York: Springer-Verlag.

Environment Canada. 1975. Combustion Technology for the Disposal and Utilization of Wood Residue. Economic and Technical Review Report EPS 3-AP-75- 4. Air Pollution Control Directorate, pp. 17, 91.

Erickson, P. J. III, J. J. Walton, S. J. Ghan, and J. E. Penner. 1990. Three-Dimensional Modeling of the Global Atmospheric Sulfur Cycle: A First Step. Submitted to *Atmos. Enviro.* Also Lawrence Livermore National Laboratory Report UCRL-JC-103973, Livermore, Calif.

Ernst, L. 1990. *Anwendung Eines Spruhsammlers auf das Problem der Differenzierung von Gasformigem und Partikelgebundenem Ammoniak.* Master's thesis, Johannes Gutenberg Universitat Mainz, FRG, Max Planck Institut für Chemie.

Esser, G. 1987. Sensitivity of Global Carbon Pools and Fluxes to Human and Potential Climatic Impacts. *Tellus,* 39, 245–260.

Ette, A. I., and Hoimuk, A. B. 1984. Comparative Studies of NO_3-Concentrations in Some Tropical Rains. *Atmospheric Environment,* 18, 2781–2786. Economic and Technical Review Report EPS 3-AP-75-4, Air Pollution Control Directorate, 17, 91.

Evans, L. F., N. K. King, P. R. Pockhaun, and L. T. Stephens. 1974. Ozone Measurements in Smoke from Forest Fires. *Environ. Sci. Technol.,* 8, 75–79.

Evans, L. F., I. A. Weeks, A. J. Eccleston, and D. R. Packham. 1977. Photochemical Ozone in Smoke from Prescribed Burning of Forests. *Environ. Sci. Technol.,* 11, 896–900.

FAO. 1981. *Tropical Forest Resources Assessment Project.* Rome: FAO.

FAO. 1981a. *Forest Resources of Tropical Asia.* Rome: FAO.

FAO. 1981b. *Tropical Forest Resources Assessment Project: Forest Resources of Tropical Africa Country Briefs, Part II,* 359–379. Rome: FAO.

FAO. 1982. *Tropical Forest Resources.* FAO Forestry Paper 30. Rome: FAO.

FAO. 1985. *Tropical Forestry Action Plan.* Rome: FAO.

FAO. 1986. *FAO Production Yearbook.* Rome: FAO.

FAO. 1988. *An Interim Report on the State of Forest Resources in the Developing Countries.* Publication FAO Forest Resources Division. Rome: FAO.

FAO. 1989. *Yearbook of Forest Products 1987 (1976–1987).* Rome: FAO.

FAO. 1990. *Preliminary 1990 Forest Assessment.* Rome: FAO.

FSI. 1987. *The State of Forest Report.* Dehradun: Forest Survey of India.

FSI. 1989. *The State of Forest Report.* Dehradun: Forest Survey of India.

Fabian, P. A., and P. G. Pruchniewicz. 1977. Meridional Distribution of Ozone in the Troposphere and Its Seasonal Variations. *J. Geophys. Res.,* 82, 2063–2073.

Fahnestock, G. R., and J. K. Agee. 1983. Biomass Consumption and Smoke Production by Prehistoric and Modern Forest Fires in Western Washington. *J. For.,* 81, 653.

Falesi, I. C. 1976. *Ecossistema de Pastagem Cultivada na Amaznia Brasileira.* Belem: Centro de Pesquisa Agropecuarla do Tropico Umido (CPATU).

Fastovsky, D. E., K. McSweeney, and L. D. Norton. 1989. Pedogenic Development at the Cretaceous-Tertiary Boundary, Garfield County, Montana. *J. Sedimentary Petrology,* 59, 758–767.

Faulkner, S. P., and A. A. de la Cruz. 1982. Nutrient Mobilization Following Winter Fires in an Irregularly Flooded Marsh. *J. Environ. Qual.,* 11, 129–133.

Faure, G. 1986. *Principles of Isotope Geology.* New York: John Wiley.

Fearnside, P. M. 1980. The Effects of Cattle Pasture on Soil Fertility in the Brazilian Amazon: Consequences for Beef Production Sustainability. *Tropical Ecol.,* 21(1), 125–137.

Fearnside, P. M. 1982. Deforestation in the Brazilian Amazon: How Fast Is It Occurring? *Interciencia,* (2), 82–88.

Fearnside, P. M. 1983a. Development Alternatives in the Brazilian Amazon: An Ecological Evaluation. *Interciencia,* 8(2), 65–78.

Fearnside, P. M. 1983b. Land Use Trends in the Brazilian Amazon Region as a Factor in Accelerating Deforestation. *Environmental Conservation,* 10(2), 141–148.

Fearnside, P. M. 1985a. Agriculture in Amazonia. In *Ecology Environments: Amazonia,* edited by G. T. Prance and T. E. Lovejoy. Oxford, U.K.: Pergamon Press.

Fearnside, P. M. 1985b. Brazil's Amazon Forest and the Global Carbon Problem. *Interciencia,* 10(4), 179–186.

Fearnside, P. M. 1985c. Burn Quality Prediction for Simulation of the Agricultural System of Brazil's Transamazon Highway Colonists for Estimating Human Carrying Capacity. In *Ecology and Resource Management in the Tropics,* edited by G. V. Govil. Varanasi, India: International Society for Tropical Ecology.

Fearnside, P. M. 1985d. Environmental Change and Deforestation in the Brazilian Amazon. In *Change in the Amazon Basin: Man's Impact on Forests and Rivers,* edited by J. Hemming. Manchester, U.K.: Manchester University Press.

Fearnside, P. M. 1986. Brazil's Amazon Forest and the Global Carbon Problem: Reply to Lugo and Brown. *Interciencia,* 11(2), 58–64.

Fearnside, P. M. 1986. *Human Carrying Capacity of the Brazilian Rainforest.* New York: Columbia University Press.

Fearnside, P. M. 1986. Spatial Concentration of Deforestation in the Brazilian Amazon. *Ambio,* 15(2), 72–79.

Fearnside, P. M. 1987a. Causes of Deforestation in the Brazilian Amazon. In *The Geophysiology of Amazonia: Vegetation and Climate Interactions,* edited by R. F. Dickinson. New York: John Wiley.

Fearnside, P. M. 1987b. Summary of Progress in Quantifying the Potential Contribution of Amazonian Deforestation to the Global

Carbon Problem. In *Proceedings of the Workshop on Biogeochemistry of Tropical Rain Forests: Problems for Research,* edited by D. Athie, T. E. Loveoy, and P. de M. Oyens. Universidade de So Paulo, Centro de Energia Nuclear na Agricultura (CENA), Piracicaba, São Paulo.

Fearnside, P. M. 1988. An Ecological Analysis of Predominant Land Uses in the Brazilian Amazon. *Environmentalist,* 8(4), 281–300.

Fearnside, P. M. 1989a. Climate Environment and International Security: The Case of Deforestation in the Brazilian Amazon. In *Climate and Geo-Sciences: A Challenge for Science and Society in the 21st Century,* edited by A. Berger, S. Schneider, and J. C. Duplesy. Dordrecht, Netherlands: Kluwer Academic Publishers.

Fearnside, P. M. 1989b. Extractive Reserves in Brazilian Amazonia: Opportunity to Maintain Tropical Rain Forest under Sustainable Use. *Bioscience,* 39(6), 387–393.

Fearnside, P. M. 1989c. Forest Management in Amazonia: The Need for New Criteria in Evaluating Development Options. *Forest Ecol. and Mgmt.,* 27(1), 61–79.

Fearnside, P. M. 1989d. A occupaao humana de rondina: Impactos limites e planeamento. *Programa POLONOROESTE Relatorio de Pesquisa No. 5.* Brasilia: Conselho Nacional de Desenvolvimento Cientifico e Tecnologico (CNPq).

Fearnside, P. M. 1989e. A Prescription for Slowing Deforestation in Amazonia. *Environ.* 31(4), 16–20, 39–40.

Fearnside, P. M. 1990a. Deforestation in Brazilian Amazonia. In *The Earth in Transition: Patterns and Processes of Biotic Impoverishment,* edited by G. M. Woodwell, 211–236. New York: Cambridge University Press.

Fearnside, P. M. 1990b. Fire in the Tropical Rain Forest of the Amazon Basin. In *Fire in the Tropical and Subtropical Biota,* edited by J. G. Goldammer, 106–116. Heidelberg, FRG: Springer-Verlag.

Fearnside, P. M. 1990c. Practical Targets for Sustainable Development in Amazonia. In *Maintenance of the Biosphere: Proceedings of the Third International Conference on the Environmental Future,* edited by N. Polunin and J. Burnett, 167–174. Edinburgh, U.K.: Edinburgh University Press.

Fearnside, P. M. 1990. *Deforestation in Brazilian Amazonia as a Source of Greenhouse Gases.* Paper presented at the Chapman Conference on Global Biomass Burning: Atmospheric, Climatic, and Biospheric Implications, Williamsburg, Va., 19–23 March (this volume, Chapter 11).

Fearnside, P. M. 1990. The Rate and Extent of Deforestation in Brazilian Amazonia. *Environmental Conservation,* 17, 213–216.

Fearnside, P. M., N. L. Filho, P. M. L. A. Graca, G. L. Ferreira, R. A. Custodio, and F. J. A. Rodrigues. nd-a. Pasture Biomass and Productivity in Brazilian Amazonia, in preparation.

Fearnside, P. M., N. L. Filho, F. J. A. Rodrigues, P. M. L. A. Graca, and J. M. Robinson. nd-b. Tropical Forest Burning in Brazilian Amazonia: Measurements of Biomass, Combustion Efficiency and Charcoal formation at Altamira, Para, in preparation.

Fearnside, P. M., M. Keller, N. L. Filho, and F. M. Fernandes. 1990. *Rainforest Burning and the Global Carbon Budget: Biomass, Combustion Efficiency and Charcoal Formation in the Brazilian Amazon.* Unpublished manuscript.

Ferrare, R. A., R. S. Fraser, and Y. J. Kaufman. 1990. Satellite Measurements of Large-Scale Air Pollution: Measurements of Forest Fire Smoke. *J. Geophys. Res.,* 95, 9911–9925.

Fillery, I. R. P., and Vlek, F. L. G. 1982. The Significance of Denitrification of Applied Nitrogen in Fallow and Cropped Rice Soils under Different Flooding Regimes. *Plant Soil,* 65, 153–116.

Finlayson-Pitts, B. J., J. N. Pitts. 1986. *Atmospheric Chemistry: Fundamentals and Experimental Techniques.* New York: John Wiley.

Fishman, J. 1988. Tropospheric Ozone from Satellite Total Ozone Measurements in Tropospheric Ozone. In *Tropospheric Ozone: Regional and Global Scale Interactions,* edited by I. S. A. Isaksen, 111–123. Dordrecht, Netherlands: Reidel.

Fishman, J., G. L. Gregory, G. W. Sachse, S. M. Beck, and G. F. Hill. 1987. Vertical Profiles of Ozone, Carbon Monoxide, and Dew Point Temperature Obtained during GTE/CITE 1, October–November 1983. *J. Geophys. Res.,* 92, 2083–2094.

Fishman, J., and J. C. Larsen. 1987. The Distribution of Total Ozone and Stratospheric Ozone in the Tropics: Implications for the Distribution of Tropospheric Ozone. *J. Geophys. Res.,* 92, 6627–6634.

Fishman, J., P. Minnis, and H. G. Reichle, Jr. 1986. Use of Satellite Data to Study Tropospheric Ozone in the Tropics. *J. Geophys. Res.,* 92, 14,451–14,465.

Fishman, J., P. Minnis, M. Taylor. 1985. *Ozone Emissions from Tropical Forest and Savannah Fires Deduced from Satellite Observations.* Eighth Conference on Fire and Forest Meteorology, Society of American Foresters, 73–80.

Fishman, J., V. Ramanathan, P. J. Crutzen, and S. C. Liu. 1979. Tropospheric Ozone and Climate, *Nature.* 282(5741), 818–820.

Fishman, J., C. E. Watson, and J. C. Larsen. 1989. The Distribution of Total Ozone, Stratospheric Ozone, and Tropospheric Ozone at Low Latitudes Deduced from Satellite Data Sets. In *Ozone in the Atmosphere,* edited by R. D. Bojkov and P. Fabian, pp. 411–414. Hampton, Va.: Deepak.

Fishman, J., C. E. Watson, J. C. Larsen, and J. A. Logan. 1990. Distribution of Tropospheric Ozone Determined from Satellite Data. *J. Geophys. Res.,* 95, No. D4, 3599–3617.

Flagan, R. C., and D. D. Taylor. 1981. Laboratory Studies of Submicron Particles from Coal Combustion. *18th Symposium (International) on Combustion,* 1227–1237. Pittsburgh, Pa.: Combustion Institute.

Fleeter, R. D., F. E. Fendell, L. M. Cohen, N. Gat and A. B. Witte. 1984. Laboratory Facility for Wind Aided Firespread along a Fuel Matrix. *Combustion and Flame,* 57, 289–311.

Flenley, J. 1979. *The Equatorial Rain Forest: A Geological History.* London: Butterworth's.

Flint, E. P., and J. F. Richards. 1991. Historical Analysis of Changes in Land Use and Carbon Stock of Vegetation in South and Southeast Asia. *Canadian J. Forest Res.,* in press.

Flocke, F. 1989. Unpublished results, KFA Julich, Institut für Chemie II, FRG.

Flohn, H. 1981. Scenarios of Cold and Warm Periods of the Past. In *Climatic Variations and Variability: Facts and Theories,* 689–698. Dordrecht, Netherlands: Reidel.

Florida Sugar Cane League. 1978. *Water Quality Studies in the Everglades Agricultural Area of Florida.* Clewiston, Fla.: FSCL.

Forsbergh, E. D. 1969. On the Climatology, Oceanography and Fisheries of the Panama Eight. *Bull. Inter-Am. Trop. Tuna Com.,* 14, 49–385.

Francis, W. 1961. *Coal, Its Formation and Composition.* London: Arnold.

Fraser, R. S., Y. J. Kaufman, and R. L. Mahoney. 1984. Satellite Measurements of Aerosol Mass and Transport. *J. Atmos. Environ.,* 18, 2577–2584.

Fraser, P. J., R. A. Rasmussen, J. W. Creffield, J. R. French, and M. A. K. Khalil. 1986. Termites and Global Methane—Another Assessment. *J. of Atmos. Chemistry,* 4, 295–310.

French, B. M. 1984. Impact Event of the Cretaceous-Tertiary Boundary: A Possible Site. *Science,* 226, 353.

Freyer, H. D. 1986. Interpretation of the Northern Hemispheric Record of the 13C/12C Trends of Atmospheric CO_2 in Tree Rings. In *The Changing Carbon Cycle: A Global Analysis,* edited by J. R. Trabalka and D. E. Reichle, 126–150. New York: Springer-Verlag.

Friedli, H., H. Lotscher, H. Oeschger, U. Siegenthaler, and B. Stauffer. 1986. Ice Core Record of the 13C/12C Ratio of Atmospheric CO_2 in the Past Two Centuries. *Nature,* 24, 237–238.

Fuchs, F. A. 1988. Fire Protection Project in China. *Fire Mgmt. Notes,* 40, 3–7.

Fuchs, F. A. 1964. *The Mechanics of Aerosols.* New York: Pergamon Press.

GTC. 1986. *Global Tropospheric Chemistry,* University Corporation for Atmospheric Research, P.O. Box 3000, Boulder, Colo.

Gadgil, A., and M. Modera. 1990. *Cookstoves and Global Warming.* Paper presented at the 12th Annual IAEE Conference, New Delhi, 4–6 January.

Galbally, I. E. 1985. The Emission of Nitrogen to the Remote Atmosphere: Background Paper. In *Biogeochemical Cycling of Sulfur and Nitrogen in the Remote Atmosphere,* edited by J. H. Galloway, R. J. Charlson, M. O. Andreae, and H. Rodhe, 27–53. Hingham, Mass.: Reidel.

Gallagher, J. L., R. J. Reimold, R. A. Linthurst, and V. J. Pfeiffer. 1980. Aerial Production, Mortality, and Mineral Accumulation-Export Dynamics in Spartina Alterniflora and Juncus Roemerianus Plant Stands in a Georgia Salt Marsh. *Ecology,* 61, 303–312.

Galloway, J. N., G. E. Likens, W. C. Keene, and J. M. Miller. 1982. The Composition of Precipitation in Remote Areas of the World. *J. Geophys. Res.,* 87, 8771–8786.

Gammon, R. H., W. D. Komhyr, and J. T. Peterson. 1986. The Global Atmospheric CO_2 Distribution 1968–1983: Interpretation of the Results of the NOAA/GMCC Measurement Program. In *The Changing Carbon Cycle: A Global Analysis,* edited by J. R. Trabalka and D. E. Reichle, 1–15. New York: Springer-Verlag.

Ganapathy, R., S. Gartner, and M. Jiang. 1981. Iridium Anomaly at the Cretaceous-Tertiary Boundary in Texas. *Earth and Planetary Science Letters,* 54, 393–396.

Garcia, O. 1985. *Atlas of Highly Reflective Clouds for the Global Tropics: 1971–1983.* Boulder, Colo.: NOAA Environmental Research Laboratories.

Garg, B. M. L. 1989. Improved Chulhas (Cookstoves) Programme in India. In *Stoves for People,* edited by Roberto Caceres, Jamuna Ramakrishna, and Kirk R. Smith, 97–100. London: IT Publications.

Garstang, M., B. E. Kelbe, G. D. Emmitt, and W. B. London. 1987. Generation of Convective Storms over the Escarpment of Northeastern South Africa. *Mon. Wea. Rev.,* 115, 429–443.

Garstang, M., J. Scala, S. Greco, R. Harriss, S. Beck, E. Browell, G. Sachse, G. Gregory, G. Gill, J. Simpson, W.-K. Tao, and A.

Torres. 1988. Trace Gas Exchanges and Convective Transports over the Amazonian Rain Forest. *J. Geophys. Res.,* 93, 1528–1550.

Gaudichet, A., J. R. Petit, R. Lefevre, and C. Lorius. 1986. An Investigation by Analytical Transmission Electron Microscopy of Individual Insoluble Microparticles From Antarctic (Dome C) Ice Core Samples. *Tellus,* 38B, 250–261.

Gavin, J., G. Kukla, and T. Karl. 1989. Sulphate Aerosols and Climate: Scientific Correspondence. *Nature,* 340, 438.

Gerber, H., and E. E. Hindman, ed. 1982. *Light Absorption by Aerosol Particles.* Hampton, Va.: Spectrum Press.

Germeraad, J. H., C. A. Hopping, and J. Muller. 1968. Palynology of Tertiary Sediments from Tropical Areas. *Rev. Palaeobot. Palynol.,* 6, 189–348.

Gerstle, R. W., and D. A. Kemnitz. 1967. Atmospheric Emissions from Open Burning. *JAPCA,* 17, 324–327.

Ghan, S. J., and J. E. Penner. 1991. Effects of Smoke on Climate. *Encyclopedia of Earth System Science.* Orlando, Fla.: Academic Press. Also Lawrence Livermore National Laboratory report UCRL-JC-103415, March.

Ghan, S. J., J. E. Penner, and K. E. Taylor. 1989. Sulphate Aerosols and Climate. *Nature,* 340, 438.

Ghan, S. J., K. E. Taylor, J. E. Penner, and D. J. Erickson III. 1990. Model Test of CCN-Cloud Albedo Climate Forcing. *Geophys. Res. Lett.,* also Lawrence Livermore National Laboratory report UCRL-101205, Rev. 1.

Gillon, D. 1983. The Fire Problem in Tropical Savannas. In *Tropical Savannas,* edited by F. Bourliere, 617–641. Amsterdam: Elsevier.

Gillon, D., and M. Rapp. 1989. Nutrient Losses during a Winter Low-Intensity Prescribed Fire in a Mediterranean Forest. *Plant Soil,* 120, 69–77.

Gilmour, I., and E. Anders. 1989. Cretaceous-Tertiary Boundary Event: Evidence for a Short Time Scale. *Geochim. Cosmochim. Acta,* 53, 503–511.

Gilmour, I., and F. Guenther. 1988. The Global Cretaceous-Tertiary Fire: Biomass or Fossil Carbon? *Global Catastrophes in Earth History,* Snowbird, Utah: Lunar Planet. Institute.

Glahn, H. R., and D. A. Lowry. 1972. The Use of Model Output Statistics (MOS), Objective Weather Forecasting. *J. Appl. Meteor.,* 11, 1203–1211.

Godfrey, R. K., and J. W. Wooten. 1979. Aquatic and Wetland Plants of Southeastern United States. *Monocotyledons.* Athens: University of Georgia Press.

Goldammer, J. G. 1979. Forest Problems in Germany. *Fire Mgmt. Notes,* 40, 7–10.

Goldammer, J. G. 1986a. *Feuer und Waldentwicklung in den Tropen und Subtropen.* Freiburger Waldschutz Abh. 6, 43.

Goldammer, J. G. 1986b. *Technical and Vocational Forestry and Forest Industries Training.* Burma: Forest Fire Management, FAO, BUR/81/001, Field Document 5.

Goldammer, J. G. 1987. Wildfires and Forest Development in Tropical and Subtropical Asia: Outlook for the Year 2000. *Proc. Wildland Fire.* Berkeley, Calif.: USDA Forest Service Gen. Tech. Rep. P SW-101.

Goldammer, J. G. 1988. Rural Land-Use and Wildland Fires in the Tropics. *Agroforestry Systems,* 6, 235–252.

Goldammer, J. G., ed. 1990. Fire in the Tropical Biota. *Ecosystem Processes and Global Challenges, Ecological Studies.* Berlin-Heidelberg: 84, Springer-Verlag.

Goldammer, J. G. 1990. Waldumwandlung und Waldverbrennung in den Tiefland-Regenwaldern des Amazonas-Beckens: Ursachen und Okologische Implikationen. *Amazonien: Versuch Einer Interdisziplinaren Annaherung,* edited by A. Hoppe. Ber. Naturforschende Ges. Freiburg 80.

Goldammer, J. G. 1991. *Fire Ecology of Tropical Forests.* Basel: Birkhauser-Verlag, in preparation.

Goldammer, J. G., and X. Di. 1990. Fire and Forest Development in the Northeast China—A Preliminary Model. In *Fire and Ecosystem Dynamics: Mediterranean and Northern Perspectives,* edited by J. G. Goldammer and M. J. Jenkins, 175. The Hague: SPB Academic.

Goldammer, J. G., and S. Penafiel. 1990. Fire in the Pine-Grassland Biomes of Tropical and Subtropical Asia. In *Fire in the Tropical Biota: Ecosystem Processes and Global Challenges,* edited by J. G. Goldammer. Ecological Studies 84, 45–62. Berlin-Heidelberg: Springer-Verlag.

Goldammer, J. G., and B. Seibert. 1989. Natural Rain Forest Fires in Eastern Borneo during the Pleistocene and Holocene. *Naturwissenschaften,* 76, 518.

Goldammer, J. G., and B. Seibert. 1990. The Impact of Droughts and Forest Fires on Tropical Lowland Rain Forest of East Kalimantan. In *Fire in the Tropical Biota: Ecosystem Processes and Global Challenges,* edited by J. G. Goldammer. Ecological Studies 84, 11–31. Berlin-Heidelberg: Springer-Verlag.

Goldammer, J. G., and K. F. Weiss. 1991. Global Estimate of Annually Burned Forest and Savanna Surface, Burnt Biomass, and Emitted Carbon to the Atmosphere in the Tropics and Subtropics. In *Tropical Forests in Transition: Ecology of Natural and Anthropogenic Disturbance Processes in Tropical Forest Biomes,* edited by J. G. Goldammer. Basel: Birkhäuser-Verlag, in preparation.

Goldberg, E. D. 1985. *Black Carbon in the Environment.* New York: John Wiley.

Goldemberg, J. 1989. A Amazonia e seu futuro. *A Folha de São Paulo,* 29 January.

Goldemberg, J., T. J. Johansson, A. K. N. Reddy, and R. H., Williams. 1988. Energy for a Sustainable World. New Delhi: Wiley Eastern Limited.

Goodland, R. 1980. Environmental Ranking of Amazonian Development Projects. *Brazil Environment Conservation,* 7(1).

Goreau, T. J., and W. Z. de Mello. 1987. Effects of Deforestation on Sources and Sinks of Atmospheric Carbon Dioxide, Nitrous Oxide, and Methane from Central Amazonian Soils and Biota during the Dry Season: A Preliminary Study. In *Proceedings of the Workshop on Biogeochemistry of Tropical Rain Forests: Problems for Research,* edited by D. Athie, T. E. Lovejoy, and P. de M. Oyens, 51–66. São Paulo: Universidade de São Paulo, Centro de Energia Nuclear na Agricultura (CENA), Piracicaba.

Goreau, T. J., and W. Z. de Mello. 1988. Tropical Deforestation: Some Effects on Atmospheric Chemistry. *Ambio,* 17(4), 275–281.

Gorham, E. 1988. Canada's Peatlands: Their Importance for the Global Carbon Cycle and Possible Effects of Greenhouse Warming. *Transaction of the Royal Society of Canada,* ser. 5, vol. 3, 21–23.

Goss, J. R., and G. E. Miller, Jr. 1973. *Study of Abatement Methods and Meteorological Conditions from Field Burning of Rice Straw.* Final Report, California Air Resources Board Project 1-101-1 and ARB 2113, University of California, Davis.

Goudriaan, J., and G. L. Ajtay. 1979. The Possible Effects of Increased CO_2 on Photosynthesis. In *The Global Carbon Cycle,* SCOPE 13, edited by B. Bolin, E. T. Degens, S. Kempe, and P. Ketner, 237–249. Chichester, U.K.: John Wiley.

Goudriaan, J., and P. Ketner. 1984. A Simulation Study for the Global Carbon Cycle, Including Man's Impact on the Biosphere. *Climatic Change,* 6, 167–192.

Grainger, A. 1990. Modelling Future Carbon Emissions from Deforestation in the Humid Tropics. In *Proceedings of the Conference on Tropical Forestry Response Options to Climate Change,* Report of workshop 9–11 January, São Paulo, Brazil. IPCC Secretariat, Geneva, and U.S. Environmental Protection Agency, Washington, pp. 105–119.

Graup, G., and B. Spettel. 1989. Mineralogy and Phase-Chemistry of an Ir-Enriched Pre-K/T Layer from the Lattengebirge, Bavarian Alps, and Significance for the KTB Problem. *Earth and Planetary Science Letters,* 95, 271–290.

Gray, H. A., G. R. Cass, J. J. Huntzicker, E. K. Heyerdahl, and J. A. Rau. 1984. Elemental and Organic Carbon Particle Concentrations: A Long-Term Perspective. *Sci. Total Environ.,* 36, 17–25.

Gray, H. A., G. R. Cass, J. J. Huntzicker, E. K. Heyerdahl, and J. A. Rau. 1986. Characteristics of Atmospheric Organic and Elemental Carbon Particle Concentrations in Los Angeles. *Environ. Sci. Technol.,* 20, 580–589.

Gray, J. T., and W. H. Schlesinger. 1981. Nutrient Cycling in Mediterranean Type Ecosystems, in Resource Use by Chaparral and Matorral. *Ecological Studies,* 39, 259–285. New York: Springer-Verlag.

Greenberg, J. P., P. R. Zimmerman, and R. B. Chatfield. 1985. Hydrocarbon and Carbon Monoxide in African Savanna Air. *Geophys. Res. Lett.,* 12, 113–116.

Greenberg, J. P., P. R. Zimmerman, L. Heidt, and W. Pollock. 1984. Hydrocarbon and Carbon Monoxide Emissions from Biomass Burning in Brazil. *J. Geophys. Res.,* 89, 1350–1354.

Gregoire J-M., S. Flasse, and J. P. Malingreau. 1988. *Octobre evaluation de l'action des feux de brousse, de Novembre 1987 a Fevrier 1988 dans la region frontalière Guinée-Sierra Leone.* Rome: CCR ISPRA.

Gregory, G., R. Harriss, R. Talbot, R. Rasmussen, M. Garstang, M. Andreae, R. Hinton, E. Browell, S. Beck, D. Sebacher, M. Khalil, R. Ferek, and S. Harriss. 1986. Air Chemistry over the Tropical Forest of Guyana. *J. Geophys. Res.,* 91, 8603–8612.

Gregory, G. L., C. Hudgins, and J. A. Ritter. 1987. In Situ Ozone Instrumentation for 10 Hz-Measurements: Development and Evaluation. In *Proceedings of the Sixth Symposium on Meteorological Observations and Instrumentation,* 136–139. Boston: American Meteorological Society.

Gregory, G. L., C. Hudgins, and R. A. Edahl, Jr. 1983. Laboratory Evaluation of an Airborne Ozone Instrument Which Compensates for Altitude/Sensitivity Effects. *Environ. Sci. Technol.,* 17, 100–103.

Gregory, G. L., E. V. Browell, L. S. Warren, and C. H. Hudgins. 1990. Amazon Basin Ozone and Aerosol: Wet-Season Observations. *J. Geophys. Res.* 95, 16,903–16,912.

Griffin, J. J., and E. D. Goldberg. 1975. The Fluxes of Elemental Carbon in Coastal Marine Sediments. *Limnol. Oceanogr.,* 20, 456–463.

Griffin, J. J., and E. D. Goldberg. 1979. Morphologies and Origin of Elemental Carbon in the Environment. *Science,* 206, 563–565.

Griffin, J. J., and E. D. Goldberg. 1981. Sphericity as a Characteristic of Solids from Fossil Fuel Burning in a Lake Michigan Sediment. *Geochim. Cosmochim. Acta,* 45, 763–769.

Griffin, J. J., and E. D. Goldberg. 1983. Impact of Fossil Fuel Combustion on Sediments of Lake Michigan: A Reprise. *Environ. Sci. Tech.,* 17, 244–245.

Griffith, D. W. T. 1990. *FTIR Remote Sensing of Biomass Burning Emissions of CO₂, CO, CH₄, CH₂O, NO, NO₂, NH₃ and N₂O.* Paper presented at the Chapman Conference on Global Biomass Burning: Atmospheric, Climatic, and Biospheric Implications, Williamsburg, Va., 19–23 March (this volume, Chapter 30).

Griffiths, J. 1972. Climates of Africa. *World Survey of Climatology,* Vol. 10. Amsterdam: Elsevier.

Groffman, P. M., and Tiedje, J. M. 1988. Denitrification Hysteresis during Wetting and Drying Cycles in Soil. *Soil Sci. Soc. Am. J.,* 52, 1626–1629.

Grosmaire, H. 1958. *Les Feux de Brousse au Sahel Senegalais.* CSA/CCTA Pub., 43, 1–33.

Groves, R. H. 1983. Nutrient Cycling in Australian Heath and South African Fynbos, in Mediterranean Type Ecosystems. *Ecological Studies,* 43, 179–191. New York: Springer-Verlag.

Gruell, G. E. 1983. *Fire in Vegetative Trends in the Northern Rockies: Interpretations from 1871–1982 Photographs.* U.S.D.A. Forest Service General Technical Report INT-158.

Gruell, G. E. nd. *Fire's Influence on Wildlife Habitat on the Bridger-Teton National Forest, Wyoming. Vol. 1, Photographic Record and Analysis.* U.S.D.A. Forest Service Research Paper INT-235.

Guenzi, W. D., W. E. Beard, F. S. Watanabe, S. R. Olsen, and L. K. Porter. 1978. Nitrification and Denitrification in Cattle Manure Amended Soil. *J. Environ. Qual.,* 7, 196–202.

Guillaume, J. 1959. *Le Crepuscule des Hommes.* Paris: Del Duca.

Gulmon, S. L. 1983. Carbon and Nitrogen Economy in Diplacus Aurantiacus a Californian Mediterranean-Climate Drought-Deciduous Shrub in Mediterranean Type Ecosystems. *Ecological Studies,* 43, 167–176. New York: Springer-Verlag.

Gupta A. C. 1987. Forest Fire Hazard. *Environment Today,* 5–6.

Haaland, D. M., and R. G. 1980. Easterling. Improved Sensitivity of Infrared Spectroscopy by the Application of Least Squares Methods. *Appl. Spectrosc.,* 34, 539–548.

Hackney, C. T., and A. A. de la Cruz. 1983. Effects of Winter Fire on the Productivity and Species Composition of Two Brackish Marsh Communities in Mississippi. *Intern. J. Ecology Envir. Sci.,* 9, 185–208.

Haffer, J. 1969. Speciation in Amazonian Forest Birds. *Science,* 165, 131.

Haider, K., A. Mosier, and O. Heinemeyer. 1985. Phytotron Experiments to Evaluate the Effect of Growing Plants on Denitrification. *Soil Sci. Soc. Am. J.,* 49, 636–641.

Haider, K., Mosier, A., and Heinemeyer, O. 1987. The Effect of Growing Plants on Denitrification at High Soil Nitrate Concentrations. *Soil Soc. Am. J.,* 51, 97–102.

Haines, D. A., and E. L. Kuehnast. 1970. When the Midwest Burned. *Weatherwise,* 2, 112–119.

Haines, D. A., and R. W. Sando. 1969. *Climatic Conditions Preceding Historically Created Fires in the North Central Region.* U.S.D.A. Research Paper NC-34.

Hall, C. A. S., R. P. Detwiler, P. Bogdonoff, and S. Underhill. 1985. Land Use Change and Carbon Exchange in the Tropics. I. Detailed Estimates for Costa Rica, Panama, Peru, and Bolivia. *Environ. Mgmt.,* 9, 345–354.

Hall, D. O. 1989. Carbon Flows in the Biosphere: Present and Future. *J. Geol. Soc.,* 146, 175–181.

Hall, D. O., and F. Rosillo-Calle. 1990. African Forests and Grasslands: Sources or Sinks of Greenhouse Gases? In *Global Warming and Climate Change: African Perspective,* edited by C. Juma. Nairobi: African Centre for Technology Studies.

Hall, D. O., and J. M. Scurlock. 1990. Climate Change and Productivity of Natural Grasslands. *Ann. Bot.*

Haller, T., and Stolp, H. 1985. Quantitative Estimation of Root Exudation of Maize Plants. *Plant Soil,* 86, 207–216.

Hallett, J., J. G. Hudson, and C. F. Rogers. 1989. Characterization of Combustion Aerosols for Haze and Cloud Formation. *Aerosol Sci. and Technol.,* 10, 70–83.

Hameed, S., and J. Dignon. 1988. Changes in the Geographical Distributions of Global Emissions of NO and SO from Fossil-Fuel Combustion between 1966 and 1980. *Atmos. Environ.,* 22, 441–149.

Hamilton, P. M., P. H. Varey, and M. M. Mila. 1978. Remote Sensing of Sulphur Dioxide. *Atmos. Environ.,* 12, 127–133.

Hamm, S., and P. Warneck. 1990. The Interhemispheric Distribution of Acetonitrile in the Troposphere. *J. Geophys. Res.,* 95D, 20,593–20,606.

Hanel, G. 1981. An Attempt to Interpret the Humidity Dependencies of Aerosol Extinction and Scattering Coefficients. *Atmos. Environ.,* 15, 403–406.

Hansen, A. D. A., B. A. Bodhaine, E. G. Dutton, and R. C. Schnell. 1988. Aerosol Black Carbon Measurements at the South Pole: Initial Results 1986–1987. *Geophys. Rev. Lett.,* 15, 1193–1196.

Hansen, H. J., R. Gwozdz, J. M. Hansen, R. G. Romley, and K. L. Rasmussen. 1986. The Diachronous C/T Plankton Extinction in the Danish Basin. In *Global Bio-Events,* edited by O. H. Walliser, 381–384. Berlin: Springer-Verlag.

Hansen, H. J., R. Gwozdz, and K. L. Rasmussen. 1988. High-Resolution Trace Element Chemistry across the Cretaceous-Tertiary Boundary in Denmark. *Revista Espanola de Paleontologia,* n deg Extraordinario, 21–29.

Hansen, H. J., K. L. Rasmussen, R. Gwozdz, and H. Kunzendorf. 1987. Iridium-Bearing Carbon Black at the Cretaceous-Tertiary Boundary. *Bull. Geological Society of Denmark,* 36, 305–314.

Hansen, J. M. 1977. Dinoflagellate Stratigraphy and Echinoid Distribution in Upper Maastrichtian and Danian Deposits from Denmark. *Bull. of Geological Society of Denmark,* 26, 1–26.

Hao, W. M., M. H. Liu, and P. J. Crutzen. 1990. Estimates of Annual and Regional Release of CO₂ and Other Trace Gases to the Atmosphere from Fires in the Tropics, Based on the FAO Statistics for the Period 1975–1980. In *Fire in the Tropical Biota: Ecosystem Processes and Global Challenges,* edited by J. G. Goldammer. Ecological Studies 84, 440–462. Berlin-Heidelberg: Springer-Verlag.

Hao, W. M., D. Scharffe, P. J. Crutzen, and E. Sanhueza. 1988. Production of N_2O, CH_4, and CO_2 from Soils in the Tropical Savannah During the Dry Season. *J. Atmos. Chem.*, 7, 93–105.

Hao W. M., D. Scharffe; J. M. Lobert, and P. J. Crutzen. 1990. *Biomass Burning: An Important Source of Atmospheric CO, CO_2, and Hydrocarbons.* Paper presented at the Chapman Conference on Global Biomass Burning: Atmospheric, Climatic, and Biospheric Implications, Williamsburg, Va., 19–23 March.

Hao, W. M., D. H. Scharffe, J. M. Lobert, and P. J. Crutzen. 1991. Emissions of Nitrous Oxide from the Burning of Biomass in an Experimental System. *Geophys. Res. Lett.*, 18, 999–1002.

Hao, W. M., S. C. Wofsy, M. B. McElroy, J. M. Beer, and M. A. Toqan. 1987. Sources of Atmospheric Nitrous Oxide from Combustion. *J. Geophys. Res.*, 92, 3098–3104.

Harrington, J. B. 1982. *A Statistical Study of Area Burned by Wildfire in Canada 1959–1980.* Chalk River, Ont.: Environment Canada.

Harris, G. W., T. E. Kleindienst, and J. N. Pitts, Jr. 1981. Rate Constants for the Reaction of OH Radicals with CH_3CN, C_2H_5CN and $CH_2 = CHCN$ in the Temperature Range 298–424 K. *Chem. Phys. Lett.*, 80, 479–483.

Harriss, R. C., et al. 1988. The Amazon Boundary Layer Experiment (ABLE 2A): Dry Season 1985. *J. Geophys. Res.*, 93 D2, 1351–1360.

Harriss, R. C., and D. I. Sebacher. 1981. Methane Flux in Forested Freshwater Swamps of the Southeastern United States. *Geophys. Res. Lett.*, 8, 1002–1004.

Harriss, R. C., D. I. Sebacher, and F. P. Day, Jr. 1982. Methane Flux in the Great Dismal Swamp. *Nature*, 297, 673–674.

Harriss, R. C., S. C. Wofsy, M. Garstang, E. V. Browell, L. C. B. Molion, R. J. McNeal, J. M. Hoell, R. J. Bendura, J. R. Coelho, R. L. Navarro, J. T. Riley, and R. L. Snell. 1990. The Amazonas Boundary Layer Experiment (ABLE-2B): Wet Season, 1987. *J. Geophys. Res.* 95, 16,721.

Hartmann, W. R. 1990. *Organische Sauren in der Atmosphere.* Ph. D. thesis, Johannes Gutenberg Universitat Mainz, FRG, Max Planck Institut für Chemie.

Hartung, J. B., G. A. Izett, C. W. Naeser, M. J. Kunk, and J. F. Suter. 1986. *The Manson, Iowa, Impact Structure and the Cretaceous-Tertiary Boundary Event* (abstract). Lunar and Planetary Science Conference 17, 31–32.

Harwood, D. M. 1988. Upper Cretaceous and Lower Paleocene Diatom and Silicoflagellate Biostratigraphy of Seymour Island, Eastern Antarctic Peninsula. *Geology and Paleontology of Seymour Island, Antarctic Peninsula,* edited by R. M. Feldmann and M. O. Woodburne, 55–129. Boulder, Colo.: Geological Society of America, Memoir 169.

Hastenrath, S. 1985. *Climate and Circulation of the Tropics.* Dordrecht, Netherlands: Riedel.

Hastenrath, S. 1990. The Relationship of Highly Reflective Clouds to Tropical Climate Anomalies. *J. of Climate,* 3, 353–365.

Hastings, J. and R. Turner. 1965. *The Changing Mile.* Tucson, Ariz.: University of Arizona Press.

Hawthorne, S. B., M. S. Krieger, D. J. Miller, and M. B. Mathiason. 1989. Collection and Quantitation of Methoxylated Phenol Tracers for Atmospheric Pollution from Residential Wood Stoves. *Environ. Sci. Technol.*, 23, 470–475.

Hawthorne, S. B., D. J. Miller, R. M. Barkley, and M. S. Krieger. 1988. Identification of Methoxylated Phenols as Candidate Tracers for Atmospheric Wood Smoke Pollution. *Environ. Sci. Technol.,* 22, 1191.

Hecht, S. B. 1981. Deforestation in the Amazon Basin: Magnitude, Dynamics, and Soil Resource Effects. *Studies in Third World Societies,* 13, 61–108.

Hecht, S. B., R. B. Norgaard, and G. Possio. 1988. The Economics of Cattle Ranching in Eastern Amazonia. *Interciencia,* 13(5), 233–240.

Hegg, D. A., P. V. Hobbs, and L. F. Radke. 1984. Measurements of the Scavenging of Sulfate and Nitrate in Clouds. *Atmos. Environ.,* 18, 1939–1946.

Hegg, D. A., L. F. Radke, P. V. Hobbs, and C. A. Brock. 1987. Nitrogen and Sulfur Emissions from the Burning of Forest Products Near Large Urban Areas. *J. Geophys. Res.,* 92, 14, 701–14, 709.

Hegg, D. A., L. F. Radke, P. V. Hobbs, R. A. Rasmussen, and P. J. Riggan. 1990. Emissions of Some Trace Gases from Biomass Fires. *J. Geophys. Res.,* 95, 5669–5675.

Hegg, D. A., L. F. Radke, P. V. Hobbs, R. A. Rasmussen, and P. J. Riggan. 1989. Emissions of Some Biomass Fires. In *Proceedings, 1989 National Air and Waste Management Association Meeting,* June 25–30; Anaheim, Calif., No. 089-025.003.

Hegg, D. A., L. F. Radke, P. V. Hobbs, and P. J. Riggan. 1988. Ammonia Emissions from Biomass Burning. *Geophys. Res. Lett.,* 15, 335–337.

Hegner, R. 1979. Nichtimmergrune Waldformationen der Tropen. *Kolner Geographische Arbeiten,* 37, Koln: Inst. Geography Univ.

Heintzenberg, J. 1982. Size-Segregated Measurements of Particulate Elemental Carbon and Aerosol Light Absorption at Remote Arctic Locations. *Atmos. Environ.* 1, 2461–2469.

Heintzenberg, J., H.-C. Hansson, and H. Lannefors. 1981. The Chemical Composition of Arctic Haze at Ny-Alesund, Spitzbergen. *Tellus,* 33, 162–171.

Helas, G., M. O. Andreae, J. Fontan, B. Cros, and R. Delmas. 1988. *Ozone Measurements in Equatorial Africa during DECAFE 88, Workshop on Tropospheric Ozone,* Quadrennial Ozone Symposium, Gottingen RFA, August.

Helas, G., H. Bingemer, and M. O. Andreae. 1991. Organic Acids over Equatorial Africa: Results from DECAFE 88. *J. Geophys. Res.,* in press.

Helfert, M. 1986. NASA, Johnson Space Center, Space Shuttle Earth Observations Project, Houston, Tex., personal communication.

Helfert, M. R., and K. P. Lulla. 1989. Monitoring Tropical Environments with Space Shuttle Photography. *Geocarto International* 4(1), 55–68.

Helfert, M. R., and C. A. Wood. 1989. The NASA Space Shuttle Earth Observations Office. *Geocarto International* 4(1), 15–24.

Henderson-Sellers, A., and K. McGuffie. 1989. Sulphate Aerosols and Climate: Scientific Correspondence. *Nature,* 340, 436–437.

Henderson-Sellers, A., M. F. Wilson, G. Thomas, and R. E. Dickinson. 1986. *Current Global Land-Surface Data Sets for Use in Climate-Related Studies.* Boulder, Colo.: National Center for Atmospheric Research Technical Note, NCAR/TN-272+STR.

Herberg, G. A., J. M. Harris, and B. A. Bodhaine. 1989. Atmospheric Transport during AGASP-II: The Alaskan Flights (2–10 April 1986). *Atmospheric Environment,* 23, 2521–2535.

Herring, J. R. 1977. *Charcoal Fluxes into Cenozoic Sediments of the North Pacific.* Ph.D. thesis, University of California, San Diego.

Herring, J. R. 1985. Charcoal Fluxes into Sediments of the North Pacific Ocean: The Cenozoic Record of Burning. In *The Carbon Cycle and Atmospheric CO₂: Natural Variations Archean to Present,* edited by E. T. Sundquist and W. S. Broecker, 419–442. Washington, D.C.: American Geophysical Union.

Hickey, L. J. 1981. Land Plant Evidence Compatible with Gradual, Not Catastrophic Change at the End of the Cretaceous. *Nature,* 292, 529–531.

Hildebrand, A. R., and W. V. Boynton. 1990. Proximal Cretaceous-Tertiary Boundary Impact Deposits in the Caribbean. *Science,* 248, 843–847.

Hildebrand, A. R., and W. S. Wolbach. 1989. Carbon and Chalcophiles at a Nonmarine K/T Boundary: Joint Investigations of the Raton Section, NM. *Lunar Planet. Sci.,* 20, 414–415.

Hildemann, L. M. 1990. *A Study of the Origin of Atmospheric Organic Aerosols.* Ph. D. thesis, California Institute of Technology, Pasadena, Calif.

Hingane, L. S. 1991. Some Aspects of Carbon Dioxide Exchange between Atmospheric and Indian Plant Biota. *Climatic Change,* in press.

Hinkle, C. R., P. A. Schmalzer, and T. H. Englert. 1990. *Effects of Fire on Coastal Soils in East Central Florida.* Internal report, NASA Kennedy Space Center, Fla.

Hobbie, J. E., J. J. Cole, J. L. Dungan, R. A. Houghton, and B. J. Peterson. 1984. The Controversy on the Role of the Biota in the Global CO₂ Balance. *Bioscience,* 34, 492–498.

Hobbs, N. T., D. S. Schimel, C. E. Owensby, and D. J. Ojima. 1991. Fire and Grazing in the Tallgrass Prairie: Contingent Effects on Nitrogen Budgets. *Ecology,* in press.

Hobbs, P. V., and L. F. Radke. 1969. Cloud Condensation Nuclei from a Simulated Forest Fire. *Science,* 163, 279–280.

Hobbs, P. V., J. L. Stith, and L. F. Radke. 1978. *Airborne Measurements of Emissions from the Controlled Burning of Forest Slash.* Report to the Weyerhaeuser Company, University of Washington, Seattle, Wash.

Holben, B. N., Y. J. Kaufman; D. Tanre, and D. E. Ward. 1990. *Optical Properties of Aerosol from Biomass Burning in the Tropics.* Paper presented at the Chapman Conference on Global Biomass Burning: Atmospheric, Climatic, and Biospheric Implications, Williamsburg, Va., 19–23 March (this volume, Chapter 49).

Holbrook, S. H. 1944. *Burning an Empire: The Story of American Forest Fires.* New York: Macmillan.

Holland, H. D. 1978. *The Chemistry of the Atmosphere and Oceans.* New York: Wiley-Interscience.

Hopkins, B. 1966. Vegetation of the Olokemeji Forest Reserve, Nigeria. IV. The Litter and Soil with Special Reference to Their Seasonal Changes. *Ecology,* 47, 687–703.

Hopkinson, C. S., J. G. Gosselink, and R. T. Parrondo. 1978. Above Ground Production of Seven Marsh Plant Species in Coastal Louisiana. *Ecology,* 59, 760–769.

Horibe, Y., and Craig, H. 1988. Is the Air in Amber Ancient? *Science,* 241, 720–721.

Houghton, J. T., G. J. Jenkins, and J. J. Ephraums, eds. 1990. *Climate Change: The IPCC Scientific Assessment.* Report prepared for Intergovernmental Panel on Climate Change by Working Group I, Cambridge University Press.

Houghton, R. A. 1986. Estimating Changes in the Carbon Content of Terrestrial Ecosystems from Historical Data. In *The Changing Carbon Cycle: A Global Analysis,* edited by J. R. Trabalka and D. E. Reichle, 175–193. New York: Springer-Verlag.

Houghton, R. A. 1989. *The Long-Term Flux of Carbon to the Atmosphere from Changes in Land Use.* Paper presented at the Third International Conference on Analysis and Evaluation of Atmospheric CO₂ Data Present and Past, World Meteorological Organization, Environmental Pollution Monitoring and Research Programme No. 59, Hinterzarten, 1–20. October.

Houghton, R. A. 1990. The Global Effects of Tropical Deforestation. *Environ. Sci. Technol.,* 24.

Houghton, R. A. 1991. Tropical Deforestation and Atmospheric Carbon Dioxide. *Climatic Change.* In press.

Houghton, R. A., R. D. Boone, J. R. Fruci, J. E. Hobbie, J. M. Melillo, C. A. Palm, B. J. Peterson, G. R. Shaver, G. M. Woodwell, B. Moore, D. L. Skole, and N. Meyers. 1987. The Flux of Carbon from Terrestrial Ecosystems to the Atmosphere in 1980 Due to Changes in Land Use: Geographic Distribution of the Global Flux. *Tellus,* 39B, 122–139.

Houghton, R. A., R. D. Boone, J. M. Melillo, C. A. Palm, G. M. Woodwell, N. Meyers, B. Moore III, and D. L. Skole. 1985. Net Flux of Carbon Dioxide from Tropical Forest in 1980. *Nature,* 316, 617–620.

Houghton, R. A., J. E. Hobbie, J. M. Melillo, B. Moore, B. J. Peterson, G. R. Shaver, and G. M. Woodwell. 1983. Changes in the Carbon Content of Terrestrial Biota and Soils between 1860 and 1980: A Net Release of CO₂ to the Atmosphere. *Ecol. Monogr.,* 53, 235–262.

Houghton, R. A., D. S. Lefkowitz, and D. L. Skole. 1991. Changes in the Landscape of Latin America between 1850 and 1980: 1. A Progressive Loss of Forests. *Forest Ecology and Management,* 38, 143–172.

Houghton, R. A., and D. L. Skole. 1990. The Carbon Cycle. In *The Earth As Transformed by Human Action,* edited by B. L. Turner. Cambridge: Cambridge University Press.

Houghton, R. A., D. L. Skole, and D. S. Lefkowitz. 1991. Changes in the Landscape of Latin America between 1850 and 1980: A Net Release of CO₂ to the Atmosphere. *Forest Ecology and Mgmt.,* 38, 173–199.

Houghton, R. A., and G. M. Woodwell. 1989. Global Climatic Change. *Scientific American,* 260, 306–344.

Hsu, K. J., Q. He, J. A. McKenzie, H. Weissert, K. Perch-Nielsen, H. Oberhansli, K. Kelts, J. LaBrecque, L. Tauxe, U. Krahenbuhl, S. F. Percival, Jr., R. Wright, A. M. Karpoff, N. Petersen, P. Tucker, R. Z. Poore, A. M. Gombos, K. Pisciotto, M. F. Carman, Jr., and E. Schreiber. 1982. Mass Mortality and Its Environmental and Evolutionary Consequences. *Science,* 216, 249–256.

Hsu, K. J., and J. A. McKenzie. 1985. A Strangelove Ocean in Earliest Tertiary. In *The Carbon Cycle and Atmospheric CO₂: Natural Variations Archean to Present,* edited by E. T. Sundquist and W. Broecker. *Geophysical Monograph Series,* 32, 487–492.

Hubbell, S. P. 1979. Tree Dispersion, Abundance, and Diversity in a Tropical Dry Forest. *Science,* 203, 1299.

Hubbell, S. P., and R. B. Foster. 1986. Canopy Gaps and the Dynamics in a Neotropical Forest. In *Plant Ecology,* edited by J. Crawley. Oxford: Blackwell.

Hubbs, C. L., and G. I. Roden. 1964. Oceanography and Marine Life along the Pacific Coast of Middle America. In *Handbook of Middle American Indians. Vol. 1,* edited by R. C. West, 143–186. Austin: University of Texas Press.

Huckle, H. F., H. D. Dollar, and R. F. Pendleton. 1974. *Soil Survey of Brevard County, Florida.* Washington, D.C.: USDA Soil Conservation Service.

Hudon, R. J., K. R. Drew, and L. M. Baskin (eds.). 1989. *Wildlife Production Systems: Economic Utilisation of Wild Ungulates.* Cambridge: Cambridge University Press, 469 pp.

Hudson, J. G. 1983. Effect of CCN Concentrations on Stratus Clouds. *J. Clim. Appl. Meteorol.,* 23, 480–486.

Hudson, J. G. 1989. An Instantaneous CCN Spectrometer. *J. Atmos. Oceanic Technol.,* 6, 1055–1065.

Hudson, J. G. 1990. Influence of Anthropogenic CCN on Stratus Microphysics. *Preprints, Conference on Cloud Physics.* San Francisco, Calif., 23–27 July.

Hudson, J. G., J. Hallett, and C. F. Rogers. 1990. Field Studies of Cloud-Forming Properties of Nuclei Produced from Hydrocarbon and Vegetative Burns. Unpublished manuscript.

Hudson, J. G., and C. F. Rogers. 1986. Relationship between Critical Supersaturation and Cloud Droplet Size: Implications for Cloud Mixing Processes. *J. Atmos. Sci.,* 43, 2341–2359.

Hui-hai, Wang. *The Main Characteristics and Rational Utilization of Land Resource of Xishuangbanna.* Xishuangbanna, People's Republic of China.

Hultberg, S. U. 1986. Danian Dinoflagellate Zonation, the C/T Boundary and the Stratigraphical Position of the Fish Clay in Southern Scandinavia. *J. Micropaleontology,* 5, 37–47.

Humphrey, Robert R. 1987. *90 Years and 535 Miles: Vegetation Changes along the Mexican Border.* Albuquerque: University of New Mexico Press.

Huntley, B. J., and B. H. Walker, eds. 1982. Ecology of Tropical Savannas. *Ecological Studies,* 42, Berlin-Heidelberg: Springer-Verlag.

IBGE. 1985. Brasilia: Instituto Brasileiro de Geografia e Estatistica, Departamento de Estudos Geograficos.

IGBP. 1990a. *The International Geosphere-Biosphere Program: The Initial Core Projects.* IGBP Report No. 12, Stockholm.

IGBP. 1990b. *Terrestrial Biosphere Exchange with Global Atmospheric Chemistry.* IGBP Report No. 13, Stockholm.

IBM-IRAT. 1983. Cartographie Automatique des Feux. *Rapport Multigr.*

INEMET. 1989. *Boletim Agroclimatolgico Decendial.* Nos. 24 and 25 (in Portuguese).

INPE. 1989. *Avaliasao da Alteracao da Cobertura Vegetal na Amazonia Legal Utilzando Sensoriamento Remoto Orbital,* 2nd ed. São Paulo, Brazil: Instituto de Pesquisas Espaciais, S. J. Campos.

INPE. 1989a. *Climanalise,* (8) and (9) (in Portuguese).

IPCC. 1989. *Response Strategies Working Group of the Intergovernmental Panel on Climate Change.* Draft report of the U.S.–Netherlands Expert Group on Emissions Scenarios, Bilthoven, The Netherlands.

Ilaithani, G. P., V. K. Bahuguna, and P. Lal. 1986. Forest Fire Season in Different Parts of India—A Statistical Approach. *Journal of Tropical Forestry* 2(3), 188–195.

Imort, M. 1989a. *Das Beziehungsgeflecht und die Auswirkungen der Degradationsfaktoren Feuer, Weide und Brennholzgewinnung auf die Savannenvegetation im Sahel.* Dipl. thesis, Inst. Forest Zoology, Department of Forestry, University Freiburg.

Imort, M. 1989b. The Influence of Fire, Grazing and Logging on a Tree-Shrub Savanna in Khayes Region, Mali. In *Proceedings of the Third International Symposium Fire Ecology.* Freiburg.

International Institute for Environment and Development/World Resources Institute. 1987. *World Resources 1987.* New York: Basic Books.

Isaksen, I. S. A., and O. Hov. 1987. Calculation of Trends in the Tropospheric Concentration of O_3, OH, CH_4, and NO_x. *Tellus,* 39B, 271–285.

Isichei, A. 1979. *Elucidation of Stochs and Flows of Nitrogen in Some Nigerian Savanna Ecosystems.* Unpublished Ph.D. thesis, University of Ife, Nigeria.

Isichei, A., A. J. Morton, and F. Ekeleme. 1990. Mineral Nutrient Flow from an Inselberg in Southwestern Nigeria. *J. Tropical Ecology.*

Isichei, A., and W. W. Sanford. 1980. Nitrogen Loss by Burning from Nigerian Grassland Ecosystems. In *Nitrogen Cycling in West African Ecosystems,* edited by T. Rosswall, 325–331. Stockholm: Royal Swedish Academy of Sciences.

Isichei, A. O., F. Ekeleme, and B. A. Jimoll. 1986. Changes in a Secondary Forest in Southwestern Nigeria Following a Ground Fire. *J. Tropical Ecology,* 2, 249–256.

Izett, G. A. 1987. *The Cretaceous-Tertiary (K-T) Boundary Interval, Raton Basin, Colorado and New Mexico, and Its Content of Shock-Metamorphosed Minerals: Implications Concerning the K-T Boundary Impact-Extinction Theory.* U.S. Geological Survey Open File Report 87–606. Denver, Colo.

Izett, G. A., and C. L. Pillmore. 1985. Shock-Metamorphic Minerals in the Cretaceous-Tertiary Boundary, Raton Basin, Colorado, and New Mexico, Provide Evidence for Asteroid Impact In Continental Crust. *Eos,* 66, 1149–1150.

Izlar, R. L. 1984. Some Comments on Fire and Climate in the Okefenokee Swamp-Marsh Complex. In *The Okefenokee Swamp: Its Natural History, Geology, and Geochemistry,* edited by A. D. Cohen, D. J. Casagrande, M. J. Andrejko, and G. R. Best, 70–85. Los Alamos, N.M.: Wetland Surveys.

Jackman, J. H., J. E. Frederick, and R. S. Stolarski. 1980. Production of Odd Nitrogen in the Stratosphere and Mesosphere: An Intercomparison of Source Strengths. *J. Geophys. Res.,* 85, 495–7505.

Jacobson, J. 1979. Recent Developments in Southern Asian Prehistory and Protohistory. *Ann. Rev. Anthrop.,* 8, 467.

Jaeger, P. 1986. Contribution à l'etude des forats reliques du Soudan Occidental. *Bull. Inst. Afr. Noire,* ser. A, 18, 993–1053.

Jaen Suarez, O. 1980. Cinco siglos de poblacion en el Istmo de Panama. *Loteria,* 291, 75–94.

Jaen Suarez, O. 1979. *La poblacion del Istmo de Panama.* Panama: Impresora de La Nacion.

Jaenicke, R. 1981. *Climatic Variations and Variability: Facts and Theories,* edited by A. Berger, Dordrecht: Reidel, 577–597.

James, S. R. 1989. Hominid Use of Fire in the Lower and Middle Pleistocene: A Review of the Evidence. *Current Anthro.,* 30, 1–26.

Jansson, M. B. 1982. *Land Erosion by Water in Different Climates.*

UNGI Report n 57, Department of Physical Geography, Uppsala University, Sweden.

Janzen J. 1980. Extinction of Light by Highly Non-Spherical Strongly Absorbing Particles: Spectrophotometric Determination of Volume Distributions for Carbon Blacks. *Appl. Optics*, 19, 2977–2985.

Jeffrey, D. 1989. Yellowstone: The Great Fires of 1988. *Natl. Geogr.*, 175, 255–273.

Jeffreys, M. D. W. 1951. Feux de Brousse. *Bulletin IFAN,* Tome 13, N deg 3, 683–710, Julliet.

Jensen, J. B., and R. J. Charlson. 1984. On the Efficiency of Nucleation Scavenging. *Tellus*, 36B, 367–375.

Johannessen, C. L. 1963. *Savannas of Interior Honduras.* Berkeley: University of California Press.

Johansson, C., and L. Granat. 1984. Emission of Nitric Oxide from Arable Land. *Tellus*, 36B, 26–37.

Johansson, C., H. Rodhe, and E. Sanhueza. 1988. Emission of NO in a Tropical Savanna and a Cloud Forest during Dry Season. *J. Geophys Res.*, 93D, 7180–7192.

Johansson, C., and E. Sanhueza. 1988. Emissions of NO from Savanna Soils during the Rainy Season. *J. Geophys. Res.*, 93, 14193–14198.

Johnson, B. 1984. *The Great Fire of Borneo.* London: World Wildlife Fund-UK.

Johnson, R., and C. Huntzicker. 1981. An Automated Thermal Optical Method for the Analysis of Carbonaceous Aerosols. In *Atmospheric Aerosol Sources—Air Quality Relationships,* edited by E. S. Macias and P. K. Hopke, Washington, D.C.: American Chemical Society.

Johnson, R. L., J. J. Shah, R. A. Cary, and J. J. Huntzicker. 1981. An Automated Thermal-Optical Method for the Analysis of Carbonaceous Aerosol. In *ACS Symposium Series, No. 167, Atmospheric Aerosol: Source/Air Quality Relationships,* edited by E. S. Macias and P. K. Hopke, Washington, D.C.: American Chemical Society.

Johnston, W. R., and J. C. Kang. 1971. Mechanisms of Hydrogen Cyanide Formation from the Pyrolysis of Amino Acids and Related Compounds. *J. Org. Chem.*, 36, 189–192.

Jones, E. W. 1963. The Forest Outliners in the Guinea Zone of Northern Nigeria. *J. Ecol.*, 15, 415–434.

Jones, M. J., and A. R. Bromfield. 1970. Nitrogen in the Rainfall at Samaru, Nigeria. *Nature.* (London) 227, 86.

Jones, M. J., and A. Wild. 1975. *Soils of the West African Savanna.* Harpenden, U.K.: Commonwealth Bureau of Soils.

Jones, R. 1979. The Fifth Continent: Problems Concerning Human Colonisation of Australia. *Ann. Rev. Anthro.*, 8, 445–466.

Jordan C. F. 1987. Shifting Cultivation. *Amazon Rain Forests,* Ecological Studies 60, 9–23. New York: Springer-Verlag.

Jordan, C. F., and C. E. Russell. 1983. Jari: Productividad de las plantaciones y perdida de nutrientes debido al corte y la quema. *Interciencia,* 8(5), 294–297.

Jorgensen, J. R., and C. G. Wells. 1971. Apparent Nitrogen Fixation in Soil Influenced by Prescribed Burning. *Soil Sci. Soc. Amer. Proc.*, 35, 806–810.

Joshi, V., D. R. Ahuja, and C. Venkataraman. 1989. Emissions from Burning Biofuel in Metal Cookstoves. *Environ. Mgmt.*, 13(6), 763–772.

Joshi, V., D. R. Ahuja, and C. Venkataraman. 1990. Thermal Performance and Emission Characteristics of Biomass-burning Heavy Stoves with Flues. *Pacific and Asian J. Energy.*

Joumard, R. 1987. *INRTS Colloque pollution de l'air par les transports.* Paris.

Justice, C., J. Townshend, B. Holben, and C. Tucker. 1985. Analysis of the Phenology of Global Vegetation Using Meteorological Satellite Data. *Intl. J. Remote Sen.*, 6, 1271–1318.

Kadeba, O., and E. A. Aduzyi. 1985. Litter Production, Nutrient Cycling and Litter Accumulation in Pinus Caribaea Morelet var. Hondurensis Stands in Northern Guinea Savanna of Nigeria. *Plant and Soil,* 86, 197–206.

Kanakidou, M., B. Bonsang, and G. Lambert. 1989. Light Hydrocarbona Vertical Profiles and Fluxes in a French Rural Area. *Atmos. Environ.*, 23, 921–927.

Kaplan W. A., S. C. Wofsy, I. Keller, and J. I. DeCosta. 1988. Emission of NO and Deposition of O_3 in a Tropical Forest System. *J. Geophys. Res.*, 93, 1389–1395.

Kasibhatla, P. S., H. Levy, II, W. J. Moxim, and W. L. Chamiedes. 1991. The Relative Impact of Stratospheric Production on Tropospheric NO_y Levels: A Model Study. *J. Geophys. Res.*, in press.

Kasibhatla, P. S., H. Levy, II, W. J. Moxim, and J. A. Logan. 1990. Simulated Influence Biomass Burning on Regional and Global NO_y Distributions. Paper presented at the Chapman Conference on Global Biomass Burning: Atmospheric, Climatic, and Biospheric Implications, Williamsburg, Va., 19–23 March (this volume, Chapter 37).

Kastner, M., F. Asaro, H. V. Michel, W. Alvarez, and L. W. Alvarez. 1984. The Precursor of the Cretaceous-Tertiary Boundary Clays at Stevns Klint, Denmark, and DSDP Hole 465A. *Science,* 226, 137–143.

Kaufman, Y. J. 1987. Satellite Sensing of Aerosol Absorption. *J. Geophys. Res.*, 92, 4307–4317.

Kaufman, Y. J., and C. Sendra. 1988. Satellite Mapping of Aerosol Loading over Vegetated Areas. In *Aerosol and Climate,* edited by P. V. Hobbs and M. P. McCormick, 1–67. Hampton, Va.: Deepak.

Kaufman, Y. J. et al. 1990. *The BASE-A Experiment: An Overview.* Paper presented at the Chapman Conference on Global Biomass Burning: Atmospheric, Climatic, and Biospheric Implications, Williamsburg, Va., 19–23 March.

Kaufman, Y. J., A. W. Setzer, C. Justice, C. J. Tucker, and I. Fung. 1990. Remote Sensing of Biomass Burning in the Tropics. In *Fire in the Tropical and Subtropical Biota,* edited by J. G. Goldammer. Heidelberg, FRG: Springer-Verlag.

Kaufman, Y. J., C. J. Tucker, and I. Fung. 1988. Remote Sensing of Biomass Burning. In *Current Problems in Atmospheric Radiation,* edited by J. Lenoble and J. F. Geleyn, 322–325. Hampton, Va.: Deepak.

Keeling, C. D. 1973a. The Carbon Dioxide Cycle: Reservoir Models to Depict the Exchange of Atmospheric Carbon Dioxide with the Oceans and Land Plants. In *Chemistry of the Lower Atmosphere,* edited by S. I. Rasool. New York: Plenum Press.

Keeling, C. D. 1973b. Industrial Production of Carbon Dioxide From Fossil Fuels and Limestone. *Tellus,* 25, 174–198.

Keeling, C. D. 1986. *Atmospheric CO_2 Concentrations—Mauna Loa Observatory, Hawaii, 1958–1986.* NDP-001/RI. Oak Ridge,

Tenn.: Carbon Dioxide Information Center, Oak Ridge National Laboratory.

Keeling, C. D., R. B. Bacastow, A. F. Carter, S. C. Piper, T. P. Whorf, M. Heimann, W. G. Mook, and H. Roeloffzen. 1989. A Three-Dimensional Model of Atmospheric CO_2 Transport Based on Observed Winds: 1. Analysis of Observational Data. In *Aspects of Climate Variability in the Pacific and the Western Americas,* Geophysical Monograph 55, edited by D. H. Peterson 165–236. Washington D.C.: American Geophysical Union.

Keeling, C. D., A. F. Carter, and W. G. Mook. 1984. Seasonal, Latitudinal, and Secular Variations in the Abundance and Isotopic Ratios of Atmospheric CO_2. *J. Geophys. Res.,* 89, 4615–4628.

Keene, W. C., and J. N. Galloway. 1986. Considerations Regarding Sources for Formic and Acetic Acids in the Troposphere. *J. Geophys. Res.,* 91, 14,466–14,474.

Keeney, S. R., and S. W. Nelson. 1982. Nitrogen-Inorganic Forms. In *Methods of Soil Analysis, Part 2, Chemical and Microbial Properties, Agronomy, Vol. 9,* edited by A. L. Page, L. H. Miller, and D. R. Keeney, 643–698. Madison, Wis.: American Society of Agronomy.

Keller, G. 1988. Biotic Turnover in Benthic Foraminifera across the Cretaceous/Tertiary Boundary at El Kef, Tunisia. *Palaeogeography, Palaeoclimatology, Palaeoecology,* 66, 153–171.

Keller, G. 1989. Extended Cretaceous/Tertiary Boundary Extinctions and Delayed Population Change in Planktonic Foraminifera from the Brazos River, Texas. *Paleoceanography,* 4, 287–332.

Keller, G. 1989. Extended Period of Extinctions across the Cretaceous/Tertiary Boundary in Planktonic Foraminifera of Continental-Shelf Sections: Implications for Impact and Volcano Theories. *Geological Soc. of America Bull.,* 101, 1408–1419.

Keller, G. 1988. Extinction, Survivorship and Evolution of Planktic Foraminifera across the Cretaceous-Tertiary Boundary at El Kef, Tunisia. *Marine Micropaleontology,* 13, 239–263.

Keller, G., and M. Lindinger. 1989. Stable Isotope, TOC, and $CaCO_3$ Record across the Cretaceous/Tertiary Boundary at El Kef, Tunisia. *Palaeogeography, Palaeoclimatology, Palaeocology,* 73, 243–265.

Keller, M., D. J. Jacob, S. C. Wofsy, and R. C. Harriss. 1991. Effects of Tropical Deforestation on Global and Regional Atmospheric Chemistry. *Climatic Change,* in press.

Keller, M., W. A. Kaplan, and S. C. Wofsy. 1986. Emissions of N_2O, CH_4, and CO_2 from Tropical Forest Soils. *J. Geophys. Res.* 91, 11,791–11,802.

Kellogg, W. W., and Z.-C. Zhao. 1988. Sensitivity of Soil Moisture to Doubling of Carbon Dioxide in Climate Model Experiments, Pt. 1: North America. *J. Climate,* 1, 348–366.

Kemp, E. M. 1981. Pre-Quaternary Fire in Australia. In *Fire and the Australian Biota,* edited by A. M. Gill, R. H. Groves, and I. R. Noble. Canberra; Australian Academy of Science.

Kendall, J. D., and C. O. Justice. 1990. *Remotely Sensed Observations of Biomass Burning in Central Africa and South America.* Paper presented at the Chapman Conference on Global Biomass Burning: Atmospheric, Climatic, and Biospheric Implications, Williamsburg, Va., 19–23 March.

Kesel, R. 1975. *Grassland Ecology: A Symposium, Geoscience and Man, Volume 10.* Baton Rouge: School of Geoscience, Louisiana State University.

Kessler, E. 1969. On the Distribution and Continuity of Water

Substance in Atmospheric Circulation. *Meteorol. Monogr.,* 10, No. 32. Boston Mass: American Meteorological Society.

Khalil, M. A. K., and R. A. Rasmussen. 1983. Sources, Sinks, and Seasonal Cycles of Atmospheric Methane. *J. Geophys. Res.,* 88, 5131–5144.

Khalil, M. A. K., and R. A. Rasmussen. 1984. Global Sources, Lifetimes and Mass Balances of Carbonyl Sulfide (OCS) and Carbon Disulfide (CS_2) in the Earth's Atmosphere. *Atmos. Environ.,* 18, 1805–1813.

Khalil, M. A. K., and R. A. Rasmussen. 1987. Atmospheric Methane: Trends over the Last 10,000 Years. *Atmos. Environ.,* 21, 2445–2452.

Khalil, M. A. K., and R. A. Rasmussen. 1990. Atmospheric Carbon Monoxide: Latitudinal Distribution of Sources. *Geophys. Res. Lett.,* 17, 1913–1916.

Killough, G. G. 1980. A Dynamic Model for Estimating Radiation Dose to the World Population from Releases of 14C to the Atmosphere. *Health Phy.,* 38, 269–300.

Killough, G. G., and W. R. Emanuel. 1981. A Comparison of Several Models of Carbon Turnover in the Ocean with Respect to Their Distributions of Transit Time and Age, and Responses to Atmospheric CO_2 and ^{14}C. *Tellus,* 33, 274–290.

King, A. W., W. R. Emanuel, and W. M. Post. Internal report, Environmental Sciences Division, Oak Ridge National Laboratory, P. O. Box 2008, Oak Ridge, Tenn. 37831.

Kirchhoff, V. W. J. H. 1988. Surface Ozone Measurements in Amazonia. *J. Geophys. Res.,* 93, 1469–1476.

Kirchhoff, V. W. J. H. 1990. *Biomass Burning in the Brazilian Amazon Region.* Paper presented at the Chapman Conf. on Global Biomass Burning: Atmospheric, Climatic, and Biospheric Implications, Williamsburg, Va. 19–33 March (this volume, Chapter 12).

Kirchhoff, V. W. J. H., E. V. Browell, and G. L. Gregory. 1988. Ozone Measurements in the Troposphere of an Amazonian Rain Forest Environment. *J. Geophys. Res.,* 93, 15850–15860.

Kirchhoff, V. W. J. H., and I. M. O. da Silva. 1988. Medidas de Ozonio em Floresta Amazonica na Estacao Chuvosa: Primeiros Resultados. *Rev. Bras. Meteorologia,* 3, 193–197.

Kirchhoff, V. W. J. H., I. M. O. da Silva, and E. V. Browell. 1990. Ozone Measurements in Amazonia: Dry Season vs. Wet Season. *J. Geophys. Res.* 95, 16,913–16,926.

Kirchhoff, V. W. J. H., E. Hilsenrath, A. G. Motta, Y. Sahai, and R. A. Medrano-B. 1983. Equatorial Ozone Characteristics as Measured at Natal (5.9°S, 35.2°W). *J. Geophys. Res.,* 88, 6812–6818.

Kirchhoff, V. W. J. H., and E. V. A. Marinho. 1988. A Survey of Continental Concentrations of Atmospheric CO in the Southern Hemisphere. *Atmos. Environ.,* 23, 461–466.

Kirchhoff, V. W. J. H., and E. V. A. Marinho. 1990. Surface Carbon Monoxide Measurements in Amazonia. *J. Geophys. Res.* 95, 16,933.

Kirchhoff, V. W. J. H., E. V. A. Marinho, P. L. S. Dias, R. Calheiros, R. Andre, and C. Volpe. 1989. O_3 and CO from Burning Sugar Cane. *Nature,* 339, 264.

Kirchhoff, V. W. J. H., and C. A. Nobre. 1986. Atmospheric Chemistry Research in Brazil: Ozone Measurements at Natal, Manaus, and Cuiba. *Revista Geofisica,* 24, 95–108.

Kirchhoff, V. W. J. H., and R. A. Rasmussen. 1990. Time Variations of CO and O_3 Concentrations in a Region Subject to Biomass Burning. *J. Geophys. Res.,* 95, 7521–7532.

Kirchhoff, V. W. J. H., Y. Sahai, and A. G Motta. 1981. First Ozone Profiles Measured with ECC Sondes at Natal (5.9°S, 35.2°W). *Geophys. Res. Lett.,* 8, 1171–1172.

Kirchhoff, V. W. J. H., A. W. Setzer, and M. C. Pereira. 1989. Biomass Burning in Amazonia: Seasonal Effects on Atmospheric O_3 and CO. *Geophys. Res. Lett.,* 16, 469–472.

Klemp, J. B., and R. B. Wilhelmson. 1978. The Simulation of Three-Dimensional Convective Storm Dynamics. *J. Atmos. Sci.,* 35, 1070–1096.

Klinge, H., and W. A. Rodrigues. 1974. Phytomass Estimation in a Central Amazonian Rain Forest. In *IUFRD Biomass Studies,* edited by H. E. Young, 339–350. Orono, Maine: University Press.

Klinge, H., W. A. Rodrigues, E. Brunig, and E. J. Fittkau. 1975. Biomass and Structure in a Central Amazonian Rain Forest. In *Tropical Ecological Systems: Trends in Terrestrial and Aquatic Research,* edited by F. B. Golley and E. Medina, 115–122. New York: Springer-Verlag.

Knight, H. 1966. Loss of Nitrogen from the Forest Floor by Burning. *Forestry Chronicle,* 42, 149–152.

Knudsen, D., G. A. Peterson, and P. F. Pratt. 1982. Lithium, Sodium, and Potassium. In *Methods of Soil Analysis, Part 2, Chemical and Microbial Properties, Agronomy, Vol. 9,* edited by A. L. Page, R. H. Miller, and D. R. Keeney, 225–246. Madison, Wis.: American Society of Agronomy.

Ko, M. K. W., M. B. McElroy, D. Weisenstein, and N. D. Sze. 1986. A Zonal Mean Model of Stratospheric Tracer Transport in Isentropic Coordinates: Numerical Simulations for Nitrous Oxide and Nitric Acid. *J. Geophys. Res.,* 90, 2313–2329.

Ko, M. K. W., N. D. Sze, and D. Weisenstein. 1989. The Roles of Dynamical and Chemical Processes in Determining the Stratospheric Concentration of Ozone in One-Dimensional and Two-Dimensional Models. *J. Geophys. Res.,* 94, 9889–9896.

Koeberl, C. 1989. Iridium Enrichment in Volcanic Dust from Blue Ice Fields, Antarctica, and Possible Relevance to the K/T Boundary Event. *Earth and Planetary Science Letters,* 92, 317–322.

Koeberl, C., V. L. Sharpton, A. V. Murali, and K. Burke. 1990. Kara and Ust-Tara Impact Structures (USSR) and Their Relevance to the K/T Boundary Event. *Geology,* 18, 50–53.

Kohlmaier, G. H., H. Brohl, and R. Fricke. 1987. The Biogenic Fuels: Fuelwood, Charcoal, Crop Residues and Animal Dung as a Net Source of Atmospheric Carbon Dioxide. *SCOPE/UNEP* 64, 29–43.

Kohlmaier, G. H., H. Brohl, E. O. Sire, and M. Plochl. 1987. Modelling Stimulation of Plants and Ecosystem Response to Present Levels of Excess Atmospheric CO_2. *Tellus,* 39B, 155–170.

Kohlmaier, G. H., G. Kratz, H. Brohl, and E. O. Sire. 1981. The Source-Sink Function of the Terrestrial Biota Within the Global Carbon Cycle. In *Energy and Ecological Modelling,* edited by W. J. Mitsch, R. W. Bosserman, and J. M. Klopatek, 57–68. Amsterdam: Elsevier Scientific.

Koide, M., and K. W. Bruland. 1975. The Electrodeposition and Determination of Radium by Isotopic Dilution in Seawater and in Sediments Simultaneously with Other Natural Radionuclides. *Anal. Chim. Acta,* 75, 1–19.

Koide, M., K. W. Bruland, and E. D. Goldberg. 1973. Th-228/Th-232 and Pb-210 Geochronologies in Marine and Lake Sediments. *Geochim. Cosmochim. Acta.,* 37, 1171–1188.

Komarek, E. V. 1973. Ancient Fires. In *Proceedings of the Annals Tall Timbers Fire Ecology Conference,* 12. Tallahassee, Fla.

Komarek Sr., E. V. 1972. Lightning and Fire Ecology in Africa. In *Proceedings of the Tall Timbers Fire Ecology Conference,* 11, 473–511.

Koonce, A. L., and A. Gonzalez-Caban. 1990. Social and Ecological Aspects of Fire in Central America. In *Fire in the Tropical Biota: Ecosystem Processes and Global Challenges (Ecological Studies 84),* edited by J. G. Goldammer, 135–158. Berlin-Heidelberg: Springer-Verlag.

Kouda, M. 1981. *Analyse synchronique et diachronique de la vegetation en zone semi-aride (Haute-Volta) par teledetection multispectrale.* Doctoral diss., University Paul Sabatier, Toulouse.

Kowal, N. E. 1966. Shifting Cultivation, Fire, and Pine Forest in the Cordillera Central, Luzon, Philippines. *Ecol. Monogr.,* 36, 389.

Kozlowski, T. T., and C. E. Ahlgren, eds. 1974. *Fire and Ecosystems.* New York: Academic Press.

Kroto, H. 1988. Space, Stars, C60, and Soot. *Science,* 242, 1139–1145.

Kuhlbusch, T. A. 1990. Masters thesis, Johannes Gutenberg Universitat, Mainz/Max Planck Institut für Chemie.

Kuhlbusch, T. A., J. M. Lobert, P. J. Crutzen, and P. Warneck. 1991. Molecular Nitrogen Emissions from Biomass Burning, a Denitrification Process. *Nature,* in press.

Kuhn, W. R. 1985. Photochemistry, Composition, and Climate. In *The Photochemistry of Atmospheres,* edited by J. S. Levine, 129–163. San Diego: Academic Press.

Kunk, M. J., G. A. Izett, R. A. Haugerud, and J. F. Sutter. 1989. 40Ar-39Ar Dating of the Manson Impact Structure: a Cretaceous-Tertiary Boundary Candidate. *Science,* 244, 1565–1568.

Kurz, H., and K. Wagner. 1957. *Tidal Marshes of the Gulf and Atlantic Coasts of Northern Florida and Charleston, South Carolina.* Tallahassee: Florida State University Studies. 24.

Kuslys, M., and U. Krahenbuhl. 1983. Noble Metals in Cretaceous/Tertiary sediments from El Kef. *Radiochim. Acta.,* 34, 139–141.

Kyte, F. T., Smit, J., and Wasson, J. T. 1985. Siderophile Interelement Variations in the Cretaceous-Tertiary Boundary Sediments from Caravaca, Spain. *Earth Planet. Sci. Lett.,* 73, 183–195.

Landsat DUN (anonymous). 1988. The Yellowstone Wildfires of 1988, EOSAT, 3, 3, December.

Lacaux, J. P., J. Servant; M. L. Huertas; B. Cros; R. Delmas; J. Loemba-Ndembi, and M. O. Andreae. 1987. Acid Rain Water in the Tropical Forest Off Western Africa. In *Scientific and Technical Advance: Acid Rains,* edited by Berry and Harrison, 264–270.

Lacaux, J. P., R. Delmas, B. Cros. G. Kouadio, and M. O. Andreae. 1990. Precipitation Chemistry in the Mayombe Forest of Equatorial Africa. *J. Geophys. Res.,* Special Issue DECAFE.

Lacaux, J. P., J. Servant, and J. G. R. Baudet. 1987. Acid Rain in the Tropical Forests of the Ivory Coast. *Atmos. Environ.,* 21, 2643–2647.

Lacaux, J. P., J. Servant, M. L. Huertas, B. Cros, R. Delmas, J. Loemba-Ndembi, and M. O. Andreae. 1988. Precipitation Chemistry from Remote Sites in the African Equatorial Forest. *Eos,* 69, 1069.

Lacey, C. J., J. Walker, and I. R. Noble. 1982. Fire in Australian

Tropical Savannas. *Ecology of Tropical Savannas (Ecological Studies* 42), edited by B. J. Huntly and B. H. Walker, 246–272. Berlin: Springer Verlag.

Lamb, B., A. Guenther, D. Gay, and H. Westberg. 1987. A National Inventory of Biogenic Hydrocarbon Emissions. *Atmos. Environ.*, 21, 1695–1705.

Lambert, G., P. Bristeau, and G. Polian. 1976. Emission and Enrichment of Radon Daughters from Etna Volcano Magma. *Geophys. Res. Lett.*, 3, 724–726.

Lambert, G., G. Polian, J. Sanak, B. Ardouin, A. Buisson, A. Jegou, and J. C. Le Roulley. 1982. Cycle du Radon et de ses Descendants: Application à l'Étude des Échanges Troposphère-Stratosphère. *Annales Géophys.*, 38, 497–531.

Lamotte, M., J. L. Tireford, J. R. Baudet, J. J. Bertrand, P. Pagney, and G. Riou. 1988. *Le Climat de LAMTO (Cote d'Ivoire) et sa place dans les climats de l'Ouest Africain.* Travaux des chercheurs de la station de Lamto, Paris: Rapport PIREN/CNRS, 8, 1–146.

Lamotte, M., et al. 1986. Variations Saisonnières de la Fixation et de la Liberation du CO_2 dans les Milieux de Savane de Cote d'Ivoire., Action des feux de brousse. *Programme Piren CO_2,* Juillet 1986. Paris: ENS

Lamprecht, H. 1986. *Waldbau in den Tropen.* Hamburg: Parey.

Langkamp, P. J., and M. J. Dalling. 1983. Nutrient Cycling in a Stand of Acacia Holosericea. *Aust. J. Bot.*, 31, 141–149.

Lanly, J. P. 1981. *Forest Resources of Tropical Africa.* UN 32/6/1301-78-04, Tropical Forest Resources Assessment Project, Tech. Rep. 2. Rome: Food and Agriculture Organization of the United Nations.

Lanly, J. P. 1982. *Tropical Forest Resources.* FAO Forestry Paper 30. Rome: Food and Agriculture Organization of the United Nations (FAO).

Lanly, J. P. 1985. Defining and Measuring Shifting Cultivation. *Unasylva,* 37 (147).

Lanly, J. P., and J. Clement. 1979. *Present and Future Forest and Plantation Areas in the Tropics.* Rome: FAO Report.

Lanning, F. C., and L. N. Eleuterius. 1983. Silsca and Ash in Tissues of Some Coastal Plants. *Annals of Botany,* 51, 835–850.

Lanyon, L. E., and W. R. Heald. 1982. Magnesium, Calcium, Strontium, and Barium. In *Methods of Soil Analysis. Part 2, Chemical and Microbial Properties, Agronomy, Vol. 9,* edited by. A. L. Page, R. H. Miller, and D. R. Keeney, 247–262. Madison, Wis. American Society of Agronomy.

Lashof, D. A. 1990. The Contribution of Biomass Burning to Global Warming: An Integrated Assessment. Paper presented at the Chapman Conference on Global Biomass Burning: Atmospheric, Climatic, and Biospheric Implications, Williamsburg, Va., 19–23 March (this volume, Chapter 54).

Lashof, D. A., and D. R. Ahuja. 1990. Relative Contributions of Greenhouse Gas Emissions to Global Warming. *Nature,* 344, 529–531.

Lashof, D. A., and D. Tirpak. 1989. *Policy Options for Stabilizing Global Climate.* Draft Report to Congress. Washington, D.C.: U.S. Environmental Protection Agency.

Lavenu, F. 1982. *Teledetection des feux de savanes dans le Parc National de la Comoe (Cote-d'Ivoire),* Dissertation, University of Paris.

Lavenu, F. 1984. *Teledetection et vegetation tropicale: Example du Nord-Est de la Cote d'Ivoire et du Bengladesh.* Thesis, UPS Toulouse.

Lavenu, F., M. F. Bellau, and J. Fontes. 1987. *Digited Vegetation Map of Africa, Scale 1/5,000,000.* Descriptive memoir and map, prepared for the Department of Forest Resources, FAO, Rome.

LeBel, P. J., W. R. Cofer III, J. S. Levine, S. A. Vay, and P. D. Roberts. 1988. Nitric Acid and Ammonia Emissions from a Mid-Latitude Prescribed Wetlands Fire. *Geophys. Res. Lett.*, 15, 792–795.

LeBel, P. J., J. M. Hoell, J. S. Levine, and S. A. Vay. 1985. Aircraft Measurements of Ammonia and Nitric Acid in the Lower Troposphere. *Geophys. Res. Lett.*, 12, 401–404.

LeBel, P. J., S. A. Vay, and P. D. Roberts. 1990. Simultaneous Measurements of Ammonia and Nitric Acid Emissions during Biomass Burning. Paper presented at the Chapman Conference on Global Biomass Burning: Atmospheric, Climatic, and Biospheric Implications, Williamsburg, Va., 19–23 March (this volume, Chapter 29).

Le Cloarec, M. F., G. Lambert, J. C. Le Roulley, and B. Ardouin. 1984. Echanges de Matériaux Volatils entre Phase Solide, Liquide et Gazeuse au Cours de l'Éruption de l'Etna de 1983. *C.R. Acad. Sci., Ser. 2,* 298, 805–808.

LeHouerou, H. N. 1974. Fire and Vegetation in the Mediterranean Basin. *Proceedings of the Tall Timbers Fire Ecology Conference,* 13, 237–277.

Leach, G., and R. Mearns. 1988. *Beyond the Woodfuel Crisis: People, Land and Trees in Africa.* London: Earthscan Publications.

Lean, J., and A. Warrilow. 1989. Simulation of the Redonal Climatic Impact of Amazon Deforestation. *Nature,* 342, 411–413.

Ledbetter, M. T., and R. S. J. Sparks. 1989. Duration of Large-Magnitude Explosive Eruptions Deduced from Graded Bedding in Deep-Sea Ash Layers. *Geology,* 7, 240–244.

Lee, R. C., Jr., W. P. Leenhouts, and J. E. Sasser. 1981. *Fire Management Plan, Merritt Island National Wildlife Refuge.* Titusville, Fla.: U.S. Fish and Wildlife Service, Merritt Island National Wildlife Refuge.

Leenhouts, W. P., and J. L. Baker. 1982. Vegetation Dynamics in Dusky Seaside Sparrow Habitat on Merritt Island National Wildlife Refuge. *Wildlife Soc. Bull.,* 10, 127–132.

Legris, P. 1963. *La Vegetation de l'Inde: Ecologie et flore.* Ph.D. dissertation, University of Toulouse.

Le Guern, F., J. C. Le Roulley, and G. Lambert. 1982. Condensation du Polonium dans les Gaz Volcaniques. *C.R. Acad. Sci.,* 294, 887–890.

Lepel, E. A., K. M. Stefansson, and W. H. Zoller. 1978. The Enrichment of Volatile Elements in the Atmosphere by Volcanic Activity: Augustine Volcano 1976. *J. Geophys. Res.,* 5, 245–248.

Lerbekmo, J F., A. R. Sweet, and R. M. St. Louis. 1987. The Relationship between the Iridium Anomaly and Palynological Floral Events at Three Cretaceous-Tertiary Boundary Localities in Western Canada. *Geological Soc. of America Bull.,* 99, 325–330.

Levantamento de Recursos Naturais, Vols. 1–23. DNPM, Rio de Janeiro.

Levine, J. S. 1982. The photochemistry of the paleoatmosphere. *J. Molecular Evolution,* 18, 161–172.

Levine, J. S. 1989. Photochemistry of Biogenic Gases. In *Global Ecology: Towards a Science of the Biosphere,* edited by M. B.

Rambler, L. Margulis, and R. Fester, 51–74. San Diego: Academic Press.

Levine, J. S. 1990a. Global Biomass Burning: Atmospheric, Climatic and Biospheric Implications. *Eos Transactions, American Geophysical Union,* 71, 1075–1077.

Levine, J. S. 1990b. Atmospheric Trace Gases: Burning Trees and Bridges. *Nature,* 346, 511–512.

Levine, J. S. 1991. Climate. *Encyclopedia of Earth System Science,* Volume I, Academic Press, Orlando, Florida, pp. 101–113.

Levine, J. S., W. R. Cofer III; R. P. Rhinehart, E. L. Winstead, D. I. Sebacher, S. Sebacher, C. R. Hinkle, P. A. Schmalzer, and A. M. Koller, Jr. 1990. Enhanced Biogenic Emissions of CH₄, N₂O, and NO Following Burning. Paper presented at the Chapman Conference on Global Biomass Burning: Atmospheric, Climatic, and Biospheric Implications, Williamsburg, Va., 19–23 March (this volume, Chapter 34).

Levine, J. S., W. R. Cofer III, D. I. Sebacher, R. P. Rhinehart, E. L. Winstead, S. Sebacher, C. R. Hinkle, P. A. Schmalzer, and A. M. Koller, Jr. 1990. The Effects of Fire on Biogenic Emissions of Methane and Nitric Oxide from Wetlands. *J. Geophys. Res.,* 95, 1853–1864.

Levine, J. S., W. R. Cofer III, D. I. Sebacher, E. L. Winstead, S. Sebacher, and P. J. Boston. 1988. The Effects of Fire on Biogenic Soil Emissions of Nitric Oxide and Nitrous Oxide. *Global Biogeochem. Cycles,* 2, 445–449.

Levine, J. S., W. R. Cofer III, E. L. Winstead, K. G. Hoffman, B. J. Stocks. 1990. *The Great Chinese Fire of 1987: Emission of Trace Gases to the Atmosphere.* Paper presented at the Chapman Conference on Global Biomass Burning: Atmospheric, Climatic, and Biospheric Implications, Williamsburg, Va., 19–23 March (this volume, Chapter 34).

Levine, J. S., C. P. Rinsland, and G. M. Tennille. 1985. The Photochemistry of Methane and Carbon Monoxide in the Troposphere in 1950 and 1985. *Nature,* 318, 254–257.

Levy, H., and W. J. Moxim. 1989. Simulated Global Distribution and Deposition of Reactive Nitrogen Emitted by Fossil Fuel Combustion. *Tellus,* 41B, 25–271.

Lewis, J. S., G. H. Watkins, H. Hartman, and R. G. Prinn. 1982. Consequences of major Impact Events on Earth. *Geol. Soc. Am. Spec. Pap. 190,* 21–221.

Li, S.-M., and J. W. Winchester. 1989. Geochemistry of Organic and Inorganic Ions of Late Winter Arctic Aerosols. *Atmos. Environ.,* 23, 2401–2415.

Likens, G. E., F. H. Bormann, R. S. Pierce, J. S. Eaton, and R. E. Munn. 1984. Long-Term Trends in Precipitation Chemistry at Hubbard Brook, New Hampshire. *Atmos. Environ.,* 18, 2641–2647.

Linak, W. P., J. A. McSorley, R. E. Hall, J. V. Ryan, R. K. Srivastava, J. O. Wendt, and J. B. Mereb. 1990. Nitrous Oxide Emissions from Fossil Fuel Combustion. *J. Geophys. Res.,* 95, 7533–7541.

Lobert, J. M. 1989. *Verbrennungen Pflanzlicher Biomass als Quelle Atmosphaerische Spurengase: Cyanoverbindungen, CO, CO₂, und NOₓ.* Ph. D. thesis, Johannes Gutenberg Universitat, Mainz.

Lobert, J., D. Scharffe, W. Hao, and P. Crutzen. 1990. Importance of Biomass Burning in the Atmospheric Budgets of Nitrogen-Containing Gases. *Nature,* 346, 552–554.

Lobert, J. M., D. H. Scharffe, W. M. Hao, T. A. Kuhlbusch, R. Seuwen, and P. J. Crutzen. 1991a. In preparation.

Lobert, J. M., and P. Warneck. 1991b. In preparation.

Logan, J. 1987. Harvard University, Center for Earth Planetary Physics, Cambridge, Mass. Personal communication.

Logan, J. A. 1983. Nitrogen Oxides in the Troposphere: Global and Regional Budgets. *J. Geophys. Res.,* 88, 10,785, 10,807.

Logan, J. A. 1985. Tropospheric Ozone: Seasonal Behavior, Trends, and Anthropogenic Influence. *J. Geophys. Res.,* 90, 10,463–10,482.

Logan, J. A., J. Dignon, and E. Gottlieb. 1990. *Biomass Burning and the Global Budget of CO: A Study Using a Chemical Tracer Model.* Paper presented at the Chapman Conference on Global Biomass Burning: Atmospheric, Climatic, and Biospheric Implications, Williamsburg, Va., 19–23 March.

Logan, J. A., and V. W. J. H. Kirchhoff. 1986. Seasonal Variations of Tropospheric Ozone at Natal, Brazil. *J. Geophys. Res.,* 91, 7875–7881.

Logan, J. A., M. J. Prather, S. C. Wofsy, and M. B. McElroy. 1981. Tropospheric Chemistry: A Global Perspective. *J. Geophys. Res.,* 86, 7210–7254.

Long, S. P., E. Garcia-Moya, S. K. Imbamba, A. Kamnalrut, M. T. Piedade, J. M. Scurlock, Y. K. Shen, and D. O. Hall. 1989. Primary Productivity of Natural Grass Ecosystems of the Tropics: A Reappraisal. *Plant and Soil,* 115, 155–166.

Lopez A., M. O. Bartomeuf, and M. L. Huertas. 1989. Simulation of Chemical Process Occurring in an Atmospheric Boundary Layer: Influence of Light and Biogenic Hydrocarbons on the Formation of Oxydants. *Atmos. Environ.,* 23, 7, 1465–1478.

Lopez A. and M. L. Huertas. 1991. Numerical Simulation of the Ozone Chemistry Observed over Forest Tropical Areas during DECAFE Experiments. *J. Geophys. Res.,* in press.

Loudjani, P. 1988. *Cartographie de la production primaire des zones savanicoles d'Afrique de l'Ouest a partir de données satellitaires: Comparaison avec des données de terrain.* DEA dissertation, University of Paris VI.

Loudjani, P., P. Y. Deschamps, F. Lavenu, J. C. Menaut, A. Podaire, J. P. Puyravaud, and B. Saugier. 1988. *Evaluating West African Aerial Herbaceous Primary Production with Satellite Data: Comparison With Ground Data.* Third International Congress of Rangelands, New Delhi, India, November.

Louw, C. 1987. Coordinator: National Programme for Weather, Climate and Atmospheric Research, CSIR, South Africa, Personal communication.

Lowenthal, D. H., and Rahn, K. A. 1985. Regional Sources of Pollution Aerosol at Barrow, Alaska during Winter 1979–1980 as Deduced from Elemental Tracers. *Atmos. Environ.,* 19, 2011–2024.

Lowman, P. D. 1966. The Earth from Orbit. *Natl. Geogr.,* Nov., 644–671.

Lugo, A. E., M. M. Sanchez, and S. Brown. 1986. Land Use and Organic Carbon Content of Some Subtropical Soils. *Plant and Soil,* 96, 185–196.

Luizao, F., P. Matson, G. Livingston, R. Luizao, and P. Vitousek. 1989. Nitrous Oxide Flux Following Tropical Land Clearing. *Global Biogeochemical Cycles,* 3, 281–285.

Lumley, J. L., and H. A. Panofsky. 1964. *The Structure of Atmospheric Turbulence.* New York: John Wiley.

Lutz, H. J. 1956. *Ecological Effects of Forest Fires in the Interior of Alaska.* U.S. Forest Service Technical Bulletin 1133.

Lynch, J. J. 1941. The Place of Burning in Management of the Gulf Coast Refuges. *J. Wildlife Mgmt.,* 5, 454–458.

Lyon, R. K., J. C. Kramlich, and J. A. Cole. 1989. Nitrous Oxide: Sources, Sampling, and Science Policy. *Environ. Sci. Technol.,* 23, 392–393.

Lyons, W., R. Calby, and C. Keen. 1987. The Impact of Mesoscale Convective Systems on Regional Visibility and Oxidant Distributions during Persistent Elevated Pollution Episodes. *J. Appl. Meteor. and Clim.,* 25, 1518–1531.

Mabbutt, J. A. 1989. Impacts of Carbon Dioxide Warming on Climate and Man in the Semi-arid Tropics. *Climate Change,* 15, 191–221.

MacArthur, R. H. 1972. *Geographical Ecology: Patterns in the Distribution of Species.* New York: Harper & Row.

Madge, D. S. 1965. Leaf Fall and Litter Disappearance in a Tropical Forest. *Pedobiologia,* 5, 273–288.

Maenhaut, W., P. Cornille, J. M. Pacyna, and V. Vitols. 1989. Trace Element Composition and Origin of the Atmospheric Aerosol in the Norwegian Arctic. *Atmos. Environ.,* 23, 2551–2569.

Maglaras, George. 1983. *Alaskan Temperature, Surface Wind, Probability of Precipitation, Conditional Probability of Frozen Precipitation, and Cloud Amount Guidance (FMAK1 Bulletin).* Technical Procedures Bulletin No. 329. Silver Spring, Md., National Weather Service.

Mahar, D. J. 1979. *Frontier Development Policy in Brazil: A Study of Amazonia.* New York: Praeger.

Mahar, D. J. 1988. *Government Policies and Deforestation in Brazil's Amazon Region.* Environment Department Working Paper No. 7. Washington, D.C.: World Bank.

Mahar, D. J. 1989. *Government Policies and Deforestation in Brazil's Amazon Region.* Washington, D.C.: World Bank.

Maier-Reimer, E., and K. Hasselmann. 1987. Transport and Storage in the Ocean—An Inorganic Ocean-Circulation Carbon Cycle Model. *Climate Dynamics,* 2, 63–90.

Maitani, T. 1979. A Comparison of Turbulence Statistics in the Surface Layer over Plant Canopies with Those over Several Other Surfaces. *Boundary-Layer Meteorology,* 17, 213–222.

Malingreau, J. P. 1984. Remote Sensing and Disaster Monitoring, A Review of Applications in Indonesia. *Proceedings of the 18th International Symposium on Remote Sensing of Environment,* 289–291, Paris, France, 1–5 October.

Malingreau, J. P., G. Stephens, and L. Fellows. 1985a. *The 1982–83 Forest Fires of Kalimantan and North Borneo.* New York: NASA-GSFC, NOAA and GISS. Unpublished report.

Malingreau, J. P., G. Stephens, and L. Fellows. 1985b. Remote Sensing of Forest Fires: Kilimantan and North Borneo in 1982–83. *Ambio,* 14(6), 314–321.

Malingreau, J. P., and C. J. Tucker. 1988. Large-Scale Deforestation in the Southeastern Amazon Basin of Brazil. *Ambio,* 17, 49–55.

Malingreau J. P., C. J. Tucker, and N. Laporte. 1989. AVHRR for Monitoring Global Tropical Deforestation. *International Journal of Remote Sensing,* 10, 4–5, 855–867.

Mallik, A. U., and R. W. Wein. 1986. Response of a Typha Marsh Community to Draining, Flooding, and Seasonal Burning. *Can. J. Bot.,* 64, 2136–2143.

Malone, R. C., L. H. Auer, G. A. Glatzmaier, M. C. Wood, and O. B. Toon. 1986. Nuclear Winter: Three-Dimensional Simulations Including Interactive Transport Scavenging and Solar Heat of Smoke. *J. Geophys. Res.,* 91, 1039–1053.

Maloney, B. K. 1985. Man's Impact on the Rain Forests of West Malesia: The Palynological Record. *J. Biogeogr.,* 12, 537.

Manabe, S., D. G. Hahn, and J. L. Holloway, Jr. 1974. The Seasonal Variation of the Tropical Circulation as Simulated by a Global Model of the Atmosphere. *J. Atmos. Sci.,* 31, 43–83.

Manabe, S., and J. L. Holloway, Jr. 1975. The Seasonal Variation of the Hydrologic Cycle as Simulated by a Global Model of the Atmosphere. *J Geophys. Res.,* 80, 1617–1649.

Mankin, W. G. 1978. Airborne Fourier Transform Spectroscopy of the Upper Atmosphere. *Opt. Engr.,* 17, 39–43.

Marenco A. 1986. Variations of CO and O_3 in the Troposphere: Evidence of O_3 Photochemistry. *Atmos. Environ.,* 20, 911–918.

Marenco A. 1990. *Study of Bush Fire Emissions and Ozone Formation over West Africa from Large Scale and Regional Aircraft Campaigns (STRATOZ; TROPOZ).* Paper presented at the Chapman Conference on Global Biomass Burning: Atmospheric, Climatic, and Biospheric Implications, Williamsburg, Va., 19–23 March.

Marenco, A. 1990. Personal communication.

Marenco A., M. Macaigne, and S. Prieur. 1989. Meridional and Vertical CO and CH_4 Distributions in the Background Troposphere (70 deg N-60 deg S; 0–12 km altitude) from Scientific Aircraft Measurements during the STRATOZ III Experiment (June 1984). *Atmos. Environ.,* 23, 185–200.

Marenco, A., and F. Said. 1989. Meridional and Vertical Ozone Distribution in the Background Troposphere (70 deg N-60 deg S; 0–12 km altitude) from Scientific Aircraft Measurements during the STRATOZ III Experiment (June 1984). *Atmos. Environ.,* 23, 201–214.

Margolis, S. V., J. F. Mount, E. Doehne, W. Showers, and P. Ward. 1987. The Cretaceous-Tertiary Boundary Carbon and Oxygen Isotope Stratigraphy, Diagenesis, and Paleoceanography at Zumaya, Spain. *Paleoceanography,* 2, 361–377.

Markham, B. L., and J. L. Barker. 1986. *Landsat MSS and ATM Post-Calibration Dynamic Ranges, Exoatmospheric Reflectances and At-Satellite Temperatures.* EOSAT Landsat technical notes, 1, August.

Markham, B. L., and J. L. Barker. 1987. Thematic Mapper Bandpass Solar Exoatmospheric Irradiances. *Intl. J. Remote Sensing,* 8, 3, 517–523.

Marland G. 1988. *The Prospects of Solving the CO_2 Problem through Global Reforestation.* Oak Ridge, Tenn.: U.S.: Dept. Energy, Office Energy Res., DOE/NBB-0082.

Marland, G. 1989. *Fossil Fuel CO_2 Emissions.* Oak Ridge, Tenn.: Carbon Dioxide Information Center.

Marland, G., T. A. Boden, R. C. Griffin, S. F. Huang, P. Ianciruk, and T. R. Nelson. 1989. *Estimates of CO_2 Emissions from Fossil Fuel Burning and Cement Manufacturing Using the United Nations Energy Statistics and the U.S. Bureau of Mines Cement Manufacturing Data,* ORNL/CDIAC-25, NDP-00. Oak Ridge, Tenn.: Oak Ridge National Laboratory.

Marland, G., and R. M. Rotty. Carbon Dioxide Emissions from

Fossil Fuels: A Procedure for Estimation and Results for 1950–1982. *Tellus*, 36, 232–261.

Marsden, E. 1964. Incidence and Possible Significance of Inhaled or Ingested Polonium. *Nature*, 203, 230–233.

Marshall, E. 1989. EPA's Plan for Cooling the Global Greenhouse. *Science*, 243, 1544–1545.

Martinelli, L. A., I. F. Brown, R. L. Victoria, M. Z. Morelra, C. A. C. Ferreira, and W. W. Thomas. 1990. *Estimativa de Biomassa e Produgao de CO₂ via Desmatamento em Floresta Tropical Bomida*. Usina Hidroeletrica de Samuel, Rondonia. Unpublished manuscript.

Mass, C., and A. Robock. 1982. The Short-Term Influence of the Mount St. Helens Volcanic Eruption on Surface Temperature in the Northwest United States. *Mon. Wea. Rev.*, 110, 614–622.

Matson, M., and Dozier J. 1981. Identification of Subresolution High Temperature Sources Using a Thermal IR Sensor. *Phonogrammetric Engineering and Remote Sensing*, 47, 9, 1311–1318.

Matson, M., and B. Holben. 1987. Satellite Detection of Tropical Burning in Brazil. *Int. J. Remote Sens.*, 8, 509–546.

Matson, M., S. R. Schneider, B. Aldridge, and B. Satchwell. 1984. *Fire Detection Using the NOAA-Series Satellites*. NOAA Techn. Report NESDIS-7. U.S. Department Commission, National Oceanic and Atmospheric Administration Natural Environmental Satellite Data and Information Service, National Weather Service, Washington, D.C.

Matson, M., G. Stephens, and J. Robinson. 1987. Fire Detection Using Data from the NOAA-N Satellites. *Int. J. Remote Sens.*, 8, 961–970.

Matson, P. A., and P. M. Vitousek. 1987. Cross-System Comparisons of Soil Nitrogen Transformations and Nitrous Oxide Flux in Tropical Forest Ecosystems. *Global Biogeochem. Cycles*, 1, 163–170.

Matson, P. A., P. M. Vitousek, J. J. Ewel, M. J. Mazzarino, and G. P. Robertson. 1987. Nitrogen Transformations Following Tropical Forest Felling and Burning on a Volcanic Soil. *Ecology*, 68, 491–502.

Matson, P. A., P. M. Vitousek, G. P. Livingston, and N. A. Swanberg. 1988. Nitrous Oxide Flux from Amazon Ecosystems: Fertility and Disturbance Effects. *Eos*, 69, 320.

Matthews, E. 1983. Global Vegetation and Land Use: New High-Resolution Data Bases for Climate Studies. *J. Climate and Applied Meteorol.* 22, 474–487.

Matthews, E. 1985. *Atlas of Archived Vegetation, Land-Use and Seasonal Albedo Data Sets*. NASA Tech Memo 86199. New York: Goddard Space Flight Center, Institute for Space Studies.

Matthews, E., and I. Fung. 1987. Methane Emission from Natural Wetlands: Global Distribution, Area, and Environmental Characteristics of Sources. *Global Biogeochem. Cycles*, 1, 61–86.

Matuska, P., M. Koval, and W. Seiler. 1986. *A High Resolution GC-Analysis Method for Determination of C₂–C₁₀ Hydrocarbons in Air Samples*, 557–583.

Mazurek, M. A., G. R. Cass, and B. R. T. Simoneit. 1988. *Quantification of the Source Contributions to Organic Aerosols in the Remote Desert Atmosphere—Final Report to Electric Power Research Institute*. Pasadena, Calif.: California Institute of Technology.

Mazurek, M. A., G. R. Cass, and B. R. T. Simoneit. 1989. Interpretation of High-Resolution Gas Chromatography and High-Resolution Gas Chromatography/Mass Spectrometry Data Acquired From Atmospheric Organic Aerosol Samples. *Aerosol Sci. Technol.*, 10, 408–420.

Mazurek, M. A., L. M. Hildemann, G. R. Cass, B. R. T. Simoneit, and W. F. Rogge. 1990a. *Measurement and Flux of Carbonaceous Aerosol Species from Pine and Oak Combustion*. Paper presented at the Chapman Conference on Global Biomass Burning: Atmospheric, Climatic, and Biospheric Implications, Williamsburg, Va., 19–23 March.

Mazurek, M. A., L. M. Hildemann, G. R. Cass, B. R. T. Simoneit, and W. F. Rogge. 1990b. *Methods of Analysis for Complex Organic Mixtures from Urban Sources of Particulate Carbon*. ACS Symposium on Measurement of Airborne Compounds: Sampling, Analysis, and Data Interpretation, Washington, D.C., 26–31 August, edited by E. D. Winegar and L. H. Keith. Chelsea, Mich.: Lewis Publishers.

Mazurek, M. A., L. M. Hildemann, B. R. T. Simoneit, G. R. Cass, and H. A. Gray. 1987. Quantitative High-Resolution Gas Chromatography and High-Resolution Gas Chromatography/Mass Spectrometry Analyses of Carbonaceous Fine Aerosol Particles. *Intl. J. Environ. Anal. Chem.*, 29, 119–139.

McBean, G. A. 1971. The Variations of the Statistics of Wind, Temperature and Humidity Fluctuations With Stability. *Boundary-Layer Meteorol.*, 1, 438–457.

McClenny, W. A., P. C. Gailey, R. S. Braman, and T. J. Shelley. 1982. Tungstic Acid Technique for Monitoring Nitric Acid and Ammonia in Ambient Air. *Anal. Chem.*, 54, 365–369.

McConnell, J. C., M. B. McElroy, and S. C. Wofsy. 1971. Natural Sources of Atmospheric CO. *Nature*, (GB), 233, 187–188.

McDowell, W. H. 1988. Potential Effects of Acid Deposition on Tropical Terrestrial Ecosystems. In *Acidification in Tropical Countries*, edited by H. Rodhe and R. Herrera, 117–139. Chichester: John Wiley.

McLean, D. M. 1978. A Terminal Mesozoic Greenhouse: Lessons from the Past. *Science*, 201, 401–406.

McLean, D. M. 1981. *Terminal Cretaceous Extinctions and Volcanism: A Link*. Abstract. American Association for the Advancement of Science, 147th National Meeting, Toronto, Canada.

McLean, D. M. 1982. *Discussions in K-Tec II Cretaceous-Tertiary Extinctions and Possible Terrestrial and Extraterrestrial Causes*, edited by D. A. Russell and G. Rice, Syllogeus Series 39. National Museums of Canada.

McLean, D. M. 1982. *Deccan Volcanism: The Cretaceous-Tertiary Marine Boundary Timing Event*. Abstract. Geological Society of America, 95th Annual Meeting, New Orleans, La.

McLean, D. M. 1985. Deccan Traps Mantle Degassing in the Terminal Cretaceous Marine Extinctions. *Cretaceous Res.*, 6, 235–259.

McLean, D. M. 1985. Mantle Degassing Induced Dead Ocean in the Cretaceous-Tertiary Transition. In *The Carbon Cycle and Atmospheric CO₂: Natural Variations Archean to Present*, edited by E. T. Sundquist and W. Broecker. Geophysical Monograph Series, 32, 493–503. Washington, D.C.: American Geophysical Union.

McLean, D. M. 1985. Mantle Degassing Unification of the Trans-K-T Geobiological Record. *Evolutionary Biology*, 19, 287–313.

McLean, D. M. 1988. K-T Transition into Chaos. *J. Geol. Educ.*, 36, 237–243.

McLean, E. O. 1982. Soil pH and Lime Requirement. In *Methods of Soil Analysis, Part 2, Chemical and Microbiological Properties*,

Agronomy, vol. 9, edited by A. L. Page, R. H. Miller, and D. R. Keeney, 199–224. Madison, Wis.: American Society of Agronomy.

McMahon, C. K. 1983. *Characteristics of Forest Fuels, Fires and Emissions*. Paper 83-45.1, presented at the 76th Annual Meeting of the Air Pollution Control Association, Atlanta, Ga., 19–24 June.

McMenamin, M. A. S., and D. S. McMenamin. 1987. Late Cretaceous Atmospheric Oxygen. *Science*, 235, 1561–1562.

McRae, D. 1989. *The Effects on the Atmosphere of a Major Nuclear Exchange*. Data presented at joint review meeting in Missoula, Mont., October 1989. Forestry Canada, Saulte Ste. Marie, Ont.

Medalia, A. I., and D. Rivin. 1982. Particulate Carbon and Other Components of Soot and Carbon Black. *Carbon*, 20, 481–492.

Medina, E. 1982. Physiological Ecology of Neotropical Savanna. *Ecology of Tropical Savannas*, Ecological Studies 42, 308–335, New York: Springer-Verlag.

Melosh, H. J., N. M. Schneider, K. J. Zahnle, and D. Latham. 1990. Ignition of Global Wildfires at the Cretaceous/Tertiary Boundary. *Nature*, 343, 251–254.

Menaut, J. C. 1974. Chute de feuilles et apport au sol de litière par les ligneux dans une savane preforestière de Cote d'Ivoire. *Bull. Soc. Ecol.*, 5, 27–39.

Menaut, J. C. 1979. Primary Production, *Tropical Grazing and Land Ecosystems—A State of Knowledge*. UNESCO/UNEP/FAO Report Natural Resources, 16, 122–145.

Menaut, J. C. 1983a. The Vegetation of African Savannas. In *Ecosystems of the World: Tropical Savannas*, edited by F. Bourliere, 109–149. Amsterdam: Elsevier.

Menaut, J. C. 1983b. The Vegetation of African Savannas. In *Tropical Finna*, edited by F. Bourliere, 109–149. Amsterdam: Elsevier.

Menaut, J. C., L. Abbadie, and P. Mordelet. 1990. *Impact of Burning on the Biogeochemical Process in a Humid Savanna of Ivory Coast*. Paper presented at the Chapman Conference on Global Biomass Burning: Atmospheric, Climatic, and Biospheric Implications, Williamsburg, Va., 19–23 March.

Menaut, J. C., R. Barbault, P. Lavelle, and M. Lepage. 1985. African Savannas: Biological Systems of Humification and Mineralization. In *Ecology and Management of the World's Savannas*, edited by J. C. Tothill and J. J. Mott, 14–33. Canberra: Australian Academy of Science.

Menaut, J. C., and J. Cesar. 1979. *Structure and Primary Productivity of Lamto Savannas (Ivory Coast)*. 60, 1197–1210.

Menaut, J. C., and J. Cesar. 1982. The Structure and Dynamics of a West African Savanna. In *Ecology of Tropical Savannas*, edited by B. J. Hintley and B. H. Waller, 80–100. Ecological studies 42. Berlin: Springer-Verlag.

Menaut, J. C., F. Lavenu, P. Loudjani, and A. Podaire. 1990. *Biomass Burning in West African Savannas*. Paper presented at the Chapman Conference on Global Biomass Burning: Atmospheric, Climatic and Biosheric Implications, Williamsburg, Va., 19–23 March (this volume, Chapter 17).

Menzel, W. P. 1990. *Pre-Launch Study Report of VAS-D Performance*. Report to NASA under contract NAS5-21965, Space Science and Engineering Center, University of Wisconsin, Madison.

Menzel, W. P., E. C. Cutrim, and E. M. Prins. 1990. *Geostationary Satellite Estimations of Biomass Burning in Amazonia*. Paper presented at the Chapman Conference on Global Biomass Burning: Atmospheric Climatic and Biospheric Implications, Williamsburg, Va., 19–23 March (this volume, Chapter 4).

Miller B. P., and H. E. Dunn. 1988. Orthogonal Least-Squares Line Fit with Variable Scaling. *Computers in Physics*, 2(4), 59–61.

Miller, S. L., and H. C. Urey. 1959. Organic Compound Synthesis on the Primitive Earth. *Science*, 130, 245.

Milligan, K., and B. Sule. 1982. Natural Forage Resources and Their Dietary Value, W. W. Sanford, H. M. Yesufu, and J. S. O. Ayeni, eds., *Nigerian Savanna—State of Knowledge*. New Bussa, Nigeria: Kaini Lake Research Institute.

Miner, S. 1969. *Preliminary Air Pollution Survey of Ammonia*. Report APTD-69-25, Raleigh, N.C. National Air Pollution Control Administration.

Mirov, N. T. 1967. *The Genus Pinus*. New York: Ronald Press.

Mo, E. and F. 19xx. *India's Forests 1987*. Report of the Survey and Utilization Division, Department of Environment, Forests and Wild Life, Ministry of Environment and Forests, Government of India.

Mobley, H. E., ed. 1976. *Southern Forestry Smoke Management Guidebook*, U.S.D.A. Forest Service, Southeastern Forest Experiment Station, Asheville, N.C., and Southern Forest Fire Laboratory, Macon, Ga.

Molenkamp, C. R., and M. l. Bradley. 1990. Parameterization of Aerosol Scavenging in a Convectite Cloud Model. *Proceedings of the American Meteorological Society Conference on Cloud Physics*, San Francisco, California, 23–27 July; also Lawrence Livermore National Laboratory Report IJCRL-101998, Livermore, Calif.

Molion, L. C. B. 1976. *A Climatonomic Study of the Energy and Moisture Fluxes of Amazonas Basin with Consideration of Deforestation Effects*. INPE 923-TPT/035, São Paulo: S. J. Campos.

Molion, L. C. B. 1986. Land Use and Agrosystem Management in the Humid Tropics. *Land Use and Agrosystem Management under Severe Climatic Conditions*. WMO Technical Note 184, Geneva: WMO.

Molion, L. C. B. 1987. Micrometeorology of an Amazonian Rainforest. In *The Geophysiology of Amazonia*, edited by R. E. Dickinson, 255–270. New York: John Wiley.

Moller D. 1984. Estimation of the Global Man-Made Sulphur Emission. *Atmos. Env.*, 18, 19–27.

Monnier, Y. 1968. *Les Effets des feux de brousse sur une savane preforestière de Cote-d'Ivoire*. Etudes Eburnennes, 9, 1–260.

Monnier, Y. 1981. *La Poussière et la Cendre*. Paris: ACCT.

Montagnini, F., and R. Buschbacher. 1989. Nitrification Rates in Two Undisturbed Tropical Rain Forests and Three Slash-and-Burn Sites of the Venezuelan Amazon. *Biotropica*, 21, 9–14.

Monteith, J. C. 1972. Solar Radiation and Productivity in Tropical Ecosystems. *J. Appl. Ecol.*, 9, 747–766.

Mooney, H. A., P. M. Vitousek, and P. A. Watson. 1987. Exchange of Materials between Terrestrial Ecosystems and the Atmosphere. *Science*, 238, 926–932.

Moore, B., R. D. Boone, J. E. Hobbie, R. A. Houghton, J. M. Melillo, B. J. Peterson, G. R. Shaver, C. J. Vorosmarty, and G. M. Woodwell. 1989. A Simple Model for Analysis of the Role of Terrestrial Ecosystems in the Global Carbon Budget. *Carbon Cycle Modelling*, SCOPE 16, edited by B. Bolin, 365–385. New York: John Wiley.

Morales, J. A., M. Hermoso, J. Serrano, and E. Sanhueza. 1990.

Trace Elements in the Venezuelan Atmosphere during Dry and Wet Periods, with and without Vegetation Burning. *Atmos. Environ.*, 24A, 407–414.

Moran, J. M., and R. D. Stieglitz. 1983. Tornadoes of Fire: The Tragic Story of Williamsonville, Wisconsin, October 8, 1871. *Weatherwise*, 298–300.

Morgan, C. 1987. A Contemporary Mass Extinction: Deforestation of Tropical Rainforests and Faunal Effects. *Palios*, 2, 165–171.

Morgan, W. J. 1971. Convection Plumes in the Lower Mantle. *Nature*, 230, 42–43.

Morgan, W. J. 1972a. Deep Mantle Convective Plumes and Plate Motions. *Amer. Assn. of Petrol. Geologists Bull.*, 56, 203–213.

Morgan, W. J. 1972b. Plate Motions and Deep Mantle Convection. *Geological Society of America, Memoir 132*, 7–22.

Morgan, W. J. 1981. Hotspot Tracks and the Opening of the Atlantic and Indian Oceans. In *The Oceanic Lithosphere*, edited by C. Emiliani, 443–487. New York: Wiley-Interscience.

Mori, S., and G. T. Prance. 1987. Species Diversity, Phenology, Plant-Animal Interactions, and Their Correlation with Climate, as Illustrated by the Brazil Nut Family (Lecythidaceae). In *The Geophysiology of Amazonia*, edited by R. E. Dickinson, 69–89. New York: John Wiley.

Mount, J. F., S. V. Margolis, W. Showers, P. Ward, and E. Doehne. 1986. Carbon and Oxygen Isotope Stratigraphy of the Upper Maastrichtian, Zumaya, Spain: A Record of Oceanographic and Biologic Changes at the End of the Cretaceous Period. *Palaios*, 1, 87–92.

Mroz, E., and W. H. Zoller. 1975. Composition of Atmospheric Particulate Matter from the Eruption of Heimaey, Iceland. *Science*, 190, 461–464.

Mueller-Dombois, D., and J. G. Goldammer. 1990. Fire in Tropical Ecosystems and Global Environmental Change: An Introduction. In *Fire in the Tropical Biota, Ecosystem Processes and Global Challenges*, edited by J. G. Goldammer, Ecological Studies 84, 1–10. Heidelberg-Berlin: Springer-Verlag.

Muller, J. 1959. Palynology of Recent Orinoco Delta and Shelf Sediments: Reports of the Orinoco Shelf Expedition. Vol. 5. *Micropaleontology*, 5, 1–32.

Munro, N. 1966. The Fire Ecology of Caribbean Pine in Nicaragua. *Proceedings Annual Tall Timbers Fire Ecological Conference 5*, Tallahassee, Fla.: Tall Timbers Research Station.

Murray, J. W., B. Spell, and B. Paul. 1983. The Contrasting Geochemistry of Manganese and Chromium in the Eastern Tropical Pacific Ocean. In *Trace Metals in Seawater*, edited by C. S. Wong, 643–669. New York: Plenum Press.

Muzio, L. J., and J. C. Kramlich. 1988. An Artifact in the Measurement of N_2O from Combustion Sources. *Geophys. Res. Lett.*, 15, 1369–1372.

Muzio, L. J., M. E. Teague, J. C. Kramlich, J. A. Cole, J. M. McCarthy, and R. K. Lyon. 1989. Errors in Grab Sample Measurements of N_2O from Combustion Sources. *J. Air Pollution Control Assn.* 39, 287–293.

Myers, N. 1980. *Conversion of Tropical Moist Forests*. Washington, D.C.: National Academy of Sciences Press.

Myers, N. 1981. The Hamburger Connection: How Central America's Forests Become North America's Hamburgers. *Ambio*, 10, 3–8.

Myers, N. 1984. *The Primary Source*. W. W. Norton.

Myers, N. 1989a. *Deforestation Rates in Tropical Forests and Their Climatic Implications*. London: Friends of the Earth.

Myers, N. 1989b. The Greenhouse Effect: A Tropical Forestry Response. *Biomass*, 18, 73–78.

NASA. 1988. Earth System Science Advisory Committee. *Earth System Science: A Closer View*. Boulder, Colo.: Office of Interdisciplinary Earth Studies, University Corporation for Atmosphere Research.

Natarajan, I. 1985. *Domestic Fuel Survey with Special Reference to Kerosene—Vols. I and II*. New Delhi: National Council of Applied Economic Research (NCAER).

National Academy of Science. 1984. *Global Tropospheric Chemistry: A Plan for Action*. Washington, D.C.: National Academy Press.

National Academy of Science. 1986. *Global Tropospheric Chemistry: Plans for the U.S. Research Effort*. Washington, D.C.: National Academy Press.

National Research Council. 1983. *Causes and Effects of Changes in Stratospheric Ozone: Update*. Washington, D.C.: National Academy Press.

National Research Council Committee on the Atmospheric Effects of Nuclear Explosions. 1985. *The Effects on the Atmosphere of a Major Nuclear Exchange*. Washington D.C.: National Academy Press.

Neftel, A., E. Moor, H. Oeschger, and B. Stauffer. 1985. Evidence from Polar Ice Cores for the Increase in Atmospheric CO_2 in the Past Two Centuries. *Nature*, 315, 45–47.

Nelson, D. W., and L. E. Sommers. 1982. Total Carbon, Organic Carbon, and Organic Matter. In *Methods of Soil Analysis, Part 2, Chemical and Microbial Properties, Agronomy, Vol. 9*, edited by A. L. Page, R. H. Miller, and D. R. Keeney, 539–579. Madison, Wis.: American Society of Agronomy.

Nelson, J. 1989. Fractality of Sooty Smoke: Implications for the Severity of Nuclear Winter. *Nature*, 339, 611–612, 22 June.

Nelson, R. M., Jr. 1982. *An Evaluation of the Carbon Mass Balance Technique for Estimating Emission Factors and Fuel Consumption in Forest Fires*. Res. Paper SE-231. Asheville, N.C.: U.S. Department of Agriculture, Forest Service, Southeastern Forest Experiment Station.

Nelson, R., and B. Holben. 1986. Identifying Deforestation in Brazil Using Multi-Resolution Satellite Data. *Int. J. Remote Sensing*, 7, 429–448.

Nepstad, D. C., F. H. Bormann, and C. Uhl. 1990. Deep Roots in Amazonian Ecosystem. Submitted to *Science*.

Neurath, G. 1969. Stockstoffverbindungen im Tabakrauch, *Beitr. Tabakforsch.*, 5, 115–133.

Neville, N., R. J. Quann, B. S. Haynes, and A. F. Sarofim. 1981. Vaporization and Condensation of Mineral Matter During Pulverized Coal Combustion. In *18th Symposium (International) on Combustion*, 1267–1274. Pittsburgh, PA.: Combustion Institute,

Newell, R. E., H. G. Reichle, Jr., and W. Seiler. 1989. Carbon Monoxide and the Burning Earth. *Scientific American*, 260(10), 82–88.

Newell, R., S. Shipley, V. Connors, and H. Reichle, Jr. 1988. Regional Studies of Potential CO Sources Based on Space Shuttle and Aircraft Measurements. *J. Atmos. Chem.*, 6, 61–81.

Newson, L. 1986. *The Cost of Conquest: Indian Decline in Honduras under Spanish Rule.* Boulder, Colo.: Westview Press.

Nguyen, B. C., B. Bonsang, N. Mihalopoulos, and S. Belviso. 1990. *Carbonyl Sulfide Emissions from African Savanna Burning.* Paper presented at the Chapman Conference on Global Biomass Burning: Atmospheric, Climatic, and Biospheric Implications, Williamsburg, Va., 19–23 March.

Nieuwstadt, F. T. M., and H. van Dop. 1982. *Atmospheric Turbulence and Air Pollution Modelling.* Dordrecht, Netherlands: Reidel.

Nimer, E. 1989. *Climatologia do Brasil.* Rio de Janeiro: IBGE.

Nobre, C. A., P. L. S. Dias, M. A. R. Santos, J. Cohen, PJ. da Rocha, R. Guedes, R. N. Ferreira, and I. A. Santos. 1988. Mean Large Scale Meteorological Aspects of ABLE-2B. *EOS Transactions,* 69(16).

Nordin, C. F., and R. H. Meade. 1982. Deforestation and Increased Flooding of the Upper Amazon. *Science* 215, 426.

Nye P. H. 1961. Organic Matter and Nutrient Cycles under Moist Tropical Forests. *Plant and Soil,* 13, 333–346.

Nye, P. H., and D. J. Greenland. 1960. *The Soil under Shifting Cultivation.* Technical Communication No. 51. Harpenden, U.K.: Commonwealth Agricultural Bureaux of Soils.

O'Keefe, J. D., and T. J. Ahrens. 1989. Did the Greenhouse Effect Kill the Dinosaurs? In *Planetary Geosciences—1988,* edited by M. Zuber, O. James, G. McPherson, and J. Plescia, 30–31. Washington, D.C.: NASA.

O'Keefe, J. D., and T. J. Ahrens. 1988. *Impact Production of CO₂ by the K-T Extinction Bolide, and the Resultant Heating of the Whole Earth (abstract).* Lunar and Planetary Science Conference 19, 885–886, Houston, Tex., March.

O'Keefe, J. D., T. J. Ahrens, and D. Koschny. 1988. Environmental Effects of Large Impacts on the Earth—Relation to Extinction Mechanisms. *Global Catastrophes in Earth History,* 133–134. Snowbird, Utah: Lunar Planetary Institute.

Octavio, K. H., A. Arrocha, and E. Sanhueza. 1987. Low Dispersion Capacity of Nocturnal Venezuelan Savannah Atmosphere. *Tellus,* 39B, 286–292.

Oeschger, H., and M. Heimann. 1983. Uncertainties of Predictions of Future Atmospheric CO₂ Concentrations. *J. Geophys. Res.,* 88, 1258–1262.

Oeschger, H., U. Siegenthaler, U. Schotterer, and A. Gugelmann. 1975. A Box Diffusion Model to Study the Carbon Dioxide Exchange in Nature. *Tellus,* 27, 168–192.

Officer, C. B., A. Hallam, C. L. Drake, and J. D. Devine. 1987. Late Cretaceous and Paroxysmal Cretaceous/Tertiary Extinctions. *Nature,* 326, 143–149.

Oguntala, A. B. 1983. *Analysis of the Litter Fuel Load in Some Nigerian Forests in Relation to Period of Fires.* Paper presented at the Workshop, Forest Fires: Ecology and the Environment, Forestry Research Institute of Nigeria, Ibadan. Mimeograph.

Ohtaki, E. 1980. Turbulent Transport of Carbon Dioxide over a Paddy Field. *Boundary-Layer Meteorol.,* 19, 315–336.

Ohtou, A., T. Maitani, and T. Seo. 1983. Direct Measurement of Vorticity and Its Transport in the Surface Layer over a Paddy Field. *Boundary-Layer Meteorol.,* 197–207.

Okali, D. U. U., and B. A. Ola-Adams. 1987. Tree Population Changes in Treated Rainforest at Omo Forest Reserve, Southwestern Nigeria. *J. Tropical Ecology,* 3, 291–303.

Olsen, S. R., and L. E. Sommers. 1982. Phosphorous, In *Methods of Soil Analysis, Part 2. Chemical and Microbial Properties, Agronomy, Vol. 9,* edited by A. L. Page, R. H. Miller, and D. R. Keeney, 403–430. Madison, Wis.: American Society of Agronomy.

Olson, J. S., J. A. Watts, and L. J. Allison. 1983. *Carbon in Live Vegetation of Major World Ecosystems.* Report TR004, DOE/NRB-0037. Washington, D.C.: U.S. Department of Energy,

Onochie, C. F. A. 1979. The Nigerian Rainforest Ecosystem: An Overview. In *The Nigerian Rainforest Ecosystem,* edited by D. U. Okali. Ibadan, Nigeria: Nigerian National MAB Committee.

Opik, E. J. 1958. On the Catastrophic Effects of Collisions with Celestial Bodies. *Irish Astron. J.,* 5, 34–36.

Oppenheimer, M. 1987. Stratospheric Sulphate Production and the Photochemistry of the Antarctic Circumpolar Vortex. *Nature,* 328, 702–704.

Orville, R. E., and R. U. Henderson. 1986. Global Distribution of Midnight Lightning: September 1977 to August 1978. *Mon. Wea. Rev.,* 114, 2640–2653.

Osemeobo, G. J. 1988. The Human Causes of Forest Depletion in Nigeria. *Envir. Conservation,* 15, 17–28.

Ottar, B., Y. Gotaas, O. Hov, T. Iversen, E. Joranger, M. Oehme, J. M. Pacyna, A. Semb, W. Thomas, and V. Vitols. 1986. *Air Pollutants in the Arctic.* NILU OR Rept. No. 30/86. Lillestrom, Norway: Norwegian Institute for Air Research.

Ovington J. D., and J. S. Olson. 1970. Biomass and Chemical Content of El Verde Lower Iontane Rain Forest Plants. *A Tropical Rain Forest.* Washington, D.C.: H. T. Odum, U.S. Atomic Energy Commission.

Ozygit, A., and C. C. Wilson. 1976. *Forestry and Forest Fire in Turkey.* Fire Mgmt. Note, 37, 17–20.

Pa, R. C., and Joshi, V. 1989. Improved Woodstoves for Household Energy Management—A Case Study. *Productivity,* 30(1), 53–59.

Pacyna, J. M., and B. Ottar. 1988. Vertical Distribution of Aerosols in the Norwegian Arctic. *Atmos. Environ.,* 22, 2213–2222.

Pacyna, J. M., and B. Ottar. 1989. Origin of Natural Constituents in the Arctic Aerosol. *Atmos. Environ.,* 23, 809–815.

Pacyna, J. M., B. Ottar, U. Tomza, and W. Maenhaut. 1985. Long-Range Transport of Trace Elements to Ny-Alesund, Spitzbergen. *Atmos. Environ.,* 19, 857–865.

Pacyna, J. M., V. Vitols, and J. E. Hanssen. 1984. Size Differentiated Composition of the Arctic Aerosol at Ny-Alesund, Spitzbergen. *Atmos. Environ.,* 18, 2447–2459.

Paegle, J. 1987. Interactions between Convective and Large-Scale Motions over Amazonia. In *The Geophysiology of Amazonia,* edited by R. E. Dickinson, 347–390. New York: John Wiley.

Palm, C. A., R. A. Houghton, J. M. Melillo, and D. L. Skole. 1986. Atmospheric Carbon Dioxide from Deforestation in Southeast Asia. *Biotropica,* 18, 177–188.

Panofsky, H. A., and I. A. Dutton. 1984. *Atmospheric Turbulence.* New York: John Wiley.

Parker, F. D. 1964. *The Central American Republics.* London: Oxford University Press.

Parrish, A., et al. 1991. A Comparison of the Total Reactive

Nitrogen Levels and the Partitioning among Individual Species at Six Rural Sites in Eastern North America. *J. Geophys. Res.*, in press.

Parsons, J. J. 1965. Cotton and Cattle in the Pacific Lowlands of Central America. *J. Inter-Am. Stud.*, 7, 149–159.

Parsons, J. J. 1976. Forest to Pasture: Development or Destruction? *Revista de Biologia Tropical*, 24 (Supl. 1), 121–138.

Parton, W. J., A. R. Mosier, and D. S. Schimel. 1988. Rates and Pathways of Nitrous Oxide Production in a Shortgrass Steppe. *Biogeochem.*, 6, 45–58.

Pascoe, E. H. 1964. *A Manual of the Geology of India and Burma, Volume III*. Calcutta: Government of India.

Pasquill, F. 1974. *Atmospheric Diffusion*. New York: John Wiley.

Patrick, W. H., Jr. 1982. Nitrogen Transformations in Submerged Soils. *Nitrogen in Agricultural Soils*, Agronomy Monograph 22, 449–465. Madison, Wis.: American Society of Agronomy.

Patterson, E. M., and C. K. McMahon. 1984. Absorption Characteristics of Forest Fire Particulate Matter. *Atmos. Environ.*, 18, 2541–2551.

Patterson, E. M., C. K. McMahon, and D. E. Ward. 1986. Absorption Properties and Graphitic Carbon Emission Factors of Forest Fire Aerosols. *Geophys. Res. Lett.*, 13, 129–132.

Patzold, H. 1986. *Nutzpflanzen der Tropen und Subtropen*. Leipzig: Hirzel-Verlag.

Pearce, F. 1989. *New Scientist*, 123, 37–41.

Penfound, W. T., and G. S. Hathaway. 1938. Plant Communities in the Marshland of Southeastern Louisiana. *Ecol. Monog.*, 8, 1–56.

Peng, T.-H. 1986. Uptake of Anthropogenic CO_2 by Lateral Transport Models of the Ocean Based on the Distribution of Bomb-Produced ^{14}C. *Radiocarbon*, 28, 363–375.

Peng, T.-H., W. S. Broecker, H. D. Freyer, and S. Trumbore. 1983. A Deconvolution of the Tree-Ring Based Delta ^{13}C Record. *J. Geophys. Res.*, 88, 3609–3620.

Peng, T.-H., and H. D. Freyer. 1986. Revised Estimates of Atmospheric CO_2 Variations Based on the Tree-Ring ^{13}C Record. In *The Changing Carbon Cycle: A Global Analysis*, edited by J. R. Trabalka and D. E. Reichle, 151–159. New York: Springer-Verlag.

Penner, J. E., C. S. Atherton, J. Dignon, S. J. Ghan, J. J. Walton, and S. Hameed. 1990. Tropospheric Nitrogen: A Three-Dimensional Study of Sources, Distribution, and Deposition. Submitted to *J. Geophys. Res.*, also Lawrence Livermore, National Laboratory report UCRL-102183, Rev. 1.

Penner, J. E., C. S. Atherton, J. J. Walton, and S. Hameed. 1990. The Global Cycle of Reactive Nitrogen, to be published in *The Proceedings of the International Conference on Global and Regional Environmental Atmospheric Chemistry, Beijing, China, May 3–10, 1989*, also Lawrence Livermore National Laboratory report UCRL-JC-104052.

Penner, J. E., and L. L. Edwards. 1986. Nucleation Scavenging of Smoke and Aerosol Particles in Convective Updrafts. *Proceedings of the Conference on Radar Meteorology and Conference on Cloud Phys.*, American Meteorological Society, Snowmass, Colo.

Penner, J. E., S. J. Ghan, and J. J. Walton. The Role of Biomass Burning in the Budget and Cycle of Carbonaceous Soot Aerosols and Their Climate Impact. Paper presented at the Chapman Conference on Global Biomass Burning: Atmospheric, Climatic, and Biospheric Implications, Williamsburg, Va., 19–23 March (this volume, Chapter 47).

Penner, J. E., L. C. Haselmen, Jr., and L. L. Edwards. 1986. Smoke-Plume Distributions above Large-Scale Fires: Implications for Simulations of Nuclear Winter. *J. Climate and Appl. Meteorol.*, 25, 1434–1444.

Pennisi, M., M. F. Le Cloarec, G. Lambert, and J. C. Le Roulley. 1988. Fractionation of Metals in Volcanic Emissions. *Earth Planet. Sci. Lett.*, 88, 284–288.

Perch-Nielsen, K., J. McKenzie, and Q. He. 1982. Biostratigraphy and Isotope Stratigraphy and the Catastrophic Extinction of Calcareous Nannoplankton at the Cretaceous/Tertiary Boundary. *Geological Society of America Special Paper 190*, 353–371.

Pereira, M. C. P. 1988. *Detection, Monitoring and Analysis of Some Environmental Impacts of Biomass Burning in Amazonia through NOAA and Landsat Satellites and Aircraft Data*. M.S. dissertation, in Portuguese. INPE-4503-TDL/326.

Perkin-Elmer Corporation. 1982. *Analytical Methods for Atomic Absorption Spectrophotometry*, Norwalk, Conn.: Perkin-Elmer Corporation.

Perkins, D. F. 1978. The Distribution and Transfer of Energy and Nutrients in the Agrostis Festuca Grassland Ecosystems. *Production Ecology of British Moors and Mountain Grasslands*, Ecological Studies 27, 375–393. New York: Springer-Verlag.

Perkins, D. F., V. Jones, R. O. Miller, and P. Neep. 1978. Primary Production, Mineral Nutrients and Litter Decomposition in the Grassland Ecosystem. *Production Ecology of British Moors and Mountain Grasslands*, Ecological Studies 27, 375–393. New York: Springer-Verlag.

Pernin, P. 1971. The Great Peshtigo Fire: An Eyewitness Account. *Wisconsin Magazine of History*, 54, 24272.

Perret, F. A. 1935. *The Eruption of Mt. Pelee 1929–1932*. Baltimore: Carnegie Institution of Washington.

Perry, R. H., and C. H. Chilton. 1973. *Chemical Engineer's Handbook*, 5th ed. New York: McGraw-Hill.

Peters, C., A. Gentry, and R. Mendelsohn. 1989. Valuation of an Amazonian Rainforest. *Nature* 339 (June) 655–656.

Peters, W. J., and L. F. Neuenschwander. 1988. *Slash and Burn: Farming in the Third World Forest*. Moscow: University of Idaho Press.

Peterson, B. J., and J. M. Melillo. 1985. The Potential Storage of Carbon Caused by Eutrophication of the Biosphere. *Tellus*, 37, 117–127.

Peterson, J. T., and C. E. Junge. 1971. Sources of Particulate Matter in the Atmosphere. In *Man's Impact on the Climate*, edited by W. H. Matthews, W. W. Kellogg, and G. D. Robinson, 310–320. Cambridge, Mass.: MIT Press.

Phillips, J. 1965. Fire—As Master and Servant: Its Influence in the Bioclimatic Regions of Trans-Saharan Africa. *Proceedings of the Tall Timbers Fire Ecology Conference*, 4, 7–109.

Pickett, S. T. A., and P. S. White, eds. 1985. *The Ecology of Natural Disturbance and Patch Dynamics*. Orlando: Academic Press.

Pierotti, D., and R. A. Rasmussen. 1976. Combustion as a Source of Nitrous Oxide in the Atmosphere. *Geophys. Res. Lett.*, 3, 265–267.

Pinchot, G. 1967. *The Fight For Conservation*. Seattle, Wash.: University of Washington Press, reprint of 1910 edition.

Pinto, J. P., Y. L. Yung, D. Rind, G. L. Russell, J. A. Lerner, J. E. Hansen, and S. Hameed. 1983. A General Circulation Model

Study of Atmospheric Carbon Monoxide. *J. Geophys. Res., 88,* 3691–3702.

Pittock, A. B., T. P. Ackerman, P. J. Crutzen, M. C. MacCracken, C. S. Shapiro, and R. P. Turco. 1986. *Environmental Consequences of Nuclear War. Vol. 1, Physical and Atmospheric Effects.* Scientific Committee on Problems of the Environment, 28, (SCOPE) New York: John Wiley.

Plate, E. 1982. *Engineering Meteorology.* Amsterdam: Elsevier Scientific.

Ponnamperuma, F. N. 1972. The Chemistry of Submerged Soils. *Adv. Agron., 24,* 29–96.

Poppleton, J. E., A. G. Shuey, and H. A. Sweet. 1977. Vegetation of Central Florida's East Coast: A Checklist of the Vascular Plants. *Florida Sci., 40,* 362–389.

Porter, D. M. 1973. *The Vegetation of Panama: A Review, in Vegetation and Vegetational History of Northern Latin America,* edited by A. Graham, 167–201. Amsterdam: Elsevier Scientific.

Post, W. M., W. R. Emanuel, P. J. Zinke, and A. G. Strangenberger. 1982. Soil Carbon Pools and World Life Zones. *Nature, 298,* 156–159.

Postgate, J. R. 1982. *Biological Nitrogen Fixation: Fundamentals.* Phil. Trans. R. Soc. Lond. B296, 375–385.

Powell, John Wesley. 1878. *Report on the Lands of the Arid Region of the United States.* Washington, D.C.: U. S. Government Printing Office.

Prance, G. T., ed. 1982. *Biological Diversification in the Tropics.* New York: Columbia University Press.

Prather, M. 1989. *An Assessment Model for Atmospheric Composition.* Washington, D.C.: NASA Conference Publication 3023.

Prather, M. J. 1990. Tropospheric Hydroxyl Concentrations and the Lifetimes of Hydrochlorofluorocarbons (HCFCs). *AFEAS Report: Appendix in World Meteorological Organization Global Ozone Research and Monitoring Project—Report No. 2., Scientific Assessment of Stratospheric Ozone,* Geneva.

Pratt, R., and P. Falconer. 1979. Circumpolar Measurements of Ozone, Particles, and CO from a Commercial Airliner. *J. Geophys. Res., 84,* 7876–7882.

Preisinger, A., E. Zobetz, A. J. Gratz, R. Lahodynsky, M. Becke, H. J. Mauritsch, G. Eder, F. Grass, F. Rogl, H. Stradner, and R. Surenian. 1986. *Nature, 322,* 794–799.

Prinn, R. G., D. Cunnold, R. Rasmussen, P. Simmonds, F. Alyea, A. Crawford, P. Fraser, and R. Rosen. 1987. Atmospheric Trends in Methylchloroform and the Global Average for the Hydroxyl Radical. *Science, 238,* 946–950.

Prinn, R. G., and Fegley, B., Jr. 1987. Bolide Impacts, Acid Rain, and Biospheric Traumas at the Cretaceous-Tertiary Boundary. *Earth Planet. Sci. Lett., 83,* 1–15.

Prinn, R. G., P. G. Simmonds, R. A. Rasmussen, R. D. Rosen, F. N. Alyea, C. A. Cardelino, A. J. Crawford, D. M. Cunnold, P. J. Fraser, and J. E. Lovelock. 1983. The Atmospheric Lifetime Experiment, 1. Introduction, Instrumentation and Overview. *J. Geophys. Res.* 88, 8353–8367.

Prinn, R., D. Cunnold, R. Rasmussen, P. Simmonds, F. Alyea, A. Crawford, P. Fraser, and R. Rosen. 1990. Atmospheric Emissions and Trends of Nitrous Oxide Deduced from Ten Years of ALE/GAGE Data, *J. Geophys. Res.* 95, 18, 369.

Prins, E. M. 1989. *Geostationary Satellite Detection of Biomass Burning in South America.* M. S. thesis, University of Wisconsin, Madison.

Prins, E. M., and W. P. Menzel. 1991. Geostationary Satellite Detection of Biomass Burning in South America. *Intl. J. Remote Sens.,* in press.

Progulske, D. R. 1974. *Yellow Ore, Yellow Hair, Yellow Pine: A Photographic Study of a Century of Forest Ecology.* Bulletin 616. Brookings, S. Dak.: South Dakota State University, Agriculture Experiment Station.

Prospero, J. M., R. J. Charlson, V. Mohnen, R. Jaenicke, A. C. Delany, J. Moyers, W. Zoller, and K. Rahn. 1983. The Atmospheric Aerosol System: An Overview. *Rev. of Geophys. and Space Phys., 21,* 1607–1629.

Provancha, M. J., P. A. Schmalzer, and C. R. Hinkle. 1986. Vegetation Types. John F. Kennedy Space Center, Biomedical Operations and Research Office (Maps in Master Planning format, 1:9600 scale, digitization by ERDAS, Inc.).

Pruppacher, H. R., and J. Klett. 1978. *Microphysics of Clouds and Precipitation.* Boston: Reidel.

Pueschei, R. F., J. M. Livingston, P. B. Russell, D. A. Colburn, T. P. Ackerman, D. A. Allen, B. D. Zak, and W. Einfeld. 1988. Smoke Optical Depths: Magnitude Variability and Wavelength Dependence. *J. Geophys. Res.* 93, 8388–8402.

Pyne, S. J. 1982. *Fire in America.* Princeton, N.J.: Princeton University Press.

Pyne, S. J. 1990. Fire Conservancy: The Origins of Wildland Fire Protection in British India, America and Australia. In *Fire in the Tropical Biota: Ecosystem Processes and Global Challenges,* edited by J. G. Goldammer. Ecological Studies 84, 319–336. Berlin-Heidelberg: Springer-Verlag.

Pyne, S. J. 1991. *Burning Bush: A Fire History of Australia.* New York: Henry Holt.

Pyne, S. J. 1988. *Fire in America: A Cultural History of Wildland and Rural Fire.* Princeton, N.J.: Princeton University Press.

Raabe, O. G. 1968. A General Method for Fitting Size Distributions to Multicomponent Aerosol Data Using Weighted Least-Squares. *Environ. Science and Technol.,* 12(10), 1162–1167.

Radke, L. F. 1983. Preliminary Measurements of the Size Distribution of Cloud Interstitial Aerosol. In *Precipitation Scavenging, Dry Deposition and Resuspension,* edited by H. R. Pruppacher, R. G. Semonin, and W. G. N. Slinn, 71–78. N.Y.: Elsevier.

Radke, L. F. 1989. Airborne Observations of Cloud Microphysics Modified by Anthropogenic Forcing. *Proceedings of a Symposium on the Role of Clouds in Atmospheric Chemistry and Global Climate,* American Meteorological Society, 310–315. Anaheim, Calif., 29 January–3 February.

Radke, L. F., A. S. Ackerman, D. A. Hegg, J. H. Lyons, P. V. Hobbs, and J. E. Penner. 1991. Effects of Aging on the Smoke from a Forest Fire: Implications for the Nuclear Winter Hypothesis. *J. Geophys. Res.,* in press.

Radke, L. F., D. A. Hegg, J. H. Lyons, C. A. Brock, P. V. Hobbs, R. Weiss, and R. Rasmussen. 1988. Airborne Measurements on Smokes from Biomass Burning. In *Aerosols and Climate,* edited by P. V. Hobbs and M. P. McCormick, 411–422. Hampton, Va.: Deepak.

Radke, L. F., and P. V. Hobbs. 1976. Cloud Condensation Nuclei on the Atlantic Seaboard of the United States. *Science,* 193, 999–1002.

Radke, L. F., J. H. Lyons, D. A. Hegg, P. V. Hobbs, D. V. Sandberg, and D. E. Ward. 1983. *Airborne Monitoring and Smoke Characterization of Prescribed Fires on Forest Lands in Western Washington and Oregon.* Environmental Protection Agency Report EPA 600/-83-047. Washington, D.C.: EPA.

Radke, L. F., J. H. Lyons, P. V. Hobbs, D. A. Hegg, D. V. Sandberg, and D. E. Ward. 1990. *Airborne Monitoring and Smoke Characterization of Prescribed Fires on Forest Lands in Western Washington and Oregon.* Gen. Tech. Rep. PNW-GTR-21, Portland, Ore.: U.S.D.A. Forest Service, Pacific Northwest Research Station.

Radke, L. F., J. H. Lyons, P. V. Hobbs, and R. E. Weiss. 1990. Smokes from the Burning of Aviation Fuels and Their Self-Lofting by Solar Heating. *J. Geophys. Res.,* 95, 14071.

Radke, L. F., J. L. Stith, D. A. Hegg, and P. V. Hobbs. 1978. Airborne Studies of Particulates and Gases From Forest Fires. *J. Air Pollut. Control Assoc.,* 28, 30–33.

Raison, R. J. 1979. Modification of the Soil Environment by Vegetation Fires, with Particular Reference to Nitrogen Transformations: A Review. *Plant and Soil,* 51, 73–108.

Ramage, C. S. 1983. Role of a Tropical Maritime Continent in the Atmospheric Circulation. *Mon. Wea. Rev.,* 365–370.

Ramakrishna, J., M. S. Durgaprasad, and K. R. Smith. 1989. Cooking in India: The Impact of Improved Stoves on Indoor Air Quality. *Environ. Int.* 15.

Ramakrishnan, P. S. 1987. In *Energy Flows and Shifting Cultivation in Rural Energy Planning for the Indian Himalaya,* edited by T. M. Jinodkumar and D. R. Ahuja. India: TERI and ICIMOD.

Ramanathan, V., L. Callis, R. Cess, J. Hansen, I. Isaksen, W. Kuhn, A. Lacis, F. Luther, J. Mahlman, R. Reck, and M. Schlesinger. 1987. Climate-Chemical Interactions and Effects of Changing Atmospheric Trace Gases. *Rev. Geophys.,* 25, 1441–1482.

Ramanathan V., R. J. Cicerone, H. B. Singh, and J. T. Kiehl. 1985. Trace Gas Trends and Their Potential Role in Climate Change. *J. Geophys. Res.,* 90, 5547–5566.

Ramaswamy, V., and J. T. Kiehl. 1985. Sensitivities of the Radiative Forcing Due to Large Loadings of Smoke and Dust Aerosols. *J. Geophys. Res.,* 90, 5597–5613.

Ramdahl, T. 1983. Retene—A Molecular Marker of Wood Combustion in Ambient Air. *Nature,* 306, 580–582.

Rampino, M. R., and R. C. Reynolds. 1983. Clay Mineralogy of the Cretaceous-Tertiary Boundary Clay. *Science,* 219, 495–498.

Rampino, M. R., and R. B. Stothers. 1988. Flood Basalt Volcanism during the Past 250 Million Years. *Science,* 241, 663–668.

Rampino, M. R., and T. Volk. 1988. Mass Extinctions, Atmospheric Sulphur and Climatic Warming at the K/T Boundary. *Nature,* 332, 63–65.

Rasmussen, R. A., and M. A. K. Khalil. 1981a. Atmospheric Methane (CH_4): Trends and Seasonal Cycles. *J. Geophys. Res.,* 86, 9826–9832.

Rasmussen, R. A., and M. A. K. Khalil. 1981b. Increase in the Concentration of Atmospheric Methane. *Atmos. Environ.,* 15, 883–886.

Rasmussen, R. A., and M. A. K. Khalil. 1983. Global Production of Methane by Termites. *Nature,* 301., 700–702.

Rasmussen, R. A., J. Krasnec, and D. Pierotti. 1976. N_2O Analysis in the Atmosphere via Electron Capture-Gas Chromatography. *Geophys. Res. Lett.,* 3, 615–618.

Rasmussen, R. A., L. E. Rasmussen, M. A. K. Khalil, and R. W. Dalluge. 1980. Concentration Distribution of Methyl Chloride in the Atmosphere. *J. Geophys. Res.,* 85, 7350–7356.

Ravindranath, N. H., and R. Shailaja. 1989. Dissemination and Evaluation of Fuel Efficient and Smokeless ASTRA Stove in Karnataka. *Energy Environ. Monitor,* 5(2), 48–60.

Read, D. J., and D. T. Mitchell. 1983. Decomposition and Mineralization Processes in Mediterranean-Type Ecosystems and in Heathlands of Similar Structure. *Mediterranean Type Ecosystems,* Ecological Studies 43, 167–176. New York: Springer-Verlag.

Reddy, K. R., and W. H. Patrick. 1984. Nitrogen Transformations and Loss in Flooded Soils and Sediments. *CRC Crit. Rev. Environ. Control,* 13, 273–309.

Reichle, H. G., V. S. Connors, J. A. Holland, W. D. Hypes, H. A. Wallio, J. C. Casas, B. B. Gormsen, M. S. Saylor, and W. D. Hasketh. 1986. Middle and Upper Tropospheric Carbon Monoxide Mixing Ratios as Measured by a Satellite Borne Remote Sensor during November 1981. *J. Geophys. Res.,* 91, 10,865–10,888.

Reichle, H. G., V. S. Connors, H. A. Wallio, J. A. Holland, R. T. Sherrill, J. C. Casas, and B. B. Gormsen. 1990. The Distribution of Middle Tropospheric Carbon Monoxide during Early October 1984. *J. Geophys. Res.,* 95, 9845–9856.

Reiners, W. A. 1973. A Summary of the World Carbon Cycle and Recommendations for Critical Research. In *Carbon and the Biosphere,* AEC Symposium series 30, CONF-720510, edited by G. M. Woodwell and E. V. Pecan, 368–382. Washington, D.C.: U.S. Atomic Energy Commission.

Revelle, R., and W. Munk. 1977. The Carbon Dioxide Cycle and the Biosphere. In *Energy and Climate,* National Research Council, Panel on Energy and Climate, R. Revelle, chair, 140–158, Washington, D.C.: National Academy of Sciences.

Rhoades, J. D. 1982. Soluble Salts, In *Methods of Soil Analysis, Part 2, Chemical and Microbial Properties, Agronomy, Vol. 9,* edited by A. L. Page, R. H. Miller, and D. R. Keeney, 167–179. Madison, Wis.: American Society of Agronomy.

Rich, P. V., T. H. Rich, B. E. Wagstaff, J. M. Mason, C. B. Douthitt, R. T. Gregory, and E. A. Felton. 1988. Evidence for Low Temperatures and Biologic Diversity in Cretaceous High Latitudes of Australia. *Science,* 242, 1403–1406.

Richards, J. F., J. S. Olson, and R. M. Rotty. 1983. *Development of a Data Base for Carbon-Dioxide Releases Resulting from Conversion of Land to Agricultural Uses.* ORA U/IEA-82-10(M), ORNL/TM-8801. Oak Ridge, Tenn.: Oak Ridge National Laboratory.

Richards, M. A., R. A. Duncan, and V. E. Courtillot. 1989. Flood Basalts and Hot-Spot Tracks: Plume Heads and Tails. *Science,* 246, 103–107.

Richards, P. W. 1976. *The Tropical Rain Forest.* Cambridge: Cambridge University Press.

Richey, J. E. 1989. Personal communication.

Rinsland, C. P., and J. S. Levine. 1985. Free Tropospheric Carbon Monoxide Concentrations in 1950 and 1951 Deduced from Infrared Total Column Amount Measurements. *Nature,* 318, 250–254.

Rinsland, C. P., J. S. Levine, and T. Miles. 1985. Concentration of Methane in the Troposphere Deduced from 1951 Solar Spectra. *Nature,* 318, 245–249.

Rizzo, B., and E. Wiken. 1990. Assessing the Sensitivity of Canada's Ecosystems to Climatic Change. *Climatic Change.*

Robertson, G. P., and T. Rosswall. 1986. Nitrogen in West Africa: The Regional Cycle. *Ecology Monogr.,* 56, 43–72.

Robertson, G. P., and J. M. Tiedje. 1988. Deforestation Alters Denitrification in a Lowland Tropical Rain Forest. *Nature,* 336, 756–759.

Robinson, J. M. 1987. *The Role of Fire on Earth: Review of the State of Knowledge and Systems Framework for Satellite and Ground-Based Evaluations.* Ph.D. dissertation, Department of Geography, UCSB, Santa Barbara, Calif.

Robinson, J. M. 1989. On Uncertainty in the Computation of Global Emission from Biomass Burning. *Climatic Change,* 14, 243–262.

Robinson, J. M. 1991. Fire from Space: Global Fire Evaluation Using IR Remote Sensing. *Intl. J. Remote Sensing,* 12, 3–24.

Robock A. 1988a. Surface Temperature Effects of Forest Fire Smoke Plumes. In *Aerosol and Climate,* edited by P. V. Hobbs and M. P. McCormick. Hampton, Va.: Deepak, 435–442.

Robock, A. 1988b. Enhancement of Surface Cooling Due to Forest Fire Smoke. *Science,* 24, 911–913.

Robock, A., and C. Mass. 1982. The Mount St. Helens Volcanic Eruption of 18 May 1980: Large Short-Term Surface Temperature Effects. *Science,* 216, 628–630.

Robock, A. 1989. Policy Implications of Nuclear Winter and Ideas for Solutions. *Ambio,* 18, 360–366.

Robock, A. and G. Grodzinsky. 1990. Surface Cooling from Yellowstone Forest Fire Smoke. Submitted to *Science.*

Rocchia, R., D. Boclet, P. Bonte, A. Castellarin, V. Courtillot, C. Jehanno, and F. C. Wezel. 1989. On the Existence of Several Iridium Enriched Layers at the K/T Boundary and in a Jurassic Sequence (abstract). *Lunar and Planetary Science Conference,* 20, 914–915.

Rodin, L. Y., N. I. Basilevitch, and N. N. Rozov. 1975. Productivity of the World's Main Ecosystems. *Productivity of the World Ecosystems.* Washington, D.C.: National Academy of Science.

Rogers, G. F., J. G. Hudson, J. Hallett, and J. E. Penner. 1990. Cloud Droplet Nucleation by Crude Oil Smoke and Coagulated Crude Oil/Wood Smoke Particles. *Atmos. Environ,* 25A, 2571–2580.

Rogers, G. F., J. G. Hudson, J. Hallett, B. Zielinska, J. G. Watson, and R. L. Tanner. 1991. The Potential Climatic Importance of Cloud Condensation Nuclei from Tropical Biomass Burning. *Nature,* in press.

Rogers, G. F. 1982. *Then and Now: A Photographic History of Vegetation Change in the Central Great Bagin Desert.* Salt Lake City: University of Utah Press.

Rondon, A., and E. Sanhueza. 1989. High HONO Atmospheric Concentrations during Vegetation Burning in the Tropical Savannah. *Tellus,* 41B, 474–477.

Rondon, A., and E. Sanhueza. 1990. Seasonal Variation of Gaseous HNO_3 and NH_3 at a Tropical Savannah Site, *J. Atmos. Chem.,* 11, 245–254.

Root, B. 1976. An Estimate of Annual Global Atmospheric Pollutant Emissions from Grassland Fires and Agricultural Burning in the Tropics. *Professional Geographer,* 28, 349–352.

Rose-Innes, R. 1972. Fire in West African Vegetation. *Proceedings of the Tall Timbers Fire Ecology Conference,* 11, 147–173.

Rothermel, R. C. 1972. *A Mathematical Model for Predicting Fire Spread in Wildland Fuels.* Res. Paper INT-115. Ogden, Utah: U.S.D.A. Forest Service, Intermountain Forest and Range Experiment Station.

Rothermel, R. C. 1983. *How to Predict the Spread and Intensity of Forest and Range Fires.* Gen. Tech. Rep. INT-14. Ogden, Utah: U.S.D.A. Forest Service, Intermountain Forest and Range Experiment Station.

Rothman, L. S., R. R. Gamache, A. Barbe, A. Goldman, J. R. Gillis, L. R. Brown, R. A. Todt, J.-M. Flaud, and C. Camy-Peyret. 1983. AFGL Atmospheric Absorption Line Parameters Compilation: 1982 Edition. *Appl. Optics,* 22, 2247–2256.

Rothman, L. S., R. R. Gamache, A. Goldman, L. R. Brown, R. A. Toth, H. M. Pickett, R. L. Poynter, J.-M. Flaud, C. Camy-Peyret, A. Barbe, N. Husson, C. P. Rinsland, and M. A. Smith. 1987. The HITRAN Database: 1986 Edition. *Appl. Optics,* 22, 4058–4097.

Rothman, L. S., A. Goldman, J. R. Gillis, R. R. Gamache, H. M. Pickett, R. L. Poynter, N. Husson, and A. Chedin. 1983. AFGL Trace Gas Compilation: 1982 Version. *Appl. Optics,* 22, 1616.

Rotmans, J., D. Lashof, and M. den Elzen. 1990. *Global Warming Potentials, Chapter 3.3 in Targets and Indicators of Climatic Change.* Advisory Group on Greenhouse Gases, Report of Working Group 2, edited by F. R. Rijsberman and R. J. Swart. Stockholm: Stockholm Environment Institute.

Royal Forest Department Thailand. 1988. *Forest Fire Control in Thailand.* Bangkok: Forest Fire Control Subdivision.

Rudolph J., and D. H. Ehhalt. 1981. Measurements of C_2–C_5 Hydrocarbons over the North Atlantic. *J. Geophys. Res.,* 86, 11959–11964.

Rudolph, J., and F. J. Johnen. 1990. Measurements of Light Hydrocarbons over the Atlantic in Regions of Low Biological Productivity. *J. Geophys. Res.,* 95, 20,583.

Rudolph, J., A. Khedim, B. Bonsang, G. Helas, and M. O. Andreae. 1990. *Hydrocarbon Emission from Tropical Biomass Burning and Ozone Formation.* Paper presented at the Chapman Conference on Global Biomass Burning: Atmospheric, Climatic, and Biospheric Implications, Williamsburg, Va., 19–23 March.

Rudolph J., R. Koppmann, and B. Bonsang. 1988. *Measurements of Light Hydrocarbons and Carbon Monoxide over an Equatorial Forest in Africa.* AGU Fall Meeting, San Francisco, 5–9 December.

Rundel, P. W. 1982. Fire as an Ecological Factor. *Physiological Plant Ecology I,* 12A, 501–538. New York: Springer-Verlag.

Rundel, P. W. 1983. Impact on Fire on Nutrient Cycles in Mediterranean-Type Ecosystems with Reference to Chaparral. *Mediterranean Type Ecosystems,* Ecological Studies 43, 167–176. New York: Springer-Verlag.

Russell, C. E. 1987. Plantation Forests. *Amazon Rain Forests,* Ecological Studies 60, 76–89. New York: Springer-Verlag.

Russell, D. A. 1975. Reptilian Diversity and the Cretaceous-Tertiary Transition in North America. *The Cretaceous System in the Western Interior of North America,* edited by W. G. E. Caldwell, 119–136. Geological Association of Canada Special Paper 13.

Russell, D. A. 1977. *A Vanished World: The Dinosaurs of Western Canada.* Ottawa: National Museums of Canada.

Russell, D. A. 1979a. The Cretaceous-Tertiary Boundary Problem. *Episodes,* 21–24.

Russell, D. A. 1979b. The Enigma of the Extinction of the Dinosaurs. *Ann. Rev. Earth and Planetary Sciences,* 7, 163–182.

Russell, D. A. 1982. The Mass Extinctions of the Late Mesozoic. *Scientific Amer.,* 246, 58–65.

Russell, D. A. 1984. The Gradual Decline of the Dinosaurs—Fact or Fancy. *Nature,* 307, 360–361.

Ryden, J. C. 1981. Nitrous Oxide Exchange between a Grassland Soil and the Atmosphere. *Nature,* 292, 235–237.

SCOPE. 1986. *Environmental Consequences of Nuclear War, SCOPE 28. Vol. 1, Physical and Atmospheric Effects,* edited by A. B. Pittock, T. P. Ackerman, P. J. Crutzen, M. C. MacCracken, C. S. Shapiro, and R. P. Turco. New York: John Wiley.

Sachse, G. W., G. F. Hill, L. O. Wade, and M. G. Perry. 1987. Fast-Response, High-Precision Carbon Monoxide Sensor Using a Tunable Diode Laser Technique. *J. Geophys. Res.,* 92, 2071–2081.

Sackett, W. M., C. W. Poag, and B. J. Eadie. 1974. Kerogen Recycling in the Ross Sea, Antarctica. *Science,* 185, 1045–1047.

Sadaphal, P. M., R. C. Pal, and V. Joshi. 1989. *Evaluation of Improved Chulha Programme in Tamil Nadu, Raiasthan and West Bengal.* Report submitted to the Department of Nonconventional Energy Sources, Ministry of Energy, Government of India, TERI.

Saha, A. K., J. Rai, V. Raman, R. C. Sharma, D. C. Parashar, S. P. Sen, and B. Sarkar. 1989. *Ind. J. Radio and Space Physics,* 18, 215–217.

Sahrawat, K. L., and Keeney, D. R. 1986. Nitrous Oxide Emission from Soils. *Adv. Soil Sci.,* 4, 103–148.

Salati, E. 1987. The Forest and the Hydrological Cycle. In *The Geophysiology of Amazonia,* edited by R. E. Dickinson 273–287. New York: John Wiley.

Saldarriaga, G. 1987. Recovery after Shifting Cultivation in Amazon Rain Forests. *Amazon Rain Forests,* Ecological Studies 60, 24–33. New York: Springer-Verlag.

Saldarriaga, J. G., and D. C. West. 1986. Holocene Fires in the Northern Amazon Basin. *Quat. Res.,* 26, 358.

Salisbury, H. E. 1989. *The Great Black Dragon Fire.* Boston: Little, Brown.

Salo, J. 1987. Pleistocene Forest Refuges in the Amazon: Evaluation of the Biostratigraphical, Lithostratigraphical and Geomorphological Data. *Ann. Zool. Fennici,* 24, 203.

Sanchez, P. 1989. Deforestation Reduction Initiative: An Imperative for World Sustainability in the Twenty-first Century. Paper presented at the International Conference on Soils and the Greenhouse Effect, Wageningen, the Netherlands, 14–18 August.

Sanchez, P. A. 1976. *Properties and Management of Soils in the Tropics.* New York: John Wiley.

Sandberg, D. V. 1983. Research Leads to Less Smoke from Prescribed Fires. *Proceedings 1983 Northwest Fire Council Annual Meeting,* Olympia, Wash., 21–22 November.

Sanford, R. L., J. Saldarriaga, K. E. Clark, C. Uhl, and R. Herrera. 1985. Amazon Rain-Forest Fires. *Science,* 227, 53.

Sanford, W. W. 1982a. The Effects of Seasonal Burning: A Review. In *Nigerian Savanna—State of Knowledge,* edited by W. W. Sanford, I. M. Yesufu, and J. S. Ayeni, 160–187. New Bussa, Nigeria: Kainji Lake Research Institute.

Sanford, W. W. 1982b. Savanna: A General Review. In *Nigerian Savanna,* edited by W. W. Sanford, H. M. Yefusu, and J. S. O. Ayeni, 3–22. New Bussa: Kainji Lake Research Institute.

Sanhueza, E. 1988. Atmospheric Cycles of Acidic Compounds. In *Acidification in Tropical Countries,* SCOPE 36, edited by H. Rodhe and R. Herrer. Stockholm: SCOPE.

Sanhueza, E., M. C. Arias, L. Donoso, N. Graterol, M. Hermoso, J. Romero, A. Rondon, and M. Santana. 1990. Chemical Composition of Acid Rains in the Venezuelan Savannah Climatic Region. Submitted to *Tellus.*

Sanhueza, E., G. Cuenca, M. J. Gomez, R. Herrera, C. Ishizaki, I. Marti, and J. Paolini. 1988. Characterization of the Venezuelan Environment and Its Potential for Acidification. *Acidification in Tropical Countries,* edited by H. Rodhe and R. Herrera, 197–225. Chichester, U.K.: John Wiley.

Sanhueza, E., W. Elbert, A. Rondon, M. Corina Arias, and M. Hermoso. 1989. Organic and Norganic Acids in Rain from a Remote Site of the Venezuelan Savannah. *Tellus,* 41B, 170–176.

Sanhueza, E., Z. Ferrer, J. Romero, and M. Santana. 1990. HCHO and HCOOH in Tropical Rains. *Ambio,* in press.

Sanhueza, I. B., K. Getavio, and A. Arrocha. 1985. Surface Ozone Measurements in the Venezuelan Tropical Savannah. *J. Atmos. Chem.,* 2, 377–385.

Sanhueza, E., A. Rondon, and J. Romero. 1987. Airborne Particles in the Venezuelan Savannah during Burning and Non-burning Periods. *Atmos. Environ.,* 21, 2227–2231.

Sanhueza, E., J. Serrano, and C. Bifano. 1986. Preliminary Determination of Trace Elements during the Vegetation Burning Season in the Venezuelan Savannah Atmosphere. In *Proceedings of the International Conference on Chemical in the Environment,* edited by J. N. Lester, R. Perry and R. M. Steritt, 348–353. Lisbon.

Sarin, M. 1989. *An Evaluation Survey of the Nada Chulha System in Stoves for People,* edited by R. Caceres, J. Ramakrishna, and K. R. Smith, 24–31. London: IT Publications.

Sarmiento, G., and M. Monasterio. 1975. A Critical Consideration of the Environmental Conditions Associated with the Occurrence of Savannah Ecosystems in Tropical America. *Tropical Ecological Systems* (Ecological Studies 11), edited by F. B. Golley and E. Medina, 223–250. Berlin: Springer-Verlag.

Sauer, C. O. 1969. *The Early Spanish Main.* Berkeley: University of California Press.

Savoie, D. L. 1984. Nitrate and Non-Sea-Salt Sulfate Aerosols over Major Regions of the World Ocean: Concentrations, Sources, and Fluxes. Ph.D. dissertation, University of Miami.

Savoie, D. L., and J. M. Prospero. 1989. Effect of Continental Sources on Nitrate Concentrations over the Pacific Ocean. *Nature,* 339, 687–689.

Savoie, D. L., J. H. Prospero, J. T. Merrill, and M. Uematsu. 1989. Nitrate in the Atmospheric Boundary Layer of the Tropical South Pacific: Implications Regarding Sources and Transport. *J. Atmos. Chem.,* 8, 391–415.

Savoie, D., J. M. Prospero, and E. S. Salzman. 1989. Non-Sea-Salt Sulfate and Nitrate in Trade Wind Aerosols at Barbados: Evidence for Long-Range Transport. *J. Geophys. Res.,* 94, 5069–5080.

Scala, J., M. Garstang, W.-K. Tao, K. E. Pickering, A. M. Thompson, J. Simpson, V. W. J. H. Kirchhoff, E. V. Browell, G. W. Sachse, A. L. Torres, G. L. Gregory, R. Rasmussen, and M. Khalil. 1990. Cloud Draft Structure and Trace Gas Transport. *J. Geophys. Res.,* 95, 17,015.

Schlesinger, W. H., and J. M. Melack. 1981. Transport of Organic Carbon in the World's Rivers. *Tellus,* 33, 172–187.

Schmalzer, P. A., and C. R. Hinkle. 1985. *A Brief Overview of Plant Communities and the Status of Selected Plant Species at John F. Kennedy Space Center, Florida.* Report submitted to Biomedical Office, NASA/Kennedy Space Center, Fla.

Schmalzer, P. A., and C. R. Hinkle. 1989. Responses of Oak Scrub Vegetation and Soils on John F. Kennedy Space Center to Intense Fire (abstract). In *Proceedings of the 17th Tall Timbers Fire Ecology Conference.*

Schmidt, M., R. Borchers, P. Fablan, G. Flentje, W. Matthews, A. Szabo, and S. Lal. 1984. Trace Gas Measurements during Aircraft Flights in Tropopause Region over Europe and North Africa. *J. Atmos. Chem.,* 2, 133–144.

Schmitz, B. 1985. Metal Precipitation in the Cretaceous-Tertiary Boundary Clay at Stevns Klint, Denmark. *Geochimica et Cosmochimica Acta,* 49, 2361–2370.

Schmitz, B. 1988. Origin of Microlayering in Worldwide Distributed Ir-rich Marine Cretaceous/Tertiary Boundary Clays. *Geology,* 16, 1068–1072.

Schneider, R. S., K. A. Jungbluth, D. A. Rotzol, and D. R. Johnson. 1986. *Atmospheric Pollutant Regional Transport and Dispersion: A Meteorological Perspective.* Madison: Space Science and Engineering Center, University of Wisconsin.

Schnell, R. C. 1984. Arctic Haze and the Arctic Gas and Aerosol Sampling Program (AGASP). *Geophys. Res. Lett.,* 11, 361–364.

Schoch, P., and D. Binkley. 1986. Prescribed Burning Increased Nitrogen Availability in a Mature Loblolly Pine Stand. *Forest Ecology Mgmt.,* 14, 13–22.

Scholle, P. A., and M. A. Arthur. 1980. Carbon Isotope Fluctuations in Cretaceous Pelagic Limestones; Potential Stratigraphic and Petroleum Tool. *American Asso. of Petrologists Bull.,* 64, 67–87.

Schule, W. 1990. Landscapes and Climate in Prehistory: Interaction of Wildlife, Man and Fire. In *Fire in the Tropical Biota: Ecosystem Processes and Global Challenges,* edited by J. Goldammer, 273–318. Berlin: Springer-Verlag.

Schuman, G. E., M. A. Stanley, and D. Knudsen. 1973. Automated Total Nitrogen Analysis of Soil and Plant Samples. *Soil Sci. Soc. Amer. Proc.,* 37, 480–481.

Schutz, H., A. Holzapfel-Pschorn, R. Conrad, H. Rennenberg, and W. Seiler. 1989. A 3-year Continuous Record on the Influence of Daytime, Season, and Fertilizer Treatment on Methane Emission Rates From an Italian Rice Paddy. *J. Geophys. Res.,* 94, 16405–16416.

Schwartz, S. E. 1988. Control of Global Cloud Albedo and Climate by Marine Phytoplankton: A Test by Anthropogenic Sulfur Dioxide Emissions. *Nature,* 336, 441–445.

Scott, G. A. C. 1987. Shifting Cultivation Where Land Is Limited. *Amazon Rain Forests,* Ecological Studies 60, 34–45. New York: Springer-Verlag.

Scurlock, J. M. O., and D. O. Hall. 1990. The Contribution of Biomass to Global Energy Use. *Biomass,* 21, 75–81.

Sebacher, D. I., R. C. Harriss, and K. B. Bartlett. 1985. Methane Emissions to the Atmosphere through Aquatic Plants. *J. Environ. Qual.,* 14, 40–46.

Sebacher, D. I., R. C. Harriss, K. B. Bartlett, S. M. Sebacher, and S. S. Grice. 1986. Atmospheric Methane Sources: Alaskan Tundra

Bogs, and Alpine Fen and a Subarctic Boreal Marsh. *Tellus,* 38B, 1–10.

Sedjo, R. 1989a. Forests: A Tool to Moderate Global Warming? *Environ.,* 31(1), 14–20.

Sedjo, R. 1989b. Forests to Offset the Greenhouse Effect. *J. Forestry,* July, 12–15.

Sehmel, G. A. 1984. Deposition and Resuspension. In *Atmospheric Science and Power Production,* edited by D. Randerson. Washington, D.C.: Technical Information Center, Office of Scientific and Technical Information, U.S. Dept. of Energy, DOE/ TIC-27601 (DE84005177), 533–583.

Seiler, W., A. Holzapfel-Pschorn, R. Conrad, and D. Scharffe. 1989. *J. Atmos. Chem.,* (Holland), 1, 241.

Seiler, W., and R. Conrad. 1987. Contribution of Tropical Ecosystems to the Global Budgets of Trace Gases, Especially CH_4, H_2, CO and N_2O. In *The Geophysiology of Amazonia,* edited by R. E. Dixon, New York: John Wiley.

Seiler, W., R. Conrad, and D. Scharffe. 1984. Field Studies of Methane Emission from Termite Nests into the Atmosphere and Measurements of Methane Uptake by Tropical Soils. *J. Atmos. Chem.,* 1, 171–186.

Seiler, W., and P. J. Crutzen. 1980. Estimates of Gross and Net Fluxes of Carbon between the Biosphere and the Atmosphere from Biomass Burning. *Clim. Change,* 2, 207–247.

Seiler, W., and J. Fishman. 1981. The Distribution of Carbon Monoxide and Ozone in the Free Troposphere. *J. Geophys. Res.,* 86, 7255–7266.

Selmar, D., R. Lieberei, and B. Biehl. 1988. Mobilization and Utilization of Cyanogenic Glycosides (The Linustatin Pathway). *Plant Physiol.,* 86, 711–716.

Setzer, A. Presentation given to the Environmental Protection Agency Biomass Burning Workshop, Washington, D.C., 14 Nov. 1989.

Setzer, A., M. Pereira, and A. Pereira. 1990. *AVHRR Estimates of Biomass Burning during BASE-A.* Paper presented at the Chapman Conference on Global Biomass Burning: Atmospheric, Climatic, and Biospheric Implications, Williamsburg, Va., 19–23 March.

Setzer, A. W., and M. C. Pereira. 1988. *Relatorio de Atividades do Projeto IBDF-INPE SEQE—Ano 1987,* INPE-4534-RPE/565. São Jose dos Campos, São Paulo: INPE.

Setzer, A. W., and M. C. Pereira. 1991. Amazon Biomass Burnings in 1987 and Their Tropospheric Emissions. *Ambio,* 20, 19–22.

Setzer, A. W., M. C. Pereira, A. C. Pereira, Jr. and S. A. O. Almeida. 1988. *Relatorio de Atividades do Projeto IBDF-INPE SEOE—Ano 1987.* Instituto de Pesquisas Espaciais (INPE), Pub. No. INPE-4S34-RPE/565. São Jose dos Campos, São Paulo: INPE.

Setzer, A. W., and M. C. P. Pereira. 1990. *Project SEQE Report for 1988-Remote Sensing of Biomass Burning.* In Portuguese. INPE internal report.

Seuwen, R. 1990. *Bestimmung Volatiler, Aliphatischer Amine in Rauchgasen bei der Verbrennung Pflanzlicher Biomasse.* Master's thesis, Johannes Gutenberg Universitat Mainz, FRG, Max Planck Institut für Chemie.

Shaver, G. R. 1981. Mineral Nutrient and Nonstructural Carbon Utilization. *Resource Use by Chaparral and Matorral.* Ecological Studies 39, 237–257. New York: Springer-Verlag.

Shaw, G. E. 1988. Chemical Air Mass Systems in Alaska. *Atmos. Environ.*, 22, 2239–2248.

Shea, D. J. 1986. *Climatological Atlas: 1950–1979.* NCAR Technical Note, NCAR/N-269 + STR. Boulder, Colo.: National Center for Atmospheric Research.

Shostakovitch, V. B. 1925. Forest Conflagrations in Siberia. *J. For.*, 23(4), 365–371.

Shukla, J., C. A. Nobre, and P. Sellers. 1990. Amazon Deforestation and Climate Change. *Science*, 247, 1322–1325.

Shuttleworth, W. J. 1988. Evaporation from Amazonian Rainforest. *Proc. R. Soc. Lond.*, B233, 321–346,

Siegenthaler, U. 1983. Uptake of Excess CO_2 by an Outcrop-Diffusion Model of the Ocean. *J. Geophys. Res.*, 88, 3599–3608.

Siegenthaler, U., M. Heimann, and H. Oeschger. 1978. Model Responses of the Atmospheric CO_2 Level and $^{13}C/^{12}C$ Ratio to Biogenic Input. In *Carbon Dioxide, Climate, and Society,* edited by J. Williams, 79–87. Oxford: Pergamon Press.

Siegenthaler, U., and H. Oeschger. 1987. Biospheric CO_2 Emissions during the Past 200 Years Reconstructed by Deconvolution of Ice-Core CO_2 Data. *Tellus,* 9B, 140–154.

Sillans, R. 1958. *Les Savanes de l'Afrique Centrale.* Paris: Lechevallier.

Simoneit, B. R. T., and H. R. Beller. 1987. Lipid Geochemistry of Cretaceous/Tertiary Boundary Sediments, Hole 605, Deep Sea Drilling Project Leg 93, and Stevns Klint, Denmark. In *Init. Repts. DSDP,* edited by J. E. van Hinte, et al., 93, 52, 1211–1215. Washington, D.C: U.S. Government Printing Office.

Simpson, B. B. 1982. The Refuge Theory (a review). *Science,* 217, 526.

Simpson, B. B, and J. Haffer. 1978. Speciation Patterns in the Amazonian Forest Biota. *Ann. Rev. Ecol. Syst.,* 9, 497.

Singh, G., A. P. Kershaw, and R. Clark. 1981. Quaternary Vegetation and Fire History in Australia. *Fire and the Australian Biota,* edited by A. M. Gill, R. H. Groves, and I. R. Noble. Canberra: Australian Academy of Science.

Siren, G. 1974. Some Remarks on Fire Ecology in Finnish Forestry. *Proceedings of the Tall Timbers Fire Ecology Conference,* 13, 191–209.

Slemr, K. R., and W. Seiler. 1984. Field Measurements of NO and NO_2 Emissions from Fertilized and Unfertilized Soils. *J. Atmos. Chem.,* 2, 1–2.

Slingo, A. 1990. Sensitivity of the Earth's Radiation Budget to Changes in Low Clouds. *Nature,* 343, 49–51.

Sloan, R. E., J. K. Rigby, Jr., L. M. Van Valen, and D. Gabriel. 1986. Gradual Dinosaur Extinction and Simultaneous Ungulate Radiation in the Hell Creek Formation. *Science,* 23, 629–633.

Small, R. D., and B. W. Bush. 1985. Smoke Production from Multiple Nuclear Explosions in Non-Urban Areas. *Science,* 229, 465–466.

Smit, J. 1982. Extinction and Evolution of Planktonic Foraminifera after a Major Impact at the Cretaceous/Tertiary Boundary. *Geological Society of America Special Paper 190,* 329–352.

Smit, J., and J. Hertogen. 1980. An Extraterrestrial Event at the Cretaceous-Tertiary Boundary. *Nature,* 285, 198–200.

Smith, D. M., J. J. Griffin, and E. D. Goldberg. 1975. Spec-

trometric Method for the Quantitative Determination of Elemental Carbon. *Anal. Chem.,* 47, 233–238.

Smith, K. R. 1987. *Biofuels, Air Pollution and Health: A Global Review.* New York: Plenum Publishing Co.

Smith, K. R. 1990. *Village Stoves and Global Warming: How Close the Connection.* Honolulu: East-West Center, in prep.

Smith, L. M., and J. A. Kadlec. 1984. Effects of Prescribed Burning on Nutritive Quality of Marsh Plants in Utah. *J. Wildlife Mgmt.,* 48, 285–288.

Smith, M. S., and K. Zimmerman. 1981. Nitrous Oxide Production by Nondenitrifying Soil Nitrate Reducers. *Soil Sci. Soc. Am. J.,* 45, 865–871.

Smith, W. L., P. K. Rao, R. Koffler, and W. R. Curtis. 1970. The Determination of Sea Surface Temperature from Satellite High-Resolution Infrared Window Radiation Measurements. *Mon. Wea. Rev.,* 98, 604–611.

Soderlund, R., and B. H. Svensson. 1976. Nitrogen, Phosphorus and Sulfur. *Ecol. Bull. Stockholm,* 22, 23–73.

Solomon, S. 1990. Progress Towards a Quantitative Understanding of Antarctic Ozone Depletion. *Nature,* 347, 347–354.

Sommer, A. 1976. Attempt at an Assessment of the World's Tropical Moist Forests. *Unasylva,* 28, 5–24.

Song Miao, P. S., S. C. Wofsy, P. S. Bakwin, and D. J. Jacob. 1990. Atmosphere-Biosphere Exchange of CO_2 and O_3 in the Central Amazon Forest. *J. Geophys. Res.,* 95, 16851.

Spivakovsky, C. M., R. Yevich, J. A. Logan, S. C. Wofsy, M. B. McElroy, and M. J. Prather. 1990. Tropospheric OH in a Three-Dimensional Chemical Tracer Model: An Assessment Based on Observations of CH_3CCl_3. *J. Geophys. Res.,* 95, 18441–18447.

Squires, P. 1958. The Microstructure and Colloidal Stability of Warm Clouds. Part II, The Causes of the Variations in Microstructure. *Tellus,* 10, 262–271.

Squires, P., and S. Twomey. 1960. The Relation between Cloud Droplet Spectra and the Spectrum of Cloud Nuclei, Physics of Precipitation. Washington, D.C.: Amer. Geophys. Union. *Geophys. Monogram,* 5, 211–219.

Standley, L. J., and B. R. T. Simoneit. 1987. Characterization of Extractable Plant, Resin, and Thermally Matured Components in Smoke Particles from Prescribed Burns. *Environ. Sci. Technol.,* 21, 163–169.

Standley, L. J., and B. R. T. Simoneit. 1990. Preliminary Correlation of Organic Molecular Tracers in Residential Wood Smoke with the Source of Fuel. *Atmos. Environ.,* 24B, 67–73.

Stark, N., and M. Spratt. 1977. Biomass and Nutrient Storage in Rainforest Oxisols Near San Carlos de Rio Negro. *Tropical Ecology,* 18(1), 1–9.

Statistics Canada. 1984. *Canadian Forestry Statistic.* Ottawa: Canadian Ministry of Supply and Services.

Stearns, F., and G. Guntenspergen. 1987. *Presettlement Forests of the Lake States* (map). Conservation Foundation Lake States Governor's Conference on Forestry.

Steele, L. P., P. J. Fraser, R. A. Rasmussen, M. A. K. Khalil, T. J. Conway, A. J. Crawford, R. H. Gammon, K. A. Masavie, and K. W Thoning, 1987. *J. Atmos. Chem.* 5, 125–171.

Stefanson, R. C. 1972. Soil Denitrification in Sealed Soil-Plant Systems. I, Effect of Plants, Soil Water Content and Soil Organic Matter Content. *Plant Soil,* 33, 113–127.

Steinhardt, U., and H. W. Fassbender. 1979. Caracteristicas y composicion quimica de las lluvias de los Andes Occidentales de Venezuela. *Turrialba,* 29, 175–182.

Stephens, J. C. 1969. Peat and Muck Drainage Problems. *Irri. Drain. Div. Proc. Am. Soc. Civil Eng.,* 95, 285–290.

Stith, J. L., L. F. Radke, and P. V. Hobbs. 1981. Particule Emissions and the Production of Ozone and Nitrogen Oxides from the Burning of Forest Slash. *Atmos. Environ.,* 15, 73–82.

Stock, W. D., and O. A. M. Lewis. 1986. Soil Nitrogen and the Role of Fire as a Mineralizing Agent in a South African Coastal Fynbos Ecosystem. *J. Ecology,* 74, 317–328.

Stocks, B. J. 1988. The Canada/US Cooperative Mass Fire Behavior and Atmospheric Environmental Impact Study. Northwest Fire Council Annual Meetings, Victoria, B.C., 84–88.

Stocks, B. J. 1990. *The Extent and Impact of Forest Fires in Northern Circumpolor Countries.* Paper presented at the Chapman Conference on Global Biomass Burning: Atmospheric, Climatic and Biospheric Implications, Williamsburg, Va., 19–23 March (this volume, Chapter 26).

Stocks, B. 1990. Private communication.

Stocks, B. J., and R. J. Barney. 1981. *Forest Fire Statistics for Northern Circumpolar Countries.* Sault Ste-Marie, Ontario: Department of the Environment, Canadian Forest Service, Rep. O-X-322.

Stocks, B. J. and J. Z. Jin. 1988. The China Fire of 1987: Extremes in Fire Weather and Behavior. *Proceedings of the 1988 Annual Meeting Northwest Fire Council Fire Management in a Climate of Change,* Victoria, B.C., 14–15 November 67–69.

Stocks, B. J., and D. J. McRae. 1991. The Canada/United States Cooperative Mass Fire Behavior and Atmospheric Environmental Impact Study. *Proceedings of the Eleventh Conference on Fire and Forest Meteorology,* Missoula, Mont., 16–19 April, in press.

Stott, P., J. G. Goldammer, and W. W. Werner. 1990. The Role of Fire in the Tropical Lowland Deciduous Forests of Asia. In *Fire in the Tropical Biota: Ecosystem Processes and Global Challenges,* edited by J. G. Goldammer. Ecological Studies 84. Berlin-Heidelberg: Springer-Verlag.

Stott, P. 1988. Savanna Forest and Seasonal Fire in South East Asia. *Plants Today,* 1, 196.

Stott, P. 1988. The Forest as Phoenix: Towards a Biogeography of Fire in Mainland South East Asia. *Geogr. J.,* 154, 337.

Strain, B. R., and T. V. Armentano. 1982. *Response of Unmanaged Ecosystems. Vol. 2, Part 12 of Environmental and Societal Consequences of a Possible CO$_2$-Induced Climate Change.* DOE/EV/10019-12. Washington, D.C.: U.S. Department of Energy.

Strong, C. P., R. R. Brooks, S. M. Wilson, R. D. Reeves, C. J. Orth, Mao Xue-Ying, L. R. Quintana, and E. Anders. 1987. A New Cretaceous-Tertiary Boundary Site at Flaxbourne River, New Zealand: Biostratigraphy and Geochemistry. *Geochim. Cosmochim. Acta,* 51, 2769–2777.

Struwe, S., and A. Kjoller. 1989. Field Determination of Denitrification in Water-Logged Forest Soils. *FEMS Microbiology Ecology,* 62, 71–78.

Stuart, A., and J. K. Ord. 1987. *Kendall's Advanced Theory of Statistics,* 5th ed. vol. 1. London: Charles Griffin.

Stuiver, M, R. L. Burk, and P. D. Quay. 1984. $^{13}C/^{12}C$ Ratios in Tree Rings and the Transfer of Biospheric Carbon to the Atmosphere. *J. Geophys. Res.,* 89, 11731–11748.

Stuiver, M., P. D. Quay, and H. G. Ostlund. 1983. Abyssal Water Carbon-14 Distribution and the Age of the World Oceans. *Science,* 219, 849–851.

Stutzer, David. 1990. *Temperature Anomalies and Smoke.* Unpublished paper, Department of Meteorology, University of Maryland.

Subbarao, K. V., and R. N. Sukheswala. 1981. Introduction. In *Deccan Volcanism and Related Basalt Provinces in Other Parts of the World,* edited by K. V. Subbarao and R. N. Sukheswala, v–vii. Geological Society of India, Memoir 3. Bangalore: Rashtrothana Press.

Sullivan, K. 1986. Personal communication. NASA, Johnson Space Center, Astronaut Corps, Houston, Texas.

Sullivan, R. 1987. Contribution in Science of the Natural History Museum of Los Angeles County, 391, 1–26.

Sullivan, W. T., III. 1984. Our Endangered Night Skies. *Sky and Telescope,* 412–414.

Suman, D. 1986. Charcoal Production from Agricultural Burning in Central Panama and Its Deposition in Sediments of Gulf of Panama. *Environmental Conservation,* 13, 51–60.

Suman, D. O. 1983. *Agricultural Burning in Panama and Central America: Burning Parameters and the Coastal Sedimentary Record.* Ph.D. thesis, University of California, San Diego.

Suman, D. O. 1984. The Production and Transport of Charcoal Formed during Agricultural Burning in Central Panama. *Interciencia,* 9, 311–313.

Suman, D. O. 1986. Charcoal Production from Agricultural Burning in Central Panama and Its Deposition in the Sediments of the Gulf of Panama. *Environ. Conserv.,* 13, 51–60.

Suman, D. O. 1988. The Flux of Charcoal to the Troposphere during the Period of Agricultural Burning in Panama. *J. Atmos. Chem.,* 6, 21–34.

Suome Antropologi, 1987. Special Issue on Swidden Cultivation, Jussi Raumolin, ed., 4, 12.

Susott, R., D. Ward, R. Babbitt, and D. Latham. 1990. Fire Dynamics and Chemistry of Large Fires, in preparation.

Sutton, O. G. 1960. *Atmospheric Turbulence.* New York: John Wiley.

Svensson, B. H. 1980. Carbon Dioxide and Methane Fluxes from the Ombrotropic Parts of a Subarctic Mire. *Ecol. Bull.,* 30, 235–250.

Sverdrup, H. U., M. W. Johnson, and R. H. Fleming. 1942. *The Oceans—Their Physics, Chemistry, and General Biology.* Englewood Cliffs, N.J.: Prentice-Hall.

Swaminathan, M. S. 1984. *Rice in 2000 A.D.:* Indian National Science Academy Report of National Relevance–I.

Sweet, A. R., and T. Jerzykiewicz. 1985. *The Cretaceous-Tertiary Boundary,* GEOS., 14, S-9.

Sweet, A. R., and J. F. Lerbekmo. 1988. *The Terrestrial Biota of the Cretaceous-Tertiary Boundary Interval with Emphasis on the Palynoflora of Western Canada* (abstract). 2.

Swift, M. J., J. Russell-Smith, and T. J. Perfect. 1981. Decomposition and Mineral Nutrient Dynamics of Plant Litter in Regenerating Bush Fallow in Sub-Humid Tropical Nigeria. *Journal of Ecology,* 69, 981–995.

Sze, N. D. 1977. Anthropogenic CO Emissions: Implications for the Atmospheric CO-OH-CH$_4$ Cycle. *Science, 195*, 673.

Takahashi, T., W. S. Broecker, A. E. Bainbridge, and R. F. Weiss. 1980. *Carbonate Chemistry of the Atlantic, Pacific, and Indian Oceans: The Results of the GEOSECS Expeditions, 1972–1978.* Technical Report No. 1, CV-1-80. Palisades, N.Y.: Lamont-Doherty Geological Observatory.

Talbot, R. W., M. O. Andreae, T. W. Andreae, and R. C. Harriss. 1988. Regional Aerosol Chemistry of the Amazon Basin. *J. Geophys. Res., 93*, 1499–1508.

Talbot, R. W., K. M. Beecher, R. C. Harriss, and W. R. Cofer. 1988. Atmospheric Geochemistry of Formic and Acetic Acids at a Mid-Latitude Temperate Site. *J. Geophys. Res., 93*, 1638–1652.

Talbot, R. W., K. M. Stein, R. C. Harriss, and W. R. Cofer, III. 1988. Atmospheric Geochemistry of Formic and Acetic Acids. *J. Geophys. Res., 93*, 1638–1652.

Tall Timbers Research Station: Fire in Africa. 1972. In *Proceedings of the Annual Tall Timbers Fire Ecology Conference, 11.* Tallahassee, Fla.: Tall Timbers Research Station.

Tans, P. P., I. Y. Fung, and T. Takahashi. 1990. Observational Constraints on the Global Atmospheric CO$_2$ Budget. *Science, 247*, 1431–1438.

Tardin, A. T., D. C. L. Lee, R. J. R. Santos, O. R. de Assis, M P. dos Santos Barbosa, M. L. Moreira, M. T. Pereira, D. Silva, and C. P. dos Santos Filho. 1980. Subprojeto Desmatamento, Convenio IBDFICNPq-INPE 1979. INPE 1649-RPE/103. SJ. Cunpos, São Paulo: INPE.

Tassios, S., and D. R. Packham. 1985. The Release of Methyl Chloride from Biomass Burning in Australia. *JAPCA 35*, 41–42.

Taylor, B. W. 1963. An Outline of the Vegetation of Nicaragua. *Ecology, 51*, 27–54.

Taylor, J. A., G. Brasseur, P. Zimmerman, and R. Cicerone. 1990. Study of the Sources and Sinks of Methane and Methyl Chloroform Using a Global 3-D Lagrangian Tropospheric Tracer Model. *J. Geophys. Res., 96*, 3013–3044.

Tchernia, P. 1980. *Descriptive Regional Oceanography.* Oxford: Pergamon Press.

Technicon Industrial Systems. 1973. *Nitrate and Nitrite in Water and Wastewater, Industrial Method No. 100–70W.* Tarrytown, N.Y.: Technicon Ind. Sys.

Technicon Industrial Systems. 1983a. *Ammonia in Water and Wastewater, Multi-Test Cartridge Method No. 696-82W: 1D–4D.* Tarrytown, N.Y.: Technicon Ind. Sys.

Technicon Industrial Systems. 1983b. *Nitrogen, Total Kjeldahl, in Water and Wastewater, Multi-Test Cartridge No. 696-82W:lC–5C.* Tarrytown, N.Y.: Technicon Ind. Sys.

Technicon Industrial Systems. 1983c. *Ortho-Phosphorous, Multi-Test Cartridge No. 696-82w: lB–4B.* Tarrytown, N.Y.: Technicon Ind. Sys.

Technicon Industrial Systems. 1983d. *Phosphorous, Total, in Water and Wastewater, Multi-Test Cartridge Method No. 696-82W:1A–9A.* Tarrytown, N.Y.: Technicon Ind. Sys.

Teddy, T. 1988. *Energy Data Directory and Yearbook 1988.* New Delhi: Tata Energy Research Institute.

Teran, F., and J. I. Barquero. 1964. *Geografia de Nicaragua.* Managua: Banco Central de Nicaragua.

Terry, R. E., R. L. Tate, and J. M. Duxbury. 1981. Nitrous Oxide Emissions from Drained, Cultivated Organic Soils of South Florida. *J. Air Pollut. Cont. Assoc., 31*, 1173–1176.

Thierstein, H. R. 1981. Late Cretaceous Nannoplankton and the Change at the Cretaceous-Tertiary Boundary. *Society of Economic Paleontologists and Mineralogists Special Publication 32*, 355–394.

Thierstein, H. R. 1982. Terminal Cretaceous Plankton Extinctions: A Critical Assessment. *Geological Society of American Special Paper 190*, 385–399.

Thompson, A. M., R. W. Stewart, M. A. Owens, and J. A. Herwehe. 1989. Sensitivity of Tropospheric Oxidants to Global Chemical and Climate Change. *Atmos. Environ., 23*, 519–532.

Thompson, A. M., and R. J. Cicerone. 1986. Possible Perturbations to Atmospheric CO, CH$_4$ and OH. *J. Geophys. Res., 91*, 10,853–10,864.

Thompson, D. J., and J. M. Shay. 1985. The Effects of Fire on Phragmites Australis in the Delta Marsh, Manitoba. *Can. J. Bot., 63*, 1864–1869.

Thompson, J. F., I. K. Smith, and D. P. Moore. 1970. Sulfur in Plant Nutrition. In *Sulfur in Nutrition*, edited by O. H. Muth and J. E. Oldfield, 80–96. Westport, Conn.: Avi Publishing.

Thompson, S. L. 1988. *Multi-Year Global Climatic Effects of Atmospheric Dust from Large Bolide Impacts, Global Catastrophes in Earth History.* Snowbird, Utah: Lunar Planetary Institute.

Tilton, F. 1931. The Great Fires in Wisconsin, Robinson and Kustermann Publishers, Green Bay. Reprinted in *Green Bay Historical Bulletin, 7*, 1–99.

Toon, O. B., J. B. Pollack, T. P. Ackerman, R. P. Turco, C. P. McKay, and M. S. Liu. 1982. Evolution of an Impact Generated Dust Cloud and Its Effects on the Atmosphere. *Geol. Soc. Am. Spec. Pap. 190*, 187–200.

Tournier, J. L. 1948. Sur les prairies d'altitude de la Chalne du Nimba. *Notes Africaines, 38*, 7.

Toutain, J., and G. Meyer. 1989. Iridium-Bearing Sublimates at the Hot-Spot Volcano (Piton de la Fournaise, Indian Ocean). *Geophysical Research Letters, 16*, 1391–1394.

Trexler, M., P. Faeth, and J. Kramer. 1989. Forestry as a Response to Global Warming: An Analysis of the Guatemala Agroforestry and Carbon Sequestration Project. Washington, D.C.: World Resources Institute, 68 pp.

Tschudy, R. H., C. L. Pillmore, C. J. Orh, J. S. Gilmore, and J. D. Knight. 1984. Disruption of the Terrestrial Plant Ecosystem at the Cretaceous-Tertiary Boundary, Western Interior. *Science, 225*, 1030–1032.

Tschudy, R. H., and B. D. Tschudy. 1986. Extinction and Survival of Plant Life Following the Cretaceous/Tertiary Boundary Event, Western Interior, North America. *Geology, 14*, 667–670.

Tsukada, M., and E. S. Deevey. 1967. Pollen Analyses from Four Lakes in the Southern Maya Area of Guatemala and El Salvador. In *Quaternary Paleoecology*, edited by E. J. Cushing and H. E. Wright, 303–331. New Haven: Yale University Press.

Tso, T. C., N. Harley, and L. T. Alexander. 1966. Source of Lead-210 and Polonium-210 in Tobacco. *Science, 153*, 880–882.

Tucker, C., J. Townshend, and T. Goff. 1985. African Land Cover Classification Using Satellite Data. *Science, 227*, 369–375.

Tucker, C. J., B. N. Holben, and T. E. Goff. 1984. Intensive Forest

Clearing in Rondonia, Brazil, As Detected by Satellite Remote Sensing. *Rem. Sens. Environ.,* 15, 255.

Turco, R. P.: 1985. The Photochemistry of the Stratosphere. In *The Photochemistry of Atmospheres: Earth, the Other Planets, and Comets,* edited by J. S. Levine, 77–128. Orlando, Fla.: Academic Press.

Turco, R. P., O. B. Toon, T. Ackerman, J. B. Pollack, and C. Sagan. 1983. Nuclear Winter, Global Consequences of Multiple Nuclear Explosions. *Science,* 222, 1283–1292.

Turco, R. P., O. B. Toon, T. P. Ackerman, J. B. Pollack, and C. Sagan. 1990. Climate and Smoke: An Appraisal of Nuclear Winter, *Science,* 247, 166–176.

Turco, R. P., O. B. Toon, R. C. Whitten, J. B. Pollack, and P. Hamill. 1983. The Global Cycle of Particulate Elemental Carbon: A Theoretical Assessment. In *Precipitation Scavenging, Dry Deposition, and Resuspension,* edited by H. R. Pruppacher, R. G. Semonin, and W. G. N. Slinn, 1337–1351. New York: Elsevier.

Turner, B. L. 1978. The Development and Demise of the Swidden Thesis of Maya Agriculture. In *Pre-Hispanic Mayan Agriculture,* edited by P. D. Harrison and B. L. Turner, 13–22. Albuquerque: University of New Mexico Press.

Turner, J., and M. J. Lambert. 1980. Sulfur Nutrition in Forests. In *Atmospheric Sulfur Deposition: Environmental Impact and Health Effects,* edited by D. S. Shriner, C. R. Richmond, and S. E. Lindberg, 321–333. Ann Arbor, Mich.: Ann Arbor Science Publishers.

Twomey, S. 1971. The Composition of Cloud Nuclei. *J. Atmos. Sciences,* 28, 377–381.

Twomey, S., R. Gall, and M. Leuthold. 1987. Pollution and Cloud Reflectance. *Boundary-Layer Meteorol.,* 41, 335–348.

Twomey, S., and J. Warner. 1967. Comparison of Measurements of Cloud Droplets and Cloud Nuclei. *J. Atmos. Sci.,* 24, 702.

Twomey, S., and T. A. Wojciechowski. 1969. Observations of the Geographical Variation of Cloud Nuclei. *J. Atmos. Sci.,* 2, 684–688.

Twomey, S. A. 1977. The Influence of Pollution on the Short-Wave Albedo of Clouds. *J. Atmos. Sci.,* 34, 1149–1152.

Twomey, S. A., M. Piepgrass, and T. L. Wolfe. 1989. An Assessment of the Impact of Pollution on Global Cloud Albedo. *Tellus,* 36B, 356–366.

Tyson, P. D., F. J. Kruger, and C. W. Louw. 1988. Atmospheric Pollution and Its Implications in the Eastern Transvaal Highveld. *South African National Scientific Programmes Report No. 150.* Pretoria: NPWCAR, CSIR.

U.S. Environmental Protection Agency. 1990. *Policy Options to Stabilize Global Climate.* Report to Congress by Office of Policy Analysis, U.S. Environmental Protection Agency, 3 volumes, January 1989 and February 1990 (revised).

U.S.D.A. Forest Service. 1982. *1979 Wildfire Statistics.* Washington, D.C.: U.S.D.A. Forest Service.

Uhl, C., D. Nepstad, R. Buschbacher, K. Clark, B. Kauffman, and S. Subler. 1989. Disturbance and Regeneration in Amazonia: Lessons for Sustainable Land-Use. *Ecologist,* 19(6), 235–240.

Uhl, C., and J. Saldarriaga. 1990. *The Disappearance of Wood Mass Following Slash and Burn Agriculture in the Venezuelan Amazon.* Unpublished manuscript.

Valasco, P. 1983. Assisting Portugal-Fire Handtool Training. *Fire Mgmt. Notes,* 44, 3–6.

Van Cleve, K., L. A. Viereck, and C. T. Dyrness. 1983. Dynamics of a Black Spruce Ecosystem in Comparison to Other Forest Types: A Multi-Disciplinary Study in Interior Alaska. In *Resources and Dynamics of the Boreal Zone,* edited by R. W. Wein, R. R. Riewe, and I. R. Methven, 148–166. Ottawa: Association of Canadian Universities for Northern Studies.

Van Cleve, K., and L. Viereck. 1981. Forest Succession in Relation to Nutrient Cycling in the Boreal Forest of Alaska. In *Forest Succession: Concepts and Applications,* edited by D. C. West, H. H. Shugart, and D. B. Botkin, 185–211. New York: Springer-Verlag.

Van Wagner, C. E. 1988. The Historical Pattern of Annual Burned Area in Canada. *Forestry Chronicle,* June.

Van Wagner, C. E. 1989. Personal communication.

Van de Hulst, H. C. 1957. *Light Scattering by Small Particles.* New York: John Wiley.

Van de Hulst, H. C. 1980. *Multiple Light Scattering: Tables, Formulas and Applications.* 2 vols. New York: Academic Press.

Van der Hammen, T. 1983. The Palaeoecology and Palaeogeography of Savannas. In *Tropical Savannas,* edited by F. Bourliere, 19–35. Amsterdam: Elsevier.

Van Arman, J., and R. Goodrick. 1979. Effects of Fire on a Kissimmee River Marsh. *Florida Sci.,* 42, 183–195.

Veltishchev, N. N., A. S. Ginsburg, and G. S. Golitsyn. 1988. *Climatic Effects of Mass Fire* [in Russian]. Izvestia Academy of Sciences of the USSR—Atm. and Oceanic Physics, 24, 296–304.

Venkatsan, M. I., and Dahl, J. 1989. Further Geochemical Evidence for Global Fires at the Cretaceous-Tertiary Boundary, *Nature,* 338, 57–60.

Verger, F. 1980. *Identification par la forme des zones brulées au Senegal.* Senegal: Rapport Programme TECASEN.

Vernet, P. 1988. *Une approche integrée de l'analyse de l'approvisionnement urbain.* Seminaire sur les Energies Domestiques à Base de Bois, Korhogo, Cote d'Ivoire. 10–12 November.

Vogelmann, A. M., A. Robock, and R. G. Ellingson. 1988. Effects of Dirty Snow in Nuclear Winter Simulations. *J Geophys. Res.,* 93, 5319–5332.

Vogl, R. J. 1973. Effects of Fire on the Plants and Animals of a Florida Wetland. *Am. Midl. Nat.,* 89, 334–347.

Vogt, P. R. 1972. Evidence for Global Synchronism in Mantle Plume Convection and Possible Significance for Geology. *Nature,* 240, 338–342.

Voice, M. E., and F. J. Gauntlett. 1985. The 1983 Ash Wednesday Fires in Australia. *Mon. Wea. Rev.,* 112, 584–590.

Von Rheinbaben, W., and G. Trolldenier. 1984. Influence of Plant Growth on Denitrification in Relation to Soil Moisture and Potassium Nutrition. *Z. Pflanzenern. Boden. K.,* 147, 730–738.

Vorphal, J. A., J. G. Sparrow, and E. P. Ney. 1970. Satellite Observations of Lightning. *Science,* 169, 860–862.

WMO. 1985. *Atmospheric Ozone, Global Ozone Research and Monitoring Project.* Report No. 16. Geneva: World Meteorological Organization.

Wade, D., J. Ewel, and R. Hofstetter. 1980. *Fire in South Florida Ecosystems.* USDA Forestry Service Gen. Tech. Rep. SE-17, Southeastern Forestry Expeditionary Station, Asheville, N.C.

Waggoner, A. P., R. E. Weiss, N. C. Ahlquist, D. S. Covert, S. Will, and R. J. Charlson. 1981. Optical Characteristics of Atmospheric Aerosols. *Atmos. Environ.,* 15, 1891–1909.

Wagner, P. L. 1958. *Nicoya: A Cultural Geography.* University of California Publications in Geography, 12(3), 195–250.

Wagner, P. L. 1964. Natural Vegetation of Middle America. *Handbook of Middle American Indians. Vol. 1,* edited by R. C. West, 216–264. Austin: University of Texas Press.

Walker, J., and A. N. Gillison. 1982. Australian Savannas. *Ecology of Tropical Savanna.* Ecological Studies 42, 5–24. New York: Springer-Verlag.

Wallace, G. F., and P. Barrett. 1981. *Analytical Methods Development for Inductively Coupled Plasma Spectrometry.* Norwalk, Conn.: Perkin-Elmer Corporation.

Walton, J. J., M. C. MacCracken, and S. J. Ghan. 1988. A Global-Scale Lagrangian Trace Species Model of Transport, Transformation and Removal Processes. *J. Geophys. Res.,* 8339–8354.

Ward, D. E. 1983. Source-Strength Modeling of Particulate Matter Emissions from Forest Fires. *Proceedings of the 76th Annual Meeting of the Air Pollution Control Association.* Paper No. 83-45.2, Atlanta, Ga., 19–24 June.

Ward, D. E. 1984. Particulate Matter Emissions from Forest Fires: A Comparison of Methods and Results. Paper presented at the Conference on Large Scale Fire Phenomenology, Sponsored by the Defense Nuclear Agency; Federal Energy Management Agency; and the National Bureau of Standards, Gaithersburg, Md., 10–13 September.

Ward, D. E. 1986. Field Scale Measurements of Emission from Open Fires. Technical paper presented at the Defense Nuclear Agency Global Effects Review, Defense Nuclear Agency, Washington, D.C.

Ward, D. E. 1989. Factors Influencing the Emissions of Gases and Particulate Matter from Biomass Burning. Paper presented at the Third International Symposium on Fire Ecology, Freiburg University, FRG. Available from D. E. Ward, U.S.D.A. Forest Service, Intermountain Fire Science Laboratory, Missoula, Mont. 59807.

Ward, D. E., A. Setzer, Y. Kaufman, and R. Rasmussen. 1990. *Characteristics of Smoke Emissions from Biomass Fires of the Amazon Region—BASE-A Experiment.* Paper presented at the Chapman Conference on Biomass Burning: Atmospheric, Climatic, and Biospheric Implications, Williamsburg, Va. 19–23 March (this volume, Chapter 48).

Ward, D. E., and C. C. Hardy. 1984. Advances in the Characterization and Control of Emissions from Prescribed Fires. *Proceedings of the 78th Annual Meeting of the Air Pollution Control Association,* San Francisco, Calif., Paper No. 84-363.

Ward, D. E., and C. C. Hardy. 1989. Emissions from Prescribed Burning of Chaparral. *Proceedings of the 1989 Annual Meeting of the Air and Waste Management Association,* Anaheim, Calif., June.

Ward, D. E., and C. C. Hardy. 1991. Smoke Emissions from Wildland Fires. *Environment International,* 17, 117–134.

Ward, D. E., C. C. Hardy, R. D. Ottmar, and D. V. Sandberg. 1982. A Sampling System for Measuring Emissions from West Coast Prescribed Fires. *Proceedings of the Pacific Northwest Section of the Air Pollution Control Association,* Vancouver, B.C.

Ward, D. E., C. C. Hardy, D. V. Sandberg, and T. E. Reinhardt. 1989. Part III—Emissions Characterization. In *Mitigation of Prescribed Fire Atmospheric Pollution through Increased Utilization of Hardwoods, Piled Residues, and Long-Needled Conifers,* edited by D. V. Sandberg, D. E. Ward, and R. D. Ottman. Final Report to the Bonneville Power and U.S. Department of Energy under IAG DE-AI179-85BP18509 (PNW-85-423), 15 July.

Ward, D. E., R. N. Nelson, and D. Adams. 1979. Forest Fire Smoke Documentation. *Proceedings of the 72nd Annual Meeting of the Air Pollution Control Association,* 21–29 June Cincinnati, Ohio, Paper No. 79-6.3.

Ward, D. E., D. V. Sandberg, R. D. Ottman, J. A. Anderson, G. G. Hofner, and C. K. Fitzsimmons. 1982. Measurements of Smokes from Two Prescribed Fires in the Pacific Northwest. Seventy-fifth Annual Meeting of the Air Pollution Control Association. Available from D. E. Ward, U.S.D.A. Forest Service, Intermountain Fire Science Laboratory, Missoula, Mont. 59807.

Ward, D. E., R. Susott, R. Babbitt, and C. Hardy. Properties and Concentrations of Smoke near the Ground from Biomass Field Tests. Paper presented at the Symposium on Smoke/Obscurants XIV; 17–19 April 1990, Kossiakoff Conference and Education Center, Johns Hopkins University, Laurel, Md., in press.

Warneck, P. 1988. *Chemistry of the Natural Atmosphere.* Orlando, Fla.: Academic Press.

Warner, J., and S. Twomey. 1967. The Production of Cloud Nuclei by Cane Fires and the Effect on Cloud Drop Concentrations. *J. Atmos. Sci.,* 24, 704.

Warren, S. G., C. J. Hahn, J. London, R. M. Chervin, and R. L. Jenne. 1986. *Global Distribution of Total Cloud Cover and Cloud Type Amounts over Land.* DOE/ER/60085-HI, U.S. Dept. of Energy and National Center for Atmospheric Research.

Watson, A., J. E. Lovelock, and L. Margulis. 1978. Methanogenesis, Fires and the Regulation of Atmospheric Oxygen. *Biosystems,* 10, 293–298.

Watters, R. F. 1960. The Nature of Shifting Cultivation. *Pacific Viewpoint,* 1, 59.

Watters, R. F. 1971. *Shifting Cultivation in Latin America.* FAO Foreign Development Paper. 17. Rome: FAO.

Wein, R. W., and D. A. MacLean. 1983. *The Role of Fire in Northern Circumpolar Ecosystems.* Scope 18, New York: John Wiley.

Weiss, R. E., and L. F. Radke. 1990. Optical Extinction and Absorption of Smokes from Large Fires: Absorption Techniques and Observations. *J. Atmos. Ocean Tech.*

Weiss, R. F. 1981. The Temporal and Spatial Distribution of Tropospheric Nitrous Oxide. *J. Geophys. Res.,* 86, 7185–7195.

Weiss, R. F., and H. Craig. 1976. Production of Atmospheric Nitrous Oxide by Combustion. *Geophys. Res. Lett.,* 3, 751–753.

Wellman, P., and M. W. McElhinny. 1970. K-Ar Age of the Deccan Traps, India. *Nature,* 227, 595–596.

Wells, C. G., R. E. Campbell, L. F. DeBano, C. E. Lewis, R. L. Fredrikson, E. C. Franklin, R. C. Froelich, and P. H. Dunn. 1979. *Effects of Fire on Soil.* U.S.D.A. Forestry Service Gen. Tech. Rep. W0-7.

Westberg, H., K. Sexton, and D. Flyckt. 1981. Hydrocarbon Production and Photochemical Ozone Formation in Forest Burn Plumes. *J. Air Pollut, Contr. Assoc.,* 31, 661–664.

Westphal, D. L., and O. B. Toon. 1991. Simulation of Microphysical, Radiative, and Dynamical Processes in a Continental-Scale Forest Fire Smoke Plume. *J. Geophys. Res.,* in press.

Westphal, D. L., O. B. Toon, and W. R. McKie. 1989. Atmospheric Effects of a Canadian Forest Fire Smoke Plume. *IRS '88: Current Problems in Atmospheric Radiation,* edited by J. Lenoble and J. F. Geleyn, 322–325. Hampton, Va: Deepak Publishing Co.

Wexler, I. 1950. The Great Smoke Pall—September 24–30. *Weatherwise,* December, 129–142.

Wharton, C. H. 1966. Fire and Wild Cattle in North Cambodia. *Proceedings of the Annual Tall Timbers Fire Ecology Conference 5,* 23, Tallahassee, Fla.

White, D. A., T. E. Weiss, J. M. Tropani, and L. B. Thien. 1978. Productivity and Decomposition of the Dominant Salt Marsh Plants in Louisiana. *Ecology, 59,* 751–759.

White, R. S. 1989. Igneous Outbursts and Mass Extinctions. *Eos,* 14 November, 1480, 1490–1491.

White, W. A. 1958. Some Geomorphic Features of Central Peninsular Florida. *Geol. Bull.,* 41. Tallahassee: Florida Geol. Surv.

White, W. A. 1970. The Geomorphology of the Florida Peninsula. *Geol. Bull.,* 51. Tallahassee: Florida Department of Natural Resources, Bureau of Geology.

Whitmore, T. C. 1975. *Tropical Rain Forests of the Far East.* Oxford: Clarendon.

Whittaker, R. H. 1977. Evolution of Species Diversity in Land Communities. *Evol. Biol.,* 10, 1.

Whittaker, R. H. and G. E. Likens. 1975. The Biosphere and Man. In *Primary Productivity of the Biosphere,* edited by H. Lieth and R. Whittaker, 305–328. Berlin: Springer-Verlag.

Wichman, I. S. 1983. Flame Spread in an Opposed Flow with a Linear Velocity Gradient. *Combustion and Flame,* 50, 287–304.

Wichman, I. S., F. A. Williams, and I. Glassman. 1982. *Theoretical Aspects of Flame Spread in an Opposed Flow over Flat Surfaces of Solid Fuels.* Nineteenth International Symposium on Combustion, 835–845.

Wielgolaski, F. E., S. Kjelik, and P. Kallio. 1975. Mineral Content of Tundra and Forest Tundra Plants in Fennoscandia, in Tundra Plants in Fennoscandia. *Ecological Studies,* 16, 316–332. New York: Springer-Verlag.

Wigley, T. M. L. 1989. Possible Climate Change Due to SO_2-Derived Cloud Condensation Nuclei. *Nature,* 339, 365–367.

Wilbur, R. B., and N. L. Christensen. 1983. Effects of Fire on Nutrient Availability in a North Carolina Coastal Plain Pocosin. *Am. Midl. Nat.,* 110, 54–61.

Williams, E. J., D. D. Parrish, and F. C. Fehsenfeld. 1987. Determination of Nitrogen Oxide Emissions from Soils: Results from a Grassland Site in Colorado, United States. *J. Geophys. Res. 2,* 2173–2180.

Williams, R. B., and M. B. Murdoch. 1972. Compartmental Analysis of the Production of Juncus Mrianus in a North Carolina Salt Marsh. *Chesapeake Sci.,* 13, 69–79.

Wilson, A. T. 1978. Pioneer Agriculture Explosion and CO_2 Levels in the Atmosphere. *Nature,* 273, 441.

Wilson, M. F., and A. Henderson-Sellers. 1985. A Global Archive of Land Cover and Soils Data for Use in General Circulation Climate Models. *J. Climate,* 5, 119–143.

Windhorst, H.-W. 1974. Studien zur Weltwirtschaftsgeographie. A) Das Ertragspotential der Walder der Erde; B) Wald- und Forstwirtschaft in Afrika. *Beih. Geogr. Zeitschr.,* 39, Wiesbaden.

Winstead, E. L., K. G. Hoffman, W. R. Cofer, III, and J. S. Levine. 1990. Nitrous Oxide Emissions from Burning Biomass. Paper presented at the Chapman Conference on Global Biomass Burning: Atmospheric, Climatic, and Biospheric Implications, Williamsburg, Va., 19–23 March (this volume, Chapter 45).

Wirkler, P. 1988. Surface Ozone over the Atlantic Ocean. *J. Atmos. Chem.,* 7, 73–91.

Wofsy, S. C. 1988. Biomass-Burning Emissions and Associated Haze Layers over Amazonia. *J. Geophys. Res.,* 93, 1509–1527.

Wolbach, W. S. 1989. *Carbon across the Cretaceous-Tertiary Boundary.* Ph.D. thesis, University of Chicago.

Wolbach, W. S., and E. Anders. 1989. Elemental C in Sediments: Determination and Isotopic Analysis in the Presence of Kerogen. *Geochim. Cosmochim. Acta,* 53, 1637–1647.

Wolbach, W. S., E. Anders, and M. A. Nazarov. 1990. Fires at the K-T Boundary: Carbon at the Sumbar, Turkmenia, Site. *Geochim. Cosmochim. Acta.,* in press.

Wolbach, W. S., E. Anders, and C. J. Orth. 1988. Darkness after the K-T impact: Effects of Soot. In *Global Catastrophes in Earth History,* 219–229. Snowbird, Utah: Lunar Planetary Institute.

Wolbach, W. S., I. Gilmour, E. Anders, C. J. Orth, and R. R. Brooks. 1988. Global Fire at the Cretaceous-Tertiary Boundary. *Nature,* 334, 59.

Wolbach, W. S., R. S. Lewis, and E. Anders. 1985. Cretaceous Extinctions: Evidence for Wildfires and Search for Meteoritic Material. *Science,* 230, 167–170.

Wolbach, W. S., R. S. Lewis, E. Anders, M. M. Grady, C. T. Pillinger, R. R. Brooks, C. J. Orth, and J. S. Gilmore, 1986. Carbon Isotopes and Iridium at Two Cretaceous-Tertiary (K-T) Boundary Sites in New Zealand. *Meteoritics,* 21, 541–542.

Woldendorp, J. W. 1962. The Quantitative Influence of the Rhizosphere on Denitrification. *Plant Soil,* 17, 267–270.

Wolfe, J. A. 1981. Paleoclimatic Significance of the Oligocene and Neogene Floras of Northwestern United States. In *Paleobotany, Paleoecology, and Evolution,* vol. 2, edited by K. J. Niklas, 79–101. New York: Praeger.

Wolfe, J. A., and G. R. Upchurch, Jr. 1986. Vegetation, Climatic Changes at the Cretaceous-Tertiary Boundary. *Nature,* 2, 148–152.

Wolfe, J. A., and G. R. Upchurch, Jr. 1987. North American Nonmarine Climates and Vegetation during the Late Cretaceous. *Paleogeography, Palaeoclimatology, Palaeoecology,* 61, 33–77.

Wolff, G. T., R. J. Countess, P. J. Groblicki, M. A. Ferman, S. Cadle, and J. L. Muhlbaier. 1981. *Atmospheric Environment,* 15, 2485–2502.

Wong, C. S. 1978. Atmospheric Input of Carbon Dioxide from Burning Wood. *Science,* 200, 197–200.

Wood, C. A. 1989. Geologic Applications of Space Shuttle Photography. *Geocarto International* 4(1), 49–54.

Wood, C. A., M. R. Helfert, K. P. Lulla, and R. O. Covey. 1989. Earth Observations during Space Shuttle Flight STS-26: Discovery's Mission to Earth. *Geocarto International* 4(2), 55–63.

Woods, D. C., R. L. Chuan, W. R. Cofer, III, and J. S. Levine. 1990. Aerosol Characterization in Smoke Plumes from Burning at a Florida Wildlife Refuge. Paper presented at the Chapman Conference on Global Biomass Burning: Atmospheric, Climatic, and Biospheric Implications, Williamsburg, Va., 19–23 March (this volume, Chapter 31).

Woods, D. C., and R. L. Chuan. 1982. Fire Particles in the Soufriere Eruption Plume. *Science,* 1, 1118–1119.

Woods, D. C., R. L. Chuan, and W. I. Rose. 1985. Halite Particles Injected into the Stratosphere by the 1982 El Chichon Eruption. *Science,* 170–172.

Woodwell, G. M., et al. 1984. Measurement of Changes in the Vegetation of the Earth by Satellite Imagery. In The Role of Terrestrial Vegetation, *The Global Carbon Cycle: Measurement by Remote Sensing,* edited by G. M. Woodwell, 221–240. SCOPE 23. New York: John Wiley.

World Bank. 1987. *Madagascar: Situation et Perspectives du Secteur Energetique.* World Bank Report. Washington, D.C.: World Bank.

Wright, H. A., and A. W. Bailey. 1982. *Fire Ecology: United States and Southern Canada.* New York: John Wiley.

Wyckoff, S., E. Lindholm, P. A. Wehinger, B. A. Peterson, J.-M. Zucconi, and M. C. Festou. 1989. The $^{12}C/^{13}C$ Abundance Ratio in Comet Halley. *Astrophys. J.,* 339, 488–500.

Yamate, G. 1975. *State of the Art Report on Atmospheric Aspects of Forest Fires.* Report C8269-2. Chicago; ITT Research Institute.

Ying, L., and S. P. Levine. 1989. FTIR Least-Squares Methods for the Quantitative Analysis of Multicomponent Mixtures of Airborne Vapours of Industrial Hygiene Concern. *Anal. Chem.,* 61, 677.

Zachariah, D. C., and M. T. Vu. 1988. World Population Projections, 1987–1988 Edition. Baltimore: World Bank, Johns Hopkins University Press, 440 pp.

Zimmerman, P. R., J. P. Greenberg, and J. P. E. C. Darlington. 1984. *Termites and Atmospheric Gas Production,* 1224, 86.

Zimmerman, P. R., J. P. Greenberg, S. O. Wandiga, and P. J. Crutzen. 1982. Termites: A Potentially Large Source of Atmospheric Methane, Carbon Dioxide, and Molecular Hydrogen. *Science,* 218, 563–565.

Zirang-Ziyuang Zazi. 1983. The Thirty-Year Ground Climate Data of Yunnan Province, Yunnan Meteorological Bureau. *Chinese J. of Nature Resource.*

Zoller, W. H., E. S. Gladney, and R. A. Duce. 1974. Atmospheric Concentrations and Sources of Trace Metals at the South Pole. *Science,* 183, 198–200.

Zoller, W. H., J. R. Parrington, and J. M. Kotra. 1983. Iridium Enrichment in Airborne Particles from Kilauea Volcano: January 1983. *Science,* 222, 1118–1120.

Zoltai, S. C. 1988. Ecoclimatic Provinces of Canada and Man-Induced Climatic Change. *Canadian Committee on Land Classification Newsletter,* 17, 12–15.

Zontek, F. 1966. Prescribed Burning on the St. Marks National Wildlife Refuge. In *Proceedings of the 5th Tall Timbers Fire Ecology Conference,* 195–210.

Index